COMMUNITY INDICATORS MEASU

To hold infinity in the palm of your hand and eternity in an hour – is inspiration.
- William Blake

This work is dedicated to my partner and husband in the journey of life, Jay Stein. Thank you for all the inspiration – with you, everything is possible.

Community Indicators Measuring Systems

Edited by

RHONDA PHILLIPS
University of Florida, USA

ASHGATE

Published by

Ashgate Publishing Limited
Gower House
Croft Road
Aldershot
Hants GU11 3HR
England

Ashgate Publishing Company
Suite 420
101 Cherry Street
Burlington, VT 05401-4405
USA

Ashgate website: http://www.ashgate.com

British Library Cataloguing in Publication Data
Community indicators measuring systems. - (Urban and regional planning and development)
1. Planning - Statistical methods 2. Social indicators
3. Economic indicators 4. Environmental indicators 5. Planning - Canada - Statistical methods 6. Social indicators - Statistical methods 7. Economic indicators - Statistical methods 8. Environmental methods - Statistical methods
I. Phillips, Rhonda
352.3'4'0727

Library of Congress Cataloging-in-Publication Data
Community indicators measuring systems / edited by Rhonda Phillips.
p. cm. -- (Urban and regional planning and development series)
Includes bibliographical references and index.
ISBN 0-7546-4005-1
1. Community development--Canada--Evaluation. 2. Community development--United States--Evaluation. 3. City planning--Canada. 4. City planning--United States. 5. Social indicators. 6. Economic indicators.7. Environmental indicators. I. Phillips, Rhonda. II. Urban and regional planning and development.

HN110.Z9C635 2004
307.1'416'0971--dc22

2004062399

ISBN 0 7546 4005 1

Printed and bound by Athenaeum Press, Ltd.,
Gateshead, Tyne & Wear.

Contents

List of Contributors

SUSAN BRIDGES is a graduate of the Urban and Regional Planning Department at the University of Florida. Her position at Renaissance Planning Group in Tampa, Florida is focused on bicycle- and pedestrian-supportive master planning and design guidelines, transit-oriented design, and waterborne transportation. Other interests include New Urbanism and Traditional Neighborhood Development. She previously served as a graduate research assistant with the University of Florida's Center for Building Better Communities.

JEFF CARMICHAEL completed a Ph.D. in Economics at the University of Colorado at Boulder in 1997. His background and research interests include integrated assessment and multi-criteria decision modeling, watershed and water quality modeling on non-point and point sources of pollution and climate change impacts on water resources. Jeff is currently a Research Associate managing QUEST Model Development for the Georgia Basin Future Project and is also involved in studies of climate change adaptation and of future water demand and supply scenario analysis in the Okanagan region of British Columbia, Canada.

CLIFFORD COBB is a graduate of the University of California, Berkeley, where he earned a master's degree in public policy. He is author (1992) of *Responsive Schools, Renewed Communities*, a communitarian proposal for systemic education reform, co-author (1993) of *The Green National Product,* an alternative measure of the national economy, co-editor (2001) of *The Path to Justice*, essays on the philosophy and economics of Henry George, contributor (1989, 1993) to *For the Common Good*, a text in environmental economics, and author of several articles on social indicators and environmental taxes. In 1999, he was invited to write a background paper for a planning session in Denmark on social indicators in preparation for the United Nations Summit on Social Development. In 1994, he helped establish Redefining Progress, a social policy think-tank in San Francisco, and he remains a senior fellow, doing research projects on common property resources and tax policy. In 1995, he served as a non-voting member of President Clinton's Council on Sustainable Development. He is currently researching the history of the commons and private property.

RINA GHOSE specializes in GIScience, Urban Geography and India. Her research interests include the impacts and implications of GIS upon society and the social construction of GIS, democratizing GIS among traditionally marginalized citizens through public participation GIS (PPGIS) and studying the process of

PPGIS, critical introspection of GIS, societal implications of digital technology, urban and rural gentrification and growth management. Professor Ghose has published her research papers widely in international scholarly journals such as Transactions in GIS, Progress in Human Geography, Cartography and Geographic Information Science, Journal of URISA, Journal of Urban Technology and Cartographica. Professor Ghose has undertaken various synergistic actions in PPGIS such as participating in collaborative research with community-based organizations and organizing PPGIS sessions in international conferences.

HEIDI HOERNIG Heidi Hoernig is a doctoral candidate at the School of Planning, University of Waterloo. Aside from monitoring and evaluating her progress towards her doctoral goals, Ms. Hoernig's research interests lie in the areas of multiculturalism and planning, downtown revitalization and monitoring and evaluation in planning.

WILLIAM HUXHOLD has been teaching courses on geographic information systems at the University of Wisconsin-Milwaukee since 1988. Prior to his full-time appointment at the University in 1990, he was MIS Project Director at the City of Milwaukee, having responsibility for the development of information systems for the City. He is active in national efforts related to land information systems and geographic information systems, having assisted efforts sponsored by the International Association of Assessing Officials, the National Oceanic and Atmospheric Association, and the National Center for Geographic Information and Analysis. Professor Huxhold is well known internationally in local government GIS circles, having published numerous papers, lectured at universities, conducted workshops, and delivered presentations at various professional meetings for more than 25 years. He has published four texts including *GIS County Users Guide: Laboratory Exercises in Geographic Information Systems Using ARC/INFO Version 7*, published in 2001 by URISA Press. His educational background includes a B.S. degree in Industrial Engineering from Northwestern University in 1968 and an M.S. degree in Engineering Management (with a Public Administration emphasis) from the University of Dayton in 1973.

NOEL KEOUGH is a co-founder and senior researcher with Sustainable Calgary Society and a member of the Board of Directors of PLAN:NET Ltd. Noel has 20 years' experience applying participatory methodologies to sustainable community development and international development. He has been a community activist for many years in his own community. He has also worked with communities from coast to coast in Canada. Internationally, Noel has experience in Central and South America, The Middle East, Eastern Europe and Central and South East Asia. From 1997 to 2002 Noel co-coordinated Sustainable Calgary's sustainability indicator initiative and co-authored Sustainable Calgary's 1998 and 2001 State of Our City Reports. He has also worked with the Town of Banff to create its first Citizens

Sustainability Report Card. Noel is currently participating in a national effort to create a Canadian Index of Wellbeing and is completing his Doctoral Dissertation in the Department of Geography, University of Calgary. His educational background includes B.S. in Engineering, and a Master's of Environmental Design.

NAOMI OLIVER is an Assistant Director and Assistant in Research at the Florida Center for Community Design and Research (FCCDR) at the University of South Florida. In graduate school, she was a GIS Teaching Assistant for the University of South Florida's Geography Department. In 2000, Ms. Oliver served as the GIS Coordinator of the FCCDR and was then promoted to Research Faculty (Assistant in Research I) in 2001 and then to Assistant in Research II in 2002. Ms. Oliver's areas of technical expertise include managing and creating GIS data, post-processing GPS data, community development and urban geography. She has particular interest in the concepts and practices of empowering citizens and decision-makers through the use of GIS based data and user-friendly applications. During the past four years, Ms. Oliver has managed over four-hundred thousand dollars worth of grant-funded projects for various local and regional governments related to community development, community indicators, and GIS.

RHONDA PHILLIPS has been involved in economic development since 1984, working as an economic developer with private, public, and non-profit organizations. Now a faculty member at the University of Florida, she directs the Center for Building Better Communities (CBBC) and the graduate concentration in community and economic development in the College of Design, Construction and Planning. Her work with the CBBC includes redevelopment and economic revitalization planning for communities throughout Florida, the US, and internationally. Current research focuses on technology-based economic development and community revitalization approaches. She is author of *Concept Marketing for Communities* and *Evaluating Technology-based Economic Development.*

CRAIG RIXFORD has written on indicators and community sustainability in a variety of contexts, including historical analysis. He worked with Redefining Progress beginning with its inception in 1994, a national non-profit organization focusing on community and regional well-being. While with Redefining Progress, he contributed research to the Genuine Progress Indicator, a pioneering alternative to the Gross National Product. Additionally, he co-authored the *Community Indicators* Handbook, widely used by many organizations throughout North America.

JOHN ROBINSON is Professor in the Sustainable Development Research Initiative (SDRI) and the Department of Geography at UBC. He received his Ph.D.

in Geography at the University of Toronto in 1981, on the intersection of philosophy with energy policy analysis. He then worked at the University of Waterloo for 11 years, learning something about interdisciplinary research and teaching. At SDRI he currently directs several research programs in the areas of climate change and sustainability, analyzing sustainable futures in the Georgia Basin, visualization and citizen engagement and the development of modeling and scenario analysis tools. His personal research interests include sustainability theory and practice, regional-scale participatory integrated assessment, participatory processes, business and sustainability, sustainable energy systems and the relationship between science and decision-making. Dr. Robinson is a member of the Advisory Council of the David Suzuki Foundation, on the Board of Westcoast Environmental Law, the Sustainable Cities Foundation, Panos Canada, a member of the Steering Group of HELIO International, and is on the Editorial Board of the journal *Integrated Assessment*. He has twice been a Coordinating Lead Author of the Intergovernmental Panel on Climate Change and is currently a member of the Ministerial Advisory Panel on Sustainability for Public Works and Government Services Canada.

CHRISTIANA SCHUMANN is an Assistant in Research at the Florida Center for Community Design and Research (FCCDR) at the University of South Florida. With a background in Applied Anthropology, her research interests include qualitative community research and stakeholder involvement. Since 2001, her involvement at the FCCDR has included the management of the Water Resource Atlas Oral History Program where local residents are interviewed about the history of the area's lakes and rivers, as well as management of the Hillsborough County and Seminole County Water Atlas projects, which provide public access to the water resource data available in both counties. Ms. Schumann has also been involved in developing the methodology to include stakeholders in the process of website design. She has also been involved in the development of the Community Atlas program, which seeks to use web technology to present data at the neighborhood level.

MARK SEASONS is an Associate Professor with the School of Planning, University of Waterloo. He joined the School in 1998 following twenty years of professional planning practice in Ontario, Alberta and New Zealand. His current research focuses on applications of program and policy evaluation and indicators in municipal planning departments. He has a Ph.D. in Regional Planning and Resource Development from the University of Waterloo.

MARC T. SMITH is an Associate Professor in the School of Building Construction, and Associate Director of the Shimberg Center for Affordable Housing at the University of Florida. He received his Master of City Planning degree from the University of Pennsylvania and his Ph.D. from the Ohio State

University. His funded research has included a Community Outreach Partnership Center grant from the U.S. Department of Housing and Urban Development, a series of contracts with the Florida Department of Community Affairs and Florida Housing Finance Corporation for affordable housing needs assessment, a series of studies of housing prices and housing demand for the Florida Real Estate Commission Education and Research Foundation, and market studies for Neighborhood Reinvestment Corporation and Local Initiatives Support Corporation funded organizations in Florida. Publications focus on reports on housing prices, impacts of regulation on housing markets, housing policy including the Low Income Housing Tax Credits program, office markets, and housing needs estimation.

SONIA TALWAR is a Ph.D. Candidate in Geography at the University of British Columbia. Her background is in physical geography and environmental studies with a focus on information management and geographical information systems. Sonia currently focuses her attention on science and society issues including the role of information and communication technologies and the convergence of technologies and public process. Concurrently, in her role as a researcher at the Geological Survey of Canada, she is part of a team to develop programs and services to meaningfully link geoscience knowledge issues with sustainable development for educators, planners and society.

JAMES TANSEY moved to the University of British Columbia from the United Kingdom in 1999. His background is in Environmental Sciences, with a focus on environmental sociology and social risk management. He is currently involved in a major study of the social dimensions of health in forestry communities in BC and an evaluation of approaches to public deliberation of biotechnology. In addition, he is a co-investigator in the Georgia Basin Futures Project (www.basinfutures.net) and in two Climate Change Adaptation studies in British Columbia.

Preface

Indicators don't guarantee results. But results are impossible without proper indicators. And proper indicators, in themselves, can produce results.

– Donella Meadows

The idea for this book started with projects funded by the Canadian Consulate's Faculty Research Program. While conducting the research, several colleagues at Canadian universities agreed that additional literature on the subject is needed and have joined me in developing this volume. Numerous accounts of either US or Canadian indicator projects exist, but very little that combines the experiences of both countries. There is much knowledge to be gained by exploring contributions by Canadian and US researchers.

Community indicators are increasingly attracting attention. The appeal is being able to take 'measurements' that provide information about past and current trends in a community or region. These measurements reflect the status of the interplay among all the realms impacting community, regional, or national well-being: social, environmental and economic. This 'triple bottom line' information can then be used by planners and community leaders in decisions that impact future outcomes. While indicators themselves do not 'provide a model of how a community works or how to determine planning choices, they provide information that can be used by planners and others when faced with decisions about the community'.[1]

Selecting types of indicators that represent a region, community, or neighborhood – often through a community visioning or special study process – and combining them provides the basis for development of an indicators measuring system. While approaches vary from sophisticated integration with relational databases and geographic information systems to basic data reports, the goal is often the same: to provide an on-going mechanism to assess overall well-being of an area.

Numerous systems have been developed in a variety of contexts. At the multi-national level, agencies such as the United Nations, the World Bank, and the World Health Organization have implemented indicators systems or projects. Nationally, many countries have indicator projects including the United States and Canada where system development is often driven by the issue of long-term sustainability. Numerous states, provinces and communities have undertaken indicator projects. Also, neighborhood level indicator systems have been

---

[1] Phillips, Rhonda (2003), *Community Indicators*, Planning Advisory Service Report Number 517, Chicago, Illinois: American Planning Association, p.2.

developed, spurred by such collaborative efforts as the National Neighborhood Indicators Project.[2]

Section I, Theory and Basis For Community Indicators, provides an overview of community indicators. Heidi Hoernig and Mark Seasons introduce community indicators as a multi-faceted concept, addressing types of indicators, frameworks for organizing them, and selection criteria. Further, they present the material in the context of the important topic of monitoring. The authors present a convincing argument that monitoring should be included from the onset of an indicators project, and planners should be informed about strengths, challenges and limitations.

Clifford Cobb and Craig Rixford give a comparative analysis of the historical emergence of community indicators. In this fascinating foray into the underlying theory and history of indicators, the authors argue that the history is really an ongoing interplay between descriptive and diagnostic indicators and the conflict that arises. They propose several remedies to rectify the inefficiencies of relying on descriptive versus diagnostic indicators for predictive purposes.

Section II, Relating Community Indicators to Planning and Development, contains three selections. Noel Keough's chapter examines Calgary's development of its indicator project, Sustainable Calgary. In the context of neo-liberal globalization challenges, Keough provides an in-depth analysis of many facets of sustainability, including the citizen participation process. Moreover, he provides valuable insight into the process of developing an indicators project and reporting mechanisms.

Mark Seasons then reports on the results of a survey conducted of Canada's mid-sized cities. He explores the interaction of indicators and planning in core-areas. The findings of this chapter imply that mid-sized cities tend to rely on 'traditional' economic and performance-based indicators. His chapter has important recommendations for improving indicators models, and encouraging cities to adopt approaches that more fully address complex social, economic and environmental challenges.

In the concluding chapter of this section, Rhonda Phillips and Susan Bridges explore applications for integrating indicators measuring systems with economic development planning. The authors present a case study in Florida to illustrate the need for using indicators as a means to mitigate conflict and develop common economic development goals.

Section III, Technological Dimensions, explores the application of tools and the processes of user engagement for developing community indicators measuring systems. The emphasis of the tools is on geographic information and other software-based systems. Each of the three chapters in the section incorporates a case study. In the first selection, Naomi Oliver, Christiana Schumann and Marc T. Smith discuss the Tampa Bay Community Atlas Program as an innovative example of community data systems. Additionally, various aspects of data collection and management are discussed.

---

[2] See www.urban.org/nnip for additional information.

Rina Ghose and William Huxhold use the case study of the Lindsay Heights area in the City of Milwaukee to illustrate development of neighborhood level indicators. The authors show that using different levels of geographic analysis is useful in assessing quality of life aspects of the area. The chapter also discusses responses from stakeholders and policy implications of the indicator study process.

Jeff Carmichael, Sonia Talwar, James Tansey and John Robinson focus on three aspects of community indicators systems: making them more integrated with all dimensions of complex social, economic, and environmental conditions; encouraging dynamism through the use of scenarios; and involving stakeholders in meaningful participation. The authors present a case study of the Georgia Basin Future Project, a regional assessment project to charting the future of this western Canadian region over the next forty years. The GB-QUEST software is used to illustrate applications in the Georgia Basin.

The selections in this book show the multifaceted and complex issues that surround community indicators measuring systems. It is the author's hope that the information contained in this book will help planners in Canada and the US to develop effective systems in their own communities.

Acknowledgements

First and foremost, each chapter author deserves a large measure of gratitude – without their work, this book would not exist. Next, it is highly unlikely it would have been completed without the capable assistance of Katie Habgood, graduate research assistant extraordinaire. The College of Design, Construction and Planning at the University of Florida is to be especially thanked for their generous support for research funding. The Canadian Consulate's Research Program for Faculty has been invaluable through the long process of learning and exploring this topic. They deserve accolades for wonderful support of US faculty to learn more about Canada. Finally, Melodye E. Gaskin's and Michelle Sipe's help with editing are to be noted with much appreciation.

SECTION I
THEORY AND BASIS FOR COMMUNITY INDICATORS

Chapter 1

Understanding Indicators

Heidi Hoernig and Mark Seasons

Introduction

Today's urban and regional communities struggle to address two urgent, related and often conflicting concerns. On one hand, they recognize that their economic and social health relies upon their ability to compete in the dynamic global market. On the other, the growing appreciation and fear for overwhelming societal and ecological costs of the industrial age, such as urban sprawl, deteriorating air and water quality and climate change, push communities to consider their quality of life and their ability to enhance or sustain it. These concerns, set within the context of pervasive public sector reform and the transforming knowledge-intensive economy, demand a greater role for knowledge management (Pilichowski, 2001) that during the past decade has increasingly included the monitoring of community indicators.

At the same time, the technical capacity of local and regional governments to track information has advanced greatly. The last two decades have seen the introduction of geographic information systems (GIS) and the automation of administrative data collection and management in most local authorities, as well as an accompanying shift in staff technical education and skills (Kingsley, 1998). As a result, planners have an expanding set of tools and data sources to facilitate the collection and manipulation of monitoring data. These two trends, the greater sophistication in organizational and professional capacity combined with the increased societal expectation that governments develop and use this capacity to demonstrate their effectiveness, facilitate the increased use of indicators and monitoring systems.

Indicators can greatly enhance decision-making at the policy-making and implementation levels. However, planners need to understand the strengths and limitations of indicators and the requirements of monitoring systems that are both sustainable over time and directly useful in practice. This chapter will explore several dimensions of community indicators through two sections. The first section, concepts and issues, examines definitions and functions of indicators, their relationship to monitoring and evaluation and the challenges of monitoring and evaluation for community planning practice. The second section explores the implications of these concepts and issues for practice and discusses issues related to the role of indicators in practice, establishing causality, participation in indicators development, and issues in choosing indicators and managing data.

Section I: Concepts and Issues

Indicators

An indicator is a multi-faceted concept. The verb 'to indicate' is defined by the Oxford dictionary as 'to point out; make known, be a sign of, show the presence of, state briefly and give reading or measurement' (Thompson, 1993, p.449). Innes (1990, p.5) defines an indicator as 'simply a set of rules for gathering and organizing data so they can be assigned meaning.' Many discussions of indicators use analogies to illustrate the concept. For example, the Sustainable Seattle report (Sustainable Seattle, 1995, p.3) likens indicators to 'gauges and dials of an aircraft's instrument panel'. Indicators can also be seen to fit in the 'information pyramid' model of information (see Figure 1.1) in which indicators are distinct from data (Hammond et al., 1996). The literature describing indicators frequently defines them in terms of function. It describes seven primary, overlapping functions: description, simplification, measurement, trend identification, clarification, communication and catalyst for action, as summarized in Table 1.1. Indicators are either derived from data, or representative of larger sets of data. Figure 1.2 shows the place of indicators within the information chain.

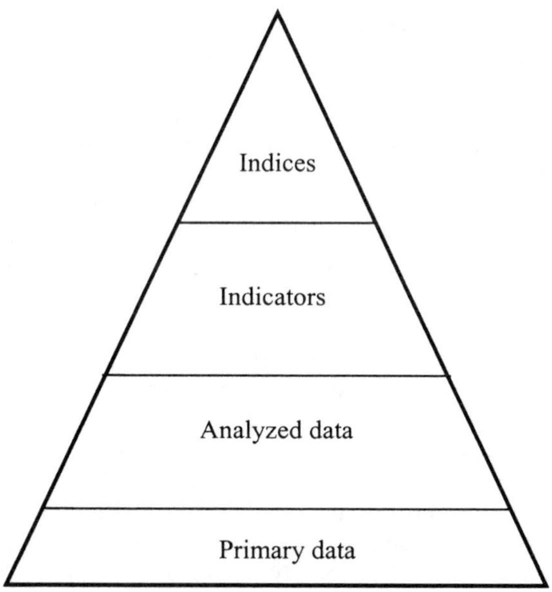

Figure 1.1 The Information Pyramid

Source: Hammond et al., 1996, p.1.

Table 1.1 Functions of Indicators

Function	**Description**	**Answers the question ...**
Description	Describe conditions or problems Increase general understanding	'What are things like?'
Simplification	Simplify complexity; provide a representative picture with significance extending to a larger phenomena of interest	'What's the big picture?'
Measurement	Measure characteristics of quality of life; measure performance of activities or services	'How much?'
Trend Identification	Establish baseline data; identify trends or patterns; show direction, improvement, disintegration, plateaus Two types: 1) Past-orientation. Indicators are chosen in light of their 'historical trend-identification properties' (MacLaren 1996, p.9) i.e. showing how dimensions of a identified phenomenon have been changing; Future-orientation. The indicator is a 'forward-looking instrument' (MacLaren 1996, p.9), used as a predictive, forecasting device.	'How did we do?' 'Where are we headed?'
Clarification	Clarify analytical issues or long-term goals; highlight areas of concern or improvement	'What's most important?'
Communication	Translate data into terms understandable by wide range of users	'How do we explain ...?'
Catalyst for Action	Stimulate public, stakeholder and political awareness, as well as interest and will to work towards change	'What next?'

Sources: Obst, 2000; International Institute of Sustainable Development, 2000; Environment Canada, 1999; Briggs, 1998; Organisation for Economic Cooperation and Development (OECD), 1998; Mitchell, 1997; Maclaren, 1996a; Maclaren, 1996b; Sustainable Seattle, 1995; Hammond, Adriaanse, Rodenburg, Bryant, and Woodward, 1996; Stewart, Belgue, Bond, L'Anglais, and Turcotte, 1975; Environment Canada and Canada Mortgage and Housing Corporation, 1999.

Indices are the aggregate forms of indicators. They are useful for providing an overall picture, but they also present special challenges of their own. During the process of generalization of data into aggregated measures, details are lost, problems masked and improvements diluted. In this way, indices can mislead policy makers from examining the complexity and subtleties of problems (Higgs and White, 2000).

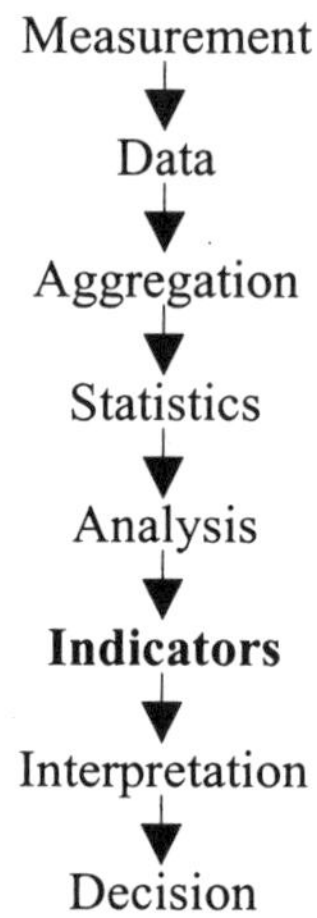

Figure 1.2 The Place of Indicators in the Information Chain

Source: Briggs, 1998 p.99.

Given the preceding overview of definitions and functions of indicators, several nuances must be kept in mind. First, indicators have been built on the positivist model of knowledge definition, production and use. In this model, knowledge production is seen as separate from its use (Innes, 1990, p.3). This has had significant implications to the policy relevance of indicators over the history of their use. Admittedly, policy relevant knowledge must be credible and stand up to criticism. To this end, the separation of data collection from users can help ensure data integrity. However, the complete separation of knowledge production and use is not necessarily an ideal situation. Indicators must be directly relevant or useful in some way to policymakers and other potential users. If not, they will not be used at all (Hammond et al., 1996). In the past, the perceived irrelevance or impracticality of indicators has contributed to their abandonment (Innes, 1990). Indicators become powerful only when they are used and when decision-makers come to anticipate and depend upon them. As such, the challenge to indicator development is simultaneously to satisfy the dual requirements of validity and usefulness.

The second aspect of concern relates to meaning attributed to collected data. Innes' (1990) simple definition of indicators as 'a set of rules' reminds us that,

> An indicator is a construct ... It is shorthand so that we can describe, discuss or manipulate our ideas about the world. It enables us to organize empirical observation, connect it with ideas and give them substance (Innes de Neufville, 1976, pp.3-4).

This social constructionist perspective on indicators illustrates that,

> [a]n indicator exists at an intersection of fact and theory. On the one hand, it purports to be an observation of fact, but on the other, it is a construct or hypothesis about what the fact looks like (Innes de Neufville, 1976, p.2).

This is a critical characteristic when applying indicators to policy making, implementation and evaluation. For example, when the gross domestic product (GDP) of a country rises or falls, it is both a statement of observed phenomenon in the world, but it is also indicative of the *theory* that changing GDP is an important reflection of the behaviour of national economic activity. In this way, '[indicators] imply a model or set of assumptions that relates the indicator to more complex phenomena' (Hammond et al., 1996, p.1). Because they are socially constructed, one may anticipate debate and disagreement over what indicators illustrate.

Disagreement over what indicators 'mean' is a debate about the validity of indicators. Indicator validity is established through the models and assumptions used to select and develop the indicator, and it extends through to data collection, analysis, interpretation and presentation. De Neufville emphasizes that the user is an integral part of indicator validation. She cautions, 'do not use the indicator in a disembodied way as if it were a thing unto itself. Recognize that an indicator is both a construct and a conceptual structure' (Innes de Neufville, 1976, p.37). She further asserts that

> [validation] is a critical part of exposing the meaning of indicators and making communication possible among people and groups who are part of the policy making. Validation of indicators depends on analysis, judgment, intuition, professionally acquired understandings and skills, and feedback from applications and social processes (Innes de Neufville, 1976, p.39).

Despite Innes' warning, indicators are frequently described as entities unto themselves. What is often implied, but not explicitly stated, is that indicators fit within some kind of monitoring strategy. Furthermore, whatever claims may be made about the capacity of indicators to perform any of the functions described in Table 1.1, indicators are not a stand-alone tool (Organisation for Economic Cooperation and Development (OECD), 1998). They require an appropriate interpretative context and they must be supplemented by other tools. Finally, the link between knowledge and action, be it for policy-making, implementation or any other decision-making process, is not automatic. In order for indicators or indeed, any monitoring or evaluative information to inform decision-making, many conditions must be met. The balance of this chapter will address these conditions.

Indicators and Monitoring

Monitoring generally refers to a definition of 'regular surveillance' (Thompson, 1993). It is a process of making systematic observations and measurements, to identify trends and patterns (Nelson and Serafin, 1994). This process begins with the establishment of baseline data, against which observations are plotted over time in order to track change. An information management system is central to a solid monitoring program in order to set up a strategy for defining, collecting, managing, analyzing and presenting information in a feasible, useful way (Patton, 1986; Rogers, 1990; Rogers, 1982).

An important distinction can be made between the monitoring of outputs and impacts or outcomes. Outputs are 'the goods, services or resources received by target groups and beneficiaries' (Mehta, 1994, p.71), whereas impacts or outcomes are the broader changes in attitudes, behaviour, development or action as a result of the intervention in question. Accordingly, recipients of outputs may be different from the beneficiaries of outcomes. For example, the output of a government's Official Plan growth management policy may be a growth management strategy, zoning by-laws or some other implementation tools which will have consequences for the development industry. The outcome of the policy could then have a broader set of effects on a municipality's environmental, social and economic character with impact on the entire community.

Table 1.2 Key Features of Monitoring Strategies

Feature	Description
Monitoring framework	The framework includes the goals and objectives of the monitoring strategy, the system or phenomena under surveillance and the participants and the users of the strategy.
Conceptual model	This model may be implicit or explicit. It provides an understanding of the relationships and networks of the system under surveillance.
Data collection and management	A set of feasible procedures and methods for addressing the practical and technical issues of data collection and management.
Data analysis and synthesis	A set of procedures and methods for analyzing and synthesizing data.
Indicators	Representative data that highlight key characteristics of the phenomena under surveillance.
Reporting strategy	Appropriate content, media, timing and target audience for presentation and reporting of results of monitoring activities.
Critical reflection	Critical reflection on the effectiveness of the strategy to provide the desired information to the users in order to affect decision-making.

Source: Adapted from Abbot and Guijt, 1998; Boyle, Kay, and Pond, 2001.

Several scholars identify key features of monitoring activity, which are summarized below in Table 1.2. These features are important considerations in the application of indicators to decision-making. The monitoring framework is the logic that links the strategy's goals and objectives, the object(s) of surveillance, the participants in the monitoring activities and the ultimate users of the information generated from the exercise. A holistic understanding of these relationships is important in order to ensure that such information can be incorporated into decision-making. The conceptual model that underlies the monitoring strategy is the set of assumptions and theories used to choose what data will be collected on the phenomena of interest. Frequently, monitoring is in place to improve this model; therefore, the model and monitoring strategy may be dynamic. Data collection, management, analysis and synthesis include sets of procedures used to make sense of the data and to generate indicators, which are those representative data that highlight key features of the phenomena under surveillance. The reporting strategy outlines the procedures through which to link information with decision-makers. Critical reflection is an important step, particularly in situations where the phenomena under study is dynamic or not well understood. Several of these features will be discussed in greater detail in the section examining implications for planning practice.

Indicators and Evaluation

A third key concept related to the topic of indicators is evaluation. There is considerable overlap in many descriptions of monitoring and evaluation. In the literature, the distinction between the two is not always clear and understandably so. The interpretation of monitoring results is an evaluative activity even when it occurs outside a formal evaluation process. The difference lies in the questions each activity answers. Monitoring collects and produces information about a general phenomenon or the implementation of the policy, program, or project. In contrast, evaluation analyzes information about the performance of the intervention in reference to a problem it was designed to address. In other words, monitoring answers the question: 'What happened, how and why?' Evaluation answers the question: 'What difference does it make?' (Mehta, 1994, p.77). Because answers to the latter question are needed to explain or justify the outcomes of planning policy and implementation, evaluation is an important complement to monitoring.

Evaluation stems from three basic goals of a public sector organization: (1) to increase knowledge; (2) to guide program or policy development and implementation; (3) and to provide accountability (Chelimsky, 1997; Mayne and Hudson, 1992). It forms the critical final link of any policy or program development cycle. Ideally, it draws together all previous stages of the development process and builds towards the beginning of another, improved cycle. Evaluation is the 'systematic gathering, analyzing and reporting of information about a program, service or intervention for use in making decisions' (Mayne and Hudson, 1992, p.2). It includes a historical analysis of the intervention (Rossi, Freeman and Lipsey, 1999), compares results to an explicit set of standards (Weiss 1984), and seeks 'reliability and usability of findings' (Organisation for Economic Cooperation and Development, 1999).

A formal and comprehensive evaluation of an entire planning policy document, (e.g. community plan, program or growth management strategy) would probably not be feasible for most local authorities. However, there are many lessons from evaluation practice that are relevant to the discussion of the effective application of monitoring to planning practice. As applied research, evaluation emphasizes the features and processes that facilitate the adoption of evaluation findings to the program or policy in question. These same elements can be used to promote the effective use of indicators in planning practice.

The evaluation literature describes several key preconditions. First, the plan or policy must be evaluable (Rutman, 1980), and it should be important enough to warrant the effort and resources required for the evaluation. Second, there must be the sufficient resources, time and skilled human resources to conduct the exercise (Guba and Lincoln, 1989; Rossi et al., 1999). Third, political and administrative support (OECD, 1999), sometimes aided by a strong champion (Seasons, 2003), are needed to conduct the evaluative activities and to carry through the recommendations provided by the results. Once these conditions are in place, an effective framework needs to be established. This framework is created by determining the type of evaluation used, the type of evaluator, the participation of key players and the process through which the evaluation will follow (OECD, 1999).

With a framework in place, a number of strategic questions must be addressed. First, the context of the evaluation must consider policy or program users, recipients, stakeholders and the users of the evaluation results. Second, the timing of the evaluation must be appropriate to the budget, program or decision-making cycle of the evaluation clients and stakeholders. Third, the research questions must be relevant in scope and topic, addressing key issues or problem areas (Rossi et al., 1999). Fourth, the data collection and analysis methods must suit the research questions and ensure adequate reliability and validity of data. At the same time, they need to be appropriate to the time, data, resources and skills available to the exercise (Julian, 1997). Sixth, there must be careful consideration given to the presentation of results, to ensure they are communicated to the right people, at the right time, in the right way. Presentations should be direct, focused, action-oriented, appropriate to the audience and use several media forms and styles (Hendricks, 1994).

Several key factors contribute to the use of evaluation findings. The encouragement and facilitation of key stakeholders' participation early on in the evaluation helps to secure their support when findings are presented later on (Weiss, 1998). Their input can be used to identify key issues and problem areas and to clarify both the internal and external context of organization, policy or program. Communication with the evaluation clients throughout the evaluation process keeps the research close to the original goals and allows adjustments to be made if necessary (Winberg, 1986). This on-going communication can also help to prepare clients for future findings, so that they may already be prepared to take action before the evaluation is even complete. Effective presentation of results uses an appropriate tone, emphasizing the strengths of the intervention and feasible recommendations for change (Hendricks, 1994).

These same considerations can be made to promote the effective use of monitoring results in planning practice by establishing facilitative preconditions, developing a monitoring framework, addressing key strategic questions, ensuring meaningful participation of key stakeholders and presenting monitoring results in a constructive fashion.

Evaluative Approaches to Monitoring Indicators

The development and application of indicators in monitoring has expanded greatly during the past decade. A wide range of organizations and government agencies have explored and applied indicators using a variety of approaches. Many of these approaches are useful to planning practice in its many dimensions. It is important to identify the conceptual frameworks behind these approaches and their implications for indicator use. Table 1.3 compares four evaluative perspectives used in monitoring indicators with relevance to planning practice. In essence, these evaluative perspectives are critiques to which planning professionals must respond, particularly those working in the public sector.

The first perspective in Table 1.3, the evaluation of community, is comprised of a body of approaches, which have coalesced around growing concerns about the current social, economic and environmental state of society. It questions the roles and actions of government and it examines the ability of government to address and balance civil society, environmental, and emancipatory goals such as social justice and environmental stewardship. There are numerous approaches related to this body of critique. Some of the better-known ones are grounded in sustainable development (World Commission on Environment and Development, 1987), the environmental movement, the public health movement and community development. A wide range of approaches to monitoring and evaluation are used in these perspectives. Key concepts in these approaches include equity, the socio-ecological model of health (Hancock, 1993; Fine and Rai, 1997; Swift, 1999), ecological integrity (Mitchell, 1997), systems (Kay, Regier, Francis and Boyle, 1999) or holistic approaches (Mitchell, 1997), participatory democracy (Zimmerman, 1986) and capacity building (Chambers, 1997). Monitoring and evaluation from these perspectives emphasize a holistic and community-based approach to measuring progress and the synthesis of social, economic and environmental goals at a community or higher order level. Monitoring systems in these approaches are frequently highly participatory and collaborative. Many community level indicator and monitoring initiatives have arisen in response to this critique (e.g. sustainability, quality of life and healthy communities indicator projects). While they draw from traditional social, economic and environmental indicators, these approaches have sought to develop indicators that can monitor agents, causes and processes of change that occur due to relationships that exist between sectors.

Table 1.3 Evaluative Perspectives Influencing/Relevant to Municipal and Regional Planning Practice

	Evaluation of Community	Evaluation of Government	Evaluation of Organization	Evaluation of Practice
Level of analysis/ focus of critique	Holistic definition of community at local, regional, or national level	Public sector	Organization	Professional municipal and regional planning practice
Key concepts	Sustainable development, quality of life, healthy communities	Performance measurement, benchmarking, New Public Management	Learning organisation, quality of work life, culture of evaluation, life-long learning	Variable: co-ordinated and integrated physical and social infrastructure, sustainable development, quality of life, new urbanism
Evaluator/ 'Client'	External: Civil society	External: Politicians, Public Internal: Senior management	Internal: Management-employee collaboration	Internal/External
Evaluation criteria	Sustainability, intergenerational equity, social and economic equity, ecological integrity, public health	Accountability, transparency, economy, efficiency, effectiveness, equity	Continuous improvement, social learning, adaptability, responsiveness, effectiveness	Land use compatibility, accessibility, connectivity, infrastructure effectiveness and efficiency, plan conformance
Monitoring systems	Community indicator projects, community report cards	Performance plans, Total Quality Management, Management by Objectives, Program Planning and Budgeting Systems	Performance plans, Total Quality Management	Five/ten year policy reviews, project, policy or program specific monitoring
Type of indicators	Environmental, social and economic. Emphasis on linkages, interaction and integration	Productivity, performance, service, cost-benefit	Productivity, performance, service	---

Source: Adapted from Hoernig, 2001.

The second theme, the evaluation of government, concerns public sector efficiency and effectiveness. This perspective has a strong financial orientation, with roots in liberal and capitalist ideology. It questions the appropriate range, level, depth and mode of government involvement and intervention. Under political and public criticism from this perspective, public sector organizations are increasingly required to show that budgets are balanced, resources are used efficiently and effectively, goals and objectives are reached, and some kind of identifiable product is produced or measurable effect is made (Chelimsky, 1997). Key concepts in this approach include efficiency (Downs and Larkey, 1986), benchmarking (Ammons, 1996 and 1999), productivity and performance measurement (Ball, 1998; Brinkerhoff and Dressler, 1990), performance management (Rogers, 1990) and public sector management (Plumptre, 1991). A strong emphasis on performance- and productivity-oriented monitoring and evaluation stems from this perspective. Several generations of monitoring systems have developed from this perspective, primarily in the private sector. Most recently, New Public Management (NPM) and Total Quality Management (TQM) have been used in the public sector.

The third perspective, the evaluation of the organization, can be categorized as a more internally critical perspective. Here the emphasis is on learning and the improvement of the organization from a self-reflexive position, rather than an external, accountability-seeking one. This body of approaches has its roots in management sciences, organizational behaviour and social psychology, with key concepts such as social learning (Hirschman, 1963), self-evaluating organization (Wildavsky, 1972), quality of the working life (Emery, 1976; Trist 1978), organizational learning (Argyris and Schoen, 1978) action research (Trist, 1981; Wright and Morley, 1989) and the reflective practitioner (Schoen, 1983). Here too there is an emphasis on performance and productivity, but the evaluative context is internally driven, participatory and learning-oriented. Many of these ideas have been incorporated into the monitoring systems such as NPM and TQM.

The final perspective, the evaluation of the planning profession, is perhaps the least developed of the four approaches. While the evaluation of planning practice is pervasive throughout academic, professional and media discourse, this has yet to lead to the systematic development of indicators of planning practice. Key evaluative concepts vary with the type of planning activity taking place. Many community plans proclaim ideals of sustainable development or quality of life. Basic evaluative criteria such as land use compatibility and accessibility are relatively standard concepts although their operationalization varies considerably across jurisdictions. While some formal review systems do exist,[1] they have not lead to the widespread development and application of monitoring systems specific to planning activities.

---

[1] e.g. Official Plan reviews are required at least every five years by the Planning Act in the province of Ontario, Canada. In the U.K., the Local Government Act 1999 requires local authorities, including planning organizations, to report upon best value performance indicators, as specified by the Audit Commission (Office of the Deputy Prime Minister, 2000).

Taken together, these four perspectives present a broad spectrum of approaches from which increased monitoring and evaluation of planning is possible. There are numerous challenges to monitoring and evaluation, discussed in the following section. Planners' choice of either a general or particular conceptual approach to monitoring and evaluation will depend upon many contextual factors. The role that monitoring can play in practice will also determine the monitoring approach taken. This will be touched upon in the section addressing implications for practice.

Challenges to Monitoring and Evaluation in Community Planning Practice

Ultimately, the application of community monitoring to planning practice concerns the evaluation of the impacts and outcomes of planning. Planners are interested in monitoring their communities because they want to develop responsive policy and implementation tools. In order to more effectively do this, planners need to be able to assess their work and its effects. Critics note that planners seldom conduct this type of evaluative activity (Baer, 1997; Talen, 1996). They point out that planners rarely determine whether or not their plans are ever actually implemented or whether or not the implementation is successful, effective or efficient.

Talen proposes that plan evaluation should perform two important roles in planning. First, it should ensure that 'planners, decision-makers and communities in general are aware of the degree to which goals have been implemented, changed or abandoned' (Talen, 1996, p.257). Second, explicit evaluation criteria can help ensure that plans address the dynamic nature of the planning environment. In this way, as goals are incorporated into plans, planners would better design the implementation of these goals if they consider, at the same time, how their implementation will be monitored and evaluated. Baer (1997, p.342) reiterates this point, stating 'focusing on plan evaluation criteria ... should allow planners to deal better with the naive enthusiasm that frequently gets incorporated into state mandates'. Therefore, as Baer (1997, p.342) later stresses, 'planners should devise their own criteria in the course of making the plan. A planner's skill should include the ability to formulate [evaluative] criteria as well as to prepare goals and objectives for a plan'. Implicit in their promotion of evaluation is the need to develop systems to monitor plan implementation.

The critique of Talen and Baer is by no means new to planning research and literature. The need to monitor and evaluate has long been recognized and explored (Seasons, 2003; Talen, 1996). However, there are numerous challenges to these endeavours due to the nature of the planning exercise. First, planning sets long-term goals, often on a 20-year horizon. The discounting period – the lifetime of the plan or policy, during which benefits will be realized and costs borne – must be considered during the policy's monitoring and evaluation. This raises the questions of how to monitor a long-term plan and at what point to evaluate whether or not it has achieved it goals.

Second, the heavy spatial focus of land-use planning makes monitoring and evaluation difficult (Bracken, 1981; Talen, 1996), as for example, in the measurement of accessibility or spatial interactions between various urban activities (e.g. accessibility to services or employment opportunities). Spatial or social impacts measurement highlights the problem of scale and sphere. Planning occurs in and has impact at many levels: city, community, neighbourhood and street (Bracken, 1981), as well as economic, social and environmental spheres. Disaggregated indicators are needed to understand the impacts of plans and policies on communities and individuals. At the same time, standard citywide aggregated indicators are also necessary to monitor the planning exercise as a whole. The monitoring and evaluation of planning must consider collection and analysis at these different levels and across different spheres, and be prepared to yield and address contradictory results. At the same time, 'what approaches are there that can be open to new information and yet not allow us to be overwhelmed by the infinity of information available?' (Wedgwood-Oppenheim, 1982, p.102).

This quotation highlights a key problem in any monitoring and evaluation exercise. Because of the broad scope of planning issues, this question of balance between inclusion and feasibility is particularly critical. It is an acute problem for general-purpose public organizations such as local authorities that have broad, generalized concerns for an area (Wedgwood-Oppenheim, 1982). Information acquisition also confronts data availability, quality and access problems (Hambleton and Thomas, 1995). These are important issues for all municipalities: local and regional, urban and rural, large and small.

Bracken (1981) also asserts that the benefits and costs to those affected by planning policies should be measured based on 'evidence of people's behaviour rather from their stated preferences' (Bracken, 1981, p.75). This is critical because social science research shows that what people say they do, is frequently different from what they do (Foddy, 1993). It may also be extremely difficult to establish cause-and-effect relationships between the impacts of plans or policies on people's behaviour. As Bracken (1981, p.81) points out,

> ... while it is fairly easy to generate quantitative indicators, which can measure the overall performance of an urban structure (such as density), it is far more difficult to generate socially meaningful indicators. For this requires an in-depth knowledge of the preferences, needs and priorities of different groups and the effects of policies upon them.

There is also the question of additional, irrelevant or invisible impacts (Hambleton et al., 1995). Additional impacts are unintended effects and can be positive or negative, direct or indirect. An irrelevant impact is 'deadweight', one which 'simply reinforced what would have happened anyway' (Hambleton et al., 1995, p.13). Impacts can also be invisible. Planning is often about preventing inappropriate development 'therefore, for planners, implementation (of strategies or land use plans), might be measured in what has not been built and in what has been conserved, protected, altered or improved' (Greed, 1996, p.2).

The measurement of impact also must consider the plurality of perceptions and experiences (Hambleton et al., 1995). Different groups throughout a community experience the outcomes of planning in different ways. Plurality requires appropriate monitoring and evaluation methods and processes to capture the spectrum of these experiences. Plurality also raises questions regarding the 'meaning of success' (Talen, 1996, p.254) and the criteria by which a plan is deemed successful (Faludi, 2000)

Talen questions 'the ability of planning to effectuate change' (1996, p.254). Given the many factors involved in the creation of urban spaces – political, economic, geographic, environmental, social, and cultural – how does one investigate the degree of impact of a plan *per se*? (Talen, 1996; Hambleton et al., 1995). Ideally, planning choreographs the co-ordination of development. Yet many organizations are involved in this dance. The ability of planning organizations to control these other actors is often limited (Faludi, 2000). This question about the extent of planners' control harkens back to the question of what is the role of planning and to theoretical questions of what planning is. Talen (1996, p.255) proposes that in many circumstances, it may not be necessary to firmly establish a cause-and-effect relationship between goals and outcomes of plans, but rather, drawing associations or connections would be sufficient. Here Talen treads on shaky methodological ground. Berke and Conroy (2000, p.31) support Talen's emphasis on performance evaluation and advocate 'focus on understanding the relationships within the continuum from state mandates (and incentives), to plans, to regulations, to standards, to outcomes'.

Faludi (2000) and others (Alexander and Faludi, 1989; Needham, Zwanikken, and Faludi, 1997) take a 'decision-centered view of planning', which focuses on interaction, negotiation and learning dimensions of the planning process. They argue that this position presents a 'more realistic' perspective which incorporates uncertainty and change. In this view, strategic plans should be monitored and judged by their ability to inform, guide and influence decisions, 'and not whether outcomes conform to intentions' (Faludi, 2000, p.316). In contrast, Minnery et al. (1993) observe that this emphasis on process has detracted from attention on the impact of planning policy and insist that planning evaluation should focus on 'congruence between policy objectives and policy impacts' (Minnery et al., 1993, p.13).

Seasons approaches monitoring and evaluation in planning from yet another perspective. He examines organizational realities i.e. 'the factors that facilitate or obstruct evaluation in municipal planning' (Seasons, 2003, p.432), a dimension frequently ignored in academic exhortation of increased and improved evaluation of planning outcomes. In his review of Ontario regional planning practice, he finds that although monitoring and evaluation are common activities, in most cases they are not strongly supported by either resource allocation or organizational culture. He lists a number of factors which facilitate effective municipal policy evaluation (Seasons, 2003). These include: 1) financial and human resources; 2) administrative and political support; 3) an organizational culture attune to learning and improvement, as well as long-range planning; 4) evaluable plans and policies; 5) monitoring of key indicators; 6) well-defined rationale for evaluation; and 7) effective communication practices.

Looking more closely at monitoring itself, Wedgwood-Oppenheim (1982) observes three traditional approaches: 1) implementation monitoring, 2) impact monitoring and 3) strategic monitoring. He highlights the important connection between the type of monitoring used and the planning environment in which the monitoring occurs (see Table 1.4). He notes several inherent difficulties in the task of monitoring planning impacts. First, monitoring the gap between reality and planned target is based on the assumption that targets exist. Targets may be appropriate in the case of 'implementation monitoring', where the task is well defined and the environment, stable and predictable. However, when the context is too uncertain, or problems are difficult to operationalize, well-defined target setting may be inappropriate. Assuming an appropriate target does exist, a second key question concerns how planners determine what is a significant divergence between the target and realized outcome. Third, planning must frequently be called upon to deal with unforeseen, unintended impacts. Citing the two examples of 1) the negative social consequences of slum clearance policies of the 1960s and 2) the increase of traffic volume that follow expanded road capacity, Wedgwood-Oppenheim (1982, p.106) points out that '... one needs therefore to be concerned with increasing responsiveness to these unintended consequences and not just measuring the achievement of intentions'.

Table 1.4 Levels of Monitoring

Monitoring type	Environment	Task	Role of monitoring
Implementation monitoring	Stable, well-understood, controllable	Clearly defined field of interest, predictable outcome	To ensure that implementation takes place according to plan
Impact monitoring	Some uncertainty and change, controllable	Clearly defined field of interest, unpredictable outcomes	To monitor for unintended outcomes To allow adjustments to be made to re-align implementation with original goals and objectives To check accuracy of forecasts
Strategic monitoring	Dynamic, uncertain, uncontrollable, imperfectly understood	Unrestricted field of interest, unpredictable outcomes	To assist in the response to new issues and in the creation of new policy tools To assess shifting policy landscape

Source: Adapted from Wedgwood-Oppenheim, 1982, pp.103-104.

Murtagh (1998) presents an approach to monitoring urban policy, which addresses many of Wedgwood-Oppenheim's observations. He presents a three-tiered approach, in which a community audit is conducted in order to establish baseline conditions. He then proposes that monitoring take place at macro, policy and delivery levels. The macro level provides a community-wide assessment, of trends, issues, needs, strengths, opportunities, weaknesses and constraints. Indicators are developed at this level. At the policy level, there is an assessment of the achievement of policy objectives, which focuses on outputs and also outcomes e.g. 'policy synergy', that is the ability of the policy to complement related policy. The third level, delivery, focuses on the community's experiences with the urban policy initiative.

Several challenges remain, even with Murtagh's comprehensive monitoring approach. Monitoring must be viewed as a full process, which includes the careful selection and definition of data to be collected, the data collection, management and analysis and their reporting and distribution. A strategic approach to reporting is necessary 'to ensure that information is well presented in a comprehensible manner to the right people at the right time' (Wedgwood-Oppenheim, 1982, p.108). Too frequently, monitoring strategies focus only on part of the process. As a result, the full benefits from the process are not realized. Or worse yet, the whole process turns into a tremendous waste of resources because the expected benefits are not forthcoming and the entire exercise is abandoned (Hoernig and Seasons, 2004).

Monitoring must also achieve a balance. On one hand, long-term monitoring requires baseline data and measurement over periods of time in order for trends and patterns to be captured and understood. On the other hand, monitoring may reveal that the original conceptual model of the system under surveillance is faulty, requiring changes to the monitoring strategy, which then affects the long-term value of the monitoring strategy itself.

Monitoring must also be able to cope with shifting circumstances and priorities, which include economic, social and political contingencies. The challenge therefore, is to develop a dynamic, responsive strategy, which has long-term benefits. There must be balance between adaptability and flexibility of a monitoring system and stability and permanence.

Monitoring attempts to apply the principles of rationality and yet is swayed and pulled in many different directions by political and social realities. It tries to construct a logical argument, which clearly demonstrates where policy works, where it fails and where improvements can be made. And yet the ground on which monitoring sits, the targets it attempts to observe, the tools by which it measures and even its observers and measurers, are all shifting by a myriad of social, economic, political, environmental and technological changes. The challenge for planners is how to harness monitoring to assist in the achievement of planning goals, a challenge well articulated by Wedgwood-Oppenheim (1982, p.102),

> Policy-making is concerned not so much with managing well-understood situations as with responding to new appreciations of the situation. Policy monitoring is concerned with the capacity to become aware that policies need to be changed because old models are no longer adequate. We need to be able to receive information that is not determined by the existing models, but which might cause them to be changed.

To summarize the preceding discussion, planners face many challenges in monitoring and evaluating their work. The extensive temporal and spatial scales involved in community planning present many possible points at which to collect information and assess results. Information acquisition and management is a challenge, in a field in which many variables and actors are involved. Other operational realities related to organizational culture, protocols and resources can greatly influence planners' ability to monitor and evaluate. Planning outcomes themselves can be unintended, additional, irrelevant or invisible. The diverse nature of communities ensures that a variety of expectations, values and experiences will affect how community members perceive and react to planning interventions. These elements highlight the contested nature of planning itself and the extent to which planners have control over their communities. Yet, these challenges to monitoring and evaluation merely reflect the complexity of the task of community planning and should not thwart planners' attempts to struggle with and improve upon their efforts to address these tensions.

Section II: Implications for Practice

This section addresses several practical issues related to incorporating the use of indicators into planning practice: 1) the role of indicators and monitoring in practice; 2) choosing indicators; 3) participation; 4) establishing causality; and 5) data management.

The Role of Indicators and Monitoring

Given the multiple functions of indicators and the several evaluative approaches that can be used to frame a monitoring strategy, community practitioners need to give consideration to the role of indicators in practice. The role will determine the types of indicators chosen and the course of action taken in their development. For example, if indicator use is to be undertaken as part of a 'legitimization' process of planning, to establish accountability, responsibility and effectiveness, then a performance measurement model might be adopted, and indicator development will follow with emphasis on the effective reporting of results.

However, if indicator use will involve a community exercise to address an issue related to community well-being, planners may draw more heavily from sustainability, quality of life or healthy communities models and work to develop strong partnership relationships with other organizations for data collection, management, analysis and reporting. As such, the role of indicators in will also determine what is needed to develop the monitoring strategy, including political and administrative support, the participants and partners, resources and skills. Table 1.5 summarizes several ways in which indicators and monitoring can be used in planning practice.

Table 1.5 Role of Indicators and Monitoring in Planning

Tool	Areas of use
Planning tool	*Land-use and servicing planning*: Infrastructure planning, development control, growth management. *Address regional concerns*: Affordable housing, downtown revitalization, water protection strategy, regional transportation.
Learning tool	*Policy review*: Assess appropriateness type and substance of policy, evaluate effectiveness and impact. *Human resource planning tool*: Link human resource planning with policy and program implementation through monitoring of inputs, outputs and outcomes. *Business plan development*: Link financial planning with policy and program implementation through monitoring of inputs, outputs and outcomes. *Benchmarking tool*: Use to compare with leading municipal practice. *Performance measurement tool*: Assess performance in core business practice through monitoring outputs and outcomes of policy and program implementation.
Communication tool	*Stakeholders*: Assist in collaboration with planning stakeholders. *Politicians*: Demonstrate effectiveness and value, justify budgeting. *Publics*: Report back to citizens, garner support, encourage feedback. *Future investors*: Attract investment and business.
Collaborative tool	*Mutual gain*: Assist in mutual problem definition and fact finding for shared concerns.

Source: Hoernig, 2001.

In determining the role of indicators, planners must also consider who will use indicators and for what purposes. Indicators developed for municipal planning could potentially be used by municipal urban planners, effected officials at all levels of government, citizen groups and other non-government organizations, interest/lobby groups, media, educators, private sector organizations and the general public. Although it may be difficult, if not impossible, to determine exactly who the users may be, it may be important to target certain kinds of users, in order to develop indicators which are understandable, poignant and practical to the user group.

Choosing Indicators

Another important issue in indicator development is choice of indicators. Choice is determined by the role of indicators in practice, which in turn determines the type of indicators needed. Indicators can be categorized in several ways. One way is to examine the function of measurement, as proposed by the following categorization of indicators by opposing pairs (Maclaren, 1996a; Dilks, 1996, p.35):

- Predictive or forward looking vs. retrospective.
- Objective (easily quantified and measured) or subjective (based on values and perceptions).
- Input or output and outcome.
- Context-specific indicators vs. comparable/global/national/regional indicators.
- Process vs. outcome indicators.

A second method of classification is to examine the conceptual frameworks underlying the monitoring exercise. In Table 1.6, Maclaren categorizes indicators in relation to specific public sector interventions or interests.

Table 1.6 Frameworks Used to Organize Indicators

Type	**Categories**
Domain Based	Environment, Economy, Society
Sectoral	Housing, Welfare, Recreation, Transportation, Environment, Economic Development
Goal Based	Carrying Capacity, Basic Human Needs, Social Well-Being, Economic Prosperity, Participation in Government *etc.*
Issue Based	Urban Sprawl, Solid Waste Management, Crime and Safety, Job Creation, Industrial Pollution, *etc.*
Causal	Conditions, Stresses, Responses

Source: Maclaren, 1996a, p.46.

In Table 1.7, indicators are organized by evaluative approach into three main sets: conventional, integrative and performance indicators. The table presents a description of the three main sets by type, focus of indicator measurement, and the main features the indicator monitors. Most of the literature on indicators also describes a generic set of selection criteria of used to choose indicators, summarized in Table 1.8.

Table 1.7 Summary of Indicators Sets and Types

Indicator sets	Characteristics of indicator set	Indicator types	Focus of measurement	Main features Monitors ...
Conventional	Discrete, single sector, often single discipline approaches	Economic	Economy	Changes in market-value activities
		Social	Societal well-being	Social well-being through surrogate concepts
		Environmental	Ecological systems	Structural and functional features of ecosystems
Integrative	Holistic, multi-sectoral, multi- and inter-disciplinary approaches	Sustainability	Sustainability	Sustainability through the integrated accounting of environmental, economic, social development factors
		Healthy Cities	Health	Health through holistic/positive health models
		Quality of Life	Quality of life	Quality of life by examining social, health, economic and environmental conditions
Performance	Performance, production or financial approaches	Performance measurement	Performance of organization, department or program	Progress towards goals and objectives
		Benchmarks		Organizational practices against 'best in the class' organizations

Source: Hoernig and Seasons, 2004.

Table 1.8 Indicator Selection Criteria

Indicator Selection Criteria
Scientifically valid; based on accurate, reliable, available and accessible data that are comparable over time
Representative; public participation addressed in selection criteria
Responsive to changes in the environment and human activities
Controllable
Relevance: to stated goals, to policy, to needs of potential users
Understandable by potential users
Comparable to thresholds or targets
Comparable to other jurisdictions
Collectable and available
Cost-effective; reasonably cost efficient and involve reasonable staff effort
Unambiguous; clear
Attractive to media
Comprehensive; integrative of environmental, social and economic factors
Includes most disadvantaged; takes account of the distribution of equity
Timeliness of feedback: will managers get information before it becomes obsolete?

Sources: Jackson, 1988; Ministry of Municipal Affairs and Housing, 1999; Hatry et al., 1992; Organisation for Economic Cooperation and Development (OECD), 1998; Maclaren, 1996a; Maclaren, 1996b; Maclaren, 2001; Dilks, 1996.

Causality

Causality in the context of indicator use and evaluation refers to establishing cause and effect relationships to the trends or patterns identified by indicators. In many cases, the number of variables and influential factors involved in any given phenomena makes this task very difficult (Pal, 1997, p.237). Causality is an important issue because perceptions of causality influence the design, collection, interpretation, reporting and policy application of indicators. However, causality must be approached with caution. Indicators are, in many ways, *only* indicators. Often, they serve as proxy or surrogate measures. They often cannot directly measure the phenomenon of interest, or even if they do, they cannot definitively establish causality. This means that interpreters and decision-makers should be careful not to take indicators too literally. They must remember that indicators provide a pragmatic short cut to dealing with complexity (Indicators for Evaluation Task Force, 1996). Continuous, full-scale, in-depth research is rarely possible; indicators provide a feasible compromise. Therefore in many cases, indicators should serve as signals or flags and should be used to instigate further investigation and action, but not to make claims of absolute truth.

Participation

There are three larger philosophical questions underlying the topic of participation. The first concerns knowledge. The key question here is – what knowledge defines, develops, monitors, analyzes and reports on indicators? It arouses debate about 'what is legitimate knowledge'; 'who decides what is "true"'; and 'who is allowed to participate in "truth production"'. It also relates to ideas of how to make knowledge understandable, accessible and relevant to stakeholders and publics. The second theme relates to governance. Three questions arise from this theme. Who is involved in making decisions? Who is chosen to represent whom? To what degree should non-bureaucrats be involved in public sector decision-making? Governance also touches upon the representation of the variety of voices and interests, particularly those of the disadvantaged and disempowered. The third theme deals with empowerment. Empowerment provides the link between knowledge and action. It follows the premise that there is a direct correlation between ability to affect change and control one's life or the course of action in a community and, the individual or community's capacity to continue to influence decisions.

Participation in monitoring and evaluation is a natural extension of efforts in recent decades to expand participation of residents in the development of their communities (Abbot and Gujit, 1998) and in the development and implementation of community plans (Hodge, 1998). Already, extensive public participation is institutionalized in various implementation processes such as zone by-law changes and community plan amendments as well as many official plan visioning processes (Hodge, 1998).

There are several potential benefits of participation, which closely relate to indicators in their 'catalyst for action' function. A study of community environmental monitoring (CEM) of 200 CEM experiences in Australia (Alexandra et al., 1996 in Abbot and Guijt, 1998) documents several advantages realized by community participation in environmental monitoring. Greater participation increased community buy-in and helped to develop a communal sense of responsibility for the continuing environmental health of the communities monitored. It also provided greater access to a wider pool of data. This was facilitated greatly by information management systems. Lastly, community participation provided greater access to other resources needed to sustain monitoring activities.

Abbot and Guijt (1998) point out that participation is a dynamic process, which changes throughout cycles of intervention. There are different levels of participation from different groups within communities and many forms of participation, 'Each person experiences a unique combination of social, economic and physical constraints and opportunities that influence their willingness and capacity to participate' (Guijt and Shah, 1998).

Guijt and Shah also address the issue of plurality and caution against the oversimplification of concepts of 'community' or the 'public'. They observe that 'communities are neither homogeneous in composition and concerns, nor necessarily harmonious in their relations' (Guijt and Shah, 1998, p.8).

Social diversity within communities also highlights the need and potential benefit for participation in monitoring activities. All people experience and evaluate communities differently. Attention to the perspectives of groups such as seniors, youth, physically and developmentally challenged people, ethnic and racial minorities can alert planners to issues of populations that are often marginalized by mainstream planning processes. Similarly, indicator development can capitalize on local knowledge bases. Although this type of knowledge may be anecdotal or less formalized than 'expert' produced knowledge, it can be rich in information about local conditions, history, relationships and interactions (Chipeniuk, 1993). In communities with ethnoculturally diverse populations, the incorporation of more than one cultural group in indicator development would raise issues related to culturally based knowledge (Sandercock, 1998). Extensive work has been done in this area with First Nations communities (Fast and Berkes, 1994; Ryan, 1995) in community development and resource management with transferability to many areas of urban planning.

Data Management

Because planning is general purpose and multi-disciplinary in nature, indicator development and monitoring in planning require extensive data sharing with other organizations and levels of government. As such, data management is a central issue in indicator and monitoring use. Key concerns are manageability and cost-effectiveness. There are technical issues such as merging databases and differences in software (Murphey, 1999) as well as social, economic and political issues related to data sharing relationships, which involve establishing rapport, cooperation and collaboration.

In their discussion of 'interoperability', Landsbergen and Wolken's comprehensive account of data management and sharing issues relates well to data issues in indicator and monitoring use. They state, 'interoperability is more than "digital plumbing" – making sure that computers talk so that bits of data flow properly. Fundamentally, interoperability is people talking and sharing information' (Landsbergen Jr. and Wolken Jr., 2001, p.206). Their research on data exchange among federal agencies identifies a number of barriers to interoperability summarized in Table 1.9.

In order to surmount these barriers, planning organizations need to equip themselves with appropriate tools and to establish the necessary social, political, economic and electronic infrastructure through which to share information and reap the benefits of the exchange. Landsbergen Jr. and Wolken Jr. (2001, p.212) identify five key elements: 1) meta-data about the information they have; 2) appropriate legislation and policy; 3) economic and budgetary mechanisms; 4) success criteria by which to monitor and evaluate the information sharing; and 5) the managerial tools to provide incentives and controls.

Table 1.9 Barriers to Interoperability

Barrier	Description
Political	*Privacy*: Privacy rights and protection need to be clear. *Statutory Authority*: Clarification is also required to establish under what circumstances data can be shared and with whom. *Openness to Public Scrutiny*: Greater data sharing and accessibility increases the public's desire to acquire data and ability to scrutinize public organizations.
Organizational	*Trust*: Establishment of trust among organizations is required. *Lack of Experience and Awareness*: Many organizations lack the experience, institution memory and knowledge 'that reflects the benefits and costs of interoperability' and recognize opportunities when they arise.
Economic	*Lack of Resources*: Resource available as well as resource sharing arrangements will impact data sharing capacity. *'Low-bid' Procurement Methods*: These influence both hardware and software purchasing decisions, often with greater long-term costs.
Technical	*Hardware or Software Incompatibility*: *Public or Private Property*: This refers to the ownership of the software system. In the cases of contractor developed software, the contractor may not be required to tell the agency how to move data in and out of the system. *Data-Sharing Standards*: This includes differing data definitions.

Source: Adapted from Landsbergen and Wolken, 2001, pp.208-10.

Summary and Conclusions

To summarize the central problems and questions in monitoring and evaluation in community planning, Table 1.10 expands upon Murtagh's initial conceptualization, to include the challenges discussed in this chapter. As one of the final stages in the plan making process, monitoring and evaluation forces planners to think through the entire planning exercise, to ask what is the very purpose and nature of the tasks which they perform. Perhaps this is why monitoring and evaluation are often set aside, because practitioners are frequently forced to solve more immediate problems, rather than deeply consider the complex nature of planning for the heterogeneous and dynamic communities.

Table 1.10 Summary of Challenges and Questions in Monitoring and Evaluating Community Planning

Category	Elements
Theoretical issues	What is the role of plans?
	What ability do plans have to effect change?
	What is the function of the evaluation?
	What is the role of values?
	Who are the clients?
	What are the criteria of success? effectiveness, efficiency, equity?
Strategic issues	Timing (frequency, point in time)
	Level (street, neighbourhood, city, region) of measurement/analysis
	Establishing baseline community conditions
Definition and measurement issues	Defining targets, operationalizing problems
	Capturing plurality of impact, both perception and experience
	Tracking unintended impacts or invisible impacts (what has been protected, what has not been built)
	Translating policy objectives into measurable indicators, *i.e.*, accessibility, interaction indicators
	How to measure people's perception of impact versus impact to their behaviour, benefits and costs to people, establishing cause-and-effect relationships of plan policy on people and their behaviour
	Sphere (social, economic, environmental, spatial) of measurement and analysis
Data management	Data manageability and feasibility of monitoring
	Data availability, quality and access
	Data analysis and synthesis
	Understanding linkages and synergies
Process issues	Establishing supportive environment for monitoring and evaluation
	Capturing the impact of policy on community capacity through participation of a cross-section of community members
	Ensuring monitoring and evaluation becomes the basis of critical self reflection and learning

Source: Adapted from Murtagh, 1998, p.133.

Practitioners considering or developing the use of indicators in community planning must be aware of the strengths, challenges and limitations of monitoring. Summary Table 1.10 is an appropriate way to conclude this chapter because not only does it provide an overview of key elements, but it also highlights the involved nature of the task. Indicators are an important tool to assist planners in many ways: to identify and understand problems and issues, to track and assess progress towards finding solutions, to educate communities and to garner and facilitate community support. Further work and research in this area will help to make monitoring indicators a standard practice for community planning.

References

Abbot, J. and Guijt, I. (1998), *Changing Views on Change: Participatory Approaches to Monitoring the Environment* (Rep. No. SARL Discussion Paper No. 2), International Institute for Environment and Development, London.

Alexander, E.R. and Faludi, A. (1989), 'Planning and Plan Implementation: Notes on Evaluation Criteria', *Environment and Planning B: Planning and Design*, vol. 16, pp.127-40.

Ammons, D. (1996), *Municipal Benchmarks: Assessing Local Performance and Establishing Community Standards*, Sage Publications, Thousand Oaks.

Ammons, D. (1999), 'The Proper Mentality for Benchmarking', *Public Administration Review*, vol. 59, pp.105-9.

Argyris, C. and Schoen, D. (1978), *Organizational Learning: A Theory of Action Perspective*, Addison-Wesley Publishing Company, Reading.

Baer, W. (1997), 'General Plan Evaluation Criteria: An Approach to Making Better Plans', *Journal of American Planning Association*, vol. 63, pp.329-44.

Ball, R. (1998), *Performance Review in Local Government*, Ashgate Publishing Ltd, Aldershot.

Berke, P.R. and Conroy, M.M (2000). Are we planning for sustainable development? An evaluation of 30 comprehensive plans. Journal of the American Planning Association 66(1), 21-33.

Boyle, M., Kay, J. and Pond, B. (2001), 'Monitoring in Support of Policy: An Adaptive Ecosystem Approach', in T. Munn (ed.), *Encyclopedia of Global Environmental Change* (Volume 5: Social and Economic Dimensions of Global Environmental Change), John Wiley and Sons, New York.

Bracken, I. (1981), *Urban Planning Methods: Research and Policy Analysis*, Methuen, London.

Briggs, D. (1998), 'State of environment reporting', in B. Nath, L. Hens, P. Compton and N. Frumkin (eds), *Environmental Management in Practice*, vol. 1, pp.90-107, Routledge, London.

Brinkerhoff, R. and Dressler, D. (1990), *Productivity Measurement: A Guide for Managers and Evaluators*, Sage Publications, Thousand Oaks.

Chambers, R. (1997), *Whose Reality Counts? Putting the First Last*, Intermediate Technology Publications, London.

Chelimsky, E. (1997), 'The Coming Transformations in Evaluation', in E. Chelimsky and W. Shadish (eds), *Evaluation for the 21st Century*, pp.1-26, Sage Publications, Thousand Oaks.

Chipeniuk, R. (1993), 'Vernacular Bio-Indicators and Citizen Monitoring of Environmental Change', in G. Nelson and R.W.G. Butler (eds), *Tourism and Sustainable Development. Department of Geography Publication Series No. 37*, pp.269-78, University of Waterloo, Waterloo.

Dilks, D. (1996), 'Measuring Urban Sustainability: Canadian Indicators Workshop', prepared on behalf of the Canada Mortgage and Housing Corporation, Ottawa.

Downs, G. and Larkey, P. (1986), *The Search for Government Efficiency: From Hubris to Helplessness*, Temple University Press, Philadelphia.

Emery, F. (1976), *Futures We Are In*, Martinus Mijhoff, Leiden.

Environment Canada (1999), *Urban Air Quality*, Environment Canada, Ottawa.

Environment Canada and Canada Mortgage and Housing Corporation (1999), Sustainable Community Indicators Program, Environment Canada and Canada Mortgage and Housing Corporation, Ottawa.

Faludi, A. (2000), 'The Performance of Spatial Planning', *Planning Practice and Research*, vol. 15, pp.299-318.

Fast, H. and Berkes, F. (1994), *Native Land Use, Traditional Knowledge and the Subsistence Economy in the Hudson Bay Bioregion*, Natural Resources Institute, Winnipeg.

Fine, R. and Rai, S. (1997), *Civil Society: Democratic Perspectives*, Cass, London.

Foddy, W. (1993), *Constructing Questions for Interviews and Questionnaires*, Cambridge University Press, Cambridge.

Greed, C. (1996), 'Implementation: Perspectives and Perceptions', in C. Greed (ed.), *Implementing Town Planning: The Role of Town Planning in the Development Process*, pp.2-15, Longman, Harlow.

Guba, E.G. and Lincoln, Y.S. (1989), *Fourth Generation Evaluation*.

Guijt, I. and Shah, M.K. (1998), 'Waking Up to Power, Conflict and Process', in I. Guijt and M.K. Shah (eds), *The Myth of Community: Gender Issues in Participatory Development*, pp.1-24, Intermediate Technology Publications Ltd, London.

Hambleton, R. and Thomas, H. (1995), 'Urban Policy Evaluation – The Contours of the Debate', in R.Hambleton and H. Thomas (eds), *Urban Policy Evaluation: Challenge and Change*, pp.1-20, Paul Chapman Publishing Ltd, London.

Hammond, A., Adriaanse, A., Rodenburg, E., Bryant, D. and Woodward, R. (1996), 'Environmental Indicators: A Systematic Approach to Measuring and Reporting on Environmental Policy Performance in the Context of Sustainable Development', World Resources Institute, Washington DC.

Hancock, T. (1993), 'The Healthy City from Concept to Application: Implications for Research', in J. Davies and M. Kelly (eds), *Healthy Cities: Research and Practice*, Routledge, London.

Hatry, H., Blair, L., Fisk, D., Greiner, J., Hall Jr., J. and Schaenman, P. (1992), *How Effective are Your Community Services? Procedures for Measuring Their Quality* (second edn.), The Urban Institute and The International City/County Management Association, Washington DC.

Hendricks, M. (1994), 'Making a Splash: Reporting Evaluation Results Effectively', in J. Wholey, H. Hatry and K. Newcomer (eds), *Handbook of Practical Program Evaluation*, pp.549-75, Jossey-Bass, San Francisco.

Higgs, G. and White, S. (2000), 'Alternatives to Census-Based Indicators of Social Disadvantate in Rural Communities', *Progress in Planning*, vol. 15, pp.1-81.

Hirschman, A. (1963), *Journey Towards Progress: Studies in Economic Policy Making in Latin America*, Twentieth Century Fund, New York.

Hodge, G. (1998), *Planning Canadian Communities: An Introduction to the Principles, Practice and Participants* (third edn.), ITP Nelson, Toronto.

Hoernig, H. (2001), *The Role of Indicators in Municipal Planning Policy Evaluation*, Unpublished Master's thesis, Waterloo: University of Waterloo.

Hoernig, H. and Seasons, M. (2004), 'Monitoring Indicators in Local and Regional Planning Practice: Concepts and Issues', *Planning, Practice and Research*, vol. 19, no.1.

Indicators for Evaluation Task Force, I.J.C. (1996), *Indicators to Evaluate Progress Under the Great Lakes Water Quality Agreement*, International Joint Commission, Ottawa.

Innes de Neufville, J. (1976), *Validating Policy Indicators*, working paper no. 258, Institute of Urban and Regional Development, University of California, Berkeley.

Innes, J.E. (1990), *Knowledge and Public Policy: The Search for Meaningful Indicators*, (second edn.), Transaction Publishers, New Brunswick.

International Institute of Sustainable Development (2000), 'Measurement and Indicators for Sustainable Development', Sustainable Development, [On-line]. Available: www.iisd.ca/measure/faqcriteria.htm.

Jackson, P. (1988). 'The Management of Performance in the Public Sector', *Public Money and Management*, winter, pp.11-16.

Julian, D. (1997), 'The Utilization of the Logic Model as a System Level Planning and Evaluation Device', *Evaluation and Program Planning*, vol. 20, pp.251-7.

Kay, J., Regier, H., Francis, G. and Boyle, M. (1999), 'An Ecosystem Approach for Sustainability: Addressing the Challenge of Complexity', *Futures*, vol. 31, pp.721-42.

Kingsley, G.T. (1998), *Neighbourhood Indicators: Taking Advantage of the New Potential*, The Urban Institute, Washington DC.

Landsbergen Jr., D. and Wolken Jr., G. (2001), 'Realizing the Promise: Government Information Systems and the Fourth Generation of Information Technology', *Public Administration Review*, vol. 61, pp.206-20.

Maclaren, V. (1996a), *Developing Indicators of Urban Sustainability: A Focus on the Canadian Experience*, Intergovernmental Committee on Urban and Regional Research, Toronto.

Maclaren, V. (1996b), 'Urban Sustainability Reporting', *Journal of American Planning Association*, vol. 62, pp.184-202.

Maclaren, V. (2001), 'Blighted or Booming? An Evaluation of Community Indicators and Their Creation', *Canadian Journal of Urban Research*, vol. 10, pp.275-91.

Mayne, J. and Hudson, J. (1992), *Program Evaluation: An Overview*, Wall and Emerson, Toronto.

Mehta, V. (1994), *The Essential Guide to Policy Development and Analysis*, Policy Planning Directorate, Canada Research Institute, Ottawa.

Ministry of Municipal Affairs and Housing (1999), *Performance Monitoring in the Planning System: A Handbook of Municipally Tested Indicators*, Ministry of Municipal Affairs and Housing, Toronto.

Minnery, J., Cameron, C., Brown, J. and Newman, P. (1993), 'Evaluation in Urban Planning', *Australian Planner*, vol. 31, pp.8-13.

Mitchell, B. (1997), *Resource and Environmental Management*, Longman, Edinburgh.

Murphey, D. (1999), 'Presenting Community-Level Data in an 'Outcomes and Indicators' Framework: Lessons from Vermont's Experience', *Public Administration Review*, vol. 59, pp.76-82.

Murtagh, B. (1998), 'Evaluating the Community Impacts of Urban Policy', *Planning Practice and Research*, vol. 13, pp.129-38.

Needham, B., Zwanikken, T. and Faludi, A. (1997), 'Strategies for Improving the Performance of Planning', *Environment and Planning B: Planning and Design*, vol. 24, p.880.

Nelson, G. and Serafin, R. (1994), 'Improving Monitoring and Assessment for Environmental Decision-Making', in J. Andrey and G. Nelson (eds), *Public Issues: A*

Geographical Perspective, pp.391-414, Department of Geography, University of Waterloo, Waterloo.

Obst, C. (2000), 'Report of the September 1999 OECD Expert Workshop on the Measurement of Sustainable Development', in *Frameworks to Measure Sustainable Development*, pp.7-18, Organisation for Economic Cooperation and Development, Paris.

Office of the Deputy Prime Minister (2000), *Performance Indicators for 2000/01*, Office of the Deputy Prime Minister, London.

Organisation for Economic Cooperation and Development (OECD) (1998), *Towards Sustainable Development: Environmental Indicators*, Organisation for Economic Cooperation and Development, Paris.

Organisation for Economic Cooperation and Development (OECD) (1999), Improving Evaluation Practice. Best Practice Guidelines for Evaluation and Background Paper, Organisation for Economic Cooperation and Development, Paris.

Pal, L.A. (1997), Beyond Policy Analysis: Public Issue Management in Turbulent Times, ITP Nelson.

Patton, M.Q. (1986), *Utilization-Focused Evaluation*, Sage Publications, Thousand Oaks.

Pilichowski, E. (2001), *Knowledge Management: Learning-by-Comparing Experiences from Private Firms and Public Organisations*, Public Management Service and Centre for Educational Research and Innovation, OECD, Paris.

Plumptre, T. (1991), *Beyond the Bottom Line: Management in Government*, Institute for Research on Public Policy, Halifax.

Rogers, S. (1982), 'The Use of Information in Policy and Management Systems', in S. Leach and J. Stewart (eds), *Approaches in Public Policy*, pp.81-101, George Allen and Unwin, London.

Rogers, S. (1990), *Performance Management in Local Government*, Longman and the Local Government Training Board, Harlow.

Rossi, P.H., Freeman, H. and Lipsey, M. (1999), *Evaluation: A Systematic Approach*, (sixth edn.), Sage Publications, Thousand Oaks.

Rutman, L. (1980), *Planning Useful Evaluations: Evaluability Assessment*, Sage Publications, Thousand Oaks.

Ryan, J. (1995), *Doing Things the Right Way: Dene Traditional Justice in Lac La Martre, N.W.T.*, The University of Calgary Press and the Arctic Institute of North America, Calgary.

Sandercock, L. (1998), *Towards Cosmopolis: Planning for Multicultural Cities*, John Wiley and Sons, Chichester.

Schoen, D. (1983), *The Reflective Practitioner: How Professionals Think in Action*, Basic Books, New York.

Seasons, M. (2003), 'Monitoring and Evaluation in Municipal Planning: Considering the Realities', *Journal of the American Planning Association*, vol. 69, pp.430-40.

Stewart, J., Belgue, D., Bond, W., L'Anglais, O. and Turcotte, H. (1975), *Urban Indicators: Quality of Life Comparisons for Canadian Cities*, Ministry of State for Urban Affairs, Ottawa.

Sustainable Seattle (1995), *Indicators of Sustainable Community: A Report to Citizens on Long-Term Trends in our Community*, Sustainable Seattle, Seattle.

Swift, J. (1999), *Civil Society in Question*, Between the Lines, Toronto.

Talen, E. (1996), 'Do Plans Get Implemented? A Review of Evaluation in Planning', *Journal of Planning Education*, vol. 10, pp.1-11.

Thompson, D. Ed. (1993), *The Oxford Dictionary of Current English* (second edn.), Oxford University Press, Oxford.

Trist, E. (1978), 'Adapting to a Changing World', *Labour Gazette*, vol. 78, pp.14-20.

Trist, E. (1981), *The Evolution of Socio-Technical Systems: A Conceptual Framework and an Action Research Program*, Ontario Quality of Working Life Centre, Toronto.

Wedgwood-Oppenheim, F. (1982), 'Monitoring and the Planning Cycle', in S. Leach and J. Stewart (eds), *Approaches in Public Policy*, pp.102-19, George Allen and Unwin, London.

Weiss, C. (1984), 'Increasing the Likelihood of Influencing Decisions', in L. Rutman (ed.), *Evaluation Research Methods: A Basic Guide*, Sage Publications, Thousand Oaks.

Weiss, C. (1998), *Evaluation* (second edn.), Prentice Hall, New Jersey.

Wildavsky, A. (1972), 'The Self-Evaluating Organization', *Public Administration Review*, vol. 32, pp.509-20.

Winberg, A. (1986), 'A Phased Approach for Conducting Program Evaluations', *Canadian Journal of Program Evaluation*, vol. 1, pp.1-11.

World Commission on Environment and Development (1987), *Our Common Future*, Oxford University Press, Oxford.

Wright, S. and Morley, D. (1989), *Learning Works: Searching for Organizational Futures*, The Adapting by Learning Group, York University, Toronto.

Zimmerman, J. (1986), *Participatory Democracy: Populism Revived*, Praeger, New York.

Chapter 2

Historical Background of Community Indicators

Clifford Cobb and Craig Rixford

Introduction

In recent decades, dozens of communities in the United States and Canada have become convinced that they can improve the quality of their lives or the performance of their governments by measuring their status and changes in it. They have amassed and adjusted data sets describing social, environmental and economic conditions and called them 'community indicators'. Often, there has been little thought at the outset regarding how the data will be used. Rather than developing a theoretical framework that would explain the connection between indicators and action, the practitioners have hoped that statistical description of social problems would make possible intelligent reforms. They have treated the value of indicators as self-evident: more public awareness of social and economic conditions will produce better policies.[1]

Yet, the secrets of how to improve society are not so quick to reveal themselves. Merely knowing that there are more homeless people, suicidal teens, or contaminants in the water supply does not automatically lead to solutions. Nor does knowledge of a social pathology inevitably mean that the public or decision-makers will be concerned about it. There are myriad claims on everyone's attention and the mere cataloging of problems is unlikely to change perceptions of their relative importance.

This is not to say that community indicators have no constructive role to play in improving public life. The question is what form they should take and how they should function. Our contention is that community indicators can provide valuable information to the extent that they shift from description to diagnosis, i.e. from a statement of current conditions to a causal model that diagnoses where

---

[1] This presupposes that the purpose of indicators is to change actual conditions within a community. In some cases, it appears that indicators programs are merely designed to promote a dialogue among members of a community, regardless of the outcome. If the process is the purpose, then it is improper to criticize them for failing to fulfill more instrumental aims, such as better schools or cleaner air. However, since we suspect that most people imagine that indicators will produce tangible outcomes in policy objectives, it is improper to claim that the process is all that matters.

intervention would be most effective. Whereas descriptive indicators are based on direct observation, diagnostic indicators require an act of scientific imagination – the conception of which points within a system of relations are most crucial in determining outcomes.

Diagnostic indicators are the ones used routinely in other fields as an aid in solving problems. Both doctors and car mechanics run diagnostic tests to evaluate the source of dysfunction. On rare occasions, the cause of an illness or mechanical failure may be obvious upon inspection. Normally, however, the cause is hidden, and the practitioner must apply a model to the situation. The model generates hypotheses about what is causing a problem, and action is then taken based on the model. If the model is generally correct, the expert may then be able to intervene successfully. If not, the practitioner must re-evaluate the model or generate a new one.

The use of diagnostic indicators could be of great value to communities. However, as in other fields of endeavor, they can be formulated only on the basis of cumulative experience among people in a discipline. For that to happen in the field of community indicators, however, would require a procedure for the accumulation of that knowledge and wisdom. As a first step in doing that, it would be helpful if practitioners had a sense of the history of their own field.

This chapter examines the history of social indicators, at both the national and community levels, as an ongoing interplay between descriptive and diagnostic indicators. These two types of indicators are directly in conflict with each other because some type of description has often been necessary to recognize problems, at which point diagnosis is required to solve them. Nevertheless, there have been various points in history when descriptive indicators were treated as substitutes for diagnostic indicators, when it was presumed that heightened awareness would be adequate to address a situation. In those situations, descriptive indicators have effectively blocked the development of diagnostic indicators and thus stood in conflict with them.

Any attempt to use descriptive indicators as problem-solving tools inevitably leads to confusion and frustration. That happened not only in the development of indicators, but in social research more generally. Researchers, journalists and citizen groups who used descriptive methods exposed the existence of problems, without offering any theoretical analysis of how or why a specific set of conditions arose. They are then puzzled when their purely inductive or descriptive methodology does not lead to the development of a theory that advances social reforms.[2]

---

[2] It appears that C. Wright Mills used the term 'abstracted empiricism' to mean something closely related to the descriptive use of statistics that we are calling historicism. (His term for what we are calling positivism is 'classic social science'.) His analysis of abstracted empiricism is similar to our view of historicism: 'The policy for progress of abstracted empiricists is very specific and quite hopeful: Let us accumulate many microscopic studies; slowly and minutely, like ants dragging many small crumbs into a great pile, we shall "build up the science"' (Mills, 1959, 127). Mills devotes an entire chapter (chapter 3) to explaining why the 'building block' methods of social inquiry employed by Paul F. Lazarsfeld, the

In this chapter, I examine the ways in which descriptive and diagnostic indicators arose historically. Before turning to that developmental process, Table 2.1 may help explain conceptually how the two approaches to indicators differ.

Table 2.1 Categorical Differences Between Diagnostic and Descriptive Indicators

	Diagnostic	**Descriptive**
Role of theory	Explicit theories about social causes guide data collection	Implicit reductionism, individual causes
Purpose of data gathering	To determine which policies will actually work, problem-solving	To determine the extent of a problem, create support for action
Bias in data collection	Conscious, purpose avowed	Unconscious, pretense of neutrality, simple collection of facts
Obstacle to good policy	Knowledge about causal relationships, resistance to counter-intuitive solutions	Political will to do what is already known to be right

Table 2.1 gives an overview of the differences between diagnostic and descriptive indicators. Since not all of the cells in the table are self-explanatory, we describe each of them below:

1) Role of theory: For diagnostic indicators, theory explicitly guides research. There is no point in collecting data until a theory has been formulated. For descriptive indictors, theory is the Holy Grail. It is something to be sought endlessly, but which may never be found. If C. Wright Mills is correct, descriptive indicators are biased toward reductionism: the belief that wholes can be explained by parts. Thus, descriptive indicators implicitly give precedence to individual behavior over social or structural factors.

2) Purpose of data gathering: Diagnostic indicators draws upon information to test whether a theory has practical significance: will a particular program actually do what it professes to do? In order to increase the problem-solving capacity of managers, the aim of diagnostic indicators is to reveal causal relationships and predict the outcome of interventions. They seek to answer the question, 'What works?' By contrast, awareness of social or economic conditions is the goal of descriptive indicators projects. Their data demonstrate the

most sophisticated spokesperson for abstracted empiricism, cannot form a coherent theoretical framework. Mills proposes that the underlying principle of abstracted empiricism is psychologism: 'an explicit metaphysical denial of social structure' that treats individuals, not social systems, as the key to understanding social phenomena (p.67). Mills summarizes his key idea succinctly: 'Social research of any kind is advanced by ideas; it is only disciplined by fact' (p.71). That could stand as the slogan of the approach we are calling positivism.

symptomatic features of a problem, i.e., to answer the question, 'Where are we?' Descriptive indicators are put forward with the belief that widespread knowledge of the existence of a housing crisis, the number of homeless people, or the frequency of farm foreclosures will help to alleviate the problems they depict. Thus, there tends to be an assumption that descriptive information is inherently valuable: if more data are available, society will be better off.

3) Bias in data collection: Diagnostic indicators are predicated on the assumption that all data gathering is done with a purpose and thus reflects some bias. The bias need not be ideological. The point is simply that data are collected to answer a particular question. It is helpful if researchers avow their theoretical purpose and make it explicit. Nevertheless, the issue is never whether data collection is unbiased, but whether the theory behind it is conscious or not. Descriptive indicators, by contrast, are based on the assumption that it is possible to collect information in a pure and unbiased manner. They treat data as neutral 'facts' that become biased only by the way in which they are used. This pretense of neutrality causes descriptive indictors to serve the status quo, because the data can only describe conditions; it cannot point toward alternative policies. By contrast, diagnostic indicators tend to challenge the status quo by showing explicitly how change is possible.

4) Chief obstacles to good policy: Descriptive indicators presuppose that experts know what to do to solve problems: break down larger problems into their components and solve each of them. The proponents of descriptive indicators consider devising solutions the easy part. They believe the difficulty lies in gathering the political will to act. Accordingly, widespread dissemination of social indicators will lead to greater public understanding of problems, which will strengthen political will to apply known remedies. The proponents of diagnostic indicators, by contrast, believe that there are many problems without any known solution and that, if there is a solution, it will be counter-intuitive. Therefore, the main obstacles to solving social problems lie in a) ignorance about which remedies will actually work and b) unwillingness to accept the counter-intuitive solutions discovered through a diagnostic methodology.

Two Strands of Indicator Development

Having summarized the differences in concepts and methods between descriptive and diagnostic indicators, we now turn to a survey of the development of each approach.

The Descriptive Approach

The descriptive approach to indicators can be traced to the concept of Staatenkunde in the 17th and 18th centuries (Lecuyer and Oberschall, 1968, pp.40-41). That term, which is the origin of the modern term 'statistics', referred to the comparative description of states: the gathering of facts that would 'illustrate the condition and prospects of society' (Eyler, 1979, p.18).[3] It is ironic that the first people to label themselves statisticians were not interested in numbers at all. In 19th century Germany, a split took place between Tabellenstatistik (the use of tables) and Staatswissenschaft (the study of the state). The former became political economy and the latter became political science and public administration. Both, however, remained highly descriptive in orientation.

In the latter part of the 19th century, the Verein fur Sozialpolitik (Social Politics Union) emerged as a reaction in Germany against the philosophy of laissez faire liberalism (Bryant, 1985). The Verein hoped to assemble vast quantities of empirical data from which it would be able to draw conclusions about the management of the economy. This was precisely the approach favored by Bismarck, who established the first model of a paternalistic welfare state. The assumption of Bismarck and his bureaucracy was that the state needed to intervene constantly in the economy to promote the well-being of citizens. Since they aimed to micromanage the economy and various institutions, they required a detailed knowledge of how they function.

This connection between the welfare state and the descriptive approach to social indicators is of great significance. Most conservative economists have been unsympathetic to any political movement that favors the compilation of social indicators. The official reason for their opposition is usually methodological. Without theory, they point out, numbers are useless. But methodology is tied to ideology. The opponents have discerned intuitively that the descriptive approach to social indicators amounts to a defense of the welfare state.

1. Objectivity, reform and the development of labor statistics The later rise of the welfare state was not foreseeable when the historical approach to indicators was just getting under way in the United States. Although one might find traces of the descriptive approach in the statistics gathered in connection with the reform movements of the 1830s, we begin with Carroll Wright and the origins of labor statistics (Leiby, 1960). Actually, Wright, as the son of a Universalist minister in Massachusetts, drew upon the moral statistics tradition that we will discuss later.[4] He believed in the goodness of all people as long as they had good information. That liberal theology taught him to consider only the outward appearance of things and not to question too deeply. Later in life, Wright distrusted any effort to use statistics to theorize about the causes of social conditions (Leiby, 1960, p.21). His faith was in the numbers themselves, not in some subtle meaning that might be derived from them. This rationalistic belief that the knowledge of social statistics would solve problems was a specific example of a more generalized faith in education as the panacea for society's ills. That was the basis of the common school ideal, another New England invention that was Protestant in spirit, yet imposed in the name of universalistic principles of tolerance and rationalism.

---

[3] The foremost representative of this discipline in the 18th century was Gottfried Achenwall, Gottingen professor of law and politics (Eyler, 1979, p.18).

[4] It is significant that theological disputes had as much bearing on the development of indicators in the United States as philosophical and political issues had in Europe.

Carroll Wright's focus on numbers as the Commissioner of Labor Statistics, first for Massachusetts and later for the federal government, allowed him to avoid taking sides in ideological disputes. When Wright took over as Commissioner in Massachusetts in 1873, he replaced a labor advocate who had used the office to gather statistics to prove theories about the exploitation of labor. Wright rejected that approach. He was determined, like his mentor, Francis A. Walker, the director of the U.S. Census in 1870, to stick to the facts and leave theorizing to others.

The underlying political philosophy of Carroll Wright and of those who followed in his footsteps in the indicators movement is captured by his view of the condition of working people in his own day. As James Leiby says, referring to Wright's theory of social problems: 'While there was no fundamental problem, there were inevitably many specific and occasional difficulties. ... The bureau's function, therefore, was not to solve one big problem, but to furnish a factual basis for the discussion of a lot of incidental ones' (Leiby, 1960, pp.67-8). Wright self-consciously eschewed any theoretical explanation of social conditions. He was content to record them and to leave theorizing to others, as if the two activities could be clearly separated.

In the midst of the labor turmoil of the time, Wright's moderate view was an essentially conservative one. Wright won support for the continuation of his bureau and later appointment as federal commissioner of labor by convincing business groups that the publication of facts had quelled agitation (Leiby, 1960, p.68).[5] In effect, Wright recapitulated in the U.S. what the German scholars did through their alliance with Bismarck: he promoted incremental micro-level reforms that would obviate deeper structural changes.

In contrast to Wright, whose ideas merely converged with the German scholars, the new American school of economics called Institutionalism was directly shaped by them. Richard Ely, for example, studied with Knies at Heidelberg and became one of the most prolific and influential American economists of his day (Gaffney, 1994, p.84, 87). He touted an inductive approach to economics. Thorstein Veblen and John R. Commons, two students of Ely's, focused attention on the historically conditioned details of institutions rather than formal theory. Although the descriptive approach underwent some changes as it crossed the Atlantic and influenced the formation of Institutional Economics, it retained its bias against deductive formulation of theory and in favor of a micro-managerial style of governance.

---

[5] The conservative nature of statistical work that describes problems rather than analyzing their causes can best be seen in comparative perspective (Lecuyer and Oberschall, 1968, p.48). In France, in the early 1840s, the moderates rejected a national survey of social problems (Coleman, 1982). But in 1848, when the socialists demanded a Ministry of Labor, the moderates proposed the same survey. The socialists, however, opposed the gathering of statistics as a bourgeois exercise. A study of textile workers had been conducted earlier that had been used to show that the conditions of workers had improved. Although liberals had also attacked the study because it revealed workers' miserable conditions, it had had little impact on law. In principle, a statistical study that merely describes events may show a horrible state of affairs, but it does not suggest a course of action. Because it encourages the treatment of symptoms, not causes, it is conservative.

2. Inductive description and the development of business cycle indicators The descriptive use of statistics became especially evident in the work of Wesley Mitchell, one of Veblen's best students. Mitchell made his reputation with the publication of *Business Cycles* in 1913. Mitchell founded the National Bureau of Economic Research (NBER) in 1920 and directed it for 25 years (Oser and Blanchfield, 1975, p.391). The focus of NBER was the compilation of mountains of data. It produced business cycle indicators that are still used today (such as leading and lagging measures). However, to the extent that the work of NBER was not guided by theory, it did not produce any useful policy recommendations.

Reacting to armchair theories of the business cycle that existed around the turn of the century, Mitchell sought to ground analysis of business cycle in empirical evidence.[6] He wanted theories to be tested by the empirical record, not by logic. Rather than developing an a priori theory and testing it, however, Mitchell set out to collect and assess thousands of economic time-series and to slowly build an understanding of the interrelationships. Eventually, he hoped that general guiding principles would emerge from the data. However, faced with the intricate interwoven nature of the business indicators, he noted that although the business cycle repeats itself there are always differences, which preclude an accurate generalized theory (Seligman, 1990, p.185).

Mitchell never explicitly articulated his theory of the business cycle. In the culminating work of his career, Mitchell (along with Arthur Burns) published *Measuring Business Cycles* in 1946. It summarized in great detail their empirical analysis of historical U.S. business cycle data. Although replete with detailed descriptions of method and intricate analysis of the indicators movements in relation to the general cycle, it offered no hypotheses that explained the business cycle.

Their approach came under attack from econometricians such as T.C. Koopmans because of its lack of emphasis on theory. Although Koopmans thought it was important to describe regularities and irregularities of the system under evaluation, he felt it was equally important to search for the laws or principles governing them. In his famous critique, 'Measurement without Theory', he suggested that with the Burns and Mitchell volume 'everything had been tossed into the statistical hopper, so that the possibility of identifying the really significant relationships had been obscured' (Seligman, 1990, p.199). Were leading indicators the cause and lagging indicators the effect? Were the coincident indicators affected by the same forces that create the general cycle? Burns and Mitchell offered no explanation of these cause and effect relationships.

At the time econometrics, based on neo-classical theory, was an emerging force in economics. Keynesian theory, which offered a 'simple' causal explanation of the business cycle, was being used to organize the national income accounts and to help formulate general economic policy. Koopmans and other critics felt that work

---

[6] Critiques of abstract theories are appropriate if the theories are not put to the test. Those who create models with ungrounded and untested economic axioms should be called to task. This should not lead to an abandonment of theory but to an insistence on testing theories and assumptions.

on business cycles needed to have some sort of analytical underpinnings to generate and test hypotheses that could be useful in understanding the business cycles.

Institutionalists responded that it is all well and good to test hypotheses, but one must first have a hypothesis to test. Mitchell and Burns, according to their defenders, were searching out patterns that could generate reasonable hypotheses. Their work was not atheoretical in nature, but was a necessary precursor to testing hypotheses. It was simply premature to generate theories since the empirical basis of the business cycle was too skimpy (Klein, p.41).[7]

Phillip Klein, an institutional economists, describes Mitchell's approach: '... business cycles are inherently complex and there is, a priori, no reason why any reasonably accurate – and therefore useful – theory attempting to explain them would not be similarly complicated' (Klein, p.44). Klein then goes on to give the example of a simple causal explanation given at the turn of the century by Henry Moore that 'attributed the modern business cycle to periodic meteorological changes that produce sunspots' (Klein, p.44). This, for Klein, is an example (albeit exaggerated) of a 'simple theory'. Descriptive statistics had initially been intended to combat that type of armchair theorizing. But the argument against armchair theorizing had no bearing on the debate with those who favored the use of empirical observation and who noted that a simple theory need not be simple-minded.

Institutionalists continued to identify and classify leading, coincident and lagging indicators, work that eventually culminated in the well know Index of Leading Economic Indicators. Klein notes that the timing relationships between the 'short list' of currently used indicators can be understood in terms of a number of business theories (pp.46-48). Although true, it is noteworthy that the theories themselves were not derived from the study of the indicators. And even though leading indicator series that have been shown to lead turning points in economic activity, they do not speak to the amplitudes of the swings nor the duration. They also tend only to be useful for prediction in the short-term (six months out) and are therefore less useful for policy.

The work of Kuznets (a student of Mitchell's and a coworker at NBER) in setting up the national income accounts was similarly driven by a desire to present facts without theory. However, in this case, the national income accounts came to have special significance in policy formation because they were relevant in testing the hypotheses of John Maynard Keynes.

3. Recent social trends The 1920s was not only the decade in which economic indicators gained popularity, it was also the period when the idea of social indicators emerged. Herbert Hoover, first as secretary of Commerce and later as President, was the chief proponent of the collection of data that could be used for social measurement (Anderson, 1988, p.161). Immediately upon taking office in the Commerce Department in 1921, he created the Conference on Unemployment. Hoover's own thinking on this subject appears to have been shaped by the historical school, particularly by the work of Mitchell. He commissioned detailed

---

[7] They also argued that neo-classical economist would trash any quantitative analysis not based on the principles of that doctrine.

descriptions (but not the development of theories) of unemployment.[8] He believed that they showed the cause of cyclical unemployment to be the waste and extravagance that arose during boom years. Hoover imagined that this was a problem that could be solved by employers, since he assumed their inefficient management practices were to blame.

Another initiative at this time was the President's Research Committee on Social Trends, which was appointed by President Hoover. In 1933, in the last days of his presidency, the committee published its report entitled *Recent Social Trends*. This work, supervised by William F. Ogburn, a sociologist at the University of Chicago, marked a milestone in the social indicators movement in the United States. It was the first official document devoted to social measurement. More than 1,500 pages in length, it covered numerous social conditions such as demographics, health, education, etc. Yet, according to Adolph Berle, a law professor and adviser to President Roosevelt in 1933 when the report was released, *Recent Social Trends* was characterized by 'barrenness of quantitative theory and statistical measurement'. It was pure description. None of the sections contained any theory about why those conditions existed. Although it was completed during the early days of the Great Depression, there was barely a hint of the extent of the crisis (Smith, 1991, p.71).[9]

Ogburn, like Wright, believed that social reports should present facts not opinions and should present the data and trends but refrain from interpretation and certainly from policy recommendations (Bulmer, 1983, p.114). He envisioned an annual social report and, in fact, edited an annual issue of the American Journal of Sociology from 1928 to 1934, which presented a statistical account of social change in the previous year (Bulmer, 1983, p.111). His approach (descriptive, inductive, pseudo-objective) became the dominant approach of academics who worked with social indicators. A number of his students, including Albert D. Biderman, Otis Dudley Duncan, Albert J. Reiss, Jr. and Eleanor Bernert Sheldon, were later active in the social indicators movement.

---

[8] As Commerce Secretary, he commissioned a study in 1921-22 that relied on a methodology that was similar in concept to the German "statisticians" of the 17th and 18th centuries. That is, the study consisted of detailed descriptions coming from hundreds of cities around the country. The reports described who was unemployed, for how long, how businesses were responding and whether any labor unrest had resulted. This amounted to 'thick' historical research and extremely 'thin' abstract theory. The intent here is not to single out Hoover for criticism. He was following the best advice of his day. The key point is that historicism so dominated statistical thought in the 1920s that no alternative seemed imaginable.

[9] New Deal 'brain trusters' like Berle may have been critical of the report, but they were also locked into a descriptive, rather than analytic, framework. The New Deal programs did not end the Depression, precisely because they were designed to micro-manage the economy in the same way that Bismarck's social welfare state had done. Even after Keynes published his theoretical work on the causes of the Depression, the policies of the Roosevelt Administration did not change much. When the economy began expanding in 1936, Roosevelt clamped the lid on it and sent the country into another depression in 1937 and 1938, a fact generally ignored by historians.

4. Change through enlightenment: the early antecedent of the community indicators movement The use of the historical method in social analysis was not limited to economists and national surveys. It was also the method of choice by social scientists who were hired by private foundations as well. The Russell Sage Foundation could be said to have initiated the development of what are now called 'community indicators' using processes that are remarkably like the ones that have been re-established in the 1990s.

Around 1910, this foundation provided a grant to the Charity Organization Society (of New York) to survey industrial conditions in Pittsburgh. Paul Kellogg, a writer and editor, was hired to carry out the study. 'The project lasted 18 months and produced a six-volume study of Pittsburgh's housing, sanitation and working conditions. Kellogg and his associates wanted a "human measure" of social conditions that would prod the government of Pittsburgh to solve the city's problems' (Smith, 1991, pp.40-41). After the study was released in 1914, the Russell Sage Foundation was besieged with requests to fund similar studies in other cities. Since it did not have the funds to do that, the Foundation provided technical advice instead. Partly as a result of this initiative, more than 2,000 local surveys were taken on education, recreation, public health, crime, or general social conditions.

Just as community indicator projects today are often associated with some existing organization, the surveys of this period were 'conducted under the supervision of citizens' committees, church federations, chambers of commerce, or civic improvement associations' (Smith, 1991, p.41). The process by which this information was expected to affect decision-making was not much different from today either: 'These groups then relayed the findings of the technical experts to the public who, enlightened by the facts, were expected to mobilize public opinion and press for appropriate reforms' (Smith, 1991, p.41).

The Russell Sage Foundation supported this precursor to community indicators work because it believed the surveys would yield results. Like many others of the day, the officers of the Foundation were guided by the metaphors of public health: they wanted to treat root causes, not merely alleviate symptoms. They did not merely want to observe society; they wanted to improve it.

Yet the methodology employed was guaranteed to fail: the collection of factual information could *only* yield observations about symptoms. In the absence of theories, the accumulation of facts cannot offer any explanation of causes. Referring to the data gathered by the surveys of living conditions, James Allen Smith notes:

> In these and countless other measurements, the professionals presumed that there were causal relationships and, thus, implicit remedies for social problems. But the surveys usually explained much less than met the eye. In reality, they were less an instrument for testing hypotheses and designing reforms than for arousing a community's conscience and "quickening community forces" for reform, as one staff member of the foundation put it (Smith, 1991, p.42).

That is why 'the actual political results [of the surveys] seldom lived up to the organizers' expectations' (Smith, 1991, p.41).

This explanation of the failure of the social surveys at the beginning of the century to lead to social change applies with equal force to current social indicators efforts that have their roots in the social indicators movement of the 1960s. Unless the methodology used to develop indicators goes beyond 'arousing a community's conscience', it is unlikely ever to lead to reforms.

5. Social indicators movement By the early 1960s, as a result of the increasingly successful management of economic policy, the deductive or analytic approach to policy-making had effectively triumphed over the inductive or descriptive method favored by the students of Ogburn. The success of the Kennedy tax cut of 1964 and the apparent accuracy of econometric predictions of its effect on the economy made economists quite influential forces in guiding public policy.

Professional economists in government (such as the Council of Economic Advisors) and in policy institutes (such as Brookings) became a model for applying social science to government policy. Critics charged that economic considerations were given undue priority in policy decisions. If social theory and planning were given equivalent institutional support, they surmised that social policy could be rationalized in the way economic policy had been. Thus, the success of economic indicators was one spur to the birth of the social indicators movement in the 1960s and early 1970s. As social crises mounted during the 1960s some politicians and social scientists began to champion the development of systems of social statistics comparable to the existing economic ones.

The event that signified the launching of the social indicators movement in the United States was the publication in 1966 of *Social Indicators*, a project sponsored by NASA (Bauer, 1966).[10] Raymond Bauer, Albert Biderman and Bertram Gross, the primary authors and influential forces in the budding social indicators movement, argued for increased collection of statistics that would be published as a social report. They also advocated the development of a system of social accounts which could help guide policy decisions.

A second influential report, *Toward a Social Report*, was issued by the Department of Health and Welfare (HEW) in early 1969, on the last day of the Johnson administration (HEW, 1969). It called for the establishment of an annual social report of the type advocated earlier by Bertram Gross in *Social Indicators*. According to Kenneth Land the HEW report 'firmly established the link of social indicators to the idea of systematic reporting on social issues for the purpose of public enlightenment' (Land, 1992, p.1844).

The Great Society, the program sponsored by President Johnson to provide an economic floor for all members of society, was in many ways a replay of the social welfare programs that were first devised in Bismarck's Germany. For every need, a new program was developed. This way of approaching policy is based on the descriptive model: create detailed solutions based on a detailed knowledge of events.

---

[10] NASA was interested in determining the second-order consequences of the space program for American society.

The Russell Sage Foundation (which had funded the early 'community indicators movement' described above) was also instrumental in keeping alive the descriptive tradition in social indicators. In 1968 it published *Indicators of Social Change*, a volume edited by Wilbert Moore and Eleanor Bernert Sheldon, a program officer at the foundation. This volume was a successor to the *Recent Social Trends* volume of the 1930s in both content and spirit. According to Sheldon, no work had been done on indicators from the publication of *Recent Social Trends* in 1933 until the mid-1960s (Sheldon, 1968, p.78).

The Russell Sage Foundation was not alone in bringing the descriptive tradition in social indicators to life again in the 1960s. The authors of *Social Indicators* were also descriptive in their approach. However, since the contributors to that book saw indicators more as tools of policy analysis, they are less clearly descriptive than the work done under the auspices of Russell Sage.[11]

In the meantime, Senator Walter Mondale, along with Senators Kennedy and Harris, put forth legislation from 1967 to 1973 to set up a Council of Social Advisors, comparable to the Council of Economic Advisors (CEA) (Booth, pp.380-5). The CSA was to issue an annual 'Social Report' like the *Economic Report of the President*. Underlying this effort was the belief that the creation of the CEA had institutionalized the use of economic information and the power of economists. Creating a comparable institution to address social problems seemed like a logical next step.

Sheldon and others at the Social Science Research Council (SSRC) opposed the creation of the CSA or any other premature application of social indicators to social policy. Instead, they argued, the pressing needs were better data series and basic research. (Sheldon et al., 1983, p.79). Since she was affiliated with the prestigious SSRC, her views had an impact.

She believed their claims for social indicators were over-enthusiastic. Following the academic approach she had inherited from Ogburn, she believed that the development of a theoretical framework for indicators was premature. Social indicators could not follow in the footsteps of economic modeling since: 1) social goals were more ambiguous than the economic ones, 2) social problems were less clearly understood than economic ones and 3) the theoretical foundations of economics were much clearer than those underlying the analysis of social problems (Land, 1983). Sheldon (1968, 1970) argued that a purely inductive approach was needed: first gather data, then develop the categories that would allow meaningful generalization.

Although Sheldon correctly pointed out that a body of social theory did not exist that was comparable to that of economic theory, she was incorrect in her portrayal of how that economic theory came into being. It was not the result of

---

[11] As Sheldon says comparing the work funded by Russell Sage with the Bauer volume: 'The Foundation's program was somewhat different in that its primary motivation was to establish a theoretical base, data and measurement techniques for understanding social change. It was not that we eschewed the importance of "policy analysis" but rather that we considered our effort as parallel – or even a precursor – to it' (Sheldon et al., p.79).

laws emerging after the careful collection and interpretation of data but of a theory based on prior insights.

Work on social indicators flourished in the United States in the 1970s with thousands of relevant articles and books being published. In the United States, Social Indicators, Volumes I, II and III were published (Office of Management and Budget, 1973; Census Bureau, 1977, 1980). The original intent of the reports was to include critical analysis and judgments about causal relationships. Political forces within the Nixon Administration turned them into 'neutral' chartbooks, full of isolated facts, with little to no interpretation. (Innes, 1990, pp.99-100).[12] There was accompanying interpretation of the 1976 and 1980 reports, but it was done in the Annals of the American Academy of Political and Social Science, an academic journal. Even then, the analysis was highly technical, evoking no political controversy.

But this early enthusiasm for social indicators was not to be long-lived. The movement in the United States was effectively over by the early 1980s although it continued in the form of annual statistical reports issued by government agencies. Some data series that were started in part because of the movement also continued to be collected. Numerous explanations have been offered that account for the early demise of the movement (e.g., Andrews, 1990; Noll and Zapf, 1994; Bulmer, 1990). The most compelling of these reasons is that the indicators were of limited use to policy makers and so few examples could be given of using indicators to bring about some reform. It was not simply the lack of political will (which the indicators were supposed to help create) or the inadequate theories of social change (that the inductive study of trends would help elucidate) that caused the demise of social indicators. It was the inability of the advocates of social indicators to offer causal explanations of social trends.

While the U.S. government worked on social indicators sporadically during the 1970s, several European nations, notably Great Britain, France, Germany and the Netherlands went further than the U.S. and institutionalized social reporting. Examples of European reports include *Donnes Sociales* in France, the *Social and Cultural Report* in the Netherlands and *Social Trends* in the U.K. Among the preconditions that enabled this were 'an articulated welfare-state program of social policy, an interventionist orientation of government, innovative statistical agencies and geographical centrality' (Noll and Zapf, 1994, p.5).

The excitement generated by the social indicators movement also had effects on international agencies. In addition, international agencies began developing indicators as a part of their mission. In the early 1970s, the OECD established the Working Party on Social Indicators which continued work into the early 1980s until the publication of *Living Conditions in OECD Countries* in 1986. At the United Nations, the relationship between economic development and human

---

[12] There was accompanying interpretation of the 1976 and 1980 reports, but it was done in the Annals of the American Academy of Political and Social Science, an academic journal. Even then, the analysis was highly technical, posing no political threat. The foremost representative of this discipline in the 18th century was Gottfried Achenwall, Gottingen professor of law and politics (Eyler, 1979, p.18).

development was explored via the Human Development Index. The World Health Organization (WHO) emphasis on human health led to the Healthy Cities Movement that developed indicators to assess improvements in the state of cities.

In a similar way, the development of environmental indicators was inspired by the social indicators movement. In the United States, the Council on Environmental Quality and the Environmental Protection Agency both began to monitor environmental trends. Similar work was begun at OECD in the 1980s. In addition, policy institutes such as Worldwatch and World Resources Institute began producing annual books describing and analyzing environmental trends. With the notion of sustainable development as highlighted by the Brundtland report and later by the Rio conference, a new framework for indicators developed in the early 1990s. Sustainability indicators attempt to describe the interrelations of economic, environmental and social concerns. These frameworks or sets of sustainability indicators offer conceptual models that illustrate the interrelationships of social, economic and environmental indicators, but they don't necessarily offer an analytical understanding of the relationships.

Other efforts characteristic of the social indicators movement in the 1980s are attempts to create indexes of social welfare that can be used as alternatives to the GDP. Examples of indicators of this variety include the Index of Social Health developed by Marc Miringoff (1997) of the Fordham Institute and the Genuine Progress Indicator (GPI) developed by Redefining Progress. Although the GPI is useful as a rhetorical tool to point out the limitations of the GDP as a measure of well-being (and demonstrates the difficulty and arbitrariness of developing any such general measure) it does not lead to any specific policies. This is the crucial distinction between descriptive and analytical work.

The social indicators movement has experienced a revival in the 1990s. The sources of that renewed energy are perhaps somewhat different than in previous decades. There is far less impetus from federal agencies or major national institutions like the SSRC. Instead, the focus in the 1990s has been on community indicators, similar to the work catalyzed by the Russell Sage Foundation around 1910. It is also characterized by consideration of economic, social and environmental conditions. However, it is similar in a critical way – the descriptive approach.

Those who have worked for years to develop better indicators have been frustrated by the lack of success at achieving social change or even institutionalizing social reporting. Much emphasis has been placed on the enlightenment or agenda-setting role of indicators and descriptive indicators are often employed effectively in the public debate. Our concern is that advocates, especially at the community level, then wonder what comes next, how they can actually effect change in what they are describing. There may be an important rhetorical or persuasive role to be played by descriptive indicators in 'raising awareness' but one can not expect those same indicators to effect change in the conditions. It is our belief that if indicators are to be a tool in bringing about change in conditions then one must approach the indicators work with a diagnostic perspective.

The Diagnostic Approach

The diagnostic approach has a tradition as long as the descriptive approach.[13] Whereas descriptive indicators presuppose that change will automatically result from publicizing facts about social conditions, diagnostic indicators are concerned with *how* effective policies can be formulated. (From the latter perspective, many of the policies formulated under the aegis of descriptive indicators are merely symbolic expressions that are directed toward symptoms rather than causes.) Diagnostic indicators aid in the development of theories or models that specify causal relationships. Those models, if correctly specified, can provide relatively accurate predictions about the consequences of policies. They can, thus, be used to solve problems.

The ultimate test of whether a model is diagnostic or not is the extent to which it is used to make predictions. The use of numbers or algebraic analysis does not necessarily qualify work as diagnostic. Statistical analysis is diagnostic if it is based on the testing of deductive hypotheses about relationships that are not directly observable. The key difference is this: descriptive methodology seeks to understand by examining relationships that can be seen; diagnostic methodology seeks to predict by drawing inferences from relationships that are unseen.

1. Petty's deductive, problem-solving approach In social analysis, one of the first applications of the diagnostic methodology was made by William Petty in the late 1600s (Studenski, 1958, pp.26-30). Petty's innovation was to bring theory and data together in understanding social and economic conditions. Before him, writers had believed that economic health depended on increasing exports of goods and imports of gold because that increased the national treasury. (They assumed that what helped the government also improved the nation.) Petty developed a theory that reversed that relationship. He argued that wealth derived from production, not from hoarding gold and that trade should therefore be taxed and regulated lightly. To help the crown, one must first improve the economic conditions of society.

Petty also theorized that England had the capacity to fight a war with France, but that Holland was rapidly growing in its capacity to wage war (due to its liberal trade policies). In order to test his hypotheses, he made calculations of the populations of France, England and Holland and of England's national income. (Gregory King estimated the national incomes of all three countries after Petty's

---

[13] The term 'positivism' has been used in many different ways in the last century. As Bryant (1985) shows, the French, German, English and American uses of the terms have varied considerably. We use the term to mean an approach to indicators based on the priority of theory over data, analysis over description and problem-solving over consciousness-raising. Indicators are relevant in the positivist framework as part of a process of testing hypotheses to determine valid relationships. Those relationships can then be used to predict the effects of policies put in place to remedy problems. We do not share the utopian hopes that some positivists have displayed historically, and we do not share the reductionist and empiricist orientation found in most Anglo-American positivists. Nevertheless, we regard this analytic, problem-solving approach as more fruitful than the purely descriptive approach of the historicists.

death, using generally similar methods.) Petty estimated that France had a larger population than England, but that England's higher productivity made the two countries equally matched in wealth (and war potential). He determined that an income tax could be levied in England that would meet the needs of the state in peace and in war. Finally, Petty used vital statistics, which he also originated along with John Graunt, to disprove the theory circulating in his day that England's population was declining.

2. The search for natural laws of society Petty is generally remembered for his initiation of national income accounting and for his testing of hypotheses about society with statistics rather than pure reason.[14] It is especially important to understand that, unlike the German statisticians of the 17th century, Petty saw statistics as a scientific enterprise involving a search for hidden causes. Petty was guided by the prevailing expectation at the Royal Society (England's preeminent institution, of which Petty was a founding member) that society, like the stars and planets, would operate in harmony with natural law.[15] Thus, all of Petty's work was shaped by deductive hypotheses rather being an inductive attempt to generate theories. He made predictions based on general principles, then sought the evidence to confirm or deny his hypotheses.

Although efforts to apply 'political arithmetick' to economic issues lay dormant throughout the 18th century, efforts to look for underlying patterns in social phenomena continued as a search for 'divine order' or natural law in human affairs (Westergaard, 1932). Astronomers (Halley and Wargentin) and other European scientists (Huygens, Neumann and Sⁿssmilch) were drawn into the investigation of demographic questions such as population size and distribution by the search for mathematical order.[16] One question they particularly wanted to

---

[14] 'Petty was in fact the most significant theorist of social statistics in Britain, at least until the heroic decade of the 1830s, and even then no single figure stood out with his overall vision' (Cullen, 1975, p.5).

[15] German intellectual traditions, in contrast to English ones, have been shaped by a strong belief in the dichotomy between the study of physical phenomena (Naturwissenschaften) and the study of human society (Geisteswissenschaften).

[16] It is interesting to compare the work of Petty and statistically-oriented scientists with the later work of Adam Smith, Malthus, Say and other classical economists. The economists were rationalists, who promulgated a theory of social equilibrium, which they were not especially interested in testing. The political arithmeticians (to use Petty's phrase to describe his work) were interested in how well theory conformed to facts. Thus, whereas Thomas Malthus (an early classical economist) asserted that life expectancies had been relatively constant throughout history, the political arithmeticians considered that a testable hypothesis. The conclusion about the merits of this idea were of great practical significance at the end of the 18th century. The economists of the day (or at least Malthus and Say) were content to ignore public health issues. Say also opposed efforts to improve social conditions in early 19th century France on the same grounds. They believed that whatever health or social improvements took place would be counterbalanced elsewhere, preserving the general condition of misery. The statisticians, on the other hand, thought that the smallpox vaccine and other public health measures could save lives on balance and that they should be introduced.

answer was the degree of imbalance in the population of men and women. Another idea circulating at the time was that the human body was in danger every seven years and that human infants were in peril during weeks that were multiples of seven and nine. Thus, some of the hypothesis-testing of early statisticians was to quell superstitious folk beliefs. They did so by contrasting population statistics with false inferences drawn from anecdotes. (The fact that legislatures still draw more on anecdotes than statistics shows that the early scientists were not entirely successful.)

By offering tantalizing conclusions based on the limited evidence available through parish records, these scientists convinced the leaders of a number of nations to initiate a national census. Sweden undertook the first modern census in 1748. In its reports, the Tabullar Commission of that country not only reported the data on population size; it also offered an explanation for the low fertility rate that was causing what they regarded as underpopulation. When the United States began the first regular (decennial) census, it was also did so to answer questions of practical importance: how were representation and taxes to be apportioned among the states.

Another 18th century development that permitted the later creation of social statistics was the refinement of life tables for the insurance business. The great regularity of mortality in large groups of people demonstrated to the mathematically-skilled scholars of the time (of which there were few) that social phenomena operated within the bounds of fixed laws that were not evident in the behavior of individuals. A life table made it possible to make predictions about the number of men and women who would die in a given year, even though it was not possible to predict the fate of particular individuals. This observation gave rise, in turn, to the belief among some statisticians that analysis of unseen regularities in society would permit leaders to shape the broad outline of the future, but not the details.

3. Public health and statistical inference in the 1830s The diagnostic approach to social indicators became much more focused in the period 1830-50 in France, Belgium, England and the United States. The interest in using statistics to reform society during this period arose in response to the strains and disruptions caused first by the Napoleonic wars and the consequent reductions in trade and later by industrialization and large-scale urbanization. In addition, the concept of public health emerged during this period, both as a product of statistical thought and as a catalyst for it. Much of the technical work on social indicators was carried out by physicians and then used by social reform groups. Applications of the theory of probability to large populations were refined during this period, so that it became possible to make more precise inferences about social phenomena. That is to say, analytical tools advanced considerably, without which useful inferences would not have been possible. In the 1830s, statistics came to be thought of in its modern sense as the science of society rather than as compilations of numbers.

Adolphe Quetelet, a Belgian mathematician, was perhaps the most important pioneer in the application of probability and statistics to social issues.[17] He was not only instrumental in developing a new way of understanding society in statistical terms, he also set up the first International Statistical Congresses starting in 1853 that brought together researchers in this field from various countries.

One of the reasons Quetelet and others were attracted to statistics in the early 19th century was that it seemed to offer a way of understanding society that was beyond ideology. Statistics revealed an underlying order in society that persisted through periods of political instability and revolution. In addition, the 19th century marked an era of increased availability of data, although much of the growth in data collection was in response to the demands of social reformers.

Quetelet hoped to understand society and its subgroupings (especially as defined by social class) by the use of statistical means, 'the average person', and variance around those means.[18] The statistical understanding of behavior gave birth to a new mode of thought that Quetelet called 'moral statistics' or 'social physics'.[19] The former term simply referred to statistics dealing with socially significant behavior, while the latter referred to the similarity between the order in society and the order found in the natural world.

Quetelet began this course of social investigation after examining the first French judicial statistics in 1827. He was surprised by the regularity of age-specific crime rates. He was later struck by the relative constancy of marriage and suicide rates from year to year. These regularities in behavior convinced Quetelet that human behavior is governed by laws that transcend individual choice, despite appearances to the contrary. In effect, he was arguing that crime (or suicide or alcoholism) is 'a property of the community, not of the malevolence of certain individuals' (Gigerenzer et al., 1989, p.47). Quetelet understood that the attribution of the causes of pathology to social forces reduced personal responsibility. He did

---

[17] The intellectual ancestry of Quetelet can traced to Poisson, Laplace and Fourier, the French mathematicians who had already been working on the application of probabilistic concepts to social events (Landau and Lazarsfeld, 1968, p.248). Quetelet was sent by his government's minister of education to Paris in the 1820s to meet with scientists and mathematicians in order to learn from them how to establish an astronomical observatory in Brussels. As in the 18th century, those interested in celestial harmonies were again the leaders in thinking about social harmonies that could be discovered mathematically.

[18] The concept of the average person represented a sharp break from the traditional concept of 'the rational person', which had previously been understood as the norm for thinking about society. Whereas one might ask what the rational person would do in a given situation, one could ask what the average person actually does.

[19] Auguste Comte, often credited as the originator of positivism, 'was annoyed because that "mere" statistician Quetelet had appropriated his "social physics" and obliged him to coin a new term, sociology ...' (Gigerenzer, 1989, et al., p.46). In addition, Comte opposed quantification of social phenomena, although most later positivists did not share that prejudice. Comte believed in natural laws of historical development, which presumably were hidden from 'mere' statistics. 'Quetelet's statistical approach was the purest form of positivism, ... too positivistic ... for the founder of positivism himself, Auguste Comte' (p.42).

not deny the existence of free will, but he believed it operated within a much narrower framework than a more purely individualist philosophy allowed. Although everyone's freedom is constrained to some extent, social reformers and political leaders were not so limited by the statistical regularities. They were relatively free to determine the conditions that would either raise or lower the frequency of social pathology.

According to Quetelet, statistical laws showed that order was universal, that irrationality had its reasons, that even crime was subject to law. The findings of statistics implied that antisocial acts were products of the social condition and that these social maladies could be cured, or at least alleviated, by the 'scientific reformer' (Gigerenzer et al., 1989, p.41).

The analysis of statistical regularities would enable a benign government to intervene successfully in solving social problems – but only if the government focused on the condition of society, not the behavior or background of particular individuals. The state could not change which choices individuals would make, but it could (in principle) change social conditions and thereby influence outcomes on a statistical level.

Louis-René Villermé independently came to many of the same conclusions as Quetelet, perhaps because they were also derived from the mathematical ideas of Laplace and Fourier. Villermé, a physician, sought the cause of differential mortality in Paris in the 1820s (Coleman, 1966). Like Quetelet, he discovered that social conditions were to blame.[20] By combining data from a survey of vital statistics with information on the distribution of wealth, he showed that there was a statistical correlation between poverty and high death rates. He showed the poor had higher birth rates, higher infant mortality rates and lower average age of death. He also discovered that the rich abandoned unwanted infants more often than the poor did. In other words, he found not only differential mortality, but differential morality as well.[21]

In England at the same time, a reform movement arose that was based on the idea that statistics could guide the development of policy. William Farr, one of the leading figures of this movement, developed a statistical approach to social problems that was similar to the models used by Quetelet and Villermé (Eyler,

---

[20] In the case of medicine, the alternative explanation was not so much individual behavior as physical causes in the surrounding environment. Villermé was first able to disprove the long-standing assumption (based on Hippocratic theories) that illness was caused by environmental factors such as dampness and bad air (Latin: mal aria). He also found no statistical correlation between death rates in different sectors of Paris and the density of population, contrary to another popular theory at the time. Thus, he proved, by the standards of his time, that nature is not the cause of differential mortality. In doing so, he was as much responsible for the development of public health as William Farr and Edwin Chadwick in England. It is important to recognize, however, that physicians interested in public health were in a minority at this time. The statistical approach to medicine was considered an affront to the ethics of caring for each individual as a separate case.

[21] Villermé also later pioneered in the statistical study of working conditions through his analysis of textile workers.

1979). That is due, in part, to his study of medicine in Paris starting in 1829.[22] In Paris, Farr acquired the image of public hygiene as the model of social improvement. Farr became aware from military records that clinical interventions with individual patients had virtually no effect on their death rates, but that public hygiene measures did reduce mortality substantially. This background in public health had a life-long influence on his work in statistics and social reform. Many of the other men who were active in social reform through statistical analysis in Victorian England were also physicians, in part because medical conditions and mortality were easier to quantify than many other social conditions.

Farr belonged to the first generation of career statisticians in England.[23] Although he maintained a very practical view of statistical work, he was, nevertheless, concerned with theory (Eyler, 1979). He rejected the mindless garnering of numbers and had nothing but scorn for the 'empiric who throws heaps of tables in our faces ...'. In a medical editorial in 1837 he wrote: 'Facts, however numerous, do not constitute a science. Like innumerable grains of sand on the sea shore, single facts appear isolated, useless, shapeless; it is only when compared, when arranged in their natural relations, when crystallized by the intellect, that they constitute the eternal truths of science' (Eyler, 1979, p.29).

Thus, he eschewed mere numerical description in favor of testing hypotheses and theories. Without theories, it is not possible to solve problems. This became the standard view of the statistical societies of the time. 'Statistics ... was to serve as a link between the other sciences and practical affairs. ... The statistician, or "statist," as he was called, was to be the scientific expert in matters of policy, the one who spoke from a knowledge of facts, not from mere conviction' (Eyler, 1979, p.16). The key point is that statistical inference played a special role that was relevant for public policy. There was an implicit recognition that conclusions drawn from direct experience (with individual criminals or patients in a hospital) might be contradicted by statistical inferences. The latter were held to be more reliable as a basis for public health and other policy measures.

It would be wrong to infer that the diagnostic-descriptive distinction was clear to anyone at the time or that it corresponded in a simple way with the differences between England and Germany. Influenced by German studies, English statistical societies conducted social surveys of English cities to measure social welfare. They were trying to answer the 'condition of England question' with descriptive statistics (Eyler, 1979, p.17). Nevertheless, some English thinkers saw that it was possible to go beyond description. Farr and others with a background in

---

[22] Farr was also influenced by 'the British empirical medical tradition, a tradition represented by Thomas Syndenham and the latter's eighteenth century descendants' (Eyler, 1979, p.10). The British physicians, including Farr, were not as far advanced in statistics as the French.

[23] Farr himself was both the chief statistician of the General Register Office from 1837 to the 1870s and an active member of the Statistical Society of London (later the Royal Statistical Society) as well as the statistical section (Section F) of the British Association for the Advancement of Science. Some of the other professional statisticians were Poor Law Commissioners.

public health used the data in the surveys not simply to write a detailed report, decry existing conditions, or create programs aimed at symptomatic relief (the response of academicians, social reformers and bureaucrats, respectively). Instead, they laid the groundwork for a new kind of scientific understanding, reform and government policy by analyzing causes and pointing to systemic remedies. They were particularly interested in determining the causes of epidemics and other threats to social well-being.[24]

Statistical work has never been neutral. The biases of the statistician inevitably play a role in formulating hypotheses and in choosing which facts to gather. In early Victorian England, the reformers who used statistical techniques to analyze cause and effect were shaped by the interests of their social class.[25] The Manchester Statistical Society and other provincial societies argued that the suffering of the working class was due more to urbanization than to working conditions in factories (Cullen, 1975, pp.106-8, 128).[26] The statisticians sought to deflect industrial reform measures and other potential attacks on property by channeling reform efforts into education and sanitation. Farr's conclusion (contradicted by Quetelet and especially by Villermé) that disease is caused by environmental factors (miasma) more than by social factors (such as income differentials) was almost certainly affected by his laissez-faire ideology and middle class background.

4. Hygiene and temperance: moral statistics in the U.S. There were no individual figures in the U.S. at this time comparable to Quetelet, Villermé, or Farr who offered intellectual contributions to the development of the diagnostic approach to

---

[24] The prevalence of the statistical philosophy in solving social problems in the 1830s does not mean that all of the supporters of the statistical movement were competent in the use of statistics. Nassau Senior, a member of the House of Lords and the head of the Poor Law Commission, prevented an alteration of the Poor Law Amendment Bill in 1834 in the following manner: 'Lord Ellenborough wished to make it possible for the authorities to grant relief at their discretion to those over 60. This would, in Nassau Senior's opinion, turn the aged into paupers, so he looked up John Rickman's life-tables for Essex in the period 1813-30. They showed that out of those who survived to age 20, nearly half survived to age 60. Hence, concluded Senior, nearly half of the adult population were over 60' (Cullen, 1975, p.84). On the basis of that confusion, he convinced Ellenborough to withdraw his proposal.

[25] Francis Galton, independently wealthy cousin of Charles Darwin and the person who coined the term 'eugenics' in 1883, took over the chair of the Anthropometric Committee of the British Association for the Advancement of Science from Farr in 1880 (Eyler, 1979, p.28; Gould, 1981, p.75). Galton represented the conservative school of thought that all characteristics were inherited. Farr and the other liberals believed that most observed characteristics were caused by environmental factors. Social reform ultimately lost out because of the dominance of the hereditarians who believed that social reforms would not help those they were intended to improve.

[26] It may in fact be true that urbanization was a worse culprit than the factory system since those working at home (in 'cottages') also had high mortality rates. In retrospect, it would appear that the Enclosure Movements, which had forced millions off their land, was the underlying cause of the misery of many members of the working class.

indicators. Nevertheless, there were physicians in the U.S. who were also intrigued by the new ideas about public health. From the 1830s onward, reform groups in the U.S. adopted both the metaphors and the statistical orientation of the public health movement and applied them to various social ills.

Reformers in the United States were primarily intent on purifying the morality of the nation. This was a concern of reformers in Victorian England as well, but it was a more pronounced theme in the United States than hygienic reform, which dominated the Old World. As a result, the statistical movement in America concentrated on studies of alcohol consumption, crime and prostitution. (The pro- and anti-slavery forces also used statistics in their arguments.) In some cases, these studies involved gathering statistics that were intended to shock; in other words, they were descriptive indicators.[27] Still, many of the studies went beyond description and formulated a hypothesis.

The most common theory in mid-19th century America was that alcohol was the source of social evil. Almost every vice in American society was traced to 'demon rum' and affiliated liquids. In particular Samuel Chipman concluded in 1836 from his surveys of poorhouses and prisons that heavy consumption of alcohol was a major factor in both poverty and crime (Davis, 1972, p.172; Cohen, 1982, pp.211-2).[28] The desire to analyze statistically the consequences of intemperance was not isolated to a few individuals. In 1833, a national temperance convention passed a resolution calling for the collection of accurate statistics on alcohol consumption and on 'the number of deaths, paupers and crimes attributable to drink' (Cohen, 1982, p.171).

5. Race: the enduring shadow in American society The temperance debate was overshadowed from 1840 to 1860 by the national debate over slavery. After the census of 1840 the debate shifted from the intrinsic morality or immorality of the 'peculiar institution' to its statistically verifiable consequences. That census marked one of the world's first efforts to gather national social statistics, including data on insanity and 'idiocy' (Anderson, 1988, p.28; Cohen, 1982, pp.176-8).[29] When the data became available in 1842, Edward Jarvis, a Massachusetts doctor specializing in the study of insanity, discovered major anomalies: 'the returns showed that the black population in the North suffered from insanity at a rate more than ten times greater than blacks in the South' (Cohen, 1982, p.192). They showed that the rate of 'idiocy' and insanity among free African-Americans in the

---

[27] For example, the movement to eliminate prostitution used statistics in an attempt to alarm the public. A publication of the New York Moral Reform Society claimed that there were 12,000 brothels in the U.S., 75,000 to 120,000 prostitutes and 500,000 licentious men. They also said that 20,000 women died each year from prostitution (Cohen, p.207-08). These statistics were not aimed at increasing understanding of the causes of prostitution. They were merely intended to shock the sensibilities of small-town readers.

[28] The same (incorrect) inference has been drawn in the 1980s and 1990s that 'drugs' are a primary cause of poverty and crime.

[29] Cohen (1982, p.178) explains that statistics on insanity were collected in 1840 at the behest of 'physicians who had been making assiduous private efforts to find correlations that would explain the rising insanity rate ...'.

North was higher (one in 162) than among the slaves in the South (one in 1,558). The rates ranged from a high of one in 14 in Maine to a low of one in 5,650 in Louisiana. The farther north one went, the higher the rate of insanity among African-Americans. (The rate for whites was fairly constant: one in 970 in the North; one in 945 in the South.) John C. Calhoun and other proponents of slavery cited this as evidence that slavery was indeed a beneficial institution for those who were enslaved. (The figures also suggest that whites in the South would have benefited from being enslaved as well, but no one seems to have drawn that inference.)

The only way to refute the conclusions was to engage in esoteric debates about methodology. Jarvis found some errors in the raw census results that helped explain what had gone wrong.[30] Yet, neither he nor anyone else at the time was able to discover the true cause of the error, in particular the simple fact that the larger number of African-Americans in the South (the denominator of the proportions) diluted the effects of enumeration errors more than in the North.[31] No official investigation was made because of a lack of apparent desire to determine the truth.

The truth, we are told, is the first casualty of war. It does not survive political conflict very well either. The misuse of statistical evidence by all sides has been the mainstay of politics ever since the census of 1840. Descriptive statistics are the mainstay of politics because they more malleable and evocative. But when a diagnostic model of causality fits the ideology of one side or another in political debate, it is an even more powerful tool.

The conflict over race since the Civil War has been heavily influenced by diagnostic statistics and indicators. In the 19th century and much of the 20th century, the racist use of statistical models was the norm. Stephen Jay Gould has traced that sad history. In the 1840s, Samuel George Morton provided 'evidence' of the intellectual inferiority of various people of color by demonstrating that their cranial capacity was smaller than that of Indo-Europeans. The errors that Morton made that biased his estimates were not conscious. As Stephen Jay Gould says about Morton, 'All I can discern is an a priori conviction about racial ranking so powerful that it directed his tabulations along preestablished lines' (Gould, 1981, p.69). Robert Bennett Bean published similar brain-size measurements in 1906. His mentor at Johns Hopkins, Franklin P. Mall, demonstrated the error in Bean's procedures, 'but not before a leading journal had recommended that blacks be

---

[30] Jarvis was able to determine that there were large populations of whites in the asylums of Massachusetts who had been recorded on the census as black. He also discovered counties in Maine that contained not a single African-American, yet which were listed in the census as having several insane blacks.

[31] The error was caused by several factors (Cohen, 1982, pp.202-4): the confusing manner of printing the census forms, the likelihood of classifying older, white, senile people as lunatic blacks and the much larger population of African-Americans in the South, which raised the denominator when calculating the rates of insanity among blacks in those states. Ultimately, the problem stemmed from a badly designed census form, not a conspiracy. As Cohen explains, 'Each schedule contained 74 columns, with headings in microscopic type printed across the tops of two pages' (p.186). Under those circumstances, mistakes were inevitable.

barred from voting as a consequence of their innate stupidity' (Gould, 1981, p.82). This pattern begun by Morton and Bean continued for decades in the work of respected scientists. As recently as 1951, the Hall of Man at the American Museum of Natural History still displayed the characters of human races by linear arrays running from apes to whites. Standard anatomical illustrations, until this generation, depicted a chimp, a Negro and a white, part by part in that order – even though variation among whites and blacks is always large enough to generate a different order with other individuals: chimp, white, black (Gould, 1981, p.88).

Gould also shows how a confused hereditary theory, combined with misuse of test scores as true measures of general, innate intelligence, led to continued pseudo-scientific validation of racist and nativist theories. In 1923, C. C. Brigham published *A Study of American Intelligence*, drawing on a report by Robert M. Yerkes on the results of testing thousands of army recruits.[32] Brigham used the flawed data and incorrect inference to conclude that Jews and other recent immigrants were of subnormal intelligence and that immigration of certain nationalities or ethnicities should be restricted on eugenic grounds. Cornelia James Cannon argued in the *Atlantic Monthly* in 1922 that the test results also demonstrated the wisdom of 'separate but equal' schools in the South. Thus, the indicators of intelligence that were devised by a few scientists destroyed the lives of millions of people, including many who died in the Holocaust – because Jews were denied entry into the U.S. As Gould concludes: 'The paths to destruction are often indirect, but ideas can be agents as sure as guns and bombs' (p.233).

By the 1930s, however, the tide began to turn. Anthropologists provided statistical evidence countering the earlier assessments of differences in intelligence by race. As a result of that work, the Supreme Court declared in 1954 in Brown v. Board of Education that laws imposing 'separate but equal' schools were unconstitutional. The court explicitly cited the work of Kenneth Clark (1950) and others, which showed the deleterious effects of segregation on school performance among African-American children. More than a century after the first statistical justifications of racism had appeared a universalistic approach had been given legal status.[33]

Racist uses of statistical analysis continue to plague us. The serious treatment of The Bell Curve (Herrnstein and Murray, 1984), which treats group differences in intelligence test scores as evidence of innate differences in ability,

---

[32] In 1930, Brigham publicly modified his views, pointing out the biases in the tests from which he had drawn his conclusions, but by then it was too late. The country was in the throes of a nativist spasm, and his recantation was little noted.

[33] It might seem that the 'Coleman Report' (Coleman et al., 1966) and later follow-up statistical analyses perpetuated the justification of segregation by finding that school integration, by itself, made no significant difference in the educational outcomes of African-American children. These studies also found that changes in school spending had little effect on school performance, contrary to conventional wisdom (Jencks, 1972, p.194). But those studies did not confirm racial inferiority. Instead, they demonstrated that schools cannot carry the entire burden of equalizing opportunity and that segregated job and housing markets are greater obstacles to equality of opportunity than segregated schools. More than 30 years later, Americans are still resisting the clear implications of the statistical results.

demonstrates that the issue will not go away. It might seem, therefore, that diagnostic statistical analysis has little value if it enables pernicious ideas to maintain credibility. Yet, consider that in the early 19th century, many abolitionists were convinced that Africans were innately inferior to Europeans, whereas a minority of people in America believe that today. Analytic, not descriptive, uses of statistics were largely responsible for that paradigmatic change in an entire culture. Because the debate was over the status of groups, not of individuals, the defeat of pseudo-scientific racism depended upon statistical arguments.

6. The Negative Income Tax and the Welfare Reform Debate of the 1970s The demonstration research projects conducted in the United States in the 1970s are an interesting application of the diagnostic approach to analyzing social problems. In the 1960s and 1970s, concern about the rising costs of the welfare system led to a number of proposals to reform welfare. One proposal, the negative income tax, was favored by some within the Nixon Administration.[34] A negative income tax essentially allows those on welfare to work and retain a certain percentage of their welfare benefits based on their earned income.

Rather than focus on overall trends of inputs (such as welfare costs) and outputs (such as changes in household income or poverty levels) to assess the overall effects of the welfare system, the demonstration research project used indicators to measure the effectiveness of a particular policy proposal. While it is true that a descriptive approach might analyze the overall effects on subgroups in order to make more accurate generalizations, the diagnostic approach employed in the research project used indicators to test a specific hypothesis about welfare reform.

The research project designed to test the idea of the negative income tax was formulated during the Johnson administration and carried out during Nixon's. The first demonstration project took place in New Jersey. Subsequent studies took place in Denver and Seattle. Advocates of the negative income tax believed that the policy would increase the incentive to work and decrease the overall cost of welfare by allowing people to retain a percentage of their welfare benefits while they earned additional income. The catch was that those who were not on welfare could earn the same overall income while working less by obtaining part of their income from welfare payments. The question, therefore, was whether there would be a net increase or decrease of aggregate welfare payments. Although there is some controversy over the interpretation of results, the general consensus is that the negative income tax does not work to reduce the total cost of welfare (Nathan, 1988, pp.49-60). This surprised many advocates of this approach who had expected the results to be positive.

Another interesting case was the 'supported-work' demonstration project conducted by the Manpower Demonstration Research Corporation (MRDC). Supported-work represents a different hypothesis about how to reform welfare.

---

[34] Meanwhile, in California, Governor Ronald Reagan was developing the 'conservative' vision of welfare reform (simply eliminate benefits) that culminated in the recent welfare reform in the United States.

This approach advocates the use of services (such as counseling, job placement services, vocational training and education,) as a way of helping those dependent on welfare to increase their opportunities to work. This 'service approach' to reforming welfare, as opposed to the 'income maintenance' strategy described above, was not just of interest to those interested in welfare reform. Those interested in increasing employment amongst youth or helping ex-addicts in reentering the workforce were interested in using the same approach in their work (Nathan, 1988, pp.102-3).

The results from the supported work demonstration project showed some interesting results. The supported work approach was most helpful for single-mother AFDC recipients, as it both increased their income and reduced the total cost of welfare. It did not make as much difference for the other groups in the project (Nathan, 1988, pp.106-10). This implied that the most effective 'supported-work' program would target single mother AFDC recipients and that other ideas were needed to address the problems faced by the other groups in the study. Another interesting finding was that the AFDC 'participants with serious problems tended to register relatively larger gains than more experienced and better-qualified workers' (Nathan, 1988, p.119). This finding is counter to the commonplace notion that such programs tend be more useful to those who are already at a relative advantage.

Based on his involvement as a user and producer of public policy research, Nathan advocates the use of large-scale demonstration research projects that are interdisciplinary in design, use both quantitative and qualitative methods and address how to do something rather than what is to be done.[35] In his view, applied social science research should move away from the description of conditions and trends (with its implied goal setting) and onto testing whether a policy (based on some causal hypothesis) works to ameliorate the problem (Nathan, 1988, pp.16-7).

A diagnostic approach to understanding social problems relies on testing hypotheses about what causes problems and what we can do to change social conditions. If the purpose of indicators is to help us improve social conditions and quality of life, we will likely have more success if we employ a diagnostic approach to the work. Although indicators do not play a pivotal role in demonstration research projects described above, the studies illustrate the potential benefits trying to test alternative ideas about what will change conditions and trends, instead of trying to more accurately describe them.

---

[35] Demonstration research amounts to a social experiment to test a hypothesis about whether a policy will work or not. Ideally the projects use random assignment to create a control group to allow comparisons with the test group. There are many technical, administrative and ethical difficulties in conducting large scale experiments discussed at length in Nathan's account of his experience with the Manpower Demonstration Research Corporation. Although this work does not specifically utilize indicators, it is illustrative of the type of positivist approach that Demonstration research evaluates pilot programs, Evaluation research examines the effects of programs already in place.

Conclusion

Although our brief history might suggest that historicism and positivism have run along parallel tracks, they have intersected frequently and contentiously. In the 19th century, a key dispute was over the relative weight to assign social and individual factors in illness. Another debate was about whether to adopt a liberal (laissez-faire) approach to economic life or a managerial (welfare state) approach. In the 20th century, there was a conflict between those who favored the use of Keynesian macro-economic tools and those who hoped to discover the mysteries of the business cycle by examining it in intricate detail.

The influence of Keynesian models and of the diagnostic paradigm grew until the mid-1960s, when faith in them began to evaporate. Since that time, policy has been driven by multi-sector models that rely on masses of data rather than relatively simple models based on theoretical principles. Thus, even within the field of national economic policy, positivism has been largely eclipsed by historicism. (Micro-economic policies are still diagnostic in orientation, but even there, the descriptive framework has creeps in through case studies, anecdotal evidence and the like.)

Much of what passes for analysis today in the social sciences is still only a matter of complex description. A great deal of systems theory now ignores the original diagnostic, cybernetic models (with a few control variables) and substitutes detailed descriptions of a multitude of interactions.

The reliance on descriptive methods is especially true of social indicators. Since descriptive indicators fit more closely with common sense, they have always dominated social indicators research. There would be no harm in that if planners and practitioners expected nothing more from those indicators than impressionistic results. However, many who favor the descriptive approach also hope that it will lead to real social change if they try hard enough. They are doomed to failure.

The more reasonable response to this condition is a displacement of goals. Since descriptive indicators cannot be used as predictive variables, they can become the framework for consciousness-raising and consensus-building processes. Nevertheless, this is not likely to be fully satisfying. Few people want the results of their work on indicators to be purely symbolic or affective.

Yet, if the indicators movement does not transcend the descriptive approach to social phenomena, there is a danger that it will flourish once again for a few years and then die. That happened in the 1870s, 1920s and 1970s when national indicators gained prominence. It also happened in the 1910s when the precursors of community indicators blossomed under the leadership of the Russell Sage Foundation.

Recognition of the contribution offered by positivism could deepen, transform and sustain the social indicators movement. The analytic and predictive orientation of the diagnostic approach would improve all aspects of indicators development.

First, it would point toward more fruitful data collection. Data are relevant to social reform only when they help prove or disprove a hypothesis. The descriptive approach begins with data and hopes to find a pattern in it. The diagnostic approach begins with a hypothesis about what the pattern ought to look like and gathers data to test it.

Second, the diagnostic approach offers causal models for testing. It begins with a relevant question about some social relationship that is perceived as problematic. By contrast, the descriptive approach begins with facts and tries to discover a framework that will make them meaningful. The purpose of diagnostic indicators is an integral part of them, not something added. As a result, if an indicators group followed a diagnostic model in developing social indicators, there is no danger that it would end up by asking, 'What should we do with what we have?' or, 'How can we convince the public that our indicators are significant?' If a group devised indicators that would lead to more effective policies, the value of the results would be self-evident. What is less self-evident is the value of simply collecting statistics that show how terrible conditions are.

Third, a diagnostic approach clarifies the relationship between factual disputes and value formation. Whereas the descriptive approach treats values as fixed formations that can be applied to changing circumstances, positivists regard values as arenas of conflict that emerge from the discovery of new factual premises. Positivists are interested less in general issues (e.g., 'quality health care' as a value) on which there is consensus than in specific issues which are disputed (e.g., single payer vs. managed care vs. health-care vouchers). Diagnostic indicators are of use in part because they make clear the nature of particular value conflicts.

While descriptive indicators enable a consensus to emerge about the existence of problems, they offer no guidance about how to solve them. Descriptive indicators promote the mistaken notion that to know a problem and have the will to solve it is enough. If those who promote descriptive indicators are to be part of the solution and not part of the problem, they will need to examine how they might learn from the diagnostic tradition about the use of indicators as predictive variables.

References

Anderson, M.J. (1988), *The American Census: A Social History*, Yale University Press, New Haven.

Andrews, F.M. (1990), 'The Evolution of a Movement', *Journal of Public Policy*, vol. 9(4), pp.401-5.

Bauer, R.A. (ed.) (1966), *Social Indicators*, MIT Press, Cambridge.

Booth, T. (1992), 'Social Indicators and the Mondale Initiative', *Knowledge, Creation, Diffusion, Utilization*, vol. 13(4), pp.371-98.

Bulmer, M. (1983), 'The Methodology of Early Social Indicator Research: William Fielding Ogburn and "Recent Social Trends", 1933', *Social Indicators Research*, vol. 13, pp.109-30.

Bulmer, M. (1990), 'Problems of Theory and Measurement', *Journal of Public Policy*, vol. 9(4), pp.407-12.

Burns, A.F. and Wesley C. Mitchell. (1946), *Measuring Business Cycles*, NEBR, New York.

Bryant, C.G.A. (1985), *Positivism in Social Theory and Research*, Macmillan, London.

Clark, K.B. (1950), *Effect of Prejudice and Discrimination on Personality Development*, Mid-century White House Conference on Children and Youth, City (cite incomplete).

Cohen, P.C. (1982), *A Calculating People: The Spread of Numeracy in Early America*, University of Chicago Press, Chicago.

Coleman, J.S., et al. (1966), *Equal Educational Opportunity*, U.S. Government Printing Office, Washington DC.

Coleman, W. (1982), *Death is a Social Disease: Public Health and Political Economy in Early Industrial France*, University of Wisconsin Press: Madison.

Cullen, M.J. (1975), *The Statistical Movement in Early Victorian Britain: The Foundations of Empirical Social Research*, Barnes and Noble, New York.

Davis, R.C. (1972), 'The Beginnings of American Social Research', in G. H. Daniels (ed.), *Nineteenth-century American Science: A Reappraisal*, Northwestern University Press, Evanston.

Eyler, J.M. (1979), *Victorian Social Medicine: The Ideas and Methods of William Farr*, Johns Hopkins University Press, Baltimore.

Gaffney, M. (1994), *The Corruption of Economics*, Shepheard-Walwyn, London.

Gigerenzer, G., Swijtink, Z., Porter, T., Daston, L., Beatty, J. and Kruger, L. (1989), *The Empire of Chance: How Probability Changed Science and Everyday Life*, Cambridge University Press, Cambridge.

Gould, S.J. (1981), *The Mismeasure of Man*, W.W. Norton, New York.

Herrnstein, R.J. and Murray, C. (1994), *The Bell Curve: Intelligence and Class Structure in American Life*, Free Press, New York.

Innes, J.E. (1990), *Knowledge and Public Policy: The Search for Meaningful Indicators*, Transaction Publishers, New Brunswick.

Jencks, C. (1972), *Inequality*, Basic Books, New York.

Klein, Philip A. (1992), 'The theoretical basis of cyclical indicators', in Karl Heinrich Oppenlander (ed.), *Business Cycle Indicators*, Avebury, Aldershot and Brookfield, Vermont.

Landau, D. and Lazarsfeld, P.F. (1968), 'Quetelet, Adolphe' in *International Encyclopedia of the Social Sciences*, vol. 13, pp.247-57, Macmillan, New York.

Lecuyer, B. and Oberschall, A.R. (1968), 'Sociology: Early History of Social Research', in *International Encyclopedia of the Social Sciences*, vol. 15, pp.36-53, Macmillan, New York.

Leiby, J. (1960), *Carroll Wright and Labor Reform: The Origin of Labor Statistics*, Harvard University Press, Cambridge.

Mills, C.W. (1959), *The Sociological Imagination*, Oxford University Press, London.

Miringoff, M.L. (1997), *The Index of Social Health 1996: Monitoring the Social Well-Being of the Nation*, Fordham Institute for Innovation in Public Policy, Tarrytown, New York.

Nathan, R.P. (1988), *Social Science in Government: Uses and Misuses*, Basic Books, New York.

Noll, H. and Zapf, W. (1994), 'Social Indicators Research: Societal Monitoring and Social Reporting', in I. Borg and P. Mohler (eds), *Trends and Perspectives in Empirical Social Research*, Walter de Gruyter, Berlin.

Oser, J. and Blanchfield, W.C. (1975), *The Evolution of Economic Thought* (third edn.), Harcourt, Brace Jovanovich, Inc, New York.

President's Research Committee on Social Trends (1933), *Recent Social Trends in the United States*, McGraw Hill, New York.

Seligman, B. (1990), *Main Currents in Modern Economics: Economic Thought Since 1870*, Transaction Publishers, New Brunswick, originally published in 1962 by The Free Press of Glencoe.

Sheldon, E.B. and Moore, W.E. (eds) (1968), *Indicators of Social Change*, Russell Sage Foundation, New York.

Sheldon, E.B. and Freeman, H.E. (1970), 'Notes on Social Indicators', *Policy Sciences*, vol. 1, pp.97-111.

Sheldon, E.B., Parke, R. and Aborn, M. (1983), 'Recollections and Views of Key Figures in the Social Indicators Program', *Items*, vol. 37(4), pp.78-89.

Smith, J.A. (1991), *The Idea Brokers: Think Tanks and the Rise of the New Policy Elite*, Free Press, New York.

Studenski, P. (1958), *The Income of Nations. Theory, Measurement and Analysis: Past and Present*, New York University Press, New York.

U.S. Bureau of the Census, Office of Federal Statistical Policy and Standards (1977), *Social indicators, 1976: Selected Data on Social Conditions and Trends in the United States*, U.S. Government Printing Office, Washington DC.

U.S. Bureau of the Census (1980), *Social Indicators III: Selected Data on Social Conditions and Trends in the United States*, U.S. Government Printing Office, Washington DC.

U.S. Department of Health, Education and Welfare (HEW). (1969), *Toward a Social Report*, U.S. Government Printing Office, Washington DC.

U.S. Office of Management and Budget, Statistical Policy Division (OMB) (1973), *Social Indicators, 1973: Selected Data on Social Conditions and Trends in the United States*, U.S. Government Printing Office, Washington DC.

Westergaard, H. (1932), *Contributions to the History of Statistics*, King, London.

SECTION II
RELATING COMMUNITY INDICATORS TO PLANNING AND DEVELOPMENT

Chapter 3

The Sustainable Calgary Story: A Local Response to a Global Challenge

Noel Keough

Introduction

Since 1996, the Sustainable Calgary Society has been a catalyst for education, research and action on issues of sustainability. This chapter chronicles Sustainable Calgary's experience in identifying, researching and reporting on a set of social, ecological and economic indicators of sustainability. In telling this story, three sets of questions are often asked. First, people ask, 'Where did sustainability indicator reporting come from; that is, what motivated this initiative?' Second, people want to understand why community sustainability reporting is important in the larger scheme of things; 'Why does it matter, how does it advance our thinking on the concept of sustainability? Third, people ask, What's next; after we produce a report where do we go from there; can indicator processes actually promote action to make our communities sustainable? This chapter will address these questions by providing a contextualized account of Sustainable Calgary's indicator project.

In short, Sustainable Calgary emerged out of a particular historical context largely defined over the past 50 years by the emergence of globalization and neo-liberal economics but countered by the rise of the environmental movement and ecological thinking as well as an increasing call for more participatory forms of democracy. Part One provides a brief characterization of globalization as well as an historical overview of the evolution of sustainability and the indicator concept.

From this backdrop of great forces of social change, a modest local initiative emerged in Calgary. Part Two tells this story. Sustainable Calgary is, in large part, a response to the dominance of neo-liberal globalization pursued at the expense of social and ecological sustainability. A small group of citizens came together in 1996 to identify a set of indicators for the community. The inclusive and participatory process used to define, research and report on the indicators was as important as the outcomes. Involving a broad cross-section of citizens in the process has shown to be an effective way to arrive at a truly meaningful set of indicators that reflect the community. The small businessperson begins to understand the ecological impacts of packaging choices, while the social worker sees new linkages among jobs, poverty, and habitat preservation. The Sustainable Calgary process explicitly sought to engage citizens rather than interest groups, stakeholders or sector representatives. The process was consciously designed to harness the diversity of knowledge, experience and expertise represented in the community rather than rely on experts.

In my experience, indicator projects are not an end in themselves but one important piece of the sustainability puzzle. They are an entre into critical discussions and explorations of what kind of a world we want. The debates they spark, the questions they raise and the opportunities for education they present are as important as the indicator reporting itself. The indicators challenge us to re-examine social, ecological and economic assumptions that are taken for granted. This is the focus of Part Three. Starting from Sustainable Calgary's work on a select number of what we call 'indicators in progress', I will illustrate this point by discussing four important issues that I think are central to the challenge of community sustainability. These issues have come to the forefront during my decade of working with sustainability indicators – what does a transformational learning process look like; how do we envision the relationship of humans to the rest of nature; how do we create, nurture and reclaim a sense of local community, in a world enamoured with globalization; and how do we 'problematize' economic growth.

Having achieved our immediate objective by completing two *State of Our City Reports*, we at Sustainable Calgary stepped back to ask ourselves, 'What next; how do indicators contribute to the ultimate goal – a sustainable Calgary?' I take up this question in Part Four. From Jacksonville to Seattle to Hamilton, people involved in indicators processes are asking the same question – we have our indicators report, now what? How de we move from the margins to the centre? Finally, then, this chapter will discuss strategies and initiatives already underway and those still being formulated and debated in Calgary with respect to achieving this ultimate goal. This requires simultaneous bottom-up and top-down strategies – a transformational education process to reach an ever-broader segment of the community as well as effective political, policy and planning mechanisms to mainstream indicators and sustainability.

Part One: The Challenge of Neo-Liberal Globalization

Globalization is a complex social, cultural and economic process. Leading social theorist Manuel Castells brought us the image of a globally networked society shaped by revolutionary communications technologies. He has documented the social problems as well as the promise of this new reality (Castells, 1999). Anthropologists Ulf Hannerz and Arjun Appadurai have written about the complex, novel, cultural mutations that are evolving from this global network of human, material and information flows and commodified relations (Hannerz, 1992; Appadurai, 1996). Each author wants to remind us that globalization is not monolithic as human and cultural agency plays a part in shaping it. Still, within the postmodernist condition described by geographer David Harvey (1989), the prime driver of globalization remains the capitalist economic system. The capitalist system harnesses new communication, transportation and production process technologies as tools to increase and concentrate capital. Harvey has demonstrated how the experience of capitalist development is qualitatively different today due to the compression of this process over space and time.

It must also be acknowledged that neo-liberal globalization has a spatial centre of gravity located in the western developed market economies, which are dominated to a large extent by the United States. As such, the specific form globalization has taken has been greatly influenced by the exhaustion of Fordism and the Keynesian welfare state in the developed market economies as well as the subsequent introduction of a neo-liberalist economic regime de-coupled from any significant system of state regulation (Peck and Tickell, 1994). The hallmarks of neo-liberal globalization include the negotiation of free trade agreements, deregulation of global financial markets, privatization of state enterprises and services, investment liberalization and international integration of production processes.

Most of the debates about globalization focus on the extent and inevitability of the transformation and whether or not the costs outweigh the benefits. The mainstream view of globalization – the one absorbed from CNN and globalization's gurus – is of a borderless world where free markets work their magic through consumer preference, human freedom is enhanced as the nation state fades away and economic man is emancipated (Ohmae, 1996; Freidman, 2000). Opponents point out that neo-liberal globalization has resulted in negative outcomes including a corporatized global economy whereby economic control has shifted substantially to transnational corporations; increased environmental degradation; and created greater gaps between the haves and have-nots (Korten, 1995). Opponents have also outlined in some detail the process of localization as an alternative to globalization (Korten, 1997).

The environment has not escaped the juggernaut of globalization either. James O'Connor is a leading theorist of this environment-globalization nexus. From a Marxist perspective, he theorizes global environmental degradation within what he calls the second contradiction of capital, that is, the inevitable depletion of the natural resource base (the first contradiction being the overproduction of capital) upon which the capitalist system depends (O'Connor, 1998). From this perspective, it could be argued that modern environmentalism, usually traced to the publishing of Rachel Carson's *Silent Spring* in 1962, arose as a response to the crisis induced by the second contradiction. *Silent Spring* was indicative of the growing concern over localized pollution caused by industrialization in the developed world (Carson, 1962). In 1972, the first international environmental conference in Stockholm began to address pollution at an international level. The *Limits to Growth* reports of the mid-1970s heralded a shift from concern over pollution to the effects of the industrial economy on global resource depletion. Since the 1990s, environmental concern has shifted to the effects of industrial growth on the global commons including issues such as global warming, ozone depletion, ocean fisheries depletion, deforestation and loss of biodiversity.

Although processes of globalization have been evolving since humans first populated the earth, the present situation is unique. For the first time in history the vast majority of the peoples of the world live under a capitalist economic system. Technological developments and fallout (e.g. communications, transportation, production technologies and genetic engineering) have created an unprecedented paradox of opportunity and risk (Beck, 2000). Human population and per capita resource demand are bringing humans up against global ecological limits of resource availability and assimilative and regenerative capacity. Overall, the critics make a case that though neo-liberal globalization is real and damaging, it is not monolithic; there are alternatives and the nation state and individuals retain a significant opportunity for agency within, and transformation of, the current global structures and processes.

Sustainability

For most of the 20th century, the term sustainability has been commonly associated with ecological science. From the early 1900s, maximum sustainable yield (MSY) became the guiding principle of resource management particularly as applied to forest and fishery resources[1] (Larkin, 1977). MSY represented the simple principle that a resource could not be extracted at a faster rate than it could be replaced. The term 'sustainable development' was first coined by the World Conservation Strategy (WCS) in the late 1970s. The WCS adapted the concept of sustainability to global ecological systems. While maintaining an ecological focus, the WCS recognized the need to resolve the conflicts between human systems and ecological systems. Sustainable development was popularized by the Brundtland Commission Report, e.g. the United Nations Report on Environment and Economy. The Brundtland report attempted a grand synthesis of environmental protection and human development arguing that they are not incompatible (World Commission on Environment and Development, 1987).

But sustainable development is not without its critics. Wolfgang Sachs (1992) argues that the development paradigm itself, launched as the post-war global economic vision of the United States, is problematic. In his view, development, in its many reincarnations, has always been a rather one-sided conversation intended to bring the global South within the sphere of the modern industrial economy on the global North's terms. From Sach's perspective, since the early 1970s, the environmental problem has increasingly been framed as an economically, socially and culturally undifferentiated human impact on the biosphere. The focus has shifted from the fundamental questioning from the grassroots level of our economic system toward the technical fix of global environmental management designed to maintain the current economic system (Sachs, 1999). Yet despite the continuing debate, the Brundtland concept of sustainable development has captured the imagination of environmentalists, developmentalists and capitalists alike.

As the struggle to define the concept has played itself out, conceptual models of sustainable development have proliferated. Some models depict

---

1 Bell and Morse (2000) point out that *Agenda 21* still calls for the deployment of MSY for international fisheries management.

sustainable development as the common ground between society, economy and the environment (World Commission on Environment and Development, 1987). Other models depict the economy as a sub-set of society and society as a sub-set of the environment (Rees, 1996). Much of the literature positions contending models of sustainable development along an anthropocentric-ecocentric continuum. Neo-liberal accounts presume an unlimited resource base and tout market solutions for all problems. Technocentric models put their faith in the power of technology and expert systems to resolve and mediate the problems of unsustainability. Ecological modernization places great faith in technology while recognizing the need for social equity and advocating more democratic means of decision-making to achieve the desired ends (Mason, 1999). For each of these three models, economic growth remains a central premise of sustainable development. Ecocentric models question economic growth, the role of technology and who holds the power to privilege what kinds of knowledge. For ecocentrisits, social equity is at the heart of the sustainability debate (Eckersley, 1992).

Indicators

Clifford Cobb traces the first attempts to systematically identify indicators of quality of life in the U.S. to research begun in 1910 (Cobb, 2000).[2] In large part, the current wave of quality of life and community sustainability indicator reporting is a response to what many see as the inappropriate reliance on a narrow set of economic indicators – the System of National Accounts (specifically the Gross National Product/Gross Domestic Product), major stock market indices and currency valuations. In her book *Counting for Nothing: What Men Value and What Women Are Worth*, Marilyn Waring traces the birth of national accounts to as far back as 1665. In 1925, Canada and Russia became the first national governments to produce national accounts. Thereafter, motivated by the need to manage the war effort and working from a framework based largely on the work of Milton Keynes, the United States and Great Britain shaped the modern United Nations System of National Accounts (UNSNA). Concurrent with the emergence of the environmental movement in the 1960s, critiques of the UNSNA from outside of the economics profession began to find voice. Leading this movement was Hazel Henderson. At the time it was published, her book *Creating Alternative Futures* was considered one of the most lucid critiques of GNP and the current economic order (Henderson, 1978).

In the North American context, the creation of local sustainability indicator projects has also been a result of grassroots action to revitalize communities. Jacksonville, Florida's Initiative, begun in 1985 and inspired by the work of Hazel Henderson, is perhaps the longest running indicator project. The next significant advance for quality of life indicator development occurred in the aftermath of the Rio Earth Summit. The Summit mandated that the United Nations

[2] In 1914, with funding from the Russell Sage Foundation, the Charity Organization Society of New York produced a study of education, recreation, public health, crime and general social conditions in Pittsburgh.

establish a set of indicators of sustainable development to measure progress on *Agenda 21*, a major component of which was the call for local engagement in sustainable development (Bell, 2000). That mandate has inspired a proliferation of local sustainability indicator projects around the world. One of the most well-known initiatives, Sustainable Seattle, was a result of consultations undertaken in response to the Rio Earth Summit (Atkisson, 1999). Sustainable Calgary's initiative was also largely influenced by the preparations for and aftermath of the Rio Earth Summit.

Agenda 21 has been the catalyst for the creation of environmental and sustainability indicator reporting at the national level as well. Many countries have produced State of the Environment reports.[3] Internationally, the most high profile indicator work has been the United Nations Development Programme's Human Development Index (HDI) (United Nations Development Programme, 2002). The HDI's intellectual inspiration has been the work of Nobel economist Amartya Sen.[4] Sen argues against a utilitarian approach to indicator development, that is, counting the things a person has. Sen takes a rights-based approach that goes beyond the negative freedoms propositions of John Rawls (e.g. the right to avoid discrimination) and asserts the need to examine positive rights or capabilities (e.g. freedom from hunger, the ability to read) (Sen, 1999).

Cobb has investigated the application of Sen's capabilities approach to community quality of life indicator initiatives (Cobb, 2000). Many think indicators are a short cut around conflict and politics – a neutral tool. One of the most interesting assertions of Cobb's work is that indicator development proceeds within and reflects the outcome of human relations mediated by power (Cobb, 2000). Within these power-mediated human relations, individual and group interests are defined from more or less inclusive or exclusive perspectives. Therefore, an analysis of indicator initiatives must include an assessment of the political nature of the sponsors of the indicator initiatives.

A dominant theme in sustainability indicator work has been the application of systems thinking to sustainability indicator processes and analysis (Atkisson, 1999; Bossel, 1999; Bell, 2000; Cobb, 2000). In the past 20 years, some of the most interesting theoretical advances in broad-based indicator development have been the promotion of a capabilities approach; the synthesis of economic, quality of life and environmental indicators under the banner of sustainability; and experimentation with participatory methodologies.

In summary, neo-liberal globalization is part of the ongoing process of capitalist economic expansion. The capitalist process, by its very nature, creates winners and losers. It is based upon the appropriation of the fruits of the labour and

---

[3] To date, Australia is the only study country with a continuing sustainability reporting process. Although Canada's 1996 *State of the Environment Report* was discontinued, the National Roundtable on Environment and Economy has been charged with the creation of a set of sustainability indicators that will eventually include a social and ecological account in the System of National Accounts.

[4] One of Sen's most popular books is *Development As Freedom*, published by Anchor Books.

natural endowments of some individuals, communities and nations by others and the transformation of the natural wealth of the planet into marketable commodities. The concept of sustainable development provides a theoretical and moral wedge to challenge neo-liberal globalization. Sustainability indicators are one tool within the sustainable development toolbox.

Part Two: The Sustainable Calgary Story

Calgary is the fifth largest city in Canada. It was built on the strength of the agriculture, cattle ranching, and oil and gas industries. It sits on the northwest extremity of the great plains of North America at the edge of the foothills of the Rocky Mountains. It is home to nearly 1,000,000 people within a land area of almost 800 square kilometers. The city is generally considered to have benefited tremendously from trade liberalization, particularly the Canada-US Free Trade Agreement (FTA) and the North American Free Trade Agreement (NAFTA). The oil and gas sector is in a boom period with growing exports to the United States, and there is anticipation of growth in the Information Technology sector. Over the past decade, upwards of 25,000 people a year have moved to Calgary, many of them contributing to the growing immigrant population. Calgary is the centre of conservative politics in Canada, but rapid population and economic growth is changing the nature of that conservatism. The combination of unprecedented economic growth and the influx of new Calgarians is challenging the sustainability of this community on many fronts. Even former Conservative Premier Peter Lougheed, one of the champions of the FTA, has recently expressed doubts about the outcome of the free trade regime in terms of impacts on our communities and the sovereignty of the Canadian economy. Meanwhile, the environmental movement has come into its own in Calgary. An ongoing community process included initiatives such as the creation of the Calgary Eco-Centre and Healthy Calgary (affiliated with the Healthy Cities network) and a 1995 workshop entitled 'Reclaiming Our Communities from the Global Economy'.

In 1996, against this backdrop, a core group of citizens came together to form Sustainable Calgary under the banner of The Arusha Centre, a community development centre with a 30-year history in Calgary. The Centre originated as a development education resource centre dedicated to bringing issues of international development and social justice to Calgarians. In the mid 1990s, funding difficulties were a catalyst for the Arusha Centre to refocus on local community development including sustainability, cultural diversity, anti-racism and community economic development. In 1996, a group of Arusha volunteers learned about the Sustainable Seattle initiative and became excited about the prospects for a similar process in Calgary. Sustainable Calgary was founded after the Arusha Centre hosted a series of public forums that were held to introduce the idea of an indicator project and gauge public support. The idea struck a chord with attendance at three public forums exceeding expectations. Sustainable Calgary was founded by a group of educators, university students, engineers, planners, social workers, environmentalists and small business owners, eventually becoming a non-profit society in its own right. Since 1996, more than 2,000 Calgarians have been involved in the process of creating the 1998 and 2001 *State of Our City Reports* (Sustainable Calgary Society 1998, 2001). A third report is being prepared for 2004.

Getting Our House in Order

For the first year, the sustainability indicator working group focused its efforts on researching community indicator projects and the relevant literature; networking with a diverse group of individuals and organizations throughout the city; mapping out a public process to help realize a first *State of Our City Report*; and defining our mission, objectives, principles and sectoral groupings.

The working group identified Sustainable Calgary's mission as the promotion, encouragement and support of community-level discussion, actions and initiatives that move Calgary toward a sustainable future. The working group also identified five major objectives in support of this mission. It would create a focal point for discussion of sustainability issues in general and raise these issues to a higher level of public debate. It would provide an educational tool for teachers, community organizations and private and public sector decision-makers. It would monitor issues, actions and policies that impact Calgary's sustainability and quality of life. It would provide a basis for action and influence policy, planning and community processes. Finally, it would highlight the linkages among social, economic and ecological spheres of activity.

The work of Sustainable Calgary was guided by four sustainability principles. First, a sustainable community maintains or enhances its ecological integrity. It lives in harmony with the natural world. It protects the air, water, soil, flora, fauna and ecosystems on which it depends for its survival. These are the life support systems for all human communities. Second, a sustainable community promotes social equity. Each and every citizen is afforded access to the benefits and opportunities that a community has to offer without social or economic discrimination. Third, a sustainable community provides the opportunity for meaningful work and livelihood for all citizens. A strong, resilient and dynamic local economy is essential for community sustainability. Fourth, a sustainable community encourages democratic participation of all citizens. We live in a democracy. The bedrock of a democracy is citizen participation in the functioning, planning and decision-making of society. In a sustainable community, participation is both a right and a responsibility and should be available to every citizen. Subsequent to a 2001 indicators report, the working-class community of Dover, in Calgary's southeast, completed a local indicators project with the assistance of Sustainable Calgary. The Dover project steering committee added a fifth principle of sustainability – a sustainable community acts responsibly in its relations with all other communities (Dover Community Association, 2002).

Following a review of current literature and practice, the indicator project team compiled a list of sustainability indicators used or suggested by others as a starting point for the selection process. In addition, eight indicator selection criterion were adapted from other processes and the literature on sustainability as follows. First, does the indicator link economic, social and/or ecological factors? Second, will people understand and care about this indicator? Third, will this indicator trigger action? Fourth, is this indicator responsive to interventions? Fifth, is there a way to accurately measure this indicator? Sixth, is the data for this indicator cost effective to collect? Seventh, is this indicator comparable to other reference points and standards? Eighth, is the indicator consistent with Sustainable Calgary's sustainability principles?

We also defined what we mean by a sustainability indicator. An indicator provides an understanding of where you are, which way you are going and how far you are from where you want to be. A good indicator is an early warning of an emerging problem and helps you recognize what needs to be done to fix it. One distinguishing feature of a sustainability indicator is its ability to illuminate the linkages among indicators – a direct or indirect relationship between two or more indicators, where changes in one affect the status of another.

Invoking sustainable development begs the question of what is it one wants to sustain? Over time, it was realized that we needed to provide a response to that question in order to clarify what we were trying to accomplish in this process. To respond to the question, the ends desired and the means chosen to achieve those ends had to be understood. The goal of a sustainable community is to achieve a good quality of life including love, comfort, health, education, physical sustenance, meaningful work, spiritual meaning and a sense of belonging. In a sustainable community, one seeks to attain these qualities through the most efficient and wise use of time, effort and resources. For a long time now, economic growth has been the means chosen to achieve good quality of life. Sustainability reporting helps us examine whether economic growth is the appropriate means to achieve our desired ends. Perhaps a more fitting model, reflecting the natural world, would be a state of dynamic equilibrium where change, innovation and development are possible and desirable, but are not dependent on constant growth. Key to sustainability is the relationship between lifestyle and quality of life. Most Calgarians enjoy a high quality of life. High levels of resource consumption characterize the particular lifestyle that supports our quality of life. Sustainability reporting challenges the community to examine whether this lifestyle is sustainable for the long term and, if not, what changes can be made to create a sustainable lifestyle that can deliver an equal or greater quality of life for our children, grandchildren and future generations.

Finally, prior to launching the public process at the beginning of the second year of this work, we defined a sectoral grouping of indicators we thought made sense for Calgary. Each indicator project creates a unique way of organizing its indicators. We wanted a grouping that was intuitive and meaningful to people at first glance. We avoided a systematized approach used, for example, in the pressure-state-response model; we did not feel the infrastructure-inspired groupings such as housing, transportation, energy, etc., were appropriate for our process as they lean toward dependence on experts. Our 2001 process used six sector groupings as outlined in Table 3.1.

Introducing the Issue of Citizenship

A key element of our public process was the invitation for people to participate in the indicator project as citizens. John Ralston Saul (1995) has written extensively on the importance of acting as citizens in what he describes as a very corporate world, where individual action, corporate allegiance or identity group is the norm. The following passage was included in the materials distributed to indicator workshop participants:

> In this process we are asking people to participate as citizens of Calgary. We are not asking you to represent a particular sector or stakeholder. The issue at stake is the sustainability of our community, not a particular sector or interest. While expertise in a particular area is valuable and a necessary contribution, the process works best when you step out of your own area of expertise and contribute with the big picture in mind, with the good of the community as a whole in mind. Each sectoral think tank should ideally be made up of people with a wide variety of expertise – business people in the community sector, health care professionals in the economic sector, etc. Each person comes to the process with particular indicators they think are important. These views need to be heard and you should be passionate about making your case as to why they are important. In the end we will all be asked, as citizens, to judge what we think are the best indicators to include in the 2001 Report. We should all bear in mind that everybody's choices may not be selected in that process, but it will be a collective decision arrived at through a fair and open process (Saul, 1995, p.1).

The Public Participation Process

What has made the work of Sustainable Calgary effective is our determination to involve as many people as possible through a participatory democratic process. This process is informed by many of the indicator project working group's experience with participatory methodologies including participatory action research, participatory appraisal, popular education and popular theatre techniques. Our process can be broken down into five distinct steps:

1. Process Initiation and Indicator Nomination In 1997, the public process was launched with a city-wide workshop attended by over 100 people. The sustainability indicator concept was introduced, and the process for selection, research and reporting on the indicators was presented and discussed. At the end of the workshop, participants volunteered for one of the five sector working groups, or, as they were called, indicator think tanks. After completing the 1998 report, we identified segments of our community who were underrepresented in the process. These included youth, business, ethnic communities and the disabled. The 2001 process was adjusted to correct this shortcoming. One adjustment was to not rely solely on people coming to our events as in 1998, but to offer to bring the process to them. As a result, the initial phase of the 2001 *State of Our City Report* consisted of a series of over 50 workshops. Some of these were open workshops held at the public library. Others were hosted by particular organizations. The host organizations were a diverse group including the Chamber of Commerce, the Rotary Club, the Developmental Disabilities Resource Centre, Calgary Aldermen and staff, University classes, the Calgary Mennonite Centre for Newcomers and neighbourhood community associations. In these workshops, the 24 indicators documented in the 1998 report were presented and participants were asked to comment on the appropriateness of the current indicators and nominate new indicators for the 2001 report. Participants were aided in their deliberations by the indicator criteria and the sustainability principles. In addition, Sustainable Calgary sponsored a city-wide Youth Creative Writing and Art Contest to engage youth in our process. More than 50 youth submitted their visions of what a sustainable Calgary would look like in the year 2020.

2. Indicator Think Tanks The next phase was to hold a series of three indicator think tank sessions over a six-week period. The think tank sessions were open to all those who had participated in a workshop. Participants self-selected into one of the five or six think tanks. Each think tank reviewed all of the new indicator nominations that had come out of the public workshops and brainstormed additional indicators. The indicator criteria were used as a ranking tool in the process. After lively brainstorming, debate and discussion, each sectoral group nominated five indicators (10 in the 1998 process) to be considered for inclusion in the 2001 Report in addition to the original 24 indicators from the 1998 Report.

3. "Dotmocracy" in Action These 30 nominations (50 in 1998) were presented by the think tank teams to a plenary "Dotmocracy" session. The sector teams presented and justified their nominations. With all of the indicator nominations before us, there was one final opportunity for further nominations from the floor. Then all those present were given five 'sticky' dots and asked to place their dots beside the indicators they thought were the most important indicators to include in the report. All of the votes were tallied, and the top 12 indicators were added to the original 24 from the 1998 report. Thus, we arrived at a list of 36 indicators for the 2001 report.

4. Community-Based Research Following the selection of indicators, volunteers were again invited to join the process as community researchers. The selected indicators were a diverse assemblage of sub-sectors (e.g. crime and safety), indicators (e.g. healthy birthweight babies) and measures (e.g. airborne particulates). The task of the community researchers was to clarify or clearly define the indicator and gather the information required to report on the indicator. The task involved investigative reporting, sleuthing, interviewing and collecting data from local social development agencies, research institutes and government agencies. The community researchers were asked to complete a draft indicator page using the template provided. It included information on indicator definition and current value; its trend over time; why the indicator is important for sustainability; what linkages the indicator had with other indicators; individual and collective actions that could be taken to improve the indicator; and the source of the information collected. In practice, information was often inconsistent, nonexistent or expensive to purchase. Often local information was not available and information aggregated at provincial or national levels had to be used.

5. Indicator in Progress Task Forces The 1998 report had identified several 'indicators in progress' for which no suitable measure or data source could be found. In preparation for the 2001 report, a series of Task Forces were convened to study in more detail the 'indicators in progress'. Research was commissioned with the collaboration of various funders, and workshops were held with individuals with interest, experience and/or expertise in the indicators in progress. Valuing cultural diversity is one example of an indicator in progress. Citizens were not just interested in reporting on the statistical diversity of Calgary's population. They wanted to know if Calgarians value the diversity in our community. We were not satisfied with the measure contained in the 1998 report, so we convened a Task Force including immigrant serving agencies, The City of Calgary diversity program, diversity funders such as The United Way of Calgary, and University of Calgary researchers. After examination of several potential measures, the group came to the following consensus regarding the measure as the diversity represented in positions of power and influence in Calgary – the rationale being that if Calgarians valued the cultural diversity of the city, that diversity would be represented in positions of power and influence. Sustainable Calgary volunteers gathered data on a group of 220 positions representing elected representatives, corporate boards, NGO agency boards and media to measure representation of women, visible minorities and aboriginal people in this group as compared with their representation in the general population.

Is Calgary Sustainable?

One of the first questions asked when Sustainable Calgary is introduced is, of course, "Is Calgary sustainable?" The question has to be answered in a meaningful way that tells a coherent and compelling story and is true to the documented research. In our process, the project team deliberated and arrived at consensus on the sustainability of each of the indicators. We had chosen not to establish targets for the indicators so there was no unequivocal benchmark from which to judge them. The deliberations were qualitative, subjective, informed assessments of each indicator. Each indicator was judged unsustainable, sustainable or neutral based on an assessment of the value of the indicator, that is, whether it was moving toward or away from sustainability and how rapidly it was changing. The results of that ranking are included in Table 3.1. Our process did not include quantitative aggregated indicators by sector as many indicator sets do. Our reports included an assessment by sector using a narrative aggregation.

Our analysis revealed that there are signs of wear in the fabric of our community. Tentative steps are being taken to address resource consumption, and we enjoy a relatively healthy natural environment. Our education and health systems are still among the best in the world but are showing signs of stress. Economically, our prosperity still rests on a lucrative but finite resource and masks intolerable inequities. Given global trends represented by indicators such as

Table 3.1 Calgary's 36 Community Sustainability Indicators

Community Indicators	
Crime Rate and Rate of Victimization	Happy Face
Leisure Activity	Happy Face
Membership in Community Associations	Neutral Face
Number of and Attendance at Public Festivals	Happy Face
Sense of Community	Neutral Face
Valuing Cultural Diversity	Sad Face
Volunteerism	Neutral Face
Economy Indicators	
Housing Affordability	Sad Face
Economic Diversification – Oil and Gas Reliance	Neutral Face
Food Bank Usage	Neutral Face
Hours Required to Meet Basic Needs at Minimum Wage	Sad Face
Income Equity: Gap between Rich and Poor	Sad Face
Unemployment Rate	Happy face
Education Indicators	
Grade Three Achievement Scores	Happy Face
Adult Literacy	Sad Face
Daycare Worker Salaries and Turnover	Sad Face
Lifelong Learning – Library Use	Happy Face
Teacher/Student Ratios	Sad Face
Natural Environment Indicators	
Air Quality	Neutral Face
Bird Population Surveys	Neutral Face
Food Grown Locally	Happy Face
Pesticide Use	Sad Face
Surface Water Quality	Happy Face
Water Consumption	Happy Face
Resource Use Indicators	
Ecological Footprint	Sad Face
Domestic Waste	Happy Face
Energy Use	Sad Face
Population Density	Sad Face
Transportation Infrastructure Spending	Neutral Face
Transit Usage for Work Trips	Sad Face
Wellness Indicators	
Childhood Asthma Hospitalization Rate	Sad Face
Support for the Most Vulnerable	Sad Face
Healthy Birth Weight Babies	Neutral Face
Access to Primary and Alternative Health Resources	Sad Face
Self-Rated Health	Sad Face
Youth Wellness	Neutral Face

ecological footprint (Rees, 1996) and living planet index (World Wildlife Fund, 2002) and our current energy and resource-intensive lifestyle, we judged that Calgary's current trajectory cannot be considered sustainable.

Our community indicators suggest a mixed message. As Calgarians we pride ourselves on our volunteerism and attachment to our city. The prevailing mood seems to be a concern that our quality of life is a little more tenuous than it was five years ago. The area of most concern in this sector is how we value cultural diversity. The leadership among institutions of power and influence in Calgary does not reflect the cultural diversity of the city. On the positive side, crime has been decreasing for several years. In addition, our survey of major public festivals in Calgary found that in the past 10 years the number and diversity of festivals available to Calgarians has increased significantly.

Contrary to most conventional economic analyses of Calgary, our assessment of the sustainability of our economy is troubling. Despite being in a boom period, housing affordability has decreased, homelessness is at an historic high, and individuals or families who make minimum wage must work long hours to meet their basic needs. On the positive side, food bank usage has leveled off over the past two years and employment trends are still strong. There are positive though tentative steps toward diversification. But the oil and gas sector has allowed us to create a high-maintenance city and an affluent lifestyle that a post oil and gas economy will not likely support.

Results are also mixed in the education sector. Grade three achievement scores in language arts continue to improve, and the public library is still a pillar of the community and a vital resource for lifelong learning. Perhaps the most urgent problem is the lack of attention to early childhood education, where Alberta lags far behind the rest of the country. We do not pay workers in this sector a living wage, and, as a consequence, turnover rates are very high and the quality of care is in question.

The natural environment sector shows the most positive signs. The increased environmental awareness of the past 10 years seems to be paying off in a generally healthy natural habitat. Air quality remains relatively good. Bird populations are for the most part healthy, water quality shows signs of improvement and our consumption of water continues to decline. There is a small but growing trend toward growing our own food and supporting local producers through farmers' markets. The biggest red flag in this sector is the continuing use of chemical pesticides on public land and even more so around our own homes.

The resource use sector presents us with one of our biggest challenges. Although most of the indicators have stabilized and there are signs that we have reached a turning point, our levels of consumption are still among the highest in the world. Energy use is still way above that of even European countries, and we actually produce more greenhouse gases per unit of energy consumed today than we did 10 years ago. However, we have begun a slow shift to renewable energy sources. Transit usage for getting to and from work has stalled at about 15 per cent of commuters. The sprawl of our city is a deterrent to more sustainable modes of transportation, but with new subdivision and inner-city developments in the works, opportunity exists to design and build more compact communities. Although we are on the verge of turning the corner in this regard, much creativity and resolve will be needed to put the city on a sustainable course.

The wellness sector reveals an interesting dichotomy. While our access to health care is still the envy of most of the world, the trend for most of our wellness indicators is away from sustainability. Childhood asthma hospital visits have decreased over the past five years but remain very high. One area of concern is the lack of movement toward a preventive care model in our health care system. Alternative health care is growing in popularity, but it is not covered within our universal health care system. One of the most troubling new indicators is the meager resources we make available for the most vulnerable in our society.

The *State of Our City Reports* contained explicit discussion of Calgary in the global village. Our assessment of the sustainability of Calgary is set within the context of relevant global economic, environmental and social development indicators. Ecological footprint analysis suggests that our lifestyle is not sustainable. If every individual on the earth consumed as many resources as the average Canadian, we would need four more planets to provide for everyone. As a result of this over-consumption, the Living Planet Index (compiled each year by the World Wildlife Fund for Nature) warns that every major ecosystem type on the planet is in decline and that since 1960, there has been a 33 per cent reduction in the productivity of these ecosystems (World Wildlife Fund, 2002). Even though Calgarians enjoy a comparatively high quality of life, as the United Nations Human Development Index demonstrates, these global indicators suggest it is at least in part, at the expense of other communities in the global village.

Achievements to Date

Identification and documentation of a set of sustainability indicators for Calgary is an immediate output of the Sustainable Calgary indicator initiative. But obviously it does not stop there. Ultimately the purpose of the initiative is to catalyze actions that move the indicators in a positive direction and to create a truly sustainable community. Beyond the immediate results (outputs), a successful indicator process will see mid-term results (outcomes) and long-term results (impacts).

The immediate outputs are, of course, the easiest to identify. Two *State of Our City Reports* have been published. Two thousand copies of each report have been distribution. More than 2,000 people have been directly involved in the process resulting in a greater awareness and understanding of sustainability among that group.

Mid-term outcomes are also readily identifiable. The indicator initiative has brought a high degree of recognition of Sustainable Calgary and the 36 indicators. Sustainable Calgary is regularly requested to participate in other forums where social, ecological and economic issues are debated and where planning and policy are formulated. Recognition from local media is another outcome of the initiative. The release of our second report was a significant media event in the city, with week-long coverage on television, radio and in newspapers. Sustainable Calgary is recognized by the media as a credible source of information and commentary on issues related to the social, economic and ecological health of

Calgary. Recognition has also come in the form of awards honoring the indicator reporting and the participatory process by which the indicators were derived and documented. The indicator initiative has also been a catalyst for new community research – particularly related to the 'indicators in progress'. Most notably, Sustainable Calgary convened a working group including the City of Calgary, the Calgary Health Authority, the United Way of Calgary and the Calgary Foundation to create a survey instrument to measure sense of community. This is a unique undertaking, and its outcome will have relevance beyond Calgary. The indicator initiative has also spawned several complementary Sustainable Calgary projects including a Calgary Green Map, an Ecological Footprint Educational initiative, the application of community based social marketing to sustainable behavoiur change and an Eco-Expo. Each of these outcomes helps create an enabling environment toward achieving sustainability.

Other outcomes signify an enabling structure or politic. These outcomes include new policies, planning, programs, and practices. While there has undoubtedly been progress in each of these areas, it is difficult to attribute this progress solely, decisively, or even significantly to the indicator initiative. This is not to say that the initiative has not contributed in these ways, but there has not been sufficient tracking of the process to make such attributions. Any attribution has been general and anecdotal. For example, the Calgary Homeless Foundation has used the 2001 report to inform its planning and educational activities.

Ultimately, the impact sought as a result of the indicator initiative is improvement in the status of each indicator and the realization of a dynamic, sustainable community. Although between the 1998 and 2001 reports there were improvements in some of the indicators, there has been no concerted effort to identify the contribution the indicator initiative has made to those positive changes. In all likelihood, any contribution would be minimal at this early stage in the initiative. Likewise, the reporting process is too young and the changes to indicators too small and contradictory (some improving, others worsening) to make conclusive statements about any contribution the indicator initiative has made to an overall trend or indeed if there is a discernable trend.

Part Three: From Action to Theory

The second question posed in the introduction was how does this indicator initiative advance our thinking on the concept of sustainability. In this section that question is addressed in response to four others that arise from an analysis of the indicators, particularly the indicators in progress: 1) what are the key elements of a transformational learning process; 2) how do we envision the relationship of humans to the rest of nature; 3) how do we create, nurture and reclaim a sense of local community in a world enamoured with globalization; and 4) how do we problematize economic growth?

Transformational Learning and Environmental Democracy

The key to the success of the indicator initiative was broad-based participation. Humans are social animals. It is through social interaction that we learn. For each individual, the objective of learning may simply be improved technique; learning to replicate current social, cultural and economic norms; or the transformation of those norms. The process of transformational learning is central to the transition to sustainability. In the context of the sustainability indicator initiative, another key element of participation are the mechanisms that facilitate transformational learning at a societal level – the forms of organization that allow for more participatory democratic processes to flourish. In the sustainability literature, this is often referred to as ecological or environmental democracy. This section will explore both of these fundamental aspects of participation.

The educational field of transformational learning is indebted to the work of Paulo Freire, a Brazilian literacy educator who worked predominantly with the poor and dispossessed of the developing world. But his theory and practice have been widely influential in developed countries as well. Even thirty years after the English-language publication of *Pedagogy of the Oppressed*, Freire (1996) remains the touchstone for the contemplation of, and theorizing and action for transformational learning. Transformational learning presumes a subject who acts in the world rather than an object who is acted upon by the world. For Freire, the emancipatory goal of education was social transformation to a more just society. His key concept, conscientization, refers to 'learning to perceive social, political and economic contradictions and to take action against the oppressive elements of reality' (1996, p.17) – in order to transform it. Friere emphasized that conscientization is a dialectical process, popularized as the action-reflection-action learning spiral.

Over the past decade there has been a dynamic evolution of thinking in the North American adult education literature. The initial exploration of Freire's work in a North American context engaged the critical theory of Jurgen Habermas (see, for example, Henry Giroux, *Critical Theory and Educational Practice*, 1983) and took a postmodern turn in the 1990s to engage questions of difference and culture (see for example, Henry Giroux, *Border Crossings: Cultural Workers and the Politics of Education*, 1992). Most recently a questioning of the postmodern path, in light of what radical educators see as its ambivalence to the ravages of what Peter MacLaren calls the 'multi-scalar, multi-sited, and market driven globalization process' (Holst, 2002, p.xvi), has led to the re-emergence of traditional Marxist analysis (see for example Allman, 1994; Mayo, 1999; McLaren, 2000). Others continue to apply the critical theory of Jurgen Habermas to the question of transformational education in social movements (Welton, 1995).

In *Learning in Social Action: A Contribution to Understanding Informal Education*, Griff Foley (1999) explores the lived experience of social movements as sites for emanicipatory learning. Foley describes processes of emancipatory learning and action as complex, contradictory, ambiguous and contested. For Foley, 'some of the most interesting and significant learning occurs informally and incidentally, in people's everyday lives. And some of the most powerful learning occurs as people struggle against oppression, as they struggle to make sense of what is happening to them and to work out ways of doing something about it.' (1999, p.1). Foley observes that 'learning in social movements is often incidental – tacit and embedded in action and often not recognized for what it is' (1999, p.3). For Foley, two important processes can enhance the learning process: 1) narrative storytelling of what people or communities experience directly, and 2) critical analysis of the dominant economic context of their direct experience – the capitalist system. The Sustainable Calgary indicator initiative was, of course, an incidental tacit learning process, embedded in action.

What does this body of knowledge tell us of the relevance to sustainability of the question of social transformation? I would suggest that there are four essential foundations of human nature that define transformational learning. First is dialectical process. By this I mean that human beings and the social systems and institutions we create are interrelated and simultaneously transform one another. The second is the recognition of subjectivity or human agency embedded in that dialectical process; in other words, consciousness and a free will. Third is social justice or the trajectory of human agency. The human impulse is to seek ethical relations with one another. Fourth, Foley's work reminds us that learning is a process that defines human beings. It is by no means confined to formal educational venues. The learning process is integral to human sustenance, growth and creativity.

If we move from fundamental principles to principles of practice, a distillation of the literature on transformational learning (e.g. Hall, 1981; Tandon, 1988; Kemmis and McTaggart, 1990) reveals at least five principles of practice that can be applied to transformational learning processes, including sustainability indicator initiatives:

1. Contextualize learning socially, politically, economically, culturally and ecologically.
2. Treat learners as subjects of the learning process rather than objects of it.
3. Create a process that encourages each individual to share the roles of teacher and learner.
4. Recognize that there are multiple ways of knowing and understanding the world.
5. Encourage critical and creative learning approaches and techniques.

Turning now to the social organization aspects of transformational learning, Michael Mason, in his book *Environmental Democracy*, seeks to identify a situation wherein a participatory process in support of sustainability is genuinely inclusive, transparent and able to resolve the potential contradictions between environmental protection and social equity (1999). Mason defines such a situation as environmental democracy – 'a participatory and ecologically rational form of collective decision-making: it prioritizes judgments based on long-term generalizable interests, facilitated by communicative political procedures and a radicalization of existing liberal rights' (1999, p.1) to include participatory democratic forms and economic democracy. He arrives at his definition via an exploration of Jurgen Habermas's discourse theory. Within discourse theory, Habermas defines a discourse principle wherein decisions can only be made through debate and discussion with the participation of all those affected by the decision. Habermas also defines an ideal speech situation as existing when the participants in a debate and discussion have access to the same knowledge and can interact in the absence of any form of coercion. Based on his understanding of human nature, Habermas asserts that in such a situation people will seek to express themselves accurately in light of known facts, legitimately in light of ethical norms and to give a truthful account of their feelings and interpretation of their experience. Mason refers to this as a truth discourse. Mason believes that mainstream sustainable development practice falls short of this ideal in that participation is tempered by the market (your economic power), exclusionary expert based decision-making and the difficulty in being inclusive beyond national borders when clearly environmental effects cross borders.

Mason argues that a logical step in the direction of environmental democracy would be toward decentralization of decision-making to the lowest most effective level possible. In the European context, this issue is engaged through the concept of subsidiarity – that decisions should be taken at the lowest effective public authority (Lafferty, 2001). He singles out the notion of quality of life and the tool of sustainability indicators as useful for inviting people to expand their self-interest into notions of community interest and individual as community member as opposed to individual as consumer.

Implicit in Mason's theoretical treatment of participatory process are several practical principles that would facilitate transformational learning: 1) every affected person should be involved in decision-making; 2) every participant requires access to the same information; 3) the process should be free of political manipulation and participants should be free of economic or political coercion and 4) decisions should be taken at the lowest effective level.

Ecological Footprint and Our Relationship with the Rest of Nature

The relationships between Calgarians, other citizens around the world and the rest of nature are complex, especially in this period of accelerating global trade. Ecological footprinting is an evolving methodology and what we referred to in the *State of Our City Reports* as an indicator in progress. The ecological footprint model aspires to quantify the amount of resources an individual, community or nation consumes and how much land area is required to provide those resources on a sustainable basis. The model obliges us to struggle with the question of how much of the earth should be left untouched for the rest of nature. Some ecologists are calling for 12 per cent of the land base to be protected from human activity, others argue for more or less. We estimated that in 2000, on average, Calgarians required nearly 23 acres per person in order to provide the materials, food and energy resources s/he consumes, for a total area of 31,000 square miles (Sustainable Calgary, 2001). The city's physical footprint occupied approximately 164 square miles. If every individual on the earth consumed as many resources as the average Calgarian, we would need at least four more planets to provide for everyone. Put another way, each year human society consumes 30 per cent more of the earth's natural capital than is regenerated (Rees, 1996). Ecological footprint analysis suggests that our current lifestyle is not sustainable and obliges us to reexamine our relationship with other communities and with the rest of nature in order to inquire into the potential of other ways of relating with nature that could maintain or even improve our quality of life. Not unexpectedly, the literature on this issue is monumental.

Capitalism is the dominant economic model of organization in the world today. The mode of human-nature relations in capitalism is quite familiar to us – an instrumental exploitation by capital of the rest of nature for our own species' purposes. Fortunately, a wide spectrum of perspectives, including cosmological, indigenous, eco-philosophical, phenomenological, human ecological, poststructural and historical materialist conceptions can help us craft an alternative to the instrumental relationship.

In *The Universe Story*, Thomas Berry and Brian Swimme's (1992) project is to begin a new story, a modern creation story, congruent with a wide array of current thought in the humanities, social theory and the nature sciences. For Berry 'eventually there is only one story; the story of the universe. Every form of being is integral with this comprehensive story. Nothing is itself without everything else' (Berry, 2000, p.268). Berry and Swimme tell us that without an 'entrancement' toward the world that 'comes from the immediate communion of the human with the natural world, a capacity to appreciate the ultimate subjectivity and spontaneities within every form of natural being...it is unlikely that the human community will have the psychic[5] energy needed for the renewal of the Earth' (p.268).

From the field of human ecology, Paul Shepard has constructed a unique evolutionary perspective on the human-nature relationship (1996). Shepard's vocation has been to demonstrate how animals, 'the Others', made us human. He identifies the Pleistocene era as a pivotal epoch when the essential features of human bodies and consciousness developed as we moved out of the forests onto the grasslands and became omnivorous. Shepard makes the case that our spatial abilities, our capacity for abstract thought and our identities all came about in large measure as a result of our interrelationship with the others. He refers to this as 'the ecological doorway to symbolic thought' (p.15). Similarly, in *Person and Nature in Native American World Views*, Anna Peterson (2001) demonstrates that 'Amerindians view of their world was fundamentally relational...The continuous, extensive and varied social interactions among humans and between humans and non-humans help define, even constitute, all parties involved' (p.120).

The political ecology literature argues the social construction of nature, that is, humans have already altered most of nature so in essence nature is nothing more than what humans define it to be (Haraway, 1991; Escobar, 2001). Carolyn Merchant has chronicled the historical evolution of the social construction of nature through the Enlightenment, Renaissance and Modern Eras with particular

[5] The notion of an original psychic universe comes from Pierre Teilhard de Chardin's *The Phenomenon of Man* (1959).

attention to its collusion with the oppression of women (1980). In *Social Nature: Theory, Practice and Politics*, Noel Castree (2001) argues that post-structuralism has given us the conceptual tools to realize that no conception of nature is human-value free. In *The Social Creation of Nature*, Neil Evernden (1992) argues that social constructionism does not have to be the last word. He makes the case that phenomenological inquiry is a way out of the impasse of dualism created by the Renaissance and the Enlightenment. According to Evernden, if we abandon, or more precisely, set aside the culturally constructed concepts of 'nature...we can experience the world and ourselves as an integral part of it' (p.110).[6]

Sense of Community and the Art of Place-Making

Over the past several years, through its extensive community consultations and workshops, Sustainable Calgary has tapped into a strong desire for a sense of belonging or sense of community in Calgary (Sustainable Calgary 1998, 2001). Our indicator, 'sense of community', acknowledges that desire. In the *State of Our City Reports*, we draw from surveys by the Calgary Health Authority that report the percentage of Calgarians who feel they have people they can call on for help in case of emergency. Sense of community is designated as an indicator in progress because we are not satisfied with the current measures available. In order to improve understanding of this issue, Sustainable Calgary convened a working group to create a survey instrument to measure sense of community.

This indicator in progress illuminates another key concept in the theoretical literature, that of place. I believe place is a strategic rallying point for sustainability because perhaps more than any other concept it is meaningful to people's everyday experience. Until recently the concept of place was thought to be an anachronism. In this age of globalization, abstract and infinite space is the dominant concept. Locality, or local place, is generally devalued (Harvey, 1996). But recently, dynamic and engaging discussions of place have been taken up by anthropologists, geographers, and political ecologists. The re-emergence of place has also been a result of the observation that in a globalizing, fast-paced world, people are seeking security in place in many different ways, some healthy, some not so healthy.[7] So, how do we as humans construct the places we inhabit? Geographers Yi Fu Tuan (1974), Anne Buttimer (1980), Edward Relph (1976) and Nicholas Entrikin (1991) all recognize that place-making is a dialectic process. It captures what Tuan considers a predisposition of the ecology of the human brain to organize phenomena in binary oppositions – 'security and adventure', 'place and journey', 'home and horizons of reach', 'closed and open', 'being (static structure) and becoming (dynamic process)'. The very act of perception – a 'reaching out to the world', (Tuan, 1974, p.16) – embodies this dialectic.

[6] In *Justice, Nature and the Geography of Difference*, David Harvey (1996) alludes to this experienced processual world when he writes that 'process, flux, and flow should be given a certain ontological priority in understanding the world' (p.8).

[7] Appadurai (1996), for example, laments the misreading of ethnic unrest in terms of tribalism rather than as an outcome of the culturally destabilizing processes of globalization.

To help understand place, I distinguish three very distinct kinds of place-making, (1) capitalist; (2) cosmopolitan; and (3) ecological. Capitalist place-making, predominantly an inauthentic[8] exogenous phenomenon, consists of two distinct processes. The first is uneven development – the process whereby capitalist production creates the conditions (e.g. scarcity of resources, labour organizing) for its own demise and the necessity to move to the next best location for exploitation (Smith, 1984). As it establishes itself in a new location, capital reshapes the new location to its own logic. The second process is considered a hallmark of the condition of postmodernity – time-space compression. According to David Harvey (1989), as a result of time-space compression there are fewer anchors of identity in places, in our minds or even in our biology. The pace, breadth and depth of infiltration of the processes of globalization into our lifeworlds seems to have reached the threshold of 'our capacity to grapple with the realities unfolding around us [as] it becomes harder and harder to react accurately to events' (Harvey, 1989, p.306). The technological and external mediation of place-making overpowers our impulse for individual and collective agency.

Cosmopolitan place-making is also premised on an understanding that the compression of time and space has qualitatively altered our experience of the world. The networks and complexities of social interaction and interconnection have become so fluid that particular places are constantly dissolving and reforming, along with individual identities and cultures (Castells, 1999). Some theorists argue that this is not necessarily a negative thing. Certainly a relatively small cosmopolitan elite comfortably circulate in these fluid networks, but exclusion from them defines marginalization for many. Cosmopolitan place-making retains only the slimmest of connections to geographic place and only a marginal recognition of the role the rest of nature plays in the process of place-making.

Ecological place-making maintains that geographic place, the rest of nature and humans are indispensable partners in place-making. Ecological place-making invokes human ecology to question the sustainability of the rapid dissolution and reconstitution of place that is characteristic of the modern world. Ecological place is imagined as a global tapestry of unique and relatively stable place-based communities where change, while not isolated from external influence, is largely driven by internal dynamics. In creating the world we want, it is important to understand the kind of place-making that dominates our communities and to promote debate about what kind of place we want – capitalist, cosmopolitan or ecological – and what kind of place is sustainable.

Nurturing a sense of community or sense of place is dependent on our ability to detach from inauthentic capitalist places and engage in authentic place-making. Strategically we can do four things: 1) We begin by engaging people in local, personal experiences of place-making. In Calgary there are abundant opportunities for authentic place-making. Calgary Dollars (2003), the

---

[8] Use of the term 'authentic', of course, raises the question of how it is defined. Although I will not venture a definition in the limited space available, the term centres around two concepts – wholism and agency.

Neighbourhood Grants Program of The Calgary Foundation (2003) and Prairie Sky Co-housing (2003) are examples of initiatives, that while small, have proliferated over the past decade. 2) We engage in 'global-local' networks of resistance to neo-liberal globalization. The International Council for Local Environmental Initiatives is potentially one such network. 3) We engage in the flows, networks and circuits of capital with the view of reclaiming the space they occupy. An example of this strategy was the defeat of the Multi-lateral Agreement on Investment. 4) We tell our story. Nicholas Entrikin (1991) writes that narrative is a means of inscribing meaning into place and culture. Thomas Berry maintains that meaning expressed in the story of who we are is at the centre of the human project. The story or narrative is also at the heart of the human ecological (Shepard, 1996), feminist (Peterson, 2001), native (Peterson, 2001) and phenomenological (Evernden, 1992) prescriptions for authentic place-making. Each community can begin to honour its own authentic 'story' and create an alternative to the dominant story of neo-liberal globalization. Through strategies such as these, we can gradually expand the field of authentic place-making and contract the field of inauthentic occupation of place. With reference to sustainability indicators, the challenge is to refine the measures that will allow us to determine whether we are expanding the field of authentic place-making.

Economic Diversification and Growth – An Issue of Scale

Another indicator that made its presence felt in the *State of Our City Reports* by virtue of its absence is economic growth – a seemingly unassailable virtue of capitalist economies. Capitalism requires that the exploitation of both labour and nature proceed at an ever-increasing pace as measured by economic growth. Technological advance enables this process (O'Connor, 1998). Conventional wisdom is that economic growth and technological advance are key to the health of our current economic system. An assessment of Calgary's sustainability indicators challenges that conventional wisdom. Further investigation into the role these twin processes play leads to a consideration of the concept of scale – a topic that has recently generated dynamic debate in the academic literature.

As the oil capital of Canada, Calgary hosts one of the paradigmatic symbols of capitalism. Calgary's business and political leaders make much of the strong growth this city has enjoyed. From 1995 to 2001, Calgary's economy grew five per cent annually – almost double the Canadian average (City of Calgary, 2002). However, during this period of strong economic growth, homelessness and foodbank usage increased; housing affordability declined; salaries below the living wage resulted in the highest turnover of daycare workers in the country; government support for the most vulnerable remained below poverty levels; hours of work required at minimum wage increased; and the income gap widened (Sustainable Calgary, 2001). A recent report of the Parkland Institute in Alberta highlighted the fact that a decade of economic growth in Alberta has not resulted in improved household incomes. In fact many households have a reduced real income through this period (Parkland Institute, 2001). The Pembina Institute in Alberta released a report highlighting an alternative to the GDP, our most quoted proxy for national well-being (Pembina Institute, 2001). The Genuine Progress Indicator (GPI) is an alternative to the Gross National Product (GNP). For more than a decade now, economists from around the world have recognized the inadequacy of the GNP (Daly and Cobb, 1989). The GNP counts expenditures for such things as environmental cleanups, repair bills for automobile accidents and legal bills for divorce proceedings as benefits to society. In contrast, the GPI subtracts such negative costs and includes real but uncounted benefits such as unpaid household work to give a more realistic picture of how we are progressing. Even though we have an undeniably high quality of life in Calgary, as confirmed in the Human Development Index, the calculation of Alberta's GPI suggests our quality of life is deteriorating (Pembina Institute, 2001). Taken together these indices suggest that economic growth is not increasing our well-being. In fact, a strong case can be made that it is contributing to a reduction in our well-being at this point in our history. In this full world, economic growth seems to be depleting both natural capital and social capital.

Understanding the concept of scale is key to understanding the ambiguous role played by economic growth. The present interest in scale is in large part attributable to the efforts of radical geographers of the historical materialist tradition to map the contours of neo-liberal globalization, and in particular counter the claim of neo-liberal globalization's boosters, that the scales that have been constructed – imposed – for their agenda, are in no way natural.[9] Geographers such as Harvey (1996), Massey (1994), Smith (1984), Swyngedouw (2001) and Lake (2002) have begun to tease out the social construction of scale. As with the discussion of transformational learning and place, we find in the literature analysis of scale as a dialectic process manifest by variably stable configurations over time (Harvey, 1996). At any point in time particular scales dominate. Geographic analysis has demonstrated that any particular scalar configuration is neither 'natural', static nor immutable. Scale is constituted within a field of cultural, social, economic and physical forces.

Economic growth advocates tend to ascribe to the economy of scale argument that asserts 'bigger is better' while environmentalists often cite E.F Schumacher's 'small is beautiful' dictum (Schumacher, 1974). Both argue for their version of a natural scale of things. Scalar theory challenges sustainable development advocates to think through issues of scale more clearly. Questions such as, "Are there natural human or ecological scales?" require more theoretical consideration through the synthesis of ecological, social constructionist and phenomenological accounts of the human-nature relationship. Though scale is in large measure socially constructed and highly malleable to our interventions, there are likely fundamental ecological, physical and processual rigidities. I would hypothesize that there are 'natural' ecological, economic, technological, cultural, psychic and inter-subjective scales each with their own range of stability.

Sustainable development is usually conceptualized in its social, ecological and economic dimensions. Scalar analysis can help us understand the variable

---

[9] An excellent expose of the fallacy of naturalness in the current scalar order focusing in Canada is Linda McQuaig's *The Cult of Impotence* (1996).

configurations, rigidities and malleabilities in each of these dimensions. Each dimension manifests a unique web of relationships. The totality of all the webs of relationships, although constituted at different scales, itself constitutes a web of relationship. Transformation over time in each dimension creates variable tensile and compressive forces that help define the stability of the totality. Exploring how far these webs can be stretched before they break would be a vital inquiry for sustainability research. In this context, to ask what is the optimum size of our economy is a legitimate question.

Part 4: Beyond Indicator Reporting

As discussed in the introduction, a universal challenge to sustainability initiatives in general and indicator initiatives in particular is how do we take the next steps toward realizing the goal of a dynamic sustainable community? Sustainable Calgary has embarked upon that journey with the vision of a Sustainable Calgary Action Forum. While the exact character of the forum is still emerging, the Sustainable Calgary experience suggests several important roads that must be traveled. From our vantage point education, collaborative research, intervention in the political and policy arenas, creating workable models of sustainable governance and showcasing early adapters and success stories are key elements in the strategy for action.

Education has always been at the heart of the Sustainable Calgary mission. Subsequent to the indicator initiative, Sustainable Calgary has created several initiatives – each with a core objective to increase sustainability literacy. These include the Calgary Green Map, community-based social marketing, ecological footprint and the Eco-Expo. The educational front will be strengthened through an understanding of the process of learning for social transformation.

The *State of Our City Reports* have demonstrated that there are significant gaps in our knowledge about community sustainability including, but not restricted to, the indicators in progress. The nature of sustainability requires holistic collaborative approaches to research that allow multiple fields of expertise, as well as local knowledge, to be harnessed towards complimentary goals. The research must enhance our understanding of social, ecological and economic processes and their relationship to sustainability. The research should also be action-oriented. Applied research through pilot projects would be one vehicle to achieve these ends. Participatory research methodologies would also be consistent with the sustainability principles identified in Part Two of this chapter.

As stressed by Cobb, sustainability indicator initiatives are not politically neutral. It matters whether an initiative is conceived of as triple bottom line, sustainable development or sustainability. The question then is whether indicator initiatives are pursued in ignorance of their political character or consciously engage the political dimension of sustainability. Furthermore, the work of Sustainable Calgary has highlighted the reality that most of the indicators we have chosen are not neatly contained within any particular political boundaries. Given these political and scalar realities, one political strategy is to contribute to building a constituency for sustainability at higher levels of organization – regionally, nationally and globally. As Sustainable Seattle was for us, (Sustainable Seattle, 1995) Sustainable Calgary has become a valuable resource for other communities contemplating or embarking upon similar initiatives. Banff (Town of Banff, 2001), Canmore and Lethbridge, Alberta; and Nelson, British Columbia are examples of communities that have sought to learn from the Sustainable Calgary experience. If we acknowledge the global nature of the sustainability challenge, then building these constituencies enhances the chances of success locally and globally.

A logical next step from indicator reporting is the creation of policies designed to mobilize resources to improve the indicators. While intervention in the mainstream policy-making process is critical, an equally important task is to promote bottom-up policy-making – what one might call creating a people's agenda. The same participatory principle and citizen-based methodology employed in identifying and researching Sustainable Calgary's indicators could be employed in the policy arena. An outstanding example of such an initiative is the city of Porto Alegre in Brazil (Porto Alegre, 2004). Each year as many as 50,000 citizens engage in a year-long participatory budget process where spending priorities for the city are established. Another example is the *Dialogue With The City* initiative in Perth, Australia (Perth, 2004). More than 1,000 people came together for a day-long discussion of what kind of a city they wanted in 2030.

A step towards policy design was taken in the 2001 *State of Our City Report*. We asked the question, "What initiatives would have a significant positive impact on the sustainability of our community?" After much debate and reflection four priority actions were highlighted. 1) Create a sense-of-community audit for assessing all major developments. Our analysis of the indicator research suggested the need to come to a better understanding of what contributes to a strong sense of community and what detracts from it. Such an understanding would be applied to all major proposed developments. 2) Reduce greenhouse gas emissions by 50 per cent over 30 years through reduced energy consumption and a shift to renewable sources. The indicators presented in the report demonstrated that reduction in our energy consumption and a shift to non-polluting forms of energy has the potential to significantly improve our health, our environmental and our economy. 3) Integrate green economic instruments into municipal decision-making. In assessing the indicators, we came to the conclusion that we are trying to build a 21st century society with 20th century economic tools. Green economic instruments would include green tax mechanisms, full-cost accounting procedures, ecological footprint analysis and the genuine progress indicator. 4) Support and promote a culture of simplicity. Our indicator research clarified for us that we need to redefine our quality of life based on healthy living, a spiritual sense of purpose, voluntary simplicity and a balance of work, family and community. One way we could support and promote a culture of simplicity at the municipal government and neighbourhood levels would be to create policies that support people to make this shift in lifestyle and eliminate policies that block or discourage the shift (Sustainable Calgary, 2001).

Another key element that has been identified by sustainability practitioners and theorists is the challenge of transforming hierarchical, vertical, command and control management organizations into horizontal, collaborative, holistic organizations. This is often referred to as sustainable governance. There are at least four important elements of sustainable governance. First, governance, as opposed to government, involves the participation of all stakeholders from the public and private sectors and civil society. Second, the principle of subsidiarity – that management and decision-making should be mandated at the lowest practical level of organization. Third, that cross-sectoral multi-stakeholder coordination is required at the highest levels of organization. Fourth, sustainable governance requires a learning organizational philosophy. At Sustainable Calgary, we are attempting to enshrine these principles into the concept of a Sustainable Calgary Action Forum.

Sustainability will require actions that change the concrete material reality of our lifestyles. Certainly change will have to happen at a large scale to turn the tide of community development toward sustainability, but often an impediment to any action is a sense of hopelessness or uncertainty that there are actually alternatives. To that end, the 2001 report included five success stories highlighting the activities of early adapters, or those we called sustainability pioneers. Calgarians identify strongly with the pioneer spirit that built the city in the 20th century. We invoked that spirit to introduce the new pioneers who are leading the transition to sustainability in the 21st century. The pioneer stories we told included organic farmers, wind energy entrepreneurs, eco-building managers, local currency promoters and collective kitchen organizers. The sustainability pioneers are proof that there are alternatives and that small scale modeling of sustainability is already occurring. Pilot projects are another step toward adoption of full-scale sustainable systems and technologies.

Conclusion

The movement of local sustainability indicator reporting is consistent with a growing realization that cities, regions, nations and the global community all rely upon a solid foundation of sustainable local communities. As citizens engaged in the indicator process, we wrestle with issues like how do we measure the value we place on cultural diversity; how do we measure the spiritual health of our community; what makes a diverse sustainable economy; how do we know we are protecting the biodiversity of life within our urban environment; and what effect does urban sprawl have on our economy, our health and our environment? What we are learning from the process is that we need to move beyond the narrow set of economic indicators we currently employ to make our decisions. We need to pay attention to the qualitative as well as the quantitative aspects of community life. We need to understand the linkages across ecological, social and economic boundaries and to develop the skill to bring a broad range of indicators into our decision-making process without having to reduce everything to dollars and cents.

At its grandest, sustainability can be a link between the practical problems of the world we inhabit in the present and the big questions of meaning and the ultimate fate of humanity. It can serve to expand our field of vision from the shallow horizons of instantaneous stock trading, quarterly earnings reports and four year election cycles towards a field of caring and a meaningful engagement with our own life's work, the seventh generation and what Thomas Berry calls The Great Work – coming to terms with our relationship with the rest of nature (Berry, 2000).

Sustainability reporting is one small but important piece of the sustainability puzzle. I hope I have been able to demonstrate that the practical hands-on nature of indicator initiatives has proven an effective tool for engaging people in the big questions about the world we want and to shed some light on the challenge of moving to the next stage – from reporting to action.

References

Allman, P. (1994), 'Paulo Freire's Contributions to Radical Adult Education', *Studies in the Education of Adults*, vol. 26(2).

Appadurai, A. (1996), *Modernity at Large: Cultural Dimensions of Globalization*, University of Minnesota Press, Minnesota.

AtKisson, A. (1999), *Believing Cassandra: An Optimist Looks At A Pessimist's World*, Chelsea Green, White River Junction.

Beck, U. (2000), 'Risk Society Revisited: Theory, Politics and Research Programmes' in Adam, Barbara, U. Beck and J. Van Loon (eds), *The Risk Society and Beyond: Critical Issues for Social Theory*, Sage Publications, Thousand Oaks.

Bell, S. and Morse, S. (2000), *Sustainability Indicators: Measuring the Immeasurable*, Earthscan, London and Sterling.

Berry, T. (2000), *The Great Work: Our Way into The Future*, Bell Tower, New York.

Berry, T. and Swimme, B. (1992), *The Universe Story: From the Primordial Flaring Forth to the Ecozoic Era – A Celebration of the Unfolding of the Cosmos*, Harper, San Francisco.

Bossel, H. (1999), 'Indicators for Sustainable Development: Theory, Method, Applications: A Report to the Balaton Group', International Institute for Sustainable Development, Winnipeg, Manitoba, Canada.

Buttimer, A. and Seamon D. (eds) (1980), *The Human Experience of Space and Place*, St. Martin's Press, New York.

Calgary Dollars (2003), available online: www.calgarydollars.ca (June 13, 2003).

Calgary Foundation, The (2003), available online: www.thecalgaryfoundation.org (June 13, 2003).

Carson, R. (1962), *Silent Spring*, Houghton Mifflin, New York.

Castells, M. (1999), *Information Technology, Globalization and Social Development*, United Nations Research Institute for Social Development (UNRISD) DP114, UNRISD, Geneva.

Castree, N. (2001), 'Socializing Nature: Theory, Practice, and Politics', in N. Castree and B. Braun (eds), *Social Nature: Theory, Practice, and Politics*, Blackwell, Oxford.

City of Calgary (2002), 'Socio-Economic Outlook 2002-2007: Calgary and Region', City of Calgary, Calgary, Alberta.

Cobb, C. (2000), *Measurement Tools and the Quality of Life*, Redefining Progress, San Francisco.

Daly, H.E. and Cobb, Jr., J.B. (1989), *For the Common Good: Redirecting the Economy Towards Community, the Environment, and a Sustainable Future*, Beacon Press, Boston.

Dover Community Association (2001), *Dover Community Sustainability Project 2001*, available online: www.sustainablecalgary.ca (April, 2004).

Eckersley R. (1992), *Environmentalism and Political Theory: Toward an Ecocentric Approach*, UCL Press, London.

Entrikin, N. (1991), *The Betweenness of Place: Toward a Geography of Modernity*, Macmillan, Hampshire, England.

Escobar, A. (2001), 'Culture Sits in Places: Reflections on Globalism and Subaltern Strategies of Localization', *Political Geography*, vol. 20, pp.139-74.

Evernden, N. (1992), *The Social Creation of Nature*, Johns Hopkins University Press, Baltimore.

Foley, G. (1999), *Learning in Social Action: A Contribution to Understanding Informal Education*, Zed Books, London.

Freire, P. (1996), *Pedagogy of the Oppressed*, Seabury Press, New York.

Friedman, T. (2000), *The Lexus and the Olive Tree*, Anchor Books, New York.

Giroux, H.A. (1983), *Critical Theory and Educational Practice*, Deakin University, Australia.

Giroux, H.A. (1992), *Border Crossings: Cultural Workers and the Politics of Education*, Routledge, New York.

Hall, B. (1981), 'Participatory Research, Popular Knowledge and Power: A Personal Reflection', *Convergence*, vol. 14 (3), pp.6-19.

Hannerz, U. (1992). *Cultural Complexity: Studies in the Social Organization of Meaning*, Columbia University Press, New York.

Haraway, D. (1991) *Simians, Cyborgs and Women: The Reinvention of Nature*, Routledge, London.

Harvey, D. (1996), *Justice, Nature and the Geography of Difference*, Blackwell, Oxford.

Harvey, D. (1989), *The Condition of Post-modernity*, Blackwell, Oxford.

Henderson, H. (1978), *Creating Alternative Futures*, Berkeley, New York.

Holst, J. (2002), *Social Movements, Civil Society, and Radical Adult Education*, Bergin and Harvey, Westport and London.

International Union for the Conservation of Nature and Natural Resources, WWF and UNEP (1980), The World Conservation Strategy: Living Resource Conservation for Sustainable Development, IUCN, Gland, Switzerland.

Kemmis, S. and McTaggart R. (eds) (1990), *The Action Research Reader*, Deakin University Press, Victoria, Australia.

Korten, D. (1995), *When Corporations Rule The World*, Kumerian Press, West Hartford.

Korten, D. (1997), *Post-Corporate World: Life After Capitalism*, Kumerian Press, West Hartford.

Lafferty, W. (2001), *Sustainable Communities in Europe*, Earthscan, London.

Lake, R.W. (2002), 'Exclusionary Environmentalism, Local Self-Sufficiency, and Dilemmas of Scale in Urban Environmental Politics', presented at the Annual Meeting of the Association of American Geographers.

Larkin, P.A. (1977), 'An Epitaph for the Concept of Maximum Sustainable Yield', *Transactions of the American Fisheries Society*, vol. 106(1), pp.1-11.

Mason, M. (1999), *Environmental Democracy*, Earthscan, London.

Massey, D. (1994), *Space, Place and Gender*, Polity Press, Oxford.

Mayo, P. (1999), *Gramsci, Freire and Adult Education: Possibilities for Transformative Action*, Zed Books, London.

McLaren, P. (2000), *Che Guevara, Paulo Freire and the Pedagogy of Revolution*, Rowman and Littlefield, Oxford.

McQuaig, L. (1996), *The Cult of Impotence*, McClelland and Stewart, Toronto.

Merchant, C. (1980), *The Death of Nature: Women, Ecology and the Scientific Revolution*, Harper and Row, San Francisco.

O'Connor, J. (1998), *Natural Causes*, Guilford, New York and London

Ohmae, K. (1996), *The End of the Nation State: The Rise of the Regional Economies*, Simon and Schuster, New York.

Parkland Institute (2001), available online: www.ualberta.ca/~parkland.

Peck, J. and Tickell A. (1994), 'Searching for a New Institutional Fix: The After-Fordist Crisis and the Global-Local Disorder', in A. Amin, (ed.) *Post-Fordism: A Reader*, Blackwell, Oxford.

Pembina Institute (2001), *Charting A Sustainable Future: The Alberta GPI Accounts*, The Pembina Institute, Pembina, Alberta.

Perth (2004), 'Dialogue with the City', available online: www.dpi.wa.gov.au/dialogue/plan.html (March 11, 2004).

Peterson, A. (2001), *Being Human: Ethics, Environment, and Our Place in the World*, University of California Press, Berkeley.

Porto Alegre (2004), 'The Participatory Budget', available online: www.portoalegre.rs.gov.br/op/ (March, 11, 2004).

Prairie Sky Co-Housing Calgary (2003), available online: www.cohousing.org/news/news.html (June 13, 2003)

Rees, W. and Wackernagel, M. (1996), *Our Ecological Footprint: Reducing Human Impact on the Earth*, New Society Publishers, Gabriola Island, B.C.

Relph, E. (1976), *The Phenomenological Foundation of Geography*, Discussion paper No. 21, Department of Geography, University of Toronto, Toronto.

Sachs, W. (1992) 'Environment', in W. Sachs (ed.) *The Development Dictionary: A Guide To Knowledge As Power*, Zed Books, Toronto.

Sachs, W. (1999), *Planet Dialectics: Explorations in Environment and Development*, Zed Books, Toronto.

Saul, John Ralston. (1995), *The Unconscious Civilization: From Corporatism to Democracy*, CBC Enterprises, Toronto.

Schumacher, E.F. (1974), *Small is Beautiful: A Study of Economics As If People Mattered*, Abacus, London.

Sen, A. (1999), *Development As Freedom*, Anchor Books, New York.

Shepard, P. (1996), *The Others: How Animals Made Us Human*, Island Press, Washington DC.

Smith, N. (1984), *Uneven Development: Nature, Capital and the Production of Space*, Basil Blackwell, New York.

Sustainable Calgary Society (1998), *State of Our City Report*, Sustainable Calgary Society, Calgary.

Sustainable Calgary Society (2001), *State of Our City Report*, Sustainable Calgary Society, Calgary.

Sustainable Seattle (1995), available online: www.sustainablesesattle.org (April, 2004).

Swyngedouw, E. (2001), 'Scaled Geographies: Nature, Place and the Politics of Scale', forthcoming in R. McMaster and E. Sheppard (eds), *Scale and Geographic Inquiry: Nature, Society and Method*, Blackwell, Cambridge.

Tandon, R. (1988), 'Social Transformation and Participatory Research', *Convergence*, vol. 21(2/3).

Town of Banff (2001), 'Banff Community Sustainability Indicators 2001: A Citizens Report', Town of Banff, Banff.

Tuan, Y.F. (1974), *Topophilia: A Study of Environmental Perception, Attitudes and Values*, Prentice-Hall, New Jersey.

UNDP (United Nations Development Programme) (2002), 'Human Development Report 2002', Oxford University Press, Oxford.

Waring, M. (1999), *Counting For Nothing: What Men Value and What Women Are Worth*, University of Toronto Press, Toronto and Buffalo.

Welton, M. (ed.) (1995), *In Defense of the Life World: Critical Perspectives on Adult Learning*, State University of New York Press, Albany.

World Commission on Environment and Development (WCED) (1987), *Our Common Future*, Oxford University Press, Oxford.

World Wide Fund for Nature (WWF) (2002), 'Living Planet Index Report', available online: www.panda.org/news_facts/publications/general/livingplanet/index.cfm.

Chapter 4

Indicators and Core Area Planning: Applications in Canada's Mid-Sized Cities

Mark Seasons

Introduction

Urban core areas are complex places that present multiple challenges to planners. In many cities, core areas are stressed, and their viability is affected by changing retail trends, suburbanisation, and myriad social, environmental and economic forces. These areas are a source of concern to local politicians and community residents. There are calls for action, significant investments of public monies are made, and expectations of success are high. In this difficult context, planners responsible for core areas need a clear understanding of the forces at play, and of the impacts of their decisions. Monitoring and evaluation processes, supported by a foundation of indicators, can meet these needs.

This chapter explores indicator use when planning in core areas of Canada's mid-sized cities. Mid-sized cities (50,000–500,000 population) have received little attention from the research community, which tends to focus on large metropolitan centres (Robertson, 1999). As a consequence, we know very little about planning practice in mid-sized cities. The evolution of indicators is reviewed, and the context for planning core areas in mid-sized cities is discussed. The chapter then reports on findings from a survey of indicator use by planners responsible for core areas[1] of mid-sized cities.

Monitoring, Evaluation and Indicators

Planners are acutely aware of the consequences of poorly informed decision-making. These consequences include missed opportunities, misallocation or waste of limited resources, and damage to political and professional reputations. According to Minnery *et al.* (1993), planners need a realistic and objective evaluation of planning alternatives. They also need an improved sense of outcomes

[1] Core areas are defined as the traditional downtown or central business district, plus adjacent neighbourhoods.

and impacts of their interventions in communities (Murtagh, 1998). Planners need clearer definitions of 'success' or 'failure' in specific contexts (Talen, 1996a, 1996b). Answers can be provided by regular and consistent monitoring and evaluation processes, and by appropriate indicators. The combination of significant resource limitations and high public visibility, combined with the general trend towards less and smaller government, means that municipal governments are adopting a corporate model of organizational management and decision making that emphasises efficiency (Seasons, 2001, 2002). Monitoring and evaluation have achieved new importance because of this growing interest in performance, value for money and calls for accountability.

Monitoring implies a continuous evaluation or assessment of activities in policies, programmes, processes or plans. This involves the collection and interpretation of data on a regular basis. Rossi *et al.* (1999) interpret monitoring as the systematic documentation of aspects of performance that indicate whether or not activities are functioning as intended or according to some appropriate standard. Weiss defines *evaluation* as: '... the systematic assessment of the operation and/or outcomes of a program or policy, compared to a set of explicit or implicit standards, as a means of contributing to the improvement' (1998, p.4) of the activity.

Indicators can be described as 'a set of rules for gathering and organizing data so they can be assigned meaning' (Innes, 1990, p.5). They can perform many functions: description, simplification, measurement, trend identification, clarification, communication and instigation (Hoernig, 2001). Indicators fit into a larger information pyramid which is composed of primary and secondary data, and they can be further aggregated into indices. Indicators provide the foundation of information that is monitored on a continual basis to identify trends and patterns, and is then analysed through a formal process of evaluation.

Evolution of Indicators

The literature on indicators is vast. Indeed, there is no shortage of indicators and indicator models from which to choose. There are dozens of groups that develop and apply indicator models, producing hundreds of indicators.[2] Much of the research on indicators emerged in the 1950s, 1960s and 1970s (Warren, 1965; Bauer, 1966; Maslove, 1973; Organisation for Economic Cooperation and Development (OECD), 1973; Stewart *et al.*, 1975) and expanded in the 1980s and 1990s with the arrival of sustainable development.

Traditional indicator categories include economic indicators, social indicators and environmental indicators. Hoernig (2001) notes that economic indicators have been used in the United States since the 1920s. Social indicators achieved prominence in the 1960s. Environmental indicators evolved during the 1970s as the environment became a mainstream issue, and governments responded

[2] Information on dozens of Canadian and international experiences with sustainable indicators may be found at: http://www.rprogress.org/resources/cip/links/cips web.html.

with environmental assessment legislation and processes. Typically, the traditional indicator groupings were discrete – developed and applied in isolation. This changed in the 1980s with the arrival of two powerful, integrative models: sustainable development and healthy communities. Table 4.1 portrays the evolution of indicators.

Table 4.1 Changing Nature of Indicators

	From	**To**
Period	1900-1985	1985-current
Character	Traditional	Broadly defined
	Discrete	Integrative
Type	Economic	Economic
	Social	Social
	Environmental	Environmental
		Sustainable
		Quality of Life
		Performance based
Scale	International	International
	National	National
	State/provincial	State/provincial
Large Municipalities	Regional	
		Large municipalities
		Community
		Neighborhood
Data	Quantitative	Quantitative
		Qualitative
Actors	Technocrats	Multiple stakeholders
Process	Top down	Collaborative
Sponsors	Governments	Governments
		Non-government organizations
		Communities

Source: Seasons (2001).

Sustainable-development-related indicators are now commonly used at the national, regional and local levels in Canada, the United States and many nations overseas (Dilks, 1996; Maclaren, 1996a, 1996b; Flood, 1997). The experiences of large urban centres such as Seattle, WA (Sustainable Seattle) and Hamilton, ON (Vision 2020) come readily to mind; there are many other examples. The healthy community model continues to frame analysis, although it seems to have been eclipsed since the late 1990s by the quality-of-life model.

Recent interpretations of sustainability have been broadened to include measures of governance and the ecological footprint. We see alternatives to traditional economic indicators such as the Genuine Progress Indicator and related indices. There has also been a shift of emphasis to sub-municipal, community or neighbourhood-based indicator models (see Urban Institute, 1999; Baum, 2001). These models focus on issues such as quality of life in the neighbourhood context. Table 4.2 provides examples of indicators, demonstrating the evolution from traditional to more integrative indicators.

The Mid-Sized City Context

Mid-sized cities can be classified according to different population sizes or other criteria (e.g. role in space economies). Researchers at the University of Waterloo's Centre for Core Area Research and Design, and Mid-Size City Research Centre have categorised cities in the 50,000–500,000 population range as mid-sized. This classification reflects the Canadian urban settlement pattern which, in many evaluative categories, has evolved differently when compared with similar-sized cities in the United States (Mercer and England, 2000; Vey and Forman, 2002). As McCann and Simmons (2000, p.84) note, 'the Canadian urban system comprises a majority of small cities, with only a few large cities'. The 50,000–500,000 population range captures all but the largest cities, and represents 23% of the Canadian population.

According to 2001 national census data, there are 84 mid-sized cities in Canada in this population range (Statistics Canada, 2002). The majority of these cities are located in the provinces of Ontario, Quebec and British Columbia. In Ontario, most mid-sized cities are within the Greater Toronto Area or within an hour's commute from Toronto. A similar pattern of concentration of mid-sized cities exists in the Vancouver and Montreal regions. Mid-sized cities in these large metropolitan regions can be the central city in an urban region, or a suburban mid-sized city. However, the experience is different in other parts of the country, where smaller provincial populations mean that cities that would be classified as mid-sized in Ontario are the dominant urban centres. This is the case in the Prairie provinces and Atlantic Canada, where mid-sized cities tend to be stand alone in nature, outside the sphere of large metropolitan centres. In Canada's metropolitan city-regions, this urban form reflects increased suburbanisation and limited population growth in the established urban centers (Bourne, 2000). Significant growth has been experienced in mid-sized cities that are within the commuter-shed of the metropolitan city-regions of Toronto, Vancouver and Montreal. However, the census data indicate that spatial proximity to large metropolitan areas does not guarantee significant population or economic growth.

Multiple and inter-connected factors such as structural economic change, perceived quality of life (in all its dimensions), and the presence or absence of comparative economic advantage can affect location and investment decisions (Filion and Rutherford, 2000). Mid-sized city social issues often parallel those of large metropolitan centres, with the possible exception of the impacts of large-scale immigration experienced in Canada's top-tier cities which are the preferred destination for most new Canadians (Mercer and England, 2000).

Table 4.2 Examples of Community-Level Indicator Models

Type	Examples
Traditional	
Economic	Gross Domestic Product (GDP)
	Income per capita
	Unemployment rate
Social	Highest level of education
	Marital status
	Literacy level
	Population trends
Environmental	Air quality
	Water quality
	Soil quality
Integrative	
Sustainable development	Renewable resource use
	Extent of recycling
	Ration renewable/non-renewable energy use
	Diversity of employment
	Local financial resilience
	Ability to participate in democratic process
Healthy community	Availability of health services
	Accessibility to educational opportunities
	Type, quality and distribution of hard services
Quality of life	Community affordability
	Population resources
	Quality of employment
	Quality of housing
	Community stress
	Health of community
	Community safety
	Community participation
Genuine progress indicator	Ecological footprint
	Non-renewable subsoil assets
	Income distribution
	Debt levels
	Health care
	Costs of crime
Performance measurement	Efficient use of government resources
	Client satisfaction levels
	Meet performance targets/standards

Sources: GPI Atlantic (2001), Hamilton (2001), International Council for Local Environmental Initiatives (ICLEI) (2001b), Pembina (2001).

Depending on the historical evolution, location and asset base of the community, mid-sized cities can face significant economic development challenges presented by global, national or regional economic change – whether structural and/or cyclical in nature. Urban environmental problems such as the rehabilitation and redevelopment of brownfield sites are common in mid-sized cities (Tyler, 2000). Infrastructure (roads, water and waste management) in mid-sized cities is aging and is often obsolete, and many inner-city communities must contend with school closures. Finally, mid-sized cities must deal with urban/rural fringe development conflicts (Bryant *et al.*, 2000).

Clearly, many complicated social, economic and environmental issues are manifest and must be addressed at the municipal government level. There are factors and forces that demand a careful handling of issues in cities, perhaps more so in mid-sized cities that often lack the diversified economic base and own-source revenues to address fiscal, socio-cultural and other challenges. Indeed, urban planners in mid-sized cities have as much need, for many of the same reasons as their large-city colleagues, to understand whether, and how well, their plans and policies have performed. These places address complex planning challenges that are similar in every respect, except perhaps scale, to those encountered in large metropolitan centres.

Planning for Core Areas

Planning for core areas in cities represents a relevant application of monitoring, evaluation and indicators. As Birch notes, 'downtowns ... are ever-changing places' (2002, p.17). Core areas play many roles, and they serve many interests in mid-sized communities. They are often the symbolic and functional heart of communities, proving an identity and sense of place (Robertson, 1995, 1999; Paradis, 2000). Core areas act as a barometer of the overall health and stability of the community. These areas are typically home to the primary business district, and the location of considerable public investment (Robertson, 1995). A thriving core area attracts residents, tourists and investment. A deteriorated core area has adverse impacts on the community's ability to attract economic development, and its overall quality of life (Filion and Rutherford, 2000).

To date, planning for core areas has proven frustrating and expensive, and revitalisation difficult to achieve. Certainly, there are success stories and considerable talk of a renaissance in core areas, especially in the American urban context (Moulton, 1999; Sohmer and Lang, 2001). Recent research by Filion *et al.* (2001) has identified mid-sized American and Canadian cities considered to have successful core areas. These are communities that build on their assets; they carry out critical analyses of their comparative strengths and weaknesses, and develop strategies that are context-specific (Nunn, 2001). Often, this involves a shift in role from traditional retail/commercial activities to a more diverse set of uses and activities.

Levers and incentives such as preferential taxation policies and public–private partnerships are used to create a positive investment climate. Successful core areas often benefit from a stable, major institutional presence, and the emergence of a strong quaternary economic sector. These are places with a high level of urban amenities, and high-density levels of development and activity. They feature a pleasing urban realm with a pedestrian-friendly environment and a lively street-level atmosphere that is attractive to tourists and residents alike. These communities have undertaken effective restoration of the core's cultural landscape, and adaptive re-use of heritage buildings. Core areas in these places are perceived as safe. There is a mix of commercial and residential land uses; retail does not predominate, although it is typically quite strong (see Broadway, 1994; Robertson, 1995, 1999; Bunting and Millward, 1998a, 1998b; Jones, 2000; Birch, 2002).

However, these successful core areas seem the exception. The core areas of many cities in Canada and the United States are in decline. This is not a new process, according to Robertson (1995, p.430), who argues that American downtowns have declined in importance since the 1920s. A similar phenomenon is identified in Canadian cities by Gad and Matthew, who note that as early as the 1920s, many of the higher-order functions associated with core areas had begun to 'scatter over a much larger area' (2000, p.253).

Contributing factors are as diverse as a changing retail landscape (Bunting and Millward, 1998; Jones, 2000), abandoned factory buildings caused by the loss of traditional manufacturing activities due to structural economic change, property tax burdens and the relative attractions of suburban and 'edge' cities (Filion and Rutherford, 2000, p.367). People and businesses have been moving out of the core for decades in many mid-sized cities. The traditional retail/commercial role of core areas has been assumed by suburban malls or 'big-box' retail complexes on the urban fringe (Jones, 2000; Mercer and England, 2000, pp.65-66). Largescale retail projects ('mega projects') in the core have failed in many mid-sized cities, leaving behind a wasteland of empty stores and unleased malls (Winter, 2003).

As Robertson notes, a lack of confidence in downtown 'tends to scare away prospective development' (1995, p.274). A vicious cycle ensues. The resultant challenges faced by vulnerable core areas in cities are considerable from symbolic and economic perspectives. They can include issues of health and safety, social problems such as homelessness and poverty (Peressini and McDonald, 2000), high rates of office and store vacancies, poor building condition, poor aesthetics and design, economic instability, substandard physical infrastructure and a deteriorated natural environment (Broadway and Jesty, 1998; Gad and Matthew, 2000; Grant, 2000).

These are challenges that are shared by many mid-sized cities. In addition, according to Bunting and Filion (1991), mid-sized cities in Canada have an urban form that is dispersed, rather than concentrated around the central business district or core. As a consequence, commercial, retail and institutional development occurs throughout the community rather than in the traditional core area (Gad and Matthew, 2000). This makes urban transit and viable core areas difficult to achieve because population and urban activity are not concentrated in the critical mass necessary to support viable systems.

Finally, mid-sized cities often lack the fiscal resources to intervene to guide the future evolution of their core areas. This combination of social, economic and environmental problems makes planning core areas in mid-sized cities a difficult proposition. Core areas are not simply another residential neighbourhood, nor are they industrial zones or institutional enclaves. Rather, core areas are multi-faceted, complex areas in mid-sized cities. As the preceding discussion demonstrates, core area fortunes wax and wane. Many core areas in mid-sized cities are delicately balanced between success and failure. Few mid-sized cities can afford the adverse fiscal, symbolic or functional costs of a declining core area. In this demanding context, mid-sized cities need sound information to guide their planning of core areas.

Indicators can provide the foundation for core area decision making and planning, monitoring and evaluation. Since indicators can play such an important role, we need to identify indicators that would be appropriate when planning core areas in mid-sized cities. This process of inquiry begins with the establishment of baseline data and information about indicator use in the context of mid-sized city core areas. In this first stage of research, we need to understand how and why indicators are used when planning mid-sized city core areas.

Research Strategy

The research methods comprised a review of the literature on core areas, specifically in mid-sized cities; monitoring and evaluation practice generally, and indicators specifically. This information provided the basis for a structured questionnaire[3] that was designed to identify the nature and extent of indicator use when planning core areas in mid-sized cities. Traditional paper-based surveys, sent through a traditional mail service, were considered too expensive to distribute and easy to ignore. A Web- and email-based survey tool, the *Zoomerang* survey package, was selected because of its ability to design, deliver and manage surveys in a timely and accurate manner. The *Zoomerang* software formatted the questions in an easily understood response format, sent introductory emails, and stored responses, with anonymity guaranteed, on its corporate server. In the first stage of survey launch, 55 municipalities in English-speaking Canada were sent the survey by email.[4] To properly target survey respondents, the research team acquired a current list of Canadian mid-sized municipal contacts from the Federation of Canadian Municipalities. This list identified the Chief Administrative Officer for each municipality, who was contacted by email to identify the individual appropriate recipient for the survey. Responses were slow to the first round of emails, so a second round of reminder emails was sent 10 days later. This strategy boosted responses to 48/55, or 85% of total possible respondents – a very positive response.

---

[3] The questionnaire received ethics approval from the University of Waterloo's Office of Research.

[4] The survey is in translation for later distribution to the remaining 29 mid-sized cities in Canada's French-speaking province of Quebec.

Survey Design

The survey was exploratory in nature. The objective was to determine whether, and how, indicators are used when planning core areas in mid-sized cities. A review of the World Wide Web using the search words 'downtown' and 'indicators' produced thousands of references. This search identified American and Canadian cities that collected indicators that pertained to core areas. This search found that core area indicators tend to be performance based, with more attention paid to economic indicators than social or environmental indicators. A sample of these indicators was incorporated as prompts in the questionnaire. Typical examples of economic indicators included: vacancy rates in office or commercial premises; employment or unemployment rates; building permit information; development activity; and data on attendance rates at downtown events and festivals. Social indicators seemed less prevalent and included factors such as: use of health and social services; crime statistics; and, to a lesser extent, data regarding homelessness.

Environmental factors associated with core areas were rare. Indeed, social and environmental indicators were cited in cities that embraced community-wide sustainability with its triad of economic, social and environmental concerns and goals. As a result, core areas did not receive special attention, but were considered the same as neighbourhoods. This review of the Web and literature identified issues that required further research. In particular, the mid-sized city experience with indicators was difficult to determine. Thus, the survey was designed to elicit responses to basic, fact-finding questions about indicators practice. Respondents were asked to respond to several question themes based on the literature review and Web search. Recent work (Seasons, 2001, 2002) examined factors that seem to impede or facilitate monitoring, evaluation and indicators in regional planning organisations.

Survey respondents were asked to identify, through rank of importance, the factors that facilitate indicator use. Multiple categories of response were possible, and simple percentage data were generated. The questions in the survey included: reasons for indicator use; influences for indicator choice; types of indicators used; frequency of data collection and analysis; responsibility for collecting indicators; responsibility for analysing indicators; data sources; communication of findings; target audiences for indicator reporting; and factors that facilitate use of indicators.

Reasons for Indicator Use

When asked whether they used indicators when planning and managing core areas, 36/48 (75%) responded in the affirmative. Survey respondents demonstrated that indicators are collected and analysed to inform decision makers about trends in core areas, followed by the need to monitor and evaluate municipal policies. These responses are consistent with our understanding that municipal planners are expected to account for policy, programme or policy performance through monitoring and evaluation. Interestingly, few respondents were obliged to collect and interpret indicators because of statutory obligations. This suggests that monitoring and evaluation processes, and supportive indicators, have been developed to meet local municipal needs, rather than provincial laws.

Influences on Indicator Choice

Personal professional experience was cited as the primary basis for indicator choice, followed by knowledge of existing indicator models. This latter finding is encouraging because it suggests that planners in mid-sized cities are conversant with the great variety of available models. The responses also demonstrated that municipal policy guidelines and community-based interventions are important considerations when selecting indicators. The former finding may reflect the emerging performance-based municipal government culture, while the latter suggests the increasing role played by community groups in the selection of indicators.

Types of Indicators Used

As expected, the majority of mid-sized cities use economic indicators when planning and managing core areas. Table 4.3 suggests that the dominant indicators

Table 4.3 Economic Indicators

Indicator	Number of responses ($N = 36$)	Response ratio (%)
Economic indicators use	34	94
Building permit statistics	30	88
Office vacancy rates	27	79
Retail/commercial vacancy rates	26	76
Residential vacancy rates	22	65
Average housing prices	20	59
Development permit statistics	18	53
Property valuation statistics	17	50
Employment by industry	16	47
Unemployment rates	14	41
Business start-ups/closures	13	38
Office absorption rates	12	35
Retail sales/capita	7	21

were building permits, followed by retail/commercial vacancy rates, office vacancy rates and housing prices. These are traditional indicators of the state of retail and commercial activity in core areas. The comparative under-use of data such as retail sales/capita and business start-ups/closures was intriguing. These findings suggest that mid-sized cities may have trouble collecting and analyzing these data, or they have overlooked the significance of these indicators of business performance and investment climate when planning core areas.

When asked whether social indicators were used when planning core areas, a majority of respondents replied positively. This suggests a broadened perspective about the texture and complexity of core areas. The responses to a follow-up question about the types of social indicators used produced some predictable responses, as well as interesting findings. Table 4.4 suggests that data regarding total population by age and sex were collected by most respondents, followed by highest level of education and crime rates in order of frequency. These are common and expected data sets. However, many cities collect data on complex social indicators such as homelessness, usage statistics for social services and ethnicity. This suggests that mid-sized cities must contend with social problems that are similar in type, if not extent, to those of larger metropolitan areas.

Table 4.4 Social Indicators

Indicator	Number of responses ($N = 34$)	Response ratio (%)
Use of social indicators	31	91
Total population by sex and age	28	88
Highest level of education	19	59
Crime rates	13	41
Usage rates-social services	13	41
Homeless statistics	12	38
Ethnicity	12	38
City parks/recreation expenditures per capita	9	28

A slight majority of respondents used environmental indicators. The findings in Table 4.5 are consistent with the literature review and search of the World Wide Web that suggested environmental indicators were under used. The respondents tracked environmental indicators such as amount of natural areas and quality of natural areas. Other top-ranked environmental indicators included water quality and amount of permeable land surfaces. Air quality, which is a high-profile issue in many parts of Canada, was tracked by a minority of respondents. The relative under use of environmental indicators probably reflects the small scale of core areas. Many environmental indicators are collected on a city-wide basis (e.g. air and water quality), thus core-area-specific indicators may be difficult to collect.

Table 4.5 Environmental Indicators

Indicators	**Number of responses (*N* = 34)**	**Response ratio (%)**
Use of environmental indicators	18	53
Miscellaneous: amount of previous surfaces, soil contamination, amount of riparian vegatatio, noise pollution, combined sewer overflows	16	80
Amount of natural areas	14	70
Quality of natural areas	13	65
Air quality	3	15

Data Collection Responsibility

Table 4.6 indicates that urban planners take the lead in indicator-based data collection concerning core areas. Where two-tier government exists (regional/local municipal), regional staff are responsible for data collection. Next in priority are

Table 4.6 Data Collection Responsibility

Responsibility	**Number of responses (*N* = 36)**	**Response ratio (%)**
Planning department staff	32	89
Economic development staff	14	39
Regional government staff	13	36
Business organisations	13	36
Transportation department staff	11	31
Miscellaneous: business improvement areas, national census	11	31
Community organisations	7	19
Stakeholders	3	8
Local elected officials	2	6
City clerk's office	1	3

economic development officials, followed by transportation staff. These findings reflect the typical roles of these departments in municipal organisations. Some non-municipal government groups also collect indicator-based data. For example, business organisations take this role according to some respondents, while community organisations are less involved. These findings suggest that non-governmental stakeholders can take an interest in, and help collect, core area indicators. The findings are similar for the analysis of data.

Data Sources

Responses to a question about sources for indicators data were fairly predictable and consistent with the literature and its emphasis on economic indicators. Municipal databases, such as building permit statistics, were used by a majority of respondents, closely followed by municipal assessment roles and national census data. Municipal census data and special-purpose surveys were used by a bare majority of respondents.

The results in Table 4.7 indicated that ease of access to data, which includes issues of affordability, was a determining factor when developing data collection strategies. Internally generated data, drawn from development applications and assessment databases, are relatively inexpensive to access and manage. While conventional economic analysis is supported with traditional data (e.g. national census), the more elusive social and environmental trends may not be tracked with sufficient rigour because these data are either difficult to disaggregate to the core area level, or may not exist.

Table 4.7 Data Sources

Data sources	**Number of responses ($N = 36$)**	**Response ratio (%)**
Municipal databases (e.g. building permits)	31	91
Municipal assessment roles	29	85
National census	28	82
Provincial or local census	21	62
Municipal special-purpose surveys	19	56
Miscellaneous: departmental databases, industry databases, provincial or federal statistics, NGOs, special-purpose bodies	14	41

Communication of Findings

Most respondents report indicator analyses to the municipal council, followed by reports to committees of council and departmental monitoring reports. These are conventional and common communication methods. A very small number of respondents used report cards to council, a method that is widely used to explain local trends and performance. Communications media include web-based information, published reports and presentations to stakeholders. Brochures and newspaper inserts were rarely used by mid-sized cities.

Target Audiences

As might be expected in municipal governments, the target audiences for a strong majority of respondents are local elected officials. These were closely followed by data and analysis shared with other city staff, and then by community organisations and business organisations. This municipal-corporate orientation is appropriate, yet suggests that core area information needs to be distributed more widely in the interests of enhanced understanding of core area issues and successes.

Factors that Facilitate Indicator Use

Respondents were asked to rank the relative importance of various factors associated with indicator data collection and analysis. When identifying the highest ranked factor, they noted that it was most important to have sufficient fiscal resources when collecting and analysing indicators. This was followed by having staff with sufficient time and the expertise to collect and analyse data, and having plans that lend themselves to evaluation. Other factors, such as a supportive political culture, senior administrative support and community-based support, were considered less important. Table 4.8 summarises these findings.

Table 4.8 Facilitative Factors

Factors	Number of responses ($N = 36$)	Response ratio (%)
Sufficient fiscal resources	11	29
Sufficient time	8	21
Evaluable plans and policies	6	16
Staff with expertise	6	16
Community-based support	5	13
Senior administrative support	4	11
Commitment to monitoring and evaluation	4	11
Political support	3	8
Supportive corporate culture	3	8

Summary and Lessons Learned

The findings suggest that traditional economic and performance-based indicators remain the mainstay of evaluation regarding mid-sized city core areas. This is development indicators, quality of life and healthy community models. Ideally, mid-sized cities should adopt indicator models that address the complex social, economic and environmental challenges encountered in mid-sized city core areas. Our understanding of the complexities associated with planning core areas tells us that indicators must respond to the core area context, and compare and contrast core area trends with community-wide patterns.

The survey of Canadian mid-sized cities suggests that planners are responsible for monitoring and evaluating core area activities, and for the collection and analysis of indicator-based data. This may be recognition of the generalist research skills of well-trained planners. However, constraints of time, funding and expertise are real concerns for planners in mid-sized cities. This means that monitoring and evaluation processes must be efficient as well as effective – simple to manage, yet generating relevant information – and be perceived as such. It also means that indicators must be meaningful and robust, as well as easily understood, collected and analysed. Data availability, accessibility and affordability are significant obstacles created by government statistical agencies which regard census data as profit-generating products. Indicator use would increase if data were more accessible and affordable, and were also disaggregated at the core area level. Otherwise, mid-sized cities will continue to rely on data that are available, but not necessarily appropriate, to meet the community's monitoring and evaluation needs.

Trial and error are important aspects of the indicator selection process. Mid-sized cities should experiment with various indicators, and test their suitability through application. The indicator selection and refinement process should be shared by city planners and community stakeholders to ensure that indicators are relevant and useful, and perceived as such. The primary clients for indicator analysis remain municipal councils and staff in mid-sized cities; other, external stakeholders expect and deserve a greater role in monitoring and evaluation processes. Communication with municipal government politicians and bureaucrats is necessary but not sufficient in communities that expect transparency of process and decision making; community stakeholders should be involved in the indicator selection process.

Mid-sized cities need advice regarding appropriate indicators. This could be provided through the development of models, frameworks or scenarios that identify indicators that are suitable in specific situations. Information needs for core areas are different from those in other parts of mid-sized cities. Table 4.9 provides an illustration of indicators that would be appropriate in mid-sized city core areas.

Table 4.9 Appropriate Core Area Indicators

	Generic community-wide and *core-area-specific* indicators
Economic	Income per capita Income distribution Unemployment rate Labour force participation rate Number and value of building permits Property values Property tax revenue trends Value of public sector investment Affordability of housing Affordability of public transit *Diversity of employment by economic sector* *Occupancy/vacancy rates* *Lease-rental rates/square foot* *Retail sales/square foot* *Event attendance rates* *Number of demolition permits*
Social	Literacy levels Community safety (crime statistics) Accessibility to social services Availability of community recreation facilities Accessibility to educational opportunities Availability of community-oriented retail services Population trends Demographic trends (age, sex, language) *Homeless statistics* *Availability and use of health services* *Multicultural/immigrant services* *Number and type of cultural facilities*
Environmental	Ambient air quality Water quality (e.g. river systems) Amount of public realm versus private property Amount of publicly accessible green space Quality of housing stock Variety of housing types Infrastructure condition (hard services) *Streetscape design (facades, signage, street furniture, etc.)* *Noise pollution* *Number of vacant building sites/parking lots* *Number and contamination level of brownfield sites*

Acknowledgements

This article has been reprinted with permission from the original source of publication, Taylor & Francis Ltd., http://www.tandf.co.uk/journals, 'Indicators and Core Area Planning: Applications in Canada's Mid-Sized Cities', by Mark Seasons, *Planning Practice and Research*, Volume 18, pages 63-80, 2003. The financial support of this research by the University of Waterloo–Social Sciences and Humanities Research Council of Canada, and the UW–Centre for Core Area Research and Design is gratefully acknowledged.

References

Baer, W.C. (1997), 'General plan evaluation criteria: an approach to making better Plans', *Journal of the American Planning Association*, 63(3), pp.329-344.

Bauer, R. (ed.) (1966), *Social Indicators*, Cambridge, MA, The MIT Press.

Baum, H. (2001), 'How should we evaluate community initiatives?', *Journal of the American Planning Association*, 67(2), pp.147-157.

Birch, E. (2002), 'Having a longer view on downtown living', *Journal of the American Planning Association*, 68(1), pp.5-21.

Bourne, L. (2000), 'Urban Canada in transition to the twenty-first century: trends, issues and visions', in: T. Bunting and P. Filion (eds) *Canadian Cities in Transition*, 2nd edn., Toronto, Oxford University Press, pp.26-52.

Bryant, C., Coppack, P. and Mitchell, C. (2000), 'The city's countryside', in: T. Bunting and P. Filion (eds), *Canadian Cities in Transition*, 2nd edn., Toronto, Oxford University Press, Bunting and Filion (1991).

Dilks, D. (1996), *Measuring Urban Sustainability: Canadian Indicators Workshop 19–21 June 1995, Proceedings*, Ottawa, State of the Environment Directorate, Environment Canada and Centre for Futures Studies in Housing and Living Environments, Canada Mortgage and Housing Corporation.

Filion, P. and Bunting, T. (1991), 'Introduction: perspectives on the Canadian city', in Banting, T. and Filion, P. (eds), *Canadian Cities in Transition*, Toronto: Oxford University Press, pp.1-22.

Filion, P., Hoerning, H. and Saneb, G. (2003), 'The successful few: healthy downtowns of mid-size cities', *Journal of the American Planning Association*, forthcoming.

Filion, P. and Rutherford, T. (2000), 'Employment transitions in the city', in: T. Bunting and P. Filion (eds), *Canadian Cities in Transition*, 2nd edn., Toronto, Oxford University Press.

Flood, J. (1997), 'Urban and housing indicators', *Urban Studies*, 34(10), pp.1635-1665.

Gad, G. and Matthew, M. (2000), Central and suburban downtowns, in: T. Bunting and P. Filion (eds), *Canadian Cities in Transition*, 2nd edn., Toronto, Oxford University Press, pp.248-273.

GPI Atlantic (2001), *Genuine Progress Index for Atlantic Canada*, available at: www.gpiatlantic.org.

Grant, J. (2000), Planning Canadian cities: context, continuity and change, in: T. Bunting and P. Filion (eds), *Canadian Cities in Transition*, 2nd edn., Toronto, Oxford University Press, pp.443-461.

Hamilton (2001), *Vision 2020*, available at: www.city.hamilton.on.ca.

Hoernig, H. (2001), 'The role of indicators in municipal planning policy evaluation', unpublished Master's Thesis, Waterloo, ON, School of Planning, University of Waterloo.

Institute for Local Governance and Regional Growth (2000), *State of the Region: Performance Indicators for the Buffalo–Niagara Region in the 21st Century*, Buffalo, University at Buffalo.

International Council for Local Environmental Initiatives (2001a), *Planning Elements for Local Agenda 21: Community-Based Issue Analysis and Priority Setting*, available at: www.iclei.org/la2l/troyan.html.

International Council for Local Environmental Initiatives (2001b), *Cities 21 Core Indicators*, available at: www.iclei.org/cities21/showind.html.

Jones, K. (2000), 'Dynamics of the Canadian retail environment', in: T. Bunting and P. Filion (eds), *Canadian Cities in Transition*, 2nd edn., Toronto, Oxford University Press, pp.404–422.

Maclaren, V.W. (1996a), 'Urban sustainability reporting', *Journal of the American Planning Association*, 62(6), pp.184-202.

Maclaren, V.W. (1996b), *Developing Indicators of Urban Sustainability: A Focus on the Canadian Experience*, Toronto, Intergovernmental Committee on Urban and Regional Research–ICURR Press.

Maslove, A.M. (1973), *Urban Indicators: A Theoretical and Empirical Analysis*, Discussion Paper No. 5, Ottawa, Economic Council of Canada.

McCann, L. and Simmons, J. (2000), 'The core–periphery structure of Canada's urban system', in: T. Bunting and P. Filion (eds), *Canadian Cities in Transition*, 2nd edn., Toronto, Oxford University Press, pp.76–96.

Mercer, J. and England, K. (2000), 'Canadian cities in continental context: global and continental perspectives on Canadian urban development', in: T. Bunting and P. Filion (eds), *Canadian Cities in Transition*, 2nd edn., Toronto, Oxford University Press, pp.55–75.

Minnery, J., Cameron, C., Brown, J. and Newman, P. (1993), 'Evaluation in urban planning: a framework for analysis', *Australian Planner*, 31(1), pp.8-13.

Moulton, J. (1999), *Ten Steps to a Living Downtown*, Washington, DC, The Brookings Institution Center on Urban and Metropolitan Policy.

Murtagh, B. (1998), 'Evaluating the community inputs of urban policy', *Planning, Practice and Research*, 93(2), pp.129–138.

Nunn, S. (2001), 'Planning for inner-city retail development: the case of Indianopolis', *Journal of the American Planning Association*, 67(2), pp.159-172.

Organization for Economic Cooperation and Development (1973), *The OECD Social Indicators Program*, Paris, Organization for Economic Cooperation and Development.

Paradis, T. (2000), 'Main street transformed: community sense of place for non-metropolitan tourism business districts', *Urban Geography*, 21(7), pp.609-639.

Pembina Institute (2001), *Study Measures the Sustainability of Alberta's Progress*, available at: www.pembina.piad.ab.ca/news/press/2001/2001-04-23.ph.

Peressini, T. and McDonald, L. (2000), Urban homelessness in Canada, in: T. Bunting and P. Filion (eds), *Canadian Cities in Transition*, 2nd edn., Toronto, Oxford University Press.

Robertson, K. (1995),Downtown redevelopment strategies in the United States: an end-of-the-century assessment, *Journal of the American Planning Association*, 61(4), pp.429-437.

Robertson, K. (1999), 'Can small-city downtowns remain viable?', *Journal of the American Planning Association*, 65(3), pp.270-283.

Rossi, P.H., Freeman, H. and Lipsey, M. (eds) (1999), *Evaluation: A Systematic Approach*, 6th edn., Thousand Oaks, CA, Sage.

Seasons, M. (2000), 'Monitoring and evaluation: problems and prospects', unpublished research report, Waterloo, ON, University of Waterloo.

Seasons, M. (2001), 'Indicators and Regional Planning: Theory and Practice', paper presented at the Association of Collegiate Schools of Planning Annual Conference, Cleveland, OH, November 8.

Seasons, M. (2002), 'Monitoring and evaluation in urban planning: practice and prospects', *Canadian Journal of Program Evaluation*, 17(1), pp.43-71.

Sohmer, R. and Lang, R. (2001), *Downtown Rebound*, Washington, DC, Fannie Mae Foundation and the Brookings Institution Center on Urban and Metropolitan Policy.

Statistics Canada (2002), *Population and Dwelling Counts for Canada and Census Subdivisions (Municipalities), 2001 and 1996 Censuses*, available at: www12.statcan.ca/english/census01/prod.../table-csdn.cfm?

Stewart, J., Belgue, D., Bond, W., L'Anglais, O. and Turcotte, H. (1975), *Urban Indicators: Quality of Life Comparisons for Canadian Cities*, Ottawa, Ministry of State for Urban Affairs.

Talen, E. (1996b), 'Do plans get implemented? A review of evaluation in planning', *Journal of Planning Literature*, 10(3), pp.248-259.

Talen, E. (1997), 'Success, failure, and conformance: an alternative approach to planning evaluation', *Environment and Planning B: Planning and Design*, 24, pp.573-587.

Tyler, M.E. (2000), The ecological structuring of urban form, in: T. Bunting and P. Filion (eds), *Canadian Cities in Transition*, 2nd edn., Toronto, Oxford University Press, pp.481-501.

Urban Institute (1999), *Building and Operating Neighbourhood Indicator Systems: A Guidebook*, Washington, DC, The Urban Institute.

Vey, J. and Forman, B. (2002), *Demographic Change in Medium-Sized Cities: Evidence from the 2000 Census*, Washington, The Brookings Institution Center on Urban and Metropolitan Policy.

Warren, R.L. (1965), *Studying Your Community*, New York, The Free Press.

Weiss, C. (1998), *Evaluation*, 2nd edn., Upper Saddle River, NJ, Prentice-Hall.

Winter, J. (2003), 'Unintended consequences for well-intentioned plans', *Ontario Planning Journal*, 17(6), pp.6-7.

Chapter 5

Integrating Community Indicators with Economic Development Planning

Rhonda Phillips and Susan Bridges

Emergence of Community Indicators Systems

Community indicators provide information about past and current trends, as well as assisting planners and community leaders in decisions that impact future outcomes. Community indicators represent an immensely exciting and refreshing approach to evaluation, one that incorporates citizen involvement and participation by its very nature. It is, essentially, a 'balance sheet' check for a community. Community indicators are sets of data used to measure the progress of an area over time. According to Besleme and Mullin (1997, p.47), indicators are 'simply measurements that reflect the status of larger systems'.

How did the use of community indicators evolve, and what were major influences? Conceived during the 1960s era of social examination and unrest, indicators began as a way to understand more about communities than traditional economic indicators revealed, by incorporating such measures as demographics, quality of life factors, and environmental quality.

Several local and state governments embraced the idea of this new, more comprehensive measurement system. The idea began to grow, and by the 1970s, a few systems were apparent. One example at the local level was the 1973 *Scorecard Project* by the Fund for the City of New York; at the state level, California's Office of Planning and Research published *Putting Social Indicators to Work: An Annotated Bibliography* in 1977 (Center for Building Better Communities 2001, p.6). In the late 1980s and early 1990s, the attention of indicators measuring systems turned to quality of life and overall community measures, incorporating 'well-being from multiple perspectives, economic, social, and environmental' (Sawicki and Flynn 1996).

During the 1990s another activity emerged that established community indicators' role in planning, policy, and citizen participation activities. The ideas and applications of community sustainability have been called a 'virtual social movement', one in which citizens and other stakeholders, private foundations, public agencies, nonprofit national and international organizations have embraced quickly (Innes and Booher, 2000). Indeed, some recent articles on community indicators contribute their development to the emergence of sustainable development approaches, although indicators have been in use for a number of decades, including the most commonly recognized economic indicators. The interest in sustainable development has been encapsulated, in many ways, into the concepts and applications of community indicators. The idea of incorporating a balanced approach, one in which environmental, social, and economic variables are included, is immensely appealing based on the volume of activities at the community and regional levels.

Essentially, community indicators are bits of information that when combined, provide a picture of what is happening in a local system. They provide insight into the direction of a community: improving or declining, forward or backwards, increasing or decreasing, staying the same (Andrews, 1996; Cobb, 2000; Redefining Progress, 1997). Indicators are gauges for a community, similar to the Dow Jones Industrial Average that indicates the health of the whole stock market although it only includes a small selection of stocks. Combining indicators provides a measuring system to provide clear and honest information about past trends, current realities, and future direction, in order to aid decision-making (Hart, 2001; Oleari, 2000). Community indicators can also be thought of as a report card of community well being. They are often an outcome of a community visioning process and represent the logical next step for setting goals.

A representative of Sustainable Seattle, a non-profit community organization, best described the power of indicators with the following statement:

> The indicators a society chooses to report to itself are surprisingly powerful. They reflect collective values and inform collective decisions. A nation that keeps a watchful eye on its salmon runs or the safety of its streets makes different choices than does a nation that is only paying attention to its GNP. The idea of citizens choosing their own indicators is something new under the stars – something intensely democratic (Donella Meadows, 1995, p.5).

Other communities have found the benefits of indicators projects to include increased awareness of needs within the community. For example, Martin County, Florida's community indicators project focuses on measuring the environmental 'footprint' and sustainability of the County in direct relation to its natural, economic, and social resources. There are dozens of examples of communities throughout the U.S. applying the concepts of community indicators measuring systems; one of the best sources for more information on this is www.sustainable.org.

Indicators are used to generate a variety of information, including annual reports to the citizens, website data, and other types of summaries. The City of Gainesville, Florida in conjunction with Sustainable Alachua County, a non-profit community group, issues intermittent reports that reflect the data collected in their indicators. It is presented in the context of sustainability and relation to the elements of the comprehensive plan. They also issue a different version, a report to the citizens that presents data differently in the context of the overall community. Other public agencies are finding ways to integrate indicators into overall service performance evaluation and monitoring (Ammons, 1996; Bjornlund, 2000).

Research by Innes (1998) implies a more profound implication of designing and implementing a community indicators measuring system: indicators become more powerful as they are embedded into institutions (communities) practices and thinking. The underlying idea is that indicators can represent shared knowledge and can only influence if represents 'a socially constructed and shared understanding created in the community of policy actors' (Innes, 1998, p.8). As such, community indicators measuring systems seem to hold much potential not only as effective evaluation and monitoring systems but also as mechanisms for effective social change.

The following section examines the need for evaluation in economic development planning. Indicators systems represent a potential method for enabling more effective evaluation for economic development planning purposes. For this reason, the basic issues need to be explored. After this section, a case illustrating an approach for integrating community indicators with economic development evaluation is presented.

The Need for Economic Development Evaluation

Economic development planning is a multifaceted issue, with many variables influencing directions and outcomes. It is a highly competitive and complex activity. Cities, counties, states, and even entire regions routinely compete with each other to attract and retain corporate and public investments, tourists, and new residents. Given this competitive environment, communities periodically should review, analyze, and modify programs and resources that affect economic development outcomes.

Hundreds of millions of dollars are spent in the U.S. each year on economic development programs and incentives. Yet, many states and localities do not know how effective these efforts really are, as evaluations are often not conducted (National Conference of State Legislatures, 1998). Uncertainty abounds regarding results of economic development practices, with many states and localities implementing programs and policies based on perceived average successes, rather than 'verified' marginal successes (Phillips, 1996). In other words, states and localities emulate the behavior of other jurisdictions by implementing similar economic development programs and policies. The end result is that most states and localities offer the same type of programs, with outcomes not clearly evaluated.

Why haven't many comprehensive evaluation studies been conducted? Generally, the nature of economic development projects implies difficult methodological problems, which impede evaluation (Felsenstein, et al. 1997; James, 1991; Reese and Fasenfest, 1999). Mowery (1990) described several limitations, which make it difficult to conduct evaluations of economic development programs: the inability to isolate influential factors; and the inability to construct a counterfactual case to test what would have happened in the absence of the program under evaluation. Schmandt (1991, pp.172-173) provided a slightly different set of explanations for the impediments to evaluation of economic development programs: (1) difficulty in distinguishing between effects induced by the programs and those induced by market forces; (2) most programs have not been operational long enough to evaluate the impacts; and (3) most programs are not utilizing performance measures (patents, number of jobs created, private funds leveraged by state funding, enterprise formations, etc.) on a regular basis to provide benchmarks.

Bartik (1994) described another reason why evaluations may not be conducted: resistance to evaluation by government and nonprofit entities. Some of the reasons for resistance include: (1) high costs of conducting comprehensive evaluations; (2) the benefits of evaluation may not accrue to the entity which pays for it; and (3) fear of negative political consequences (Bartik, 1994, p.99; Giloth, 1992).

Given the inherent difficulties in evaluating programs and the resistance to evaluation, are comprehensive evaluations necessary? For a number of year, many researchers in the field promote evaluation of economic development programs as essential (Accordino, 1994; Bartik, 1994; Bearse, 1993; Feller, 1988; Reese and Fasenfest, 1997; Wolkoff, 1990). The rationale for their support of evaluations is usually found in the following arguments: information generated through evaluation efforts can oftentimes influence the status of funding and support for programs when used to justify funding; and information gathered from evaluations can be used to steer subsequent practice. Livengood (1993, p.10) summarized the benefits of conducting local and state economic development evaluation as: evaluation provides a rational basis for the allocation of scare resources; areas for management improvements are identified; feedback form the customers of the programs is obtained and can be used for program planning; the effectiveness of the economic development strategy behind the programs can be assessed; and information regarding the utility of the programs can be used for marketing.

Evaluations typically include measurement of *performance outcomes* and *process outcomes*, which are described as:

Performance Outcomes – These measurements are the most common to evaluation efforts and indicate activities such as number of contacts, number of technologies transferred, number of patents secured (Feller, 1988, p.239), number of firms and jobs created/retained with the program; and in general, service outcomes and quality (Hatry et al., 1990).

Process Outcomes – These measurements are typically qualitative in nature and strive to assess such activities as formation of new partnerships between and among educational institutions, government, and the private sector (Feller, 1988, p.235). These measures are used less frequently because of the difficulty in assessing, yet probably represent one of the more valuable impacts of technology-based economic development programs.

The lack of evaluation at the local and state levels may be indicative of a lack of accurate techniques to assess the outcomes of economic development planning programs and policies. The availability of evaluation options may not be readily apparent. While the need for evaluation is generally accepted by both academicians and practitioners, workable and feasible evaluations systems are not used pervasively at the 'planning' level. One of the more promising developments in economic development planning evaluation techniques has been the re-emergence of community indicators. The application of community indicator measuring systems to economic development planning holds much promise as an evaluation tool.

What makes community indicators any different from other measures of economic development, such as job growth or changes in per capita income? The key is developing an integrative approach – one that considers the impacts of development not only in economic terms, but also the social and environmental dimensions. It is the idea of indicators reflecting *collective values* that provide a more powerful evaluative tool than simply considering the 'economics' of change and growth. By integrating economic development planning into the overall community or region's planning, it will be easier to evaluate. The community indicators measuring system approach also incorporates both ideas of performance and process outcomes, which serves to further facilitate evaluation.

The potential positive effects of integrating economic development planning into an overall community indicators measuring system go beyond evaluation of development outcomes. In some communities, economic development is considered a separate activity, and may be performed primarily by non-profit organizations with little substantive interaction with local governments. There are some situations in which economic development is perceived as a negative activity and a threat to existing quality of life. These perceptions lead to discord between local governments land use planning functions and the private business sector, placing economic development planners in an 'adversarial' role. Integrating economic development planning with overall community planning via an indicators measuring system represents a potentially valuable method to help mitigate this situation. The following case is an example of a community in which economic development planning and any associated activities were perceived as a negative activity; developing a community indicators system was used as a mechanism for overcoming some of the adversity.

Case in Point: Hernando County, Florida

Hernando County is positioned within the Southwestern coastal region of Florida, an area that has experienced rapid growth. Since 1980, Hernando County has grown at a faster rate than Florida and the United States, experiencing a 125% increase in population between 1980 and 1990, and a 30% increase from 1990 to 2000 (Woods and Poole, 2000). It is projected to continue to experience a dramatic population increase, with an estimated rate of 37% from 2000 to 2015. Its population is older than either Florida or the United States, with a median age of 50.88 in 2000 and a projected median age of 60.44 by 2025. The county's 2000 per capita income at $22,008 is lower than both Florida ($27,999) and the United States ($28,469), and is projected to continue this trend through 2025. This is reflective of its low prevailing wages in the County as well as its high retiree population, many of whom are on fixed incomes. The majority of employment opportunities are in the services sector (32.6%), followed by retail trade (24.5%) and state and local government (12.1%). Other sectors are not as large, with manufacturing accounting for only 4.3% and wholesale trade for only 3.1% of the total, less than Florida or U.S. averages.

Hernando County's 'economic' indicators reveal its character as a bedroom community for the Tampa metropolitan area, attracting residents for its lower costs of living. It also is a destination for retirees, which is reflected in its unusually high median age. Another factor is influencing the character too: Hernando County government has expressed the need to diversify its revenue base in order to keep up with the increasing demand for public services. Given the combination of low wages, rapid growth, and increasing demand for public services, Hernando County needs to instate economic development planning. With the accompanying benefits and costs of rapid development, the need to guide the direction of growth increases.

Compounding the problem was that past efforts at economic development planning had been unsuccessful and had received damaging negative publicity. Evaluation mechanisms had not been used, resulting in questionable outcomes that could not be 'proven' either in economic or other terms. The resulting public distrust of economic development activities instigated a chasm between the public and private sectors that has impeded effective communications between the sectors of the community. The negative connotations of economic development planning, or anything associated with it, had led to adversity in relationships and dissolution of trust between citizens, government, non-profit organizations, and private businesses.

The Hernando County government, in conjunction with civic leaders in the community, decided that they needed a guidance system in place in order to direct future growth, rather than just manage the outcomes of whatever growth happens to come their way. While the need for economic development planning is paramount, they also realized that economic activities would have to be integrated into overall community activities, or it would fail as past efforts had. This led to the exploration and application of a community indicators measuring system. Thus, the desire to evaluate and benchmark, through community indicator research, the goals, objectives, and performance of past and current economic development activities, as well as establish targets for future outcomes was the impetus for the project described in this article. The first step in designing an integrative approach for an economic development measurement system in Hernando County was selection of the indicators, followed by integration with a geographic information systems (GIS) system to facilitate use. Both these elements are described in further detail in the following sections.

Selecting the Indicators

There are literally hundreds of indicators a community can chose to incorporate into a measuring system. Generally, indicators are divided into three major categories: economic, social, and environmental. Rather than arbitrarily select commonly used indicators in each of the categories, Hernando County participated in a wide-scale community assessment. The purpose of the assessment activities was to identify major issues, translate these into goals and objectives, and then select indicators to gauge progress towards the goals. This was accomplished via the following activities:

1. A historical and current economic conditions analysis;
2. An assessment of development factors and resources;
3. A strength and weaknesses assessment via personal interviews of 35 government, business, and civic leaders in the community;
4. Telephone surveys to solicit input from a randomly selected sample of 400 citizens;
5. Two citizen forums utilizing a strategic planning approach to collectively identify issues and goals where hundreds of citizens participated; and
6. A mail survey of 100 businesses in Hernando County to gauge attitudes towards development and related issues.

The motivating factor was to identify major issues of concern to the community, and translate these into goals and actions. Indicators were then selected that provide information to gauge progress towards resolving issues. The intended use of the indicators includes performance evaluation of policies, plans, and programs aimed at community economic development, as well as informing the citizens about the important trends in the community. It is also anticipated that the information generated by the set of indicators will be considered in policy setting and adjusting activities by the local government.

Seventy-two indicators in three categories were selected as feasible, measurable, and important to the overall community effort to resolve common issues. These are described below, followed by a summary of how the community indicators have been integrated into a (GIS) to further its usefulness.

Economic Indicators

Gauging the economic health of a community is vital. The ability of a local economy to provide job opportunities and income to residents is paramount, and impacts the remaining community factors such as social and environmental aspects. The economic indicators for Hernando County are divided into six areas of county revenue sources, population, employment, income, poverty, housing and real estate. Table 5.1 provides a summary of these indicators, the data sources used to calibrate the indicators, and the reasons for including each.

Table 5.1 Economic Indicators for Hernando County

	Indicator	Data source	Update	Reason
County Revenue Sources	Property Tax (taxable real estate value)	County property appraiser	Yearly	Reveals the value of homes in the area, which indicates economic prosperity of residents.
	Franchise Fees	Budget office, or Accounting and Finance Department	Yearly	Reveals the types of companies operating in the area, and hence the diversity of the economic base.
	User Fees (Impact Fees) and Service Charges	Budget office, or Accounting and Finance Department. Also Fl. Legs. Committee on Intergovernmental Relations Date Depository, http://fcn.state.fl.us/acir	Varies	Reveals commercial waste production and service consumption.
	Utility Fees	Most municipalities collect a Public Service tax for telecom, electric, and water services	Monthly	This provides a link to sustainability, by revealing the usage rates (through taxes paid) of electricity and water. Reveals commercial energy consumption.
Population	Incorporated and unincorporated	Florida Office of Economic and Demographic Research, www.state.fl.us/edr/population/citypop2000.pdf. Also available at http://fcn.state.fl.us/lcir/databank/popdata.html	Yearly	Shows where the majority and minority of people live, and may determine allocation of services. Also establishes trend of growth and migration for the city. When linked to other indicators, can determine why population grew or not.

	Indicator	Data source	Update	Reason
	By race, gender and age	www.state.fl.us/edr/population/citypop2000.pdf. Also available at http://fcn.state.fl.us/lcir/databank/popdata.html	Yearly	Determines level of diversity in community. Can link this to education, and social equity indicators.
Employment	Total employment and total number of jobs	Hernando has 1995 numbers www.webcoast.com/dhernand.htm. Current information available at bureau of labor statistics http://stats.bls.gov/news.release/cpi.toc.htm	Yearly/ Quarterly	Strong indicator of economic health of a community.
	County employment by industry	US Dept. Of Commerce, Bureau of Economic Analysis (Hernando has 1995 info)	Yearly	Determine rate of economic diversity.
	# of Business Licenses	From Hernando Code enforcement	Monthly	Reveals diversity of economic base, as well type of jobs available.
	Quarterly Employment and Wages	Agency for Workforce Innovation, through U.S. Department of Labor, Bureau of Labor Statistics	Quarterly	Reveal what industry most jobs are in, and how much those industry workers are making. This indicates economic diversity and spending capacity.
Income	Median, household, personal Income	Florida Research Economic Database, http://fred.labormarketinfo.com/. Also BLS http://stats.bls.gov/blshome.htm	Yearly	Indicator of economic health of community.
	Income Growth Rates	Florida Research Economic Database, http://fred.labormarketinfo.com/. Also BLS http://stats.bls.gov/blshome.htm	Yearly	Indicates progress. If incomes remain stagnant over several years, could indicate lack of economic development capacity.

	Indicator	Data source	Update	Reason
	Per Capita Personal Income	Florida Research Economic Database, http://fred.labormarketinfo.com/. Also BLS http://stats.bls.gov/blshome.htm	Yearly	
	Per Capita Personal Income Growth % of Fl.	Florida Research Economic Database, http://fred.labormarketinfo.com/. Also BLS http://stats.bls.gov/blshome.htm	Yearly	Helps to determine how Hernando compares to state averages.
	Florida Price Index	Florida Research Economic Database, http://fred.labormarketinfo.com/. Also BLS http://stats.bls.gov/blshome.htm	Yearly	Suggests where Hernando falls in its personal spending capabilities in Florida.
Poverty	% of pop on public assistance	U.S. Census		If large amount on public assistance may indicate a lack of jobs, or declining economic base.
	Number of residents living below federal poverty levels, and by ethnicity	U.S. Census		If large amount on public assistance may indicate a lack of jobs, or declining economic base.
Housing & real estate	Total number of housing units (low income, single family)	Local builders association	Yearly	
	# of low income homes as % of total homes	Housing Division, Habitat for Humanity, Local Builders Association	Yearly	Reveals commitment to social equity, and identifies the number of residents who live in certain types of housing.

Indicator	Data source	Update	Reason
Percent owner v. Percent renter	Property Appraiser's Office	Yearly	Increased levels of home ownership suggest the presence of a more stable tax base. This translates into stability of public services (education).
Building Permits Issued for new Construction	County Code Enforcement. For revenues assessed from these contact County Budget Office	Monthly	Indicates growth and economic vitality. Increased construction suggests more people are buying new homes, also indicates increase of revenue for county, through ad valorem taxes.
Housing Sales and Rental Affordability	Local Builder's Association	Yearly	Determines what type of income levels can purchase and rent homes in Hernando County.

Table 5.2 Social Indicators for Hernando County

	Indicator	Data Source	Update	Reason
Education	School population	School Board/Local Government	Yearly	Student population data helps to determine if overcrowding is an issue. This data also helps in comparing Hernando County with other counties in Florida and throughout the United States. Hernando County can determine where it stands and perhaps recognize some reasons for its successes and determine areas that need some additional attention.
	Student-teacher ratio	School Board/Local Government	Yearly	
	Drop out rates	School Board/Local Government	Semester	
	Percent of adults with college degrees	School Board/Local Government	Yearly	
	Degrees awarded from county colleges	School Board/Local Government	Yearly	
	Teacher qualifications	School Board/Local Government	Yearly	
	Test scores	Princeton Education Testing Services	Yearly	
	College admissions	School Board/Local Government	Yearly	
	Dollars spent per student	School Board/Local Government	Yearly	
	Pre-school and day care	School Board/Chamber of commerce	Yearly	Quality and type of early childhood care and education. Certifications of local programs.
	Parks and facilities available	Department of Parks and Recreation	Yearly	Types of programs available for youth, for enrichment, continuing education and after-school activities

	Indicator	Data Source	Update	Reason
Crime	Number of violent crimes reported	Police Department/Census Data	Yearly	
	Dollars spent on security measures	Police Department/Census Data	Yearly	
	Ratio of police officers per 1000 people	Police Department/Census Data	Yearly	
	Rehabilitation programs	Police Department/Census Data	Yearly	
	Number of drug related arrests	Police Department/Census Data	Yearly	
Health and Wellness	Number of hospital beds per capita or per 1000	Local Government	Yearly	A key element in the quality of life for any community is the quality and availability of health care. These indicators give some idea of how Hernando County ranks with other communities with regards to health care. This will also help determine whether there is sufficient health care capacity to provide high quality service to citizens. Finally, these indicators provide information about the effectiveness of emergency services.
	Rates of occurrence of various diseases	Local Government	Yearly	
	Top 5 causes of death	Local Government/Census Data	Yearly	
	Low birth weight rate	Local Government/Census Data	Yearly	
	Available/use of emergency rescue services	Local Fire Department/Paramedics	Yearly	
	Availability of medical specialists	Local Hospitals	Yearly	
	Emergency call response time	Emergency services providers	Yearly	

	Indicator	Data Source	Update	Reason
Transportation	Time spent commuting	Survey of residents	Yearly	
	Quality of existing roads	DOT	Yearly	
	Dollars budgeted for road construction/rep.	County Commissioners	Yearly	
	Number of auto accidents/deaths	Police Department/hospitals	Yearly	
	Number of autos registered	County records	Yearly	
	Demand for public transit	Surveys	Yearly	
Organizations	Active volunteer organizations	Survey/Chamber of Commerce	Yearly	This will provide information on how active citizens are in their community. This will also give some evidence as to the amount of disposable income available and the types of activities on which residents choose to spend their free time.
	Rate of volunteerism	Survey	Yearly	
	Contributions to charitable organizations	Survey	Yearly	
Government	Number of registered voters	Board of Elections	Yearly	These indicators will provide a view of how involved citizens are with their government institutions and processes.
	Percent of voter turnout	Board of Elections	Election years	
	Community attendance at public meetings	Survey	Yearly	
	Number of people running for public office	Survey	Election years	

Social Indicators

Social indicators selected include a variety of areas, and reflect the issues identified during the community assessment activities. There are 72 indicators, divided into the six categories of education, crime, health and wellness, transportation, organizations, and government (see Table 5.2). Social indicators attempt to determine how the community is changing in a variety of contexts by providing information to such questions as: is crime increasing? Is the quality of the educational system improving? Are transportation needs being met?

These indicators are vital to overall community health and well-being and as such, provide the underlying basis for economic development outcomes. It is difficult to consider economic development planning goals and activities without concurrently considering the social impacts of proposed actions.

Environmental Indicators

Environmental indicators are a vital and necessary component of an overall community evaluation system. Too often, environmental factors are pushed to the forefront after an issue reaches crisis status. It is a better approach for communities to integrate monitoring of their environmental health, resources, and status into overall planning efforts, rather than waiting until an issue has to be addressed.

Four categories of indicators, with 13 total indicators, have been selected for use in the community indicators framework for Hernando County (see Table 5.3). These reflect 'benchmarks' for gauging the impact on the environment of the rate and type of growth. The four categories of water, solid waste, air, and land cover several measures within each and when combined, provide a comprehensive picture of the state of the environment and use of resources in the area.

Integrating Community Indicators with GIS

Combining community indicators with a geographic information system (GIS) interface creates a powerful tool that planners and citizens alike may use to create, and track progress towards a comprehensive vision for the future of the community. It is difficult sometimes to rely solely on quantitative data, such as lists of community indicators, to compel people to action. However, integrating those same indicators with visual aids, in the form of maps, can provide a stimulus that encourages immediate response and action (O'Looney, 1997).

The GIS system constructed for use with the community indicators consists of a series of layers based on spatial datasets, using the ArcView software. Some features, such as parcels and land use types, stand-alone. Others, such as roads, hazards, and conservation areas, are buffered with priority zones in 100- or 1000-foot increments. The user may overlay as many layers as needed, and use the combined layers to narrow down desired regions where development is optimal. For example, a business owner in search of an appropriate site to relocate may

Table 5.3 Environmental Indicators for Hernando County

	Indicator	Data Source	Update	Reasons
Water	Water Use	Utility organizations	Yearly	Reveals the usage over time
	Total number of customers	Utility organizations	Yearly	Reveals the usage over time
	Water quality measures	Utility organizations	Yearly	Indicates needed protection measures
	Wastewater disposal	Utility organizations	Yearly	Indicates needs versus capacity and effectiveness
Solid/hazardous waste	Tons of waste	Landfill records	Yearly	Reveals the amount of solid waste and whether recycling is influential, etc.
	Hazardous waste	Landfill records	Yearly	Gauges the level and types of hazardous waste, correlate with overall environmental quality
Air	Average daily vehicle miles traveled	Transportation department records	Yearly	Measure of road use and relates to emissions
	Average levels of air pollution	Pollutant studies	Yearly	Measures common pollutants and reveals changes in air quality over time
Land	Major bodies of water	Local maps	Yearly	Protect from pollution
	Wetlands	Local maps	Yearly	Protect from pollution
	Sinkholes	Local maps	Yearly	Protect from pollution/consider development
	Storm surge	Local maps	Yearly	Consider development
	Acres of land in ‘conservation’	Local maps	Yearly	Amenities for future/protects environment

combine the buffered roads layer (for visibility and proximity to traffic) with the airport and its vicinity (for importing materials), then intersect this layer with the layer containing vacant commercial parcels. These parcels may then be color-coded based on their proximity to *both* roads and the airport, and the resulting map will show only the parcels in which the business owner is interested. In terms of integrating with overall community indicators, the system will provide those sites that are zoned appropriately and are a reflection of the overall development goals of the community.

On a larger scale, the system may be used by county planners to determine future land use designations, special districts, and other planning priorities. Integrating the visual aspects of the community with the indicators provides a clearer picture of where and what type of development should occur. Another important feature of the system is its flexibility. Any spatial dataset in ArcView format may be imported into the system and included in the analysis. This allows the users to customize the system, and include the most up-to-date information. Even simple maps that illustrate basic development outcomes, such as existing and future patterns of development can be used for highly effective impact when interacting with citizen and other groups. For example, the following figure shows the increase in future residential areas of Hernando County.

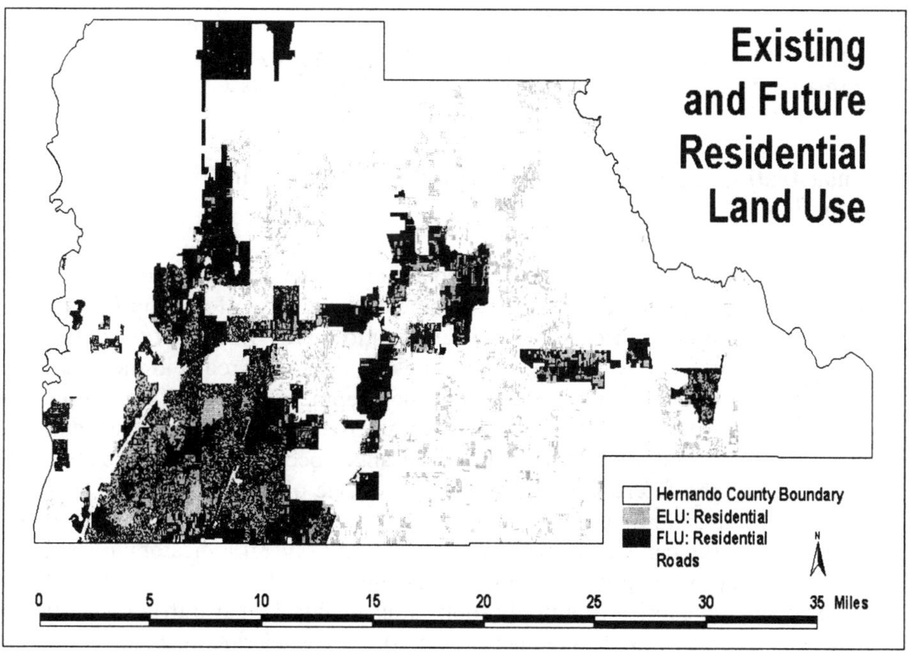

Figure 5.1 Land Use Map for Illustrating Effects of Growth

The ability to show dramatic changes, due to current policies, can help build support for proposed changes or for effective policies already in place. For example, this map shows that while existing residential land uses are spread throughout the county, future residential is more concentrated. While concentrated residential development generally represents a more desirable development pattern, Hernando County has to carefully integrate policy regarding commercial development to support it. For example, focusing on neighborhood or cluster commercial activities instead of linear type development could help mitigate the issue of sprawling development patterns. Additionally, economic indicators for housing and real estate can provide additional information on the types of residential development that should occur. If the citizens truly desire less sprawling patterns of development as expressed in their forums (and identified as an issue), then policies to promote concentrated developments should be supported. Using even a simple map of this sort can help citizens better understand the outcomes of existing and proposed policies, and can further build support for working towards desirable goals.

Whenever possible, the data in the GIS system matches the community indicators described in the previous section; however, the indicators are not always available as spatial datasets that may be displayed in ArcView. Thus, the interface with the community indicators is limited in the sense that only those indicators that can be converted into spatial representation are used. It seems as if the best use of the GIS system is to help determine where the county wishes to focus its development efforts, in conjunction with the community indicators that provide information to determine what should comprise those efforts.

Demonstrating the Significance of an Indicators Measuring System for Evaluation

GIS provides a powerful visual tool in economic development planning. One of the first steps in evaluating economic development planning in Hernando County was to import land uses, both existing and future, into a GIS system. It was clear that the most intensive land uses were stretched along the major roads in the county (as expected), and clearly follow a sprawling development pattern of strip shopping centers and individual commercial sites characterized by numerous curb cuts and access points. Unless policy and regulatory changes occur, the existing pattern will continue, given the future land use map overlay. One of the major issues identified during the goal setting process was the need to control sprawling development patterns. This issue can be 'operationalized' by using several indicators to gauge the impact of policy and regulatory changes, including those found in the transportation and air quality categories (the implication of air quality is that traffic congestion is encouraged by sprawling development patterns). By considering average commuting distances, impact on air quality, and the visual monitoring of development patterns, county planners and others can use this system to plan for more compact development patterns that reduce traffic congestion and future sprawl potential.

While more concentrated residential development represents a more generally desirable development pattern, Hernando County will have to carefully integrate policy regarding commercial development to support it. For example, focusing on neighborhood or cluster commercial activities instead of linear type development could help mitigate the issue of sprawling development patterns. Additionally, economic indicators for housing and real estate can provide additional information on the types of residential development that should occur. A second map was then constructed that interfaced with some of the social indicators by including the locations of schools, libraries, golf courses, and churches in Hernando County. Each feature is surrounded by a 500-foot buffer zone, which allows the user to calculate exactly how far a specific parcel falls from a desired amenity. The location of these social amenities were recommended for consideration as part of the residential and commercial development decisions.

Considered together, these types of GIS products demonstrate a method for providing information for economic development planning evaluation. Where, what types, and how much development should occur, and does this development support or hinder progress towards overall community goals? The GIS interface provides the ability to more clearly see the impact of changes in the community, and coupled with the overall set of indicators, can provide valuable information to evaluate economic development planning.

Conclusion

The advent of community indicators measuring systems holds much relevance for communities as supporting mechanisms in efforts to create, monitor, and adjust community goals and visions, particularly in the context of economic development planning. Indicators, when taken together, provide a 'picture' for communities to gauge changes, and as such, can provide motivation for guiding change in positive and desirable directions.

Community indicators measuring systems typically include some economic measures of community change. However, by fully integrating economic development planning into overall community planning, indicator measuring systems represent a valuable and powerful evaluation approach. Economic development planning evaluation approaches need to be accessible and useable at the local, 'planning' level, and community indicators measuring systems represent such an approach. To further enhance the usability of a community development indicators measuring system, it can be integrated with a geographic information system to enable visual assessment. This provides a powerful method for continued improvement of economic development evaluation.

As research has indicated, the process of designing a community indicators measuring system is as valuable as the outcomes. Bringing together a community to envision its future, establish specific goals, and select indicators for gauging progress towards achieving that future and goals are powerful activities. The case study of Hernando County, Florida illustrates this – by collaborating on what, where, and what types of development should occur in the future, discord between the public and private sectors and the citizens of the community will be mitigated as they work towards a common future they helped identify and articulate in the set of selected indicators. Expressing agreed upon goals and monitoring progress towards those goals via a set of commonly selected indicators is projected to serve as an effective economic development planning evaluation approach.

References

Accordino, John J. (1994), 'Evaluating Economic Development Strategies', *Economic Development Quarterly*, 8(2): 218-229.

Ammons, D.N. (1996), *Municipal Benchmarks, Assessing Local Performance and Establishing Community Standards*, Thousand Oaks, California: Sage Publications.

Andrews, J.H. (1996), Planning Practice: Going by the Numbers, *Planning* (September): 14-18.

Bartik, Timothy J. (1990), 'The Market Failure Approach to Regional Economic Development Policy', Economic Development Quarterly 4:4, pp.361-370.

Bartik, Timothy J. (1994), 'Better Evaluation is Needed for Economic Development Programs to Thrive', *Economic Development Quarterly*, 8(2) 99-106.

Bearse, Peter (1993), *The Evaluation of Business Incubation Projects*, Athens, Ohio: National Business Incubation Association.

Besleme, K. and M. Mullin. (1997), 'Community Indicators and Healthy Communities', *National Civic Review*, 86(1): 43-52.

Bjornlund, L. (2000), *Beyond Data, Current Uses of Comparative Performance Measurement in Local Government*. Washington, D.C.: International City/County Management Association.

Center for Building Better Communities (2001), *Hernando County Economic Development Study, Final Report, Part III*, Gainesville, FL: University of Florida, Center for Building Better Communities.

Cobb, C.W. (2000), *Measurement Tools and the Quality of Life*, San Francisco: Redefining Progress.

Feller, Irwin (1988), 'Evaluating State Advanced Technology Programs', *Evaluation Review*, 12(3): 232-252.

Felsenstein, Daniel, Joseph Persky and Wim Wiewel (1997), 'Integrating hard-to-measure externalities into the evaluation of local economic development projects', *The Town Planning Review*, 68(1): 55-81.

Giloth, Robert P. (1992), 'Stalking Local Economic Development Benefits: A Review of Evaluation Issues', *Economic Development Quarterly*, 6(1): 80-90.

Hart, Maureen (2001), Sustainable Measures: What is an indicator of sustainability?, www.sustainablemeasures.com/Indicators/WhatIs.html.

Hatry, Harry P., Mark Fall, Thomas O. Singer and E. Blaine Liner (1990), *Monitoring the Outcomes of Economic Development Programs*, Washington, D.C.: The Urban Institute Press.

Innes, J.E. (1998), 'Information in Communicative Planning', *Journal of the American Planning Association*, 64(1), 52-63.

Innes, J.E. and Booher, D.E. (2000), 'Indicators for Sustainable Communities, A Strategy for Building on Complexity Theory and Distributed Intelligence', *Planning Theory and Practice*, 1(2), 173-186.

Meadows, D. (1998), *Indicators and Information Systems for Sustainable Development*, Hartland Four Corners, Vermont: The Sustainability Institute.

Mowery, David (1990), 'State Government Funding for Science and Technology: Lessons for Federal Programs?', in M.O. Meredith, S.O. Nelson and A.H. Teich, eds, *AAAS Science and Technology Policy Yearbook 1991*, Washington, D.C.: American Association for the Advancement of Science, 241-250.

National Council of State Legislatures (1998), *A Review of State Economic Development Policy*, Washington, D.C.: NCSL.

Office of Planning and Research, State of California (1977), *Putting Social Indicators to Work: An Annotated Bibliography*, Sacramento: State of California.

Oleari, K. (2000), 'Making Your Job Easier: Using Whole-System Approaches to Involve the Community in Sustainable Planning and Development', *Public Management*, 82(12), 12-16.

O'Looney, J. (1997), *Beyond Maps: GIS and Decision Making in Local Government*, Washington, D.C.: International City/County Management Association.

Phillips, Rhonda (1996), 'A Test of Business Growth Through Analysis of a Technology Incubator Program', Doctoral Dissertation, Atlanta, Georgia: Georgia Institute of Technology.

Redefining Progress, Tyler Norris Associates and Sustainable Seattle (1997), *The Community Indicators Handbook*, San Francisco: Redefining Progress.

Reese, Laura and David Fasenstein (1997), 'What Works Best? Values and the Evaluation of Local Economic Development Policy', *Economic Development Quarterly*, 11(3): 195-208.

Reese, Laura and David Fasenstein (1999), 'Critical Perspectives on Local Economic Development Policy Evaluation', *Economic Development Quarterly*, 13(1): 3-7.

Sawicki, David S. and Patrice Flynn (1996), 'Neighborhood Indicators: A Review of the Literature and an Assessment of Conceptual and Methodological Issues', *Journal of the American Planning Association*, 62(2): 165-183.

Schmandt, Jurgen (1991), Regional Technology Policy: A New Role for the American States, in Ulrich Hilpert (ed.), *Regional Innovation and Decentralization: High Technology, Innovation and Government Policy*, London: Routledge.

Wolkoff, Michael J. (1990), 'New Directions in the Analysis of Economic Development Policy', *Economic Development Quarterly*, 4(4): 334-344.

Woods and Poole Economics, Inc. (2000), *Hernando County*, January.

SECTION III
TECHNOLOGICAL DIMENSIONS

Chapter 6

Community Statistical Systems: Discussion of Development and Data Issues

Naomi Oliver, Christiana Schumann and Marc T. Smith

Introduction

The use of indicators as a means to assess and track the quality of life of residents of cities, states, and nations has been expanding in the United States and internationally. Indicators have been developed for education, health, urban quality of life, and children's well-being, among other measures. For example, the National Association of Planning Councils has a social indicators project; a network of health organizations has a county level community health status indicators project; United Way has the State of Caring Index; and the Annie Casey Foundation prepares a Kids Count report. The United Nations Center for Human Settlements (Habitat) has developed a standard list of indicators generally used at the metropolitan level, and the United Nations and the World Bank have also sponsored the collection of basic human indicators in 51 countries (Malpezzi and Mayo, 1997).

More recently, the increased accessibility of local administrative records and the applications of computing and geographic information systems have made it possible to develop data sets that allow tracking and analysis of indicators at the neighbourhood level (Sawicki and Craig, 1996). Through the development of neighbourhood-level data sets, comparisons can be made across neighbourhoods and the targeting of resources improved. Neighbourhood level data can also empower residents to advocate for change with better documentation of the local situation. Kingsley (1999) points to three factors that have led to expanded use of neighbourhood level information: the recognition that communities have a powerful influence; the devolution of responsibility for social welfare policy and programs; and the decentralization of computing networks and geographic information systems.[1]

[1] Presentation by Tom Kingsley of the Urban Institute at the Community Statistical Systems Conference in Tampa, Florida in March of 2002. Details of the conference can be found at www.shimberg.ufl.edu.

This trend toward neighbourhood level data and indicators manifests through the use of Community Statistical Systems. A Community Statistical System (CSS) is an organized set of statistics derived from multiple sources. The data, which are neighbourhood/community level and generally organized, can be presented geographically using geographic information systems (GIS). Data included in a CSS are a combination of U.S. Census records, surveys, and administrative records. The latter include records from property assessors, utilities, and building officials as well as socio-demographic information from schools, voting records, police and fire departments, participatory organizations, and other sources. In addition to compiling information from multiple data sets, a CSS provides access to information for groups, such as neighbourhood associations, that traditionally have not had access to such information.

The development of a CSS parallels, but is not the same as, the development of indicators in a number of cities. A CSS is a database, and, as such, it includes information that might be used to construct indicators. Indicators are a consistent set of measures compiled from data and provided for a specific purpose, such as reporting to residents and decision-makers on the current state of an area and the trend in key measures (Swain, 2002). Indicators may also be used for advocacy, as in reporting the need for environmental protection.[2]

Therefore, Community Statistical Systems form an underlying basis for use in constructing indicators. As such, it is important to understand that the subsequent quality of indicators depends upon the quality of the CSS. This chapter discusses developments in Community Statistical Systems and obstacles to the widespread application of such approaches.

The State of the Art in Community Statistical Systems

A number of cities have developed Community Statistical Systems for use in planning, evaluation, or advocacy. Data may be collected for a single purpose such as assessing health-related indicators (Benbow et. al., 1998; Schwartz, 1998), education (Lentz, 1998), or solid waste disposal (Baud et. al., 2001) Neighbourhood data sets have also been used in studies to address topics including water quality, birth rates, education levels, housing, agriculture, and economics. The set of indicators and measures used is largely dependent on the scale at which the analysis takes place (Sawicki and Flynn, 1996) and may be at the regional, county, or city level. However, studies at these levels can mask conditions existing at the neighbourhood level (Sawicki and Flynn, 1996; Kingsley, 1999).

Therefore, the focus here is on community data systems that are generally more broadly defined as addressing the quality of life in a community. Furthermore, the systems of interest here operate at the neighbourhood level and are used to assess neighbourhoods on a comparative basis or to track change over

[2] David Swain of the Jacksonville Community Council, Inc., discussed this distinction at the Community Statistical Systems Conference in Tampa, Florida in March 2002. Details of the conference can be found at www.shimberg.ufl.edu.

time. Neighbourhood data can be an important tool in planning since it emphasizes the importance of community or neighbourhood level analysis. Studies at the neighbourhood scale can contribute to community development and improvement as, for example, when data analysis shows that city funds are not targeted on the communities most in need of assistance (Hatry et al., 2003; The Urban Institute, 1999).

The leading example of neighbourhood data is the network that operates under the umbrella of the National Neighborhood Indicators Partnership (NNIP). The NNIP is a collabourative effort between The Urban Institute and institutions in a number of different cities (twenty-one currently), all of which have built neighborhood level data systems.[3] Although the organizations have different emphases, their CSS includes information on reported crimes, birth statistics, and a host of social indicators as well as information on property conditions at parcel-level detail. So while the mix of data holdings varies across cities, in all cities a range of data is provided. Each partner is taking advantage of advances in technology that make collecting and manipulating data easier and less costly than in the past. Developing and operating a CSS has the potential to become locally self-sustaining; some of the current partners have been operating for more than ten years.

Most of the NNIP partners have similar kinds of data. Although the datasets do not match up precisely, most of them contain police data on location of crimes by type as well as public assistance records of the location of residents on TANF, food stamps, and other sorts of assistance programs. Information on property characteristics is obtained through the assessor's office from data on tax arrears, building conditions, etc. Partners also have a variety of data about schools, health, business establishments, and employment at the neighbourhood level. The concept for NNIP participants' information systems is that of a 'data warehouse' which is under construction rather than designed entirely at the outset. Rather, these 'warehouses' are places to store a large variety of data, adding to the store incrementally over time. With such a warehouse, one can obtain data quickly in order to conduct an analysis when a new 'hot issue' comes up.

Many NNIP partners are non-profit, non-governmental institutions. A number of the partners are nonprofit intermediaries whose mission is to affect and improve public policy in their own city. An example of this type of partner is the Providence Plan, an institution set up outside of government to help citizens in Providence design what their city is going to become.[4] Another is the Reinvestment Fund in Philadelphia. One of the very first, and, in many ways, more famous for its community development work at the time, is the Urban Strategies Council in Oakland, California.

NNIP partners also include two community foundations – the Boston Foundation in Boston, Massachusetts, and the Piton Foundation in Denver, Colorado. The only partner based solely at a university is the Center on Urban

---

[3] Information on the National Neighbourhood Indicators Partnership can be found at www.urban.org/nnip.

[4] See www.provplan.org.

Poverty and Social Change at Case Western University in Cleveland, although there is one partnership between a United Way and a university at the Polis Center in Indianapolis, Indiana. Only one partner is a governmental agency – a branch of a state department in Florida, the Department of Families and Children.

An important reason why most of the partners are not governmental agencies is to assure the credibility of the information they provide. When data are stored with institutions that are respected and have as their central mission the improvement of their community's interest over the long term, there is more trust in both the data and analysis. Accordingly, these institutions work very hard to develop and maintain this trust. They clean the data, honour confidentiality pledges, and are careful about the manner in which the data are released. They are advocates for the use of fact-based analysis and fact-based policy making.

These practices underscore NNIP's commitment to seeing that data are used as a basis for change. A primary mission of all of the partners is to empower neighbourhood groups to accomplish their own objectives. Information can be a bridge between city level officials and people in individual neighbourhoods, providing access to data that neither side has seen before, shaking up old positions, and opening new channels for communication. For example, maps and tables showing parcel-level data on tax delinquency, code enforcement violations, and other property conditions may present opportunities that previously were not apparent.[5]

A critical element in the success of these neighbourhood data systems is multiple stakeholder input. Stakeholders include citizens, government representatives, business leaders, and non-profit organizations. Stakeholders can assist in selecting which data to include as well as the structure of its presentation. Examples of studies that collect data from a number of participants include the Providence Plan (2002), the Quality of Life in Jacksonville study (Swain, 2000), and the Baltimore Neighborhood Indicators Alliance (2002).[6]

An Example of an NNIP Partner: Center on Urban Poverty and Social Change

The Center at Case Western Reserve University in Cleveland has built a system that is predicated on the following elements:

- A neighbourhood indicators framework
- Geographically detailed data sets
- A geographic information system
- An on-line database
- Technical assistance
- Neighbourhood surveys
- Qualitative neighbourhood data

---

[5] Much of the preceding section is drawn from a talk by Thomas Kingsley at the Community Statistical Systems Conference sponsored by the Ford Foundation and the Fannie Mae Foundation and held in Tampa Florida in March 2002. A summary of his talk is at www.shimberg.ufl.edu/conference.html.

[6] See www.bnia.org, www.provplan.org.

The Center maintains a data set of U.S. Census tract and block group level data as well as administrative data including local tax assessor parcel files, building and demolition permits, housing code enforcement, public and subsidized housing records, ES202 employment records, unemployment insurance wage records, business directories, expenditures for Community Development Block Grants, municipal income tax records, police records, juvenile court filings, child abuse and neglect reports, public school records, public assistance records, voter records, membership records, automobile registration, and health records (Coulton, 1999a; Coulton, 1999b), among others. Since many of the administrative records are individual records, the Center has entered into confidentiality agreements for the use of the data in aggregate form only. In this way, individual records are used to create indicators for the community (Bania and Coulton, 1997). The data are linked in a GIS system that allows for the generation of maps and the placing of data into neighbourhoods for analysis purposes.

Other Examples

In Denver, a neighbourhood indicator system is maintained by the privately run Piton Foundation that has a mission of 'providing opportunities for families and children in Denver to move from poverty and dependence to self-reliance' (Piton Foundation, 2002). The Denver Westside Neighborhood Leadership Program established a relationship with the Piton Foundation where Piton representatives offered training sessions to emerging community leaders. Graduates from this program have gone on to do such things as create programs to assist in stabilizing lower income families with a high mobility rate; motivate expansion of recreational and sporting activities in the evening hours at the local recreation centres; provide an impetus for the funding of a youth arts recognition program; and develop programs that encourage citizens to vote and provide explanation for key issues in the community (The Urban Institute, 1999).

The Center for Social and Community Development at Rutgers University assisted the Camden Churches Organized for People in analyzing data related to vacant houses and crime. It was determined that there was a strong link between vacancy rate experienced by individual neighbourhoods and their corresponding crime rate. As word spread about this correlation, efforts were stepped up by local agencies to rehabilitate or seal-up vacant properties (The Urban Institute, 1999).

The NNIP and other CSS efforts illustrate the potential for building and utilizing statistical systems as a basis for community change as data has become increasingly accessible and more easily manipulated through technological advances and as the ability to display data has advanced through the use of geographic information systems. However, several issues/obstacles remain in the development and expansion of the use of community statistical systems and, thus, in the development of local level indicators (Hyland, 2000).

Building a Community Statistical System

Geographic Information Systems

A key tool in realizing the potential of a CSS is the use of geographic information systems (GIS). GIS allows individual records, such as property characteristics in a property appraiser file, to be tied to a specific location. These records can then be aggregated to provide summary data, such as the level of homeownership or the distribution of housing prices, for a geographic unit such as a neighbourhood. Indicators can be derived that aggregate by geographic boundary data from administrative records that could not previously be aggregated. The geographic boundaries can be changed to match other areas for which data is collected, such as school districts or police precincts. Data can also be mapped so that, for example, the location of each housing unit with a value below a certain price can be displayed. Multiple types of information, such as the location of multifamily housing, schools, and parks, can be placed on the same map. Distances can be calculated, for example, from a house to a fire station. Different levels of information, such as the percentage of households in a U.S. Census tract that are in poverty and the location of subsidized housing in that tract, can be placed on a map. Mapping also allows the display of characteristics of different neighbourhoods, so that comparisons can be made across neighbourhoods, such as average value of housing or medial income (Local Initiatives Support Corporation, 2002).

The power of maps which show by location the level of substandard housing, crime, school delinquency, poverty, illness, and other variables that provide a picture of a neighbourhood and comparisons across neighbourhoods is profound compared to tables showing the same information (Kingsley et al., 1997). GIS maps present a readily interpretable status report for an area. For local governments, GIS presents the possibility for better allocation and targeting of resources as well as enabling and even encouraging performance measurement. By providing analytical tools and the means to look at community problems beyond anecdote, GIS presents information in a useful form not just for elected officials, but also for local community groups. Neighbourhood is the key concern for most people in an area, and GIS allows analysis at that level. With GIS and the existing databases, it is possible to build a parcel-based plat and lot system; GIS can present information about parcels and about people.

As an operating system to inform community decisions, GIS is relatively new. It has only been in the past decade that businesses have embraced GIS for decision-making, and the use of GIS for decisions relating to public policy is in its infancy. Part of the consensus for the use of CSS for analytical purposes is the need for place-based strategies and analytical systems to support locally focused efforts. In addition to promoting community development and place-based strategies, GIS also has the potential to support evaluation of the impact of these strategies.

Although GIS is powerful communication tool, it is limited by the quality of data incorporated into the system. Data sources are clearly a major obstacle in building a CSS and are the focus of discussion in the following section.

Data Sources and Obstacles

The power of a CSS is its ability to utilize data collected from a variety of sources and, through GIS, display it on a map. However, the analysis presented is limited by the availability, quality, comparability, and timeliness of the data. At one extreme, U.S. Census data is collected uniformly across the country using consistent variables, categories, and techniques. However, this data is only collected once every ten years and quickly becomes outdated. Disclosure limitations based on concerns for the privacy of citizens require that information from the U.S. Census be released only aggregated to levels that avoid revealing personal information. As a result, data is only available for U.S. Census blocks, tracts, and jurisdictions, and the boundaries of such groupings may be inconsistent with the neighbourhood boundaries desired for the compilation of data. Therefore, when using U.S. Census data, allocation methods must be devised to approximate the desired boundaries under analysis.

At the other extreme of data sources is administrative data with its advantages as well as obstacles to be overcome. An advantage of some administrative records maintained by agencies in local government is that they include data on individuals and households, and it is possible to compile this data within any applicable boundary. An example of such data would be property assessor data. These data sets have information on every parcel of property in a county, including an address, and, if properly coded geographically, can be aggregated into neighbourhood boundaries and mapped.

An issue of great concern when dealing with administrative records is data privacy since the data may be collected at a geographic level where it can be easy to identify individuals. However, confidentiality is a serious, but solvable, issue. Confidentiality of individual information can be addressed through state law, statistical techniques, and policy. Most states do not distinguish between the use of administrative records for program administration, such as benefit distribution/allocation, versus grouping data together for statistical purposes. National agencies have done an enormous amount of work on disclosure avoidance through statistical techniques, many of which are in place now. Local and state agencies need not reinvent techniques for use with administrative records, but need to be aware of and address this issue (Taeuber, 2002).

As with U.S. Census data, the boundaries used by administrative agencies to collect data may not mesh with neighbourhood boundaries. Geographic units used by the administrative sources to collect data are not consistent across cities or even across agencies collecting the data. For example, in one locality crime data may be available on the police-precinct level, and, in another, it may be tracked at the U.S. Census-tract level. Some of the needed data may be available only at the postal code or school district level. Geographic boundaries may be defined as school zones, voting precincts, police districts, and other service areas, none of which are consistent with neighbourhood boundaries. Further, even the use of a neighbourhood name may be problematic as people may have differing concepts of the boundaries of a neighbourhood.

In addition, various agencies may use different definitions in their data collection efforts, for example, counting 'calls for service' versus crime reports in police reporting. Another example is school attendance policies, a problem particularly with schools in cities where students are bused. The question is whether school enrollment information details residents of a neighbourhood or attendance at a school located in or near the neighbourhood but attended by children from elsewhere, as is particularly the case with magnet schools.

Therefore, when statistics from multiple data sets are overlaid, problems may result. CSS/GIS experts are not necessarily data experts. They may not know, for example, that 'race' from administrative records and 'race' from the U.S. Census have different definitions and that, therefore, the statistics may not be comparable. Compounding this problem is that documentation of the concepts, definitions, and methods used to collect administrative data is generally lacking. Documentation of the intake forms as well as the concepts/definitions used in their completion, how the records are processed, and so forth, are essential to begin using survey information along with the summarized administrative records for policy research purposes. In terms of national government statistical systems, goals concerning data documentation include openness, accuracy, definitions, and statements about the methodology of data collection. These goals are needed for administrative records as well.[7]

In addition to discrepancies due to methodology, administrative data have other limitations. Since the data are not collected for the CSS, the administrative purpose may not mesh with CSS data needs. Administrative records exist for program management, not statistical analysis, and inform managers about issues such as program eligibility. However, the information recorded could be used for purposes other than just managing programs.[8] Also, agency staff often have extensive knowledge about their communities that is left undocumented since much of this information exists in their heads or, informally, in their desk drawers. When the employee leaves, the information is lost. An important issue is how to capture this knowledge so that others may use it.

Further complicating the use of administrative records is the time frame for which data is collected. Data may be collected for different time frames (i.e. school year, calendar year, or fiscal year) depending on the source and location, and, therefore, may be current or available only with a time lag. Varying reporting periods make it difficult to develop a comprehensive data set for a point in time for an area.

There are several other challenges to using survey data with administrative records. One involves the conversion of administrative data to statistically usable files. An even bigger issue is to make data files comparable across areas. Currently, there is a lack of coordination between systems and, as a

---

[7] See the Taeuber speech at www.shimberg.ufl.edu.

[8] Cynthia Taeuber points out the limitations and potential of administrative data sources in a presentation made at the Community Statistical Systems Conference sponsored by the Ford Foundation and the Fannie Mae Foundation and held in Tampa Florida in March 2002. Her remarks can be found at www.shimberg.ufl.edu.

result, a lack of comparability of statistics across areas. Yet in order to use a CSS for constructing indicators for use across communities and for policy evaluation, comparable data is needed. That is, if data is not created in comparable ways, it cannot be used comparably. Although it is important to be able to combine national and local data, local data is often better. In any case, it is more detailed and by using local data and not recreating the same data on the state and national level duplication is avoided. Data must be commensurate in order to make it interoperable and usable on all levels – local, state, or national government as well as not-for-profit groups – for comparative purposes. Specifically, there is a need for comparable data systems based on reliable data sets that are current, objective, and confidential.

In summary, the great potential of CSS is that multiple data sets can be developed with new technologies for organizing and communicating the information. In order to deploy the power of the next generation of community statistical systems, it is important to address issues of accessibility, interoperability, and confidentiality for community-level data. The use of comparable and standardized data is the key to the effectiveness of these new systems.[9]

Looking Towards the Future: CSS Use Barriers and a Research Agenda

Barriers to the implementation of CSS are largely related to the issues of data comparability and standardization discussed above. The challenge facing the future of Community Statistical Systems is the development of solutions to these barriers by looking at old problems in a new light with an eye towards innovation while building on existing structures. For example, the funding environment traditionally has required community groups and local organizations to paint the worst possible picture of their community. Therefore, data is often based on needs assessment models and focuses on issues such as poverty rates, infant mortality rates, and bad test scores. But over the past few years, there has been a paradigm shift where the funding and the policy environments have begun to assess neighbourhood 'strengths' and how to build off those strengths.

The main challenges related to small area data collection and subsequent analysis, especially from U.S. Census or other administrative data, include defining neighbourhoods and standardizing data as well as having an opportunity for analysis of primary data. Developing a system for creating standardized community survey forms would facilitate such data collection and allow state and community information systems to have a coordinated statistical policy.

At the national level there is a national information system with data sets that come out at regular intervals from various statistical agencies. Statistical guidelines and policies are coordinated by the Statistical Policy Office of the Office of Management and Budget, the national mechanism for coordinating

[9] See the talk by Susan Wachter at the Community Statistical Systems Conference sponsored by the Ford Foundation and the Fannie Mae Foundation and held in Tampa, Florida in March 2002. Her remarks can be found at www.shimberg.ufl.edu.

resources and meeting the needs for statistical information while maintaining confidentiality of the information and controlling the burden on the public who must answer the survey questions. It is not so obvious what that mechanism would be below the national level. A few states use their budget office for coordination although generally to a much lesser extent than at the national level. The budget office is a logical place for such coordination for many states because statistical systems need to be able to inform the questions related to budget priorities. Currently, there is no coordinated statistical policy at state or community levels and, therefore, no comparability across areas (Taeuber, 2002).

A national government initiative involving GIS and referred to broadly as the National Spatial Data Infrastructure (NSDI) brings together local, state, and national government entities, not-for-profits, and private sector partners to create a geo-spatial, one-stop portal. NSDIs are being implemented throughout the world in 50 or more countries. This one-stop portal avoids wasteful duplication of data sets and creates the comparability and the commensurability of data that are absolutely essential for analytical purposes (Wachter, 2002).

The purpose of the NSDI initiative, which is a bottoms-up effort despite the lead of the national government, is to respond to the problem that data currently are poorly documented, undiscoverable, inconsistent, incompatible, and collected to serve specific missions in different formats and standards. The initiative's goal is to produce seamless, nationally consistent geographic information that is current, accurate, accessible, and integrated. This would be achieved by using network power utilizing different websites with a shared portal that would be the source for information on all the other providers. Importantly, the NSDI is to be built from the bottom up and is to be implemented by creating public/private partnerships with shared data collections. In order to do this, it is critical to develop and implement consensus standards and specifications as well as to determine which partner would be the best to develop and steward the data.

Applying the Lessons: The Tampa Bay Community Atlas Program

Conceptualizing the Program

Developers of new CSS programs have the advantage of being able to draw on the past experiences and lessons with CSS of groups in other locations even though the purpose of these systems may vary depending on the funders, geographic area to be served, data availability, experience of the implementing agency, and type of entity developing the CSS. Universities are uniquely situated to implement a CSS because of the resources available both in computing and expertise, their location in the community, and the increased emphasis on community service within these institutions. The following case study describes the Community Atlas, a system that is under development in the Tampa Bay area of Florida. This system is being created by the Florida Center for Community Design and Research at the University of South Florida in association with a number of partners.

The basic premise for the development of the Community Atlas is that most of the data that are of use to neighbourhoods are difficult to access since the data are located at multiple agencies at the national, state, and local level. Even development and planning professionals spend an inordinate amount of time trying to locate and reuse many of the same datasets. In fact, in most communities the data are not even cataloged in a metadata repository. If a collabourative approach to community building is to succeed, stakeholders must be able to access quality-assured community data. The Community Atlas will be the tool that helps make neighbourhood-level data more accessible for the Tampa Bay area.

Prior to developing the Community Atlas, the Florida Center spent time studying the experiences other organizations have had with CSS and its data-sharing techniques. Many of these other systems had to be built from the bottom up by groups/organizations with limited examples to draw upon. However, creators of the Community Atlas will be able to examine the technology utilized by various existing CSS, drawing upon existing technology already maintained by the University of South Florida through the Water Atlas Program,[10] a system that brings together data related to water resources and presents it on web pages related to individual bodies of water. Utilizing the lessons learned from building existing technology will greatly reduce the time devoted to development and resource building. In fact, the Community Atlas will be built directly upon the GIS and database technology already in place at the Water Atlas, allowing for the creation of a system that provides information linked to neighbourhoods.

The most successful CSS programs have drawn upon insight provided by multiple stakeholders who participated in the CSS development process. This ensures that the product will serve the needs of each potential audience. As such, the Community Atlas program will draw upon various local agencies, non-profit groups, and citizens to solicit input related to the content and design of the project's databases and accompanying website interface. In so doing, the project will ensure that community relevant data are presented through an interface that is easy to understand and accessible for all end-users.

Creators of several CSS or Indicator Systems that are in existence today started out with the ambition of collecting data that covered a wide spectrum of topics and datasets. However, over time, it was discerned that certain indicators were more useful than others and some data sets were just too unwieldy to maintain. This brings up two important issues. First of all, it is probably more prudent to establish the system with a clearly defined smaller group of datasets and plan to add more datasets later. Second, CSS programs should undergo regular assessment to determine if they are utilizing the most relevant and useful data sets and be restructured accordingly. Drawing upon the previous experience of others regarding these issues, the Community Atlas program should start out with a clearly defined number and type of data sets while devising a mechanism to add or develop additional datasets at a later date.

---

[10] See www.wateratlas.org.

What is the Community Atlas Program?

Maintained by the University of South Florida, The Community Atlas will be a data warehouse with a user-friendly web interface that seeks to increase public access to locally relevant data and provide tools, such as GIS mapping, with which to better understand those data. This web site will be designed to allow users to view data at the regional, county, municipal and neighbourhood levels. Citizens, agencies, and researchers will be able to search for information related to particular neighbourhoods, download data sets for analysis, and access trend-graphing tools. The ability for users to access data at the neighbourhood level is particularly critical as it will allow people to understand the variances within a larger geographic area. Through the Community Atlas, indicators of community well-being can be tracked and monitored. All of the information will be assembled through database technology that is tied to GIS mapping capabilities. Additionally, documents and photos related to particular areas can be presented along with the statistical data about the neighbourhood. Most importantly, the Community Atlas will be designed as a prototype with the larger intent of developing technology that can be transferred to multiple communities.

The Community Atlas will have the following four major sections and additional features:

Neighbourhood Information Pages Through the use of search tools, lists of named neighbourhoods, and online map navigation, the user will be able to easily pinpoint the neighbourhood of interest and view neighbourhood information pages that will present data, such as demographics, economics, and land-use, pertinent to that particular area. These pages will also present indicators measured against county as well as neighbourhood assets (e.g. neighbourhood groups, parks, schools, places of worship).

Interactive GIS Mapping The mapping interface will allow the user to activate multiple layers of information such as population, licensed daycare centres, schools, and aerial photography. In addition, the map will provide the user with easy-to-use tools, access to data for individual points on the map, customizable maps, and a direct link to neighbourhood pages.

Research The research section will be geared towards public and non-profit agencies as well as academic researchers. One of the primary goals in creating this section is development of a simple, yet expandable, user-friendly data-mining tool. In this section, the end-user will be able to view metadata, download data into Excel spreadsheets or comma delimited text files, create custom reports, graph trends, and view research reports about issues within the county.

Document Library This feature will allow users to search for all materials on the website that are related to particular topics, such as datasets, map layers, reports, web links, and photos.

Additional Features One additional feature of the Community Atlas will be online forms that will allow the user to communicate with the agencies providing the data as well as with the developers and sponsors behind each Community Atlas website. A second feature will be pop-up boxes that accompany each data component on the individual neighbourhood data pages to help the user understand what the data means, what potential caveats exist, and how the data are derived. A third feature will be standard graphs that can be produced from many of the data components. These will help to give the user a picture of data trends. A fourth feature will be a dynamic photo library that will allow users to upload digital pictures of neighbourhoods. Community Atlas users will also be able to scroll through all of the photos available for each neighbourhood.

Assessing the Interest of Stakeholders: Community Focus Groups and Surveys

In order to better understand the needs of project stakeholders (citizens, non-profit and public organizations, and researchers in Hillsborough County), the Florida Center held public forums and conducted surveys to ascertain whether a CSS was needed for the area as well as what data might be the most important to include in it.

The lack of access to data in Hillsborough County was documented through a survey administered in 2002 by the Florida Center for Community Design and Research as well as through community forums and interviews conducted during 2001 and 2002. The 2002 survey focused on the data needs of the local non-profit and public sectors. Representatives from a wide selection of organizations involved with issues concerning healthcare, children, the elderly, business, and employment were asked to participate in the survey. This was done because of the likely diversity of the potential users of an on-line indicator and data sharing system. The organizations asked to participate in the survey were the same as those asked to be a part of the Social Infrastructure Network (S.I.N.), a group of local organizations (led by the Hillsborough County City-County Planning Commission) whose purpose is collecting and sharing social data about the residents of Hillsborough County. There were 44 surveys returned as part of the project; of these, 25 were received from non-profit agencies, 18 from public agencies, and one from a public university. The types of services provided by these agencies vary greatly.

On the survey, the first series of questions were structured to better understand the data needs of local organizations. These questions addressed what specific and general data would be most useful, how this data is currently collected, and how the data could be best presented for their organization's use. The most common type of 'general' data chosen as useful was demographic information (98 per cent of the respondents) followed by economic information (86 per cent). In the 'other' category, types of data listed included transportation, other agencies in the area and their focus, cultural arts opportunities, business and residential relocations, and family safety (see Table 6.1).

Table 6.1 General Information Chosen as Useful

Type of information	Number of positive responses	Per cent of respondents (N=44)
Demographic	43	98
Economic	38	86
Education	35	80
Housing	32	73
Public Safety (Crime)	22	50
Health	17	39
Environmental	15	34
Other	10	23

Survey respondents were then asked about specific types of data that would be useful to their organization. The most frequently requested information included data related to children, racial and ethnic groups, gender (specifically females), the elderly, and the economic situation for county residents (including employment and income levels).

Respondents were also asked about whether data that would be of use to their organization was currently available. The majority of respondents indicated that such data were 'somewhat available' from the 2000 U.S. Census as well as from information stored within organizational databases and files. However, respondents indicated concern that the more broadly based U.S. Census data might soon be outdated while individual organizational databases and files might be too limited or specific in nature due to the particular groups of people served by the organization. Furthermore, the respondents indicated that currently they have to 'dig up data' from multiple sources and that acquiring the needed information is sometimes a 'hit or miss' proposition.

Survey respondents were asked to indicate the current sources used to acquire data. They were presented with a list of possible data sources and could choose as many as applied. The most frequently cited sources include published reports from outside organizations (70 per cent), computerized records within the organization (57 per cent), and internet resources (57 per cent). Those respondents who indicated that they accessed 'other' sources of data explained that these data came from field collection conducted by their own organizations or by subcontracted employers. Other respondents answered that they purchased data from commercial databases.

Respondents were asked to rank the grouping level or spatial boundaries at which data might be most usefully presented for the organizations. The survey provided respondents the opportunity to rank their preferences (with 'one' being the most preferred and 'nine' being the least preferred) for nine different levels (e.g. county, city, etc.) at which information could be presented. Although no one grouping level emerged as the predominate favorite, the most preferred grouping level was county followed closely by city, postal code, and neighbourhood.

When respondents were asked whether an on-line data resource would be useful to their organizations, the majority (84 per cent) indicated that it would. When asked to explain why such a resource would be useful, three major themes emerged. First, many of those surveyed thought that an on-line data resource would make it easier and more efficient to obtain data. Second, respondents stated that it would help their organizations in planning efforts as well as in identifying particular areas and groups of people in need of resources. The third theme ties into the aforementioned two. Many respondents stated that an on-line resource could assist their organizations with grant applications because it would allow them to easily access data as well as demonstrate the particular area(s) of need, both of which are often required in grant applications.

A number of questions were asked to determine the level of technology available to each group. All 44 respondents had at least one computer in their organization. Of those who responded, all had some type of personal computer. Forty-two respondents (95 per cent) answered that their organizations had an internet connection. Of those respondents, 33 gave more specific information about the type of internet connection used within their organization: twenty-five (76 per cent) answered that they had high-speed internet connections and eight (24 per cent) had dial-up service.

Respondents were also asked about the type of data typically collected and how the data were maintained. The most common response was that the data collected was related to their program participants and most typically included age, race, gender, finances, test scores, and services provided. Other types of information collected included specific information of interest to each organization, such as crime data, fire hydrants, street attributes, program referrals, voter registration, election results, and capital improvements.

Respondents were then asked how data were maintained by their agencies. This was asked to determine whether they would be able to upload data to a data-sharing system if they chose to be a part of such a system. The majority of organizations maintained their data through both written and computer files (57.5 per cent), followed by organizations that maintained only computer files (37.5 per cent). Those respondents who indicated that their data were maintained electronically were asked to identify the types of software used. The respondents were given the opportunity to name all software used for this purpose. The most frequently identified software included Excel (51 per cent), followed by Word (40 per cent), and Access (40 per cent).

When survey participants were asked whether they were required to report any of their data to state or national agencies, 25 organizations (57 per cent) indicated that they were. The state and national agencies listed included the following: the Florida Department of Health; the U.S. Department of Health and Human Services; the Center for Disease Control; the U.S. Department of Housing and Urban Development; the Florida Department of Children and Families; the Florida Department of Juvenile Justice; the State of Florida Agency for Workforce Innovation; Workforce Florida, Inc.; the U.S. Department of Labor; the U.S. State Department; the Florida Department of Elder Affairs; and the Florida Department of Community Affairs.

The respondents were then asked whether their group might be willing to share their data with other organizations. The majority (67 per cent) answered that they would while the remaining (33 per cent) indicated that they might be willing to share depending on a number of factors such as legal authorization to share information and the confidential nature of particular types of information.

Finally, survey participants were asked whether their organizations would be willing to share any of their data with the public. The responses were closely divided between organizations willing to share their data with the public (55 per cent) and those that might be willing to share with the public (45 per cent) depending upon particular factors. These factors were very similar to those stated in regard to sharing data with other organizations and included the legality of sharing particular information with the public, the confidential nature of particular information, the type of information that would be shared, and the stated need for such information.

In summary, the results of this survey support the effort to develop a community indicator website in Hillsborough County that would provide up-to-date information as well as an interactive mapping feature that would allow for the mapping of specific data. Some of the key findings that support the development of a website include:

- A current lack of access to needed data about the people and communities in Hillsborough County.
- The common use of secondary and internet sources for data acquisition by public and non-profit organizations.
- A majority of positive responses to the usefulness of mapped (spatial) data.
- A majority of positive responses to the usefulness of an on-line data source.
- High usage of computers for organizational data maintenance.
- A majority of organizations willing (or would consider the possibility) to share data with other organizations and the public.

Although many of the findings support the development of a community indicator website, some potential issues will need to be considered throughout the development process. Many of these issues concern the ability of organizations to add their own datasets to the website and include:

- The legality of sharing data.
- The confidential nature of certain types of information.
- The purpose and potential use of particular data.
- The storage of data and the compatibility with uploading it to a database.
- General agreement on what data to include.

The background research that has been conducted about the needs of stakeholders in Hillsborough County will help inform the development of the Community Atlas. Once the project is underway, more detailed design processes will be undertaken with additional stakeholder involvement.

Conclusion

To be most effective in affecting change in neighbourhoods, the operators of a community statistical system must actively engage the community in the process of framing problems, identifying what data are to be used or collected to address those problems, determining how any primary research is to be carried out, analyzing the data, determining the implications of the data, and reflecting on the research process. A community's sense of ownership of the data – in a very broad sense, the democratization of the data – is crucial for program effectiveness.

The use of CSS data and, in particular, GIS maps can lead to policy changes as community advocates are empowered by presenting data to governmental entities. Although some work in establishing a CSS may be technically complex, users must also understand the potential of the system and be able to obtain information useful to their purposes. The ability of a CSS to contain up-to-date data will be a critical factor in its success; if a CSS cannot be continually updated with data as it becomes available, the indicators produced can be questioned for validity and reliability. Another issue in sustaining a CSS is understanding how to maintain good relations with government agencies that are important data sources especially when the community is in conflict with these agencies about programs, policies, and service delivery.

The focus of CSS is on small communities (neighbourhoods) where collecting information can be very difficult. Part of the consensus for the use of CSS for analytical purposes is based on the need for place-based strategies and analytical systems to support neighbourhood level outcomes. While recognizing the power and usefulness of numbers, community improvement occurs within the economic and social marketplace as well as the political arena. The goal of CSS is to make it possible for communities to focus their resources, establish their priorities, and work proactively, rather than reactively, in order to create change.

References

Baltimore Neighborhood Indicators Alliance (2002), www.bnia.org, www.provplan.org.

Bania, N. and Coulton, C.J. (1997), *A Neighborhood Information Capacity for Community Building, Policy Analysis and Evaluation*, Case Western Reserve University Center on Urban Poverty and Social Change, Cleveland.

Benbow, N., Y. Wang and S. Whitman (1998), *The Big Cities Health Inventory, Journal of Community Health*, vol. 23(6) 471.

Community Statistical Systems Conference, available online: www.shimberg.ufl.edu/conference.html.

Coulton, C.J. (1999a), *Public Assistance Records: A Source for Neighborhood Indicators*, National Neighborhood Indicators Partnership Report, The Urban Institute, Washington DC.

Coulton, C.J. (1999b), *Vital Records: A Source for Neighborhood Indicators*, National Neighborhood Indicators Partnership Report, Center on Urban Poverty and Social Change, Case Western Reserve University, Cleveland.

Hatry, H.P., Cowan, J., Weiner, K. and Lampkin, L.M. (2003), *Developing Community-wide Outcome Indicators for Specific Services*, The Urban Institute, Washington DC.

Hyland, S.E. (2000), 'Issues in Evaluating Neighborhood Change: Economic Development and Community-Building', *Cityscape: A Journal of Policy Development and Research*, vol. 5(1), pp.209-17.

Jacksonville Community Council, Inc. (2000), *Quality of Life in Jacksonville: Indicators for Progress*, Jacksonville, Florida.

Kingsley, G.T. (1998), *Neighborhood Indicators: Taking Advantage of the New Potential* (working paper), American Planning Association, Chicago.

Kingsley, G.T. (ed.) (1999), *Building and Operating Neighborhood Indicator Systems: A Guidebook*, National Neighborhood Indicators Partnership Report, The Urban Institute, Washington, DC.

Kingsley, G.T., Coulton, C.J., Barndt, M., Sawicki, D.S. and Tatian, P. (1997), *Mapping Your Community: Using Geographic Information to Strengthen Community Initiatives*, U.S. Department of Housing and Urban Development, Washington, DC.

Lentz, C. (1998), *Why Do Some Communities Pay More Than Others? The Example of Illinois Teachers, Public Administration Review*, vol. 58(2). 139-147.

Local Initiatives Support Corporation (2002), 'Mapping for Change: Using Geographic Information Systems for Community Development', Local Initiatives Support Corporation, New York.

Malpezzi, S. and Mayo, S.K. (1997), 'Housing and Urban Development Indicators: A Good Idea Whose Time Has Returned', *Real Estate Economics*, vol. 25(1), pp.1-11.

Piton Foundation (2002), www.piton.org.

Providence Plan (2002), www.provplan.org.

Sawicki, D.C. and Craig, W.J. (1996), 'The Democratization of Data: Bridging the Gap for Community Groups', *Journal of the American Planning Association*, vol. 62(3), pp.512-23.

Sawicki, D.C. and Flynn, P. (1996), 'Neighborhood Indicators: A Review of the Literature and an Assessment of Conceptual and Methodological Issues', *Journal of the American Planning Association*, vol. 62(2), pp.165-83.

Schwartz, D. (1998), *Creating a Benchmark Data Base, Health Management Technology*, vol. 19(1), 66-67.

Swain, D. (2002), 'Measuring Progress: Community Indicators and the Quality of Life', Jacksonville Community Council, available online: www.jcci.org.

Taeuber, Cindy (2002), http://web.dcp.ufl.edu/chakabry/Shimberg_Conference/Wed-TaeuberOK.htm.

The Urban Institute (1999), *Stories: Using Information in Community Building and Local Policy*, National Neighborhood Indicators Partnership, Washington, DC.

Wachter, Susan (2002), http://web.dcp.ufl.edu/chakabry/Shimberg_Conference/Wed-WachterOK.htm.

Chapter 7

Role of Multi-Scalar GIS-Based Indicators Studies in Formulating Neighbourhood Planning Policy

Rina Ghose and William Huxhold

Introduction

The adoption of Geographic Information Systems (GIS) in planning has grown steadily in North America since the 1960s (French and Wiggins, 1990; Huxhold, 1991; Wiggins, 1993; Harris and Elmes, 1993; Budic, 1994; Huxhold and Levinsohn, 1995; Nedovic-Budic, 1998, 1999). Consequently, a wide range of GIS applications in various planning tasks evolved such as land use, zoning, transportation, site suitability analysis and economic development. GIS has aided equally in managerial and decision-making tasks in planning. Growing knowledge, awareness and acceptance of GIS coupled with lower costs of hardware and software have assisted in spreading the implementation of GIS in planning organizations.

In recent years, one notable area of application in GIS concerns neighborhood planning. Neighborhood strategic planning is considered to be an important approach for cities that are undertaking revitalization strategies for their blighted central city neighborhoods. GIS is able to greatly facilitate such neighborhood strategic planning through its data analysis and visualization capabilities. Application of GIS in this arena is helpful not only in analyzing the data, but also in communicating the outcomes of such analysis to the citizens through its visualization techniques.

Such neighborhood mapping can be an important tool for facilitating citizen participation in the neighborhood revitalization process. It is widely recognized that citizen participation in neighborhood strategic planning is highly significant and can often make the planning process more effective. However, citizen participation can be badly hampered by lack of neighborhood level data, and access to technologies that enable quick access to and analysis of such data (Barndt and Craig, 1994; Pickles, 1995; Sawicki and Craig, 1996). Based on the assumption that greater access to information leads to greater opportunities for citizens to participate in planning, a range of initiatives has been undertaken by various agencies (federal, state and local) for providing citizen groups with access to GIS (Barndt, 1999, 1998a, 1998b; Obermeyer, 1998; Sawicki and Peterman,

1998). Neighborhood based grassroots community organizations in turn have shown increasing interest in working with GIS in order to address issues that concern their neighborhood (Ghose and Huxhold, 2001; Ghose, 2001; Elwood, 2000; Kellogg, 1999; Craig and Elwood, 1998; Elwood and Leitner, 1998; Ramasubramanian, 1998).

While these developments have been positive influences in the neighborhood planning process, there is still a considerable lack of available data at the local level that can pose a serious constraint. In that respect, we will argue that GIS-based neighborhood indicators studies can assist in bridging the crucial data gap for the benefit of citizens and planners alike. Through the development and use of indicators, assessment of changing social, economic and physical conditions of a neighborhood can be conducted. Such indicators can be used in identifying neighborhoods experiencing high levels of social problems in a city, targeting public resources to these identified neighborhoods, evaluating the results of such targeted investment strategies over specific time intervals. Additionally, use of neighborhood indicators across different geographic scales can result in cross-comparisons, which can be more effective than single scale analysis in identifying problems and trends and directing policymaking in an effort to resolve such problems. The visualization techniques of GIS can be combined with specific neighborhood based indicators to assess quality of life in order to effectively distribute public resources to the areas that are most in need. Such 'equity mapping' (Talen, 1998) can be beneficial in both in neighborhood revitalization activities and in public policy research. In fact, a major objective of indicators based studies is that these should assist in public policymaking. Few studies have attempted to develop GIS based indicators in order to assess the quality of urban life at the neighborhood scale, yet GIS based indicators studies can have a significant relevance to local planning and policy formulation process.

Beginning with a discussion of the concept of indicators and their importance in public policy research, this chapter explores several interlinked studies in Milwaukee, in which GIS based neighborhood indicators were developed and mapped to examine and assess the quality of life of Milwaukee's inner city at different geographic scales. The discussions also focus upon the types of indicators that were used and the types of data that were analyzed, in order to provide directions to future indicators based research. The chapter then examines the responses of local stakeholders to the latest indicators study and discusses how the stakeholders utilized the study to implement critical planning policy changes concerning the affected neighborhood. Finally, the chapter discusses critical aspects of the study and discusses whether the study was in fact beneficial in affecting citizen action and public policymaking activities concerning inner city neighborhood planning.

Use of Indicators in Assessing Neighborhood Well-Being: A Review of Past Studies

In the United States, the public demand for a better quality of life (including quality of place) has prompted studies on the well being of a place. Indicators based studies represent a way of measuring the quality of a place that can be used for policymaking purposes. We must therefore begin by reviewing the literature and examining the concept and relevance of indicators in planning.

Since the 1960s, social scientists have studied changes in physical, social and economic conditions through the use of indicators. Such indicators are developed by the statistical transformation of raw data into useful measurement tools (Sawicki and Flynn, 1996). Three major types of indicators have developed over time; these are social, urban and neighborhood indicators. Social indicators were defined by Raymond Bauer as '...statistics, statistical series, and all other forms of evidence...that enable us to assess where we stand and are going with respect to our values and goals and to evaluate specific programs and determine their impact' (Bauer, 1966, p.1). Thus, the use of social indicators was grounded in the concept of national monitoring through a system of social accounts in order to set program priorities, establish clear social goals and policies and establish a simplified system of program evaluation (Kingsley, 1999). Urban and neighborhood indicators originated from the same concept, but examined data at different geographic scales. Social indicators were used primarily to examine data at national scale, such as the Annie E. Casey Foundation's annual report of indicators (1994) that measure the welfare of children across the nation at the geographic scale of a state.

Urban indicators were developed to assess metropolitan areas and are a precursor to the concept of neighborhood indicators (Sawicki and Flynn, 1996). The 1970s saw the emergence of urban indicators studies when 'government agencies, private research organizations and academic institutions experimented with economic and social indicators at the local level. The major thrust was the collection of disaggregated information that would serve specific audiences. Indicator projects took the form of profiles, needs assessments, state-of-the-city report cards, citizen surveys and socioeconomic data' (Sawicki and Flynn, 1996, p.166). The Urban Institute's indicator project (Flax, 1978) reviewed the use of urban indicators at the metropolitan scale from 1970-77. Topics of study included housing, crime, transportation, welfare and city service delivery. There were many challenges that hindered the development of urban indicators in the 1970s. Costs of collecting and annual updating of data were substantial and difficult. While computer capacity had increased, costs of analyzing the data and running models in the computers were still high.

Since the 1980s and 1990s, urban indicators have been widely used to measure quality of life at the geographic scale of cities and states (Sawicki and Flynn, 1996). At the city level, many communities have followed the systematic approach laid out by the National Civic League in The Healthy Communities Handbook (1993) to assess the wellness of a community. Oregon's benchmarking effort (Oregon Progress Board, 1994) is a particularly prominent model that used measurable indicators to monitor the state's progress toward its strategic goals. The benchmarks are used to serve as a baseline of reference, setting priorities, allocating resources and measuring performances of the government agencies. The Healthy Cities Project, initiated by the World Health Organization in 1986, is another example of an indicators initiative that has been popular with citizens and policymakers alike across the world in an effort to create public policies that maintain and promote individual and community health.

Neighborhood indicators were developed to inspect data at smaller areas or neighborhoods within the metropolis. According to Drummond, 'neighborhood, or small area, socioeconomic indicators can be defined as measures of social welfare based upon the collection, analysis and mapping of data at the neighborhood level, where neighborhoods are understood as relatively small, relatively homogeneous areas' (Drummond, n.d.). Sawicki and Flynn define neighborhood as 'an area that is smaller than a municipality but more than a few city blocks' (1996, p.170). The use of smaller geographic units such as the neighborhood as the basis for indicators studies has been fairly uncommon. In part, this is because assembling data at a larger geographic unit (such as the citywide scale) is more convenient because much data are available at this geographic scale. Using citywide data alone can be problematic however, as it assumes that uniformities exist across the city and does not reveal disparities that exist at the neighborhood scale. Consequently, neighborhood indicators have caught the attention of those urban scholars who 'argue that the geographic place plays an active role in altering the levels of the indicators and thus in changing the lives of the people living in those neighborhoods. The neighborhood is not viewed simply as a convenient unit of analysis...The neighborhood is chosen because some believe that the neighborhood-level indicator can be a tool to change people's lives' (Sawicki and Flynn, 1996, p.166). Thus, the belief that indicators must play a strong role in bringing positive change to public policymaking and neighborhood planning is intrinsic in the use of neighborhood indicators.

Since the 1990s, there has been an increasing interest in the development and use of neighborhood indicators in various community projects, where the indicators served to measure quality of life in the neighborhoods, monitor neighborhood conditions and highlighted the directions in which a course of action is needed. The commencement of the National Neighborhood Indicators Project (NNIP) in 1996 is an indication of the growing importance of using neighborhood indicators in local policymaking and community building activities (Kingsley, 1999). The NNIP was implemented 'to develop measures of changing social, physical and economic conditions of neighborhoods in cities throughout the United States... so that community residents, public officials and civic leaders can better plan appropriate strategies to improve their communities' (Sawicki and Flynn, 1996, p.165). Another effort is the formation of the Urban Indicators Program of United Nations Center for Human Settlements from 1994 onwards, that aims 'to build national and local capacity to collect and use policy-oriented indicators as part of a strategy for the development of sustainable human settlements', specifically cities, in order to 'anticipate, recognize, measure and interpret urban problems' in order to respond to them in policy formulation (http://www.urbanobservatory.org/indicators/). The project is thus intended to 'collect useful information on urban conditions and trends, to convert that information to knowledge through appropriate analytic techniques and to apply that knowledge in formulating and modifying urban policies' (http://www.urbanobservatory.org/indicators/).

Several factors have contributed to the growing use of neighborhood indicators. The first is the technological advancements of desktop computers and a reduction of their costs, which has made desktop computing affordable and useful for organizations. The second is the development and popularity of GIS, a technology and science that is adept at the analysis and visualization of small area spatial data. The address matching capabilities of GIS software is particularly helpful in sorting and mapping address based data by neighborhood. Additionally, the costs of GIS software package have dropped considerably over the years, and a range of such software is available at affordable prices. Thus, GIS based neighborhood indicators studies are gaining popularity in policy research (Kingsley, 1999; Drummond, n.d.). The third factor is the automating of administrative data. Over the past decade, there has been an emphasis in automating the administrative records of local public agencies, greatly simplifying the task of gathering neighborhood specific data. Finally, there has been an increasing 'shift of responsibilities for social and economic welfare from the federal to the state and local level' (Sawicki and Flynn, 1996, p.166), one that puts the focus both on increasing public-private partnerships as well as on neighborhood empowerment. In order to facilitate citizen participation in neighborhood planning, there is a great need for data concerning the environmental, social and economic conditions of neighborhoods which in turn leads to the development of neighborhood based indicators studies as well.

The Local Policy Context and the Commencement of Indicators Based Studies in Milwaukee

The commencement of urban and neighborhood indicators studies in Milwaukee is deeply linked to the issues of inner city revitalization and the local planning context. In that respect, it is appropriate to first examine the local policy context of Milwaukee that influenced the development of the indicators based studies.

Milwaukee, like many mid-western manufacturing cities, experienced severe deindustrialization and disinvestment, leaving it with the all-too-familiar problems of blighted central city neighborhoods with rising rates of poverty and unemployment and crime and generally degraded standard of living. Consequently, inner city revitalization became a top priority among the departments of City of Milwaukee and the Community Block Grant Administration (CBGA) office. CBGA oversees the distribution Community Development Block Grant (CDBG) funds that it receives from the U.S. Department of Housing and Urban Development (HUD) in order to revitalize its deeply troubled inner city neighborhoods (Ghose and Huxhold, 2001). 'The agency [CBGA] relies on Neighborhood Strategic Planning as the best way to target funds effectively, because it identifies the needs of an entire neighborhood instead of basing decisions on individual agencies' budget demands' (Huxhold and Martin, 1996, p.54).

Inner city revitalization activities undertaken by CBGA then occurs through the formal Neighborhood Strategic Planning (NSP) process within the 17 NSP areas of Milwaukee, in which CBGA allocates funding to community-based organizations that seek to increase owner occupancy, reduce the overall crime rate, or prevent or reduce the decline in property values. City of Milwaukee departments such as Department of City Development (DCD) and Department of Neighborhood Services (DNS) are also deeply involved in inner-city revitalization projects in Milwaukee, but on a case-by-case basis.

Both of these processes of inner-city revitalization however require the analysis of neighborhood-based data as assessment tools. CBGA has mandated the use of neighborhood statistics, spatial analysis and thematic maps in the construction of such Neighborhood Strategic Plans. City of Milwaukee's departments on the other hand have no formal mandates, but community-based organization staff explained that in their meetings with DCD and DNS, they typically present the neighborhood's issues and concerns through formal presentations supported with neighborhood statistics, thematic maps and if possible, spatial analysis.

Thus, in order to effectively participate in the inner city revitalization process, community-based organizations of Milwaukee have felt the need for neighborhood based spatial data and their analysis in order to identify key neighborhood concerns, to monitor neighborhood conditions over time, to develop long term planning visions for their neighborhoods and to receive CDBG funds for the costs associated with neighborhood revitalization efforts. Networks of supportive institutions have attempted to satisfy such data needs for the citizens of Milwaukee (Ghose and Huxhold, 2001). Today a range of data specific to each neighborhood is available in the areas of demographic, housing, crime, employment and business and is used widely by community-based organizations (Barndt, 1998a; 1998b; 1999).

However, the increasing focus on improving the inner city neighborhoods' quality of life along with other policy agendas in Milwaukee have also created a demand for neighborhood indicators based studies that could assess the quality of life in the different neighborhoods. In particular, the City of Milwaukee's Housing and Community Development Consolidated Plan has emphasized 'increasing property values, creating home ownership opportunities, revitalizing neighbourhoods by targeting resources to make a clear and measurable impact and improving quality of life in neighbourhoods' as key agendas (www.ci.mil.wi.us/citygov/doa/admin/Planbtm.htm#CITYOFMILWAUKEE). Such priorities have motivated both the planning agencies and community organizations to seek out indicators based research in the areas of housing and neighbourhood investment.

In response, the University of Wisconsin-Milwaukee (UWM) has undertaken the tasks of initiating GIS-based indicators studies and has acted as an important provider of data to the Milwaukee community (University of Wisconsin-Milwaukee, 1997). From 1995 onwards, the UWM School of Architecture and Urban Planning has participated in various university/community partnerships in which the University team has examined the quality of life in the 17 Neighborhood Strategic Planning areas of Milwaukee. Particular attention was paid to the housing conditions of the inner city neighbourhoods, which is in accordance with the long-term agendas of the City and the CBGA. The amount of public loans that were provided to home-owners for house purchase or home improvement was compared to the amount of private mortgage financing opportunities that were present in these distressed neighbourhoods.

One particular GIS-based indicators study undertaken by the UWM presented an interesting story about the dynamics of public loans versus private investment in housing in the inner city neighbourhoods of Milwaukee (http://www.sarup.uwm.edu/cdbg97). The spatial correlation between mortgage activities and assessed values in the central city neighbourhoods of Milwaukee indicated that private-sector investments were highest in number in the areas of increasing assessed values. On the other hand, statistical analysis showed that the private sector avoided investing in neighbourhoods within areas of low property values while public funds were heavily invested in those neighbourhoods. Results of the analyses showed the following positive and negative associations in the blocks in and around the inner city areas:

- Mortgage activity was found to be between low to medium in most of the 17 Neighborhood Strategic Planning (NSP) areas. It was lowest at the central city.
- Blocks with higher private mortgage activity experienced increase in property-assessed values.
- Blocks with higher private mortgage activity contained a greater number of housing units.
- Blocks with higher public investment exhibited a higher owner-occupancy rate.
- Blocks with higher rates of crime exhibited a lower owner-occupancy rate.
- Blocks with higher public investment experienced a lower rate of property sales.
- There was no strong relationship between crime and assessed value of properties.
- Mortgage activity was found to be between low to medium in most of the 17 Neighborhood Strategic Planning (NSP) areas. It was lowest at the central city.

The study identified the worst affected neighbourhoods in the central city and examined the effects of public investment on the most blighted neighbourhoods of Milwaukee. Such evaluations are a necessary step for creating effective public policies, and the results from this study were provided to the CBGA and the City of Milwaukee in order to assist with their public policy formation. This study also verified that not only does private sector lending institutions avoid areas of Milwaukee's central city where property assessments are decreasing, but also that the City of Milwaukee compensates for this lack of investment by targeting publicly funded grants and loans to those areas. While we cannot point to specific policy changes that occurred in response to the findings of this study, we did receive comments from a former CBGA administrator stating that the study assisted the CBGA in gaining a base level of information that in turn created a greater understanding of the problems that face inner city Milwaukee neighbourhoods (Martin, 2000).

Assessing Housing and Neighbourhood Quality in the Lindsay Heights Area, CDBG Area and the City of Milwaukee

Recently a more extensive GIS-based indicators study conducted at different geographic scales was undertaken by the Housing Research Institute of UWM. This two-year, multi-pronged study aimed to support the creation of affordable housing and revitalization of existing housing stock in Milwaukee's inner city. The main objective of this study was to assess quality of life by examining housing based indicators over time (1993-97) for two specific target neighbourhoods in Milwaukee's central city.

The Lindsay Heights area was selected as the first target neighbourhood in the study. Located in the heart of inner city Milwaukee, this neighbourhood attracted the attention of UWM because of its persistently low quality of life. It was believed that the neighbourhood would benefit from a multi-scalar indicators project, which would assist the City and the stakeholders to pursue new policy strategies concerning its revitalization. The study also directly involved the participation of two major neighbourhood stakeholder organizations, including the community-based organization known as WAICO-YMCA Central City Housing Initiative (WAICO), as well as the Wisconsin Housing and Economic Development Association (WHEDA). These stakeholders have been deeply involved in neighbourhood revitalization activities of the Lindsay Heights area since early 1990s and have a strong agenda in improving housing conditions and home ownership in this neighbourhood. In accordance with the shifts in urban governance, these stakeholders have also worked with other private lenders (such as the Milwaukee branch of the Local Initiative Support Coalition or LISC) as well as with local government agencies in pursuing their goals. Because these stakeholders possess an insiders' knowledge of the neighbourhood, they became egalitarian partners in this study and assisted in determining the indicators and in shaping the goals of the study. The study results were disseminated among them in order to facilitate their neighbourhood revitalization tasks, particularly with regards to policy reformulation activities. The reactions of these stakeholders and the consequent policy changes are discussed in the later part of the chapter.

The specific indicators that were examined include: number of residential properties, assessed value of residential property, number of residential property sales, owner occupancy rate, public investment in housing, tax delinquencies and city foreclosures, mortgage and small business lending.

This study utilized a combination of City property data and public data on private mortgages and loans as attribute data for this study, while spatial data from the City's GIS were used in the project. The City's Milwaukee Property File (MPROP) was a particularly valuable source of data in the indicators study. MPROP is a unique data repository that consists of data on individual properties from the Tax Department, the Planning Department, the Building Inspection Department and the City Treasurer's Office. An annual MPROP file is archived each year so that temporal studies can be made on changes to individual properties over the years. In addition, investment data were derived from Home Investment Partnership Program data, Home Mortgage Disclosure Act (HMDA) reports, Wisconsin Housing and Economic Development Authority (WHEDA) lending data and small business lending activity reports.

A notable aspect of this study is the comparison of these indicators at different geographic scales. Each of these indicators is first examined at the parcel level for the Lindsay Heights area, the Community Development Block Grant (CDBG) area (6000 inner city blocks of Milwaukee that receive CDBG funding) and the entire City of Milwaukee (Figure 7.1). The indicators are also examined at the census tract level within the Lindsay Heights study area alone in order to provide a more detailed picture (Housing Research Institute, 1999).

This type of multi-scalar indicators analysis is helpful in assessing the quality of life of a specific troubled area in comparison to other geographic areas. While the full details of the study are discussed in the study report (Housing Research Institute, 1999) and in the associated website (http://gis.sarup.uwm.edu/huxhold/GISdayall_files/frame.htm), the following sections summarize the key findings:

Number of Residential Properties Only one per cent of the city's residential properties are found in the Lindsay Heights area (in 1993 and in 1997). In contrast, the CDBG area includes 42 per cent. Between 1993 and 1997, the Lindsay Heights area experienced a reduction in residential properties of 8.1 per cent, as compared to 1.8 per cent for the CDBG area and 0.7 per cent for the city. It appears that the Lindsay Heights area is losing residential properties at a rate much greater than the CDBG area and the city as a whole.

Assessed Value of Residential Property The Lindsay Heights area ranks close to the bottom of all neighbourhoods in the city in terms of assessed value. The average assessed value of single-family homes in 1997 ($11,006), for example, was only 17 per cent of the citywide average ($65,633) and only 30 per cent of the average assessed value of homes in the CDBG area ($36,357). Not only are the average assessed values of all residential properties in the Lindsay Heights area far below the citywide and CDBG area average, but they are also decreasing while the city and the CDBG area properties are increasing in their assessed values. From 1993 to 1997, the average assessed value of single family homes in the Lindsay Heights area decreased by 7.3 per cent, while the city experienced an increase of 19.4 per cent and the CDBG area an increase of 10.4 per cent. While a small number of residential properties with increased assessed values appear to be randomly scattered within the area, the properties with decreasing assessed values appear to be uniformly distributed throughout the neighbourhood. If this trend continues then the Lindsay Heights area properties will continue to experience further disparities with citywide and CDBG area assessed values.

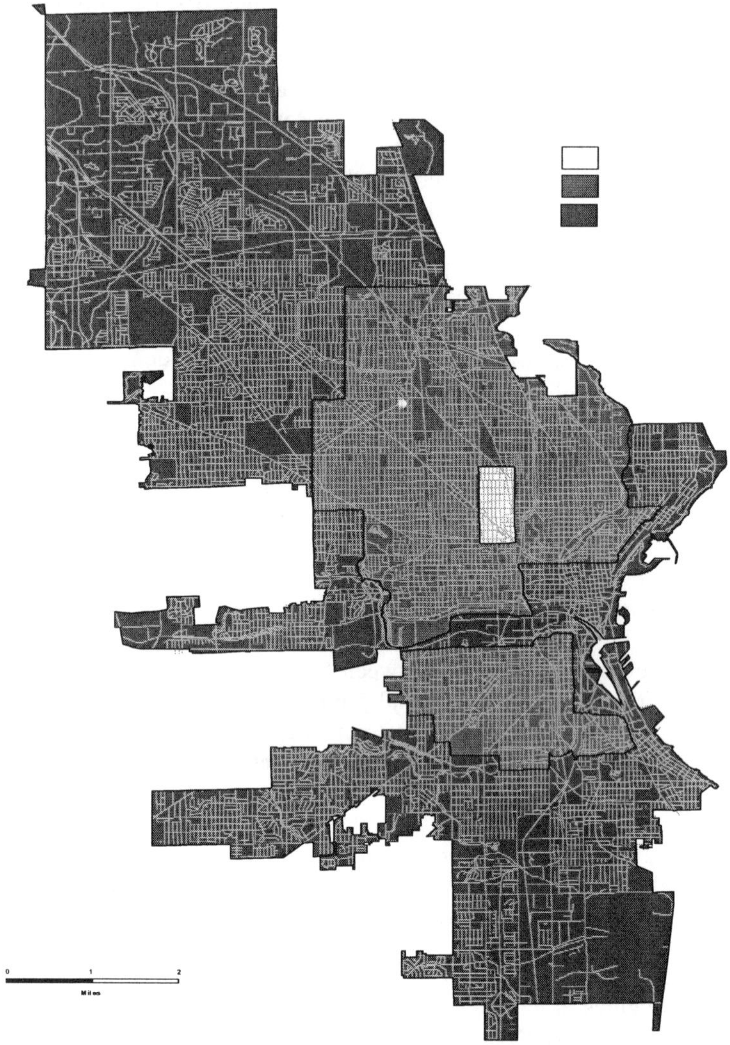

Figure 7.1 Lindsey Heights Study Area Within the Central Business District in the City of Milwaukee

Number of Residential Property Sales Property sales volume in the Lindsay Heights area considerably lags property sales in the rest of the CDBG area and the city as a whole. However, property sales in the area have increased at a higher rate than either the CDBG area or the entire city, indicating a potential turnaround in the market – especially in the single-family home market. Single-family property sales, although low in number, increased over 70 per cent in the Lindsay Heights area between the years 1993 and 1997 compared to only four per cent increase in the CDBG area and a six per cent increase citywide. This increase does not appear to be caused by an exodus of investors because, during the same time period, the per cent increase in average price of single-family properties sold in the Lindsay Heights area increased fourfold over the per cent increase in the average price of single-family properties in the city and the CDBG area. One factor contributing to the comparatively large price increases is the fact that single-family homes in the Lindsay Heights area sold for far less ($14,200) than those in the CDBG area ($40,216) and the city as a whole ($67,913). Perhaps property investors view the Lindsay Heights area as a good investment.

Owner Occupancy Rate The owner occupancy rate (number of owner-occupied properties divided by the total number of residential properties) is quite low in the Lindsay Heights area. Residents own less than 50 per cent of the residential properties in the neighbourhood, while about 64 per cent of the CDBG area residents own their homes and about 77 per cent of the residents of the entire city own their homes. In spite of this apparent negative indicator of neighbourhood quality, there is an indication that owner occupancy in the Lindsay Heights area is rather stable when compared to the apparent departure of owner occupants from their homes in the CDBG area and the entire city as well. It does not appear that owner occupants are leaving the Lindsay Heights area.

Public Investment in Housing The analysis shows that the residents of the Lindsay Heights area used Home Investment Partnership (HOME) funds more extensively than the residents of other neighbourhoods in the city. This can be an indication of the income status of the residents of the Lindsay Heights area. While the CDBG area saw a decrease in the number of properties assisted with HOME Program funds of almost 23 per cent, the Lindsay Heights area experienced an increase in properties assisted in that same time period. As a percentage of all residential properties, the Lindsay Heights area has almost two to three times as many HOME funded properties as the CDBG area, while the number of properties assisted by the HOME Program outside of the CDBG area was negligible.

Tax Delinquencies and City Foreclosures Property owners in the Lindsay Heights area appear to be having trouble keeping up with their property tax payments – more so than the rest of the city's property owners. Almost 14 per cent of all residential properties (180 properties in all) in the Lindsay Heights area were at least two years delinquent in property tax payments in 1997. This is significantly higher than the five per cent tax delinquency rate in the CDBG area and the overall 2.5 per cent rate citywide. While the Lindsay Heights area has only one per cent of the residential properties in the city, it has more than five per cent of all of the city's tax delinquent properties. In addition, the increase in the property foreclosure rate because of tax delinquency in the Lindsay Heights area was twice that of the citywide rate between 1993 and 1997. Clearly, the property owners in the Lindsay

Heights area need property tax relief or income assistance to cover property tax payments more so than the owners of property elsewhere.

Mortgage and Small Business Lending The rate of mortgage and small business lending is much lower in the Lindsay Heights area than in the CDBG area or the City of Milwaukee in general. Home purchase lending varied from less than one per 100 housing units in the Lindsay Heights area to just over three in the CDBG area and just over four in the city. Home improvement loans varied in a similar pattern. Small business lending ranged from just under three per 100 businesses in the Lindsay Heights area to more than 20 in the CDBG area and more than 40 throughout the city. Between 1993 and 1997 the rate of home purchase lending declined in the Lindsay Heights area while remaining constant elsewhere but home improvement lending increased in the Lindsay Heights area while declining elsewhere. An emerging trend, which may have both positive and negative implications, is the increasing presence of subprime lenders (lenders making loans to borrowers who have problematic credit records and do not qualify for prime market rates) in all areas.

Visual Integration

This study is an interesting example of the value of GIS-based multi-scalar analysis in the context of neighbourhood studies. While studies such as this are a more difficult proposition than constructing indicators at the city-wide scale, neighbourhood based indicators are far more effective in demonstrating the disparate conditions of housing and other quality of life factors that exist across the city. The Lindsay Heights study, for instance, is helpful in assessing quality of life in the Lindsay Heights neighbourhood in comparison to larger geographic areas and demonstrates that overall, the Lindsay Heights area fares poorly relative to the rest of the inner city neighbourhoods as well as to the city as a whole on key indicators of housing and neighbourhood quality. On the other hand, it also demonstrates that on some important measures (e.g. increase in property sales and average home prices, stability of the homeownership rate) there are clear indications of improvements.

The GIS mapping component of the indicators is an important aspect of the study. The maps in this study examine the housing quality indicators at different geographic scales (the City, the CBGA and the study area at block level, as well as the study area at census tract level) and thus enable the reader to make cross-comparisons. The maps also assist the readers to discover patterns that are otherwise not evident. For example, in examining assessed values of properties by city blocks across different geographic scales, one can observe that the Lindsay Heights area lies centrally within a larger area of the lowest assessed values in the city. When mapped by block on a citywide basis to examine the change in average assessed values spanning 1993-97, the Lindsay Heights area is found to be located within a larger area of decreasing assessed values. It should be noted however, that this area of decreasing assessed values is smaller than the area of lowest 1997 assessed values. Perhaps this is an indication of improvements in the future.

The map displaying the change in assessed values of properties within the census tracts of Lindsay Heights area shows no pattern, consequently decreasing assessed values appear to be uniformly distributed throughout the neighbourhood.

The GIS component of this study thus provides valuable visual representation of the conditions. As the stakeholders' interviews indicate below, such visual representation was particularly helpful in communication key information to the citizens and to the planning agencies (see the website for all maps associated with this study, http://www.sarup.uwm.edu/cdbg97).

Responses from Stakeholders and Policy Implications of the Indicators Study

As stated before, the Lindsay Heights area has seen the involvement of two major stakeholders in its revitalization activities. Since 1994-95, the local North-Central YMCA organization became increasingly concerned over the area's deteriorating conditions and decided to intervene in order to stabilize and improve the neighbourhoods of that area. Responding specifically to problems of poor housing quality, the organization initiated a separate housing program, naming it WAICO (Walnut Area Improvement Council) YMCA Central City Housing Initiative. This is the only major community-based organization of the Lindsay Heights area that is a stakeholder in the area's revitalization process. It participates in the Neighborhood Strategic Planning process of Milwaukee and has undertaken tasks of repairing and improving houses and bringing them up to code for very low-income residents. The organization also receives CDBG funds to acquire homes that had been boarded or vandalized and abandoned. These homes are then rehabilitated and sold to new, first time homebuyers in the low- to moderate-income category.

Wisconsin Housing and Economic Development Authority (WHEDA) is another major stakeholder in the Lindsay Heights area. It is a non-profit, quasi-state organization that is Wisconsin's housing finance authority. Its mission is to improve and increase the housing stock of Wisconsin for low- and moderate-income families. It also finances single-family homes and multi-family housing and is thus deeply involved in improving home ownership in this neighbourhood. Since 1994, it became concerned over the housing conditions in the Lindsay Heights area and began the Lindsay Heights Initiative to improve housing conditions in that area. It has sought out collaborative partnerships with major stakeholder community organizations such as WAICO and other stakeholders such as local churches, residents and smaller community groups.

Both WHEDA and WAICO became involved with the Lindsay Heights indicators study from its initial phase and participated in meetings with the university research team. For WHEDA and WAICO, the study provided 'valuable data in terms of what were the most important things to learn about those neighbourhoods, in order to measure any improvements over time' (Pukaite, 2001). As discussed by the Community Development Officer of WHEDA,

> …the study also assisted in verifying some information that WHEDA had already collected. But the study provided additional data and gave us a broader perspective of the area than what we had. It also helped us to go to other stakeholders with that data and help us sell them the value of what we are doing. We are really looking forward to the next phase [of study] report on the same indicators to see whether or not all of our energies for the last two to three years have had an impact. We believe they have, we know they have anecdotally, but we want to see it in a study done by a neutral, third party (Pukaite, 2001).

According to the community organizer from WAICO, the study had similar positive benefits:

> We found the [study] report to be very interesting. We need information like this. The study gave us a baseline of information on housing and neighbourhood quality that we did not have. Our organization's goals are to improve housing conditions in the Lindsay Heights area, construct new housing, increase property values, and increase economic development. This information is useful for our future planning efforts in housing and economic development, and it is useful to evaluate our past successes and failures. The study was also helpful in creating an increased awareness of what types of data exist, what are the sources of such data, what the data can tell us and how we can use such data to monitor conditions in our neighbourhood (Greenlee, 2001).

These stakeholders have also utilized the Lindsay Heights indicators study report to make presentations to the City of Milwaukee and believe that the study assisted in legitimizing their concerns, resulting in significant policy actions:

> I don't know if they have made any active decisions to step up their activities in that [Lindsay Heights] area because of the indicators report, but I believe the report helped us to get the attention of the City. The Department of Neighborhood Services and the Department of City Development have actively worked in the last four months with WHEDA to hammer out an expanded agreement for collaborative work that will include a Tax Incremental Finance district for the area, that will support the repair of existing homes and new home construction. We used the information from the study to make presentations to the City on the value of doing a TIF district. We were able to demonstrate to them what the status was, what had happened from 1993-97 [from the study], and then we were able to show them the changes but that information was more anecdotal. I think the people in the DCD are seeing value in what we have been doing. Now they might not attribute it directly to this study. But this study is what we have been basing all of our information and presentations [to the city and others] on. Here's where we are, here's where we are going, this is what we have already done, and if we had the collaboration of the City, this kind of positive results would be much broader.
>
> I think the information we gathered from the study helped us to demonstrate to them [the City Planning departments] that a Tax Incremental Finance district for the area would be valuable and they had never done a neighbourhood TIF district the way we have proposed that is, a very large area focusing resources on improvement of existing houses as well as new home construction. And they are about to start the process of creating the new TIF district in the next month. By the way, the City has just now begun their first

> Comprehensive Strategic Neighborhood Plan, it is being initiated in an area that is broader than Lindsay Heights, but it includes the Lindsay Heights area and they are starting in that area in large part because of the interests that's been spurred through the Lindsay Heights initiative (Pukaite, 2001).

WAICO concurs with WHEDA and comments that 'maybe this is an indirect effect [of the study], but we used the data from the study to make presentations to the City and that led to the creation of a TIF district here' (Greenlee, 2001). The creation of this TIF district or as termed by the City, the Tax Incremental District, is an important policy formulation that the City believes will 'increase residential development and increase homeownership…will provide new single family homes for residents, it also will be used to put underutilized vacant lots back into productive use which will help strengthen the fabric of the entire neighbourhood' (Lindsay Heights TID document, n.d.).

Not only did the indicators based analysis assist the stakeholders in evaluating their approaches, but such use of data also assisted them in gaining the attention of the City, leading to significant policy actions for the neighbourhood. As the Community Development officer from WHEDA commented, 'my sense is that the planners, particularly over at Department of City Development, are much more apt to be responsive to the data, to hard data. Anecdotal information is good, but only if there is hard data first' (Pukaite, 2001). This is an accurate perception, as the City of Milwaukee has a long history of using GIS and geographic information in its planning tasks and does indeed value the use of statistics and GIS in order to validate any claims (Ghose and Huxhold, 2001).

The analysis of data and its comparison at three different geographic scales was perceived to be especially helpful:

> One of the eye opening pieces, of course when I talk about getting the attention of the City, is to demonstrate that this [Lindsay Heights] area, even in relation to CDBG [the rest of the inner-city], is really very poor and very much neglected. And if there isn't a concentrated effort to change that, then in blunt terms, it's like a rot that's going to grow like a rotten apple. It's going to spread, unless we change it (Pukaite, 2001).

In addition to finding the multi-scalar analysis of neighbourhood data to be valuable, WAICO also found the visualization quality of GIS to be an excellent means of communicating the data to the neighbourhood residents:

> The maps were very useful to us. They give a visual picture of what's happening in the neighbourhood. The maps were great in presenting the information to the residents of the neighbourhood. See, the information is so easy to read from the maps. So when we were presenting the study findings to the neighbourhood residents in our monthly meeting, we used the maps heavily to explain the findings. It was easier for the residents to understand the information from the maps than from the tables, because of the visual quality and the use of colors (Greenlee, 2001).

Another important aspect to this study is the role that the University plays in neighbourhood planning activities. Both WAICO and WHEDA commented that studies such as this undertaken by the University are of great help to stakeholder organizations like them that do not have sufficient research experience or expertise or time to gather data systematically and do longitudinal comparative analysis. The community organizer from WAICO felt that such university led studies have some specific advantages:

> The university is an appropriate institution to conduct such studies. First of all, they have the human capacity and technical capacity to do this [type of research]. We don't have the staff capacity or the computer capacity to do these kinds of analysis. So it really helped us that the university did the study because this kind of study is so helpful for our planning purposes. They [the university] have active collaborations with the community; we built ties with the university that continues today. Usually there is a lot of distrust, lot of turf wars and funding wars between different organizations, so collaborations are not always possible. In our collaboration with the university, there is no such distrust and turf war. The university did not have a stake, it is a neutral party, so we trusted them [the University research team] (Greenlee, 2001).

From WHEDA's perspective, university led studies also have greater credibility in the eyes of the local government planning agencies, precisely because such agencies view the university as an impartial provider of reliable information (Pukaite, 2001).

As these interview comments indicate, the stakeholders in Lindsay Heights neighbourhood have found the indicators study to be helpful in their neighbourhood rebuilding tasks. In part, however, such success is attributed to the strong involvement of the stakeholders in the study. From WAICO's perspective, success of such community-based studies depends strongly on the level of citizen participation and input into such studies:

> A lot of studies get done and then they sit in our shelves. But this study was definitely not put into a shelf to rest. One of the reasons is because we had so much input into the study. Lot of times, studies are initiated from some outside organization, they come and ask us a few questions, then they are done. We have no input in those studies. But in this study that was not the case. From the very beginning we were involved in the meetings with other stakeholders and the university group. We had input in determining the indicators, how they can be used, what the purpose of the study is and so on. We met for one and a half year before the study began. We were even involved in the grant writing stage for the study. We understood from the beginning what was going on in the study and how the data can be useful to us for our organizational purposes. We have used the data to get City actions in new home construction, rehab of old homes, repair of homes and doing development (Greenlee, 2001).

In summary, the responses of the stakeholders indicate that this study was helpful in fulfilling several key objectives that lie behind indicators studies (Kingsley, 1999, p.56). First, it provided a baseline of information over a period of time that assisted the stakeholders in examining the conditions of the neighbourhood longitudinally, as well as evaluating the effects of their revitalization programs upon the neighbourhood. The study was helpful in democratizing information and in supporting community-building activities that address issues of persistent poverty. Second, it assisted in creating a new planning policy, whose implementation should have a significant effect upon the neighbourhood. It thus succeeded in using information as a bridge to encourage collaboration among stakeholders and in fighting for appropriate policy actions. As the outcomes indicate, this study also facilitated public-private partnerships between City of Milwaukee, WHEDA and WAICO, uniting all three in the tasks of inner-city neighbourhood revitalization. As Sawicki and Flynn (1996) have remarked, such public-private partnerships are a strong aspect of recent urban governance strategies and further justify the need of similar studies.

Conclusion

Processes of uneven development and deindustrialization have sapped the resources of many inner city neighbourhoods, transforming them into areas of urban blight and concentrations of poverty and crime. It is vital to overcome such trends and bring in positive changes into such neighbourhoods. Formulating effective strategies through specific urban policies is an obvious step in that direction. The centrality of space and place is believed to be an important aspect of urban planning. However, there is a tremendous need for data in policy making, to measure the conditions and trends, to assess current situations, to correct mistakes and to reformulate ineffective policies. Such data needs are particularly acute for neighbourhood-based community organizations and other stakeholders who need such analysis for assessing the effectiveness of their past actions and for future planning. Development and use of indicators at different geographic scales can provide a serious boost to policymaking activities. In particular, neighbourhood-based indicators studies are particularly effective in demonstrating the wide disparities that are prevalent at neighbourhood levels across the city. We believe that the Lindsay Heights study successfully demonstrates such disparities, thus enabling the stakeholders in utilizing the data from the study to create new policy formulations from the City of Milwaukee. In that respect, we believe the study fulfills the major objective of creating positive changes to policymaking and neighbourhood planning activities.

GIS, owing to its strengths in analyzing and visualizing spatial data, is an appropriate choice in such efforts. Through visualization, patterns and trends emerge that may not be evident otherwise. Such visual representations are particularly effective in communicating the results of the analysis to the community and to the policy makers. As the responses from stakeholders indicate, the visualization aspect of this study was crucial in communicating the data effectively to both government agencies and common citizens. Usually the tasks involved in gathering, compiling and analyzing indicators data in GIS can be too complex and time consuming for resource poor community-based organizations or other stakeholders involved in neighbourhood revitalization. As this chapter demonstrates, university/community partnerships are a way to 'democratize GIS', making it possible to disseminate key data and useful analysis to policy makers, community organizations and other stakeholders. As the responses to the Lindsay Heights study indicate, such partnerships can be effective in assisting neighbourhood stakeholders in formulating their strategies and gaining the attention of the City by using housing indicators over time and disseminating the results to planners, policy makers and neighbourhood-based organizations in distressed areas.

Several important issues need to be remembered in undertaking neighbourhood indicators studies. First, such studies should involve community residents, planners and stakeholders as much as possible. In the Milwaukee case, the inclusion of such partners made the study more effective. The Lindsay Heights study incorporated the input from the community in framing the research questions and in identifying the appropriate data. Follow up interviews with the stakeholders provided valuable insight to the researchers for future studies.

Second, it is best to conduct indicators studies longitudinally, during which the data can be collected first to establish a baseline of information and then be examined at different geographic scales. The Milwaukee studies have been conducted over five years (1995-2000). The first study concentrated in establishing a baseline of information across the city and identifying neighbourhoods that are in most need of attention by policymakers. This enabled the second stage of the study to begin, in which two target neighbourhoods were identified to be the focus of the study to analyze the data from 1993-97. The study continues to evolve over a second phase during which data for the two neighbourhoods will be examined from 1998-2000.

Third, the issue of geographic scales is vital to neighbourhood indicators studies. As Sawicki and Flynn (1996) have noted, indicators studies at citywide scale can create false impressions of uniformity, thus adversely affecting neighbourhood-planning tasks. The studies described in the chapter examined spatial data at different geographic scales. For planners, neighbourhood organizations and stakeholders, the multi-scalar approach of the Lindsay Heights study proved to be more effective in making cross-comparisons than a single-scale analysis would have done. Such a multi-scalar approach provided detailed monitoring of trends within a specific neighbourhood as well comparisons across the city. This approach also helped to better demonstrate needs to the City with resulting policy actions.

Acknowledgements

Credit to the first publication of a version of this article is to be given to the URISA Journal, Vol. 14, No. 2, 2002. The authors gratefully acknowledge the contributions of the members of university/community teams, the Housing Research Institute and the interview participants. The authors are particularly grateful to Joyce Witebsky for her contributions in mapping the indicators. Finally, the authors are grateful to the Fannie Mae Foundation for funding the Lindsay Heights indicators study.

References

Annie E. Casey Foundation (1994), *Kids Count Data Book: State Profiles of Child Well-Being*, The Annie E. Casey Foundation, Baltimore.

Barndt, M. (1998a), 'A Model for Evaluating Public Participation GIS Programs', Paper presented at NCGIA Specialist Meeting on Empowerment, Marginalization and Public Participation GIS, Santa Barbara, CA, available online: www.ncgia.ucsb.edu/varenius/ppgis/papers/.

Barndt, M. (1999), 'Using Spatial Information: Barriers and Solutions at the Grass Roots Level', Paper presented at the International Conference on Geographic Information and Society, University of Minnesota.

Barndt, M. (1998b), 'Public Participation GIS – Barriers to Implementation', *Cartography and Geographic Information Systems*, vol. 25(2), pp.105-12.

Barndt, M.G. and Craig, W.J. (1994), 'Data Providers Empower Community GIS Efforts', *GIS World*, vol. 7(7), pp.46-51.

Bauer, R.A. (ed.) (1966), *Social Indicators*, MIT, Cambridge.

Budic, Z.A. (1994), 'Effectiveness of Geographic Information Systems in Local Planning', *Journal of the American Planning Association*, vol. 60(2), pp.244-63.

City of Milwaukee, available online: www.ci.mil.wi.us/citygov/doa/admin/planbtm.htm#CITYOFMILWAUKEE.

Craig, W.J. and Elwood, S. (1998), 'How and Why Community Groups Use Maps and Geographic Information', *Cartography and Geographic Information Systems*, vol. 25(2), pp.95-104.

Drummond, W.J. (no date), 'The Development of GIS-Based Socio-Economic Indicators at the Neighborhood Level', draft.

Elwood, S. (2000), *Information for Change: The Social and Political Impacts of Geographic Information Technologies*, Ph.D. dissertation, University of Minnesota.

Elwood, S. and Leitner, H. (1998), 'GIS and Community-based Planning: Exploring the Diversity of Neighborhood Perspectives and Needs', *Cartography and Geographic Information Systems*, vol. 25(2), pp.77-88.

Flax, M.J. (1978), *Survey of Urban Indicator Data*, The Urban Institute, Washington DC.

French, S.P. and Wiggins, L.L. (1990), 'California Planning Agency Experiences with Automated Mapping and Geographic Information Systems', *Environment and Planning B: Planning and Design*, vol. 17(4), pp.441-50.

Ghose, R. (2001), 'Use of Information Technology for Community Empowerment: Transforming Geographic Information Systems into Community Information Systems', *Transactions in GIS*, vol. 5(2).

Ghose, R. and Huxhold, W.E. (2001), 'Role of Local Contextual Factors in Building Public Participation GIS: The Milwaukee Experience', *Cartography and Geographic Information Science* (forthcoming).

Greelee, J. (2001), Community Organizer of WAICO_YMCA, personal interview by Rina Ghose.

Harris, T. and Elmes, G. (1993), 'The Application of GIS in Urban and Regional Planning: A Review of the North American Experience', *Applied Geography*, vol. 13, pp.9-27.

Housing Research Institute (1999), *Indicators of Housing and Neighborhood Quality – The Lindsay Heights Area, Community Development Block Grant Area, and City of*

Milwaukee: 1993-1997, prepared for the Housing Research Institute, University of Wisconsin-Milwaukee, Milwaukee, available online: http://gis.sarup.uwm.edu/huxhold/GISdayall_files/frame.htm.

Huxhold, W.E. (1991), *An Introduction to Urban Geographic Information Systems*, Oxford University Press, New York.

Huxhold, W.E. and Levinsohn, A.G. (1995), *Managing Geographic Information System Projects*, Oxford University Press, New York.

Huxhold, W.E. and Martin, M. (1996), 'GIS Guides City's Neighborhood Funding Efforts', *GIS World*, vol. June, pp.54-6.

Kellogg, W. (1999), 'From the Field: Observations on Using GIS to Develop a Neighborhood Environmental Information System For Community-Based Organizations', *Journal of the Urban and Regional Information Systems Association*, vol. 11(1), pp.15-32.

Kingsley, G.T. (1999), *Building and Operating Neighborhood Indicator Systems: A Guidebook*, The Urban Institute, Washington DC.

Lindsay Heights Tax Incremental District (no date), City of Milwaukee, Wisconsin.

Martin, M. (2000), Personal Interview by Rina Ghose.

National Civic League (1993), The Healthy Communities Handbook, Denver.

National Neighborhood Indicators Partnership (1999), *Stories: Using Information in Community Building and Local Policy*, The Urban Institute, Washington DC.

Nedovic-Budic, Z. (1998), 'The Likelihood of Becoming a GIS User', *URISA Journal*, vol. 10(2), pp.6-18.

Nedovic-Budic, Z. (1999), 'Evaluating the Effects of GIS Technology: Review of Methods', *Journal of Planning Literature*, vol. 13(3), pp.284-95.

Obermeyer, N.J. (1998), 'The Evolution of Public Participation GIS', *Cartography and Geographic Information Systems*, vol. 25(2), pp.65-6.

Oregon Progress Board (1994), *Oregon Benchmarks: Standard for Measuring Statewide Progress and Institutional Performance*, report to the 1995 Legislature.

Pickles, J. (1995), 'Representations in an Electronic Age: Geography, GIS and Democracy', in J. Pickles (ed.), *Ground Truth: The Social Implications of Geographic Information Systems*, Guilford Press, New York.

Pukaite, C. (2001), Community Development Officer of WHEDA, personal interview by Rina Ghose.

Ramasubramanian, L. (1998), *Knowledge Production and Use in Community-Based Organizations: Examining the Impacts and Influence of Information Technologies*, Ph.D. Dissertation, University of Wisconsin-Milwaukee.

Sawicki, D. and Craig, W. (1996), 'Democratization of Data: Bridging the Gap for Community Groups', *Journal of the American Planning Association*, vol. 62(4), pp.512-23.

Sawicki, D. and Flynn, P. (1996), 'Neighborhood Indicators: A Review of the Literature and an Assessment of Conceptual and Methodological Issues', *Journal of the American Planning Association*, vol. 62(2), pp.165-83.

Sawicki, D. and Peterman, D.R. (1998), Understanding the Breadth and Depth of PPGIS Supply. Paper presented at NCGIA Specialist Meeting on Empowerment, Marginalization, and Public Participation GIS, Santa Barbara, CA., available online: www.ncgia.ucsb.edu/varenius/ppgis/papers/.

Talen, E. (1998), 'Visualizing Fairness: Equity Maps for Planners', *Journal of the American Planning Association*, vol. 64(1), pp.22-38.

University of Wisconsin-Milwaukee (1997), Milwaukee Community Development Block Grant Assessment Project, www.sarup.uwm.edu/cdbg97.

Urban Indicators Program of United Nations Center for Human Settlements, www.urbanobservatory.org/indicators/.

WAICO – YMCA Central City Housing Initiative (2001), Personal Interview with Community Organization Officer by Rina Ghose.

Wiggins, L.L. (1993), 'Diffusion and Use of Geographic Information Systems in Public Sector Agencies in the United States', in I. Masser and H.J. Onsrud (eds), *Diffusion and Use of Geographic Information Technologies*, Kluwer Academic Publishers, Dordrecht, Netherlands.

Wisconsin Housing and Economic Development Authority (WHEDA) (2001), Personal Interview with Community Development Officer by Rina Ghose.

Chapter 8

Where Do We Want To Be? Making Sustainability Indicators Integrated, Dynamic and Participatory

Jeff Carmichael, Sonia Talwar, James Tansey and John Robinson

The Challenge of Indicators

Social indicators have been used since the nineteenth century to both represent the key features of social systems and to create an empirical basis for social intervention. Cobb (1998) suggests that the early prison reformers in Philadelphia sought to collect data in a range of different categories that would indicate the inadequacies of the criminal justice system. The United States led the way with the development of social indicators, and, in 1869, the Massachusetts Bureau of Statistics of Labor began collecting the first official social statistics although even in those early days there was a vigorous debate about whether these data could ever be neutral and value free.

Over the course of the following century, a wider range of indicators was collected on a larger and larger scale and some authors have described a social indicators 'movement' that reached the peak of its influence in the sixties with the publication of a number of volumes including *Social Indicators* (Bauer, Biderman, et al., 1966). While the indicators movement gained a head of steam into the seventies and established an academic journal (*Social Indicators Research*), practitioners and decision-makers encountered a recurrent challenge. Simply put, the collection of a wide range of statistics did not always help to reveal the underlying causal mechanisms due to 'the inherent weakness of descriptive statistics in explaining social phenomena' (Cobb, 1998, p.11).

With the emergence of widespread public concern over environmental issues in the 1970s and 1980s, interest grew in the development of indicator systems focussed on environmental conditions. Indeed, interest in environmental indicators eclipsed interest in social indicators in the late seventies and early eighties. Since then, there has been somewhat of a convergence as a result of recognizing that environmental, social and economic impacts must all be considered simultaneously in a sustainability framework.

The rise of broader sustainability indicators has followed the increasing currency of the concept of sustainable development in the last 15 years. In Canada, sustainability indicators have been collected on a national scale since 1989 and, most recently, the National Round Table on the Environment and the Economy (NRTEE) completed a large-scale study on social and environmental indicators (National Round Table on the Environment and the Economy, 2003). Among the study's key conclusions was the observation that Canada's national information systems do not provide sufficient detail to allow for robust measures of sustainability; instead they rely on macroeconomic measures that fail to capture levels of social, human and natural capital. The committee recommended that a small list of critical indicators be incorporated into the national accounting system. This list, which focuses primarily on environmental variables including air quality measures, fresh water quality, greenhouse gas emissions, forest cover and extent of wetlands, also recommends that a central human capital indicator – educational attainment – should be measured on a national scale (National Round Table on the Environment and the Economy, 2003).

In many ways, sustainability and sustainable development are the latest rallying cries for social change. While the concept has become embedded in the everyday parlance of most stakeholders, there is a growing demand for information on the sustainability of policies, programs and activities at the community scale. In response, the development and use of community sustainability indicator systems has been growing rapidly over the past decade.

Many or most of these indicator systems have been oriented toward historical or current information on the state of the system in question in order to provide decision-makers or stakeholders with information on the current state of their community. Yet the challenge of imputing causality remains significant. The information provided by these indicators is necessary but not sufficient for determining where problems lie with past and current policies or activities. Static descriptive indicators are of even less value when the goal is to understand underlying causal mechanisms or create future-oriented strategies and goals in a rigorous and defensible fashion.

In order for sustainability indicators to be useful for planning future activities, it is necessary for them to go beyond description of current and past conditions and incorporate several features that will allow them to be used to assess future conditions. In particular, it is desirable that the development of indicator sets and sustainability indicator systems be made more integrated and dynamic. In addition, we argue that more participatory processes are needed for the development and use of indicators. We assume that there is not an elusive but ultimately objective set of sustainability indicators waiting to be discovered through the improvement of scientific and technical knowledge. Instead we focus on the development of tools that support social debate about what constitutes a desirable future for a specific region, and we argue that sustainability is an emergent property of these discussions.

The purpose of this chapter is to provide a description and some initial commentary on the development of a sustainability indicator and modeling system in the Georgia Basin, Canada, which has been designed to incorporate a high degree of integration, dynamism and participation. Following a brief description of each of these three characteristics, we will describe how they have been implemented in several related systems to date, discuss some new developments and capabilities currently under development and conclude with some thoughts about the value of such approaches.

Integration: the need for modeling

One of the most important characteristics of the concept of sustainability is its integrated nature; accordingly, it is critical to consider the interactions among complex ecological, social and economic conditions and choices (Robinson and Tinker, 1998). From this point of view, it is problematic to produce indicators which show the current or past state of single variables, such as air quality, greenhouse gas emissions, transportation fuel shares, number of vehicles, etc. Clearly, all of these indicators interact, often in complex ways. While some forms of interaction are straightforward, others are counter-intuitive. Everything else remaining equal, an increase in the number of vehicles will increase emissions, but everything else rarely remains equal and other interactions are much less obvious. Will increased urban density increase or decrease congestion or unemployment? It depends on a number of other interactions and sometimes the answer is surprising.

Given such complexity, simple indications of the current values or trends in key variables are not very useful in assessing possible future outcomes of different actions. What we really need to know is how these variables interact and how society or individuals might influence their paths over time. In other words, it is desirable to know the causal relationships that exist among the variables. This necessarily requires some form of integrated modeling of those variables and relationships. Because sustainability indicator systems encompass a large number of topics and variables, all of which interact in complex ways, formal modeling is essential to capture key relationships and present an integrated view of the systems under consideration. The ability to represent these relationships and explore the integrated view is crucial to widespread consideration of the indicators, the data that support them and the context in which they occur.

In the absence of such formal modeling, users of indicator systems are left to apply their own mental models of relationships among variables in order to try and understand the meaning of indicators.[1] Formal modeling of these relationships may allow users to learn not just the past or current state of the system, but something about how that system functions, and which key indicators may be related to which others.[2] This is crucial in developing useful policy or behavioural responses to indicator values that are deemed to be undesirable in order to affect their future states.

[1] Whether users understand and value indicators differently in the presence or absence of formal models is a research agenda of interest.

[2] As Pierre Wack reminded us many years ago, a key purpose of modeling is to change the mental models of users of the models (Wack, 1985a).

Dynamism: The Use of Scenarios

Integrated modeling of indicators allows us to better understand how individual indicators interact and influence the values we measure, like, for example, water quality or income disparity. Making indicator systems dynamic over time is one way to address the need to assess the relative effect of alternative choices and decisions about sustainability. This can be done through a variety of predictive modeling techniques or through the development of future scenarios.

Predictive modeling approaches are valuable in circumstances where the changes being modelled are within the range of previously observed behaviour (Oreskes, et al., 1994) but are often used inappropriately in the policy world resulting in an illusion of certainty (Jasanoff and Wynne, 1998). Predictive approaches focus on estimating the most likely future state of the system, using, for example, econometrically-derived economic models, or focusing on determining a most desirable state of a system, given specific goals and constraints, through optimization methods.

In contrast, scenario creation models are useful when assumptions governing current system behaviour are changed, when addressing "long" time scales, or when behaviour is not well understood. Scenario-based approaches that are based in quantitative simulation models can also support another desirable characteristic of sustainability indicator systems – the capability to explore alternative future states of the system. Some type of future orientation is required if we are to assess possible outcomes of alternative policy or behavioural decisions. If one purpose of indicators is to show us when conditions are undesirable or a remedy is required, then it is obviously helpful if such systems can also be used to indicate the possible effects associated with different remedies or responses.

All of these modeling approaches have their uses. However, since sustainability raises questions about how to achieve futures that may be very different than the past, predictive approaches based on past behaviours, or optimization approaches based on current goals and constraints, may not be very helpful. Instead, scenario-based approaches which allow us to explore alternative configurations of the future, some of which might be unlikely or non-optimal under current conditions, offer the potential to analyse how key indicators might perform across a range of possible futures (Swart, Raskin et al., forthcoming).

Participation: Involving Stakeholders

A key function of indicator systems is to indicate undesirable or unsustainable conditions and to prompt action. Indicator systems are inescapably linked to policy or behavioural questions concerning what to do to correct problems or reinforce positive outcomes. As a result, governments have typically established sustainability indicator systems in order to support policymaking. However, it has become increasingly clear that there is a gap between the intentions of policy and policy outcomes in the sustainability field (Lee and Perl, 2003). Despite a plethora of international and national scale policy development in the sustainability field, the goals and targets associated with, for example, *Agenda 21* or the Framework

Convention on Climate Change as well as national scale sustainability goals seem as elusive as ever.

One result of the growing awareness of this gap between intention and outcome has been to recognize that governments are not able to achieve sustainability on their own. Other partners' and stakeholders' involvement is needed to develop sustainability policy and decision-making. In particular, it has become clear that a major obstacle to achieving sustainability is the public's social acceptance of policies and products required to achieve sustainability.

Such social acceptance is in part a function of the knowledge that citizens and consumers have about the consequences of different policies and behaviours. This suggests that sustainability indicator systems should be accessible and meaningful to citizens as well as experts, to consumers as well as government officials. This has significant implications for the way such systems are designed, implemented and operated.

Towards Integrated, Dynamic and Participatory Indicator Systems

The preceding discussion suggests the desirability of building integration, dynamism and participation into sustainability indicator systems. This may be accomplished by creating new tools that model key relationships among key variables within a system and demonstrate to tool users the relationships between choices and indicator outcomes. This allows users to explore alternative future choices and investigate how these choices affect indicators via a system that is accessible and interesting to people who do not have specific expertise in all the components of this integrated system. In what follows, we outline several such tools that were developed in inter-related research projects located in the Georgia Basin on the west coast of Canada (Figure 8.1).

The Georgia Basin Futures Project (GBFP) is a five-year regional participatory integrated assessment project. The purpose of the GBFP is to combine public values, preferences and beliefs with expert knowledge in the production of scenarios of the future of the Georgia Basin in western Canada over the next 40 years. Key goals are to contribute to increased public involvement in the discourse about sustainability issues; to explore pathways to sustainability in the region; to create a database of public preferences, values, (un)acceptable trade-offs, etc. that can be analyzed to give a picture of how participants feel about sustainability issues and to evaluate the relationship between the use of computer-based simulation tools and the beliefs, values and behaviours of users of these tools (Tansey and Carmichael et al., 2002).

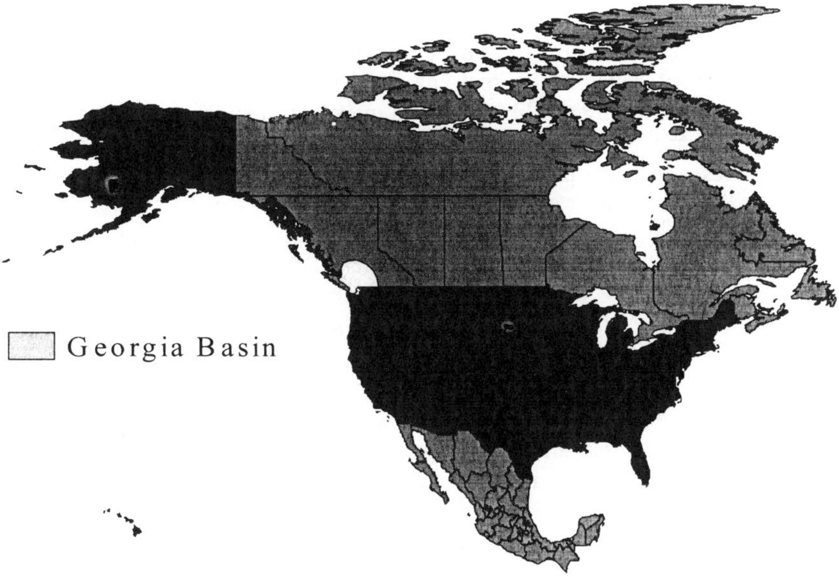

Figure 8.1 The Georgia Basin Region of Canada

The methodological core of the GBFP is the development and use of the Georgia Basin QUEST (GB-QUEST) modeling system (Rothman, 2002; Carmichael, forthcoming), the latest generation of the QUEST system which is a product of ten years of academic research and development. QUEST is a computer-based scenario generation and evaluation system designed to encourage public participation in thinking about sustainability in a regional context. Through QUEST, users explore different possible scenarios of the future in terms of their social, economic and environmental characteristics. The goal is to acquaint users with the complex realities of decision-making, specifically the uncertainties involved, necessary trade-offs and the role of subjective values. For the GB-QUEST modeling system, the geographical range encompasses the whole of the Canadian side of the Georgia Basin. The temporal scale is 40 years.

The Georgia Basin Digital Library (GBDL) project was a three-year project to develop tools to support the integration of geoscientific indicators with socio-economic factors in an environment that would be attractive to a non-expert user (Journeay, 2000). The Georgia Basin Digital Library (GBDL) itself provides access to information collections including significant geospatial information layers such as land use, transportation, various natural resources, orthophotos, etc. It also contains a network of *concepts*, organized as *ontologies* and described using *stories*, to explain what role the layers and their contents play in the sustainable development of the region. It provides a framework for the information but also possesses knowledge elements that provide context for the meaning of the information and its relationship to sustainable development.

The GBDL has been coupled to a web-based user interface called GB-Explorer that enables the contents to be viewed and explored. The interface also connects to other relevant sources of information and knowledge such as publications in other digital libraries and news stories from web-based news agencies. Furthermore, it allows individuals and groups to input their own geospatial sites and descriptions to foster the input of local knowledge, thereby stimulating dialog on issues of community or regional sustainability.

The GeoCognito project is an ongoing attempt to link some of the tools and processes developed in the GBFP and the GBDL projects. The goals are to design, test and evaluate software applications that will integrate expert and local knowledge structures of sustainability (concepts and concept relationships) with both linguistic (descriptive and narrative) and geospatial representations of these concepts in the context of a particular landscape. It focuses on the development of both web-based and workshop-oriented software applications that integrate knowledge representation and landscape visualization. The software facilitates exploration of large, integrated information/knowledge spaces and physical landscapes with an emphasis on context and focus.

The GeoCognito project is incorporating landscape visualization and information visualization into various tools, including GB-QUEST. These tools and applications are in turn being integrated into the overall architectural framework and interface design of GBExplorer for use in ongoing research and community engagement activities of the GBFP.

All of the tools developed in the GBFP, GBDL and GeoCognito projects are being used in a series of community engagement processes including exhibits in Vancouver's Science World; curriculum development for use of several of the tools in high school classrooms; case studies involving the use of tools in three municipalities in the Vancouver region; studies of the cognitive effects of playing GB-QUEST on users' mental models of sustainability, beliefs, attitudes and values and workshops with policymakers focussed on developing strategies for sustainability in the region.

The suite of interactive tools and associated processes of community engagement that are being developed in these three projects represent a rich array of attempts to develop and apply indicator and modeling systems that are integrated, dynamic and participatory. We turn now to more detailed descriptions of some of these tools and processes.

GB-QUEST

GB-QUEST Overview

GB-QUEST is an integrated model that encompasses several individual sub-models. Figure 8.2 illustrates the interactions and feedbacks among the various sub-models in GB-QUEST, showing the user's choices of external context, responses to questions and resulting outcomes (see Carmichael, forthcoming, for detailed description of GB-QUEST and the analytical techniques and procedures used to create the tool).

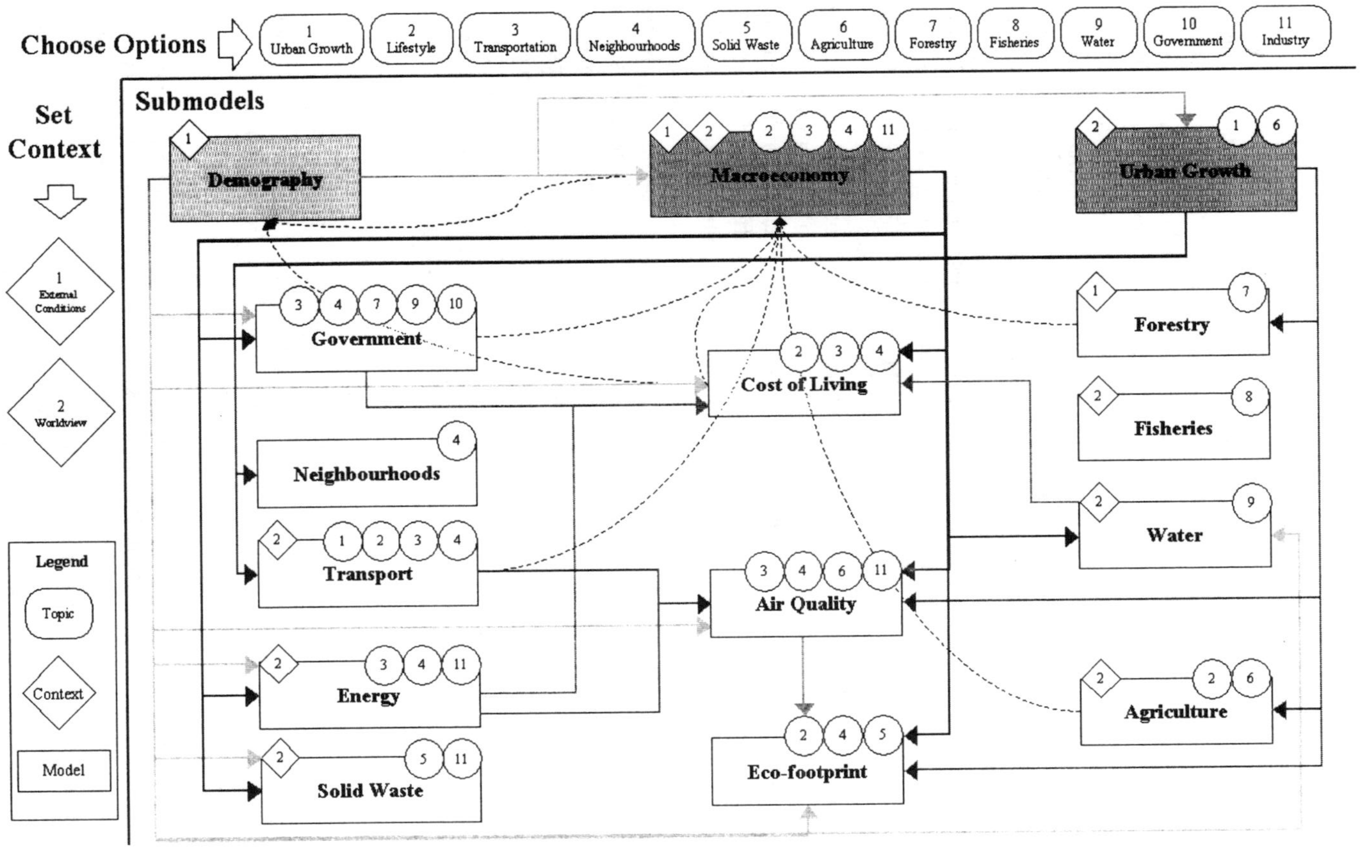

Figure 8.2 GB-QUEST Model Structure

The integrated model allows scenarios to be created that are informed by regional and cross-sectoral changes. Future air quality in the region provides a good example. Air quality is dependent on pollution from point, area and mobile sources. Point source pollution emission impacts are principally influenced by sector-specific changes in technology and resource use, both of which are uncertain and may be influenced by choices made by society. Point source impacts are also affected by economic growth and by changes in climate. Area sources are dependent on how much agricultural and forested land remains and on choices users make about how they manage that land, which in part depends on technological change and in part on societal expectations for identifying the best trade-offs between crop yields, for example, and pollution. Mobile sources depend heavily on population change and economic growth since both affect the transport of people and of goods. Mobile pollution impacts are effected by how much we travel and by what means we travel as well as how much and by what means we ship goods. Both behavioural change and technological advances play central roles. These examples illustrate the way in which GB-QUEST allows stakeholders to consider how many of the pressing regional issues interact and to examine the tradeoffs among different approaches.

GB-QUEST is not designed to be used by public audiences for detailed site- or intervention-specific policy analysis. Instead, the choices offered to the user may be thought of as desired outcomes, rather than the policies that would accomplish those outcomes. For example, a desirable scenario created in a working group may include one in which people will use more alternative means of traveling in the region, such as carpooling, public transit, walking and bicycles, instead of single-occupant cars. QUEST will illustrate some of the integrated consequences of making such choices. Users may then discuss, outside GB-QUEST, which policies and implementation strategies might be appropriate to accomplish such outcomes. Alternatively, policy models might make use of the outcomes of GB-QUEST as indications of the goals of policy and then analyse the costs and consequences of different policy measures that might be used to achieve such goals.

Stages in Using GB-QUEST

Users interact with the sub-models of GB-QUEST through the interface in three steps: Set Context, Options for the Future and Results. In the first two steps, users answer questions that affect the future scenario generated. In the last step, users explore the many results of the created scenario.

Set Context Set Context has three sections: Rank Priorities, External Conditions and How the World Works. Rank Priorities asks for the user's priorities regarding the social, economic and environmental health of the region. External Conditions provides several global scenarios that may affect the region substantially but are out of the region's control. How the World Works asks for the user's perspective on how adaptive social, technical and natural systems are to change.

(1) Rank Priorities provides the user with a choice among four simplified focuses organized in terms of value clusters. This is used to select 'front page' outcomes in the Results section. After creating a scenario, the user can compare the

four value clusters. For each cluster, GB-QUEST shows whether the nine indicators associated with that cluster improved or worsened relative to the base year. By toggling back and forth among the four clusters, the user can see that exactly the same scenario can give rise to very different conclusions about desirability depending on the values of a user. This feature is of particular value to indicator systems because our values influence both the choice of the indicators that matter most to us and our interpretation of the meaning of those indicators.

(2) External Conditions provides the user with a choice of four scenarios that provide regional interpretations of the impact of different global conditions based on global scenarios developed by the Global Scenarios Group (Raskin, 1998; Raskin, Banuri et al., 2002). These scenarios provide GB-QUEST with a set of external conditions that influence local conditions but are not within the power of the region's citizens or organizations to choose or control. Thus, this choice reflects the user's view of what external conditions are most likely. Choosing a different external condition allows the user to see its effect upon the user's preferred scenario. Each external scenario includes values for several parameters, ranging from immigration levels to timber market prices.

(3) How the World Works asks for the user's perspective on the relative speed and flexibility of three factors that influence how the region develops: Social adaptation (how quickly and how much people are willing to change their behaviour), technological innovation (the pace at which technology will change, as well as the degree of this change) and ecological resilience (the degree to which ecological systems are capable of assimilating the by-products of and adapting to changes wrought by human society). In other words, the user is asked to express their underlying beliefs about key aspects of human nature, technology and the natural world. These choices are important for two reasons. First, they are used to incorporate uncertainty regarding behavioural change and scientific processes into the models within the tool. Second, by allowing a user to view this range of uncertainty while at the same time expressing their own worldview, the user is able to see how important their underlying assumptions about reality are in determining the outcome of their scenario and thus the values of the scenario indicators.

Options for the Future The Options for the Future section provides categorized choices that are within the region's control and may affect the future of the region. The user may make choices at three different hierarchical levels ranging from detailed questions to broad decisions that encompass several detailed questions collectively. The hierarchy provides users with flexibility in terms of how deeply they explore different topics. One may choose to explore some issues in great detail yet address other issues only at the coarsest level, depending on one's interests and concerns. Questions either involve degrees of change ('a bit more, more, lots more') or involve distinct options ('dam new rivers to provide more water to cities' vs. 'focus on repairing leaks in existing systems'). The interface is intentionally qualitative, not quantitative, by design in order not to suggest spurious precision and in deference to users who are not quantitatively oriented.

Results The Results section shows the outcomes of the user's choices with a special emphasis on the priorities chosen by the user. Results are organized into two summaries of key indicator outcomes and detailed sets of results by sector.

The first Results page contains two summaries of key indicator outcomes. One summary, shown in Figure 8.3, portrays how the outcomes within the user's scenario compare to fixed targets that represent 'sustainable' activity with respect to each of nine specific outcomes in three categories: social (traffic congestion, students per teacher, hospital visits attributable to air quality related concerns), economic (government deficit, unemployment, cost of living) and ecological (eco-footprint, greenhouse gas emissions and the percentage of potentially developable land remaining in the region). Levels are shown for the beginning decade (year 2000) and ending decade (year 2040) and are compared to target levels. This allows a user to see whether each target has been reached and also to see whether the future level is closer to the target than the present level. These outcomes are portrayed each time that a scenario is generated.

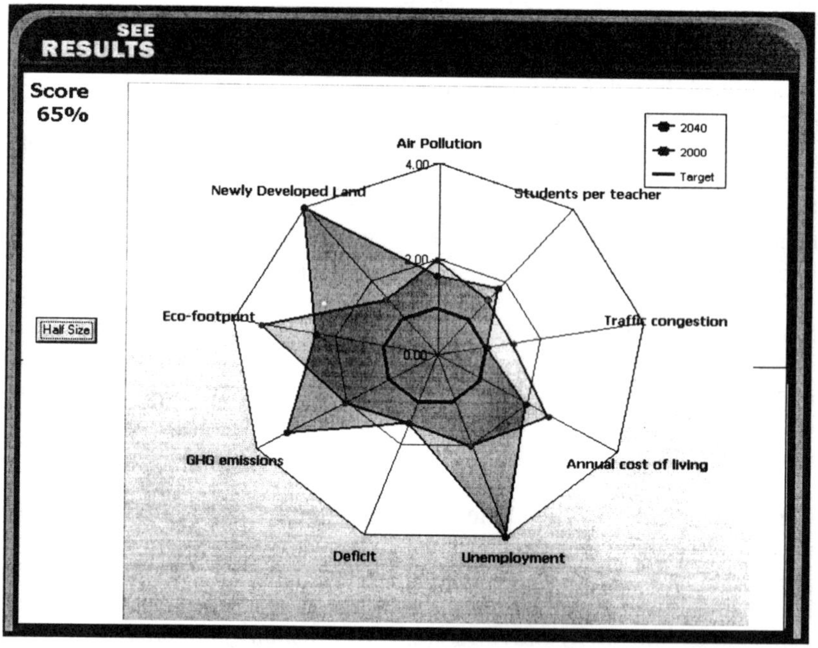

Figure 8.3 Dartboard Summary of Key Indicators

The second summary, shown in Figure 8.4, measures relative progress by illustrating how user-chosen priority outcomes (from the Rank Priorities section) change between the years 2000 and 2040. As indicated above, the user may use this feature to compare different packages of values, showing how those packages influence the desirability of the scenario outcomes.

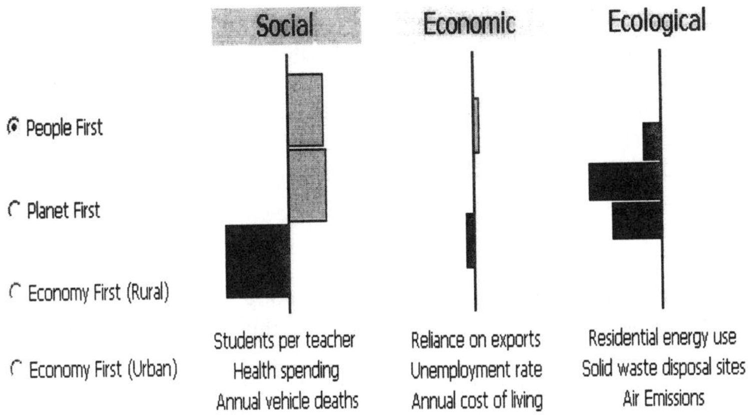

Figure 8.4 Rank Priorities Indicators

Finally, the user may examine any of the specific results within each sector. Each result is displayed separately with a list of each of the user choices that affected the outcome, ranked in order of how strongly it impacts that outcome. This helps the user understand how certain choices lead to downstream consequences. Figure 8.5 gives an air quality example, showing both the result and

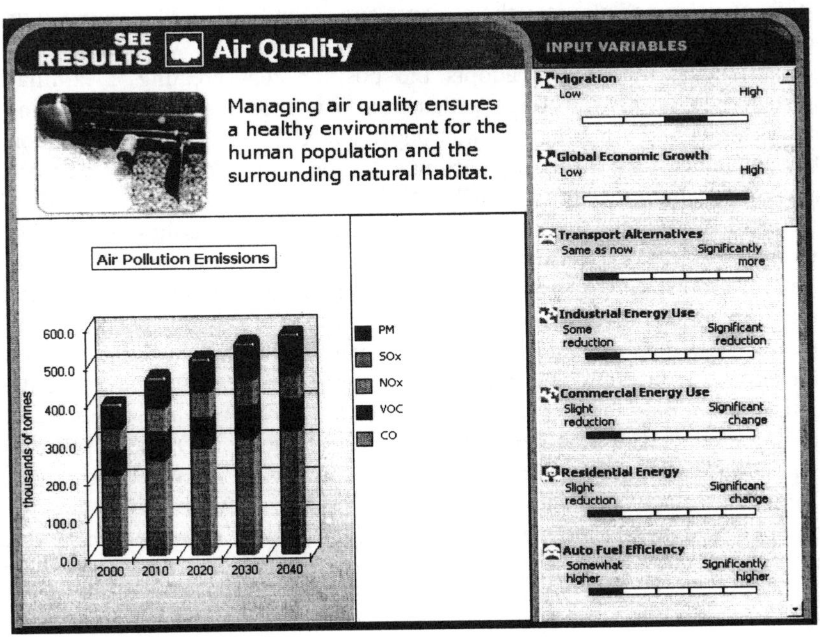

Figure 8.5 Air Quality Result and its Driving Choices

the choices that affect that result. Coupling outcomes with related choices allows users to learn about choices and consequences by exploring causal relationships within the system.

Georgia Basin Digital Library/GBExplorer

The Georgia Basin Digital Library (GBDL)

Another application that supports the interaction among variables and allows users to explore future and present conceptualizations of these variables is the Georgia Basin Digital Library (GBDL). The disaggregated nature of information, particularly across jurisdictional and disciplinary boundaries, is another barrier to representing levels of social, human and natural capital. This fragmentation of information has led to processes that fall short of embracing a holistic approach to decision making about a collective future (Journeay, 2000). In addition to being a rallying cry for social change, sustainability and the use of sustainability indicators includes a challenge of knowledge-building and integration (National Research Council, 1999). The aim of GBDL is to develop awareness and understanding of issues related to sustainability. To accomplish this, GBDL makes use of web-based geographical information systems, knowledge representation, community mapping and participatory processes. A key tenet of the approach is to support dynamic input from users to integrate their views and knowledge with pre-existing data sources. Extending beyond the view of a digital library as a static collection of digital artefacts, the GBDL adopts the position that creating a familiar and understandable frame of reference that assists in meaningfully representing and communicating the complex and interdependent ecological, social and economic systems is a necessary reality of charting a desired course for future generations. In this way, dynamic access to distributed collections and contributions of place-based local knowledge to key community sustainability issues is enabled. By providing a framework to explore integrated knowledge (represented through text, indicators, graphs and maps), the GBDL overcomes a common shortcoming of information dissemination – the lack of context (Talwar, 2003). Data and information are the building blocks to understanding; however, without a network of understood concepts, few things make sense or have meaning. Context is essential for building the mental models of sustainability that the users develop and adapt in the context of interacting with applications such as QUEST. Moreover, it assists the user in building personally relevant frames of reference from which to interpret and make indicators relevant. This recursive process helps users incorporate and assimilate new sustainability information within their existing world view, thereby potentially influencing their conceptualization of the network of relationships between sustainability indicators.

The GBExplorer Interface

GBExplorer, GBDL's interface, provides the mechanisms to examine sustainability information from different angles. GBExplorer is a collection of five applications that collectively provide a means of flexibly engaging with examples, concepts, texts, maps, images and audio on community-relevant issues. The applications include: a) Ideas and Perspectives, b) Library Collections, c) Local Stories, d) News and Information and e) Future Scenarios.

Ideas and Perspectives uses semantic networks, web-based maps and hypermedia to represent relationships between sustainability indicators. In so doing, it provides users with multiple representations of a concept or indicator such as 'Quality of Life'. A set of indicators or sustainability concepts can be captured in a semantic network of relationships (as in the top left corner of the screen). When the user clicks on one of the indicators, relevant textual or multimedia associated with that indicator is displayed so that the user can at once explore the network of relationships (context) and engage with specific examples of the indicator (focus). Points of view from government agencies, non-governmental organizations, community groups, and academics could all be represented in this framework to give a breadth of perspectives on specific sustainability indicators or issues. It lets users expand and adapt their own view in order to understand the links in someone else's, thereby simultaneously preserving their own frameworks of meaning and relating them to new ones (Berners-Lee, 1999).

Library Collections provides distributed dynamic access to data sources, particularly for geospatial information. Users can search, discover and access information through a filtered search. The search procedures comply with existing data infrastructure standards for geospatial information such as those developed by the Canadian Geospatial Data Infrastructure. This means that data sources from numerous agencies can be searched simultaneously and dynamically to return relevant metadata and data.

Local Stories is a way to incorporate issues of individual importance in the dialogue on sustainability and sustainability indicators. Merging web-based geographic information systems and narrative elements, Local Stories lets users explore their local and regional contexts spatially with map-based information and provides the ability for users to contribute stories related to, in this case, sustainability. Users interactively identify a location on the map with which to associate their story. They can identify a wide range of issues from bird or wildlife habitat, to community gardens and farmers' markets. The use of narrative to describe a user-identified issue captures often immeasurable qualities of significance such as aesthetics or inspiration. Users may also associate multimedia (images, sounds, etc.) to provide additional context to their stories.

News and Information is a quick reference source of web-based syndicated news stories related to sustainability. Using pre-assigned keywords, the application extracts relevant news articles from existing services and groups according to scale (national, regional, local) and theme (environment, people, economy). Built as an automated and dynamic service, the retrieved articles are at once current, novel and thematically relevant.

Future Scenarios will allow access to GB-QUEST in a web-environment. Imagine the ability to re-engage with a saved scenario. This would allow users to iterate through another scenario, adjusting their preferences (indicator values) to construct an alternative future to the one they had created previously. This method of interactivity helps the user understand the key indicators that drive the model and the interplay between the extensive set of indicators that support the integrated modeling in a QUEST application. Earmarked as a gateway to a web-based version of GB-QUEST, it currently provides a tutorial on GB-QUEST and background on the sub-models and model integration within GB-QUEST. Its functionality will be greatly expanded in future GeoCognito work (see below).

Use of Dynamic, Integrated and Participatory Indicators

While much of the effort expended in the Georgia Basin Futures Project and the Georgia Basin Digital Library/GeoCognito projects has been devoted to advancing the design of sophisticated, intuitive and dynamic information technology systems, a lot of time was also spent thinking about the social contexts in which they are used. For example, one could envision policy makers and politicians using GB-QUEST to represent their preferred scenario and to convince the public to support their choice. A tool like GB-QUEST or GBExplorer could also be used as an information-gathering tool with which to extract preferences from participants in the context of traditional public consultation. This form of public consultation is very common and typically informs rather than directs decisions made by appointed representatives. Finally, one could imagine GB-QUEST being used by a community group or a municipality to develop their own scenario for a region, to gather support for a range of alternatives and to support the implementation of a series of initiatives. In the latter case, this tool is being used in a more radical way to support a transformative process. As suggested earlier, it was originally imagined that the tools we proposed to develop would be used to engage the public in a series of dialogues about what they would consider to be a desirable future for the region.

These potential uses raise questions about the relationship between the research team and various audiences. We therefore preface the description of the uses of our tools with a discussion of that relationship.

A Typology of Interactivity Processes

The social scientist's dominant approach to engaging with the external world seeks to extract data from a social context, much as a chemist would extract samples from some environmental medium. The researcher strives for methodological neutrality, implying that whichever instrument he or she uses has no effect on the data that is gathered. The researcher appears to be both neutral and unbiased. In contrast, an interactive approach to social science shows greater concern for the processes of interaction between the researcher and the researched. Interactive approaches have been broadly defined as:

> a style of activity where researchers, funding agencies and "user groups" interact throughout the entire research process, including the definition of the research agenda, project selection, project execution and the application of research insights (Scott and Skea et al., 1999).

In order to make this distinction between interactive and extractive approaches more useful, we must distinguish between two different levels of interactivity. A more modest version of interactive research recognises that the quality of research products may be dependent upon the quality of the relationship between the researcher and the researched. In order to design a good data gathering process instrument, the researcher must nurture this relationship.

At a more profound level, this form of interactivity involves the commitments expressed in the definition above to include those being researched in the formulation of the research questions from the start of a project. This stronger program of interactive research fundamentally alters the purpose of the research act. For instance, in seeking to advance the use of focus groups, in-depth discussion groups, which are repeated group sessions where the goal is some form of collective change in mindset or understanding on the part of the participants, have been utilised (Burgess, Limb et al., 1988a; Burgess, Limb et al., 1988b). The explicit goal of transformative processes may include measurement of values and beliefs, but change of values and beliefs is the higher goal.

Table 8.1 describes this typology in more detail and points to the key literatures that could be used to inform the development of robust and valid processes. We distinguish between qualitative and quantitative data since much of the social science literature is divided along these lines. In contrast, we assume that both forms of data can be extremely useful and revealing. For instance, in an indicators prioritisation process using GB-QUEST, we would be interested both in the number of times a particular choice is made by a group involved in scenario development and in the reasons they express for making this choice.

The cells within the table provide examples from across a range of disciplines, and each cell implies a different form of interaction between the theories and concepts of researchers and the everyday experiences of social actors. In extractive work, the researchers seek to present the tool as a neutral device and to set his or her values apart from those of the participants in order to avoid contaminating the data. In an indicators context, this amounts to the desire to extract unbiased measures from, say, social survey data.

Of course, it is clear that there is ultimately no absolutely neutral position that any researcher can assume; however, extractive approaches generally limit unwanted and egregious bias in the design of research methodologies. Dialogical approaches take a more pragmatic approach to research and emphasise the importance of the interaction between the etic categories of researchers and the emic categories of the researched. This distinction is important since social scientists often frame social issues using terminology that is narrow and specialised. Many attempts at dialogue fail because the participants have distinct vocabularies; this can even occur within academia when participants come from different disciplines. Dialogical approaches create opportunities for enhancing mutual understanding. For instance, work in the field of integrated assessment has focused on the differences between expert and lay perceptions of the causes and impacts of climate change.

Both variants of the interactive approach seek to engage research participants much earlier in the process. Rather than present the participants with a fully formed survey instrument or indicators set, the goal is to involve them in the formulation of the key research questions and the development of specific tools. There are two critical reasons for involving research participants early and intensively in interactive research. First, it builds trust among the various research partners and gives them a stronger sense of ownership over the final product than is often the case in university initiated research projects. Second, participants often have a detailed understanding of the complex dynamics of social and natural systems that can be of great help in the design of underlying causal models.

Table 8.1 Forms of Social Science Research

	Extractive	**Interactive**	
		Dialogical	**Transformative**
Example	GB-Quest or GBExplorer as a survey tool	GB-Quest and GBExplorer as consultation/exploration tools	GB-Quest/GBDL as community planning/decision support tools
Qualitative Data	Ethnographic participant observation; usability assessment	Learning outcomes from repeated play and representation of concept relationships; consultation regarding design features	Participatory/Feminist Action Research, In-depth discussion groups, Official Community Plans, transformative community development
Quantitative Data	Questionnaire, survey, public polling, usage statistics	Focus groups leading to qualitative accounts of preferences and values; interactive workshops leading to the development of local content	Curriculum development programmes; degree of social change or participation rates; development of new consultative processes or reporting measures

Two examples from the development of GB-QUEST are illustrative. First, in the process of developing the formal funding application to the Social Sciences and Humanities Research Council, prospective partners were consulted about their research needs and interests and were asked to write letters of support for the initiative committing financial and in-kind resources to the project. Second, in the process of building the underlying models within GB-QUEST, we engaged in what has been referred to as 'participatory modelling' (van Asselt, 2001). Representatives from partner organisations were invited to a number of workshops to review the sectoral models developed by the research team and to provide input into the development of the model interface. In order to complete the latter task, we developed an Archetype Design workshop where participants engaged in a structured process to identify the choices with which users of the model would be presented on the model interface. This information was then used by the modelling team both to structure the choices offered on the interface of the GB-QUEST model and to select the main indicators in the results section of the model.

Transformative approaches to methodology must be distinguished from dialogical approaches for the simple reason that the goal of the research is quite different. The purpose of the interaction is not simply a cross-cultural dialogue but a substantive change in knowledge and behaviour. The most ambitious steps in this direction have been made in feminist research. For example, Dorothy Smith's research on the everyday lives of women in the context of broader social and economic processes has spawned the development analytic tools that analyze and sometimes confront the institutions which shape those experiences (Smith, 1986). Intensive transformative approaches pursue an interventionist therapeutic model, typically in a bounded social context where researchers are very much present.

Transformative educational programmes seek to enable behavioural and attitudinal change across a heterogenous population and, hence, they rely on an assertion that certain human characteristics are general and are subject to influence through a relatively standardised program. Such programs are likely to be evaluated according to the extent and quality of the change they engender.

Studies in the closely related fields of environmental impact assessment have also progressed beyond the dialogical model. A Finnish study demonstrates how an environmental assessment exercise around the siting of a recycling plant became a collaborative learning process where the participants' greater mutual understanding of one another and the outcome was significantly altered by the design of the process (Saarikoski, 2000). Along similar lines, a Canadian study evaluates the potential for transformative learning through legally mandated environmental assessment processes (Sinclair and Diduck, 2001). In both examples there is a fundamental tension between the goals of formal decision makers and the desire of public stakeholders to inform, influence and themselves be transformed by the process.

From an indicator point of view, the kinds of issues discussed here become particularly critical once we move to the more participatory mode of operation argued for above. Moving beyond purely extractive approaches and the ideal of producing neutral information that can be used in many ways, we are inescapably in the realm whereby the meaning of the indicators as well as the actual data itself, are co-produced by the researchers and the users. This has a significant effect on the ways in which such knowledge systems can be used.

Uses of our Information System Technologies

The preceding discussion suggests that different knowledge system tools can serve quite different purposes. More traditional extractive applications of the tools described above have produced very useful descriptive data through applications with the Georgia Basin region. For instance, the team has held countless scenario development workshops within the region and has recorded data reflecting the preferred choices of the participants. This kind of research provides useful descriptive data that indicates stakeholder and public preferences regarding development choices in the region. Ideally, users will play QUEST iteratively in order to learn about the trade-offs between the choices offered within the model. Through this kind of iterative play, it is assumed that users will move beyond *a priori* values to make more sophisticated choices.

Thus far, we have called for indicators to be interactive, dynamic and participatory. In concert with this argument, we also argue that the tools we have developed engage with users in interactive, dynamic and participatory ways. Our approach necessitates coherence between the research methods, use of the tools and the indicators themselves – this is echoed across the research process and outcomes.

The applications discussed below been developed and employed in a variety of research and test environments. The participatory modelling approach used to design the integrated QUEST models and the creation of a workshop process to envelop QUEST use, facilitate scenario creation and discussions about specific indicators. The GBDL development adopted a case study approach to design against a series of potential use situations. Its conceptual framework emerged through discussions and prototyping, all of which was manifest in the existing GBExplorer set of applications. The breadth of functionality available in GBExplorer in some measure accommodates extractive and interactive forms of social research but has a strong emphasis on supporting transformative processes.

In their current manifestations, GB-QUEST and GBExplorer engage with users in quite distinct user interaction experiences. Shiffer (1998) characterizes participation into a matrix of location and temporal coincidence (i.e. a town-hall meeting represents a same place, same time interaction of users) or separation (i.e. online discussion lists with contributions occurring at a different place and time than the recipients). With GB-QUEST, the same time, same place engagement is a key element to engaging the public in dialogues about sustainable futures. GBExplorer's interactive web environment addresses the non-coincidence, temporally or spatially, of participants. Part of the challenge and persistent question with web environments generally is how to support collective and collaborative public engagement in absence of physical proximity. We emphasize that GBExplorer has been used as one of many technologies within a larger social context for learning or planning.

Another user interaction distinction is that QUEST has a focus on collective use since it operates within a facilitated workshop environment to emphasize dialogue on model indicator inputs. This focused dialogue on desired futures provides the participants with an emphasis on the linkages between individual values and collective action and often involves taking a poll to select input choices. In the GBExplorer environment, the dominant form of interaction is that of individual use. Since GBExplorer exists in a web-environment, the user is an individual selecting the issues and navigation related to their individual preferences. By explicitly addressing different interaction experiences with GB-QUEST and GBExplorer and recognizing the value of both individual play/exploration and collective interaction, the parallel projects reciprocate experiences to inform future user experience design.

Use of GBExplorer From a GBExplorer use perspective, the Local Stories application has garnered the most attention. Developed in part to overcome a gap in the broadcast mode of web interaction (i.e. one-way from information provider to information consumer), Local Stories provides a space for local knowledge to be created and shared among community members. Although many communities have developed online information resources for its residents, few extend beyond this information provision use of the web environment. For example, the Tri-Cities region created a community portal that allowed residents to access a suite of information. This portal also included a web-based GIS that allowed residents to view spatial information (maps) and gain a range of useful information about their neighbourhood from the locations of schools, retail establishments and hiking trails to aggregated spatial views of indicators like crime rates. However, this environment did not permit users to enter their own information. As a result, they responded enthusiastically to the concept of Local Stories, and we worked with them to incorporate it within their community portal.

We also designed a workshop environment to bring people together to use Local Stories. This served three functions: 1) to explain how the application works, 2) to demonstrate its use and 3) to co-locate participants for discussion of their contributions in order to overcome the potential isolation that may occur with web-based interaction. Held in computer labs, these workshops included youth groups, seniors, writers and local government staff. Local Stories is about telling stories that hold meaning for residents in a given community. We find that this explicit emphasis on storytelling enchants users. Many quickly identify the demographics in their community and lament that the personal and community histories of their region dwindle as people age and pass away. Historical societies, libraries, writer's groups and community development agencies appreciate the contextual importance of accessible stories – meaningful and available – that Local Stories facilitates. Others have expressed interest in using Local Stories for advocacy to raise awareness about local priorities on political issues.

A key design parameter and research question in GBDL was how to make the technologies and processes surrounding its use easily adaptable to a given community or group's objectives. The adaptation and re-use of the GBExplorer prototype concepts and technologies constitutes another form of its use. The participants and communities that we have worked with often express an interest in incorporating elements from GBExplorer into their existing projects and applications. This underscores the importance for the technological development to be streamlined and flexible in order to integrate with a community's pre-existing systems. Over the past year, we have worked with two communities to modify the Local Stories application to include it within their existing portal (see example above) and supported another community in building their own Local Stories application to integrate habitat information for their region. These experiences have contributed to a technological adaptation of the five applications in GBExplorer to allow enhanced customization for our users.

GBExplorer has also been used in support of community planning processes. The community of Bowen Island, experiencing population growth and resource pressures (i.e. water supply) in their rural island community located on the edge of Vancouver, represents a microcosm of the challenge of sustainability. In particular, the library collections module provides access to community documents, maps and images, acting as a repository of documents produced by community groups (i.e. natural stewardship societies, historians and documents of public record). Thus, the library collections are a local source for information pertinent to decision-making activities related to Bowen Island. The collections also act as community memory, which can be especially helpful for volunteer groups whose memberships continually shift.

Access to these information sources is provided via the community's 'Bowen Island Geolibrary' in web and CD-ROM environments for flexibility and broad access (Journeay, 2002). The local (physical) library provides computer and web access for community members. These information resources have been used as inputs for community indicators studies. We worked with the community of Bowen Island during the design of GBExplorer and prototyped early ideas with them. Active community participants, members of municipal advisory boards and municipal staff have found it useful to have widespread access to these information resources. In addition, exploratory phases of our future research in GeoCognito (see below) have used visualization and community indicators (projected population, resource use, transportation use) in support of a community dialogue around the land use plan for a neighbourhood on the island. Preliminary indications from participants and municipal officials support the use of visualization techniques in planning processes.

As with the development and deployment of any technology, the social impacts of its use emerge over time, informing the design and deployment of future versions and uncovering unintended uses. At present, GBExplorer use occurs within social learning contexts such as education and awareness building; within deliberative processes such as planning and community visioning; and within technological transfer in the form of uptake and re-use of the applications.

Use of QUEST A simplified version of QUEST was developed for the local science centre, Science World, which attracts a half million visitors each year. The tool is used in a 200-seat decision theatre where the audience makes choices using keypads built into their seats. While the largely unmanaged nature of user interaction with this facility undermines the quality of the data, from a purely outreach perspective, it has been a great success. As part of an educational programme, the facility allows for interactive play where users can make informed choices and can select indicators that best reflect and embody their personal values. About 15,000 users, mostly schoolchildren and their teachers, have 'played' Science World QUEST. In addition, from September 2002 to June 2003, there were 30 bookings from school groups ranging from grades 4-9 with a total of 1,245 students taking part.

A number of more interactive applications of the full GB-QUEST system have been pursued in the course of our research. As part of our Strategies component, GB-QUEST has been used in a number of workshops with government policy-makers and other stakeholders. The goal in these sessions is to identify what institutional barriers exist to the implementation of a scenario that reflects the broad policy directions of the participants. The intent was to extract information from the heads of these expert informants about the political landscape of the bioregion. In this context, the model is important because it embodies integrated scientific knowledge and also offers structured choices. In a sense, the model is really a discussion tool; it represents plausible choices and realistic outcomes in a way that structures dialogue among the participants.

There are also a number of examples where the tools described above have been embedded in real decision-making processes at the municipal or regional scale and where community stakeholders have been largely responsible for directing the project. In the case of Bowen Island, GB-QUEST was used to support the planning process for the communities five-year Official Community Plan. Subsequently, GBDL and a number of specialised tools were used both to extend this planning process and to consider much more localised, site level developments. There have been two other such case studies across the region (for the City of Richmond and the Greater Vancouver Regional District), and audiences at the municipal level appear to be receptive to using tools like GBDL and GB-QUEST. As the applications are completed, the team will begin to evaluate how effective the tools have been. In particular, we are interested in the extent to which the dynamic nature of the tools helps users to make informed and sophisticated choices and which values or interests should be given priority. This data is essential for the development of meaningful and salient sets of indicators. Users of these tools have a unique opportunity to learn more about the key causal relationships affecting the indicators they deem to be most important.

Following a more conventional research design, some team members have sought to use the model to test hypotheses about learning with respect to sustainability. The so-called 'Culture and Cognition' section of the project focuses on the effect of the GB-QUEST on mental models of participants in short experimental workshops. The setting is closer to a laboratory setting, but this allows the researchers to test what the cognitive effects of GB-QUEST are in a more controlled environment. Nonetheless, this counts as interactive research since the goal is to test whether individual values can be changed through individual exposure to the tool. Similar approaches have been tried as part of GBDL. Participants are exposed to virtual reality images in a laboratory setting, and the tools are then evaluated to determine whether landscape representations enhance participants' understanding of the physical context and whether the exploration leads to the identification of new issues or raises alternatives not apparent without the immersive experience. These approaches can affect how participants consider the linkage between the physical context and policy implications.

Finally, our Teaching and Learning group has developed curriculum for the use of QUEST and another tool, the Climate Change Calculator, in high schools. This curriculum is designed to address the learning outcomes of specific high school subjects and grade levels. It has been piloted in several classroom environments by our researchers, and we hope to have teachers test it in their classrooms over the coming year.

Altogether 41 QUEST workshops have been held since July 2002. The research team is currently in the process of collating the results of these workshops and drawing conclusions as to the value and meaning of such interactive approaches to knowledge systems and futures assessment.

Future Developments

Database QUEST

Through the use of GB-QUEST with a variety of stakeholders, we have learned that people wish to explore many possible scenarios in rapid succession as a preferred means of understanding tradeoffs among scenarios. We have consequently developed a research strand to explore this and related issues, through the development of an enhanced QUEST system based upon GB-QUEST, tentatively titled 'Database QUEST'. Database QUEST will store approximately 120,000 scenario outcomes in a database allowing a user to move among scenario outcomes in real time. (In contrast, GB-QUEST requires a user to make new choices and run the underlying models for five minutes before an alternative scenario may be examined.) Database QUEST includes a new interface that will display comparative outcomes of different scenarios in a variety of ways as well as display the related set of choices that may create the scenario outcome. We intend to use Database QUEST to explore how people perceive indicators and to research the scenarios that people examine as they, through an exploratory path, choose futures that they consider most desirable.

Extending GBExplorer: Directions within GeoCognito

There is value in providing users with the ability to review and alter created scenarios in order to deeply consider the underlying causal mechanisms and indicator interplay that occurs within the QUEST modelling environment. In conjunction with the development of Database QUEST, we are also developing new means of representing and exploring the spectrum of all possible scenarios that will be required to help understand the varied landscape of desired futures. Information visualization techniques to display node and link relationships between model elements and to represent a future scenario within the context of all, or other, created scenarios for the purposes of comparison are being actively explored.

The majority of the stakeholder engagement with QUEST currently takes place in a facilitated workshop environment. This addresses collective scenario creation in a stimulating, participatory environment. The ability to enhance individual and potentially distributed play (in a web environment, for example) would address the question of uptake and learning outcomes in an individual interaction setting. To address this, a deeper understanding of how users' mental models shift throughout a QUEST session is being explored in order to inform the design of an interface that may be more adaptive to users as they play QUEST.

The ability to engage with sustainability ideas (represented as stories or indicators, for example) in the context of place offers a powerful framework to provide meaning, context and focus to the bigger picture of what sustainability can be. To this end, we are exploring the use of landscape visualization, in the form of detailed three-dimensional representations of terrain, to represent current and future scenarios. We are exploring the feasibility of overlaying a GB-QUEST future scenario on the landscape so that stakeholders can get a sense of what their desired future might actually look and feel like – even flying over it and approaching it from different angles. The landscape visualization techniques are being developed for an immersive (surrounded by three large screens, the users start to feel that they are 'in' the landscape) and a web-based environment. In relation to sustainable community-level planning, we are also investigating the issue of scale and temporality between QUEST future scenarios and a local scale planning and modelling tool, such as CommunityViz.

The five sub-applications within GBExplorer were originally developed as a tightly integrated prototype to demonstrate a reconsidered approach to developing, managing, sharing, reporting, and representing information related to sustainability. As a result of on-going engagement with stakeholders, we are redeploying these applications such that stakeholders can select and combine any of the applications together and, if appropriate, integrate them seamlessly into their existing web-based information systems. This will allow for users to customize the level of integration suitable for their needs.

Conclusions: Some Reflections on Dynamism, Integration and Participation

The challenge of indicators is to make them timely, comprehensive and meaningful. Participatory research, including interacting with user groups throughout the research process, helps entrench the relevance and impacts of sustainability indicators. Making them dynamic and integrated facilitates such participatory processes.

We have described in this chapter some of the ways we have tried to achieve this challenge in the development and use of a suite of interactive knowledge system and scenario analysis tools. In an ambitious five-year program of research, we have designed and implemented a series of such tools and tested them in different environments. While our analysis is not complete, our preliminary findings are that there is a strong interest in, and demand for, tools of this kind. While we have yet to determine the best and most appropriate uses for these tools (and have various research proposals submitted to extend out work in this respect), we believe that we have demonstrated both the feasibility and desirability of making indicator systems dynamic, integrated and participatory in the ways discussed above.

Engaging with indicators via some of the tools discussed in this chapter allows users to participate in a social debate about desired futures whilst continually refining their conceptualization of what sustainability is. This is consistent with a procedural view of sustainability that can be seen as the emergent property of a discussion about desired futures informed by some understanding of the ecological, social and economic consequences of different course of action (Robinson, 2003). In this sense, an understanding of what sustainability means, and might entail, emerges from interactive engagements among researchers and users. The indicators that result reflect the nature of that engagement and embody both scientific and value judgements.

In exploring the design and application of these tools, we have come to understand that not just the tools themselves, but also the processes of user engagement that we develop for use of the tools, must be dynamic, integrative and participatory. This discovery has led us on an extensive journey of exploration, which has taken us quite far down the path towards what we have called interactive forms of social science. The implications and appropriate approaches for such work are not yet entirely clear. What is clear is the interest and excitement associated with such research, in both the research and user community.

References

Bauer, R. and Biderman, A., et al. (1966), *Social Indicators*, NASA.

Berners-Lee, T. (1999), *Weaving the Web: The Original Design and Ultimate Destiny of the World Wide Web by Its Inventor*, HarperCollins, New York.

Burgess, J. and Limb, M., et al. (1988a), 'Exploring Environmental Values through the Medium of Small Groups: 1. Theory and Practice', *Environment and Planning A*, vol. 20, pp.308-26.

Burgess, J. and Limb, M., et al. (1988b), 'Exploring Environmental Values through the Medium of Small Groups: 2. Illustrations of a Group at Work', *Environment and Planning A*, vol. 20, pp.457-76.

Carmichael, J., Tansey, J. and Robinson, J. (forthcoming), 'Georgia Basin QUEST: An Integrated Assessment Modeling Tool', *Global Environmental Change.*

Cobb, C. and Rixford, C. (1998), *Lessons Learned from the History of Social Indicators*, Redefining Progress.

Jasanoff, S. and Wynne, B. (1998), 'Science and Decisionmaking', in S. Rayner and E. Malone (eds), *Human Choice and Climate Change*, vol 1, Battelle Press, Columbus, pp.1-87.

Journeay, M. and Dunster, J. (2002), 'The Bowen Geolibrary', CD-ROM produced by the Bowen Island Forest and Water Management Society, available online: www.bowenisland.info.

Journeay, M., Robinson, J., Talwar, S., Walsh, M., Biggs, D., McNaney, K., Kay, B., Brodaric, B. and Harrap, R. (2000), 'The Georgia Basin Digital Library: Infrastructure for a Sustainable Future', GeoCanada 2000 Conference, Calgary, Alberta.

Lee, E. and Perl, A. (eds) (2003), *The Integrity Gap – Canada's Environmental Policy and Institutions*, UBC Press, Vancouver.

Marris, C., Langford, I., Saunderson, L. and O'Riordan, T. (1997), 'Exploring the "Psychometric Paradigm": Comparisons between Aggregate and Individual Analyses', *Risk Analysis*, vol. 17(3), pp.303-13.

National Research Council (1999), *Our Common Journey: A Transition Toward Sustainability*, National Academy Press, Washington DC.

National Round Table on the Environment and the Economy (2003), 'Environment and Sustainable Development Indicators for Canada', National Round Table on the Environment and the Economy, Ottawa.

Oreskes, N. K., Shrader-Frechette, K. and Belitz, K. (1994), 'Verification, Validation and Confirmation of Numerical Models in the Earth Sciences', *Science*, vol. 263, pp.641-6.

Raskin, P. and Banuri, T. et al. (2002), *Great Transition: The Promise and Lure of the Times Ahead*, Stockholm Environment Institute, Stockholm.

Raskin, P., Gallopin, G., Hammond, A. and Swart, R. (1998), *Bending the Curve: Toward Global Sustainability*, Stockholm Environment Institute, Stockholm.

Robinson, J. (2003), 'Future Subjunctive: Backcasting as Social Learning', *Futures*, vol. 35, pp.839-56.

Robinson, J. and Tinker, J. (1998), 'Reconciling Ecological, Economic, and Social Imperatives', in J. Schnurr and S. Holtz (ed.), *The Cornerstone of Development – Integrating Environmental, Social and Economic Policies*, International Development Research Centre and Lewis Publishers, Ottawa and New York.

Rothman, D., Robinson, J. and Biggs, D. (2002), 'Signs of Life: Linking Indicators and Models in the Context of QUEST', in H. Abaza and A. Baranzini (eds), *Implementing Sustainable Development. Integrated Assessment and Participatory Decision-Making Processes*, Edward Elgar, Cheltenham (UK).

Saarikoski, H. (2000), 'Environmental Impact Assessment (EIA) as a Collaborative Learning Process', *Environmental Impact Assessment Review*, vol. 20, pp.681-700.

Scott, A. and Skea, J. et al. (1999), 'Designing "Interactive" Environmental Research for Wider Social Relevance', Brighton, Economic and Social Research Council, vol. 15.

Shiffer, M. (1998), 'Multimedia GIS for Planning Support and Public Discourse', *Cartography and Geographic Information Systems*, vol. 25(2), pp.89-94.

Sinclair, J.A. and Diduck, A.P. (2001), 'Public Involvement in EA in Canada: A Transformative Learning Perspective', *Environmental Impact Assessment Review*, vol. 21, pp.113-36.

Smith, D. (1986), 'Institutional Ethnography: A Feminist Methodology', *Resources for Feminist Research*, vol. 15, pp.6-13.

Swart, R. and Raskin, P. et al. (forthcoming), 'The Problem of the Future: Sustainability Science and Scenario Analysis', *Global Environmental Change.*

Talwar, S., Journeay, M., Harrap, R., Brodaric, B., van Ulden, J., Grant, R. and Denny, S. (2003), 'The Georgia Basin Digital Library Project: A Framework for a Place-Based Exploration of Sustainability Issues', *Geomatica*, vol. 57(2), pp.155-64.

Tansey, J. and Carmichael, J. et al. (2002), 'The Future is Not What it Used to Be: Participatory Integrated Assessment in the Georgia Basin', *Global Environmental Change*, vol. 12(2), pp.97-104.

van Asselt, M., Mellors, J., Rijkens-Klomp, N., Greeuw, S., Molendijk, K., Beers, P. and van Notten, P. (2001), 'Building Blocks for Participation in Integrated Assessment: A Review of Participatory Methods', prepared for the International Centre for Integrative Studies, Maastricht, the Netherlands.

Wack, P. (1985), 'Scenarios: Uncharted Waters Ahead', *Harvard Business Review*, vol. 5(Sept./Oct), pp.72-89.

Subject Index

WAVELET IMAGE AND VIDEO COMPRESSION

THE KLUWER INTERNATIONAL SERIES IN ENGINEERING AND COMPUTER SCIENCE

WAVELET IMAGE AND VIDEO COMPRESSION

edited by

Pankaj N. Topiwala
Sanders, A Lockheed Martin Company

KLUWER ACADEMIC PUBLISHERS
Boston / Dordrecht / London

Distributors for North, Central and South America:
Kluwer Academic Publishers
101 Philip Drive
Assinippi Park
Norwell, Massachusetts 02061 USA
Telephone (781) 871-6600
Fax (781) 681-9045
E-Mail <kluwer@wkap.com>

Distributors for all other countries:
Kluwer Academic Publishers Group
Distribution Centre
Post Office Box 322
3300 AH Dordrecht, THE NETHERLANDS
Telephone 31 78 6392 392
Fax 31 78 6546 474
E-Mail <services@wkap.nl>

Electronic Services <http://www.wkap.nl>

Library of Congress Cataloging-in-Publication Data

Wavelet image and video compression / edited by Pankaj N. Topiwala.
p.cm. —(The Kluwer international series in engineering and computer science)
Includes bibliographical references and index.
ISBN 0-7923-8182-3 (alk. paper)
1. Image processing—Digital techniques—Mathematics. 2. Image compression.
3. Wavelets (Mathematics). 4. Video compression. I. Topiwala, Pankaj N. II. Series.

TA1637 .W396 1998
621.36/7--dc21

98-175364

Printed on acid-free paper.

Printed in Great Britain by IBT Global, London

Cover photograph courtesy of the National Aeronautics and Space Administration.

This printing is a digital duplication of the original edition.

Contents

[1]Based on "Multiplication-Free Subband Coding of Color Images", by Docef, Kossentini, Chung and Smith, which appeared in the Proceedings of the Data Compression

---

Conference, Snowbird, Utah, March 1995, pp 352-361, ©1995 IEEE

[2]©1991 IEEE. Reprinted, with permission, from *IEEE Transactions of Circuits and Systems for Video Technology*, pp.86-99, March, 1991.

[3]Based on "A New Approach to Scalable Video Coding", by Chung, Kossentini and Smith, which appeared in the Proceedings of the Data Compression Conference, Snowbird, Utah, March 1995, ©1995 IEEE

1 Introduction

Pankaj N. Topiwala

1 Background

It is said that a picture is worth a thousand words. However, in the Digital Era, we find that a typical color picture corresponds to more than a million words, or bytes. Our ability to sense something like 30 frames a second of color imagery, each the equivalent of tens of millions of pixels, means that we can process a wealth of image data – the equivalent of perhaps 1 Gigabyte/second (1 GB/s or 1×10^9 B/s). The wonder and preeminent role of vision in our world can hardly be overstated, and today's digital dialect allows us to quantify this. In fact, our eyes and minds are not only acquiring and storing this much data, but processing it for a multitude of tasks, from 3-D rendering to color processing, from segmentation to pattern recognition, from scene analysis and memory recall to image understanding and finally data archiving, all in real time. Rudimentary computer science analogs of similar image processing functions can consume tens of thousands of operations *per pixel*, corresponding to perhaps 10^{13} operations/s. In the end, this continuous data stream is stored in what must be the ultimate compression system, utilizing a highly prioritized, *time-dependent* bit allocation method. Yet despite the high density mapping (which is *lossy*—nature chose lossy compression!), on important enough data sets we can reconstruct images (e.g., events) with nearly perfect clarity. While estimates on the brain's storage capacity are indeed astounding (e.g., 1.25×10^{14} B [1]), even such generous capacity must be efficiently used to permit the full breadth of human capabilities (e.g., a weekend's worth of video data, if stored "raw," would alone swamp this storage). However, there is fairly clear evidence that sensory information is not stored raw, but highly processed and stored in a kind of associative memory. At least since Plato, it has been known that we categorize objects by proximity to prototypes (i.e., sensory memory is contextual or "object-oriented"), which may be the key.

What we can do with computers today isn't anything nearly so remarkable. This is especially true in the area of pattern recognition over diverse classes of structures. Nevertheless, an exciting new development has taken place in this digital arena that has captured the imagination and talent of researchers around the globe—wavelet image compression. This technology has deep roots in theories of vision (e.g. [2]) and promises performance improvements over all other compression methods, such as those based on Fourier transforms, vectors quantizers, fractals, neural nets, and many others. It is this revolutionary new technology that we wish to present in this edited volume, in a form that is accessible to the largest readership possible. A first glance at the power of this approach is presented below in figure 1, where we achieve a dramatic 200:1 compression. Compare this to the international standard JPEG, which cannot achieve 200:1 compression on this image; at its coarsest quantization (highest compression), it delivers an interesting example of cubist art, figure 2.

(a) (b)

FIGURE 1. (a) Original shuttle image, 1k x 1k image, in full color (24 b/p); (b) Shuttle image compressed 200:1 using wavelets.

FIGURE 2. The shuttle image compressed by the international JPEG standard, using the coarsest quantization possible, giving 176:1 compression (maximum).

If we could do better pattern recognition than we can today, up to the level of "image understanding," then neural nets or some similar learning-based technology could potentially provide the most valuable avenue for compression, at least for purposes of subjective interpretation. While we may be far from that objective, wavelets offer the first advance in this direction, in that the multiresolution image analysis appears to be well-matched to the low-level characteristics of human vision. It is an exciting challenge to develop this approach further and incorporate additional aspects of human vision, such a spectral response characteristics, masking, pattern primitives, etc. [2].

Computer data compression is, of course, a powerful, enabling technology that is playing a vital role in the Information Age. But while the technologies for machine data acquisition, storage, and processing are witnessing their most dramatic developments in history, the ability to deliver that data to the broadest audiences is still often hampered by either physical limitations, such as available spectrum for broadcast TV, or existing bandlimited infrastructure, such as the twisted copper telephone network. Of the various types of data commonly transferred over networks, image data comprises the bulk of the bit traffic; for example, current estimates indicate that image data transfers take up over 90% of the volume on the Internet. The explosive growth in demand for image and video data, coupled with these and other delivery bottlenecks, mean that compression technology is at a premium. While emerging distribution systems such as Hybrid Fiber Cable Networks (HFCs), Asymmetric Digital Subscriber Lines (ADSL), Digital Video/Versatile Discs (DVDs), and satellite TV offer innovative solutions, all of these approaches still depend on heavy compression to be viable. A Fiber-To-The-Home Network could potentially boast enough bandwidth to circumvent the need for compression for the forseeable future. However, contrary to optimistic early predictions, that technology is not economically feasible today and is unlikely to be widely available anytime soon. Meanwhile, we must put digital imaging "on a diet."

The subject of digital dieting, or data compression, divides neatly into two categories: lossless compression, in which exact recovery of the original data is ensured; and lossy compression, in which only an approximate reconstruction is available. The latter naturally requires further analysis of the type and degree of loss, and its suitability for specific applications. While certain data types cannot tolerate any loss (e.g., financial data transfers), the volume of traffic in such data types is typically modest. On the other hand, image data, which is both ubiquitous and data intensive, can withstand surprisingly high levels of compression while permitting reconstruction qualities adequate for a wide variety of applications, from consumer imaging products to publishing, scientific, defense, and even law enforcement imaging. While lossless compression can offer a useful two-to-one (2:1) reduction for most types of imagery, it cannot begin to address the storage and transmission bottlenecks for the most demanding applications. Although the use of lossless compression techniques is sometimes tenaciously guarded in certain quarters, burgeoning data volumes may soon cast a new light on lossy compression. Indeed, it may be refreshing to ponder the role of lossy memory in natural selection.

While lossless compression is largely independent of lossy compression, the reverse is not true. On the face of it, this is hardly surprising since there is no harm (and possibly some value) in applying lossless coding techniques to the output of any lossy coding system. However, the relationship is actually much deeper, and lossy compression relies in a fundamental way on lossless compression. In fact, it is only a slight exaggeration to say that the art of lossy compression is to "simplify" the given data appropriately in order to make lossless compression techniques effective. We thus include a brief tour of lossless coding in this book devoted to lossy coding.

2 Compression Standards

It is one thing to pursue compression as a research topic, and another to develop live imaging applications based on it. The utility of imaging products, systems and networks depends critically on the ability of one system to "talk" with another—interoperabilty. This requirement has mandated a baffling array of standardization efforts to establish protocols, formats, and even classes of specialized compression algorithms. A number of these efforts have met with worldwide agreement leading to products and services of great public utility (e.g., fax), while others have faced division along regional or product lines (e.g., television). Within the realm of digital image compression, there have been a number of success stories in the standards efforts which bear a strong relation to our topic: JPEG, MPEG-1, MPEG-2, MPEG-4, H.261, H.263. The last five deal with video processing and will be touched upon in the section on video compression; JPEG is directly relevant here.

The JPEG standard derives its name from the international body which drafted it: the Joint Photographic Experts Group, a joint committee of the International Standards Organization (ISO), the International Telephone and Telegraph Consultative Committee (CCITT, now called the International Telecommunications Union-Telecommunications Sector, ITU-T), and the International Electrotechnical Commission (IEC). It is a transform-based coding algorithm, the structure of which will be explored in some depth in these pages. Essentially, there are three stages in such an algorithm: transform, which reorganizes the data; quantization, which reduces data complexity but incurs loss; and entropy coding, which is a lossless coding method. What distinguishes the JPEG algorithm from other transform coders is that it uses the so-called Discrete Cosine Transform (DCT) as the transform, applied individually to 8-by-8 blocks of image data.

Work on the JPEG standard began in 1986, and a draft standard was approved in 1992 to become International Standard (IS) 10918-1. Aspects of the JPEG standard are being worked out even today, e.g., the lossless standard, so that JPEG in its entirety is not completed. Even so, a new international effort called JPEG2000, to be completed in the year 2000, has been launched to develop a novel still color image compression standard to supersede JPEG. Its objective is to deliver a combination of improved performance and a host of new features like progressive rendering, low-bit rate tuning, embedded coding, error tolerance, region-based coding, and perhaps even compatibility with the yet unspecified MPEG-4 standard. While unannounced by the standards bodies, it may be conjectured that new technologies, such as the ones developed in this book, offer sufficient advantages over JPEG to warrant rethinking the standard in part. JPEG is fully covered in the book [5], while all of these standards are covered in [3].

3 Fourier versus Wavelets

The starting point of this monograph, then, is that we replace the well-worn 8-by-8 DCT by a different transform: the wavelet tranform (WT), which, for reasons to be clarified later, is now applied to the whole image. Like the DCT, the WT belongs to a class of linear, invertible, "angle-preserving" transforms called *unitary* transforms.

However, unlike the DCT, which is essentially unique, the WT has an infinitude of instances or realizations, each with somewhat different properties. In this book, we will present evidence suggesting that the wavelet transform is more suitable than the DCT for image coding, for a variety of realizations.

What are the relative virtues of wavelets versus Fourier analysis? The real strength of Fourier-based methods in science is that oscillations—waves—are everywhere in nature. All electromagnatic phenomena are associated with waves, which satisfy Maxwell's (wave) equations. Additionally, we live in a world of sound waves, vibrational waves, and many other waves. Naturally, waves are also important in vision, as light is a wave. But visual information—what is in images—doesn't appear to have much oscillatory structure. Instead, the content of natural images is typically that of variously textured objects, often with sharp boundaries. The objects themselves, and their texture, therefore constitute important structures that are often present at different "scales." Much of the structure occurs at fine scales, and is of low "amplitude" or contrast, while key structures often occur at mid to large scales with higher contrast. A basis more suitable for representing information at a variety of scales, with local contrast changes as well as larger scale structures, would be a better fit for image data; see figure 3.

The importance of Fourier methods in signal processing comes from stationarity assumptions on the statistics of signals. Stationary signals have a "spectral" representation. While this has been historically important, the assumption of stationarity—that the statistics of the signal (at least up to 2nd order) are constant in time (or space, or whatever the dimensionality)—may not be justified for many classes of signals. So it is with images; see figure 4. In essence, images have locally varying statistics, have sharp local features like edges as well as large homogeneous regions, and generally defy simple statistical models for their structure. As an interesting contrast, in the wavelet domain image in figure 5, the local statistics in most parts of the image are fairly consistent, which aids modeling. Even more important, the transform coefficients are for the most part very nearly zero in magnitude, requiring few bits for their representation.

Inevitably, this change in transform leads to different approaches to follow-on stages of quantization and even entropy coding to a lesser extent. Nevertheless, a simple baseline coding scheme can be developed in a wavelet context that roughly parallels the structure of the JPEG algorithm; the fundamental difference in the new approach is only in the transform. This allows us to measure in a sense the value added by using wavelets instead of DCT. In our experience, the evidence is conclusive: wavelet coding is superior. This is not to say that the main innovations discussed herein are in the transform. On the contrary, there is apparently fairly wide agreement on some of the best performing wavelet "filters," and the research represented here has been largely focused on the following stages of quantization and encoding.

Wavelet image coders are among the leading coders submitted for consideration in the upcoming JPEG2000 standard. While this book is not meant to address the issues before the JPEG2000 committee directly (we just want to write about our favorite subject!), it is certainly hoped that the analyses, methods and conclusions presented in this volume may serve as a valuable reference—to trained researchers and novices alike. However, to meet that objective and simultaneously reach a broad audience, it is not enough to concatenate a collection of papers on the latest wavelet

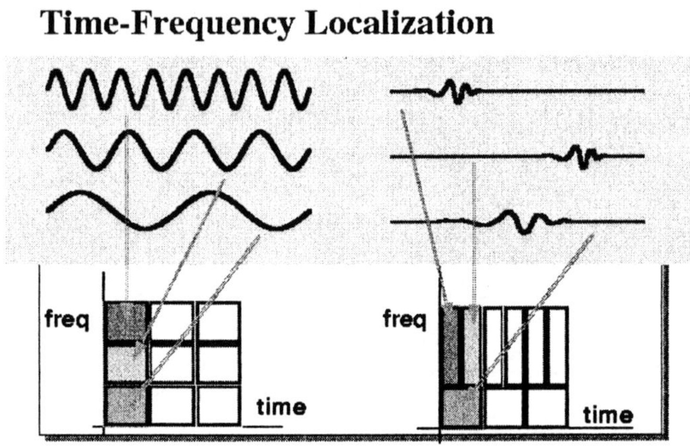

(a) (b)

FIGURE 3. The time-frequency structure of local Fourier bases (a), and wavelet bases (b).

FIGURE 4. A typical natural image (Lena), with an analysis of the local histogram variations in the image domain.

algorithms. Such an approach would not only fail to reach the vast and growing numbers of students, professionals and researchers, from whatever background and interest, who are drawn to the subject of wavelet compression and to whom we are principally addressing this book; it would perhaps also risk early obsolescence. For the algorithms presented herein will very likely be surpassed in time; the real value and intent here is to educate the readers of the methodology of creating compression

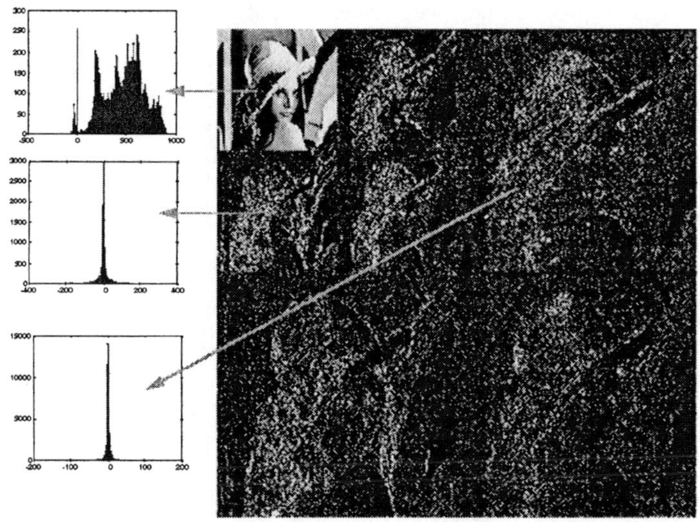

FIGURE 5. An analysis of the local histogram variations in the wavelet transform domain.

algorithms, and to enable them to take the next step on their own. Towards that end, we were compelled to provide at least a brief tour of the basic mathematical concepts involved, a few of the compression paradigms of the past and present, and some of the tools of the trade. While our introductory material is definitely not meant to be complete, we are motivated by the belief that even a little background can go a long way towards bringing the topic within reach.

4 Overview of Book

This book is divided into four parts: (I) Background Material, (II) Still Image Coding, (III) Special Topics in Image Coding, and (IV) Video Coding.

4.1 Part I: Background Material

Part I introduces the basic mathematical structures that underly image compression algorithms. The intent is to provide an easy introduction to the mathematical concepts that are prerequisites for the remaider of the book. This part, written largely by the editor, is meant to explain such topics as change of bases, scalar and vector quantization, bit allocation and rate-distortion theory, entropy coding, the discrete-cosine transform, wavelet filters, and other related topics in the simplest terms possible. In this way, it is hoped that we may reach the many potential readers who would like to understand image compression but find the research literature frustrating. Thus, it is explicitly *not* assumed that the reader regularly reads the latest research journals in this field. In particular, little attempt is made to refer the reader to the original sources of ideas in the literature, but rather the most accessible source of reference is given (usually a book). This departure from convention is dictated by our unconventional goals for such a technical subject. Part I can be skipped by advanced readers and researchers in the field, who can proceed directly to relevant topics in parts II through IV.

Chapter 2 ("Preliminaries") begins with a review of the mathematical concepts of vector spaces, linear transforms including unitary transforms, and mathematical analysis. Examples of unitary transforms include Fourier Transforms, which are at the heart of many signal processing applications. The Fourier Transform is treated in both continuous and discrete time, which leads to a discussion of digital signal processing. The chapter ends with a quick tour of probability concepts that are important in image coding.

Chapter 3 ("Time-Frequency Analysis, Wavelets and Filter Banks") reviews the continuous Fourier Transform in more detail, introduces the concepts of translations, dilations and modulations, and presents joint time-frequency analysis of signals by various tools. This leads to a discussion of the continuous wavelet transform (CWT) and *time-scale* analysis. Like the Fourier Transform, there is an associated discrete version of the CWT, which is related to bases of functions which are translations and dilations of a fixed function. Orthogonal wavelet bases can be constructed from multiresolution analysis, which then leads to digital filter banks. A review of wavelet filters, two-channel perfect reconstruction filter banks, and wavelet packets round out this chapter.

With these preliminaries, chapter 4 ("Introduction to Compression") begins the introduction to compression concepts in earnest. Compression divides into lossless and lossy compression. After a quick review of lossless coding techniques, including Huffman and arithmetic coding, there is a discussion of both scalar and vector quantization — the key area of innovation in this book. The well-known Lloyd-Max quantizers are outlined, together with a discussion of rate-distortion concepts. Finally, examples of compression algorithms are given in brief vignettes, covering vector quantization, transforms such as the discrete-cosine transform (DCT), the JPEG standard, pyramids, and wavelets. A quick tour of potential mathematical definitions of image quality metrics is provided, although this subject is still in its formative stage.

Chapter 5 ("Symmetric Extension Transforms") is written by Chris Brislawn, and explains the subtleties of how to treat image boundaries in the wavelet transform. Image boundaries are a significant source of compression errors due to the discontinuity. Good methods for treating them rely on extending the boundary, usually by reflecting the point near the boundary to achieve continuity (though not a continuous derivative). Preferred filters for image compression then have a symmetry at their middle, which can fall either on a tap or in between two taps. The appropriate reflection at boundaries depends on the type of symmetry of the filter and the length of the data. The end result is that after transform and downsampling, one preserves the sampling rate of the data exactly, while treating the discontinuity at the boundary properly for efficient coding. This method is extremely useful, and is applied in practically every algorithm discussed.

4.2 Part II: Still Image Coding

Part II presents a spectrum of wavelet still image coding techniques. Chapter 6 ("Wavelet Still Image Coding: A Baseline MSE and HVS Approach") by Pankaj Topiwala presents a very low-complexity image coder that is tuned to minimizing distortion according to either mean-squared error or models of human visual system. Short integer wavelets, a simple scalar quantizer, and a bare-bones arithmetic coder are used to get optimized compression speed. While use of simplified image models mean that the performance is suboptimal, this coder can serve as a baseline for comparision for more sophisticated coders which trade complexity for performance gains. Similar in spirit is chapter 7 by Alen Docef *et al* ("Image Coding Using Multiplier-Free Filter Banks"), which employs multiplication-free subband filters for efficient image coding. The complexity of this approach appears to be comparable to that of chapter 6's, with similar performance.

Chapter 8 ("Embedded Image Coding Using Zerotrees of Wavelet Coefficients") by Jerome Shapiro is a reprint of the landmark 1993 paper in which the concept of zerotrees was used to derive a rate-efficient, embedded coder. Essentially, the correlations and self-similarity across wavelet subbands are exploited to reorder (and reindex) the tranform coefficients in terms of "significance" to provide for embedded coding. Embedded coding means the a single coded bitstream can be decoded at any bitrate below the encoding rate (with optimal performance) by simple truncation, which can be highly advantageous for a variety of applications. Chapter 9 ("A New, Fast and Efficient Image Codec Based on Set Partitioning in Hierarchical Trees") by Amir Said and William Pearlman, presents one of the best-known wavelet image coders today. Inspired by Shapiro's embedded coder, Said-Pearlman developed a simple set-theoretic data structure to achieve a very efficient embedded coder, improving upon Shapiro's in both complexity and performance.

Chapter 10 ("Space-Frequency Quantization for Wavelet Image Coding") by Zixiang Xiong, Kannan Ramchandran and Michael Orchard, develops one of the most sophisticated and best-performing coders in the literature. In addition to using Shapiro's advance of exploiting the structure of zero coefficients across subbands, this coder uses iterative optimization to decide the order in which nonzero pixels should be nulled to achieve the best rate-distortion performance. A further innovation is to use wavelet packet decompositions, and not just wavelet ones, for enhanced performance. Chapter 11 ("Subband Coding of Images Using Classification and Trellis Coded Quantization") by Rajan Joshi and Thomas Fischer presents a very different approach to sophisticated coding. Instead of conducting an iterative search for optimal quantizers, these authors attempt to index regions of an image that have similar statistics (classification, say into four classes) in order to achieve tighter fits to models. A further innovation is to use "trellis coded quantization," which is a type of vector quantization in which codebooks are themselves divided into disjoint codebooks (e.g., successive VQ) to achieve high-performance at moderate complexity. Note that this is a "forward-adaptive" approach in that statistics-based pixel classification decisions are made first, and then quantization is applied; this is in distinction from recent "backward-adaptive" approaches such as [4] that also perform extremely well. Finally, Chapter 12 ("Low-Complexity Compression of Run Length Coded Image Subbands") by John Villasenor and Jiangtao Wen innovates on the entropy coding approach rather than in the quantization as in nearly all other chapters. Explicitly aiming for low-complexity, these authors consider the statistics of quantized and run-length coded image subbands for good statistical fits. Generalized Gaussian source statistics are used, and matched to a set of Goulomb-Rice codes for efficient encoding. Excellent performance is achieved at a very modest complexity.

4.3 Part III: Special Topics in Still Image Coding

Part III is a potpourri of example coding schemes with a special flavor in either approach or application domain. Chapter 13 ("Fractal Image Coding as Cross-Scale Wavelet Coefficient Prediction") by Geoffrey Davis is a highly original look at fractal image compression as a form of wavelet subtree quantization. This insight leads to effective ways to optimize the fractal coders but also reveals their limitations, giving the clearest evidence available on why wavelet coders outperform fractal coders. Chapter 14 ("Region of Interest Compression in Subband Coding") by Pankaj Topiwala develops a simple second-generation coding approach in which regions of interest within images are exploited to achieve image coding with variable quality. As wavelet coding techniques mature and compression gains saturate, the next performance gain available is to exploit high-level content-based criteria to trigger effective quantization decisions. The pyramidal structure of subband coders affords a simple mechanism to achieving this capability, which is especially relevant to surveillance applications. Chapter 15 ("Wavelet-Based Embedded Multispectral Image Compression") by Pankaj Topiwala develops an embedded coder in the context of a multiband image format. This is an extension of standard color coding, in which three spectral bands are given (red, green, and blue) and involves a fixed color transform (e.g., RGB to YUV, see the chapter) followed by coding of each band separately. For multiple spectral bands (from three to hundreds) a fixed spectral transform may not be efficient, and a Karhunen-Loeve Transform is used for spectral decorrelation.

Chapter 16 ("The FBI Fingerprint Image Compression Standard") by Chris Brislawn is a review of the FBI's Wavelet Scalar Quantization (WSQ) standard by one its key contributors. Set in 1993, WSQ was a landmark application of wavelet transforms for live imaging systems that signalled their ascendency. The standard was an early adopter of the now famous Daubechies9/7 filters, and uses a specific subband decomposition that is tailor-made for fingerprint images. Chapter 17 ("Embedded Image Coding using Wavelet Difference Reduction") by Jun Tian and Raymond Wells Jr. presents what is actually a general-purpose image compression algorithm. It is similar in spirit to the Said-Pearlman algorithm in that it uses sets of (in)significant coefficients and successive approximation for data ordering and quantization, and achieves similar high-quality coding. A key application discussed is in the compression of synthetic aperature radar (SAR) imagery, which is critical for many military surveillance applications. Finally, chapter 18 ("Block Transforms in Progressive Image Coding") by Trac Tran and Truong Nguyen presents an update on what block-based transforms can achieve, with the lessons learned from wavelet image coding. In particular, they develop advanced, overlapping block coding techniques, generalizing an approach initiated by Enrico Malvar called the Lapped Orthogonal Transform to achieve extremely competitive coding results, at some cost in transform complexity.

4.4 Part IV: Video Coding

Part IV examines wavelet and pyramidal coding techniques for video data. Chapter 19 ("Review of Video Coding Standards") by Pankaj Topiwala is a quick lineup of relevant video coding standards ranging from H.261 to MPEG-4, to provide appropriate context in which the following contributions on video compression can be compared. Chapter 20 ("Interpolative Multiresolution Coding of Advanced Television with Compatible Subchannels") by Metin Uz, Martin Vetterli and Didier LeGall is a reprint of a very early (1991) application of pyramidal methods (in both space and time) for video coding. It uses the freedom of pyramidal (rather than perfect reconstruction subband) coding to achieve excellent coding with bonus features, such as random access, error resiliency, and compatibility with variable resolution representation. Chapter 21 ("Subband Video Coding for Low to High Rate Applications") by Wilson Chung, Faouzi Kossentini and Mark Smith, adapts the motion-compensation and I-frame/P-frame structure of MPEG-2, but introduces spatio-temporal subband decompositions instead of 8x8 DCTs. Within the spatio-temporal subbands, an optimized rate-allocation mechanism is constructed, which allows for more flexible yet consistent picture quality in the video stream. Experimental results confirm both consistency as well as performance improvements against MPEG-2 on test sequences. A further benefit of this approach is that it is highly scalable in rate, and comparisions are provided against the H.263 standard as well.

Chapter 22 ("Very Low Bit Rate Video Coding Based on Matching Pursuits") by Ralph Neff and Avideh Zakhor is a status report of their contribution to MPEG-4. It is aimed at surpassing H.263 in performance at target bitrates of 10 and 24 kb/s. The main innovation is to use not an orthogonal basis but a highly redundant dictionaries of vectors (e.g., a "frame") made of time-frequency-scale translates of a single function in order to get greater compression and feature selectivity. The computational demands of such representations are extremely high, but these authors report fast search algorithms that are within potentially acceptable complexity costs compared to H.263, while providing some performance gains. A key objective of MPEG-4 is to achieve object-level access in the video stream, and chapter 23 ("Object-Based Subband/Wavelet Video Compression") by Soo-Chul Han and John Woods directly attempts a coding approach based on object-based image segmentation and object-tracking. Markov models are used for object transitions, and a version of I-P frame coding is adopted. Direct comparisons with H.263 indicate that while PSNRs are similar, the object-based approach delivers superior visual quality at extremely low bitrates (8 and 24 kb/s). Finally, Chapter 24 ("Embedded Video Subband Coding with 3D Set Partitioning in Hierarchical Trees (3D SPIHT)") by William Pearlman, Beong-Jo Kim and Zixiang Xiong develops a low-complexity motion-compensation free video coder based on 3D subband decompositions and application of Said-Pearlman's set-theoretic coding framework. The result is an embedded, scalable video coder that outperforms MPEG-2 and matches H.263, all with very low complexity. Furthermore, the performance gains over MPEG-2 are not just in PSNR, but are visual as well.

5 References

[1] A. Jain, *Fundamentals of Digital Image Processing,* Prentice-Hall, 1989.

[2] D. Marr, *Vision,* W. H. Freeman and Co., 1982.

[3] K. Rao and J. Hwang, *Techniques and Standards for Image, Video and Audio Coding*, Prentice-Hall, 1996.

[4] S. LoPresto, K. Ramchandran, and M. Orchard, "Image Coding Based on Mixture Modelling of Wavelet Coefficients and a Fast Estimation-Quantization Framework," DCC97, Proc. Data Comp. Conf., Snowbird, UT, pp. 221-230, March, 1997.

[5] W. Penebaker and J. Mitchell, *JPEG Still Image Data Compression Standard,* Van Nostrand, 1993.

Part I

Preliminaries

2

Preliminaries

Pankaj N. Topiwala

1 Mathematical Preliminaries

In this book, we will describe a special instance of digital signal processing. A signal is simply a function of time, space, or both (here we will stick to time for brevity). Time can be viewed as either continuous or discrete; equally, the amplitude of the function (signal) can be either continuous or discrete. Although we are actually interested in discrete-time, discrete-amplitude signals, ideas from the continuous-time, continuous amplitude domain have played a vital role in digital signal processing, and we cross this frontier freely in our exposition.

A digital signal, then, is a sequence of real or complex numbers, i.e., a vector in a finite-dimensional vector space. For real-valued signals, this is modelled on $\mathbf{R^n}$, and for complex signals, $\mathbf{C^n}$. Vector addition corresponds to superposition of signals, while scalar multiplication is amplitude scaling (including sign reversal). Thus, the concepts of vector spaces and linear algebra are natural in signal processing. This is even more so in digital image processing, since a digital image is a matrix. Matrix operations in fact play a key role in digital image processing. We now quickly review some basic mathematical concepts needed in our exposition; these are more fully covered in the literature in many places, for example in [8], [9] and [7], so we mainly want to set notation and give examples. However, we do make an effort to state definitions in some generality, in the hope that a bit of abstract thinking can actually help clarify the conceptual foundations of the engineering that follows throughout the book.

A fundamental notion is the idea that a given signal can be looked at in a myriad different ways, and that manipulating the *representation* of the signal is one of the most powerful tools available. Useful manipulations can be either linear (e.g., transforms) or nonlinear (e.g., quantization or symbol substitution). In fact, a simple change of basis, from the DCT to the WT, is the launching point of this whole research area. While the particular transforms and techniques described herein may in time weave in and out of favor, that one lesson should remain firm. That reality in part exhorts us to impart some of the science behind our methods, and not just to give a compendium of the algorithms in vogue today.

1.1 Finite-Dimensional Vector Spaces

Definition (Vector Space)

A **real** or **complex vector space** is a pair consisting of a set together with an operation ("vector addition"), $(\mathbf{V}, +)$, such that

1. (*Closure*) For any $v_1, v_2 \in \mathbf{V}$, $v_1 + v_2 \in \mathbf{V}$.
2. (*Commutativity*) For any $v_1, v_2 \in \mathbf{V}$, $v_1 + v_2 = v_2 + v_1$.
3. (*Additive Identity*) There is an element $0 \in \mathbf{V}$ such that for any $v \in \mathbf{V}, v+0 = v$.
4. (*Additive Inverse*) For any $v \in \mathbf{V}$ there is an element $(-v) \in \mathbf{V}$ such that $v + (-v) = 0$.
5. (*Scalar Multiplication*) For any $\alpha \in \mathbf{R}$ or $\mathbf{C}$, $v \in \mathbf{V}$, $\alpha v \in \mathbf{V}$,
6. (*Distributivity*) For any $\alpha \in \mathbf{R}$ or $\mathbf{C}$, $v_1, v_2 \in \mathbf{V}$, $\alpha(v_1 + v_2) = \alpha v_1 + \alpha v_2$, and
7. (*Multiplicative Identity*) For any $v \in \mathbf{V}$, $1v = v$.

Definition (Basis)

A **basis B** $= \{v_i\}$, $i = 1, \ldots, n$ is a subset of nonzero elements of **V** satisfying two conditions:

1. (*Independence*) Any equation $\sum_{i=1}^{n} \alpha_i v_i = 0$ implies all $\alpha_i = 0$, and
2. (*Completeness*) Any vector v can be represented in it, $v = \sum_{i=1}^{n} \alpha_i v_i$ for some numbers α_i.
3. The **dimension** of the vector space is n.

Thus, in a vector space **V** with a basis $\{v_i\}$, $i = 1, \ldots, n$, a vector can be represented simply as a column of numbers $v = (\alpha_1, \ldots, \alpha_n)^T$.

Definition (Maps and Inner Products)

1. A *linear transformation* is a map $L : \mathbf{V} \to \mathbf{W}$ between vector spaces such that $L(\alpha v_1 + \beta v_2) = \alpha L(v_1) + \beta L(v_2)$.
2. An *inner* (or dot) *product* is a mapping (that is, linear in each variable) $\langle , \rangle : \mathbf{V} \times \mathbf{V} \to \mathbf{R}$ **(or C)**, $\langle v, w \rangle \to \mathbf{R}$ **(or C)** which is
   (i) (*bilinear (or sesquilinear)*) $\langle \alpha v, \beta w \rangle = \bar{\alpha}\beta \langle v, w \rangle$
   (ii) (*symmetric*) $\langle v, w \rangle = \overline{\langle w, v \rangle}$, and
   (iii) (*nondegenerate*), $\langle v, v \rangle = 0 \Rightarrow v = 0$.
3. Given a basis **B**$= \{v_i\}$, $i = 1, \ldots, n$, a *dual basis* **B'**$= \{w_i\}$, $i = 1, \ldots, n$ is a basis such that $\langle v_i, w_j \rangle = \delta_{ij}$.
4. An *orthonormal basis* is a basis $\{v_i\}$, $i = 1, \ldots, n$ which is its own dual: $\langle v_i, v_j \rangle = \delta_{ij}$. More generally, the dual basis is also called a *biorthogonal* basis.
5. A *unitary* (or *orthogonal*) mapping $T : V \to W$ is one that preserves inner products: $\langle Tv_1, Tv_2 \rangle = \langle v_1, v_2 \rangle$.
6. The ℓ_p **norm** of a vector is $||v||_p = (\sum_{i=1}^{n} |\langle v, v_i \rangle|^p)^{\frac{1}{p}}$.

Examples

1. A simple example of a dot product in $\mathbf{R^n}$ or $\mathbf{C^n}$ is $\langle x, y\rangle = \sum_{i=1}^{n} \bar{x}_i y_i = x^\dagger y$. A vector space with an orthonormal basis is equivalent to the **Euclidean space $\mathbf{R^n}$** (or $\mathbf{C^n}$) with the above dot product.

2. Every basis in finite dimensions has a dual basis (also called the *biorthogonal* basis). For example, the basis $\{v_1, v_2\} = \{(1, 1/2), (1/2, 1)\}$ in $\mathbf{R^2}$ has the dual basis $\{w_1, w_2\} = \{(4/3, -2/3), (-2/3, 4/3)\}$; see figure 1.

3. The ℓ_2 norm corresponds to the usual Euclidean norm $||v||_2 = \sqrt{\langle v, v\rangle}$, and is of primary interest. Other interesting cases are $p = 1$ ($||v||_1 = \sum_{i=1}^{n} |\langle v, v_i\rangle|$), and $p = \infty$ ($||v||_\infty = \max |\langle v, v_i\rangle|$).

4. Every basis is a frame, but frames need not be bases. In $\mathbf{R^2}$, let $e_1 = (1,0), e_2 = (0,1), e_3 = (1,1)$. Then $\{e_1, e_2, e_3\}$ is a frame, $\{e_1, e_3\}$ is a basis, and $\{e_1, e_2\}$ is an orthonormal basis. Frame vectors don't need to be linearly independent, but do need to be provide a representation for any vector.

5. A square matrix A is unitary if all of its rows or columns are orthonormal. This can be stated as $A^\dagger A = AA^\dagger = I$, the identity matrix.

6. Given an orthonormal basis $\{v_i\}$, $i = 1, \ldots, n$, and $v \in \mathbf{V}$, $v = \sum_{i=1}^{n} \alpha_i v_i$ with $\alpha_i = \langle v, v_i\rangle$.

7. A linear transformation L is then represented by a **matrix** $A = (a_{ij})$, given by $Lv_i = \sum_j a_{ij} v_j$; for $v = \sum_{i=1}^{n} \alpha_i v_i$ we have $Lv = A[\alpha_1, \ldots, \alpha_n]^T$.

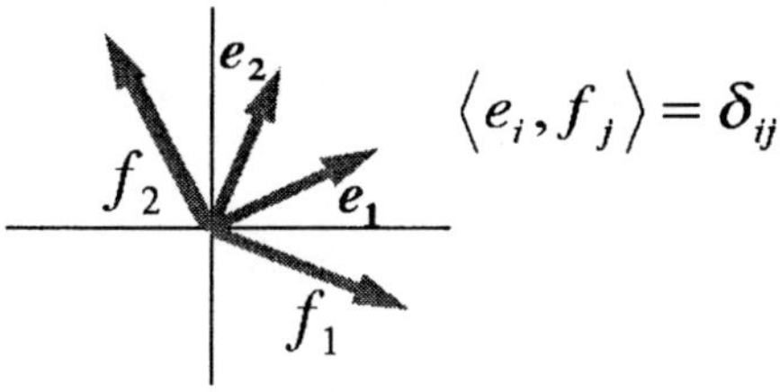

FIGURE 1. Every basis has a dual basis, also called the biorthogonal basis.

We are especially interested in unitary transformations, which, again, are maps (or matrices) that preserve dot products of vectors:

$$\langle Av, Aw\rangle = v^\dagger A^\dagger Aw = v^\dagger w = \langle v, w\rangle.$$

As such, they clearly take one orthonormal basis into another, which is an equivalent way to define them. These transformations completely preserve all the structure of a Euclidean space. From the point of view of signal analysis (where vectors are signals), dot products correspond to cross-correlations between signals. Thus, unitary transformations preserve both the first order structure (linearity) and the second order structure (correlations) of signal space. If signal space is well-modelled

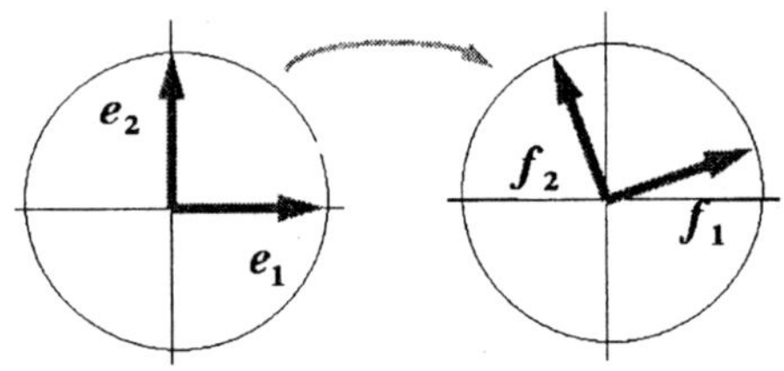

FIGURE 2. A unitary mapping is essentially a rotation in signal space; it preserves all the geometry of the space, including size of vectors and their dot products.

by a Euclidean space, then performing a unitary transformation is harmless as it changes nothing important from a signal analysis point of view.

Interestingly, while mathematically one orthonormal basis is as good as another, it turns out that in signal processing, there can be significant practical differences between orthonormal bases in representing particular classes of signals. For example, there can be differences between the distribution of coefficient amplitudes in representing a given signal in one basis versus another. As a simple illustration, in a signal space $\mathbf{V}$ with orthonormal basis $\{v_i\}$, $i = 1, \ldots, n$, the signal

$$v = \frac{1}{\sqrt{n}} \sum_i v_i$$

has nontrivial coefficients $(\frac{1}{\sqrt{n}})$ in each component. However, it is easy to see that $||v|| = 1$. In a new orthonormal basis in which v itself is one of the basis vectors, which can be done for example by a Gram-Schmidt procedure, its representation requires only one vector and one coefficient. The moral is that vectors (signals) can be well-represented in some bases and not others. That is, their expansion can be more parsimonious or compact in one basis than another. This notion of a compact representation, which is a relative concept, plays a central role in using transforms for compression. To make this notion precise, for a given $\epsilon > 0$, we will say that the representation of a vector v in one orthonormal basis, $v = \sum_i \alpha_i v_i$, is more $\epsilon-$*compact* than in another basis, $v = \sum_i \beta_i w_i$ if

$$\#\{i : |\alpha_i| > \epsilon\} \ \leq \ \#\{i : |\beta_i| > \epsilon\}.$$

The value of compact representations is that, if we discard the coefficients less than ϵ in magnitude, then we can approximate v by a small number of coefficients. If the remaining coefficients can themselves be represented by a small number of bits, then we have achieved "compression" of the signal v. In reality, instead of *thresholding* coefficients as above (i.e., deleting ones below the threshold ϵ), we *quantize* them—that is, impose a certain granularity to their precision, thereby saving bits. Quantization, in fact, will be the key focus of the compression research herein.

The surprising fact is that, for certain classes of signals, one can find fixed, well-matched bases of signals in which the typical signal in the class is compactly represented. For images in particular, wavelet bases appear to be exceptionally well-suited. The reasons for that, however, are not fully understood at this time, as no model-based optimal representation theory is available; some intuitive explanations will be provided below. In any case, we are thus interested in a class of unitary transformations which take a given signal representation into one of the wavelet bases; these will be defined in the next chapter.

1.2 Analysis

Up to now, we have been discussing finite-dimensional vector spaces. However, many of the best ideas in signal processing often originate in other subjects, which correspond to analysis in infinite dimensions. The sequence spaces

$$\ell_{\mathbf{p}} = \{(x_i)_{i=1}^{i=\infty} : \|x\|_p = (\sum_i |x_i|^p)^{\frac{1}{p}} < \infty\},$$

are infinite-dimensional vector spaces, for $p \geq 1$, which now depend on particular notions of convergence. For p=2 (only), there is a dot product available:

$$\langle (x_i), (y_i) \rangle = \sum_i \bar{x}_i y_i.$$

This is finite by the *Cauchy-Schwartz Inequality*, which says that

$$|\langle (x_i), (y_i) \rangle| \leq \|x\|_2 \|y\|_2 < \infty.$$

Morever, equality holds above if and only if $x = \alpha y$, some α.

If we plot sequences in the plane as (i, x_i), and connect the dots, we obtain a crude (piecewise linear) function. This picture hints at a relationship between sequences and functions; see figure 3. And in fact, for $p \geq 1$, there are also function space analogues of our finite-dimensional vector spaces,

$$L_p(\mathbf{R}) = \{f : \|f\|_p = (\int |f(t)|^p dt)^{1/p} < \infty\}.$$

The case $p = 2$ is again special in that the space $L_2(\mathbf{R})$ has a dot product, due again to Cauchy-Schwartz:

$$\langle f, g \rangle = \int \overline{f(t)} g(t) dt.$$

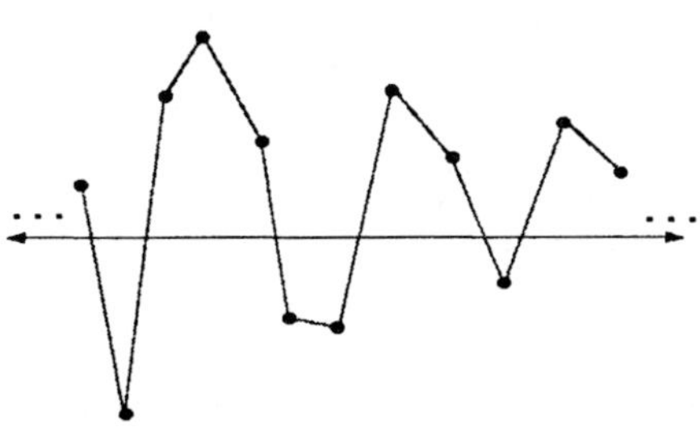

FIGURE 3. Correspondence between sequence spaces and function spaces.

We will work mainly in ℓ_2 or $L_2(\mathbf{R})$. These two notions of infinite-dimensional spaces with dot products (called *Hilbert spaces*) are actually equivalent, as we will see. The concepts of linear transformations, matrices and orthonormal bases also exist in these contexts, but now require some subtleties related to convergence.

Definition (Frames, Bases, Unitary Maps in $L_2(R)$)

1. A *frame* in $L_2(\mathbf{R})$ is a sequence of functions $\{f_i\}_{i=1}^{i=\infty}$ such that for any $f \in L_2(\mathbf{R})$, there are numbers $0 < A \le B < \infty$ such that
   $A\|f\|^2 \le \sum_i |\langle f_i, f\rangle|^2 \le B\|f\|^2$.
2. A *basis* in $L_2(\mathbf{R})$ is a frame in which all functions are linearly independent: any equation $\sum_i \alpha_i f_i = 0$ implies all $\alpha_i = 0$.
3. A *biorthogonal basis* in $L_2(\mathbf{R})$ is a basis $\{f_i\}_{i=1}^{i=\infty}$ for which there is a dual basis $\{g_i\}_{i=1}^{i=\infty} : \langle g_i, f_j\rangle = \delta_{ij}$. For any $f \in L_2(\mathbf{R})$, there are numbers α_i such that $\|f - \sum_{i=1}^{i=N} \alpha_i f_i\| \to 0$ as $N \to \infty$. An *orthonormal basis* is a special biorthogonal basis which is its own dual.
4. A *unitary* map $T : L_2(\mathbf{R}) \to L_2(\mathbf{R})$ is a linear map which preserves inner products: $\langle Tf, Tg\rangle = \langle f, g\rangle$.
5. Given an orthonormal basis $\{f_i\}_{i=1}^{i=\infty}$, there is an equivalence $L_2(\mathbf{R}) \to \ell_2$, given by $f \to \{\langle f_i, f\rangle\}^T$, an infinite column vector.
6. Maps can be represented by matrices as usual: $(a_{ij}) = \langle f_i, Tf_j\rangle$.
7. (*Integral Representation*) Any linear map on $L_2(\mathbf{R})$ can be represented by an integral, $Tf(s) = \int k(s,t)f(t)dt$.

Define the *convolution* of two functions as

$$(h * f)(x) = \int h(x-y)f(y)dy.$$

Now suppose that an operator T is *shift-invariant*, that is, for all f and all s_0,

$$\int k(s-s_0, t-s_0)f(t)dt = \int k(s,t)f(t)dt.$$

It can be shown that all such operators are convolutions, that is, its integral "kernel" $k(s,t) = h(s-t)$ for some function h. When t is time and linear operators represent systems, this has the nice interpretation that systems that are invariant in time are convolution operators — a very special type. These will play a key role in our work.

For the most part, we deal with finite digital signals, to which these issues are tangential. However, the real motivation for us in gaining familiarity with these topics is that, even in finite dimensions, many of the most important methods in digital signal processing really have their roots in Analysis, or even Mathematical Physics. Important unitary bases have not been discovered in isolation, but have historically been formulated en route to solving specific problems in Physics or Mathematics, usually involving infinite-dimensional analysis. A famous example is that of the Fourier transform. Fourier invented it in 1807 [3] explicitly to help understand the physics of heat dispersion. It was only much later that the common Fast Fourier Transform (FFT) was developed in digital signal processing [4]. Actually, even the FFT can be traced to much earlier times, to the 19th century mathematician Gauss. For a fuller account of the Fourier Transform, including the FFT, also see [6].

To define the Fourier Transform, which is inherently complex-valued, recall that the complex numbers **C** form a plane over the real numbers **R**, such that any complex number can be written as $c = a + jb$, where $j \equiv \sqrt{-1}$ is an imaginary number such that $j^2 = -1$.

1.3 Fourier Analysis

Fourier Transform (Integral)

1. The Fourier Transform is the mapping $\mathcal{F} : L_2(\mathbf{R}) \to L_2(\mathbf{R})$ given by $\mathcal{F}f(\omega) = F(\omega) = \int_{-\infty}^{\infty} f(t)e^{-2\pi j\omega t}dt = \langle e^{2\pi j\omega t}, f(t)\rangle$.
2. The Inverse Fourier Transform is the mapping $\mathcal{F}^{-1} : L_2(\mathbf{R}) \to L_2(\mathbf{R})$ given by $\mathcal{F}^{-1}F(t) = f(t) = \int_{-\infty}^{\infty} F(\omega)e^{2\pi j\omega t}d\omega = \langle e^{-2\pi j\omega t}, F(\omega)\rangle$.
3. (*Identity*) $f(t) = \langle \int e^{2\pi j\omega(s-t)}d\omega, f(s)\rangle$
4. (*Plancherel or Unitarity*) $\langle F, G\rangle = \langle f, g\rangle$.
5. (*Convolution*) $\mathcal{F}(h * f) = HF$.

These properties are not meant to be independent, and in fact item 3 implies item 4 (as is easy to see). Due to the unitary property, item 4, the Fourier Transform (FT) is a complete equivalence between two representations, one in t ("time") and the other in ω ("frequency"). The frequency-domain representation of a function (or signal) is referred to as its *spectrum*. Some other related terms: the magnitude-squared of the spectrum is called the *power spectral density*, while the integral of that is called the *energy*. Morally, the FT is a decomposition of a given function $f(t)$ into an orthonormal "basis" of functions given by complex exponentials of arbitrary frequency, $e^{2\pi j\omega t}$. Here the frequency parameter ω serves as a kind of index to the basis functions. The concept of orthonormality (and thus of unitarity) is then represented by item 3 above, where we effectively have

$$\int e^{2\pi j\omega(s-t)}d\omega = \delta(s-t),$$

the delta function. Recall that the so-called *delta function* is defined by the property that $\delta * f = f$ for all functions in $L_2(\mathbf{R})$. The delta function can be envisioned as the limit of Gaussian functions as they become more and more concentrated near the origin:

$$\delta(x) = lim_{a\to\infty}\frac{a}{\sqrt{2\pi}}e^{-a^2x^2/2}.$$

This limit tends to infinity at the origin, and zero everywhere else. That

$$\int e^{2\pi j\omega(s-t)}dt = \delta(s-t)$$

should hold can be vaguely understood by the fact that when $s = t$ above, we have an infinite integral of the constant function 1; while when $s \neq t$, the integrand oscillates and washes itself out. While this reasoning is not precise, it does suggest that item 3 is an important and subtle theorem in Fourier Analysis.

By Euler's identity, the real and imaginary parts of these complex exponential functions are the familiar sinusoids:

$$e^{2\pi j\omega t} = cos(2\pi\omega t) + jsin(2\pi\omega t).$$

The fundamental importance of these basic functions comes from their role in Physics: waves, and thus sinusoids, are ubiquitous. The physical systems that produce signals quite often oscillate, making sinusoidal basis functions an important representation.

As a result of this basic equivalence, given a problem in one domain, one can always transform it in the other domain to see if it is more amenable (often the case). So established is this equivalence in the mindset of the signal processing community, that one may even hear of the picture in the frequency domain being referred to as the "real" picture. The last property mentioned above of the FT is in fact of fundamental importance in digital signal processing. For example, the class of linear operators which are convolutions (practically all linear systems considered in signal processing), the FT says that they correspond to (complex) rescalings of the amplitude of each frequency component separately. Smoothness properties of the convolving function h then relate to the decay properties of the amplitude rescalings. (It is one of the deep facts about the FT that smoothness of a function in one of the two domains (time or frequency) corresponds to decay rates at infinity in the other [6].) Since the complex phase is highly oscillatory, one often considers the magnitude (or power, the magnitude squared) of the signal in the FT. In this setting, linear convolution operators correspond directly to positive amplitude rescalings in the frequency domain. Thus, one can develop a taxonomy of convolution operators by which portions of the frequency axis they amplify, suppress or even null. This is the origin of the concepts of lowpass, highpass, and bandpass filters, to be discussed later.

The Fourier Transform also exists in a discrete version for vectors in $\mathbf{C^N}$. For convenience, we now switch notation slightly and represent a vector $x \in \mathbf{C^N}$ as a sequence $\{x(0), \ldots, x(N-1)\}$. Let $W_N = e^{-2\pi j/N}$.

Discrete Fourier Transform (DFT)

1. The DFT is the mapping $\mathcal{F} : \mathbf{C^N} \to \mathbf{C^N}$ given by
   $X(k) = \frac{1}{\sqrt{N}} \sum_{n=0}^{N-1} x(n) W_N^{nk}$.

2. The Inverse DFT is the mapping $\mathcal{F}^{-1} : \mathbf{C^N} \to \mathbf{C^N}$ given by
   $x(n) = \frac{1}{\sqrt{N}} \sum_{k=0}^{N-1} X(k) W_N^{-nk}$.

3. (*Identity*) $x(n) = \langle \frac{1}{N} \sum_{k=0}^{N-1} W_N^{k(m-n)}, x(m) \rangle$

4. (*Plancherel or Unitarity*) $\langle X, Y \rangle = \langle x, y \rangle$.

5. (*Convolution*) $\mathcal{F}(h * x) = HX$.

While the FT of a real-variable function can take on arbitrary frequencies, the DFT of a discrete sequence can only have finitely many frequencies in its representations. In fact, the DFT has as many frequencies as the data has samples (since it takes $\mathbf{C^N}$ to $\mathbf{C^N}$), so the frequency "bandwidth" is equal to the number of samples. If the data is furthermore real, then as it turns out, the output actually has equal proportions of positive and negative frequencies. Thus, the highest frequency sinusoid that can actually be represented by a finite sampled signal has frequency equal to *half the number of samples (per unit time)*, which is known as the **Nyquist sampling theorem**.

An important computational fact is that the DFT has a fast version, the Fast Fourier Transform (FFT), for certain values of N, e.g., N a power of 2. On the face of it, the DFT requires N multiplies and $N - 1$ adds for each output, and so requires on the order of N^2 complex operations. However, the FFT can compute this in on the order of $NLogN$ operations [4]. Similar considerations apply to the Discrete Cosine Transform (DCT), which is essentially the real part of the FFT (suitable for treating real signals). Of course, this whole one-dimensional analysis carries over to higher dimensions easily. Essentially, one performs the FT in each variable separately. Among DCT computations, the special case of the 8x8 DCT for 2D image processing has received perhaps the most intense scrutiny, and very clever techniques have been developed for its efficient computation [11]. In effect, instead of requiring 64 multiplies and 63 adds for each 8x8 block, it can be done in less than one multiply and 9 adds. This stunning computational feat helped to catapult the 8x8 DCT to international standard status. While the advantages of the WT over the DCT in representing image data have been recognized for some time, the computational efficiency is only now beginning to catch up with the DCT.

2 Digital Signal Processing

Digital signal processing is the science/art of manipulating signals for a variety of practical purposes, from extracting information, enhancing or encrypting their meaning, protecting against errors, shrinking their file size, or re-expanding as appropriate. For the most part, the techniques that are well-developed and well-understood are *linear* - that is they manipulate data as if they were vectors in a vector space. It is the fundamental credo of linear systems theory that physical systems (e.g., electronic devices, transmission media, etc.) behave like linear operators on signals, which can therefore be modelled as matrix multiplications. In fact, since it is usually assumed that the underlying physical systems behave today just as they behaved yesterday (e.g., their essential characteristics are *time-invariant*), even the matrices that represent them cannot be arbitrary but must be of a very special sort. Digital signal processing is well covered in a variety of books, for example [4], [5], [1]. Our brief tour is mainly to establish notation and refamiliarize readers of the basic concepts.

2.1 Digital Filters

A digital signal is a finite numerical sequence $x = \{x(0), x(1), \ldots, x(N-1)\}$, such as the samples of a continuous waveform like speech, music, radar, sonar, etc. The *correlation* or dot product of two signals is defined as before: $\langle x, y\rangle = \sum_i \overline{x(i)}y(i)$. A digital filter, likewise, is a sequence $h = \{h(0), h(1), \ldots, h(L-1)\}$, typically of much shorter length. From here on, we generally consider real signals and real filters in this book, and will sometimes drop the overbar notation for conjugation. However, note that even for real signals, the FT converts them into complex signals, and complex representations are in fact common in signal processing (e.g., "in-phase" and "quadrature" components of signals).

A filter acts on a signal by convolution of the two sequences, producing an output signal $\{y(k)\}$, given by

$$y(k) = (h * x)(k) = \sum_i h(k-i)x(i).$$

A schematic of a digital filter acting on a digital signal is given in figure 4. In essence, the filter is clocked past the signal, producing an output value for each position equal to the dot product of the overlapping portions of the filter and signal. Note that, by the equation above, the samples of the filter are actually reversed before being clocked past the signal.

In the case when the signal is a simple *impulse*, $x = (1, 0, \ldots, 0)$, it is easy to see that the output is just the filter itself. Thus, this digital representation of the filter is known as the *impulse response* of the filter, the individual coefficients of which are called *filter taps*. Filters which have a finite number of taps are called Finite Impulse Response (FIR) filters. We will work exclusively with FIR filters, in fact usually with very short filters (often with less than 10 taps).

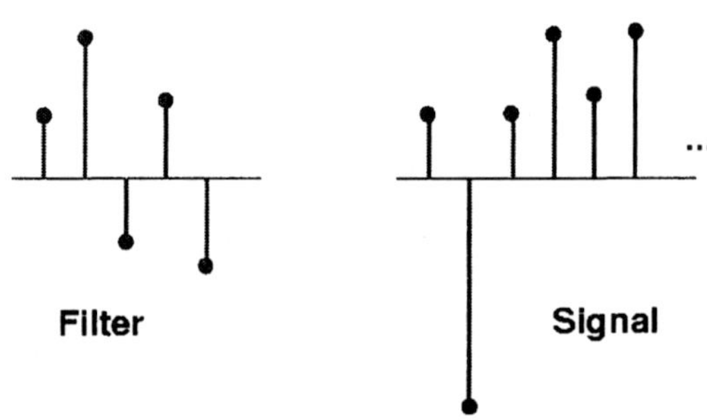

FIGURE 4. Schematic of a digital filter acting on a signal. The filter is clocked past the signal, generating an output value at each step which is equal to the dot product of the overlapping parts of the filter and signal.

Incidentally, if the filter itself is symmetric, which often occurs in our applications, this reversal makes no difference. Symmetric filters, especially ones which peak in the center, have a number of nice properties. For example, they preserve the location of sharp transition points in signals, and facilitate the treatment of signal boundaries; these points are not elementary, and are explained fully in chapter 5. For symmetric filters we frequently number the indices to make the center of the filter at the zero index. In fact, there are two types of symmetric filters: whole sample symmetric, in which the symmetry is about a single sample (e.g., $h(k) = h(-k)$), and half-sample symmetric, in which the symmetry is about two middle indices (e.g., $h(k) = h(-k-1)$). For example, $(1,2,1)$ is whole-sample symmetric, and $(1,2,2,1)$ is half-sample symmetric. In addition to these symmetry considerations, there are also issues of anti-symmetry (e.g., $(1,0,-1), (1,2,-2,-1)$). For a full-treatment of symmetry considerations on filters, see chapter 5, as well as [4], [10].

Now, in signal processing applicatons, digital filters are used to accentuate some aspects of signals while suppressing others. Often, this modality is related to the frequency-domain picture of the signal, in that the filter may accentuate some portions of the signal spectrum, but suppress others. This is somewhat analogous to what an optical color filter might do for a camera, in terms of filtering some portions of the visible spectrum. To deal with the spectral properties of filters, and for others conveniences as well, we introduce some useful notation.

2.2 *Z-Transform and Bandpass Filtering*

Given a digital signal $x = \{x(0), x(1), \ldots, x(N-1)\}$, its *z-Transform* is defined as the following power series in a complex variable z,

$$X(z) = \sum_i x(i) z^{-i}.$$

This definition has a direct resemblance to the DFT above, and essentially coincides with it if we restrict the z–Transform to certain samples on the unit circle: $z_n = e^{-2\pi jn/N}$. Many important properties of the DFT carry over to the z–Transform. In particular, the convolution of two sequences $h = \{h(0), h(1), \ldots, h(L-1)\}$ and $x = \{x(0), x(1), \ldots, x(N-1)\}$ is the product of the z–Transforms:

$$Y(z) = H(z)X(z) = \sum_k (\sum_i h(k-i)x(i)) z^{-k} = \sum_k y(k) z^{-k}.$$

Since linear time-invariant systems are widely used in engineering to model electrical circuits, transmission channels, and the like, and these are nothing but convolution operators, the z-Transform representation is obviously compact and efficient.

A key motivation for applying filters, in both optics and digital signal processing, is to pick out signal components of interest. A photographer might wish to enhance the "red" portion of an image of a sunset in order to accentuate it, and suppress the "green" or "blue" portions. What she needs is a "bandpass" filter - one that extracts just the given band from the data (e.g., a red lens). Similar considerations apply in digital signal processing. An example of a common digital filter with a notable frequency response characteristic is a *lowpass* filter, i.e., one that preserves the signal content at low frequencies, and suppresses high frequencies. Here, for a digital signal, "high frequency" is to be understood in terms of the Nyquist frequency, $\omega_{Nyq.} = \frac{\#samples/sec}{2}$. For plotting purposes, this can be generically rescaled to a fixed value, say 1/2. A complementary filter that suppresses low frequencies and maintains high frequencies is called a *highpass* filter; see figure 5.

Lowpass filters can be designed as simple averaging filters, while highpass filters can be differencing filters. The simplest of all examples of such filters are the pairwise sums and differences, respectively, called the **Haar filters**: $h_0 = \frac{1}{\sqrt{2}}(1,1)$ and $h_1 = \frac{1}{\sqrt{2}}(1,-1)$. An example of a digital signal, together with the lowpass and highpass components, is given in figure 6. Ideally, the lowpass filter would precisely pass all frequencies up to half of Nyquist, and suppress all other frequencies e.g., a "brickwall" filter with a step-function frequency response. However, it can be shown that such a brickwall filter needs an infinite number of taps, which moreover decay very slowly (like $1/n$) [4]. This is related to the fact that the Fourier transform of the rectangle function (equal to 1 on an interval, zero elsewhere) is the sinc function $(sin(x)/x)$, which decays like $1/x$ as $|x| \to \infty$. In practice, such filters cannot be realized, and only a finite number of terms are used.

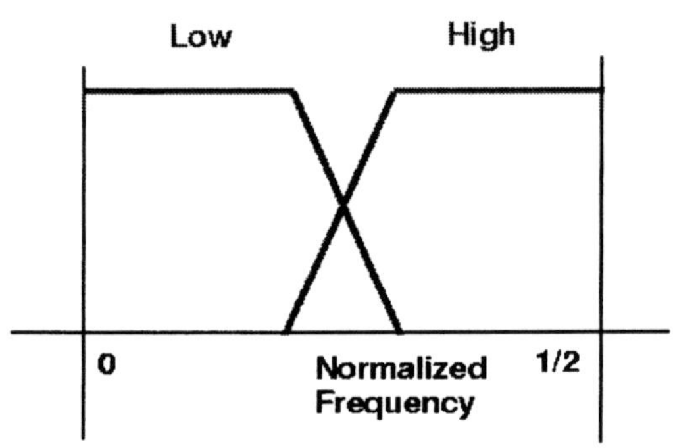

FIGURE 5. A complementary pair of lowpass and highpass filters.

If perfect band splitting into lowpass and highpass bands could be achieved, then according to the Nyquist sampling theorem since the bandwidth is cut in half the number of samples in each band needed could theoretically be cut in half without any loss of information. The totality of reduced samples in the two bands would then equal the number in the original digital signal, and in fact this transformation amounts to an equivalent representation. Since perfect band splitting cannot actually be achieved with FIR filters, this statement is really about sampling *rates*, as the number of samples tends to infinity. However, unitary changes of bases certainly exist even for finite signals, as we will see shortly.

To achieve a unitary change of bases with FIR filters, there is one phenomenon that must first be overcome. When FIR filters are used for (imperfect) bandpass filtering, if the given signal contains frequency components outside the desired "band", there will be a spillover of frequencies into the band of interest, as follows. When the bandpass-filtered signal is downsampled according to the Nyquist rate of the band, since the output must lie within the band, the out-of-band frequencies become reflected back into the band in a phenomenon called "aliasing." A simple illustration of aliasing is given in figure 7, where a high-frequency sinusoid

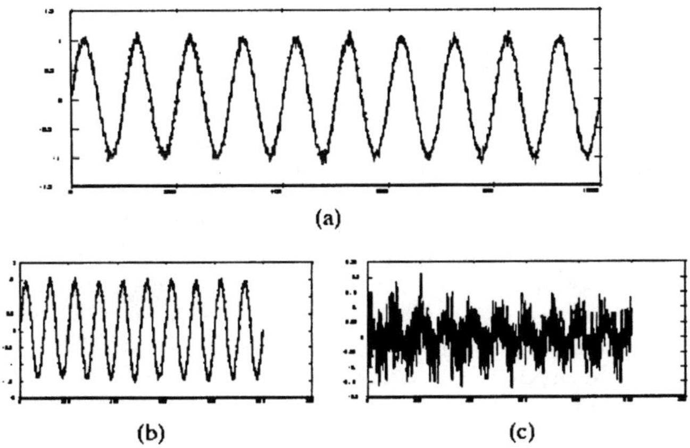

FIGURE 6. A simple example of lowpass and highpass filtering using the Haar filters, $h_0 = \frac{1}{\sqrt{2}}[1,1]$ and $h_1 = \frac{1}{\sqrt{2}}[1,-1]$. (a) A sinusoid with a low-magnitude white noise signal superimposed; (b) the lowpass filtered signal, which appears smoother than the original; (c) the highpass filtered signal, which records the high frequency noise content of the original signal.

is sampled at a rate below its Nyquist rate, resulting in a lower frequency signal (10 cycles implies Nyquist = 20 samples, whereas only 14 samples are taken).

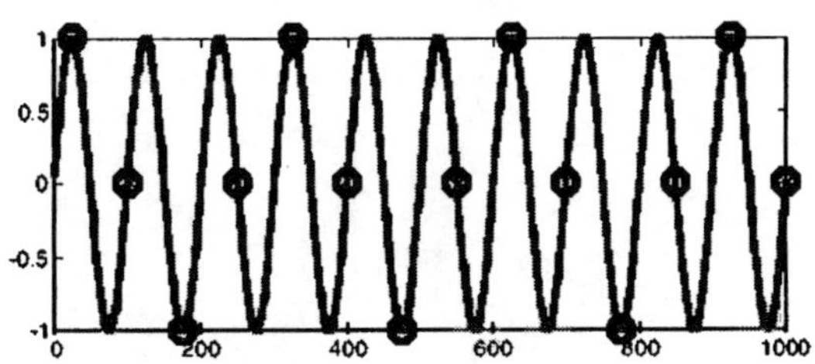

FIGURE 7. Example of aliasing. When a signal is sampled below its Nyquist rate, it appears as a lower frequency signal.

The presence of aliasing effects means it must be cancelled in the reconstruction process, or else exact reconstruction would not be possible. We examine this in greater detail in chapter 2, but for now we assume that finite FIR filters giving unitary transforms exist. In fact, decomposition using the Haar filters is certainly unitary (orthogonal, since the filters are actually real). As a matrix operation, this transform corresponds to multyplying by the block diagonal matrix in the table below.

$\ddots$	$\vdots\, 0$	
$\ldots\, 0$	$\frac{1}{\sqrt{2}}\ \frac{1}{\sqrt{2}}$	$0\, \ldots$
$\ldots\, 0$	$\frac{1}{\sqrt{2}}\ \frac{-1}{\sqrt{2}}$	$0\, \ldots$
	$0\, \vdots$	$\ddots$

TABLE 2.1. The Haar transformation matrix is block-diagonal with 2x2 blocks.

These orthogonal filters, and their close relatives the biorthogonal filters, will be used extensively throughout the book. Such filters permit efficient representation of image signals in terms of compactifying the energy. This compaction of energy then leads to natural mechanisms for compression. After a quick review of probability below, we discuss the theory of these filters in chapter 3, and approaches to compression in chapter 4. The remainder of the book is devoted to individual algorithm presentations by many of the leading wavelet compression researchers of the day.

3 Primer on Probability

Even a child knows the simplest instance of a chance event, as exemplified by tossing a coin. There are two possible outcomes (heads and tails), and repeated tossing indicates that they occur about equally often (in a *fair* coin). Furthermore, this trend is the more pronounced as the number of tosses increases without bound. This leads to the abstraction that "heads" and "tails" are two events (symbolized by "H" and "T") with equal *probability* of occurrence, in fact equal to 1/2. This is simply because they occur equally often, and the sum of all possible event (symbol) probabilities must equal unity. We can chart this simple situation in a *histogram*, as in figure 8, where for convenience we relabel heads as "1" and tails as "0".

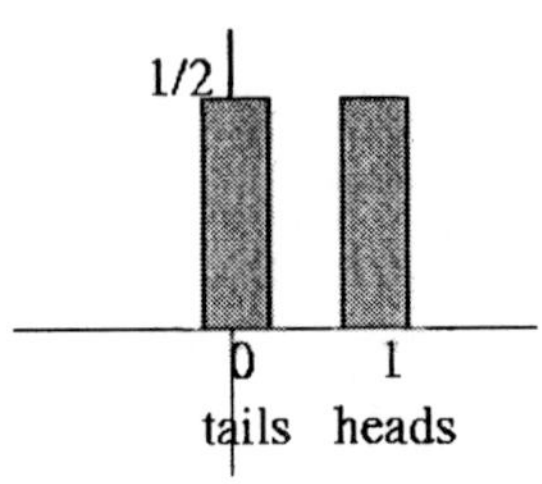

FIGURE 8. Histogram of the outcome probabilities for tossing a fair coin.

Figure 8 suggests the idea that probability is really a unit of mass, and that the points "0" and "1" split the mass equally between them (1/2 each). The total mass of 1.0 can of course be distributed into more than two pieces, say N, with nonnegative masses $p_i \geq 0, i = 0, \ldots, N-1$, such that

$$\sum_{i=0}^{N-1} p_i = 1.0.$$

The interpretation is that N different events can happen in an *experiment*, with probabilities given by p_i. The outcome of the experiment is called a *random variable.* We could toss a die, for example, and have one of 6 outcomes, while choosing a card involves 52 outcomes. In fact, any set of nonnegative numbers that sum to 1.0 constitutes what is called a discrete *probability distribution.* There is an associated histogram, generated for example by using the integers $i = 0, \ldots, N-1$ as the symbols and charting the value p_i above them.

The unit probability mass can also be spread continuously, instead of at discrete points. This leads to the concept that probability can be any function $p(x) \geq 0$ such that $\int p(x)dx = 1$ (this even includes the previous discrete case, if we allow the function to behave like a delta-function at points). The interpretation here is that our random variable can take on a continuous set of values. Now, instead of asking what is the proabability that our random variable x takes on a specific value such as 5, we ask what is the probability that x lies in some interval, $[x-a, x+b)$. For example, if we visit a kindergarten class and ask what is the likelihood that the children are of age 5, since age is actually a continuous variable (as teachers know), what we are really asking is how likely is it that their ages are in the interval $[5,6)$.

There are many probability distributions that are given by explicit formulas, involving polynomials, trigonometric functions, exponentials, logarithms, etc. For example, the Gaussian (or normal) distribution is $p(x) = \frac{1}{\sqrt{2\pi}\sigma} e^{-(x-\mu)^2/2\sigma^2}$, while the Laplace distribution is $p(x) = \frac{1}{\sqrt{2}\sigma} e^{-\sqrt{2}|x-\mu|/\sigma}$; see figure 9. The family of Generalized Gaussian distributions includes both of these as special cases, as will be discussed later.

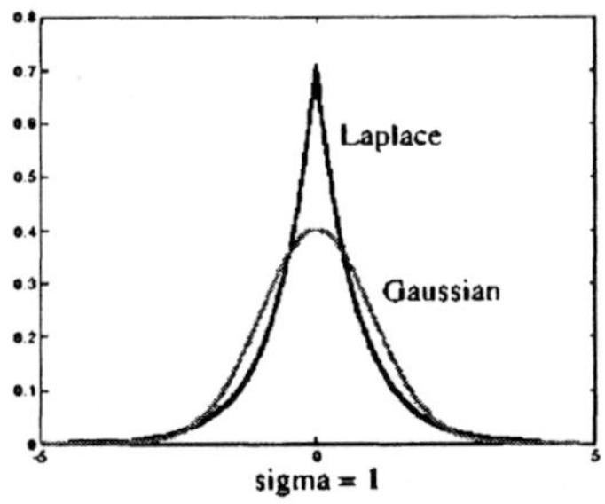

FIGURE 9. Some common probability distributions: Gaussian and Laplace.

While a probability distribution is really a continuous function, we can try to characterize it succinctly by a few numbers. The simple Gaussian and Laplace distributions can be completely characterized by two numbers, the mean or center of mass, μ, given by

$$\mu = \langle x \rangle = \int x p(x) dx$$

and the standard deviation (a measure of the width of the distribution), σ, given by

$$\sigma^2 = \langle (x-\mu)^2 \rangle = \int (x-\mu)^2 p(x) dx.$$

In general, these parameters give a first indication of where the mass is centered, and how much it is concentrated. However, for complicated distributions with many peaks, no finite number of *moments* $\langle x^n \rangle = \int x^n p(x)dx$ may give a complete description. Under reasonable conditions, though, the set of all moments does characterize any distribution.

This can in fact be seen using the Fourier Transform introduced earlier. Let

$$\phi(\omega) = \mathcal{F}(p(x)) = \int e^{-2\pi jx\omega} p(x)dx.$$

Using a well-known power series expansion for the exponential function $e^x = \sum_{n=0}^{\infty} x^n/n!$, we get that

$$\phi(\omega) = 1 + (-2\pi j\omega)\langle x \rangle + \frac{(-2\pi j\omega)^2}{2!}\langle x^2 \rangle - \ldots - \frac{(-2\pi j\omega)^n}{2!}\langle x^n \rangle + \ldots.$$

Thus, knowing all the moments of a distribution allows one to construct the Fourier Transform of the distribution (sometimes called the *characteristic function*), which can then be inverted to recover the distribution itself.

As a rule, the probability distributions for which we have formulas generally have only one peak (or *mode*), and taper off on either side. However, real-life data rarely has this type of behaviour; witness the various histograms shown in chapter 1 (figure 4) on parts of the Lena image. Not only do the histograms generally have multiple peaks, but they are widely varied, indicating that the data is very inconsistent in different parts of the picture. There is a considerable literature (e.g., the theory of stationary stochastic processes) on how to characterize data that show some consistency over time (or space); for example, we might require at least that the mean and standard deviation (as locally measured) are constant. In the Lena image, that is clearly not the case. Such data requires significant massaging to get it into a manageable form. The wavelet transform seems to do an excellent job of that, since most of the subbands in the wavelet transformed image appear to be well-structured. Furthermore, the fact that the image energy becomes highly concentrated in the wavelet domain (e.g., into the residual lowpass subband, see chap. 1, figure 5) – a phenomenon called *energy compaction*, while the rest of the data is very sparse and well-modelled, allows for its exploitation for compression.

4 References

[1] A. Jain, *Fundamentals of Digital Image Processing,* Prentice-Hall, 1989.

[2] D. Marr, *Vision,* W. H. Freeman and Co., 1982.

[3] J. Fourier, *Theorie Analytique de la Chaleur*, Gauthiers-Villars, Paris, 1888.

[4] A. Oppenheim and R. Schafer, *Discrete-Time Signal Processing*, Prentice-Hall, 1989.

[5] N. Jayant and P. Noll, *Digital Coding of Waveforms*, Prentice-Hall, 1984.

[6] A. Papoulis, *Signal Analysis*, McGraw-Hill, 1977.

[7] G. Strang and T. Nguyen, *Wavelets and Filter Banks*, Wellesley-Cambridge Press, 1996.

[8] G. Strang, *Linear Algebra and Its Applications*, Harcourt Brace Javonich, 1988.

[9] M. Vetterli and J. Kovacevic, *Wavelets and Subband Coding*, Prentice-Hall, 1995.

[10] C. Brislawn, "Classificaiton of nonexpansive symmetric extension transforms for multirate filter banks," Appl. Comp. Harm. Anal., 3:337-357, 1996.

[11] W. Penebaker and J. Mitchell, *JPEG Still Image Data Compression Standard*, Van Nostrand, 1993.

3

Time-Frequency Analysis, Wavelets And Filter Banks

Pankaj N. Topiwala

1 Fourier Transform and the Uncertainty Principle

In chapter 1 we introduced the continuous Fourier Transform, which is a unitary transform on the space of square-integrable functions, $\mathcal{F} : L_2(\mathbf{R}) \to L_2(\mathbf{R})$. Some common notations in the literature for the Fourier Transform of a function $f(t)$ are $\mathcal{F}f(\omega)$, $F(\omega)$ and $\hat{f}(\omega)$. In this chapter, we would like to review the Fourier Transform in some more detail, and try to get an intuitive feel for it. This is important, since the Fourier transform continues to play a fundamental role, even in wavelet theory. We would then like to build on the mathematical preliminaries of chapter 1 to outline the basics of time-frequency analysis of signals, and introduce wavelets. First of all, recall that the Fourier transform is an example of a unitary transform, by which we mean that dot products of functions are preserved under the transform: $\langle \mathcal{F}f, \mathcal{F}g \rangle = \langle f, g \rangle$. In particular, the norm of a function is preserved under a unitary transform, e.g., $||\mathcal{F}f||_2 = ||f||_2$.

In the three-dimensional space of common experience, a typical unitary transform could be just a rotation along an axis, which clearly preserves both the lengths of vectors as well as their inner products. Another different example is a flip in one axis (i.e., $x \to -x$ in one variable, all other coordinates fixed). In fact, it can be shown that in a finite-dimensional real vector space, all unitary transforms can be decomposed as a finite product of rotations and flips (which is called a *Givens* decomposition) [10]. Made up of only rotations and flips, unitary transforms are just changes of perspective; while they may make some things easier to see, nothing fundamental changes under them. Or does it? Indeed, if there were no value to compression, we would hardly be dwelling on them here! Again, the simple fact is that for certain classes of signals (e.g., natural images), the use of a particular unitary transformation (e.g., change to a wavlelet basis) can often reduce the number of coefficients that have amplitudes greater than a fixed threshold. We call this phenomenon energy compaction; it is one of the fundamental reasons for the success of transform coding techniques.

While unitary transforms in finite dimensions are apparently straightforward, as we have mentioned, many of the best ideas in signal processing actually originate in infinite-dimensional math (and physics). What constitutes a unitary transform in an infinite-dimensional function space may not be very intuitive. The Fourier Transform is a fairly sophisticated unitary transform on the space of square-integrable functions. Its basis functions are inherently complex-valued, while we (presumably) live in a "real" world. How the Fourier Transform came to be so commonplace in the world of even the most practical engineer is perhaps something to marvel at. In any case, two hundred years of history, and many successes, could perhaps make a similar event possible for other unitary transforms as well. In just a decade, the wavelet transform has made enormous strides.

There are no doubt some more elementary unitary transforms, even in function space; see the examples below, and figure 1.

Examples

1. (*Translation*) $T_a f(t) = f(t-a)$.
2. (*Modulation*) $M_b f(t) = e^{2j\pi bt} f(t)$.
3. (*Dilation*) $D_c f(t) = \sqrt{|c|} f(ct), \ c \neq 0$.

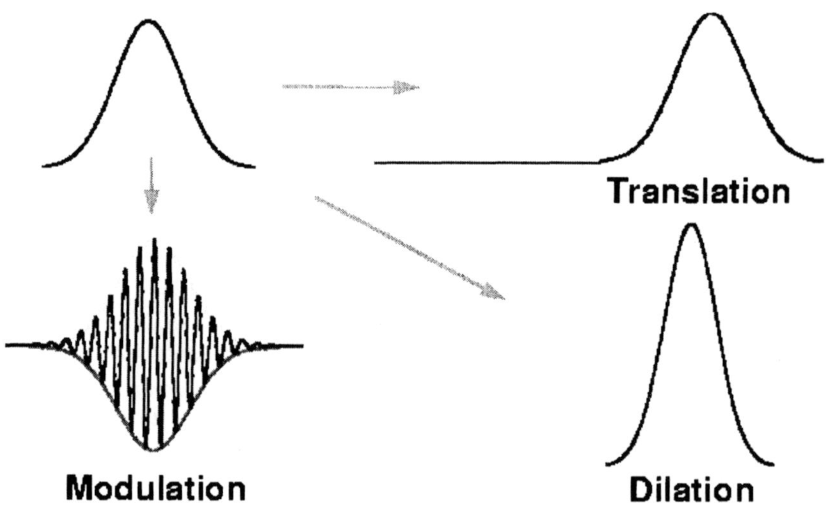

FIGURE 1. Illustration of three common unitary transforms: translations, modulations, and dilations.

Although not directly intuitive, it is in fact easy to prove by simple changes of variable that each of these transformations is actually unitary.

PROOF OF UNITARITY

1. (*Translation*) $\langle T_a f(t), T_a g(t)\rangle = \int \overline{f(t-a)} g(t-a) dt = \int \overline{f(u)} g(u) du = \langle f, g\rangle$.
2. (*Modulation*) $\langle M_a f(t), M_a g(t)\rangle = \int e^{-2j\pi at} \overline{f(t)} e^{2j\pi at} g(t) dt = \int \overline{f(t)} g(t) dt = \langle f, g\rangle$.

3. (*Dilation*) $\langle D_a f(t), D_a g(t)\rangle = |c| \int \overline{f(ct)} g(ct) dt = \int \overline{f(u)} g(u) du = \langle f, g\rangle$.

It turns out that each of these transforms also has an important role to play in our theory, as will be seen shortly. How do they behave under the Fourier Transform? The answer is surprisingly simple, and given below.

More Properties of the Fourier Transform

1. $(\mathcal{F}T_a)f(\omega) = \int e^{-2j\pi t\omega} f(t-a)dt = \int e^{-2j\pi(s-a)\omega} f(s)ds = e^{2j\pi a\omega}\mathcal{F}f(\omega) = M_a\mathcal{F}f(\omega)$.
2. $(\mathcal{F}M_a)f(\omega) = \int e^{-2j\pi t\omega} e^{2j\pi at} f(t)dt = \int e^{-2j\pi t(\omega-a)} f(t)dt = \mathcal{F}f(\omega - a) = T_a\mathcal{F}f(\omega)$.
3. $(\mathcal{F}D_a)f(\omega) = \int e^{-2j\pi t\omega}\sqrt{a}f(at)dt = \frac{1}{\sqrt{a}}\int e^{-2j\pi s\frac{\omega}{a}} f(s)ds = \frac{1}{\sqrt{a}}\mathcal{F}f(\frac{\omega}{a}) = D_{\frac{1}{a}}\mathcal{F}f(\omega)$.

Since for any nonzero L_2 function f, $0 < ||f||_2 < \infty$ by definition, we can easily rescale f (by $1/||f||_2$) so that $||f||_2 = 1$. Then $||\mathcal{F}f||_2 = 1$ also. This means that

$$\int |f(t)|^2 dt = 1 = \int |\mathcal{F}f(\omega)|^2 d\omega.$$

In effect, $|f(t)|^2$ and $|\mathcal{F}f(\omega)|^2$ are probability density functions in their respective parameters, since they are nonnegative functions which integrate to 1. Allowing for the rescaling by the positive constant $1/||f||_2$, which is essentially harmless, the above points to a relationship between the theory of L_2 functions and unitary transforms on the one hand, and probability theory and probability-preserving transforms on the other.

From this point of view, we can consider the nature of the probability distributions defined by $|f(t)|^2$ and $|\mathcal{F}f(\omega)|^2$. In particular, we can ask for their moments, or their means and standard deviations, etc. Since a function and its Fourier (or any invertible) transform uniquely determine each other, one may wonder if there are any relationships between the two probability density functions they define. For the three simple unitary tranformations above: translations, modulations, and dilations, such relationships are relatively straightforward to derive, and again are obtained by simple changes of variable. For example, translation directly shifts the mean, and leaves the standard deviation unaffected, while modulation has no effect on the absolute value square $|f(t)|^2$, and thus the associated probability density (in particular, the mean and standard deviation are unaffected). Dilation is only slightly more complicated; if the function is zero-mean, for example, then dilation leaves it zero-mean, and dilates the standard deviation. However, for the Fourier Transform, the relationships between the corresponding probability densities are both more subtle and profound.

To better appreciate the Fourier Transform, we first need some examples. While the Fourier transform of a given function is generally difficult to write down explicitly, there are some notable exceptions. To see some examples, we first must be able to write down simple functions that are square-integrable (so that the Fourier Transform is well-defined). Such elementary functions as polynomials ($f(t) = a_n t^n +$

$\ldots + a_0$) and trigonometric functions ($sin(t)$, $cos(t)$) don't qualify automatically, since their absolute squares are not integrable. One either needs a function that goes to zero quickly as $|t| \to \infty$, or just to cut off any reasonable function outside a finite interval. The Gaussian function $f(t) = 2^{\frac{1}{4}} e^{-\pi t^2}$ certainly goes to zero very quickly, and one can even use it to dampen polynomials and trig functions into being square-integrable. Or else one can take practically any function and cut it off; for example, the constant function 1 cut to an interval $[-T, T]$, called the rectangle function, which we will label $rect_{[-T,T]}(t)$.

Examples

1. (*Sinc*) Let $rect_{[-T,T]}(t) = 1, \quad t \in [-T, T]; \; 0$ else. Then

$$\begin{aligned} \mathcal{F}rect(\omega) &= \int_{-T}^{T} e^{-2\pi jt\omega} dt = \frac{-1}{2\pi j\omega}[e^{-2\pi jT\omega} - e^{2\pi jT\omega}] = 2T\frac{sin(2\pi T\omega)}{2\pi T\omega} \\ &\equiv 2T \; sinc(2\pi T\omega). \end{aligned}$$

2. (*Gaussian*) The Gaussian function $f(t) = 2^{\frac{1}{4}} e^{-\pi t^2}$ maps into itself under the Fourier transform.

$$\begin{aligned} \mathcal{F}f(\omega) &= 2^{\frac{1}{4}} \int e^{-2\pi jt\omega} e^{-\pi t^2} dt \\ &= 2^{\frac{1}{4}} \int e^{-\pi(t^2+2jt\omega)} dt = 2^{\frac{1}{4}} \int e^{-\pi(t^2+2jt\omega-\omega^2)-\pi\omega^2} dt \\ &= 2^{\frac{1}{4}} e^{-\pi\omega^2} \int e^{-\pi(t+j\omega)^2} dt = 2^{\frac{1}{4}} e^{-\pi\omega^2} \int e^{-\pi t^2} dt \\ &= 2^{\frac{1}{4}} e^{-\pi\omega^2}. \end{aligned}$$

Note that both calculations above use elements from complex analysis. Item 1 uses Euler's formula mentioned earlier, while the second to last step in item 2 uses some serious facts from integration of complex variables (namely, the so-called Residue Theorem and its use in changing contours of integration [13]); it is not a simple change of variables as it may appear. These remarks underscore a link between the analysis of real and complex variables in the context of the Fourier Transform, which we mention but cannot explore. In any case, both of the above calculations will be very useful in our discussion.

To return to relating the probability densities in time and frequency, recall that we define the mean and variance of the densities $|f(t)|^2$, $|\mathcal{F}f(\omega)|^2$ by

$$\begin{aligned} \mu_t &= \langle t \rangle = \int t \; |f(t)|^2 dt, \quad \sigma_t^2 = \langle (t-\mu_t)^2 \rangle = \int (t-\mu_t)^2 |f(t)|^2 dt, \\ \mu_\omega &= \langle \omega \rangle = \int \omega \; |\mathcal{F}f(\omega)|^2 d\omega, \quad \sigma_\omega^2 = \langle (\omega-\mu_\omega)^2 \rangle = \int (\omega-\mu_\omega)^2 |\mathcal{F}f(\omega)|^2 d\omega. \end{aligned}$$

Using the important fact that a Gaussian function maps into itself under the Fourier transform, one can construct examples showing that for a given function f with

$$f = 2^{\frac{1}{4}} e^{2\pi ibt} e^{-\pi(t-a)^2}; \quad \hat{f} = 2^{\frac{1}{4}} e^{2\pi iab} e^{2\pi ia\omega} e^{-\pi(\omega-b)^2}$$

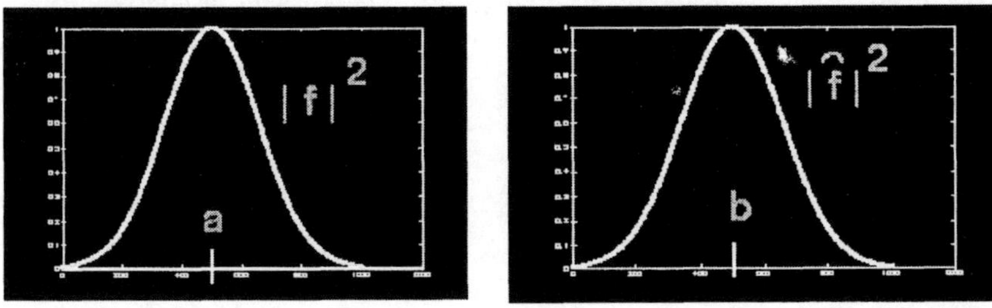

$$\left.\begin{aligned} \mu_t &= a \\ \mu_\omega &= b \end{aligned}\right\} \quad \text{arbitrary}$$

FIGURE 2. Example of a square-integrable function with norm 1 and two free parameters (a and b), for which the means in time and frequency are arbitrary.

$\|f\|_2 = 1$, the mean values in the time and frequency domains can be arbitrary, as shown in figure 2.

On the other hand, the standard deviations in the time and frequency domains cannot be arbitrary. The following fundamental theorem [14] applies for any nonzero function $f(t) \in L_2(\mathbf{R})$.

Heisenberg's Inequality

1. $\sigma_t \sigma_w \geq \frac{1}{4\pi} \|f\|_2^2$.

2. Equality holds in the above if and only if the function is among the set of translated, modulated, and dilated Gaussians,

$$f(t) = Ce^{2\pi jbt} e^{-(rt-a)^2}, \ r > 0.$$

This theorem has far-reaching consequences in many disciplines, most notably physics; it is also relevant in signal processing. The fact that the familiar Gaussian emerges as an extremal function of the Heisenberg Inequality is striking; and that translations, modulations, and dilations are available as the exclusive degrees of freedom is also impressive. This result suggests that Gaussians, under the influence of translations, modulations, and dilations, form elementary building blocks for decomposing signals. Heisenberg effectively tells us these signals are individual packets of energy that are as concentrated as possible in time and frequency. Naturally, our next questions could well be: (1) What kind of decompositions do they give? and (2) Are they good for compression? But let's take a moment to appreciate this result a little more, beginning with a closer look time and frequency shifts.

2 Fourier Series, Time-Frequency Localization

While time and frequency may be viewed as two different domains for representating signals, in reality they are inextricably linked, and it is often valuable to view signals in a unified "time-frequency plane." This is especially useful in analyzing signals that change frequencies in time. To begin our discussion of the time-frequency localization properties of signals, we start with some elementary consideration of time or frequency localization separately.

2.1 Fourier Series

A signal $f(t)$ is *time-localized* to an interval $[-T,T]$ if it is zero outside the interval. It is *frequency-localized* to an interval $[-\Omega,\Omega]$ if the Fourier Transform $\mathcal{F}f(\omega)=0$ for $|\omega|>\Omega$. Of course, these domains of localization can be arbitrary intervals, and not just symmetric about zero. In particular, suppose $f(t)$ is square-integrable, and localized to $[0,1]$ in time; that is, $f\in L_2([0,1])$. It is easy to show that the functions $\{e_n=e^{2\pi jnt}rect_{[0,1]} : n\ integer\}$ form an orthonormal basis for $L_2([0,1])$. In fact, we compute that

$$\langle e_n, e_m\rangle = \int_0^1 e^{2\pi j(m-n)t}dt = \delta_{mn}.$$

Completeness is somewhat harder to show; see [13]. Expanding a function $f(t)$ into this basis, by $f(t)=\sum_n a_n e_n$, $a_n=\langle e_n,f\rangle$, is called a **Fourier Series** expansion. This basis can be viewed graphically as a set of building blocks of signals on the interval $[0,1]$, as illustrated in the left-half of figure 3. Each basis function corresponds to one block or box, and all boxes are of identical size and shape. To get a basis for the next interval, $[1,2]$, one simply moves over each of the basis elements, $\{e_n=e^{2\pi jnt}rect_{[1,2]} : n\ integer\}$, which corresponds to an adjacent tower of blocks. In this way, we fill up the time-frequency plane. The union of all basis functions for all unit intervals $[m,m+1]$ constitutes an orthonormal basis for all of $L_2(R)$, which can be indexed by $\{e_{mn}=e^{2\pi jnt}rect_{[m,m+1]} : m,\ n\ integer\}$. This is precisely the integer translates and modulates of the rectangle function $rect_{[0,1]}$.

Observing that $rect_{[m,m+1]}=T_m rect_{[0,1]}$ and $e_n=M_n rect_{[0,1]}$, we see immediately that in fact $\{e_{mn}=e^{2\pi jnt}rect_{[m,m+1]}\}=\{M_nT_m rect_{[0,1]}\}$. This notation is now helpful in figuring out its Fourier Transform, since

$$\begin{aligned}\mathcal{F}e_{mn} &= \mathcal{F}M_nT_m rect_{[0,1]} = T_nM_m\mathcal{F}rect_{[0,1]} = T_nM_m\mathcal{F}T_{\frac{1}{2}}rect_{[-\frac{1}{2},\frac{1}{2}]}\\ &= T_nM_mM_{\frac{1}{2}}sinc(\pi\omega) = T_nM_{m+\frac{1}{2}}sinc(\pi\omega).\end{aligned}$$

So now we can expand an arbitrary $f\in L_2(\mathbf{R})$ as $f=\sum a_{mn}e_{mn}$. Note incidentally that this is a discrete series, rather than an integral as in the Fourier Transform discussed earlier. An interesting aspect of this basis expansion, and the picture in figure 3 is that if we highlight the boxes for which the amplitudes $|a_{mn}|$ are the largest, we can get a running time-frequency plot of the signal $f(t)$.

Similar to time-localized signals, one can consider frequency-localized signals, e.g., $f(t)\in L_2(\mathbf{R})$ such that $\mathcal{F}f(\omega)\in L_2([a,b])$. For example, let's take the interval to be $[-\frac{1}{2}, \frac{1}{2}]$. In the frequency domain, we know that we can get an orthonormal basis with the functions $\{f_n = e^{2\pi jn\omega} rect_{[-\frac{1}{2},\frac{1}{2}]}\} = \{M_n rect_{[-\frac{1}{2},\frac{1}{2}]}\}$. To get the time-domain picture of these functions, we must take the Inverse Fourier Transform.

$$\mathcal{F}^{-1} f_n(t) = \mathcal{F}^{-1} M_n rect_{[-\frac{1}{2},\frac{1}{2}]} = T_{-n} sinc(\pi t) = sinc(\pi t + n).$$

This follows by calculations similar to earlier ones. This shows that the functions $\{sinc(\pi t + n) : n \; integer\}$ form an orthonormal basis of the space of functions that are frequency localized on the interval $[-\frac{1}{2}, \frac{1}{2}]$. Such functions are also called *bandlimited* functions, which are limited to a finite frequency band. The fact that the integer translates of the sinc function are both orthogonal and normalized is nontrivial to see in just the time domain. Any bandlimited function $f(t)$ can be expanded into them as $f(t) = \sum a_n sinc(\pi t - n)$. In fact, we compute the expansion coefficients as follows:

$$\begin{aligned}
a_n &= \langle sinc(\pi t - n), f(t) \rangle \\
&= \langle e^{-2\pi jn\omega} rect_{[-\frac{1}{2},\frac{1}{2}]}, \mathcal{F}f(\omega) \rangle \\
&= \int_{-\frac{1}{2}}^{\frac{1}{2}} e^{2\pi jn\omega} \mathcal{F}f(\omega) d\omega \\
&= \int e^{2\pi jn\omega} e^{-2\pi jt\omega} f(t) dt d\omega \\
&= \int e^{-2\pi j\omega(t-n)} d\omega f(t) dt \\
&= \int \delta(t-n) f(t) dt \\
&= f(n).
\end{aligned}$$

In other words, the coefficients of the expansion are just the samples of the function itself! Knowing just the values of the function at the integer points determines the function everywhere. Even more, the mapping from bandlimited functions to the sequence of integer samples is a unitary transform. This remarkable formula was discovered early this century by Whittaker [15], and is known as the **Sampling Theorem** for bandlimited functions. The formula can also be used to relate the bandwidth to the sampling density in time—if the bandwidth is larger, we need a greater sampling density, and vice versa. It forms the foundation for relating digital signal processing with notions of continuous mathematics. A variety of facts about the Fourier Transform was used in the calculations above. In having a basis for bandlimited function, of course, we can obtain a basis for the whole space $L_2(\mathbf{R})$ by translating the basis to all intervals, now in frequency. Since modulations in time correspond to frequency-domain translations, this means that the set of integer translates and modulates of the sinc function forms an orthonormal basis as well.

So we have seen that integer translates and modulates of either the rectangle function $rect_{[0,1]}$ or the sinc function $sinc(\pi t)$ form an orthonormal basis. Each gives a version of the picture on the left side of figure 3. Each basis allows us to analyze, to some extent, the running spectral content of a signal. The rectangle

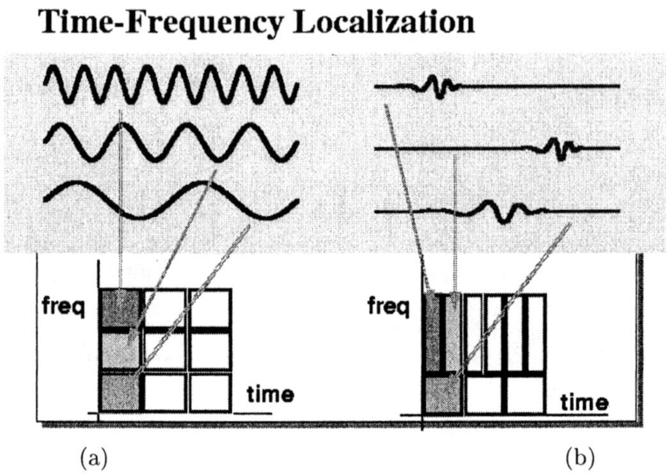

FIGURE 3. The time-frequency structure of local Fourier bases (a), and wavelet bases (b).

function is definitely localized in time, though not in frequency; the sinc is the reverse. Since the translation and modulation are common to both bases, and only the envelope is different, one can ask what set of envelope functions would be best suited for both time and frequency analysis (for example, the rectangle and sinc functions suffer discontinuities, in either time or frequency). The answer to that, from Heisenberg, is the Gaussian, which is perfectly smooth in both domains. And so we return to the question of how to understand the time-frequency structure of time-varying signals better.

A classic example of a signal changing frequency in time is music, which by definition is a time sequence of harmonies. And indeed, if we consider the standard notation for music, we see that it indicates which "notes" or frequencies are to be played at what time. In fact, musical notation is a bit optimistic, in that it demands that precisely specified frequencies occur at precise times (only), without any ambiguity about it. Heisenberg tells us that there can be no signal of precisely fixed frequency that is also fixed in time (i.e., zero spread or variance in frequency and finite spread in time). And indeed, no instrument can actually produce a single note; rather, instruments generally produce an envelope of frequencies, of which the highest intensity frequency can dominate our aural perception of it. Viewed in the time-frequency plane, a musical piece is like a footpath leading generally in the time direction, but which wavers in frequency. The path itself is made of stones large and small, of which the smallest are the imprints of Gaussians.

2.2 Time-Frequency Representations

How can one develop a precise notion of a time-frequency imprint or representation of a signal? The answer depends on what properties we demand of our time-frequency representation. Intuitively, to indicate where the signal energy is, it should be a positive function of the time-frequency plane, have total integral equal to one, and satisfy the so-called "marginals:" i.e., that integrating along lines parallel to the time axis (e.g., on the set $\{(t,\omega) : \omega = \omega_0\}$) gives $|\mathcal{F}f(\omega_0)|^2$, the probability density in frequency, while integrating along $\{(t,\omega) : t = t_0\}$ gives $|f(t_0)|^2$, the probability density in the time axis. Actually, a function on the time-frequency plane satisfying all of these conditions does not exist (for reasons related to Heisenberg Uncertainty) [14], but a reasonable approximation is the following function, called the **Wigner Distribution** of a signal $f(t)$:

$$W_f(t,\omega) = \int \overline{f(t-p/2)} f(t+p/2) e^{2\pi j p\omega} dp.$$

This function satisfies the marginals, integrates to one when $\|f\|_2 = 1$, and is real-valued though not strictly positive. In fact, it is positive for the case that $f(t)$ is a Gaussian, $\phi(t) = 2^{\frac{1}{4}} e^{-\pi t^2}$, in which case it is a two-dimensional Gaussian [14]:

$$W_\phi(t,\omega) = 2e^{-2\pi(t^2+\omega^2)}.$$

The time-frequency picture of this Wigner Distribution is shown at the center in figure 4, where we plot, for example, the 1σ spread. Since all Gaussians satisfy the equality in the Heisenberg Inequality, if we stretch the Gaussian in time, it shrinks in frequency width, and vice versa. The result is always an ellipse of equal area. As was first observed by Dennis Gabor [16], this shows the Gaussian to be a flexible, time-frequency unit out of which more complicated signals can be built.

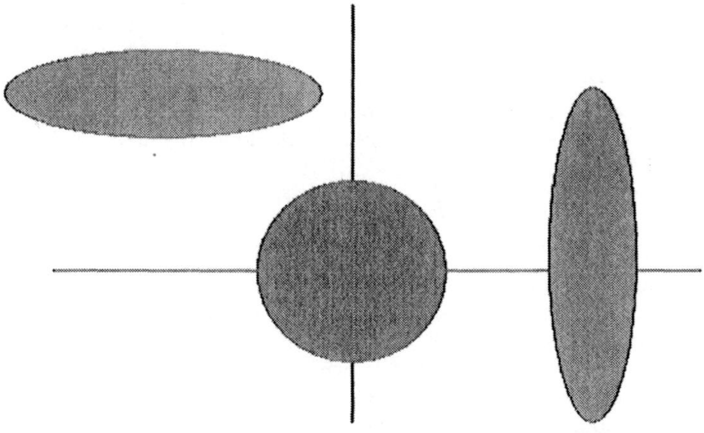

FIGURE 4. The time-frequency spread of Gaussians, which are extremals for the Heisenberg Inequality. A larger variance in time (by dilation) corresponds to a smaller variance in frequency, and vice versa. Translation merely moves the center of the 2-D Gaussian.

Armed with this knowledge about the time-frequency density of signals, and given this basic building block, how can we analyze arbitrary signals in $L_2(\mathbf{R})$? The set of translations, dilations, and modulations of the Gaussian $\phi(t) = 2^{\frac{1}{4}} e^{-\pi t^2}$ do not form an orthonormal basis in $L_2(\mathbf{R})$. For one thing, these functions are not orthogonal since any two Gaussians overlap each other, no matter where they are centered.

More precisely, by completing the square in the exponent, we can show that the product of two Gaussians is a Gaussian. From that observation, we compute:

$$\begin{aligned}\langle M_aT_b\phi(t), M_cT_d\phi(t)\rangle &= \int e^{-2\pi jt(a-c)}\phi(t-b)\phi(t-d)dt \\ &= C_1\int e^{-2\pi jt(a-c)}\phi(t-h)dt \\ &= C_2\mathcal{F}\phi(a-c) = C_2\phi(a-c) \neq 0.\end{aligned}$$

Since the Fourier Transform of a Gaussian is a Gaussian, these inner products are generally nonzero. However, these functions do form a complete set, in that any function in $L_2(\mathbf{R})$ can be expanded into them. Even leaving dilations out of it for the moment, it can be shown that analogously to the rectangle and sinc functions, the translations and modulations of a Gaussian form a complete set. In fact, let $\phi(t) = 2^{\frac{1}{4}}e^{-\pi t^2}$ and define the **Short-Time Fourier Transform** of a signal $f(t)$ to be

$$S_f(t,\omega) = \langle M_\omega T_t\phi(s), f(s)\rangle = \int e^{-2\pi j\omega s}\phi(s-t)f(s)ds.$$

Since the Gaussian is well localized around the time point $s = t$, this process essentially computes the Fourier Transform of the signal in a small window around the point t, hence the name. Because of the complex exponential, this is generally a complex quantity. The magnitude squared of the Short-Time Fourier Transform, which is certainly real and positive, is called the **Spectrogram** of a signal, and is a common tool in speech processing. Like the Wigner Distribution, it gives an alternative time-frequency representation of a signal.

A basic fact about the Short-Time Fourier Transform is that it is invertible; that is, $f(t)$ can be recovered from $S_f(t,\omega)$. In fact, the following inversion formula holds:

$$f(t) = C_\phi \int e^{2\pi j\omega t}\phi(t-s)S_f(s,\omega)dsd\omega.$$

This result even holds when ϕ is any nonzero function in $L_2(\mathbf{R})$, not just a Gaussian, although Gaussians are preferred since they extremize the Heisenberg Inequality. While the meaning of above integral, which takes place in the time-frequency plane, may not be transparent, it is the limit of partial sums of translated, modulated Gaussians (we assume ϕ is a Gaussian), with coefficients given by dot products of such Gaussians against the initial function $f(t)$. This shows that any function can be written as a superposition of such Gaussians. However, since we are conducting a two-parameter integral, we are using an uncountable infinity of such functions, which cannot all be linearly independent.

Thus, these functions could never form a basis—there are too many of them. As we have seen before, our space of square-integrable functions on the line, $L_2(\mathbf{R})$, has bases with a countable infinity of elements. For another example, if one takes the functions $\{t^ne^{-\pi t^2} : n \geq 0\}$ and orthonormalizes them using the Gram-Schmidt method, one obtains the so-called **Hermite functions** (which are again of the form polynomial times Gaussian). It turns out that, similarly, a discrete, countable set of these translated, modulated Gaussians suffices to recover $f(t)$. This is yet another type of **sampling theorem**: the function $S(s,\omega)$ can be sampled at a discrete set of points, $S(ma_0, nb_0)$, from which $f(t)$ can be recovered (and thus, the full function $S(s,\omega)$ as well). The reconstruction formula is then a sum rather than an integral, reminiscent of the Fourier Series expansions. This is again remarkable, since it says that the continuum of points in the function $S(s,\omega)$ which are not sample points provide redundant information. What constraints are there on the values, a_0, b_0? Intuitively, if we sample finely enough, we may have enough information to recover $f(t)$, whereas a coarse sampling may leak too much data to permit full recovery. It can be shown that we must have $a_0 b_0 \leq 1$ [1].

3 The Continuous Wavelet Transform

The translated and modulated Gaussian functions

$$\{\phi_{ab} = M_a T_b \phi(t)\} = \{2^{\frac{1}{4}} e^{2\pi jat} e^{-\pi j(t-b)^2}\}$$

are now called **Gabor wavelets**, in honor of Dennis Gabor [16] who first proposed them as fundamental signals. As we have noted, the functions $\{\phi_{mn} : a = ma_0,\ b = mb_0\}$ suffice to represent any L_2 function, for $a_0 b_0 \leq 1$. In fact, it is known [1] that for $a_0 b_0 < 1$, these functions form a *frame*, as we introduced in chapter 1. In particular, not only can any $f \in L_2(\mathbf{R})$ be represented as $f = \sum a_{mn}\phi_{mn}$, but the expansion satisfies the inequalities $A\|f\|_2^2 \leq \sum |\langle \phi_{mn}, f\rangle|^2 \leq B\|f\|_2^2$. Note that since the functions are not an orthonormal basis, the expansion coefficients are not given by $a_{mn} = \langle \phi_{mn}, f\rangle$. In fact, finding an expansion of a vector in a frame is a bit complicated [1]. Furthermore, in any frame, there are typically many inequivalent ways to represent a given vector (or function).

While the functions $\{\phi_{mn}\}$ form a frame and not a basis, can they nevertheless provide a good representation of signals for purposes of compression? Since a frame can be highly redundant, it may provide examplars close to a given signal of interest, potentially providing both good compression as well as pattern recognition capability. However, effective algorithms are required to determine such a representation, and to specify it compactly. So far, it appears that expansion of a signal in a frame generally leads to an increase in the sampling rate, which works against compression. Nonetheless, with "window" functions other than the Gaussian, one can hope to find orthogonal bases using time-frequency translations; this is indeed the case [1].

But for now we consider instead (and finally) the *dilations* in addition to translations as another avenue for constructing useful decompositions. First note that there would be nothing different in considering modulations and dilations, since that pair converts to translations and dilations under the Fourier Transform. Now, in analogy to the Short-Time Fourier Transform, we can certainly construct what is called the **Continuous Wavelet Transform, CWT**, as follows. Corresponding to the role of the Gaussian should be some basic function ("wavelet") $\psi(t)$ on which we consider all possible dilations and translations, and then take inner products with a given function $f(t)$:

$$CWT_f(a,b) \quad = \quad \langle D_a T_b \psi(t), f(t)\rangle, \quad a \neq 0$$

$$= \sqrt{|a|} \int \overline{\psi(at-b)} f(t)dt.$$

By now, we should be prepared to suppose that in favorable circumstances, the above data is equivalent to the given function $f(t)$. And indeed, given some mild conditions on the "analyzing function" $\psi(t)$—roughly, that it is absolutely integrable, and has mean zero—there is an inversion formula available. Furthermore, there is an analogous sampling theorem in operation, in that there is a discrete set of these wavelet functions that are needed to represent $f(t)$. First, to show the inversion formula, note by unitarity of the Fourier Transform that we can rewrite the CWT as

$$\begin{aligned}
CWT_f(a,b) &= \langle D_a T_b \psi(t), f(t)\rangle = \langle \mathcal{F} D_a T_b \psi(t), \mathcal{F} f(t)\rangle \\
&= \langle D_{\frac{1}{a}} M_b \mathcal{F}\psi, \mathcal{F} f\rangle \\
&= \frac{1}{\sqrt{|a|}} \int e^{-2\pi j b \frac{\omega}{a}} \overline{\hat{\psi}(\frac{\omega}{a})} \hat{f}(\omega) d\omega.
\end{aligned}$$

Now, we claim that a function can be recovered from its CWT; even more, that up to a constant multiplier, the CWT is essentially a unitary transform. To this end, we compute:

$$\begin{aligned}
\langle CWT_f(a,b), CWT_g(a,b)\rangle &\equiv \int \overline{CWT_f(a,b)} CWT_g(a,b) \frac{dadb}{|a|} \\
&= \int \frac{1}{|a|} e^{2\pi j \frac{b}{a}(\omega-\eta)} \overline{\hat{f}(\omega)} \hat{g}(\eta) \hat{\psi}(\frac{\omega}{a}) \overline{\hat{\psi}(\frac{\eta}{a})} d\omega d\eta \frac{dadb}{|a|} \\
&= \int e^{2\pi j \frac{b}{a}(\omega-\eta)} \frac{db}{|a|} \int \overline{\hat{f}(\omega)} \hat{g}(\eta) \hat{\psi}(\frac{\omega}{a}) \overline{\hat{\psi}(\frac{\eta}{a})} d\omega d\eta \frac{da}{|a|} \\
&= \int \delta(\omega-\eta) \overline{\hat{f}(\omega)} \hat{g}(\eta) \hat{\psi}(\frac{\omega}{a}) \overline{\hat{\psi}(\frac{\eta}{a})} d\eta d\omega \frac{da}{|a|} \\
&= \int \overline{\hat{f}(\omega)} \hat{g}(\omega) d\omega \int |\hat{\psi}(\frac{\omega}{a})|^2 \frac{da}{|a|} \\
&= \int \overline{\hat{f}(\omega)} \hat{g}(\omega) d\omega \int \frac{|\hat{\psi}(\zeta)|^2}{|\zeta|} d\zeta \\
&= C_\psi \int \overline{\hat{f}(\omega)} \hat{g}(\omega) d\omega = C_\psi \langle f, g\rangle, \\
\textit{where we assume} \quad C_\psi &\equiv \int \frac{|\hat{\psi}(\zeta)|^2}{|\zeta|} d\zeta < \infty.
\end{aligned}$$

The above calculations use a number of the properties of the Fourier Transform, as well as reordering of integrals, and requires some pondering to digest. The condition on ψ that $C_\psi < \infty$ is fairly mild, and is satisfied, for example, by real functions that are absolutely integrable (e.g., in $L_1(\mathbf{R})$) and have mean zero ($\hat{\psi}(0) = 0$). The main interest for us is that since the CWT is basically unitary, it can be inverted to recover the given function.

Of course, as before, converting a one-variable function to an equivalent two-variable function is hardly a trick as far as compression is concerned. The valuable fact is that the CWT can also be subsampled and still permit recovery. Moreover, unlike the case of the time-frequency approach (Gabor wavelets), here we can get simple orthonormal bases of functions generated by the translations and dilations of a single function. And these do give rise to compression.

4 Wavelet Bases and Multiresolution Analysis

An orthonormal basis of wavelet functions (of finite support) was discovered first in 1910 by Haar, and then not again for another 75 years. First, among many possible discrete samplings of the continuous wavelet transform, we restrict ourselves to integer translations, and dilations by powers of two; we ask, for which ψ is $\{\psi_{mn}(t) = 2^{m/2}\psi(2^m t - n) : m,\ n\ integer\}$ an orthonormal basis of $L_2(\mathbf{R})$? Haar's early example was

$$\psi_{Haar} = rect_{[0,\frac{1}{2}]} - rect_{[\frac{1}{2},1]}.$$

This function is certainly of zero mean and absolutely integrable, and so is an admissable wavelet for the CWT. In fact, it is also not hard to show that the functions $\psi_{mn}(t)$ that it generates are orthonormal (two such functions either don't overlap at all, or due to different scalings, one of them is constant on the overlap, while the other is equally 1 and -1). It is a little more work to show that the set of functions is actually complete (just as in the case of the Fourier Series). While a sporadic set of wavelet bases were discovered in the mid 80's, a fundamental theory developed that nicely encapsulated almost all wavelet bases, and moreover made the application of wavelets more intuitive; it is called *Multiresolution Analysis.* The theory can be formulated axiomatically as follows [1].

Multiresolution Analysis
Suppose given a tower of subspaces $V_k \subset L_2(\mathbf{R})$ such that

1. $V_k \subset V_{k+1}$, *all* k.
2. $\overline{\bigcup_k V_k} = L_2(\mathbf{R})$.
3. $\bigcap_k V_k = 0$.
4. $f(t) \in V_k \leftrightarrow f(2^{-k}t) \in V_0$.
5. There exists a $\phi \in V_0$ such that $\{\phi(t-n) : n\ integer\}$ is an orthonormal basis of V_0.

Then there exists a function $\psi(t) \in V_1$ such that $\{\psi_{mn}(t) = 2^{m/2}\psi(2^m t - n) : m\ , n\ integer\}$ is an orthonormal basis for $L_2(\mathbf{R})$.

The intuition in fact comes from image processing. By axioms 1 to 3 we are given tower of subspaces of the whole function space, corresponding to the full structure of signal space. In the idealized space $L_2(\mathbf{R})$, all the detailed texture of imagery can be represented, whereas in the intermediate spaces V_k, only the details up to a fixed granularity are available. Axiom 4 says that the structure of the details is the same at each scale, and it is only the granularity that changes—this is the multiresolution aspect. Finally, axiom 5, which is not directly derived from image processing intuition, says that the intermediate space V_0 and hence all spaces V_k have a simple basis given by translations of a single function. A simple example can quickly illustrate the meaning of these axioms.

Example (Haar) Let $V_0 = \{f : f|_{[n,n+1)} = constant,\ n\ integer\}$, with $\phi(t) = rect_{[0,1]}$. It is clear that $\{\phi(t-n) : n\ integer\}$ form an orthonormal basis for V_0. All the other subspaces are defined from V_0 by axiom 4. The spaces V_k, $k \to \infty$ are made up of functions that are constant on smaller and smaller *dyadic* intervals (whose endpoints are rational numbers with denominators that are powers of 2). It is not hard to prove that any L_2 function can be approximated by a sequence of step functions whose step width shrinks to zero. And in the other direction, as $k \to -\infty$, it is easy to show that only the zero-function is locally constant on intervals whose width grows unbounded. These remarks indicate that axioms 1 to 3 are satisfied, while the others are by construction. What function $\psi(t)$ do we get? Precisely the Haar wavelet, as we will show.

There is in fact a recipe for generating the wavelet from this mechanism. Let $\phi_{mn}(t) = 2^{m/2}\phi(2^m t - n)$, so that ϕ_{0n} is an orthonormal basis for V_0 for example, and ϕ_{kn} is one for V_k. Since $V_0 \subset V_1$ we must have that $\phi \in V_1$, so $\phi(t) = \sum_n h_0(n)\phi_{1n}(t) = \sqrt{2}\sum_n h_0(n)\phi(2t-n)$, for some coefficients $h_0(n)$. By orthonormality of the ϕ_{1n}, we have $h_0(n) = \langle\phi_{1n}, \phi\rangle$, and by normality of $\phi = \phi_{00}$ we have that $\sum |h_0(n)|^2 = 1$. We can also get an equation for ϕ_{0k} from above by

$$
\begin{aligned}
\phi_{0k}(t) &= \phi(t-k) = \sqrt{2}\sum_n h_0(n)\phi(2(t-k)-n) = \sqrt{2}\sum_n h_0(n)\phi(2t-(2k+n)) \\
&= \sum_m h_0(m)\phi_{1(2k+m)}.
\end{aligned}
$$

By orthonormality of the ϕ_{0k} we in fact have

$$
\begin{aligned}
\delta_{0k} &= \langle\phi_{00}, \phi_{0k}\rangle = \sum_{mn} \overline{h_0(n)}h_0(m)\langle\phi_{1n}, \phi_{1(2k+m)}\rangle \\
&= \sum_{mn} \overline{h_0(n)}h_0(m)\delta_{n(2k+m)} = \sum_m \overline{h_0(2k+m)}h_0(m).
\end{aligned}
$$

This places a constraint on the coefficients $h_0(n)$ in the expansion for ϕ, which is often called the *scaling function*; the expansion itself is called the *dilation equation*. Like ϕ, our *wavelet* ψ will also be a linear combination of the ϕ_{1n}. For a derivation of these results, see [1] or [8].

Recipe for a Wavelet in Multiresolution Analysis
$\psi = \sum_n h_1(n)\phi_{1n}, \quad h_1(n) = (-1)^n h_0(-n+1).$

Example (Haar) For the Haar case, we have $\{h_0(0), h_0(1)\} = \{\frac{1}{\sqrt{2}}, \frac{1}{\sqrt{2}}\}$ and $\{h_1(0), h_1(1)\} = \{\frac{1}{\sqrt{2}}, -\frac{1}{\sqrt{2}}\}$. This is easy to compute from the function $\phi_{00} = rect_{[0,1]}$ directly, since

$$\begin{aligned}\phi &= \phi_{00} = rect_{[0,1]} = \frac{1}{\sqrt{2}}\phi_{10} + \frac{1}{\sqrt{2}}\phi_{11} = rect_{[0,\frac{1}{2}]} + rect_{[\frac{1}{2},1]}.\\ \psi &= \frac{1}{\sqrt{2}}\phi_{10} - \frac{1}{\sqrt{2}}\phi_{11} = rect_{[0,\frac{1}{2}]} - rect_{[\frac{1}{2},1]}.\end{aligned}$$

A picture of the Haar scaling function and wavelet at level 0 is given in figure 6; the situation at other levels is only a rescaled vesion of this. A graphic illustrating the approximation of an arbitrary L_2 function by piecewise constant functions is given in figure 5. The picture for a general multiresolution analysis is similar, and only the approximating functions change.

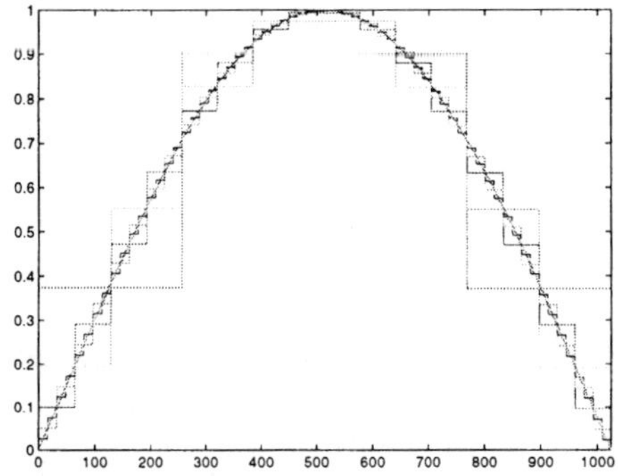

FIGURE 5. Approximation of an L_2 function by piecewise constant functions in a Haar multiresolution analysis.

For instance, the Haar example of piecewise constant functions on dyadic intervals is just the first of the set of piecewise polynomial functions—the so-called *splines*. There are in fact spline wavelets for every degree, and these play an important role in the theory as well as application of wavelets.

Example (Linear Spline) Let $\phi(t) = 1 - |t|, \ |t| \leq 1; \ 0 \ else$. It is easy to see that

$$\phi(t) = \frac{1}{2\sqrt{2}}\sqrt{2}\phi(2t-1) + \frac{1}{\sqrt{2}}\sqrt{2}\phi(2t) + \frac{1}{2\sqrt{2}}\sqrt{2}\phi(2t+1);$$

However, this $\phi(t)$ is not orthogonal to its integer translates (although it is independent from them). So this function is not yet appropriate for building a multiresolution analysis. By using a standard "orthogonalization" trick [1], one can get from it a function such that it and its integer translates are orthonormal. In this instance, we obtain $\phi^\sharp$ as

$$\hat{\phi}^\sharp(\omega) = \sqrt{\frac{3}{2\pi}} \frac{4sin^2(\omega/2)}{\omega^2(1+2cos^2(\omega/2))^{1/2}}.$$

This example serves to show that the task of writing down the scaling and wavelet functions quickly becomes very complicated. In fact, many important and useful wavelet functions do not have any explicit closed-form formulas, only an iterative way of computing them. However, since our interest is mostly in the implications for digital signal processing, these complications will be immaterial.

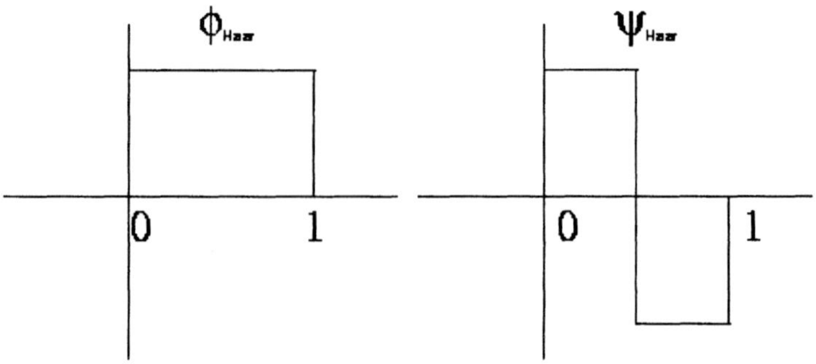

FIGURE 6. The elementary Haar scaling function and wavelet.

Given a multiresolution analysis, we can develop some intuitive processing approaches for disecting the signals (functions) of interest into manageable pieces. For if we view the elements of the spaces V_k, $k \to \infty$ as providing greater and greater detail, then we can view the projections of functions to spaces V_k, $k \to -\infty$ as providing coarsenings with reduced detail. In fact, we can show that given the axioms, the subspace $V_0 \subset V_1$ has an orthogonal complement, W_0, which in turn is spanned by the wavelets $\{\psi_{0n}\}$; this gives the orthogonal sum $V_1 = V_0 \dot{+} W_0$ (this means that any two elements from the two spaces are orthogonal). Similarly, $V_1 \subset V_2$ has an orthogonal complement, W_1, which is spanned by the $\{\psi_{1n}\}$, giving the orthogonal sum $V_2 = V_1 \dot{+} W_1$. Substituting in the previous orthogonal sum, we obtain $V_2 = V_0 \dot{+} W_0 \dot{+} W_1$. This analysis can be continued, both upwards and downwards in indices, giving

$$\begin{aligned} V_0 &= V_{-k} \dot{+} W_{-k} \dot{+} \ldots \dot{+} W_0; \\ V_k &= V_0 \dot{+} W_0 \dot{+} \ldots \dot{+} W_{k-1} = \ldots \dot{+} W_{-1} \dot{+} W_0 \dot{+} \ldots \dot{+} W_{k-1}; \\ L_2(\mathbf{R}) &= \sum_k W_k, \ all\ k. \end{aligned}$$

In these decompositions, the spaces V_k are spanned by the set $\{\phi_{kn}\}$, while the spaces W_k are spanned by the set $\{\psi_{kn}\}$. The $\{\phi_{kn}\}$ are only orthogonal for a fixed k, but have inner products for different $k's$, while the functions $\{\psi_{kn}\}$ are fully orthonormal for both indices. In terms of signal processing, then, if we start with a signal $f \in V_0$, we can decompose it as $f = f^{-1} + \delta^{-1} \in V_{-1} \dot{+} W_{-1}$. We can view f^{-1} as a first coarse resolution approximation of f, and the difference δ^{-1} as the first detail signal. This process can be continued indefinitely, deriving coarser resolution signals, and the corresponding details. Because of the orthogonal decompositions at each step, two things hold: (1) the original signal can always be recovered from the final coarse resolution signal and the sequence of detail signals (this corresponds to the first equation above); and (2) the representation $f = f^{-k} + \delta^{-k} + \ldots + \delta^{-1}$ preserves the energy of the signal (and is a unitary transform).

Starting with $f = \sum_n c_n^0 \phi_{0n} \in V_0$, how does one compute even the first decomposition, $f = f^{-1} + \delta^{-1} \in V_{-1} \dot{+} W_{-1}$? By orthonormality,

$$\begin{aligned}
f^{-1} &= \sum_m \langle \phi_{-1m}, f \rangle \phi_{-1m} = \sum_{mn} c_n^0 \langle \phi_{-1m}, \phi_{0n} \rangle \phi_{-1m} \\
&= \sum_{mn} c_n^0 \phi_{-1m} \frac{1}{\sqrt{2}} \int \overline{\phi(\frac{t}{2} - m)} \phi(t-n) dt \\
&= \sum_{mn} c_n^0 \phi_{-1m} \sqrt{2} \int \overline{\phi(s)} \phi(2s - (n-2m)) ds \\
&= \sum_{mn} c_n^0 \overline{h_0(n-2m)} \phi_{-1m} \equiv \sum_m c_m^1 \phi_{-1m}; \\
\textit{Similarly } \delta^{-1} &= \sum_n \langle \psi_{-1n}, f \rangle \psi_{-1n} \\
&= \sum_{mn} c_n^0 \overline{h_1(n-2m)} \psi_{-1m} \equiv \sum_m d_m^1 \psi_{-1m};
\end{aligned}$$

This calculation brings up the relationship between these wavelet bases and digital signal processing. In general, digital signals are viewed as living in the sequence space, l_2, while our continuous-variable functions live in $L_2(\mathbf{R})$. Given an orthonormal basis of $L_2(\mathbf{R})$, say $\{e_n(t)\}$, there is a correspondence is $\{(c_n^0)\} \to f = \sum_n c_n^0 e_n(t) \in L_2(\mathbf{R})$. This correspondence is actually a unitary equivalence between the sequence and function spaces, and goes freely in either direction. Sequences and functions are two sides of the same coin. This applies in particular to our orthonormal families of wavelet bases, $\{\psi_{mn}\}$, which arise from multiresolution analysis. Since the multiresolution analysis comes with a set of decompositions of functions into approximate and detail functions, there is a corresponding mapping on the sequences, namely

$$\{(c_n^0)\} \to \{(c_m^1), (d_m^1)\} = \{(\sum_n \overline{h_0(n-2m)} c_n^0), (\sum_n \overline{h_1(n-2m)} c_n^0)\}.$$

Since the decomposition of the functions is a unitary transform, it must also be one on the sequence space. The reconstruction formula can be derived in a manner quite similar to the above, and reads as follows:

$$\begin{aligned} c_n^0 &= \sum_m h_0(n-2m)c_m^1 + h_1(n-2m)d_m^1 \\ &= \sum_{ml} \{h_0(n-2m)\overline{h_0(l-2m)} + h_1(n-2m)\overline{h_1(l-2m)}\}c_l^0. \\ \textit{That is, } \delta_{nl} &= \sum_m \{h_0(n-2m)\overline{h_0(l-2m)} + h_1(n-2m)\overline{h_1(l-2m)}. \end{aligned}$$

The last line is actually not a separate requirement, but follows automatically from the machinery developed. The interesting point is that this set of ideas originated in the function space point of view. Notice that with an appropriate reversing of coefficient indices, we can write the above decomposition formulas as convolutions in the digital domain (see chapter 1). In the language developed in chapter 1, the coefficients $\{h_0(n)\}$ and $\{h_1(n)\}$ are filters, the first a *lowpass* filter because it tends to give a coarse approximation of reduced resolution, and the second a *highpass* filter because it gives the detail signal, lost in going from the original to the coarse signal. A second important observation is that after the filtering operation, the signal is subsampled ("downsampled") by a factor of two. In particular, if the original sequence (c_n^0) has $N = 2^M$ samples, the lowpass and highpass outputs of the decomposition, $\{(c_m^1), (d_m^1)\}$, each have $N/2 = 2^{M-1}$ samples, equalling a total of N samples. Incidentally, although this is a unitary decomposition, preservation of the number of samples is not required by unitarity (for example, the plane $\mathbf{R}^2$ can be embedded in the 3-space $\mathbf{R}^3$ in many ways, most of which give 3 samples instead of two in the standard basis). The reconstruction process is also a convolution process, except that with the indicesbuilds up fully N samples from each of the (lowpass and highpass) "channels" and adds them up to recover the original signal. The process of decomposing and reconstructing signals and sequences is described succintly in the figure 7.

$$f^0 = \sum_k c_k^0 \phi_{0k} = f^{-1} + \delta^{-1} \in V_0 = V_{-1} \dot{+} W_{-1}$$

$$f^{-1} = \sum_k c_k^1 \phi_{-1k}, \qquad \delta^{-1} = \sum_k d_k^1 \psi_{-1k}$$

FIRST APPROX. SIGNAL FIRST DETAIL SIGNAL

$$c_k^1 = \sum_n \overline{h_0(n-2k)}\, c_n^0, \qquad d_k^1 = \sum_n \overline{h_1(n-2k)}\, c_n^0,$$

$$c_n^0 = \sum_k h_0(n-2k)\, c_k^1 + h_1(n-2k)\, d_k^1.$$

FIGURE 7. The correspondence between wavelet filters and continuous wavelet functions, in a multiresolution context. Here the functions determine the filters, although the reverse is also true by using an iterative procedure. As in the function space, the decomposition in the digital signal space can be continued for a number of levels.

So we see that orthonormal wavelet bases give rise to lowpass and highpass filters, whose coefficients are given by various dot products between the dilated and translated scaling and wavelet functions. Conversely, one can show that these filters fully encapsulate the wavelet functions themselves, and are in fact recoverable by iterative procedures. A variety of algorithms have been devised for doing so (see [1]), of which perhaps the best known is called the *cascade* algorithm. Rather than present the algorithm here, we give a stunning graphic on the recovery of a highly complicated wavelet function from a single 4-tap filter (see figure 8)! More on this wavelet and filter in the next two sections.

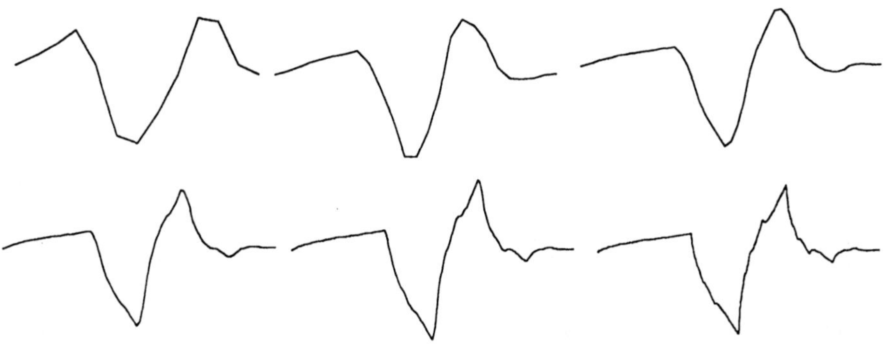

FIGURE 8. Six steps in the cascade algorithm for generating a continuous wavelet starting with a filter bank; example is for the Daub4 orthogonal wavelet.

5 Wavelets and Subband Filter Banks

5.1 Two-Channel Filter Banks

The machinery for digital signal processing developed in the last section can be encapsulated in what is commonly called a two-channel perfect reconstruction filter bank; see figure 9. Here, in the wavelet context, H_0 refers to the lowpass filter, derived from the dilation equation coefficients, perhaps with reversing of coefficients. The filter H_1 refers to the highpass filter, and in the wavelet theory developed above, it is derived directly from the lowpass filter, by $h_1(n) = (-1)^n h_0(-n+1)$. The reconstruction filters are precisely the same filters, H_0, H_1, but applied after reversing. (Note, even if we reverse the filters first to allow for a conventional convolution operation, we need to reverse them again to get the correct form of the reconstruction formula.) Thus, in the wavelet setting, there is really only one filter to design, say H_0, and all the others are determined from it.

How is this structure used in signal processing, such as compression? For typical input signals, the detail signal generally has limited energy, while the approximation signal usually carries most of the energy of the original signal. In compression processing where energy compaction is important, this decomposition mechanism is frequently applied repetitively to the lowpass (approximation) signal for a number of levels; see figure 10. Such a decomposition is called an *octave-band* or Mallat

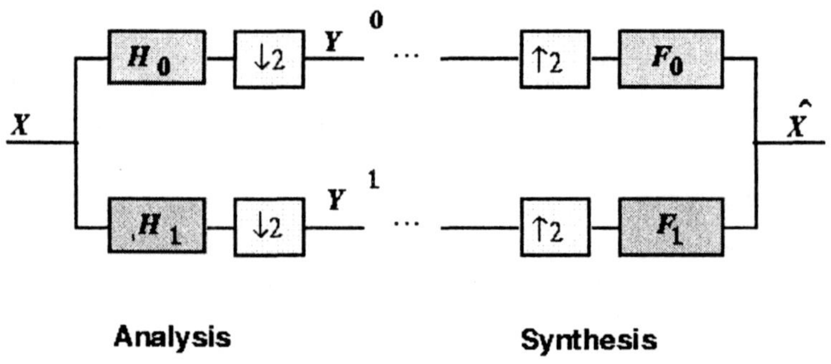

FIGURE 9. A two-channel perfect reconstruction filter bank.

decomposition. The reconstruction process, then, must precisely duplicate the decomposition in reverse. Since the transformations are unitary at each step, the concatenation of unitary transformations is still unitary, and there is no loss of either energy or information in the process. Of course, if we need to compress the signal, there will be both loss of energy and information; however, that is the result of further processing. Such additional processing would typically be done just after decomposition stage (and involve such steps as quantization of transform coefficients, and entropy coding—these will be developed in chapter 4).

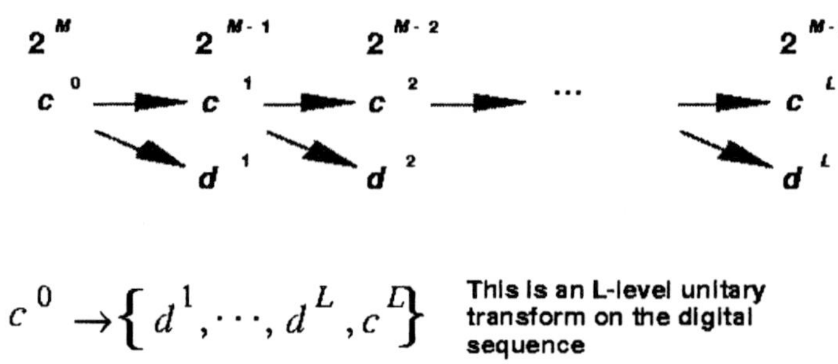

FIGURE 10. A one-dimensional octave-band decomposition using two-channel perfect reconstruction filters.

There are a number of generalizations of this basic structure that are immediately available. First of all, for image and higher dimensional signal processing, how can one apply this one-dimensional machinery? The answer is simple: just do it one coordinate at a time. For example, in image processing, one can decompose the row vectors and then the column vectors; in fact, this is precisely how FFT's are applied in higher dimensions. This "separation of variables" approach is not the only one available in wavelet theory, as "nonseparable" wavelets in multiple dimensions do exist; however, this is the most direct way, and so far, the most fruitful one.

In any case, with the lowpass and highpass filters of wavelets, this results in a decomposition into quadrants, corresponding to the four subsequent channels (low-low, low-high, high-low, and high-high); see figure 11. Again, the low-low channel (or "subband") typically has the most energy, and is usually decomposed several more times, as depicted. An example of a two-level wavelet decomposition on the well-known Lena image is given in chapter 1, figure 5.

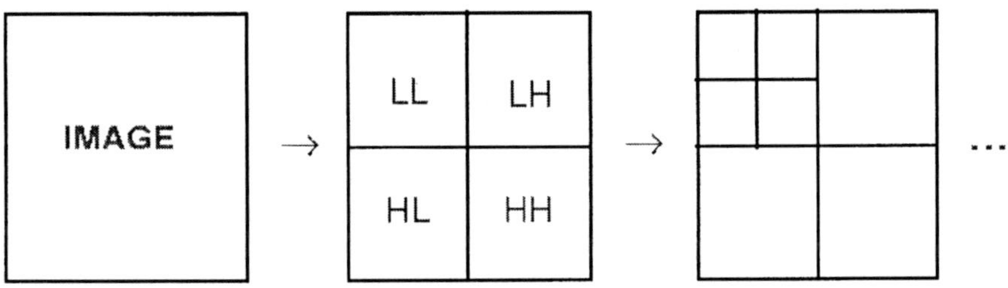

FIGURE 11. A two-dimensional octave-band decomposition using two-channel perfect reconstruction filters.

A second, less-obvious, but dramatic generalization is that, once we have formulated our problem in terms of the filter bank structure, one may ask what sets of filters $\{H_i, F_i\}$, $i = 0, 1$, satisfy the perfect reconstruction property, regardless of whether they arise from wavelet constructions or not. In fact, in this generalization, the problem is actually older than the recent wavelet development (starting in the mid-1980s); see [1]. To put the issues in a proper framework, it is convenient to recall the notation of the z-Transform from chapter 1.

A filter h acts on a signal x by convolution of the two sequences, producing an output signal $y = \{y(k)\}$, given by

$$y(k) = (h * x)(k) = \sum_i h(k-i)x(i).$$

Given a digital signal or filter, $x = \{x(0), x(1), \ldots, x(N-1)\}$, recall that its *z-Transform* is defined as the following power series in a complex variable z,

$$X(z) = \sum_i x(i)z^{-i}.$$

The convolution of two sequences $h = \{h(0), h(1), \ldots, h(L-1)\}$ and $x = \{x(0), x(1), \ldots, x(N-1)\}$ is then the product of the $z-$Transforms:

$$Y(z) = H(z)X(z) = \sum_k (\sum_i h(k-i)x(i))z^{-k} = \sum_k y(k)z^{-k}.$$

Define two further operations, *downsampling* and *upsampling*, respectively, as

$$\begin{aligned} (\downarrow x)(i) &= x(2i). \\ (\uparrow x)(2i) &= x(i), \quad (\uparrow x)(2i+1) = 0. \end{aligned}$$

$$(\uparrow\downarrow x)(2i) = x(2i), \quad (\uparrow\downarrow x)(2i+1) = 0.$$

These operations have the following impact on the z-Transform representations.

$$Y(z) = (\downarrow X)(z) = \sum_i x(2i)z^{-i};$$
$$Y(z^2) = \sum_i x(2i)z^{-2i} = \frac{1}{2}\{X(z) + X(-z)\}.$$
$$V(z) = (\uparrow X)(z) = \sum_i x(i)z^{-2i} = X(z^2).$$
$$W(z) = (\uparrow\downarrow X)(z) = \frac{1}{2}\{X(z) + X(-z)\}.$$

Armed with this machinery, we are ready to understand the two-channel filter bank of figure 9. A signal $X(z)$ enters the system, is split and passed through two channels. After analysis filtering and downsampling and upsampling, we have

$$\begin{aligned}
\uparrow Y^0(z) &= \frac{1}{2}\{H_0(z)X(z) + H_0(-z)X(-z)\}. \\
F_0(z) \uparrow Y^0(z) &= \frac{1}{2}\{F_0(z)H_0(z)X(z) + F_0(z)H_0(-z)X(-z)\}. \\
F_1(z) \uparrow Y^1(z) &= \frac{1}{2}\{F_1(z)H_1(z)X(z) + F_1(z)H_1(-z)X(-z)\}. \\
\hat{X}(z) &= \frac{1}{2}\{F_0(z)H_0(z) + F_1(z)H_1(z)\}X(z) \\
&+ \frac{1}{2}\{F_0(z)H_0(-z) + F_1(z)H_1(-z)\}X(-z).
\end{aligned}$$

In the expression for the output of the filter bank, the term proportional to $X(z)$ is the desired signal, and its coefficient is required to be either unity or at most a delay, z^{-l} for some integer l. The term proportional to $X(-z)$ is the so-called *alias* term, and corresponds to portions of the signal reflected in frequency at Nyquist due to potentially imperfect bandpass filtering in the system. It is required that the alias term be set to zero. In summary, we need

$$\begin{aligned}
\frac{1}{2}\{F_0(z)H_0(-z) + F_1(z)H_1(-z)\} &= 0. \ (\textit{Alias Cancellation (AC)}) \\
\frac{1}{2}\{F_0(z)H_0(z) + F_1(z)H_1(z)\} &= z^{-l}. \ (\textit{Perfect Reconstruction (PR)})
\end{aligned}$$

These are the fundamental equations that need to be solved to design a filtering system suitable for applications such as compression. As we remarked, orthonormal wavelets provide a special solution to these equations. First, choose $H_0(z)$, and set

$$H_1(z) = -z^{-1}H_0(-z^{-1}), \quad F_0(z) = H_0(z^{-1}), \quad F_1(z) = -zH_0(-z).$$

One can verify that the above conditions are met. In fact, the AC condition is easy, while the PR condition imposes a single condition on the filter $H_0(z)$ which turns out to be equivalent to the orthonormality condition on the wavelets; see [1] for more details.

But let us work more generally now, and tackle the AC equation first. Since we have a lot of degrees of freedom, we can, as others before us, take the easy road and simply let $F_0(z) = H_1(-z)$, $F_1(z) = -H_0(-z)$. Then the AC term is automatically zero.

Next, notice that the PR condition now reads $F_0(z)H_0(z) - H_0(-z)F_0(-z) = 2z^{-l}$. Setting $P_0(z) = F_0(z)H_0(z)$, we get

$$P_0(z) - P_0(-z) = 2z^{-l}.$$

Now, since the left-hand side is an odd function of z, l must be odd. If we let $P(z) = z^l P_0(z)$, we get the equation

$$P(z) + P(-z) = 2.$$

Finally, if we write out $P(z)$ as a series in z, we see that all even powers must be zero except the constant term, which must be 1. At the same time, the odd terms are arbitrary, since they cancel above—they are just the remaining degrees of freedom. The product filter $P(z)$, which is now called a *halfband filter* [8], thus has the form

$$P(z) = p_{-l}z^l + \ldots + p_{-1}z^1 + 1 + p_1 z^{-1} + \ldots + p_l z^{-l}.$$

Here $l = 2N + 1$, *an odd number.*

The design approach is now as follows.

Design of Perfect Reconstruction Filters

1. Design the product filter $P(z)$ as above.
2. Peel off z^l, to get $P_0(z)$.
3. Factor $P_0(z)$ as $F_0(z)H_0(z)$.
4. Set $H_1(z) = F_0(-z)$, $F_1(z) = -H_0(-z)$.

Finite length two-channel filter banks that satisfy the perfect reconstruction property are called Finite Impulse Response (FIR) Perfect Reconstruction (PR) Quadrature Mirror Filter (QMF) Banks in the literature. A detailed account of these filter banks, along with some of the history of this terminology, is available in [8, 17, 1]. We will content ourselves with some elementary but key examples. Note one interesting fact that emerges above: in step 3 above, when we factor $P_0(z)$, it is immaterial which factor is which (e.g., F_0 and H_0 can be freely interchanged while preserving perfect reconstruction). However, if there is any intermediary processing between the analysis and synthesis stages (such as compression processing), then the two factorizations may prove to be inequivalent.

5.2 *Example FIR PR QMF Banks*

Example 1 (Haar = Daub2)
$P(z) = \frac{1}{2}(z + 2 + z^{-1})$. Here $l = 1$, $P_0(z) = \frac{1}{2}(1 + 2z^{-1} + z^{-2}) = \frac{1}{2}(1 + z^{-1})^2$. This factors easily as $F_0(z) = H_0(z) = \frac{1}{\sqrt{2}}(1 + z^{-1})$. (This is the only nontrivial factorization.) Notice that these very simple filters enjoy one pleasant property: they have a symmetry (or antisymmetry).

Example 2 (Daub4)
$P(z) = \frac{1}{16}(-z^3 + 9z + 16 + 9z^{-1} - z^{-3}) = z^3 P_0(z)$. We can factor $P_0(z)$ as $P_0(z) = (1+z^{-1})^4(c - z^{-1})(\frac{1}{c} - z^{-1})$; $c = 2 - \sqrt{3}$. This polynomial has a number of nontrivial factorizations, many of which are interesting. The Daubechies orthonormal wavelet filters Daub4 choice [1] is as $F_0(z) = (1 + z^{-1})^2(c - z^{-1})$, $H_0(z) = (1 + z^{-1})^2(\frac{1}{c} - z^{-1})$. Note that with the extra factor of $(c - z^{-1})$, this filter will not have any symmetry properties. The explicit lowpass filter coefficients are $H_0 = \frac{1}{4\sqrt{2}}[(1 + \sqrt{3}), (3 + \sqrt{3}), (3 - \sqrt{3}), (1 - \sqrt{3})] \equiv [c(0), c(1), c(2), c(3)]$. The highpass filter is $H_1 = [c(0), -c(1), c(2), -c(3)]$.

Example 3 (Daub6/2)
In Example 2, we could factor $P_0(z)$ differently as $F_0(z) = (1+z^{-1})^3(c - z^{-1})(\frac{1}{c} - z^{-1})$, $H_0(z) = (1+z^{-1})$. These filters are symmetric, and work quite well in practice for compression. It it interesting to note that in practice the short lowpass filter is apparently more useful in the analysis stage than in the synthesis one. This set of filters corresponds not to an orthonormal wavelet bases, but rather to two sets of wavelet bases which are *biorthogonal*—that is, they are dual bases; see chapter 1.

Example 4 (Daub5/3)
In Example 2, we could factor $P_0(z)$ differently again, as $F_0(z) = (1 + z^{-1})^2$, $H_0(z) = (1 + z^{-1})^2(c - z^{-1})(\frac{1}{c} - z^{-1})$. These are odd-length filters which are also symmetric (symmetric odd-length filters are symmetric about a filter tap, rather than in-between filter two taps as in even-length filters). The explicit coefficients are $H_0 = \frac{1}{8}[-1, 2, 6, 2, -1]$, $F_0 = \frac{1}{2}[1, 2, 1]$. These filters also perform very well in experiments. Again, there seems to be a preference in which way they are used (the longer lowpass filter is here preferred in the analysis stage). These filters also correspond to biorthogonal wavelets.

The last two sets of filters were actually discovered prior to Daubechies by Didier LeGall [18], specifically intending their use in compression.

Example 5 (Daub2N)

$$P(z) = 2(\frac{1+z}{2})^N(\frac{1+z^{-1}}{2})^N \sum_{k=0}^{N-1} \binom{N+k-1}{k}(\frac{1-z}{2})^k(\frac{1-z^{-1}}{2})^k.$$

To get the Daubechies orthonormal wavelets of length $2N$, factor this as $P(z) = C(z)C(z^{-1})$. On the unit circle where $|z| = 1$, this equals $|C(z)|^2$, and finding the appropriate square root is called *spectral factorization* (see for example [17, 8]).

Other splittings give various biorthogonal wavelets.

6 Wavelet Packets

While figures 10 and 11 illustrate octave-band or Mallat decompositions of signals, these are by no means the only ones available. As mentioned, since the decomposition is perfectly reconstructing at every stage, in theory (and essentially in practice) we can decompose any part of the signal any number of steps. For example, we could continue to decompose some of the highpass bands, in addition to or instead of decomposing the lowpass bands. A schematic of a general decomposition is depicted in figure 12. Here, the left branch at a bifurcation indicates a lowpass channel, and the right branch indicates a highpass channel. The number of such splittings is only limited by the number of samples in the data and the size of the filters involved—one must have at least twice as many samples as the maximum length of the filters in order to split a node (for the continuous function space setting, there is no limit). Of course, the reconstruction steps must precisely traverse the decomposition in reverse, as in figure 13. This theory has been developed by Coifman, Meyer and Wickerhauser and is thoroughly explained in the excellent volume [12].

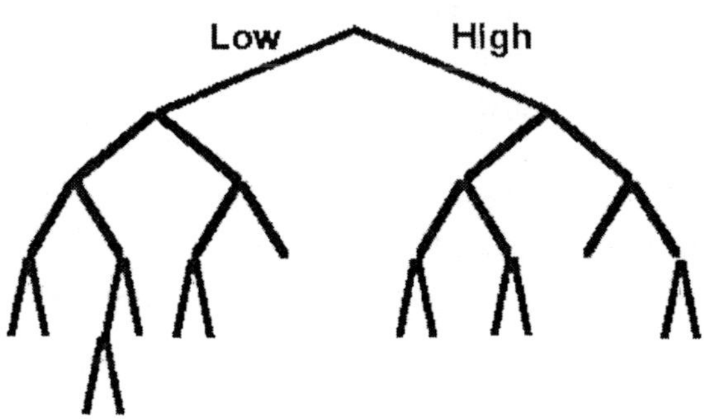

FIGURE 12. A one-dimensional wavelet packet decomposition.

This freedom in decomposing the signal in a large variety of ways corresponds in the function space setting to a large library of interrelated bases, called *wavelet packet* bases. These include the previous wavelet bases as subsets, but show considerable diversity, interpolating in part between the time-frequency analysis afforded by Fourier and Gabor analysis, and time-scale analysis afforded by traditional wavelet bases. Don't like either the Fourier or wavelet bases? Here is a huge collection of bases at one's fingertips, with widely varying characteristics that may altenerately be more appropriate to changing signal content than a fixed basis. What is more is that there exists a fast algorithm for choosing a well-matched basis (in terms of entropy extremization)—the so-called *best basis* method. See [12] for the details. In applications to image compression, it is indeed bourne out that allowing more general decompositions than octave-band permits greater coding efficiencies, at the price of a modest increase in computational complexity.

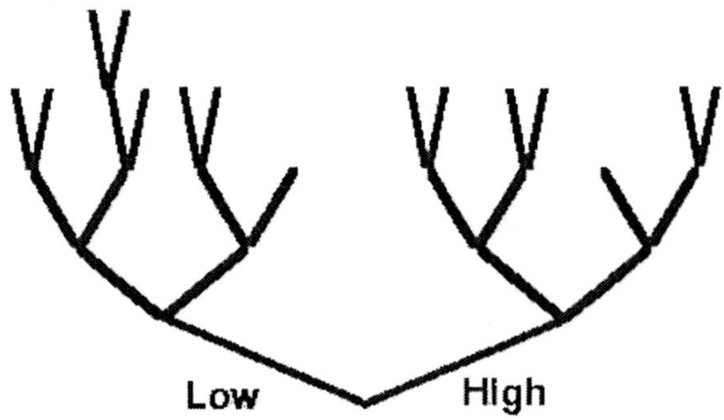

FIGURE 13. A one-dimensional wavelet packet reconstruction, in which the decomposition tree is precisely reversed.

7 References

[1] I. Daubechies, *Ten Lectures on Wavelets,* SIAM, 1992.

[2] A. Jain, *Fundamentals of Digital Image Processing,* Prentice-Hall, 1989.

[3] D. Marr, *Vision,* W. H. Freeman and Co., 1982.

[4] J. Fourier, *Theorie Analytique de la Chaleur*, Gauthiers-Villars, Paris, 1888.

[5] A. Oppenheim and R. Shafer, *Discrete-Time Signal Processing*, Prentice-Hall, 1989.

[6] N. Jayant and P. Noll, *Digital Coding of Waveforms*, Prentice-Hall, 1984.

[7] A. Papoulis, *Signal Analysis*, McGraw-Hill, 1977.

[8] G. Strang and T. Nguyen, *Wavelets and Filter Banks*, Wellesley-Cambridge Press, 1996.

[9] G. Strang, *Linear Algebra and Its Applications*, Harcourt Brace Javonich, 1988.

[10] M. Vetterli and J. Kovacevic, *Wavelets and Subband Coding*, Prentice-Hall, 1995.

[11] C. Brislawn, "Classificaiton of nonexpansive symmetric extension transforms for multirate filter banks," Los Alamos Report LA-UR-94-1747, 1994.

[12] M. Wickerhauser, *Adapted Wavelet Analysis from Theory to Software*, A K Peters, 1994.

[13] W. Rudin, *Functional Analysis*, McGraw Hill, 1972?

[14] G. Folland, *Harmonic Analysis in Phase Space*, Princeton University Press, 1989.

[15] E. Whittaker, "On the Functions which are Represented by the Expansions of Interpolation Theory," Proc. Royal Soc., Edinburgh, Section A 35, pp. 181-194, 1915.

[16] D. Gabor, "Theory of Communication," Journal Inst. Elec. Eng., 93 (III) pp. 429-457, 1946.

[17] P. Vaidyanathan, *Multirate Systems and Fitler Banks*, Prentice-Hall, 1993.

[18] D. LeGall and A. Tabatabai, "Subband Coding of Digital Images Using Symmetric Short Kernel Filters and Arithmetic Coding Techniques," *Proc. ICASSP,* IEEE, pp. 761-765, 1988.

4

Introduction To Compression

Pankaj N. Topiwala

1 Types of Compression

Digital image data compression is the science (or art) of reducing the number of bits needed to represent a given image. The purpose of compression is usually to facilitate the storage and transmission of the data. The reduced representation is typically achieved using a sequence of transformations of the data, which are either exactly or approximately invertible. That is, data compression techniques divide into two categories: lossless (or noiseless) compression, in which an exact reconstruction of the original data is available; and lossy compression, in which only an approximate reconstruction is available. The theory of exact reconstructability is an exact science, filled with mathematical formulas and precise theorems. The world of inexact reconstructability, by contrast, is an exotic blend of science, intuition, and dogma. Nevertheless, while the subject of lossless coding is highly mature, with only incremental improvements achieved in recent times, the yet immature world of lossy coding is witnessing a wholesale revolution.

Both types of compression are in wide use today, and which one is more appropriate depends largely on the application at hand. Lossless coding requires no qualifications on the data, and its applicability is limited only by its modest performance weighted against computational cost. On typical natural images (such as digital photographs), one can expect to achieve about 2:1 compression. Examples of fields which to date have relied primarily or exclusively on lossless compression are medical and space science imaging, motivated by either legal issues or the uniqueness of the data. However, the limitations of lossless coding have forced these and other users to look seriously at lossy compression in recent times. A remarkable example of a domain which deals in life-and-death issues, but which has nevertheless turned to lossy compression by necessity, is the US Justice Department's approach for storing digitized fingerprint images (to be discussed in detail in these pages). Briefly, the moral justification is that although lossy coding of fingerprint images is used, repeated tests have confirmed that the ability of FBI examiners to make fingerprint matches using compressed data is in no way impacted.

Lossy compression can offer one to two orders of magnitude greater compression than lossless compression, depending on the type of data and degree of loss that can be tolerated. Examples of fields which use lossy compression include Internet transmission and browsing, commercial imaging applications such as videoconferencing, and publishing. As a specific example, common video recorders for home use apply lossy compression to ease the storage of the massive video data, but provide reconstruction fidelity acceptable to the typical home user. As a general trend, a subjective signal such as an image, if it is to be subjectively interpreted (e.g., by human observers) can be permitted to undergo a limited amount to loss, as long as the loss is indescernible.

A staggering array of ideas and approaches have been formulated for achieving both lossless and lossy compression, each following distinct philosophies regarding the problem. Perhaps surprisingly, the performance achieved by the leading group of algorithms within each category is quite competitive, indicating that there are fundamental principles at work underlying all algorithms. In fact, all compression techniques rely on the presence of two features in the data to achieve useful reductions: *redundancy* and *irrelevancy* [1]. A trivial example of data redundancy is a binary string of all zeros (or all ones)—it carries no information, and can be coded with essentially one bit (simple differencing reduces both strings to zeros after the initial bit, which then requires only the length of the string to encode exactly). An important example of data irrelevancy occurs in the visualization of grayscale images of high dynamic range, e.g., 12 bits or more. It is an experimental fact that for monochrome images, 6 to 8 bits of dynamic range is the limit of human visual sensitivity; any extra bits do not add perceptual value and can be eliminated. The great variety of compression algorithms mainly differ in their approaches to extracting and exploiting these two features of redundacy and irrelevancy.

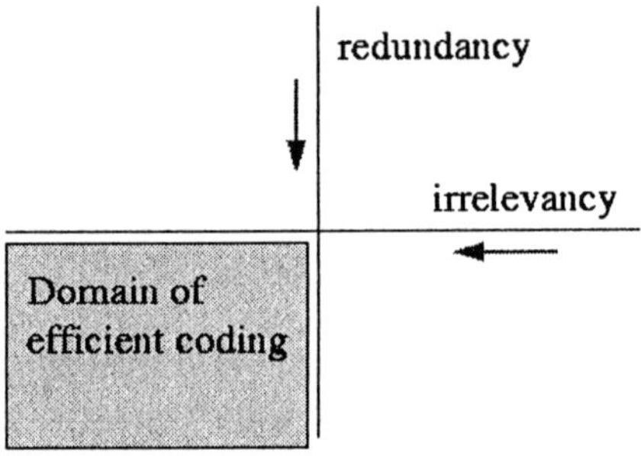

FIGURE 1. The operational space of compression algorithm design [1].

In fact, this point of view permits a fine differentiation between lossless and lossy coding. Lossless coding relies only on the redundancy feature of data, exploiting unequal symbol probabilities and symbol predictability. In a sense that can be made precise, the greater the information density of a string of symbols, the more random it appears and the harder it is to encode losslessly. Information density, lack of redundancy and lossless codability are virtually identical concepts. This subject has a long and venerable history, on which we provide the barest outline below.

Lossy coding relies on an additional feature of data: irrelevancy. Again, in any practical imaging system, if the operational requirements do not need the level of precision (either in spatio-temporal resolution or dynamic range) provided by the data, than the excess precision can be eliminated without a performance loss (e.g., 8-bits of grayscale for images suffice). As we will see later, after applying cleverly chosen transformations, even 8-bit imagery can permit some of the data to be eliminated (or downgraded to lower precision) without significant subjective loss. The art of lossy coding, using wavelet transforms to separate the relevant from the irrelevant, is the main topic of this book.

2 Resume of Lossless Compression

Lossless compression is a branch of Information Theory, a subject launched in the fundamental paper of Claude Shannon, *A mathematical theory of communication* [2]. A modern treatment of the subject can be found, for example, in [3]. Essentially, Shannon modelled a communication channel as a conveyance of endless strings of symbols, each drawn from a finite alphabet. The symbols may occur with probabilities according to some probability law, and the channel may add "noise" in the form of altering some of the symbols, again perhaps according to some law. The effective use of such a communication channel to convey information then poses two basic problems: how to represent the signal content in a minimal way (source coding), and how to effectively protect against channel errors to achieve error-free transmission (channel coding). A key result of Shannon is that these two problems can be treated individually: statistics of the data are not needed to protect against the channel, and statistics of the channel are irrelevant to coding the source [2]. We are here concerned with losslessly coding the source data (image data in particular), and henceforth we restrict attention to source coding.

Suppose the channel symbols for a given source are the letters x_i indexed by the integers $i = 1, ..., N = 2^L$, whose frequency of occurence is given by probabilities p_i. We will say then that each symbol x_i carries an amount of *information* equal to $-\log_2 p_i$ bits, which satisfies $0 \leq -\log_2 p_i$. The intuition is that the more unlikely the symbol, the greater is the information value of its occurance; in fact, $-\log_2 p_i = 0$ exactly when $p_i = 1$ and the symbol is certain (a certain event carries no information). Since each symbol occurs with probability p_i, we may define the information content of the source, or the **entropy**, by the quantity

$$H = -\sum_{i=1}^{N} p_i \log_2 p_i. \tag{4.1}$$

Such a data source can clearly be encoded by $\log_2 N = L$ bits per sample, simply by writing each integer in binary form. Shannon's *noiseless coding theorem* [2] states that such a source can actually be encoded with $H + \epsilon$ bits per sample exactly, for any $\epsilon > 0$. Alternatively, such a source can be encoded with H bits/sample, permitting a reconstruction of arbitrarily small error. Thus, H bits/sample is the raw *information rate* of the source. It can shown that

$$0 \leq H \leq \log_2 N = L.$$

Furthermore, $H = 0$ if and only if all but one of the symbols have zero probability, and the remaining symbol has unit probability; and $H = \log_2 N = L$ if and only if all symbols have equal probability, in this case $\frac{1}{N}$.

This result is typical of what Information Theory can achieve: a statement regarding infinite strings of symbols, the (nonconstructive) proof of which requires use of codes of arbitrarily long length. In reality, we are always encoding finite strings of symbols. Nevertheless, the above entropy measure serves as a benchmark and target rate in this field which can be quite valuable. Lossless coding techniques attempt to reduce a data stream to the level of its entropy, and fall under the rubric of Entropy Coding (for convenience we also include some predictive coding schemes in the discussion below). Naturally, the longer the data streams, and the more consistent the statistics of the data, the greater will be the success of these techniques in approaching the entropy level. We briefly review some of these techniques here, which will be extensively used in the sequel.

2.1 DPCM

Differential Pulse Coded Modulation (DPCM) stands for coding a sample in a time series $x[n]$ by predicting it using linear interpolation from neighboring (generally past) samples, and coding the prediction error:

$$x[n] - \sum_{i=1}^{i=k} a[i]x[n-i]. \tag{4.2}$$

Here, the $a[i]$ are the constant prediction coefficients, which can be derived, for example, by regression analysis. When the data is in fact correlated, such prediction methods tend to reduce the min-to-max (or *dynamic*) range of the data, which can then be coded with fewer bits on average. When the prediction coefficients are allowed to adapt to the data by learning, this is called *Adaptive* DPCM (ADPCM).

2.2 Huffman Coding

A simple example can quickly illustrate the basic concepts involved in Huffman Coding. Suppose we are given a message composed using an alphabet made of only four symbols, $\{A, B, C.D\}$, which occur with probabilities $\{0.3, 0.4, 0.2, 0.1\}$, respectively. Now a four-symbol alphabet can clearly be coded with 2 bits/sample (namely, by the four symbols $\{A, B, C, D\} = \{00, 01, 10, 11\}$). A simple calculation shows that the entropy of this symbol set is $H = 1.85$ bits/sample, so somehow there must be a way to shrink this representation. How? While Shannon's arguments are non-constructive, a simple and ingenious tree-based algorithm was devised by Huffman in [5] to develop an efficient binary representation. It builds a binary tree from the bottom up, and works as follows (also see figure 2).

The Huffman Coding Algorithm:

1. Place each symbol at a separate leaf node of a tree - to be grown - together with its probability.
2. While there is more than one disjoint node, iterate the following: Merge the two nodes with the smallest probabilities into one node, with the two original nodes as children. Label the parent node by the union of the children nodes, and assign its probability as the sum of the probabilities of the children nodes. Arbitrarily label the two children nodes by 0 and 1.

3. When there is a single (root) node which has all original leaf nodes as its children, the tree is complete. Simply read the new binary symbol for each leaf node sequentially down from the root node (which itself is not assigned a binary symbol). The new symbols are distinct by construction.

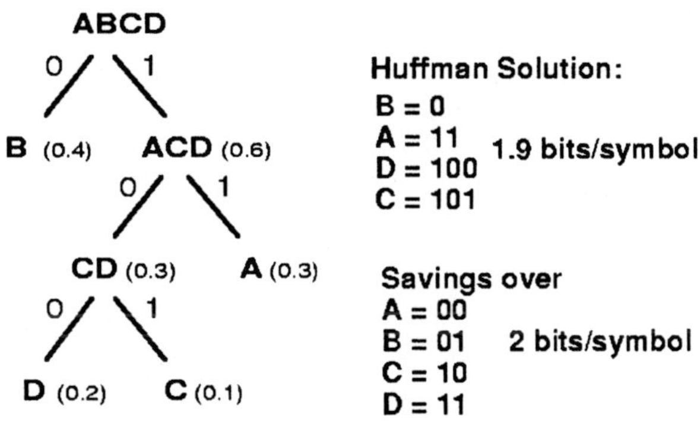

FIGURE 2. Example of a Huffman Tree For A 4-Symbol Vocabulary.

	A	B	C	D
Prob.	0.3	0.4	0.1	0.2
Symbol	11	0	101	100

TABLE 4.1. Huffman Table for a 4-symbol alphabet. Average bitrate = 1.9 bits/sample.

This elementary example serves to illustrate the type and degree of bitrate savings that are available through entropy coding techniques. The 4-symbol vocabulary can in fact be coded, on average, at 1.9 bits/sample, rather than 2 bits/sample. On the other hand, since the entropy estimate is 1.85 bits/sample, we fall short of achieving the "true" entropy. A further distinction is that if the Huffman coding rate is computed from long-term statistical estimates, the actual coding rate achieved with a given finite string may differ from both the entropy rate and the expected Huffman coding rate. As a specific instance, consider the symbol string

ABABABDCABACBBCA, which is coded as

110110110100101110111010010111.

For this specific string, 16 symbols are encoded with 30 bits, achieving 1.875 bits/sample.

One observation about the Huffman coding algorithm is that for its implementation, the algorithm requires prior knowledge of the probabilities of occurence of the various symbols. In practice, this information is not usually available *a priori* and must be inferred from the data. Inferring the symbol distributions is therefore the first task of entropy coders. This can be accomplished by an initial pass through the data to collect the statistics, or computed "on the fly" in a single pass, when it is called *adaptive* Huffman coding. The adaptive approach saves time, but comes at a cost of further bitrate inefficiencies in approaching the entropy. Of course, the longer the symbol string and the more stable the statistics, the more efficient will be the entropy coder (adaptive or not) in approaching the entropy rate.

2.3 Arithmetic Coding

Note that Huffman coding is limited to assigning an integer number of bits to each symbol (e. g., in the above example, $\{A, B, C, D\}$ were assigned $\{1, 2, 3, 3\}$ bits, respectively). From the expression for the entropy, it can be shown that in the case when the symbol probabilities are exactly reciprocal powers of two ($p_i = 2^{-k_i}$), this mechanism is perfectly rate-efficient: the i-th symbol is assigned k_i bits. In all other cases, this system is inefficient, and fails to achieve the entropy rate. However, it may not be obvious how to assign a noninteger number of bits to a symbol. Indeed, one cannot assign noninteger bits if one works one symbol at a time; however, by encoding long strings simultaneously, the coding rate per symbol can be made arbitrarily fine. In essence, if we can group symbols into substrings which themselves have densities approaching the ideal, we can achieve finer packing using Huffman coding for the strings.

This kind of idea can be taken much further, but in a totally different context. One can in fact assign real-valued symbol bitrates and achieve a compact encoding of the entire message directly, using what is called *arithmetic* coding. The trick: use real numbers to represent strings! Specifically, subintervals of the unit interval $[0, 1]$ are used to represent symbols of the code. The first letter of a string is encoded by chosing a corresponding subinterval of $[0, 1]$. The length of the subinterval is chosen to equal its expected probability of occurence; see figure 3. Successive symbols are encoded by expanding the selected subinterval to a unit interval, and chosing a corresponding subinterval. At the end of the string, a single (suitable) element of the designated subinterval is sent as the code, along with an end-of-message symbol. This scheme can achieve virtually the entropy rate because it is freed from assigning integer bits to each symbol.

2.4 Run-Length Coding

Since the above techniques of Huffman and arithmetic coding are general purpose entropy coding techniques, they can by themselves only achieve modest compression gains on natural images–typically between 2:1 and 3:1 compression. Significant compression, it turns out, can be obtained only after cleverly modifying the data, e.g., through transformations, both linear and nonlinear, in a way that tremendously increases the value of lossless coding techniques. When the result of the intermediate transforms is to produce long runs of a single symbol (e.g., '0'), a simple preprocessor prior to the entropy coding proves to be extremely valuable, and is in fact the key to high compression.

In data that has long runs of a single symbol, for example '0', interspersed with clusters of other ('nonzero') symbols, it is very useful to devise a method of simply coding the runs of '0' by a symbol representing the length of the run (e.g. a run of

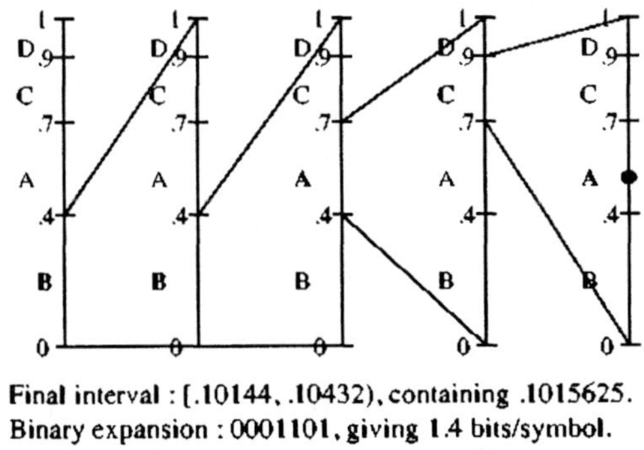

FIGURE 3. Example of arithmetic coding for the 4-ary string BBACA.

one thousand 8-bit integer 0's can be substituted by a symbol for 1000, which has a much smaller representation). This simple method is called Run-Length Coding, and is at the heart of the high compression ratios reported throughout most the book. While this technique, of course, is well understood, the real trick is to use advanced ideas to create the longest and most exploitable strings of zeros! Such strings of zeros, if they were created in the original data, would represent direct loss of data by nulling. However, the great value of transforms is in converting the data to a representation in which most of the data now has very little energy (or information), and can be practicably nulled.

3 Quantization

In mathematical terms, *quantization* is a mapping from a continuously-parameterized set V to a discrete set X, $Q : V \to X$. *Dequantization* is a mapping in the reverse direction, $D : X \to V$. The combination of the two mappings, $DQ : V \to V$, maps the continuous space V into a discrete subset of itself, which is therefore nonlinear, lossy, and irreversible. The elements of the continuously parameterized set can, for example, be either single real numbers (scalars), in which case one is representing an arbitrary real number with an integer; or points in a vector space (vectors), in which case one is representing an arbitrary vector by one of a fixed discrete set of vectors. Of course, the discretization of vectors (e.g., *vector quantization*) is a more general process, and includes the discretization of scalars (*scalar quantization*) as a special case, since scalars are vectors of dimension 1. In practice, there are usually limits to the size of the scalars or vectors that need to be quantized, so that one is typically working in a finite volume of a Euclidean vector space, with a finite number of representative scalars or vectors.

In theory, quantization can refer to discretizing points on any continuous parameter space, not necessarily a region of a vector space. For example, one may want to represent points on a sphere (e.g., a model of the globe) by a discrete set of points (e.g., a fixed set of (lattitude, longtitude) coordinates). However, in practice, the above cases of scalar and vector quantization cover virtually all useful applications, and will suffice for our work.

In computer science, quantization amounts to reducing a floating point representation to a fixed point representation. The purpose of quantization is to reduce the number of bits for improved storage and transmission. Since a discrete representation of what is originally a continuous variable will be inexact, one is naturally led to measure the degree of loss in the representation. A common measure is the mean-squared error in representing the original real-valued data by the quantized quantities, which is just the L_2 size (squared) of the error signal:

$$MSE = \frac{1}{N}||v - Q(v)||_2^2.$$

Given a quantitive measure of loss in the discretization, one can then formulate approaches to minimize the loss for the given measure, subject to constraints (e.g., the number of quantization bins permitted, or the type of quantization). Below we present a brief overview of quantization, with illustrative examples and key results. An excellent reference on this material is [1].

3.1 Scalar Quantization

A scalar quantizer is simply a mapping $Q : \mathbf{R} \to \mathbf{Z}$. The purpose of quantization is to permit a finite precision representation of data. As a simple example of a scalar quantizer, consider the map that takes each interval $[n - \frac{1}{2}, n + \frac{1}{2})$ to n for each integer n, as suggested in figure 4. In this example, the *decision boundaries* are the half-integer points $n + 1/2$, the *quantization bins* are the intervals between the decision boundaries, $[n - \frac{1}{2}, n + \frac{1}{2})$, and the integers n are the *quantization codes*, that is, the discrete set of reconstruction values to which the real numbers are mapped. It is easy to see that this mapping is lossy and irreversible since after quantization, each interval $[n - \frac{1}{2}, n + \frac{1}{2})$ is mapped to a single point, n (in particular, the interval $[-\frac{1}{2}, \frac{1}{2})$ is mapped entirely to 0). Note also that the mapping Q is nonlinear: for example, $Q(0.4 + 0.4) = Q(0.8) = 1.0 \neq 0.0 = Q(0.4) + Q(0.4)$.

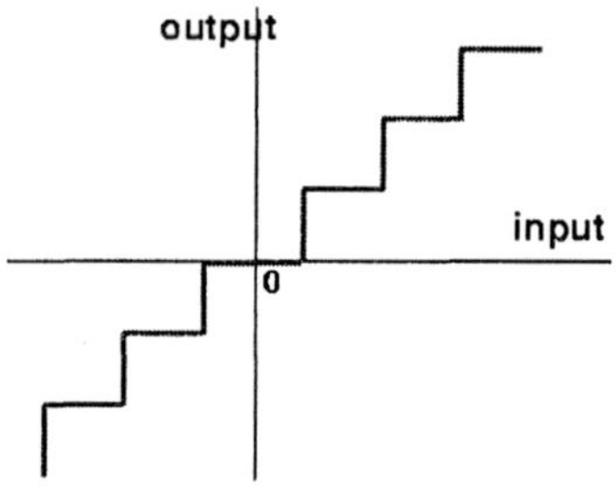

FIGURE 4. A simple example of a quantizer with uniform bin sizes. Often the zero bin is enlarged to permit more efficient entropy coding.

This particular example has two characteristics that make it especially simple. First, all the bins are of equal size, of width 1, making this a *uniform quantizer*. In practice, the bins need not all be of equal width, and in fact, nonuniform quantizers are frequently used. For example, in speech coding, one may use a quantizer whose bin sizes grow exponentially in width (called a logarithmic quantizer), on account of the apparently reduced sensitivity of listeners to speech amplitude differences at high amplitudes [1]. Second, upon *dequantization*, the integer n can be represented as itself. This fact is clearly an anomaly of the fact that our binsizes are integer and in fact all of width 1. If the binsizes were $\frac{1}{2}$, for example, then $Q(2.20) = 4$ and $D(4) = 2.0$.

As mentioned, the lossy nature of quantization means that one has to deal with the type and level of loss, especially in relation to any operational requirements arising from the target applications. A key result in an MSE context is the construction of the optimal (e.g., minimum MSE) **Lloyd-Max** quantizers. For a given probability density function (pdf) $p(x)$ for the data they compute the decision boundaries through a set of iterative formulas.

The Lloyd-Max Algorithm:

1. To construct an optimized quantizer with L bins, initialize reconstruction values by $y_i, i = 1, \ldots, L$, and set the decision points by $x_0 = -\infty, x_{L+1} = \infty, x_i = (y_i + y_{i+1})/2, i = 1, \ldots, L$.
2. Iterate the following formulas until convergence or until error is within tolerance:

$$y_i = \frac{\int_{x_{i-1}}^{x_i} xp(x)dx}{\int_{x_{i-1}}^{x_i} p(x)dx};$$

$$x_i = (y_i + y_{i+1})/2; \quad i = 1, \ldots, L.$$

These formulas say that the reconstruction values y_i should be iteratively chosen to be the centroids of the pdfs in the quantization bin $[x_{i-1}x_i]$. A priori, this algorithm converges to a local minimum for the distortion. However, it can be shown that if the pdf $p(x)$ satisfies a modest condition (that it is log-concave: $d^2p(x)/dx^2 < 0$), satisfied by Gaussian and Laplace distributions), then this algorithm converges to a global minimum. If we assume the pdf varies slowly and the quantization binsizes are small (what is called the *high bitrate approximation*), one could make the simplifying assumption that the pdf is actually constant over each bin ; in this case, the centroid is just half-way between the decision points, $y_i = (x_{i-1} + x_i)/2$. While this is only approximately true, we can use it to estimate the error variance in using a Lloyd-Max quantizer, as follows. Let p_k be the probability of being in the interval $[x_{i-1}, x_i]$ of length Δ_k, so that $p(x) = p_k/\Delta_k$ on that interval. Then

$$\sigma_Q^2 = \langle (x - Q(x))^2 \rangle = \sum_{k=1}^{k=L} \int_{x_{k-1}}^{x_k} (x - y_k)^2 p_k dx = \frac{1}{12} \sum_{k=1}^{k=L} \Delta_k^2 p_k.$$

A particularly simple case is that of a uniform quantizer, in which all quantization intervals are the same, $\Delta_k = \Delta$, in which case we have that the error variance is $\sigma_Q^2 = \Delta^2/12$, which depends only on the binsize. For a more general pdf, it can be shown that the error variance is approximately

$$\sigma_Q^2 = \langle (x - Q(x))^2 \rangle = \frac{1}{12L^2} \left\{ \int_{x_0}^{x_{L+1}} p(x)^{1/3} dx \right\}^3.$$

Uniform quantization is widely used in image compression, since it is efficient to implement. In fact, while the Lloyd-Max quantizer minimizes distortion if one considers quantization only, it can be shown that uniform quantization is (essentially) optimal if one includes entropy coding; see [1, 4].

In sum, scalar quantization is just a partition of the real line into disjoint cells C_i indexed by the integers, $\mathbf{R} = \bigcup C_i$. This setup includes the case of quantizing a subinterval I of the line, by considering the complement of the interval as a cell: $C_i = I^c$, for one i. The mapping $Q : \mathbf{R} \to \mathbf{Z}$ then is simply the map taking an element $x \in \mathbf{R}$ to the integer i such that $x \in C_i$. This point of view is convenient for generalizing to vector quantization.

3.2 Vector Quantization

Vector quantization is the process of discretizing a vector space, by partitioning it into cells, $\mathbf{V} = \bigcup C_i$ (again, the complement of a finite volume can be one of the cells), and selecting a representative from each cell. For optimality considerations here one must deal with multidimensional probability density functions. An analogue of the Lloyd-Max quantizer can be constructed, to minimize mean squared-error distortion. However, in practice, instead of working with hypothetical multidimensional pdfs, we will assume instead that we have a set of training vectors, which we will denote by $t_j, j = 1, \ldots, J$, which serve to define the underlying distribution; our task is to select a *codebook* of L vectors $y_i, i = 1, \ldots, L$ (called *codewords*) which satisfy

$$\min_{\{y_i\}_1^L \in V} \sum_j \|t_j - Q(t_j)\|^2. \tag{4.3}$$

The Generalized Lloyd-Max Algorithm:

1. Initialize reconstruction vectors $y_i, i = 1, \ldots, L$ arbitrarily.
2. Partition the set of training vectors t_j into sets S_i for which the vector y_i is the closest:
   $t_j \in S_i \iff d(y_i, t_j) \leq d(y_k, t_j)$, all k.
3. Redefine y_i to be the centroid of S_i:
   $y_i = \frac{1}{\#S_i} \sum_{x_j \in S_i} x_j$.
4. Repeat steps 2 and 3 until convergence.

In this generality, the algorithm may not converge to a global minimum for an arbitrary initial codebook (although it does converge to a local minimum) [1]; in practice this method can be quite effective, especially if the initial codebook is well-chosen.

4 Summary of Rate-Distortion Theory

Earlier we saw how the entropy of a given source—a single number—encapsulates the bitrate necessary to encode the source exactly, albeit in a limiting sense of infinitely long sequences. Suppose now that we are content with approximate coding; is there a theory of (minimal) distortion as a function of rate? It is common sense that as we allocate more bitrate, the minimum achievable distortion possible should decrease, so that there should be an inverse relationship between distortion and rate. Here, the concept of distortion must be made concrete; in the engineering literature, this notion is the standard mean squared error metric:

$$D = MSE = \frac{1}{N}||x - \tilde{x}||_2^2 = \frac{1}{N}\sum_{i=0}^{N-1} |x_i - \tilde{x}_i|^2.$$

It can be shown that for a continuous-amplitude source, the rate needed to achieve distortion-free representation actually diverges as $N \to \infty$ [1]. The intuition is that we can never approach a continuous amplitude waveform by finite precision representations. In our applications, our original source is of finite precision (e.g., discrete-amplitude) to begin with; we simply wish to represent it with fewer total bits. If we normalize the distortion as D/σ_x^2, then $0 \leq D/\sigma_x^2 \leq C$ for some finite constant C. Here, $D(0) = C\sigma_x^2$, where the constant C depends on the probability distribution of the source. If the original rate of the signal is R_0, then $D(R_0) = 0$. In between, we expect that the distortion function is monotonically decreasing, as depicted in figure 5.

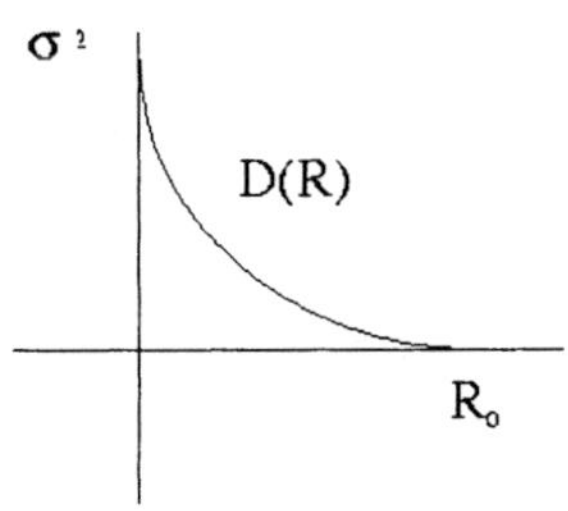

FIGURE 5. A generic distortion-rate curve for a finite-precision discrete signal.

If the data were continuous amplitude, the curve would not intersect the rate axis but approach it asymptotically. In fact, since we often transform our finite-precision data using irrational matrix coefficients, the transform coefficients may be continuous-amplitude anyway. The relevant issue is to understand the rate-distortion curve in the intermediate regime between $R = 0$ and $R = R_0$. There are approximate formulas in this regime which we indicate. As discussed in section 3.1, if there are L quantization bins $[x_{i-1},\ x_i]$ and output values y_i, then distortion is given by

$$D = \sigma_Q^2 = \sum_{i=1}^{L} \int_{x_{i-1}}^{x_i} (x - y_i)^2 p_x(x)dx,$$

where $p_x(x)$ is the probability density function (pdf) of the variable x. If the quantization stepsizes are small, and the pdf is slowly varying, one can make the approximation that it is constant over the individual bins. That is, we can assume that the variable x is uniformly distributed within each quantization bin: this is called the *high rate approximation*, which becomes inaccurate at low bitrates, though see [9]. The upshot is that we can estimate the distortion as

$$D(R) = C\sigma_x^2 2^{-2R},$$

where the constant C depends on the pdf of the variable x. For a zero-mean Gaussian variable, $C = \sqrt{3}\pi/2$; in general, $1 \leq C \leq 10$.

In particular, these theoretical distortion curves are convex; in practice, with only finite integer bitrates possible, the convexity property may not always hold. In any case, the slopes (properly interpreted) are negative, $\frac{dD(R)}{dR} < 0$. More importantly, we wish to study the case when we have multiple quantizers (e.g., pixels) x_i to which to assign bits, and are given a total bit budget of R, how should we allocate bits R_i such that $R = \sum_i R_i$ and so as to minimize total distortion? For simplicity, consider the case of only two variables (or pixels), x_1, x_2. Assume we initially allocate R_i bits to x_i, $i = 1, 2$ so that we satisfy the bit budget $R = R_1 + R_2$. Now suppose that the slopes of the distortion curves are not equal, $\frac{dD_1(R_1)}{dR} \neq \frac{dD_2(R_2)}{dR}$, with say the distortion D_1 falling off more steeply then D_2 : $\frac{dD_1(R_1)}{dR} < \frac{dD_2(R_2)}{dR}$. In this case we can incrementally steal some bitrate from x_2 to give to x_1 (e.g., set $R_1 \to R_1 + \epsilon$, $R_2 \to R_2 - \epsilon$). While the total bit budget would continue to be satisfied, we would realize a reduction in total distortion $D = D_1 + D_2$. This argument shows that the equilibrium point (optimality) is achieved when, as illustrated in figure 6, we have

$$\frac{dD_1(R_1)}{dR} = \frac{dD_2(R_2)}{dR}.$$

This argument holds for any arbitrary number of quantizers, and is treated more precisely in [10].

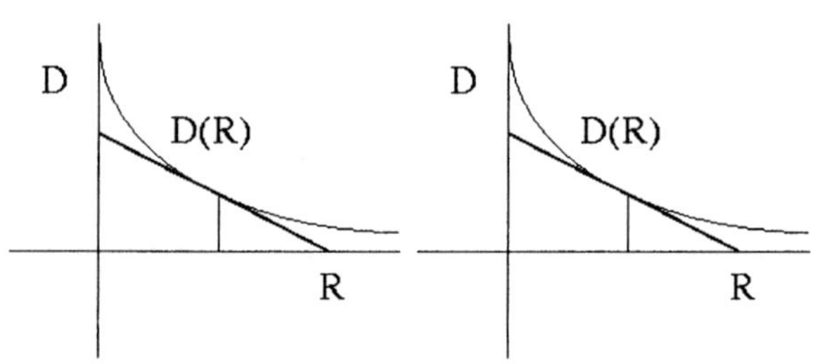

FIGURE 6. Principle of Equal Marginal Distortion: A bitrate allocation problem between competing quantizers is optimally solved for a given bit budget if the marginal change in distortion is the same for all quantizers.

5 Approaches to Lossy Compression

With the above definitions, concepts, and tools, we can finally begin the investigation of lossy compression algorithms. There are a great variety of such algorithms, even for the class of image signals; we restrict attention to VQ and transform coders (these are essentially the only ones that arise in the rest of the book).

5.1 VQ

Conceptually, after scalar quantization, VQ is the simplest approach to reducing the size of data. A scalar quantizer may take a string of quantized values and requantize them to lower bits (perhaps by some nontrivial quantization rule), thus saving bits. A vector quantizer can take blocks of data, for example, and assign codewords to each block. Since the codewords themselves can be available to the decoder, all that really needs to be sent is the index of the codewords, which can be a very sparse representation. These ideas originated in [2], and have been effectively developed since.

In practice, vector quantizers demonstrate very good performance, and they are extremely fast in the decoding stage. However their main drawback is the extensive training required to get good codebooks, as well as the iterative codeword search needed to code data—encoding can be painfully slow. In applications where encoding time is not important but fast decoding is critical (e.g., CD applications), VQ has proven to be very effective. As for encoding, techniques such as hierarchical coding can make tremendous breakthroughs in the encoding time with a modest loss in performance. However, the advent of high-performance transform coders makes VQ techniques of secondary interest—no international standard in image compression currently depends on VQ.

5.2 Transform Image Coding Paradigm

The structure of a transform image coder is given in figure 7, and consists of three blocks: transform, quantizer, and encoder. The transform decorrelates and compactifies the data, the quantizer allocates bit precisions to the transform coefficients according to the bit budget and implied relative importance of the transform coefficients, while the encoder converts the quantized data into a symbols consisting of 0's and 1's in a way that takes advantage of differences in the probability of occurence of the possible quantized values. Again, since we are particularly interested in high compression applications in which 0 will be the most common quantization value, run-length coding is typically applied prior to encoding to shrink the quantized data immediately.

5.3 JPEG

The lossy JPEG still image compression standard is described fully in the excellent monograph [7], written by key participants in the standards effort. It is a direct instance of the transform-coding paradigm outlined above, where the transform is the 8×8 discrete cosine transform (DCT). The motivation for using this transform

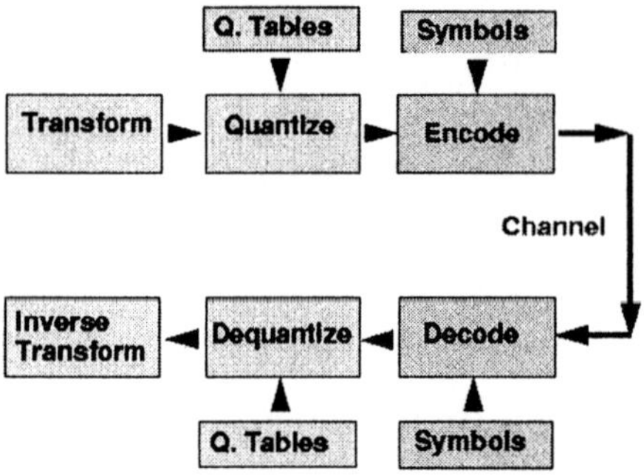

FIGURE 7. Structure of a transform image codec system.

is that, to a first approximation, we may consider the image data to be stationary. It can be shown that for the simplest model of a stationary source, the so-called autoregressive model of order 1, $AR(1)$, the best decomposition in terms of achieving the greatest decorrelation (the so-called *Karhunen-Loeve* decomposition) is the DCT [1]. There are several versions of the DCT, but the most common one for image coding, given here for 1-D signals, is

$$X(n) = \alpha(n) \sum_{k=0}^{N-1} x(k) cos(\frac{\pi(2k+1)n}{2N}),$$

$$\alpha(0) = \sqrt{\frac{1}{N}}, \quad \alpha(n) = \sqrt{\frac{2}{N}}, \; n > 0.$$

This 1-D transform is just matrix multiplication, $X = Cx$, by the matrix

$$C(n,k) = \sqrt{\frac{1}{N}}, k = 0,$$

$$= \sqrt{\frac{2}{N}} cos(\frac{\pi(2k+1)n}{2N}), k > 0.$$

The 2-D transform of a digital image $x(k,l)$ is given by $X = C\ x\ C^T$, or explicitly

$$X(m,n) = \alpha(m)\alpha(n) \sum_{k,l=0}^{N-1} x(k,l) cos(\frac{\pi(2k+1)m}{2N}) cos(\frac{\pi(2l+1)n}{2N}).$$

For the JPEG standard, the image is divided into blocks of 8×8 pixels, on which 8×8 DCTs are used. ($N = 8$ above; if the image dimensions do not divide by 8, the image may be zero-padded to the next multiple of 8.) Note that according to the formulas above, the computation of the 2-D 8×8 DCT would appear to require 64 multiplies and 63 adds per output pixel, all in floating point. Amazingly, a series of clever approaches to computing this transform have culminated in a remarkable factorization by Feig [7] that achieves this same computation in less than 1 multiply and 9 adds per pixel, all as integer ops. This makes the DCT extremely competitive in terms of complexity.

The quantization of each block of 8×8 transform coefficients is performed using uniform quantization stepsizes; however, the stepsize depends on the position of the coefficient in the 8×8 block. In the baseline model, the stepsizes are multiples of the ones in the following quantization table:

16	11	10	16	24	40	51	61
12	12	14	19	26	58	60	55
14	13	16	24	40	57	69	56
14	17	22	29	51	87	80	62
18	22	37	56	68	109	103	77
24	35	55	64	81	104	113	92
49	64	78	87	103	121	120	101
72	92	95	98	112	100	103	99

TABLE 4.2. Baseline quantization table for the 8×8 DCT transform coefficients in the JPEG standard [7]. An alternative table can also be used, but must be sent along in the bitstream.

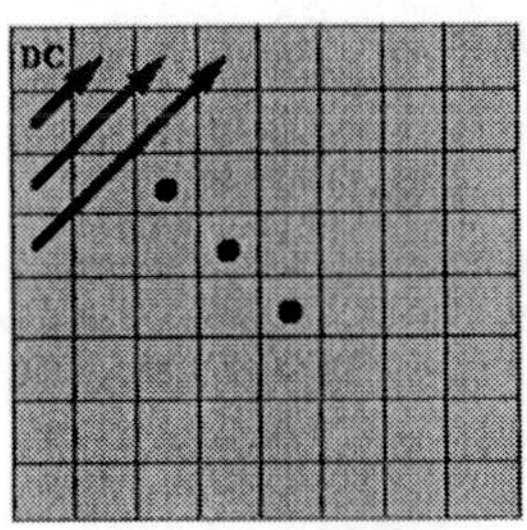

FIGURE 8. Scanning order for 8 x 8 DCT coefficients used in the JPEG Standard; alternate diagonal lines could also be reverse ordered.

The quantized DCT coefficients are scanned in a specific diagonal order, starting with the "DC" or 0-frequency component (see figure 8), then run-length coded, and finally entropy coded according to either Huffman or arithmetic coding [7]. In our discussions of transform image coding techniques, the encoding stage is often very similar to that used in JPEG.

5.4 Pyramid

An early pyramid-based image coder that in many ways launched the wavelet coding revolution was the one developed by Burt and Adelson [8] in the early 80s. This method was motivated largely by computer graphics considerations, and not directly based on the ideas of unitary change of bases developed earlier. Their basic approach is to view an image via a series of lowpass approximations which successively halve the size in each dimension at each step, together with approximate expansions. The expansion filters utilized need not invert the lowpass filter exactly.

The coding approach is to code the downsampled lowpass as well as the difference between the original and the approximate expanded reconstruction. Since the first lowpass signal can itself be further decomposed, this procedure can be carried out for a number of levels, giving successively coarser approximations of the original. Being unconcerned with exact reconstructions and unitary transforms, the lowpass and expansion filters can be extremely simple in this approach. For example, one can use the 5-tap filters $h0 = (-a, .25, .5 + 2a, .25, -a)$, where $a = 0$ or $a = .05$ are popular. The expansion filter can be the same filter, acting on upsampled data! Since $(.25, .5, .25)$ can be implemented by simple bit-shifts and adds, this is extremely efficient. The down side is that one must code the residuals, which begin at the full resolution; the total number of coefficients coded, in dimension 2, is thus approximately 4/3 times the original, leading to some inefficiencies in image coding. In dimension N the oversampling is roughly by $\frac{2^N}{2^N-1}$, and becomes less important in higher dimensions.

The pyramid of image residuals, plus the final lowpass, are typically quantized by a uniform quantizer, and run-length and entropy coded, as before.

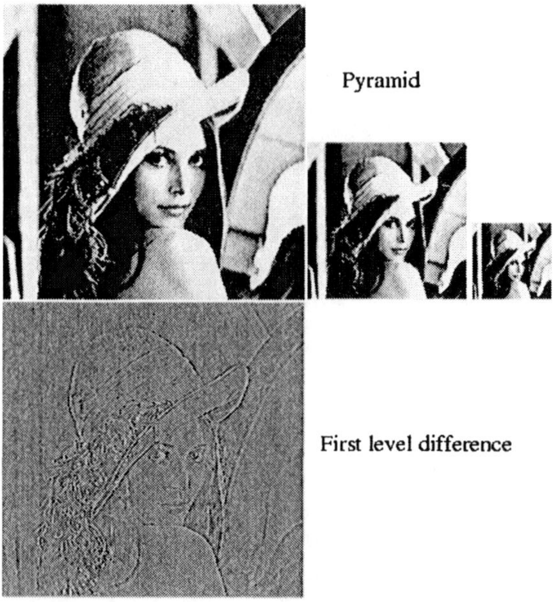

FIGURE 9. A Laplacian pyramid, and first level reconstruction difference image, as used in a pyramidal coding scheme.

5.5 *Wavelets*

The wavelet, or more generally, subband approach differs from the pyramid approach in that it strives to achieve a unitary transformation of the data, preserving the sampling rate (e.g., two-channel perfect reconstruction filter banks). Acting in two dimensions, this leads to a division of an image into quarters at each stage. Besides preserving "critical" sampling, this whole approach leads to some dramatic simplifications in the image statistics, in that all highpass bands (all but the residual lowpass band) have statistics that are essentially equivalent; see figure 10. The highpass bands are all zero-mean, and can be modelled as being Laplace distributed, $p(x) = \frac{1}{\sqrt{2}\sigma} exp(\sqrt{2}|x|/\sigma)$, or more generally as Generalized Gaussian distributed,

$$p(x) = \left\{ \frac{\nu\eta(\nu,\sigma)}{2\Gamma(1/\nu)} \right\} e^{-[\eta(\nu,\sigma)|x|]^{\nu}},$$

where

$$\eta(\nu,\sigma) = \frac{1}{\sigma} \left\{ \frac{\Gamma(3/\nu)}{\Gamma(1/\nu)} \right\}^{1/2}.$$

Here, the parameter ν is a model parameter in the family: $\nu = 1.0$ gives the Laplace density, and $\nu = 2.0$ gives the Gaussian density, $\Gamma(x)$ is the well-known gamma function, and σ is the standard deviation.

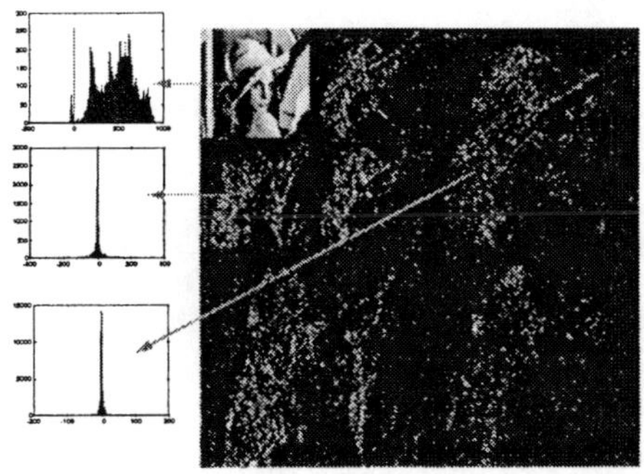

FIGURE 10. A two-level subband decomposition of the Lena image, with histograms of the subbands.

The wavelet transform coefficients are then quantized, run-length coded, and finally entropy coded, as before. The quantization can be done by a standard uniform quantizer, or perhaps one with a slightly enlarged dead-zone at the zero symbol; this results in more zero symbols being generated, leading to better compression with limited increase in distortion. There can also be some further ingenuity in quantizing the coefficients. Note, for instance, that there is considerable correlation across the subbands, depending on the position within the subband. Advanced approaches which leverage that structure provide better compression, although at the cost of complexity. The encoding can procede as before, using either Huffman or arithmetic coding. An example of wavelet compression at 32:1 ratio of an original 512x512 grayscale image "Jay" is given in figure 11, as well as a detailed comparison of JPEG vs. wavelet coding, again at 32:1, of the same image in figure 12.

Here we present these images mainly as a qualitative testament to the coding advantages of using wavelets. The chapters that follow provide detailed analysis of coding techniques, as well as figures of merit such as peak signal-to-noise. We discuss image quality metrics next.

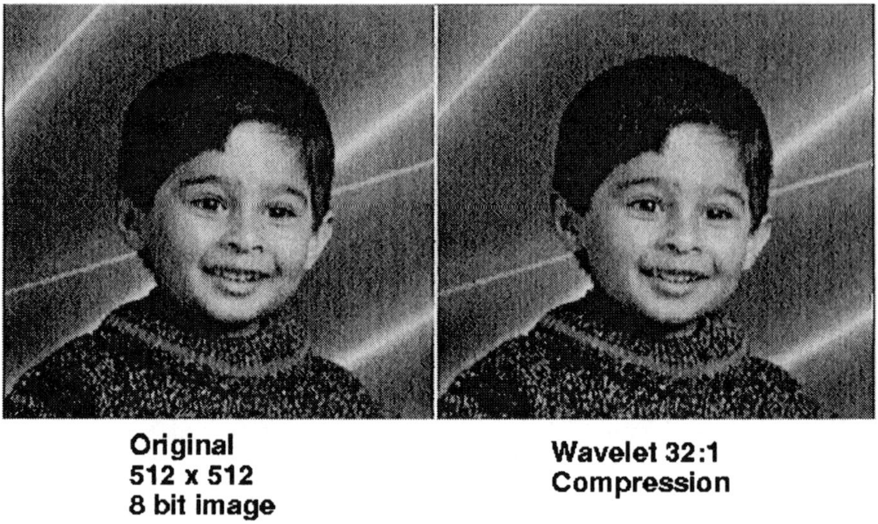

FIGURE 11. A side-by-side comparison of (a) an original 512x512 grayscale image "Jay" with (b) a wavelet compressed reconstruction at 32:1 compression (0.25 bits/pixel).

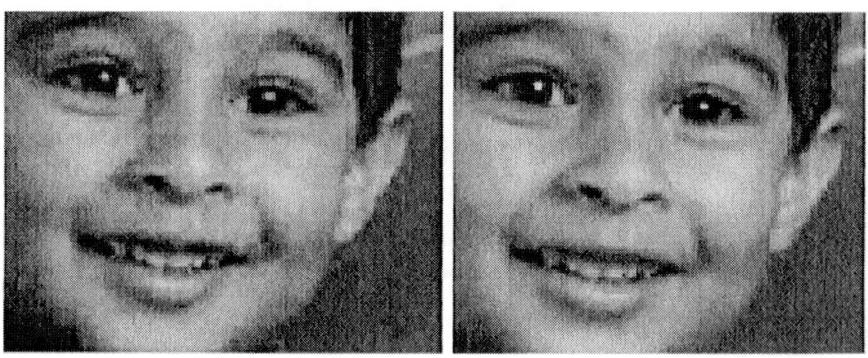

FIGURE 12. A side-by-side detailed comparison of JPEG vs. wavelet coding at 32:1 (0.25 bits/pixel).

6 Image Quality Metrics

In comparing an original signal x with a inexact reconstruction $\tilde{x} = Q(x)$, the most common error measure in the engineering literature is the mean-squared error (MSE):

$$MSE = \frac{1}{N}||x - \tilde{x}||_2^2 = \frac{1}{N}\sum_{i=0}^{N-1}|x_i - \tilde{x}_i|^2.$$

In image processing applications such as compression, we often use a logarithmic metric based on this quantity called the *peak signal-to-noise ratio* (PSNR) which, for 8-bit image data for example, is given by

$$PSNR = 10\ log_{10}\ \frac{255^2}{MSE}.$$

This error metric is commonly used for three reasons: (1) a first order analysis in which the data samples are treated as independent events suggests using a sum of squares error measure; (2) it is easy to compute, and leads to optimization approaches that are often tractable, unlike other candidate metrics (e.g., based on other L_p norms, for $p \neq 2$.); and (3) it has a reasonable though imperfect correspondence with image degradations as interpreted subjectively by human observers (e.g., image analysts), or by machine interpreters (e.g., for computer pattern recognition algorithms). While no definitive formula exists today to quantify reconstruction error as defined by human observers, it is to be hoped that future investigations will bring us closer to that point.

6.1 L_p Metrics

Besides the usual mean-squared error metric, one can consider various other weightings derived from the L_p norms discussed previously. While closed-form solutions are sometimes possible for minimizing mean-squared error, they are virtually impossible to obtain for other normalized L_p metrics, for $p \neq 2$:

$$\mathcal{E}_p = \frac{1}{N}||x - \tilde{x}||_p^p = \frac{1}{N}\sum_{i=0}^{N-1} |x_i - \tilde{x}_i|^p.$$

However, the advent of fast computer search algorithms makes other $p-$norms perfectly viable, especially if they correlate more precisely with subjective analysis. Two of these $p-$norms are often useful, e.g., $p = 1$ which corresponds to the sum of the error magnitudes, or $p = \infty$ which corresponds to the maximum absolute difference.

6.2 Human Visual System Metrics

A human observer, in viewing images, does not compute any of the above error measures, so that their use in image coding for visual interpretation is inherently flawed. The problem is, what formula should we use? To gain a deeper understanding, recall again that our eyes and minds are interpreting visual information in a highly context-dependent way, of which we know the barest outline. We interpret visual quality by how well we can recognize familiar patterns and understand the *meaning* of the sensory data. While this much is known, any mathematically-defined space of patterns can quickly become unmanageable with the slightest structure. To make progress, we have to severely limit the type of patterns we consider, and analyze their relevance to human vision.

A class of patterns that is both mathematically tractable, and for which there is considerable psycho-physical experimental data, is the class of sinusoidal patterns–visual waves! It is known that the human visual system is not uniformly sensitive to all sinusoidal patterns. In fact, a typical experimental result is that we are most sensitive to visual waves at around 5 cycles/degree, with sensitivity decreasing exponentially at higher frequencies; see figure 13 and [6]. What happens below 5 cycles/degree is not well understood, but it is believed that sensitivity decreases as well with decreasing frequency. These ideas have been successfully used in JPEG [7]. We treat the development of such HVS-based wavelet coding later in the chapter 6.

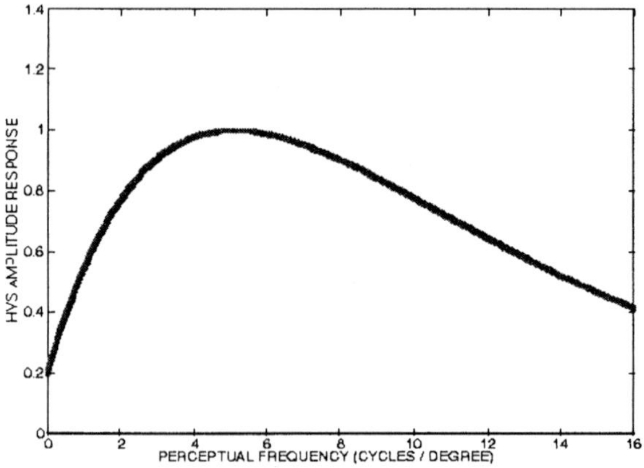

FIGURE 13. An experimental sensitivity curve for the human visual system to sinusoidal patterns, expressed in cycles/degree in the visual field.

7 References

[1] N. Jayant and P. Noll, *Digital Coding of Waveforms*, Prentice-Hall, 1984.

[2] C. Shannon and W. Weaver, *The Mathematical Theory of Communication*, University of Illinois Press, 1949.

[3] Thomas and T. Cover, *Elements of Information Theory*, Wiley, 1991.

[4] A. Jain, *Fundamentals of Digital Image Processing*, Prentice-Hall, 1989.

[5] D. Huffman, "A Method for the Construction of Minimum Redundancy Codes," Proc. IRE, 40: 1098-1101, 1952.

[6] N. Nill, "A Visual Model Weighted Cosine Transform for Image Compression and Quality Assessment," *IEEE Trans. Comm.*, pp. 551-557, June, 1985.

[7] W. Penebaker and J. Mitchell, *JPEG Still Image Data Compression Standard*, Van Nostrand, 1993.

[8] P. J. Burt and E. H. Adelson, "The Laplacian pyramid as a compact image code," IEEE Trans. Communication 31, pp. 532-540, 1983.

[9] S. Mallat and F. Falzon, "Understanding Image Transform Codes," IEEE Trans. Im. Proc., submitted, 1997.

[10] Y. Shoham and A. Gersho, "Efficient bit allocation for an arbitrary set of quantizers," IEEE Trans. Acoustics, Speech, and Signal Processing, vol. 36, pp. 1445-1453, September 1988.

[11] A. Oppenheim and R. Schafer, *Discrete-Time Signal Processing*, Prentice-Hall, 1989.

5

Symmetric Extension Transforms

Christopher M. Brislawn

1 Expansive vs. nonexpansive transforms

A ubiquitous problem in subband image coding is deciding how to convolve analysis filters with finite-length input vectors. Specifically, how does one extrapolate the signal so that the convolution is well-defined when the filters "run off the ends" of the signal? The simplest idea is to extend the signal with zeros: if the analysis filters have finite impulse response (FIR) then convolving with a zero-padded, finite length vector will produce only finitely many nonzero filter outputs. Unfortunately, due to overlap at the signal boundaries, the filter output will have *more* nonzero values than the input signal, so even in a maximally decimated filter bank the zero-padding approach generates *more* transformed samples than we had in the original signal, a defect known as *expansiveness*. Not a good way to begin a data compression algorithm.

A better idea is to take the input vector, $x(n)$, of length N_0 and form its periodic extension, $\tilde{x}(n)$. The result of applying a linear translation-invariant (LTI) filter to $\tilde{x}$ will also have period N_0; note that this approach is equivalent to circular convolution if the filter is FIR and its length is at most N_0. Now consider an M-channel perfect reconstruction multirate filter bank (PR MFB) of the sort shown in Figure 1. If the filters are applied to x by periodic extension and if the decimation ratio, M, divides N_0 then the downsampled output has period N_0/M. This means that $\tilde{x}$, and therefore x, can be reconstructed perfectly from N_0/M samples taken from each channel; we call such a transform a *periodic extension transform* (PET).

Unlike the zero-padding transform, the PET is *nonexpansive*: it maps N_0 input samples to $M \times N_0/M = N_0$ transform domain samples. The PET has two defects, however, that affect its use in subband coding applications. First, as with the zero-padding transform, the PET introduces an artificial discontinuity into the input

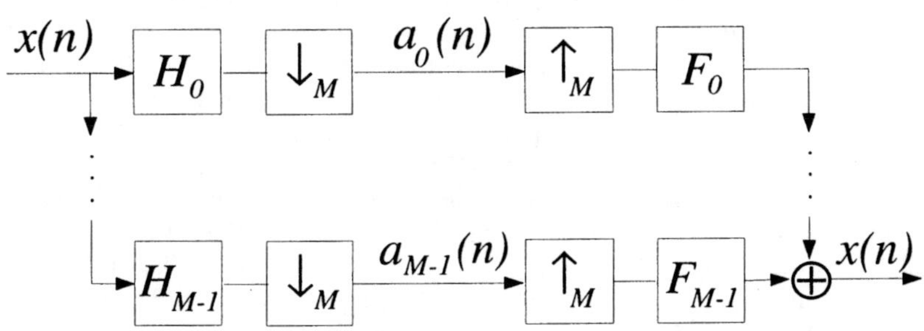

FIGURE 1. M-channel perfect reconstruction multirate filter bank.

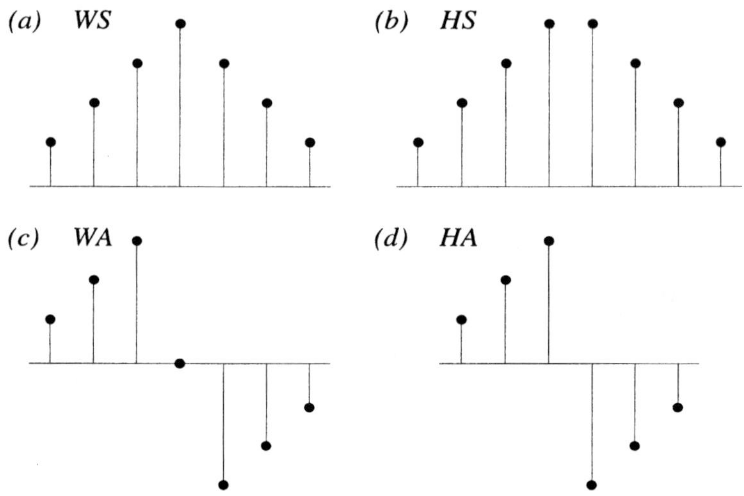

FIGURE 2. (*a*) Whole-sample symmetry. (*b*) Half-sample symmetry. (*c*) Whole-sample antisymmetry. (*d*) Half-sample antisymmetry.

signal. This generally increases the variance of highpass-filtered subbands and forces the coding scheme to waste bits coding a preprocessing artifact; the effects on transform coding gain are more pronounced if the decomposition involves several levels of filter bank cascade. Second, the applicability of the PET is limited to signals whose length, N_0, is divisible by the decimation rate, M, and if the filter bank decomposition involves L levels of cascade then the input length must be divisible by M^L. This causes headaches in applications for which the coding algorithm designer is not at liberty to change the size of the input signal.

One approach that addresses both of these problems is to transform a *symmetric* extension of the input vector, a method we call the *symmetric extension transform* (SET). The most successful SET approaches are based on the use of linear phase filters, which imposes a restriction on the allowable filter banks for subband coding applications. The use of linear phase filters for image coding is well-established, however, based on the belief that the human visual system is more tolerant of linear—as compared to nonlinear—phase distortion, and a number of strategies for designing high-quality linear phase PR MFB's have emerged in recent years. The improvement in rate-distortion performance achieved by going from periodic to symmetric extension was first demonstrated in the dissertation of Eddins [Edd90, SE90], who obtained improvements in the range of 0.2–0.8 dB PSNR for images coded at around 1 bpp. (See [Bri96] for an extensive background survey on the subject.) In Section 3 we will show how symmetric extension methods can also avoid divisibility constraints on the input length, N_0.

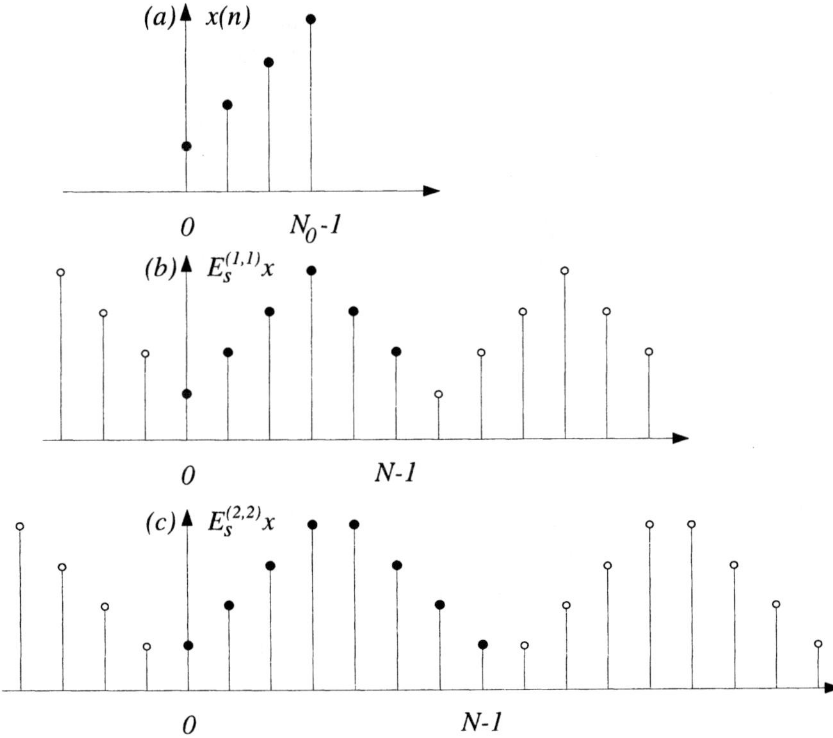

FIGURE 3. (*a*) A finite-length input vector, x. (*b*) The WS extension $y = E_s^{(1,1)}x$. (*c*) The HS extension $y = E_s^{(2,2)}x$.

2 Four types of symmetry

The four basic types of symmetry (or antisymmetry) for linear phase signals are depicted in Figure 2. Note that the axes of symmetry are integers in the whole-sample cases and odd multiples of 1/2 in the half-sample cases. In the symmetric extension approach, we generate a linear phase input signal by forming a symmetric extension of the finite-length input vector and then periodizing the symmetric extension. While it might appear that symmetric extension has doubled the length (or period) of the input, remember that the extension of the signal has not increased the number of degrees of freedom in the source because half of the samples in a symmetric extension are redundant.

Two symmetric extensions, denoted with "extension operator" notation as $E_s^{(1,1)}x$ and $E_s^{(2,2)}x$, are shown in Figure 3. The superscripts indicate the number of times the left and right endpoints of the input vector, x, are reproduced in the symmetric extension; e.g., both the left and right endpoints of x occur twice in each period of $E_s^{(2,2)}x$. We'll use N to denote the length (or period) of the extended signal; $N = 2N_0 - 2$ for $E_s^{(1,1)}x$ and $N = 2N_0$ for $E_s^{(2,2)}x$.

Using the Convolution Theorem, it is easy to see that applying a linear phase filter to a linear phase signal results in a filtered signal that also has linear phase; i.e., a filtered signal with one of the four basic symmetry properties. Briefly put, the phase of the convolution output is equal to the *sum* of the phases of the inputs. Thus, e.g., if a linear phase HS filter with group delay (= axis of symmetry) γ is applied to the HS extension, $E_s^{(2,2)}x$, then the output will be symmetric about $n = \gamma - 1/2$. Since $\gamma - 1/2$ is an integer in this case, the filtered output will be *whole*-sample symmetric (WS).

The key to implementing the symmetric extension method in a linear phase PR MFB without increasing the size of the signal (i.e., implementing it *nonexpansively*) is ensuring that the linear phase subbands remain symmetric *after* downsampling. When this condition is met, half of each symmetric subband will be redundant and can be discarded with no loss of information. If everything is set up correctly, the total number of transform domain samples that must be transmitted will be exactly equal to N_0, meaning that the transform is nonexpansive.

We next show explicitly how this can be achieved in the two-channel case using FIR filters. To keep the discussion elementary, we will not consider M-channel SET's here; the interested reader is referred to [Bri96, Bri95] for a complete treatment of FIR symmetric extension transforms and extensive references. In the way of background information, it is worth pointing out that all nontrivial two-channel FIR linear phase PR MFB's divide up naturally into two distinct categories: filter banks with two symmetric, odd-length filters (WS/WS filter banks), and filter banks with two even-length filters—a symmetric lowpass filter and an antisymmetric highpass filter (HS/HA filter banks). See [NV89, Vai93] for details.

3 Nonexpansive two-channel SET's

We will begin by describing a nonexpansive SET based on a WS/WS filter bank, $\{H_0, H_1\}$. It's simplest if we allow ourselves to consider noncausal implementations of the filters; causal implementation of FIR filters in SET algorithms is possible, but the complications involved obscure the basic ideas we're trying to get across here. Assume, therefore, that h_0 is symmetric about $\gamma_0 = 0$ and that h_1 is symmetric about $\gamma_1 = -1$. Let y be the WS extension

$$y = E_s^{(1,1)}x \quad ,$$

of period $N = 2N_0 - 2$.

First consider the computations in the lowpass (h_0) channel, as shown in Figure 4 for an even-length input ($N_0 = 4$). Since both y and h_0 are symmetric about 0, the filter output, $y * h_0$, will also be WS with axis of symmetry $n = 0$, and one easily sees that the filtered signal remains symmetric after 2:1 decimation: $b_0 \equiv \downarrow (y * h_0)$ is WS about $n = 0$ and HS about $n = 3/2$ (see Figures 4(c) and (d)). The symmetric, decimated subband has period $N/2 = N_0 - 1$ and can be reconstructed perfectly from just $N_0/2$ transmitted samples, assuming that the receiver knows how to extend the transmitted half-period, a_0, shown in Figure 4(e) to restore a full period of b_0.

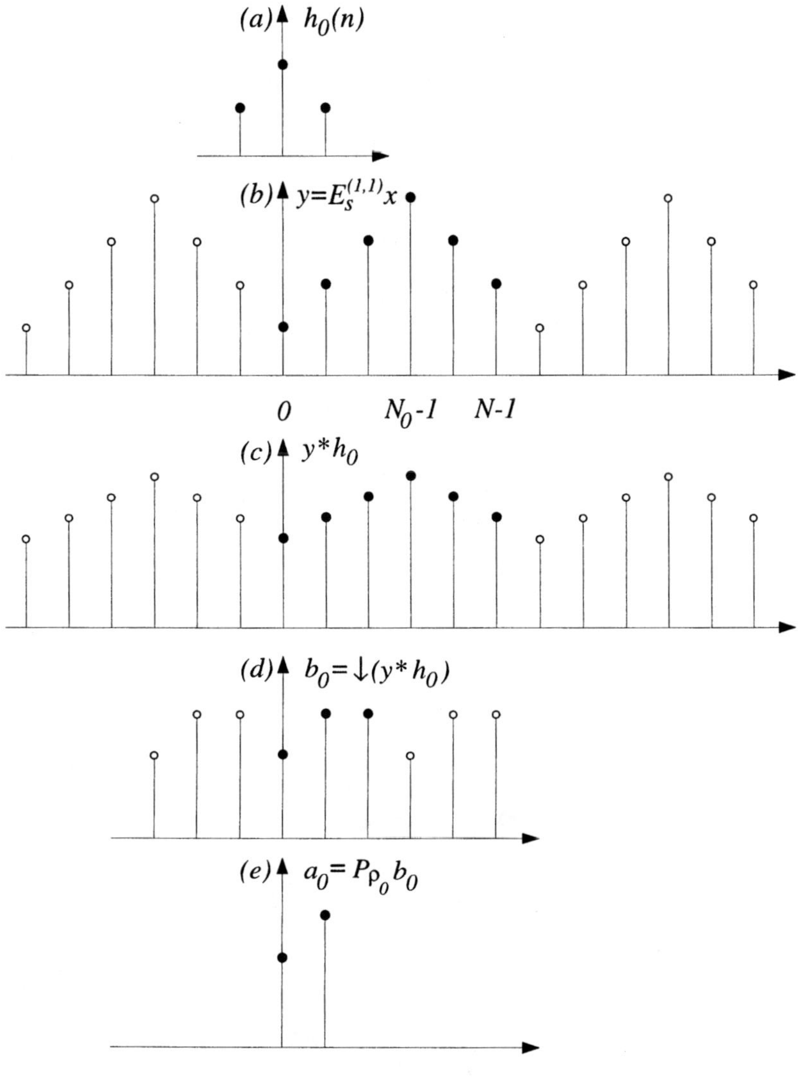

FIGURE 4. The lowpass channel of a (1,1)-SET.

In the highpass channel, depicted in Figure 5, the filter output $y * h_1$ is also WS, but because h_1 has an odd group delay ($\gamma_1 = -1$) the downsampled signal, $b_1 \equiv \downarrow (y * h_1)$, is HS about $n = -1/2$ and WS about $n = 1$. Again, the downsampled subband can be reconstructed perfectly from just $N_0/2$ transmitted samples if the decoder knows the correct symmetric extension procedure to apply to the transmitted half-period, a_1.

Let's illustrate this process with a block diagram. Let ρ_0 and ρ_1 represent the

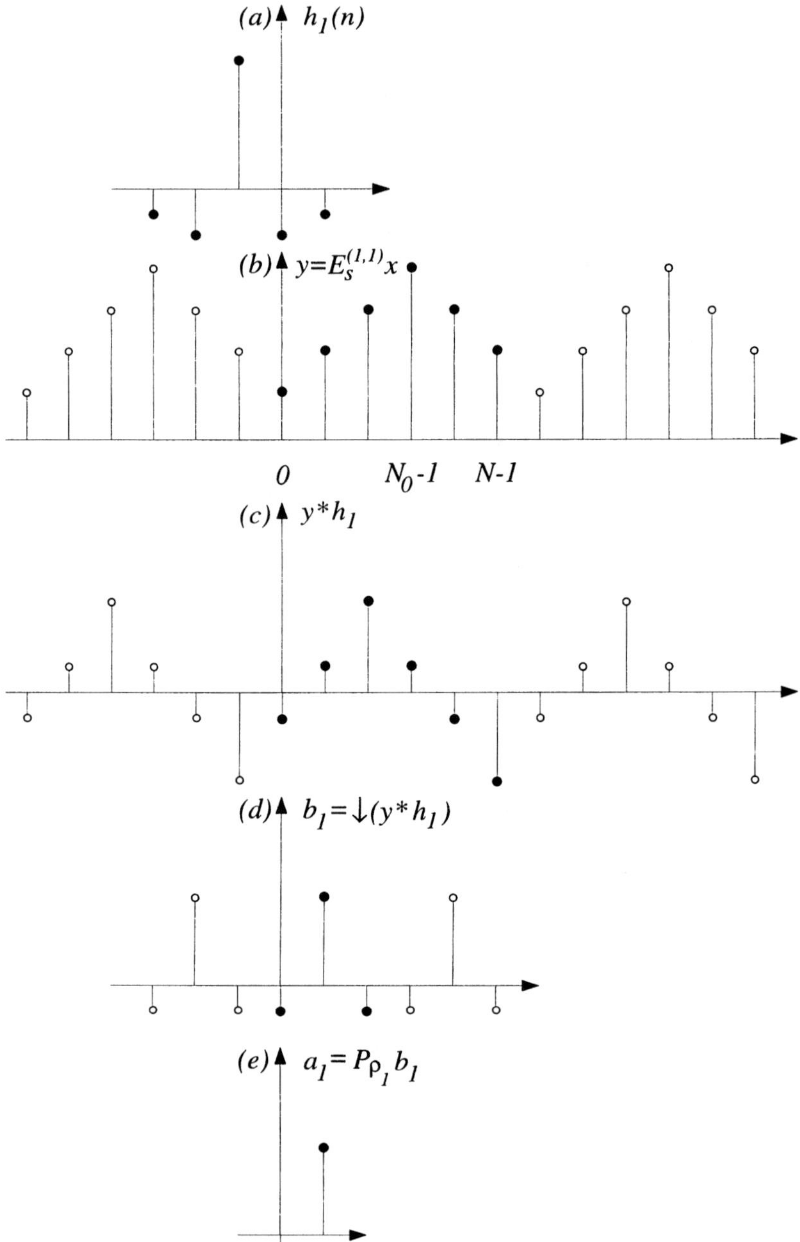

FIGURE 5. The highpass channel of a (1,1)-SET.

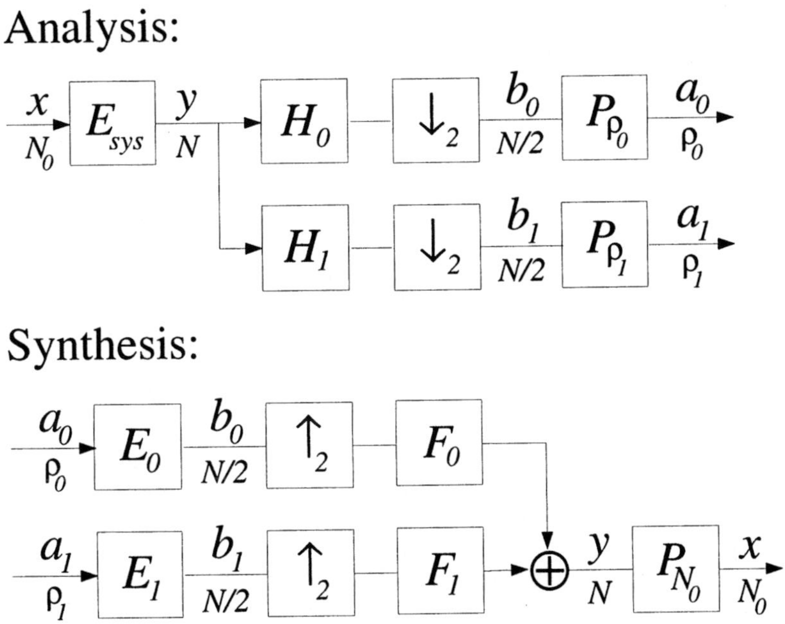

FIGURE 6. Analysis/synthesis block diagrams for a symmetric extension transform.

number of nonredundant samples in the symmetric subbands b_0 and b_1 (in Figures 4(e) and 5(e) we have $\rho_0 = \rho_1 = 2$). P_ρ will be the operator that projects a discrete signal onto its first ρ components (starting at 0):

$$P_\rho b = \{b(0), \ldots, b(\rho - 1)\} \quad .$$

With this notation, the analysis/synthesis process for the SET outlined above is represented in the block diagram of Figure 6.

The input vector, x, of length N_0 is extended by the system input extension ($E_{sys} = E_s^{(1,1)}$) to form a linear phase source, y, of period N. We have indicated the length (or period) of each signal in the diagram beneath its variable name. The extended input is filtered by the lowpass and highpass filters, then decimated, and a complete, nonredundant half-period of each symmetric subband is projected off for transmission. The total transmission requirement for this SET is

$$\rho_0 + \rho_1 = 4 = N_0 \quad ,$$

so the transform is nonexpansive.

To ensure perfect reconstruction, the decoder in Figure 6 needs to know which symmetric extensions to apply to a_0 and a_1 in the synthesis bank to reconstruct b_0 and b_1. In this example, b_0 is WS at 0 and HS at 3/2 ("(1,2)-symmetric" in the language of [Bri96]), so the correct extension operator to use on a_0 is

$$E_0 = E_s^{(1,2)}$$

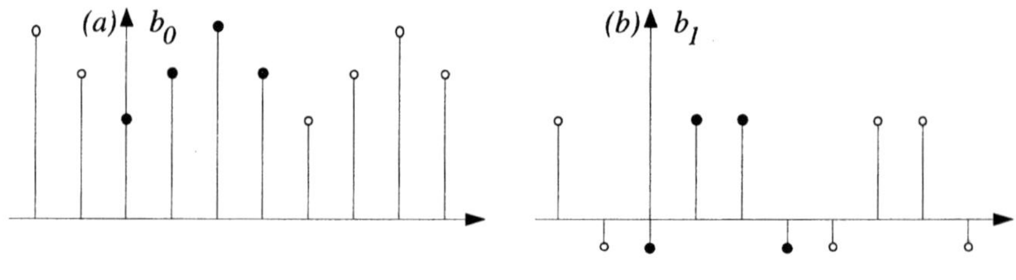

FIGURE 7. Symmetric subbands for a (1,1)-SET with odd-length input ($N_0 = 5$).

Similarly, b_1 is HS at $-1/2$ and WS at 2 ("(2,1)-symmetric"), so the correct extension operator to use on a_1 is

$$E_1 = E_s^{(2,1)} \quad .$$

These choices of synthesis extensions can be determined by the decoder from knowledge of N_0 and the symmetries and group delays of the analysis filters; this data is included as side information in a coded transmission.

Now consider an *odd*-length input, e.g., $N_0 = 5$. Since the period of the extended source, y, is $N = 2N_0 - 2 = 8$, we can still perform 2:1 decimation on the filtered signals, even though the input length was odd! This could not have been done using a periodic extension transform. Using the same WS filters as in the above example, one can verify that we still get symmetric subbands after decimation. As before, b_0 is WS at 0 and b_1 is HS at $-1/2$, but now the periods of b_0 and b_1 are $N/2 = 4$. Moreover, when N_0 is odd b_0 and b_1 have *different* numbers of nonredundant samples. Specifically, as shown in Figure 7, b_0 has $\rho_0 = 3$ nonredundant samples while b_1 has $\rho_1 = 2$. The total transmission requirement is

$$\rho_0 + \rho_1 = 5 = N_0 \quad ,$$

so this SET for odd-length inputs is also nonexpansive.

SET's based on HS/HA PR MFB's can be constructed in an entirely analogous manner. One needs to use the system input extension $y = E_s^{(2,2)}x$ with this type of filter bank, and an appropriate noncausal implementation of the filters in this case is to have both h_0 and h_1 centered at $-1/2$. As one might guess based on the above discussion, the exact subband symmetry properties obtained depend critically on the phases of the analysis filters, and setting both group delays equal to $-1/2$ will ensure that b_0 (resp., b_1) is symmetric (resp., antisymmetric) about $n = -1/2$. This means that the block diagram of Figure 6 is also applicable for SET's based on HS/HA filter banks. We encourage the reader to work out the details by following the above analysis; subband symmetry properties can be checked against the results tabulated below.

SET's based on the system input extension $y = E_s^{(1,1)}x$ using WS/WS filter banks are denoted "(1,1)-SET's," and a concise summary of their characteristics is given in Table 5.1 for noncausal filters with group delays 0 or -1. Similarly, SET's based on the system input extension $y = E_s^{(2,2)}x$ using HS/HA filter banks are called "(2,2)-SET's," and their characteristics are summarized in Table 5.2. More complete SET characteristics and many tedious details aimed at designing causal implementations and M-channel SET's can be found in [Bri96, Bri95].

4 References

[Bri95] Christopher M. Brislawn. Preservation of subband symmetry in multirate signal coding. *IEEE Trans. Signal Process.*, 43(12):3046–3050, December 1995.

[Bri96] Christopher M. Brislawn. Classification of nonexpansive symmetric extension transforms for multirate filter banks. *Appl. Comput. Harmonic Anal.*, 3:337–357, 1996.

[Edd90] Steven L. Eddins. *Subband analysis-synthesis and edge modeling methods for image coding*. PhD thesis, Georgia Institute of Technology, Atlanta, GA, November 1990.

[NV89] Truong Q. Nguyen and P. P. Vaidyanathan. Two-channel perfect-reconstruction FIR QMF structures which yield linear-phase analysis and synthesis filters. *IEEE Trans. Acoust., Speech, Signal Process.*, 37(5):676–690, May 1989.

[SE90] Mark J. T. Smith and Steven L. Eddins. Analysis/synthesis techniques for subband image coding. *IEEE Trans. Acoust., Speech, Signal Process.*, 38(8):1446–1456, August 1990.

[Vai93] P. P. Vaidyanathan. *Multirate Systems and Filter Banks*. Prentice Hall, Englewood Cliffs, NJ, 1993.

TABLE 5.1. $(1,1)$-Symmetric extension transforms.

Signal x:	Filter h:		Subband b:		Ext.
N_0	Sym.	Center γ	Sym., Center	Size ρ	E
Even	WS	0	WS, 0	$N_0/2$	$E_s^{(1,2)}$
		-1	HS, $-1/2$	$N_0/2$	$E_s^{(2,1)}$
Odd	WS	0	WS, 0	$(N_0+1)/2$	$E_s^{(1,1)}$
		-1	HS, $-1/2$	$(N_0-1)/2$	$E_s^{(2,2)}$

TABLE 5.2. $(2,2)$-Symmetric extension transforms.

Signal x:	Filter h:		Subband b:		Ext.
N_0	Sym.	Center γ	Sym., Center	Size ρ	E
Even	HS	$-1/2$	HS, $-1/2$	$N_0/2$	$E_s^{(2,2)}$
	HA	$-1/2$	HA, $-1/2$	$N_0/2$	$E_a^{(2,2)}$
Odd	HS	$-1/2$	HS, $-1/2$	$(N_0+1)/2$	$E_s^{(2,1)}$
	HA	$-1/2$	HA, $-1/2$	$(N_0-1)/2$	$E_a^{(2,1)}$

Part II

Still Image Coding

6

Wavelet Still Image Coding: A Baseline MSE and HVS Approach

Pankaj N. Topiwala

1 Introduction

Wavelet still image compression has recently been a focus of intense research, and appears to be maturing as a subject. Considerable coding gains over older DCT-based methods have been achieved, although for the most part the computational complexity has not been very competitive. We report here on a high-performance wavelet still image compression algorithm optimized for either mean- squared error (MSE) or human visual system (HVS) characteristics, but which has furthermore been streamlined for extremely efficient execution in high-level software. Ideally, all three components of a typical image compression system: transform, quantization, and entropy coding, should be optimized simultaneously. However, the highly non-linear nature of quantization and encoding complicates the formulation of the total cost function. We present the problem of (sub)optimal quantization from a Lagrange multiplier point of view, and derive novel HVS-type solutions. In this chapter, we consider optimizing the filter, and then the quantizer, separately, holding the other two components fixed. While optimal bit allocation has been treated in the literature, we specifically address the issue of setting the quantization stepsizes, which in practice is slightly different. In this chapter, we select a short high-performance filter, develop an efficient scalar MSE-quantizer, and four HVS-motivated quantizers which add value visually while sustaining negligible MSE losses. A combination of run-length and streamlined arithmetic coding is fixed in this study. This chapter builds on the previous tutorial material to develop a baseline compression algorithm that delivers high performance at rock-bottom complexity. As an example, the 512x512 Lena image can be compressed at 32:1 ratio on a Sparc20 workstation in 240 ms using only high-level software (C/C++). This pixel rate of over 10^6 pixels/s comes very close to approaching the execution speed of an excellent and widely available C-language implementation of JPEG [10], which is about 30% faster. The performance of our system is substantially better than JPEG's, and within a modest visual margin of the best-performing wavelet coders at ratios up to 32:1. For contribution-quality compression, say up to 15:1 for color imagery, the performance is indistinguishable.

2 Subband Coding

Subband coding owes its utility in image compression to its tremendous energy compaction capability. Often this is achieved by subdividing an image into an "octave-band" decomposition using a two-channel analysis/synthesis filterbank, until the image energy is highly concentrated in one lowpass subband. Among subband filters for image coding, a consensus appears to be forming that biorthogonal wavelets have a prominent role, launching an exciting search for the best filters [3], [15], [34], [35]. Among the criteria useful for filter selection are regularity, type of symmetry, number of vanishing moments, length of filters, and rationality. Note that of the three components of a typical transform coding system: transform, quantization, and entropy coding, the transform is often the computational bottleneck. Coding applications which place a premium on speed (e.g., video coding) therefore suggest the use of short, rational filters (the latter for doing fixed-point arithmetic). Such filters were invented at least as early as 1988 by LeGall [13], again by Daubechies [2], and are in current use [19]. They do not appear to suffer any coding disadvantages in comparison to irrational filters, as we will argue.

The Daubechies 9-7 filter (henceforth daub97) [2] enjoys wide popularity, is part of the FBI standard [5], and is used by a number of researchers [39], [1]. It has maximum regularity and vanishing moments for its size, but is relatively long and irrational. Rejecting the elementary Haar filter, the next simplest biorthogonal filters [3] are ones of length 6-2, used by Ricoh [19], and 5-3, adopted by us. Both of these filters performed well in the extensive search in [34],[35], and both enjoy implementations using only bit shifts and additions, avoiding multiplications altogether. We have also conducted a search over thousands of wavelet filters, details of which will be reported elsewhere. However, figure 2 shows the close resemblance of the lowpass analysis filter of a custom rational 9-3 filter, obtained by spectral factorization [3], with those of a number of leading Daubechies filters [3]. The average PSNR results for coding a test set of six images, a mosaic of which appears in figure 2, are presented in table 1. Our MSE and HVS optimized quantizers are discussed below, which are the focus of this work. A combination of run-length and adaptive arithmetic coding, fixed throughout this work, is applied on each subband separately, creating individual or grouped channel symbols according to their frequency of occurrence. The arithmetic coder is a bare-bones version of [38] optimized for speed.

Average PSNR	16:1	32:1	64:1
daub53	32.73	29.89	27.29
daub93	32.77	29.92	27.36
custom93	32.76	29.92	27.34
daub97	33.09	30.14	27.57

TABLE 6.1. Variation of filters, with a fixed quantizer (globalunif) and an arithmetic coder. Performance is averaged over six images.

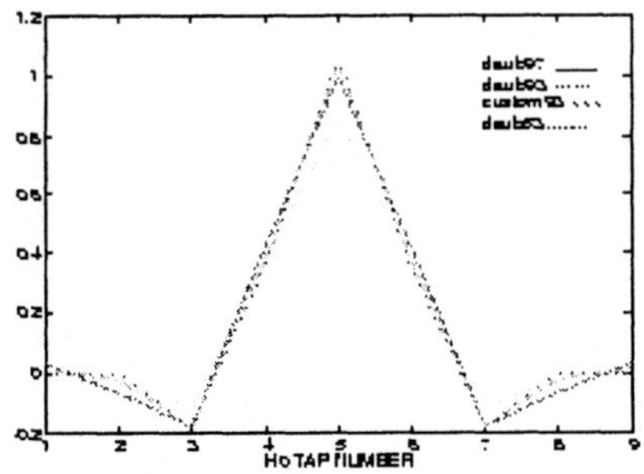

FIGURE 1. Example lowpass filters (daub97, daub93, daub53, and a custom93), showing structural similarity. Their coding performance on sample images is also similar.

Average PSNR	16:1	32:1	64:1
globalunif	33.09	30.14	27.57
logsig	32.99	30.19	27.57
hvs1	32.97	30.00	27.16
hvs2	33.05	30.21	27.56
hvs3	32.35	29.59	27.02
hvs4	32.99	30.14	27.55

TABLE 6.2. Variation of quantizers, with a fixed filter (daub97) and an arithmetic coder. Performance is averaged over six images.

3 (Sub)optimal Quantization

In designing a low-complexity quantizer for wavelet transformed image data, two points become clear immediately: first, vector quantization should probably be avoided due to its complexity, and second, given the individual nature and relative homogeneity of subbands it is natural to design quantizers separately for each subband. It is well-known [2], and easily verified, that the wavelet-transform subbands are Laplace (or more precisely, Generalized Gaussian) distributed in probability density (pdf). Although an optimal quantizer for a nonuniform pdf is necessarily nonuniform [6], this result holds only when one considers quantization by itself, and not in combination with entropy coding. In fact, with the use of entropy coding, the quantizer output stream would ideally consist of symbols with a power-of-one-half distribution for optimization (this is especially true with Huffman coding, less so with our arithmetic coder). This is precisely what is generated by a Laplace-distributed source under uniform quantization (and using the *high-rate approximation*). The issue is how to choose the quantization stepsize for each subband appropriately. The following discussion derives an MSE-optimized quantization scheme of low complexity under simplifying assumptions about the transformed data. Note that while the MSE measured in the transform domain is not equal to that in the image domain due to the use of *bi*orthogonal filters, these quantities are related

FIGURE 2. A mosaic of six images used to test various filters and quantizers empirically, using a fixed arithmetic coder.

above and below by constants (singular values of the so-called frame operator [3] of the wavelet bases). In particular, the MSE minimization problems are essentially equivalent in the two domains.

Now, the transform in transform coding not only provides energy compaction but serves as a prewhitening filter which removes interpixel correlation. Ideally, given an image, one would apply the optimal Karhunen-Loeve transform (KLT) [9], but this approach is image-dependent and of high complexity. Attempts to find good suboptimal representations begin by invoking stationarity assumptions on the image data. Since for the first nontrivial model of a stationary process, the AR(1) model, the KLT is the DCT [9], this supports use of the DCT. While it is clear that AR(1)-modeling of imagery is of limited validity, an exact solution is currently unavailable for more sophisticated models. In actual experiments, it is seen that the wavelet transform achieves at least as much decorrelation of the image data as the DCT.

In any case, these considerations lead us to assume in the first instance that the transformed image data coefficients are independent pixel-to-pixel and subband-to-subband. Since wavelet-transformed image data are also all approximately Laplace-distributed (this applies to all subbands except the residual lowpass component, which can be modelled as being Gaussian-distributed), one can assume further that the pixels are identically distributed when normalized. We will reflect on any residual interpixel correlation later.

In reality, the family of Generalized Gaussians (GGs) is a more accurate model of transform-domain data; see chapters 1 and 3. Furthermore, parameters in the Generalized Gaussian family vary among the subbands, a fact that can be exploited for differential coding of the subbands. An approach to image coding that is tuned to the specific characteristics of more precise GG models can indeed provide superior performance [25]. However, these model parameters have then to be estimated accurately, and utilized for different bitrate and encoding structures. This *Estimation-Quantization (EQ)* [25] approach, which is substantially more compute-efficient that several sophisticated schemes discussed in later chapters, is itself more complex than the baseline approach presented below.

Note that while the identical-distribution assumption may also be approximately available for more general subband filtering schemes, wavelet filters appear to excel in prewhitening and energy compaction. We review the formulation and solution of (sub)optimal scalar quantization with these assumptions, and utilize their modifications to derive novel HVS-based quantization schemes.

Problem ((Sub)optimal Scalar Quantization) Given independent random sources $x_1, \ldots, x_k$ which are identically distributed when normalized, possessing rate-distortion curves $D_i(R_i)$ and given a bitrate budget R, the problem of optimal joint quantization of the random sources is the variational problem

$$\delta \int L = 0, \quad L = \sum L_i,$$

$$L_i = D_i(R_i) + \lambda R_i, \quad i = 1, \ldots, k,$$

$$R_i \geq 0, \quad \sum R_i \leq R,$$

where λ is an undetermined Lagrange multiplier.

It can be shown that, by the independence of the sources and the additivity of the distortion and bitrate variables, the following equations obtain [36], [26]:

$$\frac{\partial L_i}{\partial R_i} = \frac{\partial D_i(R_i)}{\partial R_i} + \lambda = 0, \quad i = 1, \ldots, k.$$

This implies that for optimal quantization, all rate-distortion curves must have a constant slope, $-\lambda$. This is the variational formulation of the argument presented in chapter 3 for the same result. λ itself is determined by satisfying the bitrate budget.

To determine the bit allocations for the various subbands necessary to satisfy the bitrate budget R, we use the "high-resolution" approximation [8] [6] in quantization theory. This says that for slowly varying pdf's and fine-grained quantization, one may assume that the pdf is constant in each quantization interval (i.e., the pdf is *uniform* in each bin). By computing the distortion in each uniformly-distributed bin (and assuming the quantization code is the center of each quantization bin), and summing according to the probability law, one can obtain a formula for the distortion [6]:

$$D_i(R_i) = C_i \sigma_i^2 2^{-2R_i}.$$

Here C_i is a constant depending on the normalized distribution of x_i (thus identically constant by our assumptions), and σ_i^2 is the variance of x_i. These approximations lead to the following analytic solution to the bit-allocation problem:

$$R_i = \overline{R} + \frac{1}{2} log_2 \frac{\sigma_i^2}{\rho^2},$$

$$\overline{R} = R/k, \quad \rho^2 = (\prod \sigma_i^2)^{1/K}.$$

Note that this solution, first derived in [8] in the context of block quantization, is idealized and overlooks the fact that bitrates must be nonnegative and typically integral — conditions which must be imposed a posteriori. Finally, while this solves the bit-allocation problem, in practice we must set the quantization stepsizes instead of assigning bits, which is somewhat different.

The quantization stepsizes can be assigned by the following argument. Since the Laplace distribution does not have finite support (and in fact has "fat tails" compared to the Gaussian distribution; this holds even more for the Generalized Gaussian model), and whereas only a finite number of steps can be allocated, a tradeoff must be made between quantization granularity and saturation error — error due to a random variable exceeding the bounds of the quantization range. Since the interval $[-2\sigma_i, 2\sigma_i]$ accounts for nearly 95% of the values for zero-mean Laplace-distributed source x_i, a reasonable procedure would be to design the optimal stepsizes to accommodate the values that fall within that interval with the available bitrate budget. In practice, values outside this region are also coded by the entropy coder rather than truncated. Dividing this interval into

$$N_i = 2^{R_i} = 2^{\overline{R}} \frac{\sigma_i}{\rho}$$

steps yields a stepsize of

$$Q_i = \frac{4\sigma_i}{N_i} = \frac{2^{2-\overline{R}}}{\rho} = Constant$$

to achieve an overall distortion of

$$D = kC\rho^2 2^{-2\overline{R}}.$$

Note that this simple procedure avoids calculating variances, is independent of the subband, and provides our best coding results to date in an MSE sense; see table 1. It is interesting that our assumptions allow such a simple quantization scheme. It should be noted that the constant stepsize applies equally to the residual lowpass component. A final point is how to set the ratio of the zero binwidth to the stepsize, which could itself be subband-dependent. It is common knowledge in the compression community that, due to the zero run-length encoding which follows quantization, a slightly enlarged zero-binwidth is valuable for improved compression with minimal quality loss. A factor of 1.2 is adopted in [5] while [24] recommends a factor of 2, but we have found a factor of 1.5 to be empirically optimal in our coding.

One bonus of the above scheme is that the information-theoretic apparatus provides an ability to specify a desired bitrate with some precision. We note further that in an octave-band wavelet decomposition, the subbands at one level are approximately lowpass versions of subbands at the level below. If the image data is itself dominated by its "DC" component, one can estimate (where $H_0(0)$ is the zero-frequency response of the lowpass analysis filter)

$$\sigma_{i+1} = |H_0(0)|^2 \sigma_i = 2\sigma_i.$$

This relationship appears to be roughly bourne out in our experiments; its validity would then allow estimating the distortion at the subband level.

It would therefore seem that this globally uniform quantizer (herein "globalunif") would be the preferred approach. Another quantizer, developed for the FBI application [5], herein termed the "logsig" quantizer (where is a global quality factor) has also been considered in the literature, e.g. [34]:

$$Q_i = \frac{C_i q}{log_2 \sigma_i^2}.$$

An average PSNR comparison of these and other quantizers motivated by human visual system considerations is given in table 6.2.

4 Interband Decorrelation, Texture Suppression

A significant weakness of the above analysis is the assumption that the initial sources, representing pixels in the various subbands, are statistically independent. Although wavelets are excellent prewhitening filters, residual interpixel and especially interband correlations remain, and are in some sense entrenched – no fixed length filters can be expected to decorrelate them arbitrarily well. In fact, visual analysis of wavelet transformed data readily reveals correlations across subbands (see figure 3), a point very creatively exploited in the community [22], [39], [27], [28], [29], [20]. These issues will be treated in later chapters of this book. However, we note that exploiting this residual structure for performance gains comes at a complexity price, a trade we chose not to make here.

At the same time, the interband correlations do allow an avenue for systematically detecting and suppressing low-level textural information of little subjective value. While current methods for further decorrelation of the transformed image appear computationally expensive, we propose a simple method below for texture suppression which is in line with the analysis in [Knowles]. A two-channel subband coder, applied with levels in an octave-band manner, transforms an image into a sequence of subbands organized in a pyramid

$$(a_{ij}) \to P = (LOW_L, H_L, V_L, D_L, \ldots, H_1, V_1, D_1).$$

Here LOW_L refers to the residual lowpass subband, while the H_i, V_i, D_i are respectively the horizontal, vertical, and diagonal subbands at the various levels. The

FIGURE 3. Two-level decomposition of the Lena image, displaying interpixel and interband dependencies.

pyramidal structure implies that a given pixel in any subband type at a given level corresponds to a 2x2 matrix of pixels in the subband of the same type, but at the level immediately below. This parent-to-children relationships persists throughout the pyramid, excepting only the two ends.

In fact, this is not one pyramid but a multitude of pyramids, one for each pixel in LOW_L. The logic of this framework led coincidentally to an auxiliary development, the application of wavelets to region-of-interest compression [32]. Every partition π of LOW_L gives rise to a sequence of pyramids $P_1, \ldots, P_{|\pi|}$, with $|\pi|$ the number of regions in the partition π. Each of these partitions can be quantized separately, leading to differing levels of fidelity. In particular, image pixels outside regions of interest can be assigned relatively reduced bitrates. We subsequently learned that Shapiro had also developed such an application [23], using fairly similar methods but with some key differences. In essence, Shapiro's approach is to preprocess the image by heightening the regions of interest in the pyramid before applying standard compression on the whole image, whereas we suppress the background first. These approaches may appear to be quite comparable; however, due to the limited dynamic range of most images (e.g, 8 bits), the region-enhancement approach quickly reaches a saturation point and limits the degree of preference given to the regions. By contrast, since our approach suppresses the background, we remove the dynamic range limitations in the preference, which offers coding advantages.

Returning to optimized image coding, while interband correlations are apparently hard to characterize for nonzero pixel values, Shapiro's advance [22] was the observation that a zero pixel in one subband correlates well with zeros in the entire pyramid below it. He used this observation to develop a novel approach to coding wavelet transformed quantized data. Xiong et al [39] went much further, and developed a strategy to code even the nonzero values in accordance to this pyramidal structure. But this approach involves nonlinear search strategies, and is likely to be computationally intensive. Inspired by these ideas, we consider a simple sequential thresholding strategy, to be applied to prune the quantized wavelet coefficients in order to improve compression or suppress unwanted texture. Our reasoning is that if edge data represented by a pixel and its children are both weak, the children can be sacrificed without significant loss either to the image energy or the information content. This concept leads to the rule:

$$\begin{aligned} if\ |parent_{i+1}| &< K_{i+1}\ and \\ max\{|children_i|\} &< K_i, \\ then\ set\ children_i &= 0. \end{aligned}$$

Our limited experimentation with this rule indicates that slight thresholding (i.e., with low threshold values) can help reduce some of the ringing artifacts common in wavelet compressed images.

5 Human Visual System Quantization

While PSNRs appear to be converging for the various competitive algorithms in the wavelet compression race, a decisive victory could presumably be had by finding a quantization model geared toward the human visual system (HVS) rather than MSE. While this goal is highly meritorious in abstraction, it remains challenging in practice, and apparently no one has succeeded in carrying this philosophy to a clear success. The use of HVS modelling in imaging systems has a long history, and dates back at least to 1946 [21]. HVS models have been incorporated in DCT-based compression schemes by a number of authors [16], [7], [33], [14], [17], [4], [9], and recently been introduced into wavelet coding [12], [1]. However, [12] uses HVS modelling to directly suppress texture, as in the discussion above, while [1] uses HVS only to evaluate wavelet compressed images. Below we present a novel approach to incorporating HVS models into wavelet image coding.

The human visual system, while highly complex, has demonstrated a reasonably consistent modulation transfer function (MTF), as measured in terms of the frequency dependence of contrast sensitivity [9], [18]. Various curve fits to this MTF exist in the literature; we use the following fit obtained in [18]. First, recall that the MTF refers to the sensitivity of the HVS to visual patterns at various perceived frequencies. To apply this to image data, we model an image as a positive real function of two real variables, take a two-dimensional Fourier transform of the data and consider a radial coordinate system in it. The HVS MTF is a function of the radial variable, and the parametrization of it in [18] is as follows (see "hvs1" in figure 4):

$$HVS(r) = (0.2 + 0.45r)e^{-0.18r}.$$

This model has a peak sensitivity at around 5 cycles/degree. While there are a number of other similar models available, one key research issue is how to incorporate these models into an appropriate quantization scheme for improved coding performance. The question is how best to adjust the stepsizes in the various subbands, favoring some frequency bands over others ([12] provides one method). To derive an appropriate context in which to establish the relative value placed on a signal by a linear system, we consider its effect on the relative energetics of the signal within the frequency bands. This leads us to consider the gain at frequency r to be the squared-magnitude of the transfer function at r:

$$|HVS(r)|^2.$$

This implies that to a pixel in a given subband with a frequency interval $[r_1, r_2]$, the human visual system confers a "perceptual" weight which is the average of the squared MTF over the frequency interval.

Definition Pixels in a subband decomposition of a signal are weighted by the human visual system (HVS) according to

$$weight = \frac{1}{r_2 - r_1} \int_{r_1}^{r_2} |HVS(r)|^2 dr.$$

This is motivated by the fact that in a frequency-domain decomposition, the weight at a pixel (frequency) would be precisely $|HVS(r)|^2$. Since wavelet-domain subband pixels all reside roughly in one frequency band, it makes sense to average the weighting over the band. Assuming that $HVS(r)$ is a continuous function, by the fundamental theorem of calculus we obtain in the limit, meaningful in a purely spectral decomposition, that

$$\lim_{r_2 \to r_1} \frac{1}{r_2 - r_1} \int_{r_1}^{r_2} |HVS(r)|^2 dr = |HVS(r)|^2.$$

This limiting agreement between the perceptual weights in subband and pointwise (spectral) senses justifies our choice of perceptual weight.

To apply this procedure numerically, a correspondence between the subbands in a wavelet octave-band decomposition, and cycles/degree, must be established. This depends on the resolution of the display and the viewing distance. For CRT displays used in current-generation computer monitors, a screen resolution of approximately $80 - 100$ pixels per inch is achieved; we will adopt the 100 pixel/inch figure. With this normalization, and at a standard viewing distance of eighteen inches, a viewing angle of one degree subtends a field of x pixels covering y inches, where

$$\frac{y}{18} = tan(1^o); \quad x = 100y;$$

$$\to x \approx 31.41 \approx 32 pixels/degree.$$

This setup corresponds to a perceptual Nyquist frequency of 16 cycles/degree. Since in an octave-band wavelet decomposition the frequency widths are successively halved, starting from the upper-half band, we obtain, for example, that a five-level wavelet pyramid corresponds to a band decomposition into the following perceptual frequency widths (in cycles/degree):

$$[0, 0.5], \ [0.5, 1], \ [1, 2], \ [2, 4], \ [4, 8], \ [8, 16].$$

In this scheme, the lowpass band $[0, 0.5]$ corresponds to the very lowpass region of the five-level transformed image. The remaining bands correspond to the three high-frequency subbands at each of the levels 5, 4, 3, 2 and 1.

While the above defines the perceptual band-weights, our problem is to devise a meaningful scheme to alter the quantization stepsizes. A theory of weighted bit allocation exists [6], which is just an elaboration on the above development. In essence, if the random sources $\{x_1, \ldots, x_k\}$ are assigned weights $\{h_1, \ldots, h_k\}$ with geometric mean H, the optimal weighted bitrate allocation is given by

$$R_i = \overline{R} + \frac{1}{2} log_2 \frac{\sigma_i^2}{\rho^2} + \frac{1}{2} log_2 \frac{h_i}{H},$$

$$\overline{R} = R/k, \quad \rho^2 = (\prod \sigma_i^2)^{1/k}, \quad H = (\prod h_i)^{1/k}.$$

Finally, these rate modifications lead to quantization stepsize changes as given by previous equations. The resulting quantizer, labeled by "hvs1" in table 1, does appear subjectively to be provide an improvement over images produced by the pure means-squared error method "globalunif".

Having constructed a weighted quantizer based on one perceptual model, we can envision variations on the model itself. For example, the low-frequency information, which is the most energetic and potentially the most perceptually meaningful, is systematically suppressed in the model hvs1. This also runs counter to other strategies in coding (e.g., DCT methods such as JPEG) which typically favor low-frequency content. In order to compensate for that fact, one can consider several alterations of this curve-fitted HVS model. Three fairly natural modifications of the HVS model are presented in figure 4, each of which attempts to increase the relative value of low-frequency information according to a certain viewpoint. "hvs2" maintains the peak sensitivity down to dc, reminiscent of the spectral response to the chroma components of color images, in say, a "Y-U-V" [10] decomposition, and also similarly to a model proposed in [12] for JPEG; "hvs3" takes the position that the sensitivity decreases exponentially starting at zero frequency; and "hvs4" upgrades the y-intercept of the "hvs1" model.

$$hvs2(r) = (0.45r)e^{-0.18r}, \ r \geq 5.1111; \quad hvs2(r) = 1.0, \ r < 5.1111.$$

$$hvs3(r) = e^{-0.18r}$$

$$hvs4(r) = (1 + 0.45r)e^{-0.18r}.$$

While PSNR is not the figure of merit in the analysis of HVS-driven systems, it is interesting to note that all models remain very competitive with the MSE-optimized quantizers, even while using substantially different bit allocations. And visually, the images produced under these quantization schemes appear to be preferred to the unweighted image. An example image is presented in figure 5, using the "hvs3" model.

6 Summary

We have considered the design of a high-performance, low-complexity wavelet image coder for grayscale images. Short, rational biorthogonal wavelet filters appear preferred, of which we obtained new examples. A low-complexity optimal scalar

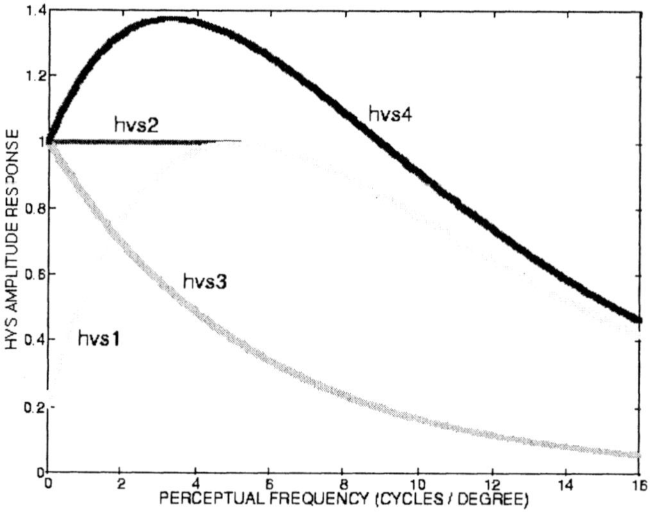

FIGURE 4. Human visual system motivated spectral response models; hvsl refers to a psychophysical model from the literature ([18]).

quantization scheme was presented which requires only a single global quality factor, and avoids calculating subband-level statistics. Within that context, a novel approach to human visual system quantization for wavelet transformed data was derived, and utilized for four different HVS-type curves including the model in [18]. An example reconstruction is provided in figure 5 below, showing the relative advantages of HVS coding. Fortunately, these quantization schemes provide improved visual quality with virtually no degradation in mean-squared error. Although our approach is based on established literature in linear systems modeling of the human visual system, the information-theoretic ideas in this paper are adaptable to contrast sensitivity curves in the wavelet domain as they become available, e.g. [37].

7 References

[1] J. Lu, R. Algazi, and R. Estes, "Comparison of Wavelet Image Coders Using the Picture Quality Scale," UC-Davis preprint, 1995.

[2] M. Antonini et al, "Image Coding Using Wavelet Transform," *IEEE Trans. Image Proc.*, pp. 205-220, April, 1992.

[3] I. Daubechies, *Ten Lectures on Wavelets,* SIAM, 1992.

[4] B. Chitraprasert, and K. Rao, "Human Visual Weighted Progressive Image Transmission," *IEEE Trans. Comm.*, vol. 38, pp. 1040-1044, 1990.

[5] "Wavelet Scalar Quantization Fingerprint Image Compression Standard," Criminal Justice Information Services, FBI, March, 1993.

[6] A. Gersho and R. Gray, *Vector Quantization and Signal Compression,* Kluwer, 1992.

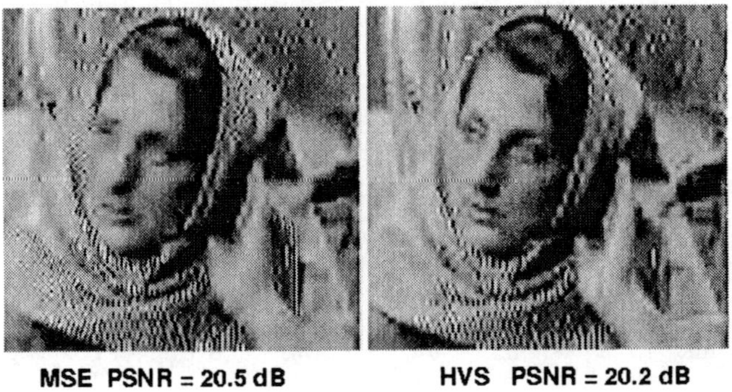

FIGURE 5. Low-bitrate coding of the facial portion of the Barbara image (256x256, 8-bit image), 64:1 compression. (a) MSE-based compression, PSNR=20.5 dB; (b) HVS-based compression, PSNR=20.2 dB. Although the PSNR is marginally higher in (a), the visual quality is significantly superior in (b).

[7] N. Griswold, "Perceptual Coding in the Cosine Transform Domain," *Opt. Eng.*, v. 19, pp. 306-311, 1980.

[8] J. Huang and P. Schultheiss, "Block Quantization of Correlated Gaussian Variables," *IEEE Trans. Comm.*, CS-11, pp. 289-296, 1963.

[9] W. Pennebaker and J. Mitchell, *JPEG Still Image Data Compression Standard,* Van Nostrand, 1993.

[10] T. Lane et al, The Independent JPEG Group Software. ftp:// ftp.uu.net /graphics/jpeg.

[11] R. Kam and P. Wong, "Customized JPEG Compression for Grayscale Printing," . Data Comp. Conf. DCC-94, IEEE, pp. 156-165, 1994.

[12] A. Lewis and G. Knowles, "Image Compression Using the 2-D Wavelet Transform," *IEEE Trans. Image Proc.,* pp. 244-250, April. 1992.

[13] D. LeGall and A. Tabatabai, "Subband Coding of Digital Images Using Symmetric Short Kernel Filters and Arithmetic Coding Techniques," *Proc. ICASSP,* IEEE, pp. 761-765, 1988.

[14] N. Lohscheller, "A Subjectively Adapted Image Communication System," *IEEE Trans. Comm.*, COM-32, pp. 1316-1322, 1984.

[15] E. Majani, "Biorthogonal Wavelets for Image Compression," *Proc. SPIE,* VCIP-94, 1994.

[16] J. Mannos and D. Sakrison, "The Effect of a Visual Fidelity Criterion on the Encoding of Images," *IEEE Trans. Info. Thry,* IT-20, pp. 525-536, 1974.

[17] K. Ngan, K. Leong, and H. Singh, "Adaptive Cosine Transform Coding of Images in Perceptual Domain," *IEEE Trans. Acc. Sp. Sig. Proc.*, ASSP-37, pp. 1743-1750, 1989.

[18] N. Nill, "A Visual Model Weighted Cosine Transform for Image Compression and Quality Assessment," *IEEE Trans. Comm.*, pp. 551-557, June, 1985.

[19] A. Zandi et al, "CREW: Compression with Reversible Embedded Wavelets," *Proc. Data Compression Conference 95,* IEEE, March, 1995.

[20] A. Said and W. Pearlman, "A new, fast, and efficient image codec based on set partitioning in hierarchical trees," IEEE Trans. Circuits and Systems for Video Technology, vol. 6, pp. 243-250, June 1996.

[21] E. Selwyn and J. Tearle, em Proc. Phys. Soc., Vo. 58, no. 33, 1946.

[22] J. Shapiro, "Embedded Image Coding Using Zerotrees of Wavelet Coefficients," *IEEE Trans. Signal Proc.,* pp. 3445-3462, December, 1993.

[23] J. Shapiro, "Smart Compression Using the Embedded Zerotree Wavelet (EZW) Algorithm," *Proc. Asilomar Conf. Sig., Syst. and Comp.,* IEEE, pp. 486-490, 1993.

[24] S. Mallat and F. Falcon, "Understanding Image Transform Codes," IEEE Trans. Image Proc., submitted, 1997.

[25] S. LoPresto, K. Ramchandran, and M. Orchard, "Image Coding Based on Mixture Modelling of Wavelet Coefficients and a Fast Estimation-Quantization Framework," DCC97, Proc. Data Comp. Conf., Snowbird, UT, pp. 221-230, March, 1997.

[26] Y. Shoham and A. Gersho, "Efficient Bit Allocation for an Arbitrary Set of Quantizers," *IEEE Trans. ASSP,* vol. 36, pp. 1445-1453, 1988.

[27] A. Docef et al, "Multiplication-Free Subband Coding of Color Images," *Proc. Data Compression Conference 95,* IEEE, pp. 352-361, March, 1995.

[28] W. Chung et al, "A New Approach to Scalable Video Coding," *Proc. Data Compression Conference 95,* IEEE, pp. 381-390, March, 1995.

[29] M. Smith, "ATR and Compression," Workshop on Clipping Service, MITRE Corporation, July, 1995.

[30] G. Strang and T. Nguyen, *Wavelets and Filter Banks,* Cambridge-Wellesley Press, 1996.

[31] P. Topiwala et al, *Fundamentals of Wavelets and Applications,* IEEE Educational Video, September, 1995.

[32] P. Topiwala et al, "Region of Interest Compression Using Pyramidal Coding Schemes," *Proc. SPIE-San Diego,* Mathematical Imaging, July, 1995.

[33] K. Tzou, T. Hsing and J. Dunham, "Applications of Physiological Human Visual System Model to Image Compression," *Proc. SPIE 504,* pp. 419-424, 1984.

[34] J. Villasenor et al, "Filter Evaluation and Selection in Wavelet Image Compression," *Proc. Data Compression Conference,* IEEE, pp. 351-360, March, 1994.

[35] J. Villasenor et al, "Wavelet Filter Evaluation for Image Compression," *IEEE Trans. Image Proc.,* August, 1995.

[36] M. Vetterli and J. Kovacevic, *Wavelets and Subband Coding,* Prentice-Hall, 1995.

[37] A. Watson et al, "Visibility of Wavelet Quantization Noise," NASA Ames Research Center preprint, 1996. http://www.vision.arc.nasa.gov /personnel/watson/watson.html

[38] I. Witten et al, "Arithmetic Coding for Data Compression," *Comm. ACM,* IEEE, pp. 520- 541, June, 1987.

[39] Z. Xiong et al, "Joint Optimization of Scalar and Tree-Structured Quantization of Wavelet Image Decompositions," *Proc. Asilomar Conf. Sig., Syst., and Comp.,* IEEE, November, 1993.

7
Image Coding Using Multiplier-Free Filter Banks

Alen Docef, Faouzi Kossentini, Wilson C. Chung, and Mark J. T. Smith

1 Introduction[1]

Subband coding is now one of the most important techniques for image compression. It was originally introduced by Crochiere in 1976, as a method for speech coding [CWF76]. Approximately a decade later it was extended to image coding by Woods and O'Neil [WO86] and has been gaining momentum ever since. There are two distinct components in subband coding: the analysis/synthesis section, in which filter banks are used to decompose the image into subband images; and the coding system, where the subband images are quantized and coded.

A wide variety of filter banks and subband decompositions have been considered for subband image coding. Among the earliest and most popular were uniform-band decompositions and octave-band decompositions [GT86, GT88, WBBW88, WO86, KSM89, JS90, Woo91]. But many alternate tree-structured filter banks were considered in the mid 1980s as well [Wes89, Vet84].

Concomitant with the investigation of filter banks was the study of coding strategies. There is great variation among subband coder methods and implementations. Common to all, however, is the notion of splitting the input image into subbands and coding these subbands at a target bit rate. The general improvements obtained by subband coding may be attributed largely to several characteristics—notably the effective exploitation of correlation within subbands, the exploitation of statistical dependencies among the subbands, and the use of efficient quantizers and entropy coders [Hus91, JS90, Sha92]. The particular subband coder that is the topic of discussion in this chapter was introduced by the authors in [KCS95] and embodies all the aforementioned attributes. In particular, intra-subband and inter-band statistical dependencies are exploited by a finite state prediction model. Quantization and entropy coding are performed jointly using multistage residual quantizers and arithmetic coders. But perhaps most important and uniquely characteristic of this particular system is that all components are designed together to optimize rate-distortion performance, subject to fixed constraints on computational complexity.

High performance in compression is clearly an important measure of overall value,

[1]Based on "Multiplication-Free Subband Coding of Color Images", by Docef, Kossentini, Chung and Smith, which appeared in the Proceedings of the Data Compression Conference, Snowbird, Utah, March 1995, pp 352-361,

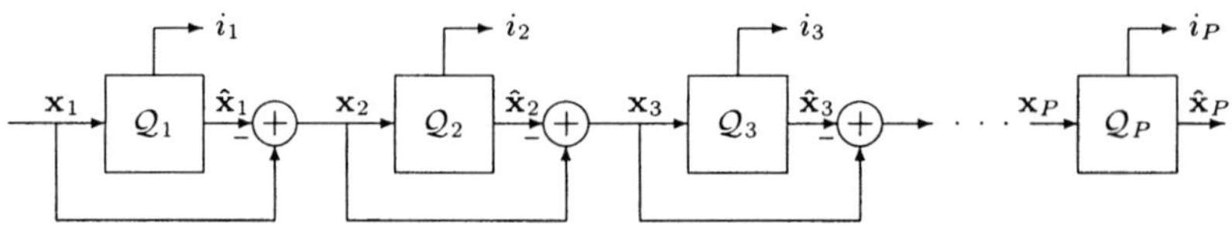

FIGURE 1. A residual scalar quantizer.

but computational complexity should not be ignored. In contrast to JPEG, subband coders tend to be more computationally intensive. In many applications, the differences in implementation complexity among competing methods, which is generally quite dramatic, can outvie the disparity in reconstruction quality. In this chapter we present an optimized subband coding algorithm, focusing on maximum complexity reduction with the minimum loss in performance.

2 Coding System

We begin by assuming that our digital input image is a color image in standard red, green, blue (RGB) format. In order to minimize any linear dependencies among the color planes, the color image is first transformed from an RGB representation into a luminance-chrominance representation, such as YUV or YIQ. In this chapter we use the NTSC YIQ representation. The Y, I, Q planes are then decomposed into 64 square uniform subbands using a multiplier-free filter bank, which is discussed in Section 4. When these subbands are quantized, an important issue is efficient bit allocation among different color planes and subbands to maximize overall performance. To quantify the overall distortion, a three component distortion measure is used that consists of a weighted sum of error components from the Y, I, and Q image planes,

$$d = w_Y d_Y + w_I d_I + w_Q d_Q,$$

where d_Y, d_I, and d_Q are the distortions associated with the Y, I, and Q color components, respectively. A good set of empirically designed weighting coefficients is $w_Y = 1.0$, $w_I = 0.3$, and $w_Q = 0.3$ [Woo91]. This color distance measure is very useful in the joint optimization of the quantizer and entropy coder. In fact, the color subband coding algorithm of [KCS95] minimizes the expected value of the three-component distortion subject to a constraint on overall entropy.

Entropy-constrained multistage residual scalar quantizers (RSQs) [KSB93, KSB94b] are used to encode the subband images, an example of which is shown in Figure 1. The input to each stage is quantized, after which the difference is computed. This difference becomes the input to the next stage and so on. These quantizers [BF93, KSB94a, KSB94b] have been shown to be effective in an entropy-constrained framework. The multistage scalar quantizer outputs are then entropy encoded using binary arithmetic coders. In order to reduce the complexity of the entropy coding step, we restrict the number of quantization levels in each of the RSQs to be 2, which simplifies binary arithmetic coding (BAC) [LR81, PMLA88]. This also simplifies the implementation of the multidimensional prediction.

A multidimensional prediction network is used that exploits intra-subband and inter-subband dependencies across the color coordinate planes and RSQ stages. Prediction is implemented by using a finite state model, as illustrated in Figure 2. The images shown in the figure are multistage approximations of the color coordinate planes. Based on a set of conditioning symbols we determine the probabilities to be used by the current arithmetic coder. The symbols used for prediction are pointed to by the arrows in Figure 2. The three types of dependencies are distinguished by the typography: the dashed arrows indicate inter-stage dependencies in the RSQ, the solid arrows indicate intra-subband dependencies, and the dotted arrows indicate inter-subband dependencies. A separate finite state model is used for each RSQ stage p, in each subband m, and in each color coordinate plane n. The state model is described by

$$u = F(s_0, s_1, \ldots, s_k), \tag{7.1}$$

where $u \in \{0, 1, \ldots, U\}$ is the state, U is the number of conditioning states, F is a mapping function, and k is the number of conditioning symbols $s_0, s_1, \ldots, s_k$. The output indices j of the RSQ are dependent on both the input $\boldsymbol{x}$ and the state u, and are given by $j = E(\boldsymbol{x}, u)$, where E is the fixed-length encoder. Since N, the stage codebook size is fixed to $N = 2$, the number of conditioning states is a power of 2. More specifically, it is equal to 2^R, where R is the number of conditioning symbols used in the multidimensional prediction. Then, the number of conditional probabilities for each stage is 2^{R+1}, and the total number of conditional probabilities T can be expressed as an exponential function of R [KCS95]. The complexity of the overall entropy coding process, expressed here in terms of T, increases linearly as a function of the number of subbands in each color plane and the number of stages for each color subband. However, because of the exponential dependence of T on R, the parameter R is usually the most important one. In fact, T can grow very large even for moderate values of R. Experiments have shown that, in order to reduce the entropy significantly, as many as $R = 8$ conditioning symbols should be used. In such a case, $2^8 = 256$ conditioning states are required, which implies that 196,608 conditional probabilities must be computed and stored.

Assuming that we are constrained by a limit on the number of conditional probabilities in the state models, then we would like to be able to find the number k and location of conditioning symbols for each stage such that the overall entropy is minimized. Such an algorithm is described in [KCS93, KCS96]. So far, the function F in equation (7.1) is a one-to-one mapping because all the conditioning symbols are used to define the state u. Complexity can be further reduced if we use a many-to-one mapping instead, in which case F is implemented by a table that maps each internal state $(s_0, s_1, \ldots, s_k)$ to a state in a reduced set of states. To obtain the new set of states that minimizes the increase in entropy, we employ the PNN algorithm [Equ89] as described below.

At each stage, we look at all possible pairs of conditioning states and compute the increase in entropy that results when the two states are merged. If we want to merge the conditioning states u_i and u_k, we can compute the resulting increase in entropy as

$$\begin{aligned} \Delta_{i,k} &= pr(u_i)H(J|u_i) + pr(u_k)H(J|u_k) \\ &\quad -pr(u_i, u_k)H(J|u_i, u_k), \end{aligned} \tag{7.2}$$

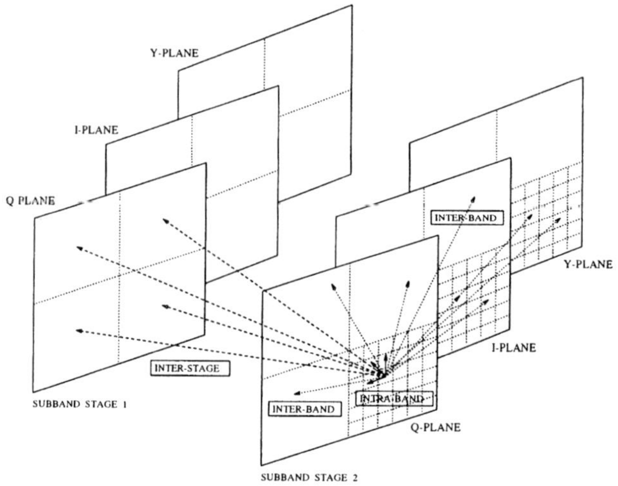

FIGURE 2. Inter-band, intra-band, and inter-stage conditioning

where J is the index random variable, pr denotes probability, and H denotes conditional entropy. Then we merge the states u_i and u_k that correspond to the minimum increase in entropy $\Delta_{i,k}$. We can repeat this procedure until we have the desired number of conditioning states. Using the PNN algorithm we can reduce the number of conditioning states by one order of magnitude with an entropy increase that is smaller than 1%.

The PNN algorithm is used here in conjunction with the generalized BFOS algorithm [Ris91] to minimize the overall complexity-constrained entropy. This is illustrated in Figure 3. A tree is first built, where each branch is a unary tree that represents a particular stage. Each unary tree consists of 2^k nodes, with each node corresponding to a pair (C, H), where C is the number of conditioning states and H is the corresponding entropy. The pair (C, H) is obtained by the PNN algorithm. The BFOS algorithm is then used to locate the best numbers of states for each stage subject to a constraint on the total number of conditional probabilities.

3 Design Algorithm

The algorithm used to design the subband coder minimizes iteratively the expected distortion, subject to a constraint on the complexity-constrained average entropy of the stage quantizers, by jointly optimizing the subband encoders, decoders, and entropy coders. The design algorithm employs the same Lagrangian parameter λ in the entropy-constrained optimization of all subband quantizers, and therefore requires no bit allocation[KCS96].

The encoder optimization step of the design algorithm usually involves dynamic M-search of the multistage RSQ in each subband independently. The decoder optimization step consists of using the Gauss-Seidel algorithm [KSB94b] to minimize iteratively the average distortion between the input and the synthesized reproduction of all stage codebooks in all subbands. Since actual entropy coders are not used

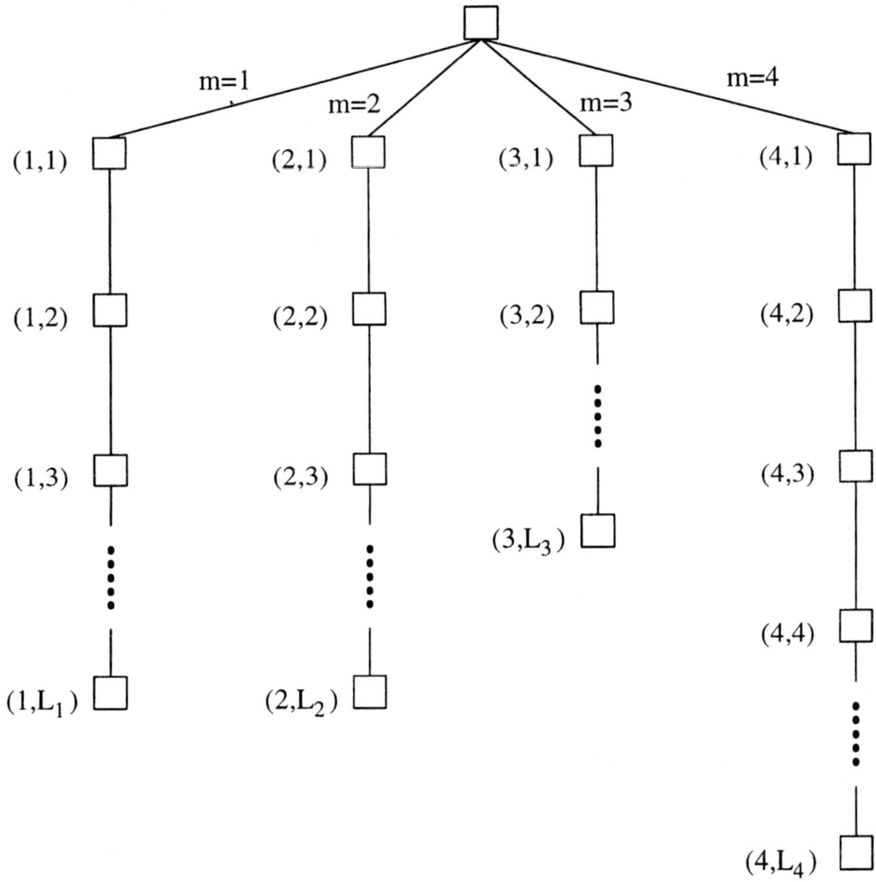

FIGURE 3. Example of a 4-ary unary tree.

explicitly in the design process, the entropy coder optimization step is effectively a high order modeling procedure. The multistage residual structure substantially reduces the large complexity demands, usually associated with finite-state modeling, and makes exploiting high order dependencies much easier by producing multiresolution approximations of the input subband images.

The distortion measure used by this coder is the absolute distortion measure. Since it only requires differences and additions to be computed, and its performance is shown empirically to be comparable to that of the squared-error measure, this choice leads to significant savings in computational complexity. Besides the actual distortion calculation, the only difference in the quantizer design is that the cell centroids are no longer computed as the average of the samples in the cell, but rather as their median. Entropy coder optimization is independent of the distortion measure.

Since multidimensional prediction is the determining factor in the high performance of the coder, one important step in the design is the entropy coder optimization procedure, consisting of the generation of statistical models for each stage and image subband. We start with a sufficiently large region of support as shown in Figure 2, with R conditioning stage symbols. The entropy coder optimization consists then of four steps. First we locate the best $K < R$ conditioning symbols for each stage (n, m, p). Then we use the BFOS algorithm to find the best orders $k < K$ subject to a limit on the total number N_1 of conditional probabilities. Then we generate a sequence of (state-entropy) pairs for each stage by using the PNN algorithm to merge conditioning states. Finally we use the generalized BFOS algorithm again to find the best number of conditioning states for each stage subject to a limit on the total number $N_2 << N_1$ of conditional probabilities.

4 Multiplierless Filter Banks

Filter banks can often consume a substantial amount of computation. To address this issue a special class of recursive filter banks was introduced in [SE91]. The recursive filter banks to which we refer use first-order allpass polyphase filters. The lowpass and highpass transfer functions for the analysis are given by

$$H_0(z) = \frac{p_0 + z^{-2}}{1 + p_0 z^{-2}} + z^{-1}\frac{p_1 + z^{-2}}{1 + p_1 z^{-2}},$$

and

$$H_1(z) = \frac{p_0 + z^{-2}}{1 + p_0 z^{-2}} + z^{-1}\frac{p_1 + z^{-2}}{1 + p_1 z^{-2}}.$$

Since multiplications can be implemented by serial additions, consider the total number of additions implied by the equations above. Let n_0 and n_1 be the number of serial adds required to implement the multiplications by p_0 and p_1. The number of adds per pixel for filtering either the rows or columns is then $(n_0+2)+(n_1+2)+1 = n_0+n_1+5$. If the binary representation of p_i has d_i nonzero digits, then $n_i = d_i - 1$ and the overall number of adds per pixel is $d_0 + d_1 + 3$. This number can be reduced by a careful design of the filter bank coefficients.

Toward this goal we employ a signed-bit representation proposed in [Sam89, Avi61] which, for a fractional number, has the form:

$$p = s_0.s_1s_2s_3\ldots = \sum_{k=0}^{\infty} s_k 2^{-k}$$

where $s_k \in \{-1, 0, 1\}$. The added flexibility of having negative digits allows for representations having fewer nonzero digits. The complexity per nonzero digit remains the same since additions and subtractions have the same complexity. The canonic signed-digit representation (CSD) of a number is the signed-digit representation having the minimum number of nonzero digits for which no two nonzero digits are adjacent. The CSD representation is unique, and a simple algorithm for converting a conventional binary representation to CSD exists [Hwa79].

To design a pair (p_0, p_1) that minimizes a frequency-domain error function and only requires a small number of additions per pixel, we use the CSD representation

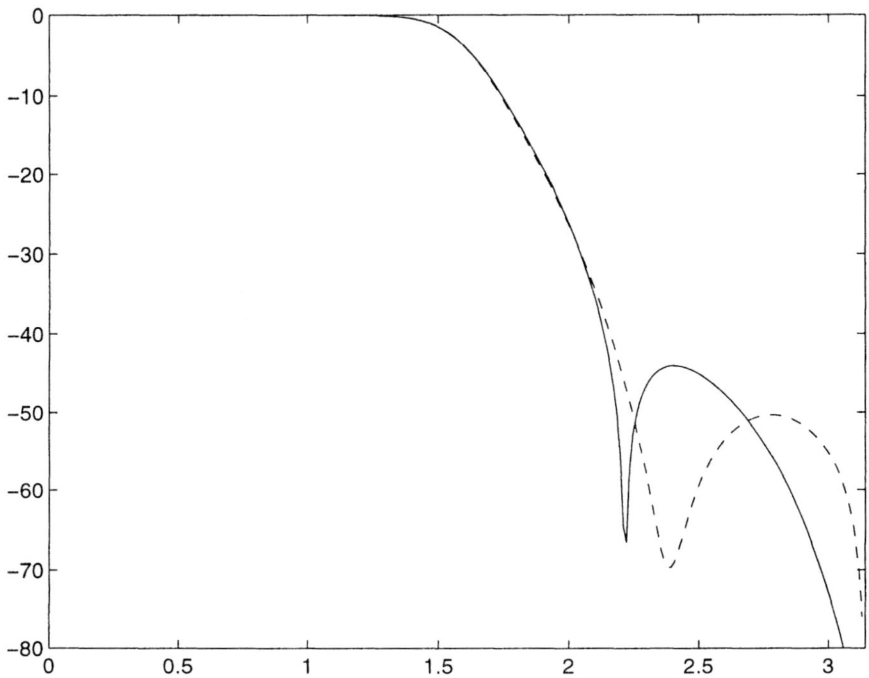

FIGURE 4. Frequency response of the lowpass analysis filter: floating-point (solid) and multiplier-free (dashed).

of the coefficients. We start with a full-precision set of coefficients (p_0^{FP}, p_1^{FP}), like the ones proposed in [SE91]. Then we compute exhaustively the error function over the points (p_0, p_1) having a finite conventional binary representation inside the rectangle $[p_0^{FP} - \Delta_0, p_0^{FP} + \Delta_0] \times [p_1^{FP} - \Delta_1, p_1^{FP} + \Delta_1]$. For all these points, we compute the complexity index as $d_0 + d_1 + 3$, where d_i is the number of nonzero digits in the CSD representation of p_i. If we are restricted to using n adds per pixel, then we choose the pair (p_0, p_1) that minimizes the error function subject to the constraint $d_0 + d_1 + 3 \leq n$.

To highlight the kinds of filters possible, Figure 4 shows a first-order IIR polyphase lowpass filter in comparison with a multiplier-free filter. As can be seen, the quality in cutoff and attenuation is exceptionally high, while the complexity is exceptionally low.

5 Performance

This system performs very well and is extremely versatile in terms of the classes of images it can handle. In this section a set of examples is provided to illustrate the system's quality at bit rates below 1 bit/pixel.

For general image coding applications, we design the system by using a large set

Bit rate	0.5bpp	0.25bpp	0.125bpp
Optimized SBC	37.19	34.07	30.73
Shapiro	36.28	33.17	30.23
Said and Pearlman	37.25	34.15	31.11
JPEG	34.8	31.55	21.93

TABLE 7.1. Peak SNR for several image coding algorithms (all values in dB).

of images of all kinds. This results in a generic finite model that tends to work well for all images. As mentioned earlier in the section on design, the training images are decomposed using a 3-level uniform tree-structure IIR filter bank, resulting in $(3)(64) = 192$ color image subbands. Interestingly, of the subbands in the Y plane many do not have sufficient energy to be considered for encoding in the low bit rate range. Similarly, only about 15% of the chrominance subbands have significant energy. These properties are largely responsible for the high compression ratios subband image coders are able to achieve.

This optimized subband coding algorithm (optimized for generic images) compares favorably with the best subband coders reported in the literature. Table 7.1 shows a comparison of several subband coders and the JPEG standard for the 512×512 test image Lena. Generally speaking, all of the subband coders listed above seem to perform at about the same level for the popular natural images. JPEG, on the other hand, performs significantly worse, particularly at the lower bit rates.

Let us look at the computational complexity and memory requirements of the optimized subband coder. The memory required to store all codebooks is only 1424 bytes, while that required to store the conditional probabilities is approximately 512 bytes. The average number of additions required for encoding 4.64 per input sample. We can see that the complexity of this coder is very modest.

First, the results of a design using the LENA image are summarized in Table 7.2 and compared with the full-precision coefficients results given in [SE91]. When the full-precision coefficients are used at a rate of 0.25 bits per pixel, the PSNR is 34.07 dB. The loss in peak SNR due to the use of finite precision arithmetic relative to the floating-point arithmetic analysis/synthesis filters is given in the last column. The first eight rows refer to an "A11" filter bank, the ninth row to an "A01" and the tenth row to an "A00" filter bank. The data show that little performance loss is incurred by reducing the complexity of the filter banks up until the very end of the table. Therefore the filter banks with three to seven adds as shown are reasonable candidates for the optimized subband coder.

Although instead of the mean square error distortion measure used to compute the PSNR we use the absolute distortion measure, the gap in PSNR between our coder and JPEG is large. In terms of complexity, the JPEG implementation used requires approximately 7 adds and 2 multiplies (per pixel) for encoding and 7 adds and 1 multiply for decoding. This is comparable to the complexity of the optimized coder, which requires approximately 11 to 16 adds for encoding (including analysis), and 10 to 15 adds for decoding (including synthesis). However, given the big differences in performance and relatively small differences in complexity, the

Decimal representation		CSD representation		Adds/ pixel	Decrease in PSNR
p_0	p_1	p_0	p_1		
0.141279	0.589817	–	–	–	0
0.14501953125	0.595703125	.00100101001	$.1010\underline{1}0001$	11	0.00
0.1435546875	0.59375	$.001001010\underline{1}$	$.1010\underline{1}$	10	0.00
0.140625	0.58984375	.001001	$.1010\underline{1}00\underline{1}$	9	0.00
0.15625	0.609375	.00101	$.10100\underline{1}$	8	-0.01
0.125	0.5703125	.001	.1001001	7	-0.01
0.125	0.5625	.001	.1001	6	-0.01
0.125	0.5	.001	.1	5	0.05
–	0.5	–	.1	3	0.40
–	–	–	–	1	3.24

TABLE 7.2. Performance for different complexity indices. ($\underline{1}$ is a -1 digit.)

optimized subband coder is attractive.

Perhaps the greatest advantage of the optimized subband coder is that is can be customized to a specific class of images to produce dramatic improvements. For example, envision a medical image compression scenario where the task is to compress X-ray CT images or magnetic resonance images, or envision an aerial surveillance scenario in which high altitude aerial images are being compressed and transmitted. The world is filled with such compression scenarios involving specific classes of images. The optimized subband coder can take advantage of such a context and yield performance results that are superior to those of a generic algorithm. In experiments with medical images, we have observed improvements as high as three decibels in PSNR along with noticeable improvements in subjective quality.

To illustrate the subjective improvements one can obtain by class optimization, consider compression of synthetic aperture radar images (SAR). We designed the coder using a wide range of SAR images and then compared the coded results to JPEG and the Said and Pearlman subband image coder. Visual comparisons are shown in Figure 5, where we see an original SAR image next to three versions of that image coded at 0.19 bits/pixel. What is clearly evident is that the optimized coder is better able to capture the natural texture of the original and does not display the blocking artifacts seen in the JPEG coded image or the ringing distortion seen in Said and Pearlman results.

6 References

[Avi61] A. Avizienis. Signed digit number representation for fast parallel arithmetic. *IRE Trans. Electron. Computers*, EC-10:389–400, September 1961.

[BF93] C. F. Barnes and R. L. Frost. Vector quantizers with direct sum codebooks. *IEEE Trans. on Information Theory*, 39(2):565–580, March 1993.

[CWF76] R. E. Crochiere, S. A. Weber, and F. L. Flanagan. Digital coding of speech in sub-bands. *Bell Syst. Tech. J.*, 55:1069–1085, Oct. 1976.

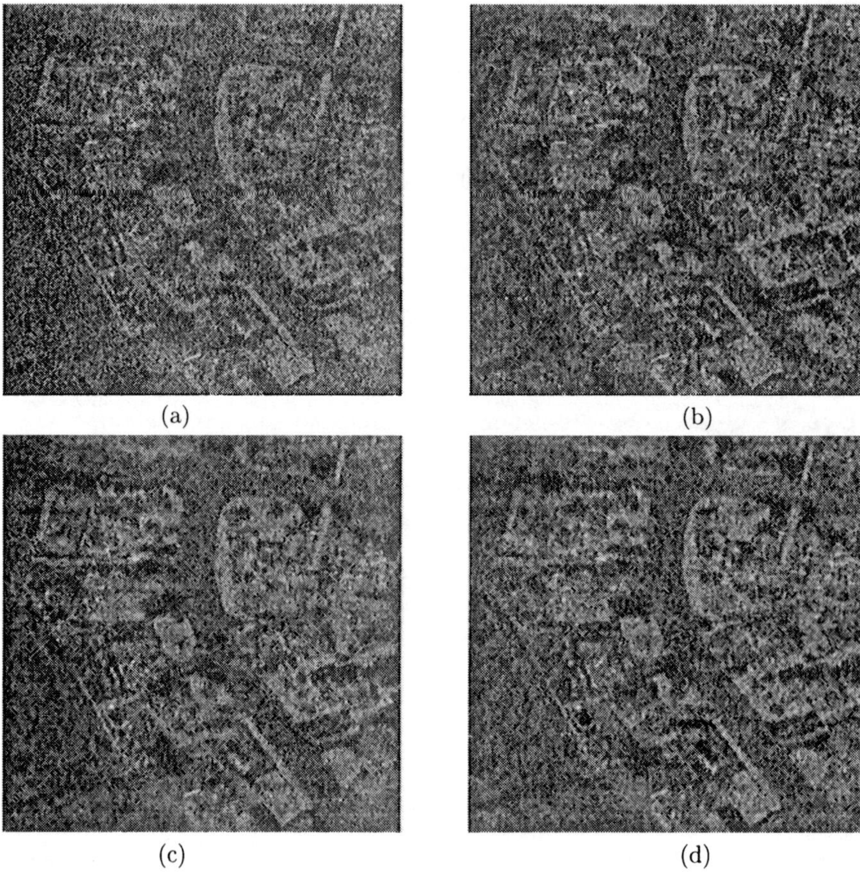

FIGURE 5. Coding results at 0.19 bits/pixel: (a) Original image; (b) Image coded with JPEG; (c)Image coded with Said and Pearlman's subband coder; (d) Image coded using the optimized subband coder.

[Equ89] W. H. Equitz. A new vector quantization clustering algorithm. *IEEE Trans. on Acoustics, Speech, and Signal Processing*, ASSP-37:1568–1575, October 1989.

[GT86] H. Gharavi and A. Tabatabai. Subband coding of digital image using two-dimensional quadrature mirror filtering. In *SPIE Proc. Visual Communications and Image Processing*, pages 51–61, 1986.

[GT88] D. Le Gall and A. Tabatabai. Subband coding of digital images using short kernel filters and arithmetic coding techniques. In *IEEE International Conference on Acoustics, Speech, and Signal Processing*, pages 761–764, April 1988.

[Hus91] J.H. Husøy. Low-complexity subband coding of still images and video.

Optical Engineering, 30(7), July 1991.

[Hwa79] K. Hwang. *Computer Arithmetic: Principles, Architecture and Design.* Wiley, New York, 1979.

[JS90] P. Jeanrenaud and M. J. T. Smith. Recursive multirate image coding with adaptive prediction and finite state vector quantization. *Signal Processing*, 20(1):25–42, May 1990.

[KCS93] F. Kossentini, W. Chung, and M. Smith. Subband image coding with intra- and inter-band subband quantization. In *Asilomar Conf. on Signals, Systems and Computers*, November 1993.

[KCS95] F. Kossentini, W. Chung, and M. Smith. Subband coding of color images with multiplierless encoders and decoders. In *IEEE International Symposium on Circuits and Systems*, May 1995.

[KCS96] F. Kossentini, W. Chung, and M. Smith. A jointly optimized subband coder. *IEEE Trans. on Image Processing*, August 1996.

[KSB93] F. Kossentini, M. Smith, and C. Barnes. Entropy-constrained residual vector quantization. In *IEEE International Conference on Acoustics, Speech, and Signal Processing*, volume V, pages 598–601, April 1993.

[KSB94a] F. Kossentini, M. Smith, and C. Barnes. Finite-state residual vector quantization. *Journal of Visual Communication and Image Representation*, 5(1):75–87, March 1994.

[KSB94b] F. Kossentini, M. Smith, and C. Barnes. Necessary conditions for the optimality of variable rate residual vector quantizers. *IEEE Trans. on Information Theory*, 1994.

[KSM89] C. S. Kim, M. Smith, and R. Mersereau. An improved sbc/vq scheme for color image coding. In *IEEE International Conference on Acoustics, Speech, and Signal Processing*, pages 1941–1944, May 1989.

[LR81] G. G. Langdon and J. Rissanen. Compression of black-white images with arithmetic coding. *IEEE Trans. on Communications*, 29(6):858–867, 1981.

[PMLA88] W. B. Pennebaker, J. L. Mitchel, G. G. Langdon, and R. B. Arps. An overview of the basic principles of the Q-coder adaptive binary arithmetic coder. *IBM J. Res. Dev.*, 32(6):717–726, November 1988.

[Ris91] E. A. Riskin. Optimal bit allocation via the generalized BFOS algorithm. *IEEE Trans. on Information Theory*, 37:400–402, March 1991.

[Sam89] H. Samueli. An improved search algorithm for the design of multiplierless fir filters with powers-of-two coefficients. *IEEE Trans. on Circuits and Systems*, 36:1044–1047, July 1989.

[SE91] M.J.T. Smith and S.L. Eddins. Analysis/synthesis techniques for sub-band image coding. *IEEE Trans. on Acoustics, Speech, and Signal Processing*, 38(8):1446–1456, August 1991.

[Sha92] J. M. Shapiro. An embedded wavelet hierarchical image coder. In *IEEE International Conference on Acoustics, Speech, and Signal Processing*, pages 657–660, March 1992.

[Vet84] M. Vetterli. Multi-dimensional sub-band coding: some theory and algorithms. *Signal Processing*, 6:97–112, February 1984.

[WBBW88] P. Westerink, J. Biemond, D. Boekee, and J. Woods. Sub-band coding of images using vector quantization. *IEEE Trans. on Communications*, 36(6):713–719, June 1988.

[Wes89] P. H. Westerink. *Subband Coding of Images.* PhD thesis, T. U. Delft, October 1989.

[WO86] J. W. Woods and S. D. O'Neil. Subband coding of images. *IEEE Trans. on Acoustics, Speech, and Signal Processing*, ASSP-34:1278–1288, October 1986.

[Woo91] John W. Woods, editor. *Subband Image Coding.* Kluwer Academic publishers, Boston, 1991.

8

Embedded Image Coding Using Zerotrees of Wavelet Coefficients

Jerome M. Shapiro

ABSTRACT [1]
The Embedded Zerotree Wavelet Algorithm (EZW) is a simple, yet remarkably effective, image compression algorithm, having the property that the bits in the bit stream are generated in order of importance, yielding a fully embedded code. The embedded code represents a sequence of binary decisions that distinguish an image from the "null" image. Using an embedded coding algorithm, an encoder can terminate the encoding at any point thereby allowing a target rate or target distortion metric to be met exactly. Also, given a bit stream, the decoder can cease decoding at any point in the bit stream and still produce exactly the same image that would have been encoded at the bit rate corresponding to the truncated bit stream. In addition to producing a fully embedded bit stream, EZW consistently produces compression results that are competitive with virtually all known compression algorithms on standard test images. Yet this performance is achieved with a technique that requires absolutely no training, no pre-stored tables or codebooks, and requires no prior knowledge of the image source.
The EZW algorithm is based on four key concepts: 1) a discrete wavelet transform or hierarchical subband decomposition, 2) prediction of the absence of significant information across scales by exploiting the self-similarity inherent in images, 3) entropy-coded successive-approximation quantization, and 4) universal lossless data compression which is achieved via adaptive arithmetic coding.

1 Introduction and Problem Statement

This paper addresses the two-fold problem of 1) obtaining the best image quality for a given bit rate, and 2) accomplishing this task in an embedded fashion, i.e. in such a way that all encodings of the same image at lower bit rates are embedded in the beginning of the bit stream for the target bit rate.

The problem is important in many applications, particularly for progressive transmission, image browsing [25], multimedia applications, and compatible transcoding in a digital hierarchy of multiple bit rates. It is also applicable to transmission over a noisy channel in the sense that the ordering of the bits in order of importance leads naturally to prioritization for the purpose of layered protection schemes.

1.1 Embedded Coding

An embedded code represents a sequence of binary decisions that distinguish an image from the "null", or all gray, image. Since, the embedded code contains all lower rate codes "embedded" at the beginning of the bit stream, effectively, the bits are "ordered in importance". Using an embedded code, an encoder can terminate the encoding at any point thereby allowing a target rate or distortion metric to be met exactly. Typically, some target parameter, such as bit count, is monitored in the encoding process. When the target is met, the encoding simply stops. Similarly, given a bit stream, the decoder can cease decoding at any point and can produce reconstructions corresponding to all lower-rate encodings.

Embedded coding is similar in spirit to binary finite-precision representations of real numbers. All real numbers can be represented by a string of binary digits. For each digit added to the right, more precision is added. Yet, the "encoding" can cease at any time and provide the "best" representation of the real number achievable within the framework of the binary digit representation. Similarly, the embedded coder can cease at any time and provide the "best" representation of an image achievable within its framework.

The embedded coding scheme presented here was motivated in part by universal coding schemes that have been used for lossless data compression in which the coder attempts to optimally encode a source using no prior knowledge of the source. An excellent review of universal coding can be found in [3]. In universal coders, the encoder must *learn* the source statistics as it progresses. In other words, the source model is incorporated into the actual bit stream. For lossy compression, there has been little work in universal coding. Typical image coders require extensive training for both quantization (both scalar and vector) and generation of non-adaptive entropy codes, such as Huffman codes. The embedded coder described in this paper attempts to be universal by incorporating all learning into the bit stream itself. This is accomplished by the exclusive use of adaptive arithmetic coding.

Intuitively, for a given rate or distortion, a non-embedded code should be more efficient than an embedded code, since it is free from the constraints imposed by embedding. In their theoretical work [9], Equitz and Cover proved that a successively refinable description can only be optimal if the source possesses certain Markovian properties. Although optimality is never claimed, a method of generating an embedded bit stream with no apparent sacrifice in image quality has been developed.

1.2 Features of the Embedded Coder

The EZW algorithm contains the following features:

- A discrete wavelet transform which provides a compact multiresolution representation of the image.

- Zerotree coding which provides a compact multiresolution representation of *significance maps*, which are binary maps indicating the positions of the significant coefficients. Zerotrees allow the successful prediction of insignificant coefficients across scales to be efficiently represented as part of exponentially growing trees.

- Successive Approximation which provides a compact *multiprecision* representation of the significant coefficients and facilitates the embedding algorithm.
- A prioritization protocol whereby the ordering of importance is determined, in order, by the precision, magnitude, scale, and spatial location of the wavelet coefficients. Note in particular, that larger coefficients are deemed more important than smaller coefficients regardless of their scale.
- Adaptive multilevel arithmetic coding which provides a fast and efficient method for entropy coding strings of symbols, and requires no training or prestored tables. The arithmetic coder used in the experiments is a customized version of that in [31].
- The algorithm runs sequentially and stops whenever a target bit rate or a target distortion is met. A target bit rate can be met *exactly*, and an operational rate vs. distortion function (RDF) can be computed point-by-point.

1.3 Paper Organization

Section II discusses how wavelet theory and multiresolution analysis provide an elegant methodology for representing "trends" and "anomalies" on a statistically equal footing. This is important in image processing because edges, which can be thought of as anomalies in the spatial domain, represent extremely important information despite the fact that they are represented in only a tiny fraction of the image samples. Section III introduces the concept of a *zerotree* and shows how zerotree coding can efficiently encode a significance map of wavelet coefficients by predicting the absence of significant information across scales. Section IV discusses how successive approximation quantization is used in conjunction with zerotree coding, and arithmetic coding to achieve efficient embedded coding. A discussion follows on the protocol by which EZW attempts to order the bits in order of importance. A key point there is that the definition of importance for the purpose of ordering information is based on the magnitudes of the uncertainty intervals as seen from the viewpoint of what the decoder can figure out. Thus, there is no additional overhead to transmit this ordering information. Section V consists of a simple 8×8 example illustrating the various points of the EZW algorithm. Section VI discusses experimental results for various rates and for various standard test images. A surprising result is that using the EZW algorithm, terminating the encoding at an arbitrary point in the encoding process does not produce any artifacts that would indicate where in the picture the termination occurs. The paper concludes with Section VII.

2 Wavelet Theory and Multiresolution Analysis

2.1 Trends and Anomalies

One of the oldest problems in statistics and signal processing is how to choose the size of an analysis window, block size, or record length of data so that statistics computed within that window provide good models of the signal behavior within that window. The choice of an analysis window involves trading the ability to analyze "anomalies", or signal behavior that is more localized in the time or space domain and tends to be wide band in the frequency domain, from "trends", or signal behavior that is more localized in frequency but persists over a large number of lags in the time domain. To model data as being generated by random processes so that computed statistics become meaningful, stationary and ergodic assumptions are usually required which tend to obscure the contribution of anomalies.

The main contribution of wavelet theory and multiresolution analysis is that it provides an elegant framework in which both anomalies and trends can be analyzed on an equal footing. Wavelets provide a signal representation in which some of the coefficients represent long data lags corresponding to a narrow band, low frequency range, and some of the coefficients represent short data lags corresponding to a wide band, high frequency range. Using the concept of *scale*, data representing a continuous tradeoff between time (or space in the case of images) and frequency is available.

For an introduction to the theory behind wavelets and multiresolution analysis, the reader is referred to several excellent tutorials on the subject [6], [7], [17], [18], [20], [26], [27].

2.2 Relevance to Image Coding

In image processing, most of the image area typically represents spatial "trends", or areas of high statistical spatial correlation. However "anomalies", such as edges or object boundaries, take on a perceptual significance that is far greater than their numerical energy contribution to an image. Traditional transform coders, such as those using the DCT, decompose images into a representation in which each coefficient corresponds to a fixed size spatial area and a fixed frequency bandwidth, where the bandwidth and spatial area are effectively the same for all coefficients in the representation. Edge information tends to disperse so that many non-zero coefficients are required to represent edges with good fidelity. However, since the edges represent relatively insignificant energy with respect to the entire image, traditional transform coders, such as those using the DCT, have been fairly successful at medium and high bit rates. At extremely low bit rates, however, traditional transform coding techniques, such as JPEG [30], tend to allocate too many bits to the "trends", and have few bits left over to represent "anomalies". As a result, blocking artifacts often result.

Wavelet techniques show promise at extremely low bit rates because trends, anomalies and information at all "scales" in between are available. A major difficulty is that fine detail coefficients representing possible anomalies constitute the largest number of coefficients, and therefore, to make effective use of the multiresolution representation, much of the information is contained in representing the *position* of those few coefficients corresponding to significant anomalies.

The techniques of this paper allow coders to effectively use the power of multiresolution representations by efficiently representing the positions of the wavelet coefficients representing significant anomalies.

2.3 A Discrete Wavelet Transform

The discrete wavelet transform used in this paper is identical to a hierarchical subband system, where the subbands are logarithmically spaced in frequency and represent an octave-band decomposition. To begin the decomposition, the image is divided into four subbands and critically subsampled as shown in Fig. 1. Each coefficient represents a spatial area corresponding to approximately a 2×2 area of the original image. The low frequencies represent a bandwidth approximately corresponding to $0 < |\omega| < \frac{\pi}{2}$, whereas the high frequencies represent the band from $\frac{\pi}{2} < |\omega| < \pi$. The four subbands arise from separable application of vertical and horizontal filters. The subbands labeled LH_1, HL_1, and HH_1 represent the finest scale wavelet coefficients. To obtain the next coarser scale of wavelet coefficients, the subband LL_1 is further decomposed and critically sampled as shown in Fig. 2. The process continues until some final scale is reached. Note that for each coarser scale, the coefficients represent a larger spatial area of the image but a narrower band of frequencies. At each scale, there are 3 subbands; the remaining lowest frequency subband is a representation of the information at all coarser scales. The issues involved in the design of the filters for the type of subband decomposition described above have been discussed by many authors and are not treated in this paper. Interested readers should consult [1], [6], [32], [35], in addition to references found in the bibliographies of the tutorial papers cited above.

It is a matter of terminology to distinguish between a transform and a subband system as they are two ways of describing the same set of numerical operations from differing points of view. Let $\mathbf{x}$ be a column vector whose elements represent a scanning of the image pixels, let $\mathbf{X}$ be a column vector whose elements are the array of coefficients resulting from the wavelet transform or subband decomposition applied to $\mathbf{x}$. From the transform point of view, $\mathbf{X}$ represents a linear transformation of $\mathbf{x}$ represented by the matrix $\mathbf{W}$, i.e.,

$$\mathbf{X} = \mathbf{W}\mathbf{x}. \tag{8.1}$$

Although not actually computed this way, the effective filters that generate the subband signals from the original signal form basis functions for the transformation, i.e, the rows of $\mathbf{W}$. Different coefficients in the same subband represent the projection of the entire image onto translates of a prototype subband filter, since from the subband point of view, they are simply regularly spaced different outputs of a convolution between the image and a subband filter. Thus, the basis functions for each coefficient in a given subband are simply translates of one another.

In subband coding systems [32], the coefficients from a given subband are usually grouped together for the purposes of designing quantizers and coders. Such a grouping suggests that statistics computed from a subband are in some sense representative of the samples in that subband. However this statistical grouping once again implicitly de-emphasizes the outliers, which tend to represent the most significant anomalies or edges. In this paper, the term "wavelet transform" is used because each wavelet coefficient is individually and deterministically compared to the same set of thresholds for the purpose of measuring significance. Thus, each coefficient is treated as a distinct, potentially important piece of data regardless of its scale, and no statistics for a whole subband are used in any form. The result is that the small number of "deterministically" significant fine scale coefficients are not obscured because of their "statistical" insignificance.

The filters used to compute the discrete wavelet transform in the coding experiments described in this paper are based on the 9-tap symmetric Quadrature Mirror filters (QMF) whose coefficients are given in [1]. This transformation has also been called a QMF-pyramid. These filters were chosen because in addition to their good localization properties, their symmetry allows for simple edge treatments, and they produce good results empirically. Additionally, using properly scaled coefficients, the transformation matrix for a discrete wavelet transform obtained using these filters is so close to unitary that it can be treated as unitary for the purpose of lossy compression. Since unitary transforms preserve L_2 norms, it makes sense from a numerical standpoint to compare all of the resulting transform coefficients to the same thresholds to assess significance.

3 Zerotrees of Wavelet Coefficients

In this section, an important aspect of low bit rate image coding is discussed: the coding of the positions of those coefficients that will be transmitted as non-zero values. Using scalar quantization followed by entropy coding, in order to achieve very low bit rates, i.e., less than 1 bit/pel, the probability of the most likely symbol after quantization - the zero symbol - must be extremely high. Typically, a large fraction of the bit budget must be spent on encoding the *significance map*, or the binary decision as to whether a sample, in this case a coefficient of a 2-D discrete wavelet transform, has a zero or non-zero quantized value. It follows that a significant improvement in encoding the significance map translates into a corresponding gain in compression efficiency.

3.1 Significance Map Encoding

To appreciate the importance of significance map encoding, consider a typical transform coding system where a decorrelating transformation is followed by an entropy-coded scalar quantizer. The following discussion is not intended to be a rigorous justification for significance map encoding, but merely to provide the reader with a sense of the relative coding costs of the position information contained in the significance map relative to amplitude and sign information.

A typical low-bit rate image coder has three basic components: a transformation, a quantizer and data compression, as shown in Fig. 3. The original image is passed through some transformation to produce transform coefficients. This transformation is considered to be lossless, although in practice this may not be the case exactly. The transform coefficients are then quantized to produce a stream of symbols, each of which corresponds to an index of a particular quantization bin. Note that virtually all of the information loss occurs in the quantization stage. The data compression stage takes the stream of symbols and attempts to losslessly represent the data stream as efficiently as possible.

The goal of the transformation is to produce coefficients that are decorrelated. If we could, we would ideally like a transformation to remove all dependencies between samples. Assume for the moment that the transformation is doing its job so well that the resulting transform coefficients are not merely uncorrelated, but statistically independent. Also, assume that we have removed the mean and coded it separately so that the transform coefficients can be modeled as zero-mean, independent, although perhaps not identically distributed random variables. Furthermore, we might additionally constrain the model so that the probability density functions (PDF) for the coefficients are symmetric.

The goal is to quantize the transform coefficients so that the entropy of the resulting distribution of bin indices is small enough so that the symbols can be entropy-coded at some target low bit rate, say for example 0.5 bits per pixel (bpp.). Assume that the quantizers will be symmetric mid-tread, perhaps non-uniform, quantizers, although different symmetric mid-tread quantizers may be used for different groups of transform coefficients. Letting the central bin be index 0, note that because of the symmetry, for a bin with a non-zero index magnitude, a positive or negative index is equally likely. In other words, for each non-zero index encoded, the entropy code is going to require at least one-bit for the sign. An entropy code can be designed based on modeling probabilities of bin indices as the fraction of coefficients in which the absolute value of a particular bin index occurs. Using this simple model, and assuming that the resulting symbols are independent, the entropy of the symbols H can be expressed as

$$H = -p\log_2 p - (1-p)\log_2(1-p) + (1-p)[1 + H_{NZ}], \tag{8.2}$$

where p is the probability that a transform coefficient is quantized to zero, and H_{NZ} represents the conditional entropy of the absolute values of the quantized coefficients conditioned on them being non-zero. The first two terms in the sum represent the first-order binary entropy of the significance map, whereas the third term represents the conditional entropy of the distribution of non- zero values multiplied by the probability of them being non-zero. Thus we can express the true cost of encoding the actual symbols as follows:

$$\text{Total Cost} = \text{Cost of Significance Map} + \text{Cost of Non-Zero Values}. \tag{8.3}$$

Returning to the model, suppose that the target is $H = 0.5$. What is the minimum probability of zero achievable? Consider the case where we only use a 3-level quantizer, i.e. $H_{NZ} = 0$. Solving for p provides a lower bound on the probability of zero given the independence assumption.

$$p_{min}(H_{NZ} = 0, H = 0.5) = 0.916. \tag{8.4}$$

In this case, under the most ideal conditions, 91.6% of the coefficients must be quantized to zero. Furthermore, 83% of the bit budget is used in encoding the significance map. Consider a more typical example where $H_{NZ} = 4$, the minimum probability of zero is

$$p_{min}(H_{NZ} = 4, H = 0.5) = 0.954. \tag{8.5}$$

In this case, the probability of zero must increase, while the cost of encoding the significance map is still 54% of the cost.

As the target rate decreases, the probability of zero increases, and the fraction of the encoding cost attributed to the significance map increases. Of course, the independence assumption is unrealistic and in practice, there are often additional dependencies between coefficients that can be exploited to further reduce the cost of encoding the significance map. Nevertheless, the conclusion is that no matter how optimal the transform, quantizer or entropy coder, under very typical conditions, the cost of determining the positions of the few significant coefficients represents a significant portion of the bit budget at low rates, and is likely to become an increasing fraction of the total cost as the rate decreases. As will be seen, by employing an image model based on an extremely simple and easy to satisfy hypothesis, we can efficiently encode significance maps of wavelet coefficients.

3.2 Compression of Significance Maps using Zerotrees of Wavelet Coefficients

To improve the compression of significance maps of wavelet coefficients, a new data structure called a *zerotree* is defined. A wavelet coefficient x is said to be *insignificant* with respect to a given threshold T if $|x| < T$. The zerotree is based on the hypothesis that if a wavelet coefficient at a coarse scale is insignificant with respect to a given threshold T, then *all* wavelet coefficients of the same orientation in the same spatial location at finer scales are likely to be insignificant with respect to T. Empirical evidence suggests that this hypothesis is often true.

More specifically, in a hierarchical subband system, with the exception of the highest frequency subbands, every coefficient at a given scale can be related to a set of coefficients at the next finer scale of similar orientation. The coefficient at the coarse scale is called the *parent*, and all coefficients corresponding to the same spatial location at the next finer scale of similar orientation are called *children*. For a given parent, the set of all coefficients at all finer scales of similar orientation corresponding to the same location are called *descendants*. Similarly, for a given child, the set of coefficients at all coarser scales of similar orientation corresponding to the same location are called *ancestors*. For a QMF-pyramid subband decomposition, the parent-child dependencies are shown in Fig. 4. A wavelet tree descending from a coefficient in subband $HH3$ is also seen in Fig. 4. With the exception of the lowest frequency subband, all parents have four children. For the lowest frequency subband, the parent-child relationship is defined such that each parent node has three children.

A scanning of the coefficients is performed in such a way that no child node is scanned before its parent. For an N-scale transform, the scan begins at the lowest frequency subband, denoted as LL_N, and scans subbands HL_N, LH_N, and HH_N, at which point it moves on to scale $N-1$, etc. The scanning pattern for a 3-scale QMF-pyramid can be seen in Fig. 5. Note that each coefficient within a given subband is scanned before any coefficient in the next subband.

Given a threshold level T to determine whether or not a coefficient is significant, a coefficient x is said to be an element of a *zerotree* for threshold T if itself and *all* of its descendents are insignificant with respect to T. An element of a zerotree for threshold T is a *zerotree root* if it is not the descendant of a previously found zerotree root for threshold T, i.e. it is not *predictably insignificant* from the discovery of a zerotree root at a coarser scale at the same threshold. A zerotree root is encoded with a special symbol indicating that the insignificance of the coefficients at finer scales is completely predictable. The significance map can be efficiently represented as a string of symbols from a 3-symbol alphabet which is then entropy-coded. The three symbols used are 1) zerotree root, 2) isolated zero, which means that the coefficient is insignificant but has some significant descendant, and 3) significant. When encoding the finest scale coefficients, since coefficients have no children, the symbols in the string come from a 2-symbol alphabet, whereby the zerotree symbol is not used.

As will be seen in Section IV, in addition to encoding the significance map, it is useful to encode the sign of significant coefficients along with the significance map. Thus, in practice, four symbols are used: 1) zerotree root, 2) isolated zero, 3) positive significant, and 4) negative significant. This minor addition will be useful for embedding. The flow chart for the decisions made at each coefficient are shown in Fig. 6.

Note that it is also possible to include two additional symbols such as "positive/negative significant, but descendants are zerotrees" etc. In practice, it was found that at low bit rates, this addition often increases the cost of coding the significance map. To see why this may occur, consider that there is a cost associated with partitioning the set of positive (or negative) significant samples into those whose descendents are zerotrees and those with significant descendants. If the cost of this decision is C bits, but the cost of encoding a zerotree is less than $C/4$ bits, then it is more efficient to code four zerotree symbols separately than to use additional symbols.

Zerotree coding reduces the cost of encoding the significance map using self-similarity. Even though the image has been transformed using a decorrelating transform the *occurrences* of insignificant coefficients are not independent events. More traditional techniques employing transform coding typically encode the binary map via some form of run-length encoding [30]. Unlike the zerotree symbol, which is a single "terminating" symbol and applies to all tree-depths, run-length encoding requires a symbol for each run-length which much be encoded. A technique that is closer in spirit to the zerotrees is the end-of-block (EOB) symbol used in JPEG [30], which is also a "terminating" symbol indicating that all remaining DCT coefficients in the block are quantized to zero. To see why zerotrees may provide an advantage over EOB symbols, consider that a zerotree represents the insignificance information in a given orientation over an approximately square spatial area at all finer scales up to and including the scale of the zerotree root. Because the wavelet transform is a hierarchical representation, varying the scale in which a zerotree root occurs automatically adapts the spatial area over which insignificance is represented. The EOB symbol, however, always represents insignificance over the same spatial area, although the number of frequency bands within that spatial area varies. Given a fixed block size, such as 8×8, there is exactly one scale in the wavelet transform in which if a zerotree root is found at that scale, it corresponds to the same spatial area as a block of the DCT. If a zerotree root can be identified at a coarser scale, then the insignificance pertaining to that orientation can be predicted over a larger area. Similarly, if the zerotree root does not occur at this scale, then looking for zerotrees at finer scales represents a hierarchical divide and conquer approach to searching for one or more smaller areas of insignificance over the same spatial regions as the DCT block size. Thus, many more coefficients can be predicted in smooth areas where a root typically occurs at a coarse scale. Furthermore, the zerotree approach can isolate interesting non-zero details by immediately eliminating large insignificant regions from consideration.

Note that this technique is quite different from previous attempts to exploit self-similarity in image coding [19] in that it is far easier to predict insignificance than to predict significant detail across scales. The zerotree approach was developed in recognition of the difficulty in achieving meaningful bit rate reductions for significant coefficients via additional prediction. Instead, the focus here is on reducing the cost of encoding the significance map so that, for a given bit budget, more bits are available to encode expensive significant coefficients. In practice, a large fraction of the insignificant coefficients are efficiently encoded as part of a zerotree.

A similar technique has been used by Lewis and Knowles (LK) [15], [16]. In that work, a "tree" is said to be zero if its energy is less than a perceptually based threshold. Also, the "zero flag" used to encode the tree is not entropy-coded. The present work represents an improvement that allows for embedded coding for two reasons. Applying a deterministic threshold to determine significance results in a zerotree symbol which guarantees that no descendant of the root has a magnitude larger than the threshold. As a result, there is no possibility of a significant coefficient being obscured by a statistical energy measure. Furthermore, the zerotree symbol developed in this paper is part of an alphabet for entropy coding the significance map which further improves its compression. As will be discussed subsequently, it is the first property that makes this method of encoding a significance map useful in conjunction with successive-approximation. Recently, a promising technique representing a compromise between the EZW algorithm and the LK coder has been presented in [34].

3.3 Interpretation as a Simple Image Model

The basic hypothesis - if a coefficient at a coarse scale is insignificant with respect to a threshold then all of its descendants, as defined above, are also insignificant - can be interpreted as an extremely general image model. One of the aspects that seems to be common to most models used to describe images is that of a "decaying spectrum". For example, this property exists for both stationary autoregressive models, and non-stationary fractal, or "nearly-$1/f$" models, as implied by the name which refers to a generalized spectrum [33]. The model for the zerotree hypothesis is even more general than "decaying spectrum" in that it allows for some deviations to "decaying spectrum" because it is linked to a specific threshold. Consider an example where the threshold is 50, and we are considering a coefficient of magnitude 30, and whose largest descendant has a magnitude of 40. Although a higher frequency descendant has a larger magnitude (40) than the coefficient under consideration (30), i.e. the "decaying spectrum" hypothesis is violated, the coefficient under consideration can still be represented using a zerotree root since the whole tree is still insignificant (magnitude less than 50). Thus, assuming the more common image models have some validity, the zerotree hypothesis should be satisfied easily and extremely often. For those instances where the hypothesis is violated, it is safe to say that an *informative*, i.e. unexpected, event has occurred, and we should expect the cost of representing this event to be commensurate with its self-information.

It should also be pointed out that the improvement in encoding significance maps provided by zerotrees is specifically *not* the result of exploiting any linear dependencies between coefficients of different scales that were not removed in the transform stage. In practice, the correlation between the values of parent and child wavelet coefficients has been found to be extremely small, implying that the wavelet transform is doing an excellent job of producing nearly uncorrelated coefficients. However, there is likely additional dependency between the *squares* (or magnitudes) of parents and children. Experiments run on about 30 images of all different types, show that the correlation coefficient between the square of a child and the square of its parent tends to be between 0.2 and 0.6 with a strong concentration around 0.35. Although this dependency is difficult to characterize in general for most images, even without access to specific statistics, it is reasonable to *expect* the magnitude of a child to be smaller than the magnitude of its parent. In other words, it can be reasonably conjectured based on experience with real-world images, that had we known the details of the statistical dependencies, and computed an "optimal" estimate, such as the conditional expectation of the child's magnitude given the parent's magnitude, that the "optimal" estimator would, with very high probability, predict that the child's magnitude would be the smaller of the two. Using only this mild assumption, based on an inexact statistical characterization, given a fixed threshold, and conditioned on the knowledge that a parent is insignificant with respect to the threshold, the "optimal" estimate of the significance of the rest of the descending wavelet tree is that it is entirely insignificant with respect to the same threshold, i.e., a zerotree. On the other hand, if the parent is significant, the "optimal" estimate of the significance of descendants is highly dependent on the details of the estimator whose knowledge would require more detailed information about the statistical nature of the image. Thus, under this mild assumption, using zerotrees to predict the insignificance of wavelet coefficients at fine scales given the insignificance of a root at a coarse scale is more likely to be successful in the absence of additional information than attempting to predict significant detail across scales.

This argument can be made more concrete. Let x be a child of y, where x and y are zero-mean random variables, whose probability density functions (PDF) are related as

$$p_x(x) = ap_y(ax), \quad a > 1. \tag{8.6}$$

This states that random variables x and y have the same PDF shape, and that

$$\sigma_y^2 = a^2\sigma_x^2. \tag{8.7}$$

Assume further that x and y are uncorrelated, i.e.,

$$E[xy] = 0. \tag{8.8}$$

Note that nothing has been said about treating the subbands as a group, or as stationary random processes, only that there is a similarity relationship between random variables of parents and children. It is also reasonable because for intermediate subbands a coefficient that is a child with respect to one coefficient is a parent with respect to others; the PDF of that coefficient should be the same in either case. Let $u = x^2$ and $v = y^2$. Suppose that u and v are correlated with correlation coefficient ρ. We have the following relationships:

$$E[u] = \sigma_x^2 \tag{8.9}$$

$$\begin{aligned} E[v] &= \sigma_y^2 && (8.10)\\ \sigma_u^2 &= E[x^4] - \sigma_x^4 && (8.11)\\ \sigma_v^2 &= E[y^4] - \sigma_y^4. && (8.12) \end{aligned}$$

Notice in particular that

$$\sigma_v^2 = a^4 \sigma_u^2. \qquad (8.13)$$

Using a well known result, the expression for the best linear unbiased estimator (BLUE) of u given v to minimize error variance is given by

$$\begin{aligned} \hat{u}_{BLUE}(v) &= E[u] - \rho \frac{\sigma_u}{\sigma_v}(E[v] - v) && (8.14)\\ &= \frac{1-\rho}{a^2}\sigma_y^2 + \rho \frac{v}{a^2} && (8.15) \end{aligned}$$

If it is observed that the magnitude of the parent is below the threshold T, i.e. $v = y^2 < T^2$, then the BLUE can be upper bounded by

$$\hat{u}_{BLUE}(v|v < T^2) < \frac{1-\rho}{a^2}\sigma_y^2 + \rho \frac{T^2}{a^2}. \qquad (8.16)$$

Consider two cases a) $T \geq \sigma_y$ and b) $T < \sigma_y$. In case (a), we have

$$\hat{u}_{BLUE}(v|v < T^2) \leq \frac{T^2}{a^2} < T^2, \qquad (8.17)$$

which implies that the BLUE of x^2 given $|y| < T$ is less than T^2, for *any* ρ, including $\rho = 0$. In case (b), we can only upper bound the right hand side of (8.16) by T^2 if ρ exceeds the lower bound

$$\rho \geq \frac{1 - \frac{a^2 T^2}{\sigma_y^2}}{1 - \frac{T^2}{\sigma_y^2}} \triangleq \rho_0. \qquad (8.18)$$

Of course, a better non-linear estimate might yield different results, but the above analysis suggests that for thresholds exceeding the standard deviation of the parent, which by (8.6) exceeds the standard deviation of all descendants, if it is observed that a parent is insignificant with respect to the threshold, then, using the above BLUE, the estimates for the magnitudes of all descendants is that they are less than the threshold, and a zerotree is expected regardless of the correlation between squares of parents and squares of children. As the threshold decreases, more correlation is required to justify *expecting* a zerotree to occur. Finally, since the lower bound $\rho_0 \to 1$ as $T \to 0$, as the threshold is reduced, it becomes increasingly difficult to expect zerotrees to occur, and more knowledge of the particular statistics are required to make inferences. The implication of this analysis is that at very low bit rates, where the probability of an insignificant sample must be high and thus, the significance threshold T must also be large, expecting the occurrence of zerotrees and encoding significance maps using zerotree coding is reasonable without even knowing the statistics. However, letting T decrease, there is some point below which the advantage of zerotree coding diminishes, and this point is dependent on the specific nature of higher order dependencies between parents and children. In particular, the stronger this dependence, the more T can be decreased while still retaining an advantage using zerotree coding. Once again, this argument is not intended to "prove" the optimality of zerotree coding, only to suggest a rationale for its demonstrable success.

3.4 Zerotree-like Structures in Other Subband Configurations

The concept of predicting the insignificance of coefficients from low frequency to high frequency information corresponding to the same spatial localization is a fairly general concept and not specific to the wavelet transform configuration shown in Fig. 4. Zerotrees are equally applicable to quincunx wavelets [2], [13], [23], [29], in which case each parent would have two children instead of four, except for the lowest frequency, where parents have a single child.

Also, a similar approach can be applied to linearly spaced subband decompositions, such as the DCT, and to other more general subband decompositions, such as wavelet packets [5] and Laplacian pyramids [4]. For example, one of many possible parent-child relationship for linearly spaced subbands can be seen in Fig. 7. Of course, with the use of linearly spaced subbands, zerotree-like coding loses its ability to adapt the spatial extent of the insignificance prediction. Nevertheless, it is possible for zerotree-like coding to outperform EOB-coding since more coefficients can be predicted from the subbands along the diagonal. For the case of wavelet packets, the situation is a bit more complicated, because a wider range of tilings of the "space-frequency" domain are possible. In that case, it may not always be possible to define similar parent-child relationships because a high-frequency coefficient may in fact correspond to a larger spatial area than a co-located lower frequency coefficient. On the other hand, in a coding scheme such as the "best-basis" approach of Coifman *et. al.* [5], had the image-dependent best basis resulted in such a situation, one wonders if the underlying hypothesis - that magnitudes of coefficients tend to decay with frequency - would be reasonable anyway. These zerotree-like extensions represent interesting areas for further research.

4 Successive-Approximation

The previous section describes a method of encoding significance maps of wavelet coefficients that, at least empirically, seems to consistently produce a code with a lower bit rate than either the empirical first-order entropy, or a run-length code of the significance map. The original motivation for employing successive-approximation in conjunction with zerotree coding was that since zerotree coding was performing so well encoding the significance map of the wavelet coefficients, it was hoped that more efficient coding could be achieved by zerotree coding more significance maps.

Another motivation for successive-approximation derives directly from the goal of developing an embedded code analogous to the binary-representation of an approximation to a real number. Consider the wavelet transform of an image as a mapping whereby an amplitude exists for each coordinate in scale-space. The scale-space coordinate system represents a coarse-to-fine "logarithmic" representation of the domain of the function. Taking the coarse-to-fine philosophy one-step further, successive-approximation provides a coarse-to-fine, multiprecision "logarithmic" representation of amplitude information, which can be thought of as the range of the image function when viewed in the scale-space coordinate system defined by the wavelet transform. Thus, in a very real sense, the EZW coder generates a representation of the image that is coarse-to-fine in both the domain and range simultaneously.

4.1 Successive-Approximation Entropy-Coded Quantization

To perform the embedded coding, successive-approximation quantization (SAQ) is applied. As will be seen, SAQ is related to bit-plane encoding of the magnitudes. The SAQ sequentially applies a sequence of thresholds $T_0, \ldots, T_{N-1}$ to determine significance, where the thresholds are chosen so that $T_i = T_{i-1}/2$. The initial threshold T_0 is chosen so that $|x_j| < 2T_0$ for all transform coefficients x_j.

During the encoding (and decoding), two separate lists of wavelet coefficients are maintained. At any point in the process, the *dominant* list contains the *coordinates* of those coefficients that have not yet been found to be significant in the same relative order as the initial scan. This scan is such that the subbands are ordered, and within each subband, the set of coefficients are ordered. Thus, using the ordering of the subbands shown in Fig. 5, all coefficients in a given subband appear on the initial dominant list prior to coefficients in the next subband. The *subordinate* list contains the *magnitudes* of those coefficients that have been found to be significant. For each threshold, each list is scanned once.

During a *dominant pass*, coefficients with coordinates on the dominant list, i.e. those that have not yet been found to be significant, are compared to the threshold T_i to determine their significance, and if significant, their sign. This significance map is then zerotree coded using the method outlined in Section III. Each time a coefficient is encoded as significant, (positive or negative), its magnitude is appended to the subordinate list, and the coefficient in the wavelet transform array is set to zero so that the significant coefficient does not prevent the occurrence of a zerotree on future dominant passes at smaller thresholds.

A dominant pass is followed by a *subordinate pass* in which all coefficients on the subordinate list are scanned and the specifications of the magnitudes available to the decoder are refined to an additional bit of precision. More specifically, during a subordinate pass, the *width* of the effective quantizer step size, which defines an uncertainty interval for the true magnitude of the coefficient, is cut in half. For each magnitude on the subordinate list, this refinement can be encoded using a binary alphabet with a "1" symbol indicating that the true value falls in the upper half of the old uncertainty interval and a "0" symbol indicating the lower half. The string of symbols from this binary alphabet that is generated during a subordinate pass is then entropy coded. Note that prior to this refinement, the width of the uncertainty region is exactly equal to the current threshold. After the completion of a subordinate pass the magnitudes on the subordinate list are sorted in decreasing magnitude, to the extent that the decoder has the information to perform the same sort.

The process continues to alternate between dominant passes and subordinate passes where the threshold is halved before each dominant pass. (In principle one could divide by other factors than 2. This factor of 2 was chosen here because it has nice interpretations in terms of bit plane encoding and numerical precision in a familiar base 2, and good coding results were obtained).

In the decoding operation, each decoded symbol, both during a dominant and a subordinate pass, refines and reduces the width of the uncertainty interval in which the true value of the coefficient (or coefficients, in the case of a zerotree root) may occur. The reconstruction value used can be anywhere in that uncertainty interval. For minimum mean-square error distortion, one could use the centroid of the uncertainty region using some model for the PDF of the coefficients. However, a practical approach, which is used in the experiments, and is also MINMAX optimal, is to simply use the center of the uncertainty interval as the reconstruction value.

The encoding stops when some target stopping condition is met, such as when the bit budget is exhausted. The encoding can cease at any time and the resulting bit stream contains all lower rate encodings. Note that if the bit stream is truncated at an arbitrary point, there may be bits at the end of the code that do not decode to a valid symbol since a codeword has been truncated. In that case, these bits do not reduce the width of an uncertainty interval or any distortion function. In fact, it is very likely that the first L bits of the bit stream will produce exactly the same image as the first $L + 1$ bits which occurs if the additional bit is insufficient to complete the decoding of another symbol. Nevertheless, terminating the decoding of an embedded bit stream at a specific point in the bit stream produces exactly the same image would have resulted had that point been the initial target rate. This ability to cease encoding or decoding anywhere is extremely useful in systems that are either rate-constrained or distortion-constrained. A side benefit of the technique is that an operational rate vs. distortion plot for the algorithm can be computed on-line.

4.2 Relationship to Bit Plane Encoding

Although the embedded coding system described here is considerably more general and more sophisticated than simple bit-plane encoding, consideration of the relationship with bit-plane encoding provides insight into the success of embedded coding.

Consider the successive-approximation quantizer for the case when all thresholds are powers of two, and all wavelet coefficients are integers. In this case, for each coefficient that eventually gets coded as significant, the sign and *bit position* of the most-significant binary digit (MSBD) are measured and encoded during a dominant pass. For example, consider the 10-bit representation of the number 41 as 0000101001. Also, consider the binary digits as a sequence of binary decisions in a binary tree. Proceeding from left to right, if we have not yet encountered a "1", we expect the probability distribution for the next digit to be strongly biased toward "0". The digits to the left and including the MSBD are called the *dominant bits*, and are measured during dominant passes. After the MSBD has been encountered, we expect a more random and much less biased distribution between a "0" and a "1", although we might still expect $P(0) > P(1)$ because most PDF models for transform coefficients decay with amplitude. Those binary digits to the right of the MSBD are called the *subordinate bits* and are measured and encoded during the subordinate pass. A zeroth-order approximation suggests that we should expect to pay close to one bit per "binary digit" for subordinate bits, while dominant bits should be far less expensive.

By using successive-approximation beginning with the largest possible threshold, where the probability of zero is extremely close to one, and by using zerotree coding, whose efficiency increases as the probability of zero increases, we should be able to code dominant bits with very few bits, since they are most often part of a zerotree.

In general, the thresholds need not be powers of two. However, by factoring out a constant mantissa, M, the starting threshold T_0 can be expressed in terms of a threshold that is a power of two

$$T_0 = M 2^E, \tag{8.19}$$

where the exponent E is an integer, in which case, the dominant and subordinate bits of appropriately scaled wavelet coefficients are coded during dominant and subordinate passes, respectively.

4.3 Advantage of Small Alphabets for Adaptive Arithmetic Coding

Note that the particular encoder alphabet used by the arithmetic coder at any given time contains either 2, 3, or 4 symbols depending whether the encoding is for a subordinate pass, a dominant pass with no zerotree root symbol, or a dominant pass with the zerotree root symbol. This is a real advantage for adapting the arithmetic coder. Since there are never more than four symbols, all of the possibilities typically occur with a reasonably measurable frequency. This allows an adaptation algorithm with a short memory to learn quickly and constantly track changing symbol probabilities. This adaptivity accounts for some of the effectiveness of the overall algorithm. Contrast this with the case of a large alphabet, as is the case in algorithms that do not use successive approximation. In that case, it takes many events before an adaptive entropy coder can reliably estimate the probabilities of unlikely symbols (see the discussion of the Zero-Frequency Problem in [3]). Furthermore, these estimates are fairly unreliable because images are typically statistically non-stationary and local symbol probabilities change from region to region.

In the practical coder used in the experiments, the arithmetic coder is based on [31]. In arithmetic coding, the encoder is separate from the model, which in [31], is basically a histogram. During the dominant passes, simple Markov conditioning is used whereby one of four histograms is chosen depending on 1) whether the previous coefficient in the scan is known to be significant, and 2) whether the parent is known to be significant. During the subordinate passes, a single histogram is used. Each histogram entry is initialized to a count of one. After encoding each symbol, the corresponding histogram entry is incremented. When the sum of all the counts in a histogram reaches the maximum count, each entry is incremented and integer-divided by two, as described in [31]. It should be mentioned, that for practical purposes, the coding gains provided by using this simple Markov conditioning may not justify the added complexity and using a single histogram strategy for the dominant pass performs almost as well (0.12 dB worse for Lena at 0.25 bpp.). The choice of maximum histogram count is probably more critical, since that controls the learning rate for the adaptation. For the experimental results presented, a maximum count of 256 was used, which provides an intermediate tradeoff between the smallest possible probability, which is the reciprocal of the maximum count, and the learning rate, which is faster with a smaller maximum histogram count.

4.4 *Order of Importance of the Bits*

Although importance is a subjective term, the order of processing used in EZW implicitly defines a precise ordering of importance that is tied to, in order, *precision*, *magnitude*, *scale*, and *spatial location* as determined by the initial dominant list.

The primary determination of ordering importance is the numerical *precision* of the coefficients. This can be seen in the fact that the uncertainty intervals for the magnitude of all coefficients are refined to the same precision before the uncertainty interval for any coefficient is refined further.

The second factor in the determination of importance is *magnitude*. Importance by magnitude manifests itself during a dominant pass because prior to the pass, all coefficients are insignificant and presumed to be zero. When they are found to be significant, they are all assumed to have the same magnitude, which is greater than the magnitudes of those coefficients that remain insignificant. Importance by magnitude manifests itself during a subordinate pass by the fact that magnitudes are refined in descending order of the center of the uncertainty intervals, i.e. the decoder's interpretation of the magnitude.

The third factor, *scale*, manifests itself in the *a priori* ordering of the subbands on the initial dominant list. Until, the significance of the magnitude of a coefficient is discovered during a dominant pass, coefficients in coarse scales are tested for significance before coefficients in fine scales. This is consistent with prioritization by the decoder's version of magnitude since for all coefficients not yet found to be significant, the magnitude is presumed to be zero.

The final factor, *spatial location*, merely implies that two coefficients that cannot yet be distinguished by the decoder in terms of either precision, magnitude, or scale, have their relative importance determined arbitrarily by the initial scanning order of the subband containing the two coefficients.

In one sense, this embedding strategy has a strictly non-increasing operational distortion-rate function for the distortion metric defined to be the sum of the widths of the uncertainty intervals of all of the wavelet coefficients. Since a discrete wavelet transform is an invertible representation of an image, a distortion function defined in the wavelet transform domain is also a distortion function defined on the image. This distortion function is also not without a rational foundation for low-bit rate coding, where noticeable artifacts must be tolerated, and perceptual metrics based on just-noticeable differences (JND's) do not always predict which artifacts human viewers will prefer. Since minimizing the widths of uncertainty intervals minimizes the largest possible errors, artifacts, which result from numerical errors large enough to exceed perceptible thresholds, are minimized. Even using this distortion function, the proposed embedding strategy is not optimal, because truncation of the bit stream in the middle of a pass causes some uncertainty intervals to be twice as large as others.

Actually, as it has been described thus far, EZW is unlikely to be optimal for *any* distortion function. Notice that in (8.19), dividing the thresholds by two simply decrements E leaving M unchanged. While there must exist an optimal starting M which minimizes a given distortion function, how to find this optimum is still an open question and seems highly image dependent. Without knowledge of the optimal M and being forced to choose it based on some other consideration, with probability one, either increasing or decreasing M would have produced an embedded code which has a lower distortion for the same rate. Despite the fact that without trial and error optimization for M, EZW is provably sub-optimal, it is nevertheless quite effective in practice.

Note also, that using the width of the uncertainty interval as a distance metric is exactly the same metric used in finite-precision fixed-point approximations of real numbers. Thus, the embedded code can be seen as an "image" generalization of finite-precision fixed-point approximations of real numbers.

4.5 Relationship to Priority-Position Coding

In a technique based on a very similar philosophy, Huang *et. al.* discuss a related approach to embedding, or ordering the information in importance, called Priority-Position Coding (PPC) [10]. They prove very elegantly that the entropy of a source is equal to the average entropy of a particular ordering of that source plus the average entropy of the position information necessary to reconstruct the source. Applying a sequence of decreasing thresholds, they attempt to sort by amplitude all of the DCT coefficients for the entire image based on a partition of the range of amplitudes. For each coding pass, they transmit the significance map which is arithmetically encoded. Additionally, when a significant coefficient is found they transmit its value to its full precision. Like the EZW algorithm, PPC implicitly defines importance with respect to the magnitudes of the transform coefficients. In one sense, PPC is a generalization of the successive-approximation method presented in this paper, because PPC allows more general partitions of the amplitude range of the transform coefficients. On the other hand, since PPC sends the value of a significant coefficient to full precision, its protocol assigns a greater importance to the least significant bit of a significant coefficient than to the identification of new significant coefficients on next PPC pass. In contrast, as a top priority, EZW tries to reduce the width of the largest uncertainty interval in all coefficients before increasing the precision further. Additionally, PPC makes no attempt to predict insignificance from low frequency to high frequency, relying solely on the arithmetic coding to encode the significance map. Also unlike EZW, the probability estimates needed for the arithmetic coder were derived via training on an image database instead of adapting to the image itself. It would be interesting to experiment with variations which combine advantages of EZW (wavelet transforms, zerotree coding, importance defined by a decreasing sequence of uncertainty intervals, and adaptive arithmetic coding using small alphabets) with the more general approach to partitioning the range of amplitudes found in PPC. In practice, however, it is unclear whether the finest grain partitioning of the amplitude range provides any coding gain, and there is certainly a much higher computational cost associated with more passes. Additionally, with the exception of the last few low-amplitude passes, the coding results reported in [10] did use power-of-two amplitudes to define the partition suggesting that, in practice, using finer partitioning buys little coding gain.

5 A Simple Example

In this section, a simple example will be used to highlight the order of operations used in the EZW algorithm. Only the string of symbols will be shown. The reader interested in the details of adaptive arithmetic coding is referred to [31]. Consider the simple 3-scale wavelet transform of an 8×8 image. The array of values is shown in Fig. 8. Since the largest coefficient magnitude is 63, we can choose our initial threshold to be anywhere in $(31.5, 63]$. Let $T_0 = 32$. Table 8.1 shows the processing on the first dominant pass. The following comments refer to Table 8.1:

1. The coefficient has magnitude 63 which is greater than the threshold 32, and is positive so a positive symbol is generated. After decoding this symbol, the decoder knows the coefficient is in the interval $[32, 64)$ whose center is 48.

2. Even though the coefficient -31 is insignificant with respect to the threshold 32, it has a significant descendant two generations down in subband $LH1$ with magnitude 47. Thus, the symbol for an isolated zero is generated.

3. The magnitude 23 is less than 32 and all descendants which include (3,-12,-14,8) in subband $HH2$ and all coefficients in subband $HH1$ are insignificant. A zerotree symbol is generated, and no symbol will be generated for any coefficient in subbands $HH2$ and $HH1$ during the current dominant pass.

4. The magnitude 10 is less than 32 and all descendants (-12,7,6,-1) also have magnitudes less than 32. Thus a zerotree symbol is generated. Notice that this tree has a violation of the "decaying spectrum" hypothesis since a coefficient (-12) in subband $HL1$ has a magnitude greater than its parent (10). Nevertheless, the entire tree has magnitude less than the threshold 32 so it is still a zerotree.

5. The magnitude 14 is insignificant with respect to 32. Its children are (-1,47,-3,2). Since its child with magnitude 47 is significant, an isolated zero symbol is generated.

6. Note that no symbols were generated from subband $HH2$ which would ordinarily precede subband $HL1$ in the scan. Also note that since subband $HL1$ has no descendants, the entropy coding can resume using a 3-symbol alphabet where the IZ and ZTR symbols are merged into the Z (zero) symbol.

7. The magnitude 47 is significant with respect to 32. Note that for the future dominant passes, this position will be replaced with the value 0, so that for the next dominant pass at threshold 16, the parent of this coefficient, which has magnitude 14, can be coded using a zerotree root symbol.

During the first dominant pass, which used a threshold of 32, four significant coefficients were identified. These coefficients will be refined during the first subordinate pass. Prior to the first subordinate pass, the uncertainty interval for the magnitudes of all of the significant coefficients is the interval $[32, 64)$. The first subordinate pass will refine these magnitudes and identify them as being either in interval $[32, 48)$, which will be encoded with the symbol "0", or in the interval $[48, 64)$, which will be encoded with the symbol "1". Thus, the decision boundary

Comment	Subband	Coefficient Value	Symbol	Reconstruction Value
(1)	LL3	63	POS	48
	HL3	-34	NEG	-48
(2)	LH3	-31	IZ	0
(3)	HH3	23	ZTR	0
	HL2	49	POS	48
(4)	HL2	10	ZTR	0
	HL2	14	ZTR	0
	HL2	-13	ZTR	0
	LH2	15	ZTR	0
(5)	LH2	14	IZ	0
	LH2	-9	ZTR	0
	LH2	-7	ZTR	0
(6)	HL1	7	Z	0
	HL1	13	Z	0
	HL1	3	Z	0
	HL1	4	Z	0
	LH1	-1	Z	0
(7)	LH1	47	POS	48
	LH1	-3	Z	0
	LH1	-2	Z	0

TABLE 8.1. Processing of first dominant pass at threshold $T = 32$. Symbols are POS for positive significant, NEG for negative significant, IZ for isolated zero, ZTR for zerotree root, and Z for a zero when their are no children. The reconstruction magnitudes are taken as the center of the uncertainty interval.

Coefficient Magnitude	Symbol	Reconstruction Magnitude
63	1	56
34	0	40
49	1	56
47	0	40

TABLE 8.2. Processing of the first subordinate pass. Magnitudes are partitioned into the uncertainty intervals $[32, 48)$ and $[48, 64)$, with symbols "0" and "1" respectively.

is the magnitude 48. It is no coincidence, that these symbols are exactly the first bit to the right of the MSBD in the binary representation of the magnitudes. The order of operations in the first subordinate pass is illustrated in Table 8.2.

The first entry has magnitude 63 and is placed in the upper interval whose center is 56. The next entry has magnitude 34, which places it in the lower interval. The third entry 49 is in the upper interval, and the fourth entry 47 is in the lower interval. Note that in the case of 47, using the center of the uncertainty interval as the reconstruction value, when the reconstruction value is changed from 48 to 40, the the reconstruction error actually increases from 1 to 7. Nevertheless, the uncertainty interval for this coefficient decreases from width 32 to width 16. At the conclusion of the processing of the entries on the subordinate list corresponding to the uncertainty interval $[32, 64)$, these magnitudes are reordered for future subordinate passes in the order (63,49,34,47). Note that 49 is moved ahead of 34 because from the decoder's point of view, the reconstruction values 56 and 40 are distinguishable. However, the magnitude 34 remains ahead of magnitude 47 because as far as the decoder can tell, both have magnitude 40, and the initial order, which is based first on importance by scale, has 34 prior to 47.

The process continues on to the second dominant pass at the new threshold of 16. During this pass, only those coefficients not yet found to be significant are scanned. Additionally, those coefficients previously found to be significant are treated as zero for the purpose of determining if a zerotree exists. Thus, the second dominant pass consists of encoding the coefficient -31 in subband $LH3$ as negative significant, the coefficient 23 in subband $HH3$ as positive significant, the three coefficients in subband $HL2$ that have not been previously found to be significant (10,14,-13) are each encoded as zerotree roots, as are all four coefficients in subband $LH2$ and all four coefficients in subband $HH2$. The four remaining coefficients in subband $HL1$ (7,13,3,4) are coded as zero. The second dominant pass terminates at this point since all other coefficients are predictably insignificant.

The subordinate list now contains, in order, the magnitudes (63,49,34,47,31,23) which, prior to this subordinate pass, represent the three uncertainty intervals $[48, 64)$, $[32, 48)$ and $[16, 31)$, each having equal width 16. The processing will refine each magnitude by creating two new uncertainty intervals for each of the three current uncertainty intervals. At the end of the second subordinate pass, the order of the magnitudes is (63,49,47,34,31,23), since at this point, the decoder could have identified 34 and 47 as being in different intervals. Using the center of the uncertainty interval as the reconstruction value, the decoder lists the magnitudes as (60,52,44,36,28,20).

The processing continues alternating between dominant and subordinate passes and can stop at any time.

6 Experimental Results

All experiments were performed by encoding and decoding an actual bit stream to verify the correctness of the algorithm. After a 12-byte header, the entire bit stream is arithmetically encoded using a single arithmetic coder with an adaptive model [31]. The model is initialized at each new threshold for each of the dominant and subordinate passes. From that point, the encoder is fully adaptive. Note in particular that there is no training of any kind, and no ensemble statistics of images are used in any way (unless one calls the zerotree hypothesis an ensemble statistic). The 12-byte header contains 1) the number of wavelet scales, 2) the dimensions of the image, 3) the maximum histogram count for the models in the arithmetic coder, 4) the image mean and 5) the initial threshold. Note that after the header, there is no overhead except for an extra symbol for end-of-bit-stream, which is always

# bytes	R	Compression	MSE	PSNR (dB)	signif. coef
32768	1.0	8:1	7.21	39.55	39446
16384	0.5	16:1	15.32	36.28	19385
8192	0.25	32:1	31.33	33.17	9774
4096	0.125	64:1	61.67	30.23	4950
2048	0.0625	128:1	114.5	27.54	2433
1024	0.03125	256:1	188.2.7	25.38	1253
512	0.015625	512:1	281.7	23.63	616
256	0.0078125	1024:1	440.2	21.69	265

TABLE 8.3. Coding results for 512 × 512 Lena showing Peak-Signal-to-Noise (PSNR) and the number of wavelet coefficients that were coded as non-zero.

maintained at minimum probability. This extra symbol is not needed for storage on computer medium if the end of a file can be detected.

The EZW coder was applied to the standard black and white 8 bpp. test images, 512 × 512 "Lena" and the 512 × 512 "Barbara", which are shown in Fig. 9a and 11a. Coding results for "Lena" are summarized in Table 8.3 and Fig. 9. Six scales of the QMF-pyramid were used. Similar results are shown for "Barbara" in Table 8.4 and Fig. 10. Additional results for the 256 × 256 "Lena" are given in [22].

Quotes of PSNR for the 512 × 512 "Lena" image are so abundant throughout the image coding literature that it is difficult to definitively compare these results with other coding results [2]. However, a literature search has only found two published results where authors generate an actual bit stream that claims higher PSNR performance at rates between 0.25 and 1 bit/pixel [12] and [21], the latter of which is a variation of the EZW algorithm. For the "Barbara" image, which is far more difficult than "Lena", the performance using EZW is substantially better, at least numerically, than the 27.82 dB for 0.534 bpp. reported in [28].

The performance of the EZW coder was also compared to a widely available version of JPEG [14]. JPEG does not allow the user to select a target bit rate but instead allows the user to choose a "Quality Factor". In the experiments shown in Fig. 11 "Barbara" is encoded first using JPEG to a file size of 12866 bytes, or a bit rate of 0.39 bpp. The PSNR in this case is 26.99 dB. The EZW encoder was then applied to "Barbara" with a target file size of exactly 12866 bytes. The resulting PSNR is 29.39 dB, significantly higher than for JPEG. The EZW encoder was then applied to "Barbara" using a target PSNR to obtain exactly the same PSNR of 26.99. The resulting file size is 8820 bytes, or 0.27 bpp. Visually, the 0.39 bpp. EZW version looks better than the 0.39 bpp. JPEG version. While there is some loss of resolution in both, there are noticeable blocking artifacts in the JPEG version. For the comparison at the same PSNR, one could probably argue in favor of the JPEG.

Another interesting figure of merit is the number of significant coefficients re-

[2]Actually there are multiple versions of the luminance only "Lena" floating around, and the one used in [22] is darker and slightly more difficult than the "official" one obtained by this author from RPI after [22] was published. Also note that this should not be confused with results using only the Green component of an RGB version which are also commonly cited.

# bytes	R	Compression	MSE	PSNR (dB)	signif. coef
32768	1.0	8:1	19.92	35.14	40766
16384	0.5	16:1	57.57	30.53	20554
8192	0.25	32:1	136.8	26.77	10167
4096	0.125	64:1	257.1	24.03	4522
2048	0.0625	128:1	318.5	23.10	2353
1024	0.03125	256:1	416.2	21.94	1259
512	0.015625	512:1	546.8	20.75	630
256	0.0078125	1024:1	772.5	19.54	291

TABLE 8.4. Coding results for 512 × 512 Barbara showing Peak-Signal-to-Noise (PSNR) and the number of wavelet coefficients that were coded as non-zero.

tained. DeVore *et. al.* used wavelet transform coding to progressively encode the same image [8]. Using 68272 bits, (8534 bytes, 0.26 bpp.), they retained 2019 coefficients and achieved a RMS error of 15.30 (MSE=234, 24.42 dB), whereas using the embedded coding scheme, 9774 coefficients are retained, using only 8192 bytes. The PSNR for these two examples differs by over 8 dB. Part of the difference can be attributed to fact that the Haar basis was used in [8]. However, closer examination shows that the zerotree coding provides a much better way of encoding the positions of the significant coefficients than was used in [8].

An interesting and perhaps surprising property of embedded coding is that when the encoding or decoding is terminated during the middle of a pass, or in the middle of the scanning of a subband, there are no artifacts produced that would indicate where the termination occurs. In other words, some coefficients in the same subband are represented with twice the precision of the others. A possible explanation of this phenomena is that at low rates, there are so few significant coefficients, that any one does not make a perceptible difference. Thus, if the last pass is a dominant pass, setting some coefficient that might be significant to zero may be imperceptible. Similarly, the fact that some have more precision than others is also imperceptible. By the time the number of significant coefficients becomes large, the picture quality is usually so good that adjacent coefficients with different precisions are imperceptible.

Another interesting property of the embedded coding is that because of the implicit global bit allocation, even at extremely high compression ratios, the performance scales. At a compression ratio of 512:1, the image quality of "Lena" is poor, but still recognizable. This is not the case with conventional block coding schemes, where at such high compression ratios, their would be insufficient bits to even encode the DC coefficients of each block.

The unavoidable artifacts produced at low bit rates using this method are typical of wavelet coding schemes coded to the same PSNR's. However, subjectively, they are not nearly as objectionable as the blocking effects typical of block transform coding schemes.

7 Conclusion

A new technique for image coding has been presented that produces a fully embedded bit stream. Furthermore, the compression performance of this algorithm is competitive with virtually all known techniques. The remarkable performance can be attributed to the use of the following four features:

- a discrete wavelet transform, which decorrelates most sources fairly well, and allows the more significant bits of precision of most coefficients to be efficiently encoded as part of exponentially growing zerotrees,
- zerotree coding, which by predicting insignificance across scales using an image model that is easy for most images to satisfy, provides substantial coding gains over the first-order entropy for significance maps,
- successive-approximation, which allows the coding of multiple significance maps using zerotrees, and allows the encoding or decoding to stop at any point,
- adaptive arithmetic coding, which allows the entropy coder to incorporate *learning* into the bit stream itself.

The precise rate control that is achieved with this algorithm is a distinct advantage. The user can choose a bit rate and encode the image to *exactly* the desired bit rate. Furthermore, since no training of any kind is required, the algorithm is fairly general and performs remarkably well with most types of images.

Acknowledgment

I would like to thank Joel Zdepski who suggested incorporating the sign of the significant values into the significance map to aid embedding, Rajesh Hingorani who wrote much of the original C code for the QMF-pyramids, Allen Gersho who provide the original "Barbara" image, and Gregory Wornell whose fruitful discussions convinced me to develop a more mathematical analysis of zerotrees in terms of bounding an optimal estimator. I would also like to thank the editor and the anonymous reviewers whose comments led to a greatly improved manuscript.

As a final point, it is acknowledged that all of the decisions made in this work have been based on strictly numerical measures and no consideration has been given to optimizing the coder for perceptual metrics. This is an interesting area for additional research and may lead to improvements in the visual quality of the coded images.

8 References

[1] E. H. Adelson, E. Simoncelli, and R. Hingorani, "Orthogonal pyramid transforms for image coding", *Proc. SPIE*, vol. 845, Cambridge, MA, pp. 50-58, Oct. 1987.

[2] R. Ansari, H. Gaggioni, and D. J. LeGall, "HDTV coding using a nonrectangular subband decomposition," in *Proc. SPIE Conf. Vis. Commun. Image Processing*, Cambridge, MA, pp. 821-824, Nov. 1988.

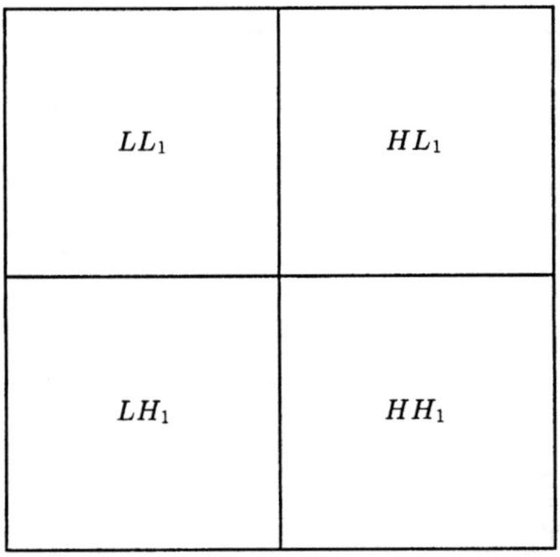

FIGURE 1. First Stage of a Discrete Wavelet Transform. The image is divided into four subbands using separable filters. Each coefficient represents a spatial area corresponding to approximately a 2 × 2 area of the original picture. The low frequencies represent a bandwidth approximately corresponding to $0 < |\omega| < \frac{\pi}{2}$, whereas the high frequencies represent the band from $\frac{\pi}{2} < |\omega| < \pi$. The four subbands arise from separable application of vertical and horizontal filters.

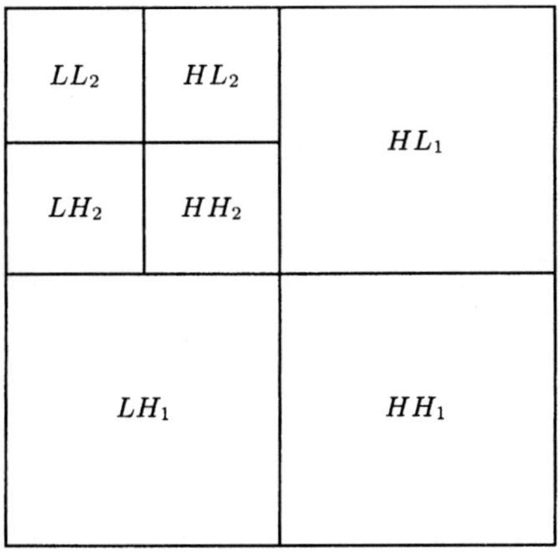

FIGURE 2. A Two Scale Wavelet Decomposition. The image is divided into four subbands using separable filters. Each coefficient in the subbands LL_2, LH_2, HL_2 and HH_2 represents a spatial area corresponding to approximately a 4 × 4 area of the original picture. The low frequencies at this scale represent a bandwidth approximately corresponding to $0 < |\omega| < \frac{\pi}{4}$, whereas the high frequencies represent the band from $\frac{\pi}{4} < |\omega| < \frac{\pi}{2}$.

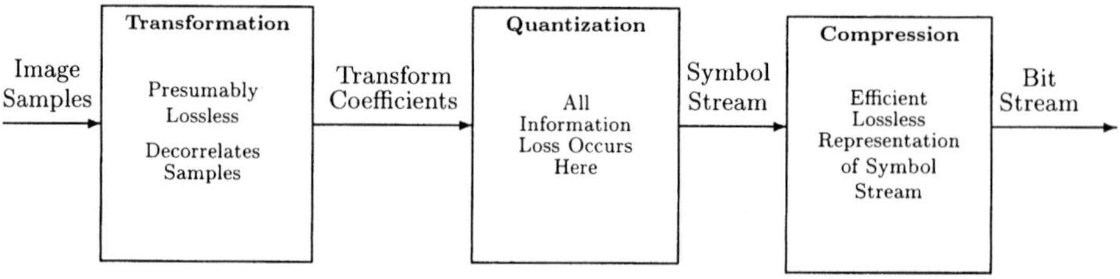

FIGURE 3. A Generic Transform Coder.

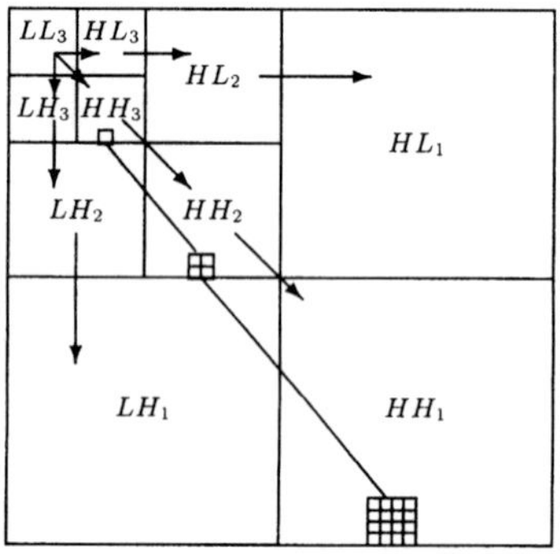

FIGURE 4. Parent-Child Dependencies of Subbands. Note that the arrow points from the subband of the parents to the subband of the children. The lowest frequency subband is the top left, and the highest frequency subband is at the bottom right. Also shown is a wavelet tree consisting of all of the descendents of a single coefficient in subband $HH3$. The coefficient in $HH3$ is a zerotree root if it is insignificant and *all* of its descendants are insignificant.

[3] T. C. Bell, J. G. Cleary, and I. H. Witten, *Text Compression*, Prentice-Hall: Englewood Cliffs, NJ 1990.

[4] P. J. Burt and E. H. Adelson, "The Laplacian pyramid as a compact image code", *IEEE Trans. Commun.*, vol. 31, pp. 532-540, 1983.

[5] R. R. Coifman and M. V. Wickerhauser, "Entropy-based algorithms for best basis selection", *IEEE Trans. Inform. Theory*, vol. 38, pp. 713-718, March, 1992.

[6] I. Daubechies, "Orthonormal bases of compactly supported wavelets", Comm. Pure Appl. Math., vol.41, pp. 909-996, 1988.

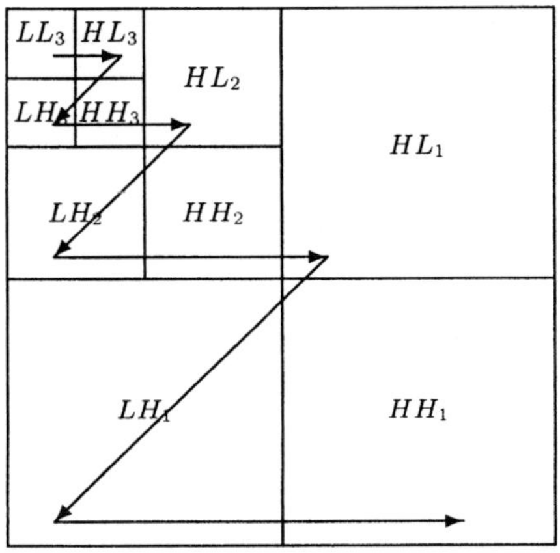

FIGURE 5. Scanning order of the Subbands for Encoding a Significance Map. Note that parents must be scanned before children. Also note that all positions in a given subband are scanned before the scan moves to the next subband.

[7] I. Daubechies, "The wavelet transform, time-frequency localization and signal analysis", *IEEE Trans. Inform. Theory*, vol. 36, pp. 961-1005, Sept. 1990.

[8] R. A. DeVore, B. Jawerth, and B. J. Lucier, "Image compression through wavelet transform coding", *IEEE Trans. Inform. Theory*, vol. 38, pp. 719-746, March, 1992.

[9] W. Equitz and T. Cover, "Successive refinement of information", *IEEE Trans. Inform. Theory*, vol. 37, pp. 269-275, March 1991.

[10] Y. Huang, H. M. Driezen, and N. P. Galatsanos "Prioritized DCT for Compression and Progressive Transmission of Images", *IEEE Trans. Image Processing*, vol. 1, pp. 477-487, Oct. 1992.

[11] N.S. Jayant and P. Noll, *Digital Coding of Waveforms*, Prentice-Hall: Englewood Cliffs, NJ 1984.

[12] Y. H. Kim and J. W. Modestino, "Adaptive entropy coded subband coding of images", *IEEE Trans. Image Process.*, vol 1, pp. 31-48, Jan. 1992.

[13] J. Kovačević and M. Vetterli, "Nonseparable multidimensional perfect reconstruction filter banks and wavelet bases for $\mathcal{R}^n$", *IEEE Trans. Inform. Theory*, vol. 38, pp. 533-555, March, 1992.

[14] T. Lane, Independent JPEG Group's free JPEG software, available by anonymous FTP at uunet.uu.net in the directory /graphics/jpeg. 1991.

[15] A. S. Lewis and G. Knowles, "A 64 Kb/s Video Codec using the 2-D wavelet transform", *Proc. Data Compression Conf.*, Snowbird, Utah, IEEE Computer Society Press, 1991.

[16] A. S. Lewis and G. Knowles, "Image Compression Using the 2-D Wavelet Transform", *IEEE Trans. Image Processing* vol. 1 pp. 244-250, April 1992.

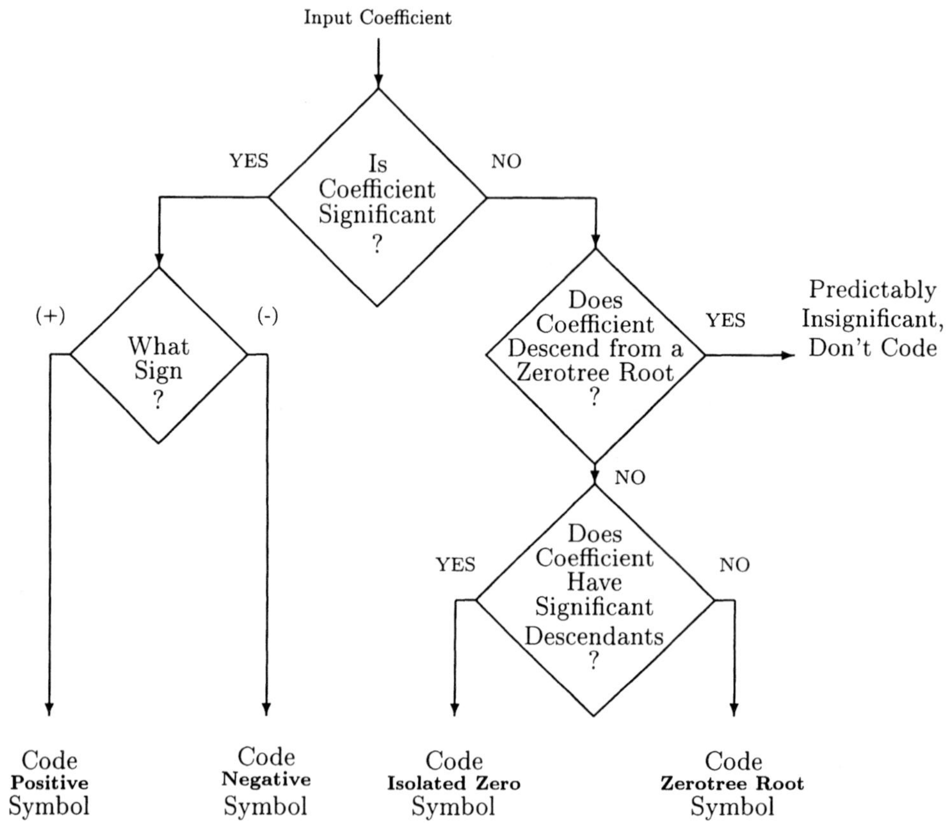

FIGURE 6. Flow Chart for Encoding a Coefficient of the Significance Map.

[17] S. Mallat, "A theory for multiresolution signal decomposition: The wavelet representation", *IEEE Trans. Pattern Anal. Machine Intell.* vol. 11 pp. 674-693, July 1989.

[18] S. Mallat, "Multifrequency channel decompositions of images and wavelet models", *IEEE Trans. Acoust. Speech and Signal Processing*, vol. 37, pp. 2091-2110, Dec. 1990.

[19] A. Pentland and B. Horowitz, "A practical approach to fractal-based image compression", *Proc. Data Compression Conf.*, Snowbird, Utah, IEEE Computer Society Press, 1991.

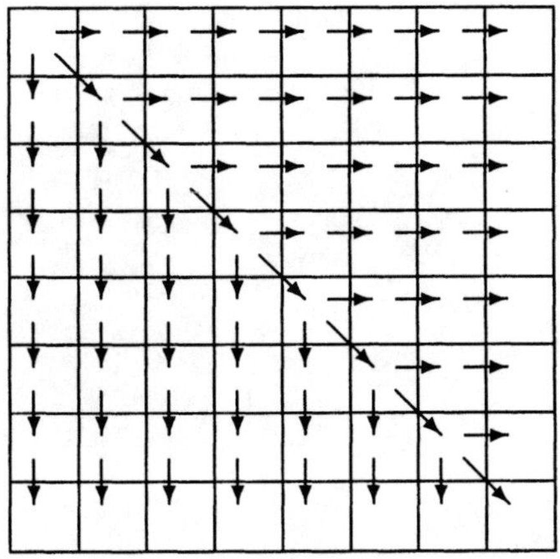

FIGURE 7. Parent-Child Dependencies For Linearly Spaced Subbands Systems such as the DCT. Note that the arrow points from the subband of the parents to the subband of the children. The lowest frequency subband is the top left, and the highest frequency subband is at the bottom right.

63	−34	49	10	7	13	−12	7
−31	23	14	−13	3	4	6	−1
15	14	3	−12	5	−7	3	9
−9	−7	−14	8	4	−2	3	2
−5	9	−1	47	4	6	−2	2
3	0	−3	2	3	−2	0	4
2	−3	6	−4	3	6	3	6
5	11	5	6	0	3	−4	4

FIGURE 8. Example of 3-scale wavelet transform of an 8×8 image.

[20] O. Rioul and M. Vetterli, "Wavelets and signal processing", *IEEE Signal Processing Magazine*, vol. 8, pp. 14-38, Oct. 1991.

[21] A. Said and W. A. Pearlman, "Image Compression using the Spatial-Orientation Tree", *Proc. IEEE Int. Symp. on Circuits and Systems*, Chicago, IL, May 1993, pp. 279-282.

[22] J. M. Shapiro, "An Embedded Wavelet Hierarchical Image Coder", *Proc. IEEE Int. Conf. Acoust., Speech, Signal Proc.*, San Francisco, CA, March 1992.

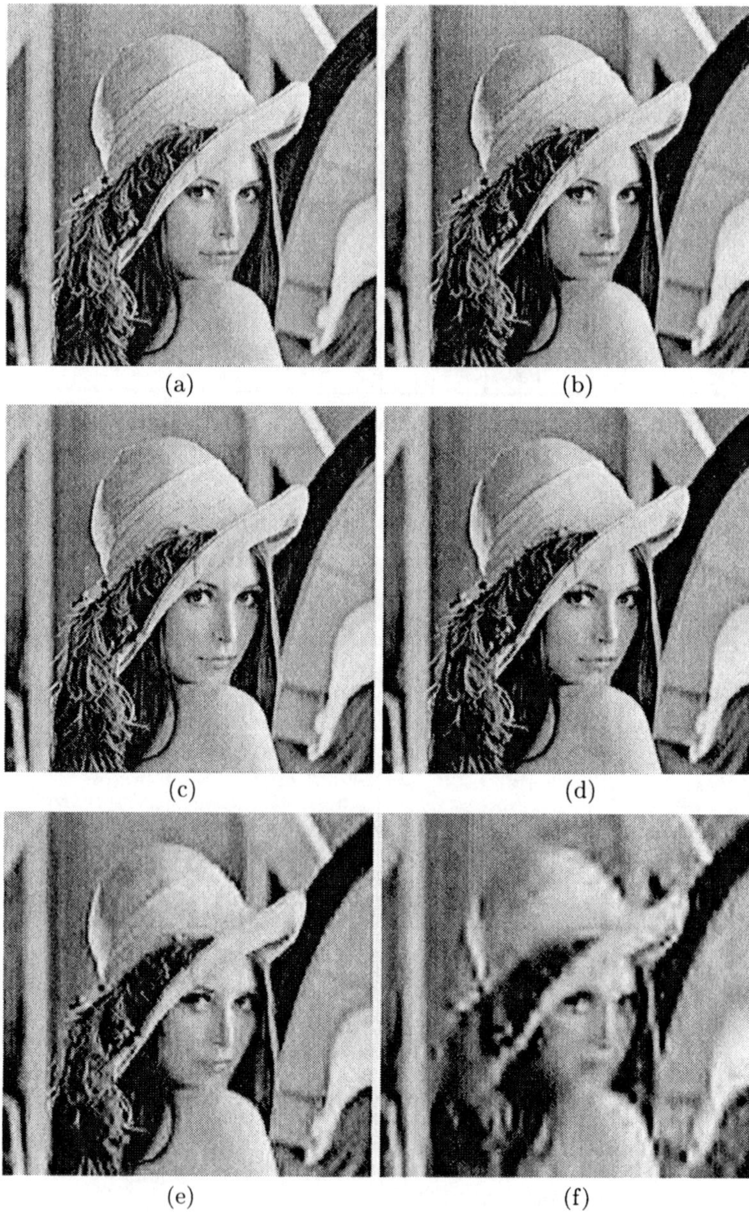

FIGURE 9. Performance of EZW Coder operating on "Lena". (a) Original 512 × 512 "Lena" image at 8 bits/pixel (b) 1.0 bits/pixel, 8:1 Compression, PSNR=39.55 dB, (c) 0.5 bits/pixel, 16:1 Compression, PSNR=36.28, (d) 0.25 bits/pixel, 32:1 Compression, PSNR=33.17 dB, (e) 0.0625 bits/pixel, 128:1 Compression, PSNR=27.54 dB, (f) 0.015625 bits/pixel, 512 :1 Compression, PSNR=23.63 dB.

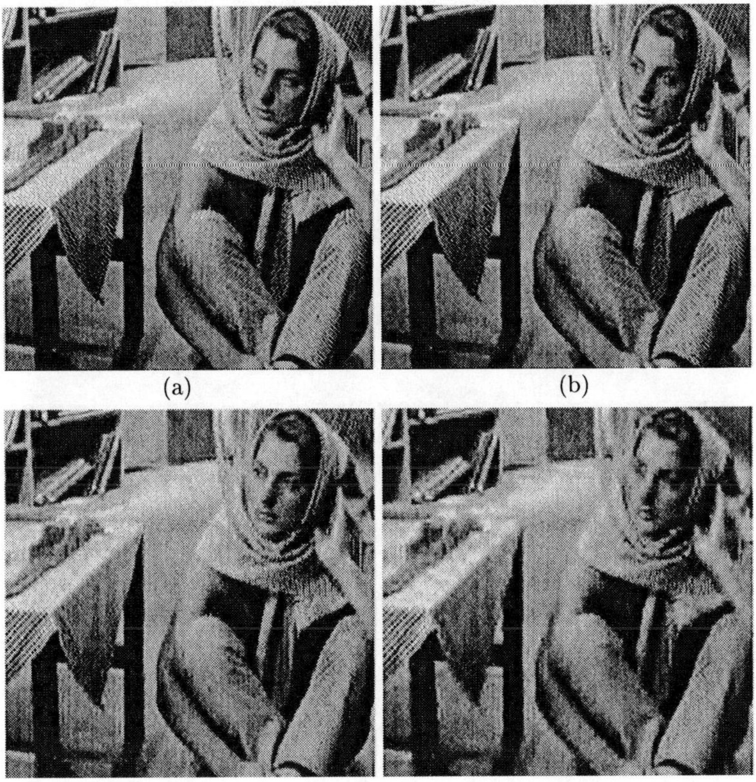

(a) (b) (c) (d)

FIGURE 10. Performance of EZW Coder operating on "Barbara" at (a) 1.0 bits/pixel, 8:1 Compression, PSNR=35.14 dB (b) 0.5 bits/pixel, 16:1 Compression, PSNR=30.53 dB, (c) 0.125 bits/pixel, 64:1 Compression, PSNR=24.03 dB, (d) 0.0625 bits/pixel , 128:1 Compression, PSNR=23.10 dB.

[23] J. M. Shapiro, "Adaptive multidimensional perfect reconstruction filter banks using McClellan transformations", *Proc. IEEE Int. Symp. Circuits Syst.*, San Diego, CA, May 1992.

[24] J. M. Shapiro, "An embedded hierarchical image coder using zerotrees of wavelet coefficients", *Proc. Data Compression Conf.*, Snowbird, Utah, IEEE Computer Society Press, 1993.

[25] J. M. Shapiro, "Application of the embedded wavelet hierarchical image coder to very low bit rate image coding", *Proc. IEEE Int. Conf. Acoust., Speech, Signal Proc.*, Minneapolis, MN, April 1993.

[26] Special Issue of *IEEE Trans. Inform. Theory*, March 1992.

[27] G. Strang, "Wavelets and dilation equations: A brief introduction", *SIAM Review* vol. 4, pp. 614-627, Dec 1989.

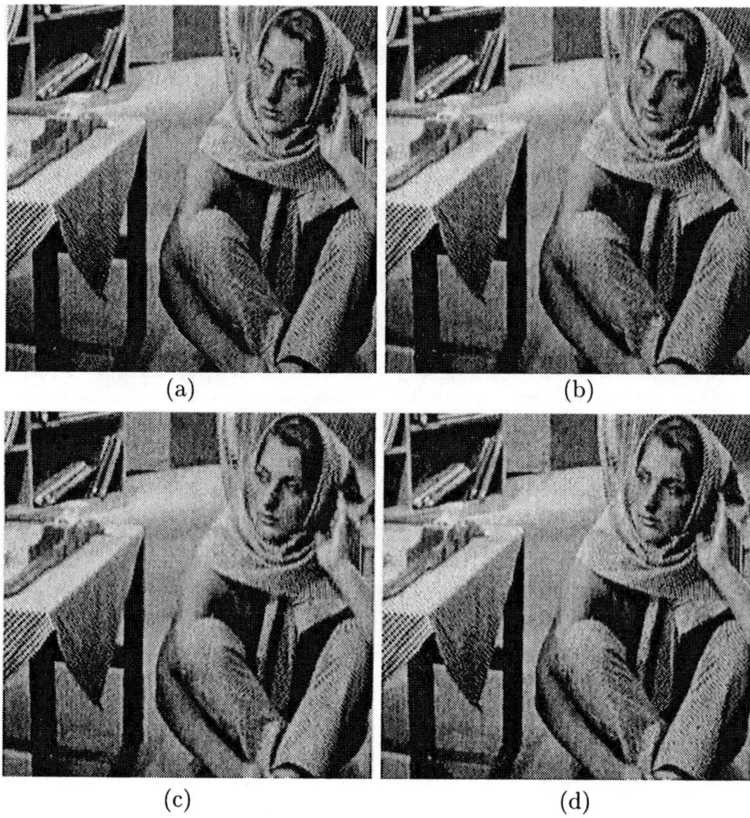

FIGURE 11. Comparison of EZW and JPEG operating on "Barbara" (a) Original 512×512 (b) EZW at 12866 bytes, 0.39 bits/pixel, 29.39 dB, (c) EZW at 8820 bytes, 0.27 bits/pixel, 26.99 dB, (d) JPEG at 12866 bytes, 0.39 bits/pixel, 26.99 dB.

[28] J. Vaisey and A. Gersho, "Image Compression with Variable Block Size Segmentation", *IEEE Trans. Signal Processing*, vol. 40, pp. 2040-2060, Aug. 1992.

[29] M. Vetterli, J. Kovačević, and D. J. LeGall, "Perfect reconstruction filter banks for HDTV representation and coding", *Image Commun.*, vol. 2, pp. 349-364, Oct 1990.

[30] G. K. Wallace, "The JPEG Still Picture Compression Standard", *Commun. of the ACM*, vol. 34, pp. 30-44, April 1991.

[31] I. H. Witten, R. Neal, and J. G. Cleary, "Arithmetic coding for data compression", *Comm. ACM*, vol. 30, pp. 520-540, June 1987.

[32] J. W. Woods ed. *Subband Image Coding*, Kluwer Academic Publishers, Boston, MA. 1991.

[33] G.W. Wornell, "A Karhunen-Loéve expansion for $1/f$ processes via wavelets", *IEEE Trans. Inform. Theory*, vol. 36, July 1990, pp. 859-861.

[34] Z. Xiong, N. Galatsanos, and M. Orchard, "Marginal analysis prioritization for image compression based on a hierarchical wavelet decomposition",*Proc. IEEE Int. Conf. Acoust., Speech, Signal Proc.*, Minneapolis, MN, April 1993.

[35] W. Zettler, J. Huffman, and D. C. P. Linden, "Applications of compactly supported wavelets to image compression", *SPIE Image Processing and Algorithms*, Santa Clara, CA 1990.

9

A New Fast/Efficient Image Codec Based on Set Partitioning in Hierarchical Trees

Amir Said and William A. Pearlman

ABSTRACT [1]

Embedded zerotree wavelet (EZW) coding, introduced by J. M. Shapiro, is a very effective and computationally simple technique for image compression. Here we offer an alternative explanation of the principles of its operation, so that the reasons for its excellent performance can be better understood. These principles are partial ordering by magnitude with a set partitioning sorting algorithm, ordered bit plane transmission, and exploitation of self-similarity across different scales of an image wavelet transform. Moreover, we present a new and different implementation, based on set partitioning in hierarchical trees (SPIHT), which provides even better performance than our previosly reported extension of the EZW that surpassed the performance of the original EZW. The image coding results, calculated from actual file sizes and images reconstructed by the decoding algorithm, are either comparable to or surpass previous results obtained through much more sophisticated and computationally complex methods. In addition, the new coding and decoding procedures are extremely fast, and they can be made even faster, with only small loss in performance, by omitting entropy coding of the bit stream by arithmetic code.

1 Introduction

Image compression techniques, especially non-reversible or lossy ones, have been known to grow computationally more complex as they grow more efficient, confirming the tenets of source coding theorems in information theory that a code for a (stationary) source approaches optimality in the limit of infinite computation (source length). Notwithstanding, the image coding technique called embedded zerotree wavelet (EZW), introduced by Shapiro [19], interrupted the simultaneous progression of efficiency and complexity. This technique not only was competitive in performance with the most complex techniques, but was extremely fast in execution and produced an embedded bit stream. With an embedded bit stream, the reception of code bits can be stopped at any point and the image can be decompressed and reconstructed. Following that significant work, we developed an alternative exposition of the underlying principles of the EZW technique and presented an extension that achieved even better results [6].

In this article, we again explain that the EZW technique is based on three concepts: (1) partial ordering of the transformed image elements by magnitude, with transmission of order by a subset partitioning algorithm that is duplicated at the decoder, (2) ordered bit plane transmission of refinement bits, and (3) exploitation of the self-similarity of the image wavelet transform across different scales. As to be explained, the partial ordering is a result of comparison of transform element (coefficient) magnitudes to a set of octavely decreasing thresholds. We say that an element is significant or insignificant with respect to a given threshold, depending on whether or not it exceeds that threshold.

In this work, crucial parts of the coding process—the way subsets of coefficients are partitioned and how the significance information is conveyed—are fundamentally different from the aforementioned works. In the previous works, arithmetic coding of the bit streams was essential to compress the ordering information as conveyed by the results of the significance tests. Here the subset partitioning is so effective and the significance information so compact that even binary uncoded transmission achieves about the same or better performance than in these previous works. Moreover, the utilization of arithmetic coding can reduce the mean squared error or increase the peak signal to noise ratio (PSNR) by 0.3 to 0.6 dB for the same rate or compressed file size and achieve results which are equal to or superior to any previously reported, regardless of complexity. Execution times are also reported to indicate the rapid speed of the encoding and decoding algorithms. The transmitted code or compressed image file is completely embedded, so that a single file for an image at a given code rate can be truncated at various points and decoded to give a series of reconstructed images at lower rates. Previous versions [19, 6] could not give their best performance with a single embedded file, and required, for each rate, the optimization of a certain parameter. The new method solves this problem by changing the transmission priority, and yields, with one embedded file, its top performance for all rates.

The encoding algorithms can be stopped at any compressed file size or let run until the compressed file is a representation of a *nearly* lossless image. We say *nearly* lossless because the compression *may* not be reversible, as the wavelet transform filters, chosen for lossy coding, have non-integer tap weights and produce non-integer transform coefficients, which are truncated to finite precision. For perfectly reversible compression, one must use an integer multiresolution transform, such as the S+P transform introduced in [14], which yields excellent reversible compression results when used with the new extended EZW techniques.

This paper is organized as follows. The next section, Section 2, describes an embedded coding or progressive transmission scheme that prioritizes the code bits according to their reduction in distortion. In Section 3 are explained the principles of partial ordering by coefficient magnitude and ordered bit plane transmission and which suggest a basis for an efficient coding method. The set partitioning sorting procedure and spatial orientation trees (called zerotrees previously) are detailed in Sections 4 and 5, respectively. Using the principles set forth in the previous sections, the coding and decoding algorithms are fully described in Section 6. In Section 7, rate, distortion, and execution time results are reported on the operation of the coding algorithm on test images and the decoding algorithm on the resultant compressed files. The figures on rate are calculated from actual compressed file sizes and on mean squared error or PSNR from the reconstructed images given by the decoding algorithm. Some reconstructed images are also displayed. These results are put into perspective by comparison to previous work. The conclusion of the paper is in Section 8.

2 Progressive Image Transmission

We assume that the original image is defined by a set of pixel values $p_{i,j}$, where (i, j) is the pixel coordinate. To simplify the notation we represent two-dimensional arrays with bold letters. The coding is actually done to the array

$$\mathbf{c} = \Omega(\mathbf{p}), \tag{9.1}$$

where $\Omega(\cdot)$ represents a unitary hierarchical subband transformation (e.g., [4]). The two-dimensional array $\mathbf{c}$ has the same dimensions of $\mathbf{p}$, and each element $c_{i,j}$ is called *transform coefficient* at coordinate (i, j). For the purpose of coding we assume that each $c_{i,j}$ is represented with a fixed-point binary format, with a small number of bits—typically 16 or less—and can be treated as an integer.

In a progressive transmission scheme, the decoder initially sets the reconstruction vector $\hat{\mathbf{c}}$ to zero and updates its components according to the coded message. After receiving the value (approximate or exact) of some coefficients, the decoder can obtain a reconstructed image

$$\hat{\mathbf{p}} = \Omega^{-1}(\hat{\mathbf{c}}). \tag{9.2}$$

A major objective in a progressive transmission scheme is to select the most important information—which yields the largest distortion reduction—to be transmitted first. For this selection we use the mean squared-error (MSE) distortion measure,

$$D_{\mathrm{mse}}(\mathbf{p} - \hat{\mathbf{p}}) = \frac{||\mathbf{p} - \hat{\mathbf{p}}||^2}{N} = \frac{1}{N} \sum_i \sum_j (p_{i,j} - \hat{p}_{i,j})^2, \tag{9.3}$$

where N is the number of image pixels. Furthermore, we use the fact that the Euclidean norm is invariant to the unitary transformation Ω, i.e.,

$$D_{\mathrm{mse}}(\mathbf{p} - \hat{\mathbf{p}}) = D_{\mathrm{mse}}(\mathbf{c} - \hat{\mathbf{c}}) = \frac{1}{N} \sum_i \sum_j (c_{i,j} - \hat{c}_{i,j})^2. \tag{9.4}$$

From (9.4) it is clear that if the exact value of the transform coefficient $c_{i,j}$ is sent to the decoder, then the MSE decreases by $|c_{i,j}|^2/N$. This means that the coefficients with larger magnitude should be transmitted first because they have a larger content of information.[2] This corresponds to the progressive transmission method proposed by DeVore *et al.* [3]. Extending their approach, we can see that the information in the value of $|c_{i,j}|$ can also be ranked according to its binary representation, and the most significant bits should be transmitted first. This idea is used, for example, in the bit-plane method for progressive transmission [8].

[2]Here the term *information* is used to indicate how much the distortion can decrease after receiving that part of the coded message.

BIT ROW																
	sign	s	s	s	s	s	s	s	s	s	s	s	s	s	s	s
msb	5	1	1	0	0	0	0	0	0	0	0	0	0	0	0	0
	4	→	→	1	1	0	0	0	0	0	0	0	0	0	0	0
	3	→	→	→	→	1	1	1	1	0	0	0	0	0	0	0
	2	→	→	→	→	→	→	→	→	1	1	1	1	1	1	1
	1	→	→	→	→	→	→	→	→	→	→	→	→	→	→	→
lsb	0	→	→	→	→	→	→	→	→	→	→	→	→	→	→	→

FIGURE 1. Binary representation of the magnitude-ordered coefficients.

Following, we present a progressive transmission scheme that incorporates these two concepts: ordering the coefficients by magnitude and transmitting the most significant bits first. To simplify the exposition we first assume that the ordering information is explicitly transmitted to the decoder. Later we show a much more efficient method to code the ordering information.

3 Transmission of the Coefficient Values

Let us assume that the coefficients are ordered according to the minimum number of bits required for its magnitude binary representation, that is, ordered according to a one-to-one mapping $\eta : I \mapsto I^2$, such that

$$\lfloor \log_2 |c_{\eta(k)}| \rfloor \geq \lfloor \log_2 |c_{\eta(k+1)}| \rfloor, \quad k = 1, \ldots, N. \qquad (9.5)$$

Fig. 1 shows the schematic binary representation of a list of magnitude-ordered coefficients. Each column k in Fig. 1 contains the bits of $c_{\eta(k)}$. The bits in the top row indicate the sign of the coefficient. The rows are numbered from the bottom up, and the bits in the lowest row are the least significant.

Now, let us assume that, besides the ordering information, the decoder also receives the numbers μ_n corresponding to the number of coefficients such that $2^n \leq |c_{i,j}| < 2^{n+1}$. In the example of Fig. 1 we have $\mu_5 = 2$, $\mu_4 = 2$, $\mu_3 = 4$, etc. Since the transformation Ω is unitary, all bits in a row have the same content of information, and the most effective order for progressive transmission is to sequentially send the bits in each row, as indicated by the arrows in Fig. 1. Note that, because the coefficients are in decreasing order of magnitude, the leading "0" bits and the first "1" of any column do not need to be transmitted, since they can be inferred from μ_n and the ordering.

The progressive transmission method outlined above can be implemented with the following algorithm to be used by the encoder.

ALGORITHM I

1. output $n = \lfloor \log_2 \left(\max_{(i,j)} \{|c_{i,j}|\} \right) \rfloor$ to the decoder;
2. output μ_n, followed by the pixel coordinates $\eta(k)$ and sign of each of the μ_n coefficients such that $2^n \leq |c_{\eta(k)}| < 2^{n+1}$ (**sorting pass**);

3. output the n-th most significant bit of all the coefficients with $|c_{i,j}| \geq 2^{n+1}$ (i.e., those that had their coordinates transmitted in previous sorting passes), in the same order used to send the coordinates (**refinement pass**);
4. decrement n by 1, and go to Step **2**.

The algorithm stops at the desired the rate or distortion. Normally, good quality images can be recovered after a relatively small fraction of the pixel coordinates are transmitted.

The fact that this coding algorithm uses uniform scalar quantization may give the impression that it must be much inferior to other methods that use non-uniform and/or vector quantization. However, this is not the case: the ordering information makes this simple quantization method very efficient. On the other hand, a large fraction of the "bit-budget" is spent in the sorting pass, and it is there that the sophisticated coding methods are needed.

4 Set Partitioning Sorting Algorithm

One of the main features of the proposed coding method is that the ordering data is not explicitly transmitted. Instead, it is based on the fact that the execution path of any algorithm is defined by the results of the comparisons on its branching points. So, if the encoder and decoder have the same sorting algorithm, then the decoder can duplicate the encoder's execution path if it receives the results of the magnitude comparisons, and the ordering information can be recovered from the execution path.

One important fact used in the design of the sorting algorithm is that we do not need to sort all coefficients. Actually, we need an algorithm that simply selects the coefficients such that $2^n \leq |c_{i,j}| < 2^{n+1}$, with n decremented in each pass. Given n, if $|c_{i,j}| \geq 2^n$ then we say that a coefficient is *significant*; otherwise it is called *insignificant*.

The sorting algorithm divides the set of pixels into partitioning subsets $\mathcal{T}_m$ and performs the magnitude test

$$\max_{(i,j)\in\mathcal{T}_m} \{|c_{i,j}|\} \geq 2^n \ ? \tag{9.6}$$

If the decoder receives a "no" to that answer (the *subset* is insignificant), then it knows that all coefficients in $\mathcal{T}_m$ are insignificant. If the answer is "yes" (the subset is significant), then a certain rule shared by the encoder and the decoder is used to partition $\mathcal{T}_m$ into new subsets $\mathcal{T}_{m,l}$, and the significance test is then applied to the new subsets. This set division process continues until the magnitude test is done to all single coordinate significant subsets in order to identify each significant coefficient.

To reduce the number of magnitude comparisons (message bits) we define a set partitioning rule that uses an expected ordering in the hierarchy defined by the subband pyramid. The objective is to create new partitions such that subsets expected to be insignificant contain a large number of elements, and subsets expected to be significant contain only one element.

To make clear the relationship between magnitude comparisons and message bits, we use the function

$$S_n(\mathcal{T}) = \begin{cases} 1, & \max_{(i,j)\in\mathcal{T}} \{|c_{i,j}|\} \geq 2^n, \\ 0, & \text{otherwise,} \end{cases} \tag{9.7}$$

to indicate the significance of a set of coordinates $\mathcal{T}$. To simplify the notation of single pixel sets, we write $S_n(\{(i,j)\})$ as $S_n(i,j)$.

5 Spatial Orientation Trees

Normally, most of an image's energy is concentrated in the low frequency components. Consequently, the variance decreases as we move from the highest to the lowest levels of the subband pyramid. Furthermore, it has been observed that there is a spatial self-similarity between subbands, and the coefficients are expected to be better magnitude-ordered if we move downward in the pyramid following the same spatial orientation. (Note the mild requirements for ordering in (9.5).) For instance, large low-activity areas are expected to be identified in the highest levels of the pyramid, and they are replicated in the lower levels at the same spatial locations.

A tree structure, called *spatial orientation tree,* naturally defines the spatial relationship on the hierarchical pyramid. Fig. 2 shows how our spatial orientation tree is defined in a pyramid constructed with recursive four-subband splitting. Each node of the tree corresponds to a pixel, and is identified by the pixel coordinate. Its direct descendants (offspring) correspond to the pixels of the same spatial orientation in the next finer level of the pyramid. The tree is defined in such a way that each node has either no offspring (the leaves) or four offspring, which always form a group of 2×2 adjacent pixels. In Fig. 2 the arrows are oriented from the parent node to its four offspring. The pixels in the highest level of the pyramid are the tree roots and are also grouped in 2×2 adjacent pixels. However, their offspring branching rule is different, and in each group one of them (indicated by the star in Fig. 2) has no descendants.

The following sets of coordinates are used to present the new coding method:

- $\mathcal{O}(i,j)$: set of coordinates of all offspring of node (i,j);
- $\mathcal{D}(i,j)$: set of coordinates of all descendants of the node (i,j);
- $\mathcal{H}$: set of coordinates of all spatial orientation tree roots (nodes in the highest pyramid level);
- $\mathcal{L}(i,j) = \mathcal{D}(i,j) - \mathcal{O}(i,j)$.

For instance, except at the highest and lowest pyramid levels, we have

$$\mathcal{O}(i,j) = \{(2i,2j),(2i,2j+1),(2i+1,2j),(2i+1,2j+1)\}. \tag{9.8}$$

We use parts of the spatial orientation trees as the partitioning subsets in the sorting algorithm. The set partitioning rules are simply:

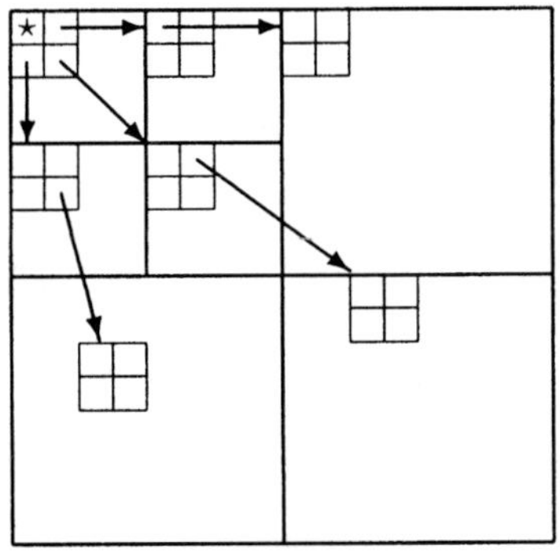

FIGURE 2. Examples of parent-offspring dependencies in the spatial-orientation tree.

1. the initial partition is formed with the sets $\{(i,j)\}$ and $\mathcal{D}(i,j)$, for all $(i,j) \in \mathcal{H}$;
2. if $\mathcal{D}(i,j)$ is significant then it is partitioned into $\mathcal{L}(i,j)$ plus the four single-element sets with $(k,l) \in \mathcal{O}(i,j)$.
3. if $\mathcal{L}(i,j)$ is significant then it is partitioned into the four sets $\mathcal{D}(k,l)$, with $(k,l) \in \mathcal{O}(i,j)$.

6 Coding Algorithm

Since the order in which the subsets are tested for significance is important, in a practical implementation the significance information is stored in three ordered lists, called *list of insignificant sets* (LIS), *list of insignificant pixels* (LIP), and *list of significant pixels* (LSP). In all lists each entry is identified by a coordinate (i,j), which in the LIP and LSP represents individual pixels, and in the LIS represents either the set $\mathcal{D}(i,j)$ or $\mathcal{L}(i,j)$. To differentiate between them we say that a LIS entry is of type A if it represents $\mathcal{D}(i,j)$, and of type B if it represents $\mathcal{L}(i,j)$.

During the sorting pass (see Algorithm I) the pixels in the LIP—which were insignificant in the previous pass—are tested, and those that become significant are moved to the LSP. Similarly, sets are sequentially evaluated following the LIS order, and when a set is found to be significant it is removed from the list and partitioned. The new subsets with more than one element are added back to the LIS, while the single-coordinate sets are added to the end of the LIP or the LSP, depending whether they are insignificant or significant, respectively. The LSP contains the coordinates of the pixels that are visited in the refinement pass.

Below we present the new encoding algorithm in its entirety. It is essentially equal to Algorithm I, but uses the set-partitioning approach in its sorting pass.

ALGORITHM II

1. **Initialization:** output $n = \left\lfloor \log_2 \left(\max_{(i,j)}\{|c_{i,j}|\} \right) \right\rfloor$; set the LSP as an empty

list, and add the coordinates $(i, j) \in \mathcal{H}$ to the LIP, and only those with descendants also to the LIS, as type A entries.

2. **Sorting pass:**

 2.1. for each entry (i, j) in the LIP do:

  2.1.1. output $S_n(i, j)$;
  2.1.2. if $S_n(i, j) = 1$ then move (i, j) to the LSP and output the sign of $c_{i,j}$;

 2.2. for each entry (i, j) in the LIS do:

  2.2.1. if the entry is of type A then
   - output $S_n(\mathcal{D}(i, j))$;
   - if $S_n(\mathcal{D}(i, j)) = 1$ then
     * for each $(k, l) \in \mathcal{O}(i, j)$ do:
       · output $S_n(k, l)$;
       · if $S_n(k, l) = 1$ then add (k, l) to the LSP and output the sign of $c_{k,l}$;
       · if $S_n(k, l) = 0$ then add (k, l) to the end of the LIP;
     * if $\mathcal{L}(i, j) \neq \emptyset$ then move (i, j) to the end of the LIS, as an entry of type B, and go to Step **2.2.2**; else, remove entry (i, j) from the LIS;

  2.2.2. if the entry is of type B then
   - output $S_n(\mathcal{L}(i, j))$;
   - if $S_n(\mathcal{L}(i, j)) = 1$ then
     * add each $(k, l) \in \mathcal{O}(i, j)$ to the end of the LIS as an entry of type A;
     * remove (i, j) from the LIS.

3. **Refinement pass:** for each entry (i, j) in the LSP, except those included in the last sorting pass (i.e., with same n), output the n-th most significant bit of $|c_{i,j}|$;

4. **Quantization-step update:** decrement n by 1 and go to Step **2**.

One important characteristic of the algorithm is that the entries added to the end of the LIS in Step 2.2 are evaluated before that same sorting pass ends. So, when we say "for each entry in the LIS" we also mean those that are being added to its end. With Algorithm II the rate can be precisely controlled because the transmitted information is formed of single bits. The encoder can also use property in equation (9.4) to estimate the progressive distortion reduction and stop at a desired distortion value.

Note that in Algorithm II all branching conditions based on the significance data S_n—which can only be calculated with the knowledge of $c_{i,j}$—are output by the encoder. Thus, to obtain the desired decoder's algorithm, which duplicates the encoder's execution path as it sorts the significant coefficients, we simply have to replace the words *output* by *input* in Algorithm II. Comparing the algorithm above to Algorithm I, we can see that the ordering information $\eta(k)$ is recovered when the coordinates of the significant coefficients are added to the end of the LSP, that is, the coefficients pointed by the coordinates in the LSP are sorted as in (9.5). But note that whenever the decoder inputs data, its three control lists (LIS, LIP, and LSP) are identical to the ones used by the encoder at the moment it outputs that data, which means that the decoder indeed recovers the ordering from the execution path. It is easy to see that with this scheme coding and decoding have the same computational complexity.

An additional task done by decoder is to update the reconstructed image. For the value of n when a coordinate is moved to the LSP, it is known that $2^n \leq |c_{i,j}| < 2^{n+1}$. So, the decoder uses that information, plus the sign bit that is input just after the insertion in the LSP, to set $\hat{c}_{i,j} = \pm 1.5 \times 2^n$. Similarly, during the refinement pass the decoder adds or subtracts 2^{n-1} to $\hat{c}_{i,j}$ when it inputs the bits of the binary representation of $|c_{i,j}|$. In this manner the distortion gradually decreases during both the sorting and refinement passes.

As with any other coding method, the efficiency of Algorithm II can be improved by entropy-coding its output, but at the expense of a larger coding/decoding time. Practical experiments have shown that normally there is little to be gained by entropy-coding the coefficient signs or the bits put out during the refinement pass. On the other hand, the significance values are not equally probable, and there is a statistical dependence between $S_n(i,j)$ and $S_n(\mathcal{D}(i,j))$, and also between the significance of adjacent pixels.

We exploited this dependence using the adaptive arithmetic coding algorithm of Witten *et al.* [7]. To increase the coding efficiency, groups of 2×2 coordinates were kept together in the lists, and their significance values were coded as a single symbol by the arithmetic coding algorithm. Since the decoder only needs to know the transition from insignificant to significant (the inverse is impossible), the amount of information that needs to be coded changes according to the number m of insignificant pixels in that group, and in each case it can be conveyed by an entropy-coding alphabet with 2^m symbols. With arithmetic coding it is straightforward to use several adaptive models [7], each with 2^m symbols, $m \in \{1, 2, 3, 4\}$, to code the information in a group of 4 pixels.

By coding the significance information together the average bit rate corresponds to a m-th order entropy. At the same time, by using different models for the different number of insignificant pixels, each adaptive model contains probabilities *conditioned* to the fact that a certain number of adjacent pixels are significant or insignificant. This way the dependence between magnitudes of adjacent pixels is fully exploited. The scheme above was also used to code the significance of trees rooted in groups of 2×2 pixels.

With arithmetic entropy-coding it is still possible to produce a coded file with the exact code rate, and possibly a few unused bits to pad the file to the desired size.

7 Numerical Results

The following results were obtained with monochrome, 8 bpp, 512×512 images. Practical tests have shown that the pyramid transformation does not have to be exactly unitary, so we used 5-level pyramids constructed with the 9/7-tap filters of [1], and using a "reflection" extension at the image edges. It is important to observe that the bit rates are not entropy estimates—they were calculated from the actual size of the compressed files. Furthermore, by using the progressive transmission ability, the sets of distortions are obtained from *the same file,* that is, the decoder read the first bytes of the file (up to the desired rate), calculated the inverse subband transformation, and then compared the recovered image with the original. The distortion is measured by the peak signal to noise ratio

$$\text{PSNR} = 10 \log_{10}\left(\frac{255^2}{\text{MSE}}\right) \text{ dB}, \tag{9.9}$$

where MSE denotes the mean squared-error between the original and reconstructed images.

Results are obtained both with and without entropy-coding the bits put out with Algorithm II. We call the version without entropy coding ***binary-uncoded.*** In Fig. 3 are plotted the PSNR versus rate obtained for the luminance (Y) component of LENA both for binary uncoded and entropy-coded using arithmetic code. Also in Fig. 3, the same is plotted for the luminance image GOLDHILL. The numerical results with arithmetic coding surpass in almost all respects the best efforts previously reported, despite their sophisticated and computationally complex algorithms (e.g., [1, 8, 9, 10, 13, 15]). Even the numbers obtained with the binary uncoded versions are superior to those in all these schemes, except possibly the arithmetic and entropy constrained trellis quantization (ACTCQ) method in [11]. PSNR versus rate points for competitive schemes, including the latter one, are also plotted in Fig. 3. The new results also surpass those in the original EZW [19], and are comparable to those for extended EZW in [6], which along with ACTCQ rely on arithmetic coding. The binary uncoded figures are only 0.3 to 0.6 dB lower in PSNR than the corresponding ones of the arithmetic coded versions, showing the efficiency of the partial ordering and set partitioning procedures. If one does not have access to the best CPU's and wishes to achieve the fastest execution, one could opt to omit arithmetic coding and suffer little consequence in PSNR degradation. Intermediary results can be obtained with, for example, Huffman entropy-coding. A recent work [24], which reports similar performance to our arithmetic coded ones at higher rates, uses arithmetic and trellis coded quantization (ACTCQ) with classification in wavelet subbands. However, at rates below about 0.5 bpp, ACTCQ is not as efficient and classification overhead is not insignificant.

Note in Fig. 3 that in both PSNR curves for the image LENA there is an almost imperceptible "dip" near 0.7 bpp. It occurs when a sorting pass begins, or equivalently, a new bit-plane begins to be coded, and is due to a discontinuity in the slope of the rate $\times$ distortion curve. In previous EZW versions [19, 6] this "dip" is much more pronounced, of up to 1 dB PSNR, meaning that their embedded files did not yield their best results for all rates. Fig. 3 shows that the new version does not present the same problem.

In Fig. 4, the original images are shown along with their corresponding reconstructions by our method (arithmetic coded only) at 0.5, 0.25, and 0.15 bpp. There are no objectionable artifacts, such as the blocking prevalent in JPEG-coded images, and even the lowest rate images show good visual quality. Table 9.1 shows the corresponding CPU times, excluding the time spent in the image transformation, for coding and decoding LENA. The pyramid transformation time was 0.2 s in an IBM RS/6000 workstation (model 590, which is particularly efficient for floating-

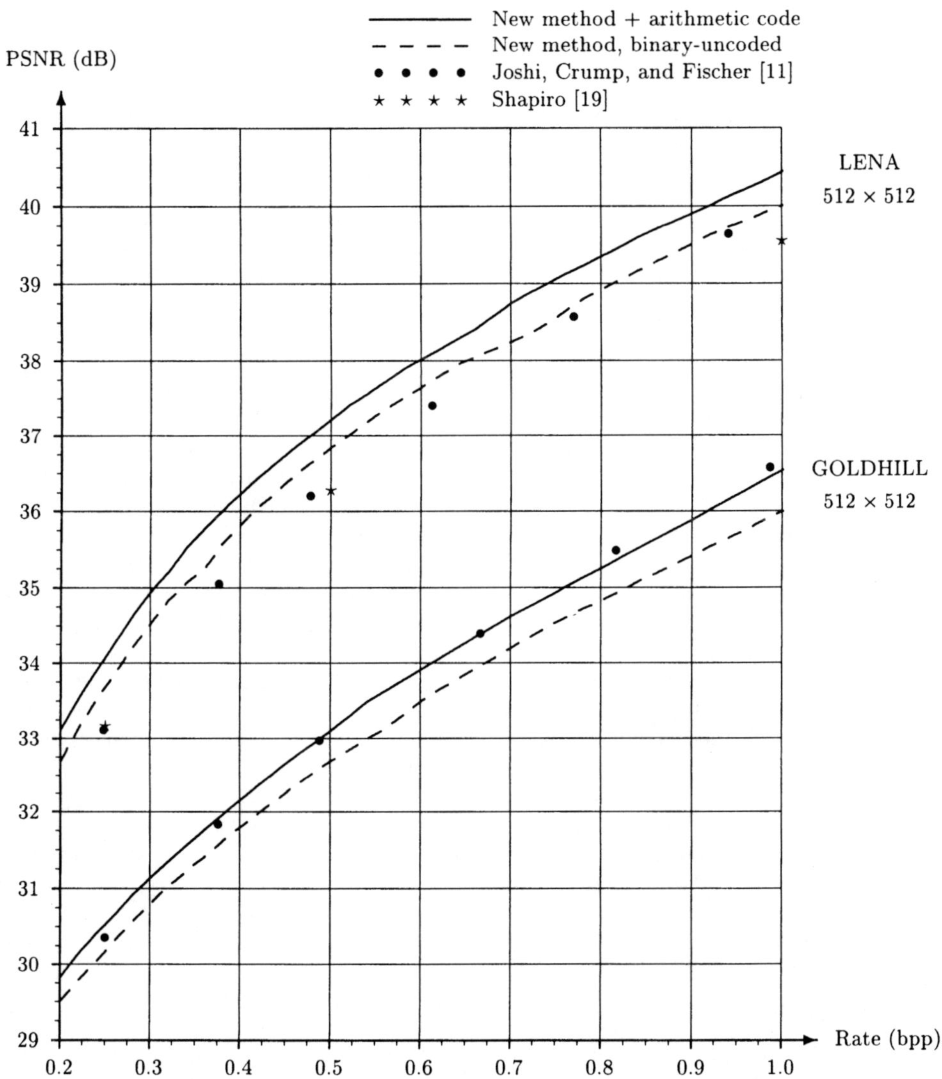

FIGURE 3. Comparative evaluation of the new coding method.

rate (bpp)	binary uncoded		arithmetic coded	
	code	decode	code	decode
0.25	0.07	0.04	0.18	0.14
0.50	0.14	0.09	0.33	0.29
1.00	0.27	0.17	0.64	0.57

TABLE 9.1. Effect of entropy-coding the significance data on the CPU times (s) to code and decode the image LENA 512 × 512 (IBM RS/6000 workstation).

point operations). The programs were not optimized to a commercial application level, and these times are shown just to give an indication of the method's speed. The ratio between the coding/decoding times of the different versions can change for other CPUs, with a larger speed advantage for the binary-uncoded version.

8 Summary and Conclusions

We have presented an algorithm that operates through set partitioning in hierarchical trees (SPIHT) and accomplishes completely embedded coding. This SPIHT algorithm uses the principles of partial ordering by magnitude, set partitioning by significance of magnitudes with respect to a sequence of octavely decreasing thresholds, ordered bit plane transmission, and self-similarity across scale in an image wavelet transform. The realization of these principles in matched coding and decoding algorithms is a new one and is shown to be more effective than in previous implementations of EZW coding. The image coding results in most cases surpass those reported previously on the same images, which use much more complex algorithms and do not possess the embedded coding property and precise rate control. The software and documentation, which are copyrighted and under patent application, may be accessed in the Internet site with URL of `http://ipl.rpi.edu/SPIHT` or by anonymous ftp to `ipl.rpi.edu` with the path `pub/EW_Code` in the compressed archive file `codetree.tar.gz`. (The file must be decompressed with the command `gunzip` and exploded with the command '`tar xvf`'; the instructions are in the file `codetree.doc`.) We feel that the results of this coding algorithm with its embedded code and fast execution are so impressive that it is a serious candidate for standardization in future image compression systems.

9 References

[1] J.M. Shapiro, "Embedded image coding using zerotrees of wavelets coefficients," *IEEE Trans. Signal Processing,* vol. 41, pp. 3445–3462, Dec. 1993.

[2] M. Rabbani, and P.W. Jones, *Digital Image Compression Techniques,* SPIE Opt. Eng. Press, Bellingham, Washington, 1991.

[3] R.A. DeVore, B. Jawerth, and B.J. Lucier, "Image compression through wavelet transform coding," *IEEE Trans. Inform. Theory,* vol. 38, pp. 719–746, March 1992.

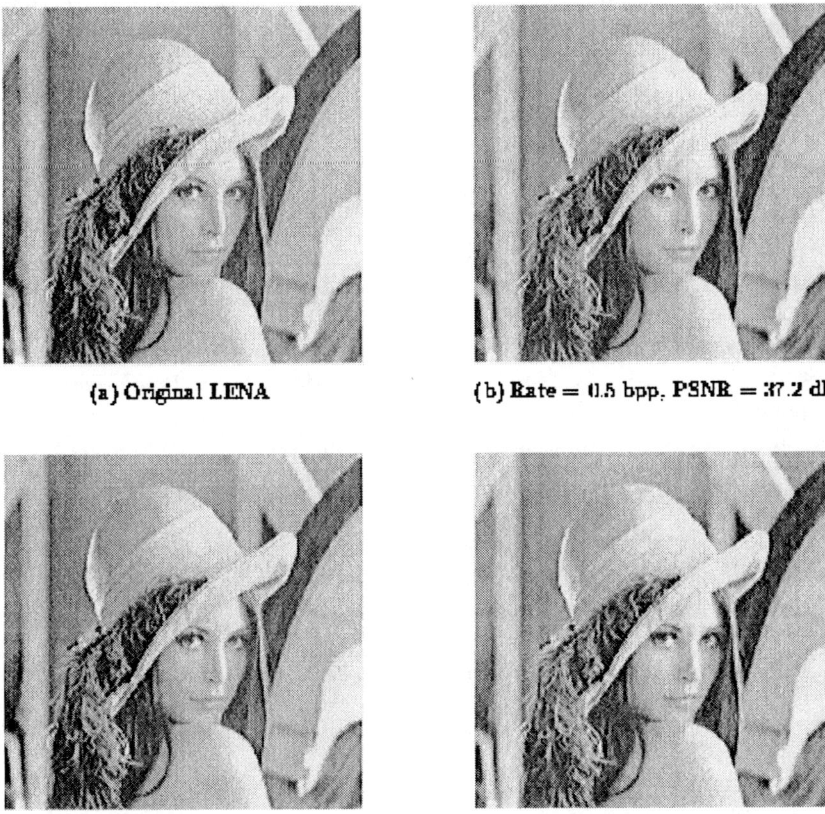

(a) Original LENA

(b) Rate = 0.5 bpp, PSNR = 37.2 dB

(c) Rate = 0.25 bpp, PSNR = 34.1 dB

(d) Rate = 0.15 bpp, PSNR = 31.9 dB

FIGURE 4. Images obtained with the arithmetic code version of the new coding method.

[4] E.H. Adelson, E. Simoncelli, and R. Hingorani, "Orthogonal pyramid transforms for image coding," *Proc. SPIE*, vol. 845 – Visual Comm. and Image Proc. II, Cambridge, MA, pp. 50–58, Oct. 1987.

[5] M. Antonini, M. Barlaud, P. Mathieu, and I. Daubechies, "Image coding using wavelet transform," *IEEE Trans. Image Processing,* vol. 1, pp. 205–220, April 1992.

[6] A. Said and W.A. Pearlman, "Image compression using the spatial-orientation tree," *IEEE Int. Symp. on Circuits and Systems,* Chicago, IL, pp. 279–282, May, 1993.

[7] I.H. Witten, R.M. Neal, and J.G. Cleary, "Arithmetic coding for data compression," *Commun. ACM*, vol. 30, pp. 520–540, June 1987.

[8] P. Sriram and M.W. Marcellin, "Wavelet coding of images using trellis coded quantization," *SPIE Conf. on Visual Inform. Process.*, Orlando, FL, pp. 238–

247, April 1992; also "Image coding using wavelet transforms and entropy-constrained trellis quantization," *IEEE Trans. Image Processing*, vol. 4, pp. 725–733, June 1995.

[9] Y.H. Kim and J.W. Modestino, "Adaptive entropy coded subband coding of images," *IEEE Trans. Image Processing*, vol. IP-1, pp. 31–48, Jan. 1992.

[10] N. Tanabe and N. Farvardin, "Subband image coding using entropy-constrained quantization over noisy channels," *IEEE J. Select. Areas in Commun.*, vol. 10, pp. 926–943, June 1992.

[11] R.L. Joshi, V.J. Crump, and T.R. Fischer, "Image subband coding using arithmetic and trellis coded quantization," *IEEE Trans. Circ. & Syst. Video Tech.*, vol. 5, pp. 515–523, Dec. 1995.

[12] R.L. Joshi, T.R. Fischer, R.H. Bamberger, "Optimum classification in subband coding of images," *Proc. 1994 IEEE Int. Conf. on Image Processing*, vol. II, pp. 883–887, Austin, TX, Nov. 1994.

[13] J.H. Kasner and M.W. Marcellin, "Adaptive wavelet coding of images," *Proc. 1994 IEEE Conf. on Image Processing*, vol. 3, pp. 358–362, Austin, TX, Nov. 1994.

[14] A. Said and W.A. Pearlman, "Reversible Image compression via multiresolution representation and predictive coding," *Proc. SPIE Conf. Visual Communications and Image Processing '93*, Proc. SPIE 2094, pp. 664–674, Cambridge, MA, Nov. 1993.

[15] D.P. de Garrido, W.A. Pearlman and W.A. Finamore, "A clustering algorithm for entropy-constrained vector quantizer design with applications in coding image pyramids," *IEEE Trans. Circ. and Syst. Video Tech.*, vol. 5, pp. 83–95, April 1995.

10

Space-frequency Quantization for Wavelet Image Coding

Zixiang Xiong
Kannan Ramchandran
Michael T. Orchard

1 Introduction

Since the introduction of wavelets as a signal processing tool in the late 1980s, considerable attention has focused on the application of wavelets to image compression [1, 2, 3, 4, 5, 6]. The hierarchical signal representation given by the dyadic wavelet transform provides a convenient framework both for exploiting the specific types of statistical dependencies found in images, and for designing quantization strategies matched to characteristics of the human visual system. Indeed, before the introduction of wavelets, a wide variety of closely related coding frameworks had been extensively studied in the image coding community, including pyramidal coding [7], transform coding [8] and subband coding [9]. Viewed in the context of this prior work, initial efforts in wavelet coding research concentrated on the promise of more effective compaction of energy into a small number of low frequency coefficients. Following the design methodology of earlier transform and subband coding algorithms, initial "wavelet-based" coding algorithms [3, 4, 5, 6] were designed to exploit the energy compaction properties of the wavelet transform by applying quantizers (either scalar or vector) optimized for the statistics of each frequency band of wavelet coefficients. Such algorithms have demonstrated modest improvements in coding efficiency over standard transform-based algorithms.

Contrasting with those early coders, this paper proposes to exploit both the frequency and spatial compaction property of the wavelet transform through the use of two very simple quantization modes. To exploit the spatial compaction properties of wavelets, we define a symbol that indicates that a spatial region of high frequency coefficients has value zero. We refer to the application of this symbol as zerotree quantization, because it will involve setting to zero a tree-structured set of wavelet coefficients. In the next section, we explain how a spatial region in the image is related to a tree-structured set of coefficients in the hierarchy of wavelet coefficients. Zerotree quantization can be viewed as a mechanism for pointing to the locations where high frequency coefficients are clustered. Thus, this quantization mode directly exploits the spatial clustering of high frequency coefficients.

For coefficients that are not set to zero by zerotree quantization, we propose to apply a common uniform scalar quantizer, independent of the coefficient's frequency band. The resulting scalar indices are coded with an entropy coder, with probabilities adapted to the statistics of each band. We select this quantization scheme for its simplicity. In addition, though we recognize that improved performance can be achieved by more complicated quantization schemes (e.g. vector quantization, scalar quantizers optimized for each band, optimized non-uniform scalar quantizers, entropy-constrained scalar quantization [10], etc.), we conjecture that these performance gains will be limited when coupled with zerotree quantization. When zerotree quantization is applied most efficiently, the remaining coefficients will be characterized by distributions that are not very peaked near zero. Consequently, uniform scalar quantization followed by entropy coding provides nearly optimal coding efficiency, and achieves nearly optimal bit-allocation among bands with differing variances. The good coding performance of our proposed algorithm provides some experimental evidence in support of this conjecture.

Though zerotree quantization has been applied in several recent wavelet-based image coders, this paper is the first to address the question of how to jointly optimize the application of spatial quantization modes (zerotree quantization) and scalar quantization of frequency bands of coefficients. In [1], Lewis and Knowles apply a perceptually-based thresholding scheme to predict zerotrees of high frequency coefficients based on low valued coefficients in a lower frequency band corresponding to the same spatial region. While this simple *ad hoc* scheme exploits interband dependencies induced by spatial clustering, it often introduces large error in the face of prediction errors. Shapiro's embedded zerotree approach [2] applies the zerotree symbol when all coefficients in the corresponding tree equal zero. While this strategy can claim to minimize distortion of the overall coding scheme, it cannot claim optimality in an operational rate and distortion sense (i.e. it does not minimize distortion over all strategies that satisfy a given rate constraint).

This paper consists of two main parts. The first part focuses on a the space-frequency quantization (SFQ) algorithm for the dyadic wavelet transform [21]. The goal of SFQ is to optimize the application of zerotree and scalar quantization in order to minimize distortion for a given rate constraint. The image coding algorithm described in the following sections is an algorithm for optimally selecting the spatial regions (from the set of regions allowed by the zerotree quantizer) for applying zerotree quantization, and for optimally setting the scalar quantizer's step size for quantizing the remaining coefficients. We observe that, although these two quantization modes are very basic, an image coding algorithm that applies these two modes in a jointly optimal manner can be competitive with (and perhaps outperform) the best coding algorithms in the literature. Consequently, we claim that the joint management of space and frequency based quantizers is one of the most important fundamental issues in the design of efficient image coders.

The second part of this paper extends the SFQ algorithm from wavelet transforms to wavelet packets (WP) [12]. Wavelet packets are a family of wavelet transforms (or filter bank structures), from which the best one can be chosen adaptively for each individual image. This allows the frequency resolution of the chosen wavelet packet transform to best match that of the input image. The main issue in designing a WP-based coder is to search for the best wavelet packet basis. Through fast implementation, we show that a WP-based SFQ coder can achieve significant gain in PSNR over the wavelet-based SFQ coder at the same bitrate for some class of images [13]. The increase of computational complexity is moderate.

2 Background and Problem Statement

2.1 Defining the tree

A wavelet image decomposition can be thought as a tree-structured set of coefficients, with each coefficient corresponding to a spatial region in the image. A spatial wavelet coefficient tree is defined as the set of coefficients from different bands that represent the same spatial region in the image. The lowest frequency band of the decomposition is represented by the root nodes (top) of the tree, the highest frequency bands by the leaf nodes (bottom) of the tree, and each parent node represents a lower frequency component than its children. Except for a root node, which has only three children nodes, each parent node has four children nodes, the 2×2 region of the same spatial location in the immediately higher frequency band.

Define a *residue tree* as the set of all descendants of any parent node in a tree. (Note: a residue tree does not contain the parent node itself.) Zerotree spatial quantization of a residue tree assigns to the elements of the residue tree either their original values or all zeros. Note the semantic distinction between residue trees and zerotrees: the residue tree of a node is the set of all its descendants, while a zerotree is an all zero residue tree. A zerotree node refers to a node whose descendants are all set to zero. Zerotrees can originate at any level of the full spatial tree, and can therefore be of variable size. When a residue tree is zerotree quantized, only a single symbol is needed to represent the set of zero quantized wavelet coefficients – our coder uses a binary zerotree map indicating the presence or absence of zerotree nodes in the spatial tree.

2.2 Motivation and high level description

The underlying theme of SFQ is that of efficiently coupling the spatial and frequency characterization modes offered by the wavelet coefficients, by defining quantization strategies that are well matched to the respective modes. The paradigm we invoke is a combination of simple uniform scalar quantization to exploit the frequency characterization, with a fast tree-structured zerotree quantization scheme to exploit the spatial characterization.

Our proposed SFQ coder has a goal of jointly finding the best combination of spatial zerotree quantization choice and the scalar frequency quantizer choice. The SFQ paradigm is conceptually simple: throw away, i.e. quantize to zero, a subset of the wavelet coefficients, and use a single simple uniform scalar quantizer on the rest. Given this framework, the key questions are obviously:

(I) what (spatial) subset of coefficients should be thrown away? and

(II) what uniform scalar (frequency) quantizer step size should be used to quantize the survivor set, i.e. the complementary set of (I)?

This paper formulates the answers to these questions invoking an operational rate-distortion optimality criterion. While the motivation is simple, this optimization task is complicated by the fact that the two questions posed above are interdependent. The reason for this is easily seen. The optimal answer to (I), i.e. the optimal spatial subset to throw away depends on the scalar quantizer choice of (II) since the zerotree pruning operation involved in (I) is driven by rate-distortion tradeoffs induced by the quantizer choice. Conversely, the scalar quantizer of (II) is applied only to the complementary subset of (I), i.e. to the population subset which survives the zerotree pruning operation involved in (I). This interplay between these modes necessitates an iterative way of optimizing the problem, which will be described in detail in the following sections.

Note that the answers to (I) and (II) are sent as side information ("map" bits) to the decoder in addition to the quantized values of the survivor coefficients ("data" bits). Since a single scalar quantizer is used for the entire image, the quantizer step size information of (II) is negligible and can be ignored. The side-information of (I) is sent as a binary zerotree map indicating whether or not tree nodes are zerotree quantized. This overhead information is not negligible and is optimized jointly with the "data" information in our SFQ coder, with the optimization being done in a rate–distortion sense. At this point we will not concern ourselves with the details of how this map is sent, but we will see later that much of this zerotree map information is actually predictable and can be inferred by the decoder using a novel prediction scheme based on the known data field of the corresponding parent band.

Finally, while it appears counter-intuitive at first glance to expect high performance from using a single uniform scalar quantizer, further introspection reveals why this is possible. The key is to recall that the quantizer is applied only to a subset of the full wavelet data, namely the survivors of the zerotree pruning operation. This pruned set has a distribution which is considerably less peaked than that of the original full set, since most of the samples populating the zero and low valued bins are discarded during the zerotree pruning operation. Thus, the spatial zerotree operation effectively "flattens" the residue density of the "trimmed" set of wavelet coefficients, endorsing the use of a single step size uniform quantizer. In summary, the motivation of resorting to multiple quantizers for the various image subbands (as is customarily done) is to account for the different degrees of "peakiness" around zero of the associated histograms of the different image subbands. In our proposed scheme, the bulk of the insignificant coefficients responsible for this peakiness are removed from consideration, rendering the bands with near flat distributions and justifying a simple single step size scalar quantizer.

2.3 Notation and problem statement

Let T denote the balanced (full) spatial tree, i.e. the tree grown to full depth (in a practical scenario, this may be restricted to four to six levels typically). Letting i denote any node of the spatial tree, T_i signifies the full balanced tree rooted at node i. Note that T is short-hand notation for the balanced full-depth tree (i.e. rooted at the origin). In keeping with conventional notation, we define a pruned subtree $S_i \preceq T_i$ as any subtree of T_i that shares its root i. Again for brevity, $S \preceq T$ refers to any pruned subtree of the full depth tree T. Note that the set of all $S \preceq T$ corresponds to the collection of all possible zerotree spatial-quantization topologies. We also need to introduce notation for residue trees. A *residue tree* U_i (corresponding to any arbitrary parent node i of T) consists of the set of all descendants of i in T but does not include i itself, i.e. $U_i = T_i - \{i\} = \cup_{j \in C_i} T_j$, where C_i is the set of children or direct descendants of i (this is a 2x2 children set for all parent nodes except the root nodes which contain only 3 children). See Fig. 1.

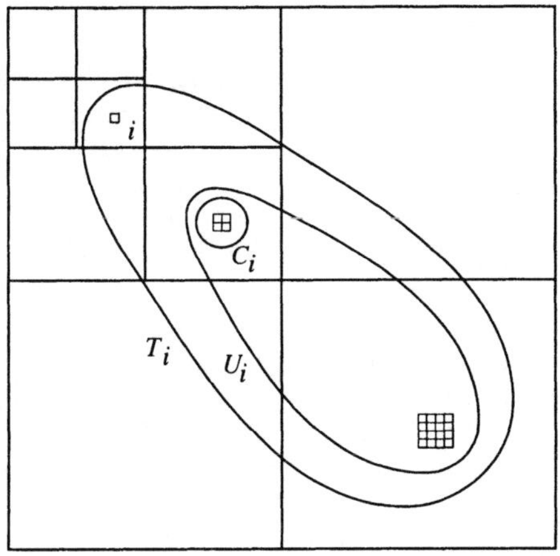

FIGURE 1. Definitions of C_i, U_i and T_i for a node i in a spatial tree. C_i is the set of children or direct descendants of i, U_i the residue tree of i, T_i the full balanced tree rooted at node i ($T_i = i \cup U_i$).

Let us now address quantization. Let Q represent the (finite) set of all admissible scalar frequency quantizer choices. Thus, the quantization modes in our framework are the spatial tree-structured quantizer $S \preceq T$ and the scalar frequency quantizer $q \in Q$ (used to quantize the coefficients of S). The unquantized and quantized wavelet coefficients associated with node i of the spatial tree will be referred to by w_i and $\hat{w}_i(q)$, with the explicit dependency on q of $\hat{w}_i$ being dropped where obvious. In this framework, we seek to minimize the average distortion subject to an average rate constraint. Let $D(q, S)$ and $R(q, S)$ denote the distortion and rate, respectively, associated with quantizer choice (q, S). We will use a squared-error distortion measure. The rate $R(q, S)$ consists of two components: tree data rate $R_{(data)}(q, S)$, measured by the first-order entropy, and tree map rate $R_{(map)}(S)$, where the superscripts will be dropped where obvious.

Then, our problem can be stated simply as:

$$\min_{\{q \in Q;\ S \preceq T\}} D(q, S) \quad \textit{subject to} \quad R(q, S) \leq R_b, \tag{10.1}$$

where R_b is the coding budget.

Stated in words, our coding goal is to find the optimal combination of spatial subset to prune (via zerotree spatial quantization) and scalar quantizer step size to apply to the survivor coefficients (frequency quantization) such that the total quantization distortion is minimized subject to a constraint on the total rate. Qualitatively stated, scalar frequency-quantizers trade off bits for distortion in proportionality to their step sizes, while zerotree spatial-quantizers trade off bits for distortion by zeroing out entire sets of coefficients but incurring little or no bitrate cost in doing so. We are interested in finding the optimal tradeoff between these two quantization modes.

2.4 Proposed approach

The constrained optimization problem of (10.1) can be converted to an unconstrained formulation using the well-known Lagrange multiplier method. That is, it can be shown [14, 15] that the solution to (10.1) is identical to the solution to the following equivalent unconstrained problem for the special case of $R(q,S) = R_b$:

$$\min_{\{q \in Q, S \preceq T\}} [J(q,S) = D(q,S) + \lambda R(q,S)], \tag{10.2}$$

where $J(q,S)$ is the Lagrangian (two-sided) cost including both rate and distortion, which are connected through the Lagrange multiplier $\lambda \geq 0$, which is the quality-factor trading off distortion for rate ($\lambda = 0$ refers to the highest attainable quality and $\lambda = \infty$ to the lowest attainable rate). Note that the entropy only or distortion only cost measure of [2] become special cases of this more general Lagrangian cost measure corresponding to $\lambda = \infty$ and $\lambda = 0$, respectively. The implication of (10.2) is that if an appropriate λ can be found for which the solution to (10.2) is (q^*, S^*) and further $R(q^*, S^*) = R_b$, then (q^*, S^*) is also the solution to (10.1). The solution of (10.2) finds points that reside on the convex-hull of the rate-distortion function, and sweeping λ from 0 to ∞ traces this convex hull. In practice, for most applications (including ours), a convex-hull approximation to the desired rate R_b suffices, and the only suspense is in determining the value of λ that is best matched to the bit budget constraint R_b. Fortunately, the search for the optimal rate-distortion slope λ^* is a fast convex search which can be done with any number of efficient methods, e.g. the bisection method [15].

Our proposed approach is therefore to find the convex-hull approximation to (10.1) by solving:

$$\overset{(c)}{\max_{\lambda \geq 0}} \; \overset{(b)}{\min_{q \in Q}} \; \overset{(a)}{\min_{S \preceq T}} \; [\; D(q,S) + \lambda(R(q,S) - R_b) \;], \tag{10.3}$$

where the innermost minimization (a) involves the search for the best spatial subtree S for fixed values of q and λ, the second minimization (b) involves search for the best scalar quantizer q (and associated $S(q)$) for a fixed value of λ, and finally the outermost optimization (c) is the convex search for the optimal value of λ that satisfies the desired rate constraint (see [15]). The solution (λ^*, q^*, S^*) to (10.3) is thus obtained in three sequential optimization steps.

Minimization (a) involves optimal tree-pruning to find the best S for a fixed q and λ and is by far the most important of the three optimization operations of (10.3). This will be described in Sections 3.1 and 3.2. Given our global framework of data + tree-map, minimization (a) can be written as:

$$\overset{(a)}{\min_{\{S \preceq T\}}} [D(q,S) + \lambda(R_{(data)}(q,S) + R_{(map)}(q,S))], \tag{10.4}$$

where $R_{(map)}(q,S)$ is the tree-map rate and $R_{(data)}(q,S)$ the tree-data rate. We seek an efficient way to jointly code the data + map information using a novel way of predicting the map information from the known data information (see Section 3.2). As this method dictates a strong coupling between the data and map field components, the chicken-and-egg problem is solved using a two-phase approach. In the first phase, (10.4) is optimized assuming that $R_{(map)}(q,S)$ is zero, or more generally that it is a fixed cost independent of the choice of q and S (this simply changes the optimal operating slope λ^* for the same target bit budget R_b).

The optimal S from the phase I tree-pruning operation is $S^*_{(data)}$, i.e.

$$S^*_{(data)} = arg \min_{\{S \preceq T\}} [D(q, S) + \lambda R_{(data)}(q, S)]. \tag{10.5}$$

In the second phase (tree-map prediction phase), the true data-map dependencies are taken into account, and the solution of phase I, $S^*_{(data)}$, is modified to reflect the globally correct choice $S^*_{(data+map)}$. Details are provided in Section 3.

At the end of phase two, for each space–frequency quantization choice, we identify a single point on the operational R-D curve corresponding to a choice of q, λ, and their best matched S. In order to find the best scalar quantizer, we search for the $q \in Q$ (in minimization (b)) which "lives" at absolute slope λ on the convex hull of the operational R-D curve. This defines the optimal combination of q and S for a fixed λ. Finally, the "correct" value of λ, λ^*, that matches the rate constraint R_b is found using a fast convex search in optimization (c).

3 The SFQ Coding Algorithm

3.1 Tree pruning algorithm: Phase I (for fixed quantizer q and fixed λ)

The proposed algorithm is designed to minimize an unweighted mean-squared error distortion measure, with distortion energy measured directly in the transform domain to reduce computational complexity. In this tree pruning algorithm, we also approximate the encoder output rate by the theoretical first order entropy, which can be approached very closely by applying adaptive arithmetic coding. Our (phase I) tree-pruning operation assumes that the cost of sending the tree map information is independent of the cost of sending the data given the tree map, an approximation which will be accounted for and corrected later in phase II. Thus, there will be no mention of the tree map rate in this phase of the algorithm, where the goal will be to search for that spatial subtree $S^*_{data} \preceq T$ whose *data* cost is minimum in the rate-distortion sense.

The lowpass band of coefficients at the coarsest scale cannot (by definition) be included in residue trees since residue trees refer only to descendants. The lowpass band quantizer operates independently of other quantizers. Therefore, we code this band separately from other highpass bands. The quantizer applied to the lowpass band is selected so that the operating slope on its R-D curve matches the overall absolute slope λ on the convex hull of the operational R-D curve for the "highpass" coder, i.e. we invoke the necessary condition that at optimality both coders operate at the same slope on their operational rate-distortion curves, else the situation can be improved by stealing bits from one coder to the other till equilibrium is established.

The following algorithm is used for a fixed value of q and λ to find the best $S \preceq T$. Note that the iteration count k is used as a superscript where needed. $n_i^{(k)}$ refers to the binary zerotree-map (at the k th iteration of the algorithm) indicating the presence ($n_i^{(k)} = 0$) or absence ($n_i^{(k)} = 1$) of a zerotree associated with node i of the tree. Recall that $n_i^{(k)} = 0$ implies that all descendants of i (i.e. elements of U_i) are set to zero at the k *th* iteration. $S^{(k)}$ refers to the (current) best subtree obtained after k iterations, with $S^{(0)}$ initialized to the full tree T. We will drop the "data" suffix from S to avoid cluttering. C_i refers to the set of children nodes (direct offspring) of node i. $J_{U_j}^*$ refers to the minimum or best (Lagrangian) cost associated with the residue tree U_j of node j, with this cost being set to zero for all leaf nodes of the full tree. $J_j^{(k)}$ is the (Lagrangian) cost of quantizing node j (with w_j and $\hat{w}_j$ denoting the unquantized and quantized values of the wavelet coefficient at node j respectively) at the k *th* iteration of the algorithm, with $D_j^{(k)}$ and $R_j^{(k)}$ referring to the distortion and rate components respectively. $p_j^{(k)}$ is the probability at the k *th* iteration, i.e. using the statistics from the set $S^{(k)}$, of the quantization bin associated with node j. We need to superscript the tree rate (and hence the Lagrangian) cost with the iteration count k because the tree changes topology (i.e. gets pruned) at every iteration, and we assume a global entropy coding scheme. Finally, we assume that the number of levels in the spatial tree is indexed by the scale parameter l, with $l = 0$ referring to the coarsest (lowest frequency) scale.

ALGORITHM I

Step 1 (Initialization):

Set $S^{(0)} \leftarrow T$; set the iteration count $k \leftarrow 0$. For all leaf nodes j of T, set $J_{U_j}^* = 0$.

$$\begin{aligned} S^{(0)} &\leftarrow T, \\ k &\leftarrow 0, \\ J_{U_j}^* &\leftarrow 0, \quad \forall j \in leaf\ nodes\ of\ T. \end{aligned}$$

Step 2 (Probability update - needed due to the use of entropy coding):

Update the probability estimates for all nodes in $S^{(k)}$, i.e. update $p_i^{(k)}, \forall i \in S^{(k)}$, where

$$p_i^{(k)} = \frac{no.\ of\ coeffs.\ quantized\ to\ bin\ no.\ \lfloor (w_i/q + 0.5) \rfloor}{no.\ of\ coeffs.\ in\ S^{(k)}}.$$

Step 3 (Zerotree pruning):

Set tree-depth count $l \leftarrow$ $maximum\ depth\ of\ S^{(k)} - 1$. For every node i at current tree-depth l of $S^{(k)}$, determine if it is cheaper to zero out or to keep its best residue tree U_i in a rate-distortion sense. Zeroing out or pruning U_i incurs a cost equal to the energy of residue tree U_i (L.H.S. of inequality (10.6)), while keeping U_i incurs the cost of sending C_i and the best residue tree representations of nodes in C_i (R.H.S. of inequality (10.6)). That is,

$$l \leftarrow maximum\ depth\ of\ S^{(k)} - 1,$$

$$\forall i \in depth\ l\ of\ S^{(k)},$$

$$\begin{array}{ll} If & \sum_{j\in U_i} w_j^2 \leq \sum_{j\in C_i}(J_j^{(k)} + J_{U_j}^*), \\ then & \{n_i^{(k)} = 0; \quad J_{U_i}^* = \sum_{j\in U_i} w_j^2\} \\ else & \{n_i^{(k)} = 1; \quad J_{U_i}^* = \sum_{j\in C_i}(J_j^{(k)} + J_{U_j}^*)\} \end{array} \tag{10.6}$$

where

$$J_j^{(k)} = D_j^{(k)} + \lambda R_j^{(k)} = (w_j - \hat{w}_j^{(k)})^2 + \lambda[-log_2(p_j^{(k)})]. \tag{10.7}$$

Step 4 (Loop bottom-up through all tree levels):

Set $l \leftarrow l - 1$ and go to **Step 2** if $l \geq 0$.

Step 5 (Check for convergence, else iterate):

Using the values of $\{n_i^{(k)}\}$ for all $i \in S^{(k)}$ found by optimal pruning, carve out the pruned subtree $S^{(k+1)}$ for the next iteration. If $S^{(k+1)} \neq S^{(k)}$ (i.e. if some nodes got pruned), then increment the iteration count $k \leftarrow (k+1)$ and go back to **Step 1** to update statistics and iterate again. Else, declare $S^*_{data} \leftarrow S^{(k+1)}$ as the converged pruned spatial tree associated with scalar quantizer choice q and rate-distortion slope λ. This uniquely defines the (locally) optimal zerotree map $\{n_i\}$ for all nodes $i \in T$.

Discussion:

1. Scalar frequency-quantization (using step size q) of all the highpass coefficients is applied in an iterative fashion. At each iteration, a fixed tree $S^{(k)}$ specifies the coefficients to be uniformly quantized, and the pruning rule of Step 2 is invoked to decide whether coefficients are worthy of being retained or if they should be killed. As the decision of whether or not to kill the descendants U_j of node j cannot be made without knowing the best representation (and associated best cost) for residue tree U_j, the pruning operation must proceed from the bottom of the tree (leaves) to the top (root).
2. Note that in Step 2, we query whether or not it is worthwhile to send any of the descendants of node i. This is done by comparing the cost of zeroing out *all* descendants of node i (assuming that zerotree quantized data incurs zero rate cost) to the best alternative associated with not choosing to do so. This latter cost is that of sending the children nodes of i together with the best cost of the residue trees associated with each of the children nodes. Since processing is done in a bottom-up fashion, these best residue tree costs are known at the time. The cheaper of these costs is used to dictate the zerotree decision to be made at node i, and is saved for future reference involving decisions to be made for the ancestors of i.
3. As a result of the pruning operation of Step 2, some of the spatial tree nodes are discarded. This affects the histogram of the surviving nodes which is recalculated in Step 1 (initially the histogram associated with the full tree is used) whenever any new node gets pruned out.
4. The above algorithm is guaranteed to converge to a local optimal choice for S^*_{data}.

Proposition 1: The above tree pruning algorithm converges to a local minimum.

Proof: See Appendix.

A plausible explanation for the above proposition is the following: Recall that the motivation for our iterative pruning algorithm is that as the trees get pruned, the probability density function (PDF) of the residue trees change dynamically. So, at each iteration we update to reflect the "correct" PDFs till the algorithm converges. The above Proposition shows that the gain in terms of Lagrangian cost comes from better representation of the PDFs of the residue trees in the iterations. In the above tree pruning algorithm, the number of non-pruned nodes is monotonically decreasing before it converges, so the above iterative algorithm converges very fast! In our simulations, the above algorithm converges in less than five iterations in all our experiments.

5. In addition to being locally optimal, our algorithm can make claims to having global merits as well. While the following discussion is not intended to be a rigorous justification, it serves to provide a relative argument from a viewpoint of image coding. After the hierarchical wavelet transform, the absolute values of most of the highpass coefficients are small, and they are deemed to be quantized to zero. The PDF of the wavelet coefficients is approximately symmetric, and sharply peaked at zero. Sending those zero quantized coefficients is not cost effective. Zerotree quantization efficiently identifies those zerotree nodes. A larger portion of the available bit budget is allocated for sending the larger coefficients which represent more energy. So, the resulting PDF after zerotree quantization is considerably less peaked than the original one. Suppose the algorithm decided to zerotree quantize residue tree U_i at the k th iteration, i.e. deemed the descendants of i to be not worth sending at the k th iteration. This is because the cost of pruning U_i which is identical to the energy of U_i (L.H.S. of inequality (10.6)), is less than or equal to the cost of sending it (R.H.S. of inequality (10.6)), or

$$\sum_{j \in U_i} w_j^2 \leq \sum_{j \in C_i} (J_j^{(k)} + J_{U_j}^*).$$

The decision to zerotree quantize the set U_i is usually because U_i consists mainly of insignificant coefficients (with respect to the quantizer step size q). Then, in the $(k+1)$th iteration, due to the trimming operation, the probability of "small" coefficients becomes smaller, i.e. the cost of sending residue tree U_i becomes larger at the $(k+1)$ th iteration, $J_j^{(k+1)}$ goes up (since $D_j^{(k+1)} = D_j^{(k)}$ while $R_j^{(k+1)} > R_j^{(k)}$ in (10.7)), thus reinforcing the wisdom in killing U_i at the kth iteration. That is, if inequality (10.6) holds at the k th iteration, then with high probability it is also true at the $(k+1)$ th iteration. Thus our algorithm's philosophy of "once pruned it is pruned forever" which leads to a fast solution is likely to be very close to the globally optimal point as well. Of course, for residue trees that only have a few large coefficients, zerotree quantization in an early iteration might affect the overall optimality. However, we expect that the probability of such subtrees to be relatively small for natural images. So, our algorithm is efficient globally.

3.2 Predicting the tree: Phase II

Recall that our coding data-structure is a combination of a zerotree map indicating which nodes of the spatial tree have their descendants set to zero, and the quantized data stream corresponding to the survivor nodes. The side information needed to send the zerotree map is obviously a key issue in our design. In the process of formulating the optimal pruned spatial-tree representation, as described in Section 3.1, we did not consider the cost needed to send the tree-description $\{n_i\}$. This is tantamount to assuming that the tree map is free, or more generally that the tree-map cost is independent of the choice of tree (i.e. all trees cost the same regardless of choice of tree). While this is certainly feasible, it is not necessarily an efficient coding strategy. In this section, we describe a novel way to improve the coding efficiency by using a prediction strategy for the tree *map* bits of the nodes of a given band based on the (decoded) *data* information of the associated parent band. We will see that this leads to a way for the decoder to *deduce* much of the tree-map information from the data field (sent top-down from lower to higher frequency bands) leaving the encoder to send zerotree bits only for nodes having unpredictable map information.

Due to the tight coupling between data and map information in our proposed scheme, the "best" tree representation, as found through Algorithm I assuming that the map bits are decoupled from the data bits, needs to be updated to correct for this bias, and zerotree decisions made at the tree nodes need to be re-examined to check for possible reversals in decision due to the removal of this bias. In short, the maxim that "zerotree maps are not all equally costly" needs to be quantitatively reflected in modifying the spatially pruned subtree obtained from Algorithm I to a tree description that is the best in the global (data + map) sense. In this subsection, we describe how to accomplish this within the framework of predictive spatial-quantization while maintaining overall rate–distortion optimality. The basic idea is to predict the significance/insignificance of a residue tree from the energy of its parent.

The predictability of subtrees depends on two thresholds output from the spatial-quantizer. This is a generalization of the prediction scheme of Lewis and Knowles [1] for both efficiently encoding the tree, and modifying the tree to optimally reflect the tree encoding. The Lewis and Knowles technique is based on the observation that the variance of a parent block centered around node i usually provides a good prediction of the energy of coefficients in the residue tree U_i. Their algorithm eliminates tree information by completely relying on this prediction. In order to improve performance, we incorporate this prediction in our algorithm by using it to represent the overall spatial tree information in a rate-distortion optimal way, rather than blindly relying on it to completely avoid sending tree-map information, which is in general suboptimal.

First, the variance of each (parent) node i is calculated as the energy of a 3×3 block centered at the corresponding wavelet coefficient of the node[1]. Note also that, for decodability or closed–loop operation, all variances should be calculated based on the *quantized* wavelet coefficients. We assume zero values for zerotree quantized

[1]This "low-pass" filtering is needed to more accurately reflect the level of activity at the node.

coefficients.

Then, the variances of the parent nodes of each band are ordered in decreasing magnitude, and the zerotree map bits corresponding to these nodes, n_i, are listed in the same order. Two thresholds T_h and T_l are sent per band as (negligible) overhead information to assist the encoding.[2] Nodes whose parents have variances above T_h are assumed to be significant (i.e. n_i is assumed to be 1), thus requiring no tree information. Similarly, nodes with parents having energy below T_l are assumed to be insignificant (i.e. n_i is assumed to be 0), and they too require no side information. Tree-map information is sent only for those nodes whose parents have a variance between T_h and T_l. The algorithm is motivated by the potential for the Lewis and Knowles predictor to be fairly accurate for nodes having very high or very low variance, but to perform quite poorly for nodes with variance near the threshold.

This naturally leads to the question of optimization of the parameters T_h and T_l within the pruned spatial-tree quantization framework. We now address their optimal design. Clearly, T_h should be at least as small as the variance of the highest insignificant node (or the first node from the top of the list for which $n_i = 0$), since setting T_h any higher would require sending redundant tree information for residue tree which could be inferred via the threshold. Likewise, T_l should be at least as large as the variance of the smallest significant node (or the first node from the bottom of the list for which $n_i = 1$).

Now let us consider if T_h should be made smaller in an optimal scenario. Let z_h denote the index of the node with variance equal to T_h (n_{z_h} must be 0), and suppose the number of 0 nodes down to n_{z_h} in the variance-ordered list is h. Note, if we shall reduce T_h at all, then, $h > 1$, since, T_h should made at least as small as the variance of the next node $n_{z_{h+1}}$ with $n_{z_{h+1}} = 0$.

Let b_h be the position difference between n_{z_h} and $n_{z_{h+1}}$, then, this change to T_h saves us $\sum_{i=1}^{h} b_i$ bits, equal to the number of positions we move T_h down in the list (we assume that these binary map symbols have an entropy of one bit per symbol). Thus, changing n_{z_i} from 0 to 1 for all i from $i = 1$ to h, decreases the map rate by

$$\Delta R_{map,h} = \sum_{i=1}^{h} b_i \tag{10.8}$$

Of course we know that reversing the map bits for the n_{z_i} nodes from 0 to 1 increases the *data* cost (in the rate-distortion or Lagrangian sense) as determined in the pruning phase of Algorithm I. So in performing a complete analysis of data + map, we re-examine and, where globally profitable, reverse the "data-only" based zerotree decisions output by Algorithm I. The rule for reversing the decisions for the nodes n_{z_i} is clear: weigh the "data cost loss" versus the "map cost gain" associated with the reversals and reverse only if the latter outweighs the former. As we are examining rate-distortion tradeoffs, we need to use Lagrangian costs in this comparison.

It is clear that in doing the tree-pruning of Algorithm I, we can store (in addition to the tree map information $\{n_i\}$) the winning and losing Lagrangian costs cor-

[2] T_h and T_l are actually sent indirectly by sending the coordinates of two parent nodes which have variances T_h and T_l respectively.

responding to each node i, where the winning cost corresponds to that associated with the optimal binary decision n_i (i.e. $J^*_{U_i}$) and the losing cost corresponds to that associated with the losing decision $\overline{n_i}$ (i.e. the larger side of inequality (10.6). Denote by $\Delta J_{data,i}$ the magnitude of this Lagrangian cost difference for node i (note the inclusion of the data subscript for emphasis); i.e. $\Delta J_{data,i}$, for every parent node $i \in T$ is the absolute value of the difference between the two sides of the inequality (10.6) after convergence of Algorithm I. Then, the rule for reversing the n_{z_i} decisions from 0 to 1 for all nodes z_i from $i = 1$ to h is clearly:

$$If \quad \lambda \Delta R_{map,h} = \lambda \sum_{i=1}^{h} b_i > \sum_{i=1}^{h} \Delta J_{data,z_i}, \tag{10.9}$$

$$Then \quad reverse\ phase\ I\ decision: \quad n_{z_i} \leftarrow 1. \tag{10.10}$$

If inequality (10.9) is not true, no decision shall be made until we try to move T_h to the next 0 node. In this case, h is incremented until inequality (10.9) *is* satisfied for some larger h, whereupon n_{z_h} is reversed to 1 for all i from $i = 1$ to h. Then h is reset to 1 and the whole operation repeated until the entire list has been exhausted.

We summarize the design of T_h as follows:

ALGORITHM II

Step 1 Order the variance of each parent node in decreasing magnitude, and list the zerotree map bits associated with these nodes in the same order.

Step 2 Identify all the **zero** nodes n_{z_h} in the list, and record b_h, the difference in list position entry between the h th and the $(h + 1)$ th **zero** nodes.

Step 3 Set $h = 1$.

Step 4 Check if inequality (10.9) is satisfied for this value of h. If it is not, increment h, if possible, and go to **Step 4**. Else, reverse the tree map bits n_{z_i} from 0 to 1 for all i from $i = 1$ to h, and go to **Step 2**.

T_h will point to the first zero node on the final modified list. It is obvious that Algorithm II optimizes the choice of T_h, using a global data + map rate-distortion perspective. A similar algorithm is used to optimize the choice of T_l. As a result of the tree prediction algorithm, the optimal pruned subtree S^*_{data} output by Algorithm I (based on data only) is modified to $S^*_{data+map}$.

3.3 Joint Optimization of Space-Frequency Quantizers

The above fast zerotree pruning algorithm tackles the innermost optimization (a) of (10.3) , i.e. finds $S^*(q)$ for each each scalar quantizer q (and λ, which is implied). As stated earlier, for a fixed quality factor λ, the optimal scalar quantizer is the one with step size q^* that minimizes the Lagrangian cost $J(q, S^*(q))$, i.e. lives at absolute slope λ on the composite distortion-rate curve. That is, from (10.3), we have:

$$q^* = arg \min_{q \in Q} \{J(q, S^*(q))\} = arg \min_{q \in Q} \{D(q, S^*(q)) + \lambda R(q, S^*(q))\}.$$

While faster ways of reducing the search time for the optimal q exist, in this work, we exhaustively search for all choices in a finite admissible list. Finally, the optimal slope λ^* is found using the convex search bisection algorithm as described in [14]. By the convexity of the pruned-tree rate-distortion function [14], starting from two extreme points in the rate-distortion curve, the bisection algorithm successively shrinks the interval in which the optimal operating point lies until it converges. The convexity of the pruned-tree rate-distortion function guarantees the convergence of the optimal space-frequency quantizer.

3.4 Complexity of the SFQ algorithm

The complexity of the SFQ algorithm lies mainly in the iterative zerotree pruning stage of the encoder, which can be substantially reduced with fast heuristics based on models rather than actual R-D data which is expensive to compute. A good complexity measure is the running time on a specific machine. On a Sun SPARC 5, it takes about 20 seconds to run the SFQ algorithm for a fixed q and λ pair. On the same machine, Said and Pearlman's coder [SP96] takes about 4 seconds to run on encoding. Although our SFQ encoder is slower than that of Said and Pearlman's, the decoder is much faster because there are only two quantization modes used in the SFQ algorithm with the classification being sent as side-information. Our coder is suitable for applications such as image libraries, CD-ROM's and centrally stored databases where asymmetric coding complexity is preferred.

4 Coding Results

Experiments are performed on standard 512×512 grayscale Lena and Barbara images to test the proposed SFQ algorithm at several bitrates. Although our analysis of scalar quantizer performance assumed the use of orthogonal wavelet filters, simulations showed that little is lost in practice from using "nearly" orthogonal wavelet filters that have been reported in the literature to produce better perceptual results. We use the 7-9 biorthogonal set of linear phase filters of [6] in all our experiments. We use a 4-scale wavelet decomposition with the coarsest lowpass band having dimension 32×32. This lowest band is coded separately from the remaining bands, and the tree node symbols are also treated separately. For decodability, bands are scanned from coarse to fine scale, so that no child node is output before its parent node. The scalar quantization step size q takes values from the set $\{q\text{: } q=7.5+0.1*k,\ k=1,2,...,245\}$. An adaptive arithmetic coding [17, 18] is used to entropy code the quantized wavelet coefficients. All reported bitrates, which include both the data rate and the map rate, are calculated from the "real" coded bitstreams. About 10% of the total bitrate is spent in coding the zerotree map. The original Lena image is shown in Fig. 3 (a), while the decoded Lena images at the bitrate of 0.25 b/p is shown in Fig. 3 (b), respectively. The corresponding PSNR's, defined as $10log_{10}[255^2/mse]$, at different bitrates for all three images are tabulated in Table 1.1.

	Rate (b/p)	0.25	0.50	1.00
Lena	PSNR (dB)	34.33	37.36	40.52
Barbara	PSNR (dB)	28.29	32.15	37.03

TABLE 10.1. Coding results of the SFQ algorithm at various bitrates for the standard 512 × 512 Lena and Barbara images.

To illustrate the zerotree pruning operation in our proposed SFQ algorithm, we show in Fig. 2 the space-frequency quantized 4-level wavelet decomposition of the Lena image (white regions represent pruned nodes). The target bitrate is 1 b/p, and the scalar quantization step size for all highpass coefficients is 7.8.

FIGURE 2. The space-frequency quantized 4-level wavelet decomposition of the Lena image (white regions represent pruned nodes). The target bitrate is 1 b/p, and the scalar quantization step size for all highpass coefficients is 7.8.

Finally, we compare the performance of our SFQ algorithm in Fig. 4 with some of the high performance image compression algorithms by Shapiro [2], Said and Pearlman [SP96] and Joshi, Crump and Fisher [24]. Our simulations show that the SFQ algorithm is competitive with the best coders in the literature. For example, our SFQ-based coder outperforms 0.2 dB and 0.7 dB better in PSNR over the coder in [SP96] for Lena and Barbara, respectively.

5 Extension of the SFQ Algorithm from Wavelet to Wavelet Packets

We now cast our SFQ scheme in an adaptive transform coding framework by extending the SFQ paradigm from wavelet transform to the signal-dependent wavelet packet transform. In motivating our work, we observe that, although the wavelet representation offers an efficient space-frequency characterization for a broad class of natural images, it remains a "fixed" decomposition that fails to fully exploit

(a) (b)

FIGURE 3. Original and decoded 512 × 512 Lena images. (a) Original image. (b) Bitrate=0.25 b/p, PSNR=34.33 dB.

space-frequency distribution attributes that are specific for each individual image. Wavelet packets [12], or arbitrary subband decomposition trees, represent an elegant generalization of the wavelet decomposition. They can be thought of as adaptive wavelet transforms, capable of providing arbitrary frequency resolution. The wavelet packet framework, combined with their efficient implementation architecture using filter bank tree-structures, has opened a new avenue of wavelet packet image coding.

The problem of wavelet packet image coding consists of considering all possible

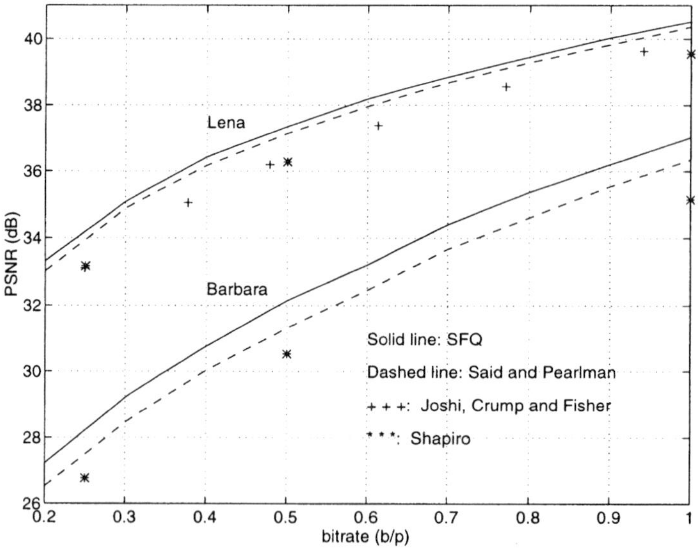

FIGURE 4. Performance Comparison of SFQ and other high performance coders.

wavelet packet bases in the library, and choosing the one that gives the best coding performance. A fast single tree algorithm with an operational rate-distortion measure was proposed in [15] to search for the best wavelet packet transform. Substantial coding improvements were reported of the resulting wavelet packet coder over wavelet coders for different classes of images, whose space-frequency characteristics were not well captured by the wavelet decomposition. However, the simple scalar quantization strategy used in [15] does not exploit spatial relationships inherent in the wavelet packet representation. Furthermore, the results in [15] were based on first order entropy. In this work, we introduce a new wavelet packet SFQ coder that undertakes the *joint* optimization of the wavelet packet transform and the powerful SFQ scheme. Experiments verify the high performances of our new coder.

Our current work represents generalizations of previous frameworks. It solves the joint transform and quantizer design problem by means of the dynamic programming based fast single tree algorithm, thus significantly lowering the computational complexity over that of an exhaustive search approach. To further reduce the complexity of the joint search in the wavelet packet SFQ coder, we also propose a much faster, but suboptimal heuristic in a practical coder with competitive coding results. The practical coder still possesses the capability of changing the wavelet packet transform to suit the input image and the space-frequency quantizer to suit the transform for maximum coding gain.

The extension of the SFQ algorithm to spatially-varying wavelet packets was done in [23]. More generalized joint time-frequency segmentation algorithms using wavelet packets appeared in [24, 25] with improved performance in speech and image coding.

6 Wavelet packets

One practical way of constructing wavelet bases is to iterate a 2-channel perfect reconstruction filter bank over the lowpass branch. This results in a dyadic division of the frequency axis that works well for most signals we encounter in practice. However, for signals with strong stationary highpass components, it is often argued that wavelet basis won't perform well, and improvements can be generally found if we search over all possible binary trees for a particular filter set, instead of using the fixed wavelet tree. Expansions produced by these arbitrary subband trees are called wavelet packet expansions, and mathematical groundwork of wavelet packets was laid by Coifman *et al* in [12]. Wavelet packet transforms are obviously more general than the wavelet transform, they also offer a rich library of bases from which the best one can be chosen for a fixed signal with a fixed cost function. The advantage of the wavelet packet framework is thus its universality in adapting the transform to a signal without training or assuming any statistical property of the signal.

The extra adaptivity of the wavelet packet framework is obtained at a price of added computation in searching for the best wavelet packet basis, so an efficient fast search algorithm is the key in applications involving wavelet packet. The problem of searching for the best basis from the wavelet packet library was examined in [12, 15], and a fast single tree algorithm was described in [15] using a rate-distortion optimization framework for image compression.

The basic idea of the single tree algorithm is to prune a full subband tree (grown to certain depth) in a recursive greedy fashion, starting from the bottom of the tree using a composite Lagrangian cost function $D + \lambda R$. The distortion D is the mean-squared quantization error, and rate R is either estimated from the first order entropy of the quantization indices or obtained from the output rate of a real coder. The Lagrange multiplier λ is a balancing factor between R and D, and the best λ can be searched by an iterative bisectional algorithm [14] to meet a given target bitrate.

7 Wavelet packet SFQ

We now extends the concept of SFQ to arbitrary wavelet packet expansions. Central to the idea of wavelet packet SFQ is the generalization of the spatial coefficient tree structure from wavelet to wavelet packet transform coefficients.

The spatial coefficient tree structure defined for the wavelet case was designed to capture the spatial relationships of coefficients in different frequency bands. Parent-child relationships reflect common spatial support of coefficients in adjacent frequency bands. For the generalized wavelet packet setting, we can still define a spatial coefficient tree as the set of coefficients in different bands that correspond to a common spatial region of the image. Fig. 5 gives examples of spatial coefficient trees corresponding to several different wavelet packet transforms. The parent-child relationship for the wavelet case can be imported to the wavelet packet case as well. We designate nodes in a lower band of a coefficient tree as parents of those in the next higher band. As the frequency bands (or wavelet packet tree) can now be arbitrary, the parent-child relationship could involve multiple or single children of each parent node and vice versa.

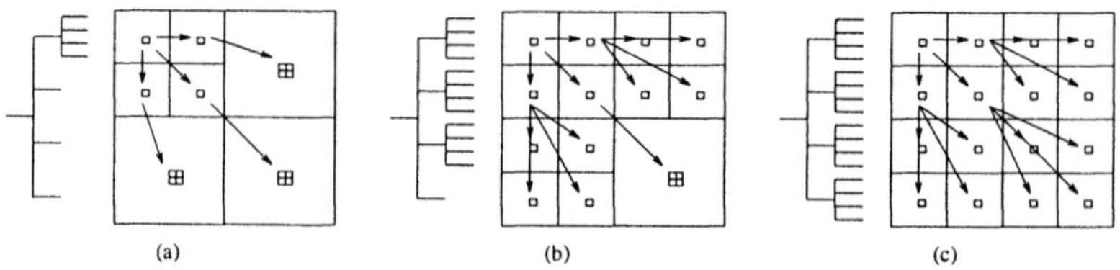

FIGURE 5. Examples of spatial coefficient trees corresponding to different wavelet packet transforms. (a) Wavelet transform. (b) An arbitrary wavelet packet transform. (c) Full subband transform. Arrows identify the parent-child dependencies.

We now propose a wavelet packet SFQ coder by allowing joint transform and quantizer design. We use the single tree algorithm to search for the best wavelet packet transform, and the SFQ scheme to search for the best quantization for each transform. Since the single tree algorithm and the SFQ scheme both rely on the rate-distortion optimization framework, the combination of these two is a natural choice for joint transform and quantizer design, where the Lagrangian multiplier (or rate-distortion equalizer) λ plays the role of connecting the transform and quantizer together.

In wavelet packet SFQ, we start by growing the full subband tree. We then invoke the extended definition of spatial trees for wavelet packet, as illustrated in Fig. 5, and use the SFQ algorithm to find an optimally pruned spatial tree representation for a given rate-distortion tradeoff factor λ and scalar quantization stepsize q. This is exactly like the wavelet-based SFQ algorithm, except run on a wavelet packet transform[3]. The minimum SFQ cost associated with the full wavelet packet tree is then compared with those associated with its pruned versions, where the leaf nodes are merged into their respective parent nodes. Again we keep the smaller cost as the winner and prune the loser with higher cost. This single tree wavelet packet pruning criterion is then used recursively at each node of the full wavelet packet tree. At the end of this single tree pruning process, when the root of the full wavelet packet tree is reached, the optimal wavelet packet basis from the entire family and its best SFQ are therefore found for fixed values of λ and q. As in SFQ, the best scalar quantization stepsize q is searched over a set of admissible choices to minimize the Lagrangian cost, and the optimal λ found bisectionally to meet the desired bitrate budget. The block diagram of the wavelet packet SFQ coder is shown in Fig. 6.

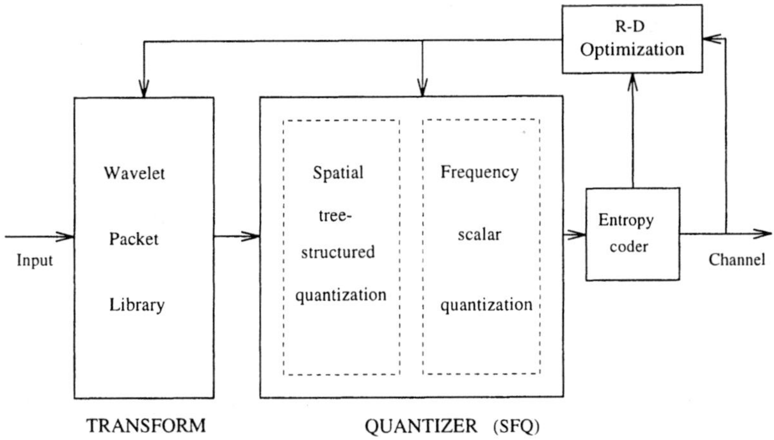

FIGURE 6. Block diagram of the wavelet packet SFQ coder.

8 Wavelet packet SFQ coder design

Although the extension from wavelet-based SFQ to wavelet packet SFQ is conceptually simple, in the practical coder design, computational complexity must be taken into account. It is certainly impractical to design a SFQ coder for each basis in the wavelet packet library in order to choose the best one. In this section, we first describe a locally optimal wavelet packet SFQ coder. It involves the joint application of the single tree algorithm and SFQ. The complexity of this approach grows exponentially with respect to the wavelet packet tree depth. We then propose a practical coder in which the designs of transform and quantizer are decoupled,

[3]A practically equivalent implementation is to rearrange the wavelet packet transform coefficients in a wavelet-transform-like fashion, and apply SFQ as designed for wavelet transforms.

and the single tree and the SFQ algorithms are applied sequentially, thus achieving computational savings.

8.1 *Optimal design: Joint application of the single tree algorithm and SFQ*

In the optimal design of the wavelet packet SFQ coder, we first grow a full subband tree, and then start pruning the full tree using the single tree algorithm, with the cost at each node generated by calling the SFQ algorithm. In other words, at each stage of the tree pruning, we first call modified versions of the SFQ algorithm (based on the wavelet packet tree in the current stage) to generate the minimum costs for the current tree and its pruned version with four leaf nodes merged into one parent node. We then make a decision on whether to keep the current tree or its pruned version based on which one gives smaller cost. This single tree pruning decision is made recursively at each node of the wavelet packet tree, starting from the leaf nodes of the full tree. Due to the greedy nature of the single tree pruning criterion, we are guaranteed at each stage of the pruning to have the minimum cost so far and its associated best wavelet packet tree. When the tree pruning reaches the root node of the full wavelet packet tree, we have exhausted the search, and found the optimal pruned wavelet packet tree to use for the input image, together with its best SFQ.

8.2 *Fast heuristic: Sequential applications of the single tree algorithm and SFQ*

The optimal wavelet packet SFQ design is computationally costly because the SFQ algorithm has to be run at each stage of the single tree pruning. To further reduce the computational complexity of the optimal wavelet packet SFQ design, we now propose a practical coder that is orders of magnitude faster than the optimal one. The basic idea is to decouple the joint transform and quantizer design, and to optimize the transform and quantization operators sequentially instead.

In the practical coder, for fixed values of λ and q, we first choose a near-optimal wavelet packet basis using the single tree algorithm and the Lagrangian cost function $D + \lambda R$. This is done by using MSE of scalar quantization as distortion D and first order entropy of the scalar quantized wavelet packet coefficients as rate R. Note that we only apply scalar quantization to the transform coefficients. No zerotree quantization is applied at this stage. We then apply the SFQ scheme on the decomposition given by the above near-optimal wavelet packet basis. We use the same values of λ and q in both the single tree algorithm and the SFQ algorithm, which guarantee the same quality factor (λ) is used in both the transform and the quantizer designs. The best scalar quantization stepsize q is searched within the admissible set Q and the best λ is found using the bisection algorithm.

Essentially, the suboptimal heuristic can be considered as the applications of the single tree algorithm in [15] and a modified SFQ algorithm in tandem. The overall complexity of the practical wavelet packet coder using the suboptimal heuristic is thus the sum of the complexities of the two algorithms.

9 Experimental Results

We test our wavelet packet SFQ algorithm on the Lena and Barbara images using the 7/9 biorthogonal filters of [6]. Adaptive arithmetic coding [18] is used to code the quantized wavelet packet coefficients. All reported bitrates, which include 85 bits that are needed to inform the decoder of the best wavelet packet tree (of maximum depth-4), are calculated from the "real" coded bitstreams.

9.1 Results from the joint wavelet packet transform and SFQ design

We first compare the performance differences of wavelet-based SFQ and wavelet packet SFQ for the 512 × 512 Barbara image. Numerical PSNR values of wavelet-based SFQ and wavelet packet SFQ at several bitrates are tabulated in Table 1.2, together with those obtained from the EZW coder of [2]. We see that wavelet packet SFQ performs much better than wavelet-based SFQ for Barbara. For example, at 0.5 b/p, the wavelet packet SFQ achieves a PSNR that is one dB higher than the wavelet-based SFQ.

	Rate (b/p)	0.25	0.50	1.00
EZW	PSNR (dB)	26.77	30.53	35.14
WT+SFQ	PSNR (dB)	28.29	32.15	37.03
WP&SFQ	PSNR (dB)	29.37	33.13	37.70

TABLE 10.2. Optimal wavelet packet SFQ vs. wavelet-based SFQ for the Barbara image (scale 4) in bitrate and PSNR (Results from Shapiro's wavelet-based coder are also included for comparisons).

9.2 Results from the sequential wavelet packet transform and SFQ design

We also compare the performance of the sequential wavelet packet SFQ design with the optimal wavelet packet SFQ design for the Barbara image. Numerical PSNR values obtained from the sequential wavelet packet SFQ design are tabulated in Table 1.3.

	Rate (b/p)	0.25	0.50	1.00
Lena	PSNR (dB)	34.35	37.40	40.55
Barbara	PSNR (dB)	29.25	33.12	37.69

TABLE 10.3. Coding results of the practical wavelet packet SFQ coder at various bitrates for the standard 512 × 512 Lena and Barbara images.

The original Barbara image together with the wavelet packet decomposed and decoded Barbara images at 0.5 b/p are shown in Fig. 4. By comparing the coding results in Table 1.2 and 1.3, we see that the performance difference between the joint optimal wavelet packet SFQ design and the suboptimal heuristic is at most 0.12 dB at comparable bitrates. The computational savings of the suboptimal heuristic, however, is quite impressive. In our experiments, we only need to run the SFQ algorithm once in the practical coder, but 85 times in the optimal coder! Based on the near-optimal performance of the practical coder, we henceforth use the practical coder exclusively in our experiments for other images. Results of the wavelet packet SFQ coder for the 512 × 512 Lena are also tabulated in Table 1.3.

To demonstrate the versatility of our wavelet packet SFQ coder, we benchmark against a 768 × 768 (eight bits resolution) fingerprint image using the FBI Wavelet Scalar Quantization (WSQ) standard [26]. Brislawn *et al* reported a PSNR of 36.05 dB when coding the fingerprint image using the WSQ standard at .5882 b/p (43366 bytes) [26]. Using the wavelet packet SFQ algorithm, we get a PSNR of 37.30 dB at the same rate. The original fingerprint together with the wavelet packet decomposed and decoded fingerprint images at 0.5882 b/p are shown in Fig. 8.

10 Discussion and Conclusions

The SFQ algorithm developed in this paper tests the hypothesis that high performance coding depends on exploiting both frequency and spatial compaction of energy in a space-frequency transform. The two simple quantization modes used in the SFQ algorithm put a limit its overall performance. This can be improved by introducing sophisticated schemes such as trellis-coded quantization and subband classification [20, 21] to exploit "packing gain" in the scalar quantizer, a type of gain quite separate and above anything considered in this paper. Zerotree quantization can be viewed as providing a mechanism for spatial classification of wavelet coefficients into two classes: "zerotree" pruned coefficients and non-pruned coefficients. The tree-structure constraint of the SFQ classification permits us to efficiently implement RD-optimization, but produces suboptimal classification for coding purposes. I.e. if our classification procedure searched over a richer collection of possible sets of coefficients (e.g. including some non-tree-structured sets), algorithm complexity would be increased, but improved results could be realized. In fact, performance improvements in other zerotree algorithms have been realized by expanding the classification to consider sets of coefficients other than purely tree-structured sets [SP96].

While the SFQ algorithm focuses on the quantization aspect of wavelet coefficients, the WP-based SFQ algorithm exploits the "transform gain", which was not considered in any of the above-mentioned coding schemes. Although the improvement of using the best wavelet packet transform over the fixed dyadic wavelet transform is image dependent, WP-based SFQ provides a framework for joint transform and quantizer design.

It is interesting to compare the SFQ algorithm with the rate-distortion optimized version of JPEG proposed in [22]. These two algorithms are built around very similar rate-distortion optimization frameworks, with the algorithms of [22] using block DCT's instead of the wavelet transform, and using runlengths of zeros instead of zerotrees. The R-D optimization provides a large gain over standard JPEG (0.7 dB at 1 b/p for Lena), but the final PSNR results (e. g. 39.6 dB at 1 b/p for Lena) remain about 0.9 dB below SFQ for the Lena image at a bitrate of 1 b/p. We interpret these results as reflecting the fact that the block DCT is not as effective as the wavelet transform at compacting high frequency energy around edges. (I.e. blocks containing edges tend to spread high frequency energy among many coefficients.)

APPENDIX

Proof of Proposition 1: Algorithm I converges to a local optimum

We will show that $J(S^{(k+1)}) \leq J(S^{(k)})$ thereby establishing the proposition, since the number of iterations is guaranteed to be finite, given that the tree is always being pruned.

Since $S^{(k+1)}$ is a pruned version of $S^{(k)}$, let us refer to the set of tree coefficients pruned out in the $(k+1)$ th iteration as $\Delta S^{(k+1)}$, i.e.

$$\Delta S^{(k+1)} = S^{(k)} \setminus S^{(k+1)} \textit{ or } S^{(k)} = S^{(k+1)} \cup \Delta S^{(k+1)}. \qquad (10.11)$$

Let us now evaluate the costs $J(S^{(k+1)})$ and $J(S^{(k)})$:

$$\begin{aligned} J(S^{(k+1)}) &\overset{(a)}{=} \sum_{i \in S^{(k+1)}} J_i^{(k+1)} + \sum_{j \in T \setminus S^{(k+1)}} w_j^2 \\ &\overset{(b)}{=} \sum_{i \in S^{(k+1)}} J_i^{(k+1)} + \sum_{j \in \Delta S^{(k+1)}} w_j^2 + \sum_{j \in T \setminus S^{(k)}} w_j^2, \end{aligned} \qquad (10.12)$$

where (a) above follows from the cost of $S^{(k+1)}$ being the sum of the Lagrangian cost $J_i^{(k+1)}$ of all nodes in $S^{(k+1)}$ plus the cost of zerotree quantizing the set $T \setminus S^{(k+1)}$, and (b) follows from (10.11).

Similarly, we have:

$$\begin{aligned} J(S^{(k)}) &\overset{(a)}{=} \sum_{i \in S^{(k)}} J_i^{(k)} + \sum_{j \in T \setminus S^{(k)}} w_j^2 \\ &\overset{(b)}{=} \sum_{i \in S^{(k+1)}} J_i^{(k)} + \sum_{j \in \Delta S^{(k+1)}} J_j^{(k)} + \sum_{j \in T \setminus S^{(k)}} w_j^2, \end{aligned} \qquad (10.13)$$

where (a) above follows from the definition of $J(S^{(k)})$ and (b) above follows from (10.11).

We therefore have:

$$\begin{aligned} & J(S^{(k)}) - J(S^{(k+1)}) \\ = & \sum_{i \in S^{(k+1)}} (J_i^{(k)} - J_i^{(k+1)}) + \sum_{j \in \Delta S^{(k+1)}} (J_j^{(k)} - w_j^2) && (10.14) \\ \geq & \sum_{i \in S^{(k+1)}} (J_i^{(k)} - J_i^{(k+1)}) && (10.15) \\ = & \lambda \sum_{i \in S^{(k+1)}} (R_i^{(k)} - R_i^{(k+1)}) && (10.16) \\ = & \lambda |S^{(k+1)}| \sum_{bin=-\infty}^{+\infty} -p^{(k+1)}(bin) \log_2 [\frac{p^{(k)}(bin)}{p^{(k+1)}(bin)}] && (10.17) \\ \geq & \ 0, && (10.18) \end{aligned}$$

where (10.14) follows from (10.12) and (10.13); (10.15) follows because the second summation of (10.14) is greater than or equal to zero (else $\Delta S^{(k+1)}$ would not have been pruned out in the $k+1$ th iteration, by hypothesis); (10.16) follows since $J_i = D_i + \lambda R_i$, with $D_i^{(k)} = D_i^{(k+1)} = (w_i - \hat{w}_i)^2$; (10.17) expresses the rates for the k th and $(k+1)$ th iterations in terms of their first-order entropies induced by the distributions $\{p^{(k)}(bin)\}$ and $\{p^{(k+1)}(bin)\}$ respectively (note that bin is the histogram bin number and $|S^{(k+1)}|$ refers to the cardinality or number of elements in $S^{(k+1)}$); and (10.18) follows from the fact that the summation in (10.17) is the Kullback-Leibler distance or relative entropy between the distributions $\{p^{(k)}(bin)\}$ and $\{p^{(k+1)}(bin)\}$ which is nonnegative.

11 References

[1] A. Lewis and G. Knowles, "Image compression using the 2-D wavelet transform," *IEEE Trans. Image Processing*, vol. 1, pp. 244-250, April 1992.

[2] J. M. Shapiro, "Embedded image coding using zerotrees of wavelet coefficients," *IEEE Trans. Signal Processing*, vol. 41, pp. 3445-3463, December 1993.

[3] J. W. Woods and S. O'Neil, "Subband coding of images," *IEEE Trans. Acoustics, Speech, and Signal Processing*, vol. 34, pp. 1278-1288, October 1986.

[4] P. Westerink, "Subband coding of images," Ph.D. dissertation, The Delft University of Technology, October 1989.

[5] Y. H. Kim and J. W. Modestino, "Adaptive entropy coded subband coding of images," *IEEE Trans. Image Processing*, vol. 1, pp. 31-48, January 1992.

[6] M. Antonini, M. Barlaud, P. Mathieu, and I. Daubechies, "Image coding using wavelet transform," *IEEE Trans. Image Processing*, vol. 1, pp. 205-221, April 1992.

[7] P. J. Burt and E. H. Adelson, "The Laplacian pyramid as a compact image code," *IEEE Trans. Communication*, vol. 31, pp. 532-540, 1983.

[8] R. J. Clarke, *Transform Coding of Images.* Orlando, FL: Academic Press, 1985.

[9] *Subband Image Coding*, J. W. Woods, Ed. Norwell, MA: Kluwer Academic, 1991.

[10] N. Farvardin and J. Modestino, "Optimum quantizer performance for a class of non-Gaussian memoryless sources," *IEEE Trans. Inform. Theory*, vol. 30, pp. 485-497, May 1984.

[11] Z. Xiong, K. Ramchandran, and M. T. Orchard, "Space-frequency quantization for wavelet image coding," *IEEE Trans. Image Processing*, 1997.

[12] R. Coifman and V. Wickerhauser, "Entropy-based algorithms for best basis selection," *IEEE Trans. Inform. Theory*, vol. 38, pp. 713-718, March 1992.

[13] Z. Xiong, K. Ramchandran, M. T. Orchard, and K. Asai, "Wavelet packets-based image coding using joint space-frequency quantization," Proc. ICIP'94, Austin, Texas, November, 1994.

[14] Y. Shoham and A. Gersho, "Efficient bit allocation for an arbitrary set of quantizers," *IEEE Trans. Acoustics, Speech, and Signal Processing*, vol. 36, pp. 1445-1453, September 1988.

[15] K. Ramchandran and M. Vetterli, "Best wavelet packet bases in a rate-distortion sense," *IEEE Trans. Image Processing*, vol. 2, pp. 160-176, April 1993.

[16] A. Said and W. A. Pearlman, "A new, fast, and efficient image codec based on set partitioning in hierarchical trees," *IEEE Trans. Circuits and Systems for Video Technology*, vol. 6, pp. 243-250, June 1996.

[17] G. Langdon, "An introduction to arithmetic coding," *IBM J. Res. Develop.*, 28, pp. 135-149, 1984.

[18] I. Witten, R. Neal, and J. Cleary, "Arithmetic coding for data compression," *Communications of the ACM*, 30, pp. 520-540, 1987.

[19] R. L. Joshi, V. J. Crump, and T. R. Fisher, "Image subband coding using arithmetic and trellis coded quantization," *IEEE Trans. Circuits and Systems for Video Technology*, vol. 5, pp. 515-523, December 1995.

[20] M. W. Marcellin and T. R. Fischer, "Trellis coded quantization of memoryless and Gaussian-Markov sources," *IEEE Trans. Communications*, vol. 38, no. 1, pp. 82-93, January 1990.

[21] R. L. Joshi, H. Jafarkhani, J. H. Kasner, T. R. Fisher, N. Farvardin, M. W. Marcellin, and R. H. Bamberger, "Comparison of different methods of classification in subband coding of images," submitted to *IEEE Trans. Image Processing*, 1995.

[22] M. Crouse and K. Ramchandran, "Joint thresholding and quantizer selection for decoder-compatible baseline JPEG," *Proc. of ICASSP'95*, Detroit, MI, May 1995.

[23] K. Ramchandran, Z. Xiong, K. Asai, and M. Vetterli, "Adaptive transforms for image coding using spatially-varying wavelet packets," *IEEE Trans. Image Processing*, vol. 5, pp. 1197-1204, July 1996.

[24] D. P. Bertsekas, *Dynamic Programming: Deterministic and Stochastic Models.* Englewood Cliffs, NJ: Prentice Hall, 1987.

[25] C. Herley, Z. Xiong, K. Ramchandran and M. T. Orchard, "Joint Space-frequency Segmentation Using Balanced Wavelet Packets Trees for Least-cost Image Representation", *IEEE Trans. on Image Processing*, May 1997.

[26] J. Bradley, C. Brislawn, and T. Hopper, "The FBI wavelet/scalar quantization standard for gray-scale fingerprint image compression," *Proc. VCIP'93*, Orlando, FL, April 1993.

(a)

(b)

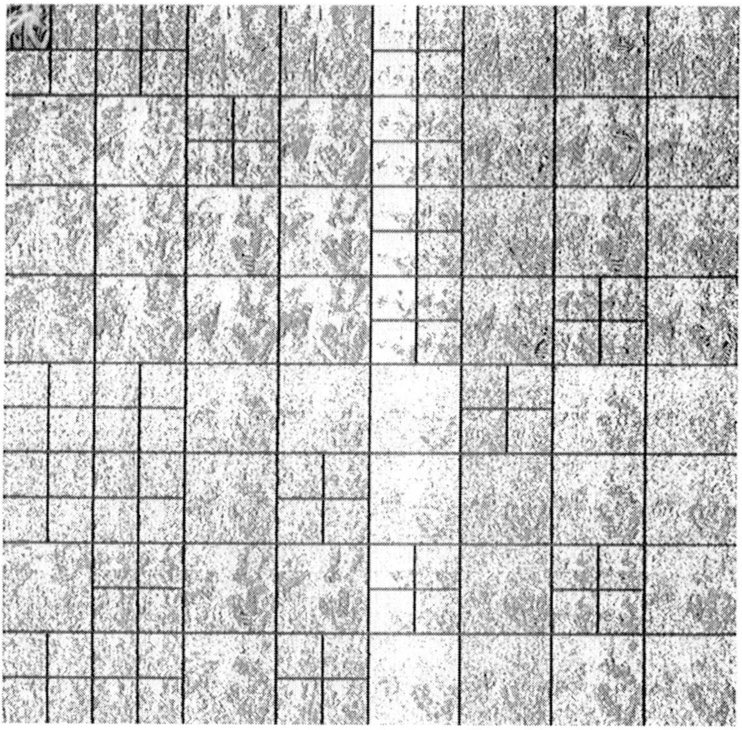

(c)

FIGURE 7. Original, wavelet packet decomposed and decoded 512×512 Barbara image. (a) Original Barbara. (b) Decoded Barbara from (c) using SFQ. Rate=0.5 b/p, PSNR=33.12 dB. (c) Wavelet packet decomposition of Barbara at 0.5 b/p. Black lines represent frequency boundaries. Highpass bands are processed for display.

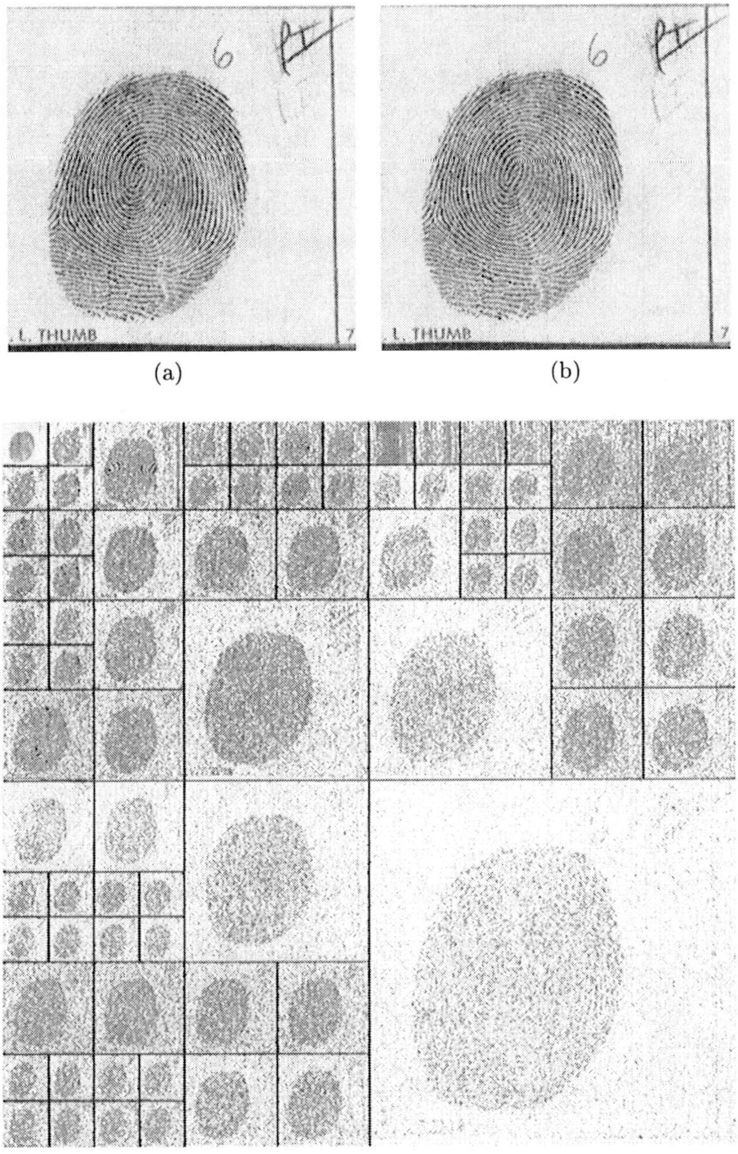

FIGURE 8. Original, wavelet packet decomposed and decoded 768 × 768 FBI fingerprint images. (a) Original image. (b) Decoded fingerprint from (c) using SFQ. Rate=0.5882 b/p, PSNR=37.30 dB. The decoded image using the FBI WSQ standard has a PSNR of 36.05 dB at the same bitrate. (c) Wavelet packet decomposition of the fingerprint image at .5882 b/p (43366 bytes). Black lines represent frequency boundaries. Highpass bands are processed for display.

11

Subband Coding of Images Using Classification and Trellis Coded Quantization

Rajan L. Joshi and Thomas R. Fischer

1 Introduction

Effective quantizers and source codes are based on effective source models. In statistically-based image coding, classification is a method for identifying portions of the image data that have similar statistical properties so that the data can be encoded using quantizers and codes matched to the data statistics. Classification in this context is, thus, a method for identifying a set of source models appropriate for an image, and then using the respective model parameters to define the quantizers and codes used to encode the image.

This chapter describes an image coding algorithm based on block classification and trellis coded quantization. The basic principles of block classification for coding are developed and several classification methods are outlined. Trellis coded quantization is described, and a formulation presented that is appropriate for arithmetic coding of the TCQ codeword indices. Finally, the performance of the classification-based subband image coding algorithm is presented.

2 Classification of blocks of an image subband

Figure 1 shows the frequency partition induced by a 22-band decomposition. This decomposition can be thought of as a 2-level uniform split (16-band uniform) followed by a 2-level octave split of the lowest frequency subband. The 2-level octave split exploits the correlation present in the lowest frequency subband [36]. Typically, the histogram of a subband is peaked around zero. One possible approach to model the subbands is to treat them as memoryless generalized Gaussian (GG) sources [24]. Several researchers [15], [35], [36] have demonstrated that using quantizers and codes based on such models can yield good subband coding performance.

It is well-understood, however, that image subbands are often not memoryless. Typically, the bulk of the energy in high frequency subbands (HFS's) is concentrated around edge or texture features in the image. Possible approaches to deal with this regional energy vatiation in image subbands include spatially adaptive quantization, spatially-varying filter banks [2], [7] and wavelet packets [11], [30].

Spatially adaptive quantization methods can be further classified as forward-adaptive or backward-adaptive. Recent work by LoPresto et al. [23], and Chrysafis and Ortega [6] are examples of backward-adaptive methods. In this chapter, we present analysis of forward-adaptive methods and also discuss the trade-off between coding gain and side information penalty. The reader is referred to [18], [19] for further details. Other state-of-the-art subband image coders such as the SPIHT coder [32] and the SFQ coder [43] exploit the inter-band energy dependence in the form of zero-trees [33].

Chen and Smith [4] pioneered the use of classification in discrete cosine transform (DCT) coding of images. Their approach is to classify the DCT blocks according to block energy and adapt the quantizer to the class being encoded. Woods and O'Neil [41] propose use of a similar classification method in subband coding. Subsequently, a number of researchers [19], [22], [27], [37] have investigated the use of classification in subband coding. There is a common framework behind these different approaches to classifying image subbands. A subband is divided into a number of blocks of fixed size. Then the blocks are assigned to different classes according to their energy and a different quantizer is used to encode each class. For each block, the class index has to be sent to the decoder as side information.

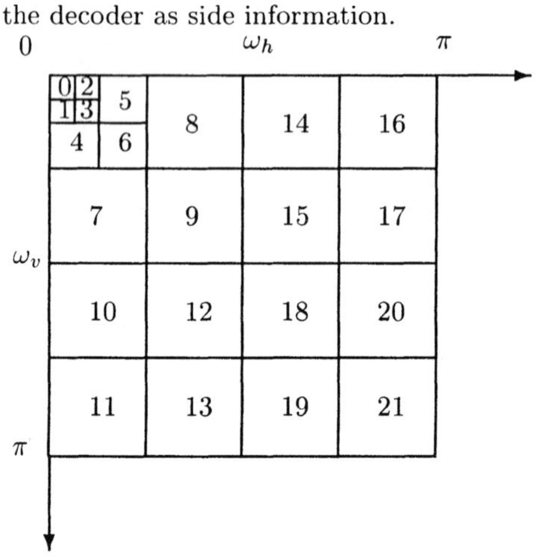

FIGURE 1. Frequency partition for a 22-band decomposition. Reprinted with permission from "Comparison of different methods of classification in subband coding of images," R. L. Joshi, H. Jafarkhani, J. H. Kasner, T. R. Fischer, N. Farvardin, M. W. Marcellin and R. H. Bamberger, © IEEE.

2.1 *Classification gain for a single subband*

Consider a subband X of size $r \times c$. Divide the subband into rectangular blocks of size $m \times n$. Assume that the subband size is an integral multiple of the block size, in both row and column directions. Let $N = (rc)/(mn)$ be the total number of blocks in subband X. Assume that the subband has zero mean and let the variance of the subband be σ_x^2.

Let each block be assigned to one of J classes. Group samples from all the blocks assigned to class i, into source X_i (for $1 \leq i \leq J$). Let the total number of blocks assigned to source X_i be N_i. Let σ_i^2 be the sample variance of source X_i and p_i be the probability that a sample from the subband belongs to source X_i. Then $p_i = N_i/N$, $1 \leq i \leq J$.

The mean-squared error distortion for encoding source X_i at a rate R_i is of the form [13]

$$D_i(R_i) = \epsilon_i^2(R_i)\,\sigma_i^2 2^{-2R_i}. \tag{11.1}$$

The distortion-rate performance for the encoding of the subband samples, X, is assumed to have a similar general form. The factor ϵ_i depends on the rate, the density of the source and the type of encoding used (for example, whether or not entropy coding is being used with the quantization). Under the high-rate assumption, ϵ_i can be assumed to be independent of the encoding rate. Then

$$D_i(R_i) = \epsilon_i^2 \sigma_i^2 2^{-2R_i}. \tag{11.2}$$

Let the side rate for encoding the classification information be R_s. If R_T bits per sample are available to encode source X, then only R bits per sample can be used for encoding sources X_i, $1 \leq i \leq J$, where $R = R_T - R_s$.

The rate allocation problem for classification can be defined as

$$\min \sum_{i=1}^{J} p_i D_i(R_i), \tag{11.3}$$

subject to the constraint

$$\sum_{i=1}^{J} p_i R_i = R. \tag{11.4}$$

Using Lagrangian multiplier techniques and assuming $R_i > 0$ for $1 \leq i \leq J$, the solution of the above problem is

$$R_i = R + \frac{1}{2}\log_2 \frac{\epsilon_i^2 \sigma_i^2}{\prod_{j=1}^{J} (\epsilon_j^2 \sigma_j^2)^{p_j}}, \tag{11.5}$$

and

$$D_{opt}(R) = \prod_{j=1}^{J} (\epsilon_j^2 \sigma_j^2)^{p_j} 2^{-2R}. \tag{11.6}$$

Thus the classification gain is

$$G_c = \frac{\epsilon_x^2 \sigma_x^2}{\prod_{j=1}^{J} (\epsilon_j^2 \sigma_j^2)^{p_j}} 2^{-2R_s}. \tag{11.7}$$

If the subband samples X and the classified sources X_i, $1 \leq i \leq J$, have the same distribution ($\epsilon_x^2 = \epsilon_1^2 = \epsilon_2^2 = \ldots = \epsilon_J^2$) and ignoring the side information ($R_T = R$), the expression for classification gain simplifies to

$$G_c = \frac{\sigma_x^2}{\prod_{j=1}^{J} (\sigma_j^2)^{p_j}}. \tag{11.8}$$

2.2 Subband classification gain

Consider an exact reconstruction filter bank which splits an image into M subbands. Suppose that each subband is classified into J classes as outlined in the previous section. In this subsection, we examine the overall coding gain due to subband decomposition and classification. Let f_i be the decimation factor for subband i. The notation is as before, with the addition that a subscript (ij) refers to the jth class from the ith subband. Assuming that the analysis and synthesis filters form an orthogonal filter bank, the overall squared distortion satisfies

$$D_x = \sum_{i=1}^{M} f_i \left(\sum_{j=1}^{J} p_{ij} D_{ij} \right). \tag{11.9}$$

We can solve the problem of allocating rate R to MJ sources (J classes each in M subbands) using the Lagrangian multiplier technique used in the previous subsection. Assuming that all rates are positive, the solution is given by

$$R_{ij} = R + \frac{1}{2} \log_2 \frac{\epsilon_{ij}^2 \sigma_{ij}^2}{\prod_{m=1}^{M} \left(\prod_{n=1}^{J} (\epsilon_{mn}^2 \sigma_{mn}^2)^{p_{mn}} \right)^{f_m}}, \tag{11.10}$$

Thus, the overall subband classification gain is

$$G_{sc} = \frac{\epsilon_x^2 \sigma_x^2}{\prod_{m=1}^{M} \left(\prod_{n=1}^{J} (\epsilon_{mn}^2 \sigma_{mn}^2)^{p_{mn}} \right)^{f_m}} 2^{-2R_s}. \tag{11.11}$$

For a fixed amount of side information, a lower geometric mean results in a higher subband classification gain. In other words, a greater disparity in class variances corresponds to a higher subband classification gain. Also, it should be noted that there is an inverse relationship between subband classification gain and side information.

If biorthogonal filters are being used, the distortion from each subband has to be weighted differently [42]. It should be noted that the classification gain in Equation (11.11) is implicitly based on the assumption of large encoding rate. If the encoding rate is small, then the term ϵ_i^2 in Equation (11.2) should be modified to the form $\epsilon_i^2(R_i)$ and the optimum rate allocation should be re-derived as in [3]. However, in our simulations, we found that regardless of the rate, there is a monotonic relationship between G_c and actual classification gain; that is, a higher value of G_c corresponds to a higher classification gain.

2.3 Non-uniform classification

Dividing the blocks (DCT or subband) into equally populated classes is, generally, not optimal. For example, consider a subband where 80% of the blocks have low energy and 20% of the blocks have high energy. Suppose it is estimated that the subband is a mixture of a low variance source and a high variance source. If the blocks are divided into two equally populated classes, then one of the classes will again contain a mixture of blocks from high and low variance sources. This is clearly not desirable. By classifying the subband into 5 equally-populated classes, we can ensure that each class contains samples from only one source. But in that case, the low variance source is divided into 4 classes and the side information also increases by a factor of more than 2. Also, from a rate-distortion point of view, the choice of equally-populated classes does not necessarily maximize the classification gain.

The capability of allowing different number of blocks in each class can, potentially, improve classification. The problem is that of determining the number of blocks in each class. Two different approaches have been proposed in the literature to solve this problem.

1. Maximum gain classification: This approach was proposed in [14]. The main idea is to choose the number of blocks in each class so as to maximize the classification gain given by equation (11.8). This also maximizes the overall subband classification gain G_{sc} given by equation (11.11).

2. Equal mean-normalized standard deviation (EMNSD) classification: This approach was proposed in [12]. The main idea is to have blocks with similar statistical properties within each class. The 'coefficient of variation' (defined by standard deviation divided by mean) of block energies in a class is used as the measure of dispersion and the number of blocks in each class is determined so as to equalize the dispersion in each class.

2.4 The trade-off between the side rate and the classification gain

In a classification-based subband image coder, the classification map for each subband has to be sent as side information. Thus, if each subband is classified independently, the side rate can be a significant portion of the overall encoding rate at low bit rates. A possible approach to reducing the side information is to constrain the classification maps in some manner. Sending a single classification map for the subbands having the same frequency orientation [16], [20] is an example of this approach. This has the effect of reducing the side information and (possibly) the complexity of the classification process, albeit at the cost of some decrease in classification gain. As reported in [19], in general, these methods perform slightly worse compared to the case when each subband is classified independently, but they can provide flexibility in terms of computational complexity.

Another approach is to reduce the side information by exploiting the redundancies in the classification maps. Figure 2 plots the maximum classification gain maps of subbands for a 2 level (16-band uniform) split of the luminance component of the 512×512 'Lenna' image (using $J = 4$ classes per subband). It can be seen that there is a dependence between the classification maps of subbands having the same frequency orientation. In addition, there is also some dependence between the classification indices of spatially adjacent blocks from the same subband. Any one or both of these dependencies can be exploited to reduce the side information required for sending the classification maps. This is accomplished by using conditional entropy coding for encoding the classification maps. The class index of the previous block from the same subband, and the class index of the spatially corresponding block from the lower frequency subband having the same frequency orientation, are used as conditioning contexts. It was reported in [19] that side rate reduction of $15 - 20\%$ can be obtained using this method.

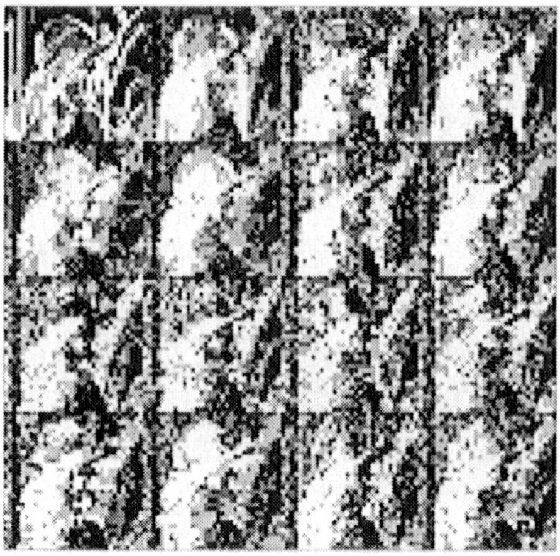

FIGURE 2. Classification map for 'Lenna' for a 16-band uniform decomposition (maximum classification gain method; $J = 4$). Reprinted with permission from "Comparison of different methods of classification in subband coding of images," R. L. Joshi, H. Jafarkhani, J. H. Kasner, T. R. Fischer, N. Farvardin, M. W. Marcellin and R. H. Bamberger, © IEEE.

3 Arithmetic coded trellis coded quantization

In subband image coding, generalized Gaussian densities (GGD's) have been a popular tool for modeling subbands [15], [36] or sources derived from subbands [14]. Thus improved quantization methods for encoding generalized Gaussian sources have the potential of improving subband image coding performance. In this section, we present the arithmetic coded trellis coded quantization (ACTCQ) system which has a better rate-distortion performance compared to optimum entropy-constrained scalar quantization (ECSQ). Description of the ACTCQ system is based on [15],[18].

Marcellin and Fischer [25] introduced the method of trellis coded quantization (TCQ). It is based on mapping by set-partitioning ideas introduced by Ungerboeck [38] for trellis coded modulation. The main idea behind trellis coded quantization is to partition a large codebook into a number of disjoint codebooks. The choice of codebook at any given time instant is constrained by a trellis. One advantage of trellis coded quantization is its ability to obtain granular gain at much lower complexity when compared to vector quantization. Entropy-constrained TCQ (ECTCQ) was introduced in [9] and [26]. The ECTCQ uses the Lagrangian multiplier approach, first formulated by Chou, Lookabaugh and Gray [5], for entropy-constrained vector quantization. Although the ECTCQ systems perform very well for encoding generalized Gaussian sources, the implementation is cumbersome because separate TCQ codebooks, Lagrangian multipliers, and noiseless codes must be pre-computed and stored for each encoding rate, both at the TCQ encoder and decoder.

The ACTCQ system described in this section overcomes most of these drawbacks. The main feature of this method is that the TCQ encoder uses uniform codebooks. This results in a reduced storage requirement at the encoder (but not at the decoder), as it is no longer necessary to store TCQ codebooks at each encoding rate.

Since TCQ requires multiple scalar quantizations for a single input sample, using uniform thresholds speeds up scalar quantization. Also, the system does not require *iterative* design of codebooks, based on long training sequences for each different rate, as in the case of ECTCQ[1]. Thus the uniform threshold TCQ system has certain advantages from a practical standpoint. Kasner and Marcellin [21] have presented a modified version of the system which reduces the storage requirement at the decoder as well and requires no training.

There is an interesting theoretical parallel between ACTCQ and uniform threshold quantization (UTQ). The decision levels in a UTQ are uniformly spaced along the real line, but the reconstruction codewords are the centroids of the decision regions. Farvardin and Modestino [8] show that UTQ, followed by entropy coding, can perform almost as well as optimum ECSQ at all bit rates and for a variety of memoryless sources. This section demonstrates that this result extends to trellis coded quantization also.

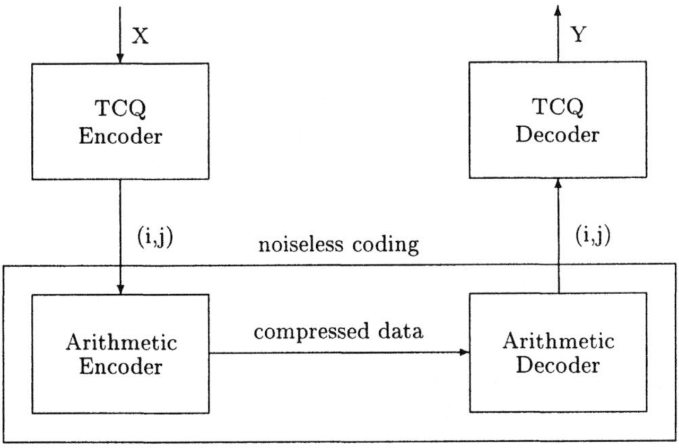

FIGURE 3. Block diagram of the ACTCQ system. Reprinted with permission from "Image subband coding using arithmetic coded trellis coded quantization," Rajan L. Joshi, Valerie J. Crump, and Thomas R. Fischer, © IEEE 1995.

The block diagram of the ACTCQ system is shown in Figure 3. The system consists of 4 modules, namely, TCQ encoder, arithmetic encoder, arithmetic decoder and TCQ decoder. The process of arithmetic encoding of a sequence of source symbols, followed by arithmetic decoding, is noiseless. Thus the output of the TCQ encoder is identical to the input of the TCQ decoder. Hence, it is convenient to separate the description of the TCQ encoder-decoder from that of arithmetic coding.

[1]Training is necessary for calculating centroids at the decoder, but it is based on generalized Gaussian sources of relatively small size and is not iterative.

3.1 Trellis coded quantization

Operation of the TCQ encoder

The TCQ encoder uses a rate-1/2 convolutional encoder. A rate-1/2 convolutional encoder is a finite state machine which maps 1 input bit into 2 output bits. The convolutional encoder can be represented by an equivalent N-state trellis diagram. Figure 4 shows the 8 state trellis. The ACTCQ system uses Ungerboeck's trellises [38]. These trellises have two branches entering and leaving each state, and possess certain symmetry.

The encoding alphabet is selected as the scaled integer lattice (with scale factor α) as shown in Figure 5. The codewords are partitioned into subsets, from left to right, as $\ldots D_0, D_1, D_2, D_3, D_0, D_1, \ldots$, with D_0 labeling the zero codeword.

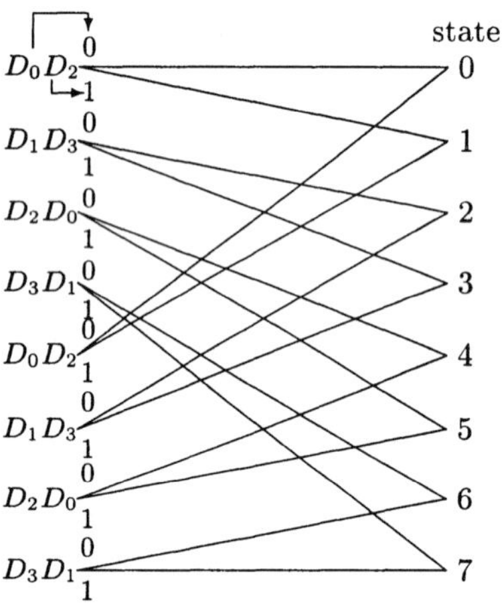

FIGURE 4. 8-state trellis labeled by 4 subsets. Reprinted with permission from "Image subband coding using arithmetic coded trellis coded quantization," Rajan L. Joshi, Valerie J. Crump, and Thomas R. Fischer, © IEEE 1995.

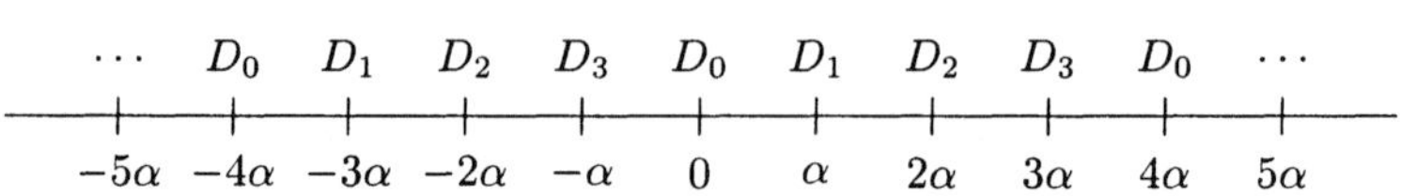

FIGURE 5. Uniform codebook for the TCQ encoder. Reprinted with permission from "Image subband coding using arithmetic coded trellis coded quantization," Rajan L. Joshi, Valerie J. Crump, and Thomas R. Fischer, © IEEE 1995.

Assume that J samples $\{x_1, x_2, \ldots, x_J\}$ from a memoryless source X are to be encoded. Let $\hat{x}_{ni}$ denote a codeword from subset D_i which is closest to input sample x_n. The Viterbi algorithm [10] is used in TCQ encoding to find a sequence of symbols $\{\hat{x}_1, \hat{x}_2, \ldots, \hat{x}_J\}$ such that

$$\sum_{n=1}^{J}(x_n - \hat{x}_n)^2, \quad \hat{x}_n \in \{\hat{x}_{ni}, 0 \le i \le 3\}$$

is minimized and the sequence of symbols satisfies the restrictions imposed by the trellis. To specify a sequence $\hat{x}_n$, the encoder has to specify a sequence of subsets and the corresponding sequence of indices of the respective codewords.

Union codebooks

Let the reproduction alphabet be divided into two union codebooks A_0 and A_1, defined by

$$A_0 = D_0 \bigcup D_2 \quad \text{and} \quad A_1 = D_1 \bigcup D_3.$$

Then the present state of the trellis limits the choice of the next codeword to only one of the union codebooks. Furthermore, the trellis structure is such that if the current state is even (odd), the next reproduction codeword must be chosen from A_0 (A_1). For example, if the current state is 0, then the next reproduction codeword has to be chosen from either D_0 or D_2, that is, from A_0.

Dropping the scale factor α from the notation, we can denote the A_0 and A_1 union codebooks as

$$A_0 = \{0, 2, -2, 4, -4, 6, -6, \cdots\} \quad \text{and} \quad A_1 = \{1, -1, 3, -3, 5, -5, \cdots\}.$$

Now consider the mapping (with integer division)

$$k = \begin{cases} j/2 & \text{for } j > 0 \\ |j| & \text{for } j \le 0 \end{cases}$$

Applying this mapping, each union codebook is mapped to the non-negative integers,

$$\tilde{A} = \{0, 1, 2, 3, \cdots\}.$$

Hence, encoding a sequence of TCQ symbols reduces to encoding a sequence of non-negative integers.

TCQ decoder

The TCQ decoder converts the non-negative integers in $\tilde{A}$ into a sequence of indices from A_0 and A_1 with the help of the trellis diagram. We denote the decoded symbol as a union codebook-codeword pair, (i, j), where i indexes the union codebook, and j indexes the respective codeword. This requires sequential updating of the trellis state as each symbol is decoded.

In general it is suboptimal to use the same reproduction alphabet at the TCQ decoder as is used for the uniform threshold encoding. Let B_{ij} be the set of all input samples which are mapped to the union codebook-codeword pair (i, j). Then, it can be shown that the choice of reproduction codeword $\hat{y}(i, j)$ which minimizes $\sum_{x \in B_{ij}} (x - y(i, j))^2$ is the centroid

$$\hat{y}(i, j) = \frac{\sum_{x \in B_{ij}} x}{\|B_{ij}\|}$$

where $||B_{ij}||$ is the cardinality of set B_{ij}. In practice, $\hat{y}(i,j)$ can be estimated from training data and pre-stored at the decoder or can be sent as side information [21]. For large encoding rates, the centroids are reasonably well approximated by the TCQ encoder codeword levels. For low encoding rates, using centroids provides a slight improvement in signal-to-noise ratio.

3.2 Arithmetic coding

Arithmetic coding [29], [31] is a variable rate coding technique in which a sequence of symbols is represented by an interval in the range $[0,1)$. As each new symbol in the sequence is encoded, the existing interval is narrowed according to the probability of that particular symbol. The more probable the symbol, the less the interval is narrowed. As the input symbol sequence grows longer, the interval required to represent it becomes smaller, requiring greater precision, or equivalently, more bits.

The TCQ encoder outputs a sequence of TCQ symbols, each symbol a non-negative integer in $\tilde{A}$. The arithmetic coder performs variable-length encoding of the trellis symbols, providing an average codeword length approaching the first-order entropy of the source. Since the symbols in $\tilde{A}$ are drawn from A_0 and A_1, it is generally necessary to use context-based arithmetic coding for good performance.

Use of two contexts in arithmetic coding of TCQ symbols

In context-based arithmetic coding, one model (probability table) is maintained for each context. Given an interval which represents the symbol sequence up to the present time and a context, for a new symbol the interval is narrowed using the model for that particular context. For example, if the input source X_n is a first order Markov source, x_{n-1} is the context for encoding x_n and the conditional probability mass function $p(x_n|x_{n-1})$ is used as the model (context-dependent) for narrowing the interval. It is necessary that the decoder has knowledge of the context or is able to derive it from the information received previously.

Since the allowable symbols in a TCQ encoded sequence are trellis-state dependent, we define state-dependent contexts for the arithmetic coding. Recall that if the present state is even (odd), the next codeword belongs to union codebook A_0 (A_1). Marcellin [26] has found that virtually nothing is lost by using just two conditional probability distributions, one for each union codebook and representing an average over the corresponding trellis states. Since in a typical application there might be 4 to 256 states, the practical advantage of using just two probability tables is obvious. The probability tables used for the two contexts are based on conditional probabilities, $P(y(i,j)|A_0)$ and $P(y(i,j)|A_1)$. Since the arithmetic decoder can decode symbol by symbol, given the present state and the next subset information, both the TCQ encoder and decoder can find the next state based on the trellis diagram. Thus, the context information is readily available both at the encoder and decoder.

Adaptive source models

Each union codebook uses a separate adaptive source model. The source model for the appropriate union codebook is modified after each new sample, as described in [40]. The adaptation of the source model is achieved by increasing by one the count for the symbol which just occurred.

For adaptive source models, it is common practice to use a model having equal probabilities for all symbols, as the initial source model. This is quite reasonable when the symbol alphabet size is small and the number of source samples to code is large. We have found, however, that such uniform initialization can cause a significant loss of performance in some cases. The coding method selected uses adaptive source models, with the initial probability counts for the union codebooks computed from generalized Gaussian source models.

3.3 Encoding generalized Gaussian sources with ACTCQ system

The probability density function (pdf) for the generalized Gaussian distribution (GGD) is given by [8]:

$$p(x) = \left[\frac{\nu\eta(\nu,\sigma)}{2\Gamma(1/\nu)}\right] \exp(-[\eta(\nu,\sigma)|x|]^{\nu}), \tag{11.12}$$

where

$$\eta(\nu,\sigma) = \sigma^{-1}\left[\frac{\Gamma(3/\nu)}{\Gamma(1/\nu)}\right]^{1/2}, \tag{11.13}$$

ν is a shape parameter describing the exponential rate of decay, and σ is the standard deviation. Laplacian and Gaussian densities are members of the family of GGD's with $\nu = 1.0$ and 2.0, respectively. GGD parameter values of $\nu < 1.0$ have been found as appropriate for modeling image subband data [36]. If subbands are classified, then appropriate shape parameters for modeling the classes are often in the range $1.0 \leq \nu \leq 2.0$ [17].

The ACTCQ system was used to encode generalized Gaussian sources with shape parameter ν ranging from 0.5 to 2.0. Figure 6 shows the rate vs SNR performance for encoding a Laplacian source using an 8-state trellis. Also shown are the rate vs SNR performance of entropy-constrained TCQ (ECTCQ) [9], uniform threshold quantization (UTQ) [8], and the Shannon Lower Bound (SLB) to the rate-distortion function for the Laplacian source. For rates larger than about 2.0 bits/samples, the ACTCQ system performance is identical with ECTCQ[2], better than UTQ by about 1.0 dB and is within 0.5 dB of the SLB for rates greater than 2.0 bits. However, at low rates (below 1.5 bits per sample) the performance of the ACTCQ system deteriorates. This drop in performance is due to not having a zero level available in the A_1 union codebook.

Figure 7 shows a simple modification that overcomes this problem. All the positive codewords from the reproduction alphabet in Figure 5 are shifted to the left by α. Thus zero is a codeword in both D_0 and D_1. Dropping the scale factor α from the notation, the union codebooks become

$$A_0 = \{0, 1, -2, 3, -4, 5, -6, \cdots\} \quad \text{and} \quad A_1 = \{0, -1, 2, -3, 4, -5, 6, \cdots\}$$

Again, it is straightforward to map A_i to $\tilde{A}$. Note that if the source is zero mean and has a symmetric density, then the symmetry in A_0 and A_1 implies that a

[2]The slight difference in the performance of ACTCQ and ECTCQ is due to the fact that the ECTCQ results are for a 4-state trellis, while the ACTCQ results are for an 8-state trellis.

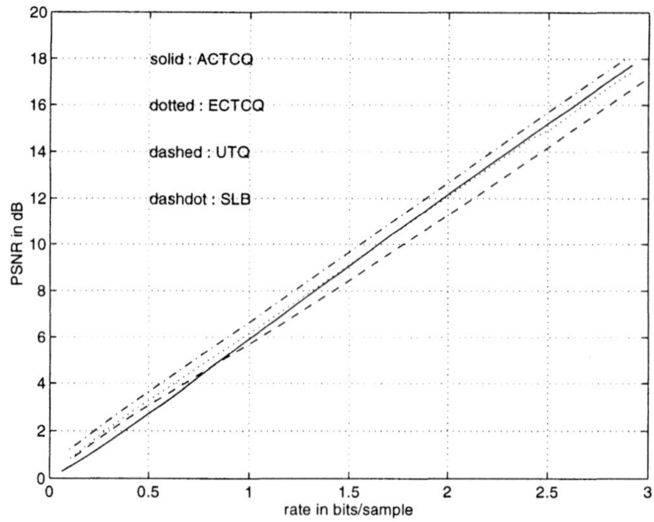

FIGURE 6. Comparison of the ACTCQ system with ECTCQ, UTQ and SLB for encoding a Laplacian Source. Reprinted with permission from "Image subband coding using arithmetic coded trellis coded quantization," Rajan L. Joshi, Valerie J. Crump, and Thomas R. Fischer, © IEEE 1995.

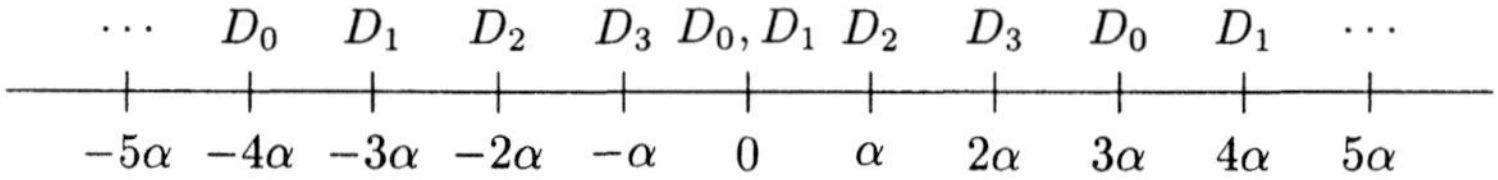

FIGURE 7. Modified codebook for low encoding rates. Reprinted with permission from "Image subband coding using arithmetic coded trellis coded quantization," Rajan L. Joshi, Valerie J. Crump, and Thomas R. Fischer, © IEEE 1995.

single context is sufficient for the arithmetic coding. In practice, we use the TCQ codebook of Figure 5 for high encoding rates, and the codebook of Figure 7 for low-rate encoding, where the switch between codebooks depends on the source [15]. The performance of the ACTCQ system for encoding a Laplacian source, using modified codebooks at low bit rates, is shown in Figure 8.

Sending the centroids as side information

In the ACTCQ system described so far, the decoder has to store the reconstruction codebook. The codebook is obtained by means of a training sequence generated from a generalized Gaussian density of appropriate shape parameter. One drawback of this is that the the ACTCQ decoder has to store codebooks corresponding to a variety of step sizes and a variety of shape parameters. This can be avoided by using universal TCQ [21]. In universal TCQ, a few codewords are trained on the actual input sequence and sent to the decoder as side information; the remaining

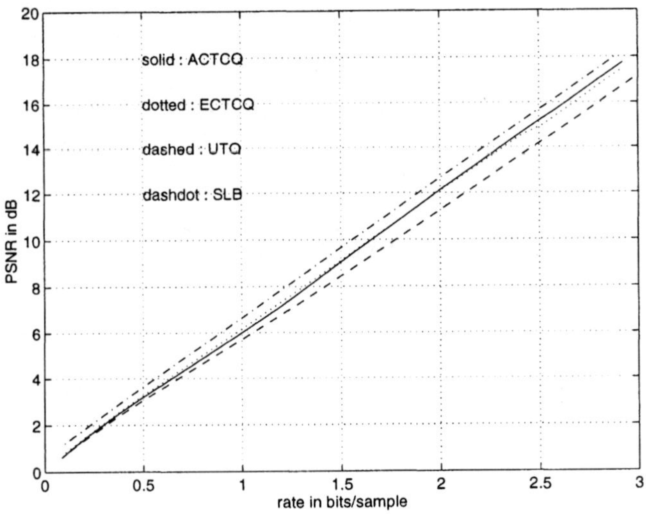

FIGURE 8. Performance of the ACTCQ system (using the modified codebook at low bit rates) for encoding a Laplacian source. Reprinted with permission from "Image subband coding using arithmetic coded trellis coded quantization," Rajan L. Joshi, Valerie J. Crump, and Thomas R. Fischer, © IEEE 1995.

decoder codewords are identical to those used for encoding. For generalized Gaussian densities, in most cases it is adequate to send fewer than 5 codewords to the decoder. The exact number of codewords sent can be fixed, or dependent on the step size and the shape parameter. The performance of universal TCQ system is very close to the ACTCQ system using trained codewords for reconstruction, and furthermore this method requires no stored codebooks. Universal TCQ was used to encode the subbands in the results presented in the next section.

4 Image subband coder based on classification and ACTCQ

In this section we present simulation results for a subband image coder based on classification and trellis coded quantization. The basic steps in the encoder can be summarized as follows :

- Decompose the image into M subbands.
- Classify each subband into J classes. Estimate the generalized Gaussian parameters ν and σ for each class.
- Optimally allocate rate between MJ generalized Gaussian sources.
- For each subband,

- Encode the classification map and other side information.
- Encode each class in the subband which is allocated non-zero rate using the universal ACTCQ encoder.

In the following subsection, each of the above steps is described in more detail. Some of the choices, such as using equally likely classification, have been influenced by considerations of computational complexity.

4.1 Description of the image subband coder

Subband decomposition

Throughout the simulations, the 22-band decomposition is used. The decomposition is implemented using a 2-D separable tree-structured filter bank using the 9-7 biorthogonal spline filters designed by Antonini et al. [1]. Symmetric extension is used at the convolution boundaries. The mean of the LFS is subtracted and sent as side information (16 bits).

Classification of subbands

Each subband is classified into 4 *equally likely* classes. The block size for subbands 0-6 is always 2×2. The block size for subbands 7-21 is dependent on the targeted rate. For targeted rates up to 0.4 b/p, the block size is 8×8, whereas for higher rates, the block size is 4×4. Each class is modeled as a generalized Gaussian source. Although a variety of sophisticated methods, such as maximum likelihood estimation and the Kolmogrov-Smirnov test, can be used for estimating the parameters ν and σ, we found that a simple method described by Mallat [24] is adequate. In this method the first absolute moment and variance of the GGD is estimated by using the sample averages. Then the shape parameter ν is estimated by solving

$$F(\hat{\nu}) = \left(\frac{\hat{E}[|X|]}{\hat{\sigma}} \right) \tag{11.14}$$

where

$$F(\alpha) = \frac{\Gamma(2/\alpha)}{\sqrt{\Gamma(1/\alpha)\Gamma(3/\alpha)}} \tag{11.15}$$

The $\hat{}$ denotes quantities estimated from the data set. In our simulations, ν was restricted to the set $\{0.5, 0.6, 0.7, 0.8, 0.9, 1.0, 2.0\}$.

Rate allocation

The rate allocation equations developed in section 2.2 can be used for rate allocation. For targeted rates of 1.0 bits/pixel and below, however, a more accurate solution can be obtained using Shoham and Gersho's algorithm [34] for optimal bit allocation for an arbitrary set of quantizers. We used a variation of this algorithm developed by Westerink et al. [39] which traces the convex hull of the composite rate-distortion curve, starting with the lowest possible rate.

The operational rate-distortion curves for GGD's encoded using the ACTCQ system are pre-computed and stored. This is done only for permissible values of

ν. Typically for each rate-distortion curve, 100 points are stored with bit rates ranging from 0 bits to about 9 bits. The points are (roughly) uniformly spaced with respect to rate. After estimating the GG parameters for various subbands and DCT coefficients, the operational rate-distortion curves for the respective parameters, suitably scaled by the variance and decimation factor, are used in rate allocation. After finding the optimum rate allocation, the actual step sizes to be used for the ACTCQ system can be found by using a look-up table.

Side information

For each subband, all the classes having zero rate are merged into a single class and the classification map is modified accordingly. If all classes are assigned positive rate, an uncompressed classification map requires 38,912 bits (for a 512×512 image), when the block size in subbands 7-21 is 4×4. The classification map can be arithmetic coded using conditional probability models. In the results to follow, only limited (one sample) intra-band dependence is used for conditioning[3].

In addition to the classification maps, the following information is sent to the decoder as side information. For each class in the subband, a 2 bit index is sent to the decoder. This index specifies the type of TCQ codebook that will be used by the ACTCQ system. A zero index means that the class is allocated zero rate. Also, for each class which is allocated nonzero rate, the variance (16 bits), the shape parameter (3 bits) and the index of the step size to be used by the ACTCQ system (8 bits) have to be sent as side information. Thus, for a class allocated non-zero rate, the amount of side information is 27 bits. Now consider a 22-band decomposition in which each subband is further classified into 4 classes. Assuming every class is allocated non-zero rate, the total side information is $((27 + 2) \times 88 + 16) = 2568$ bits[4]. This is a worst case scenario. Thus for a 512×512 image, the maximum side information is 0.001 bits/pixel. In practice, even for rates of 1.0 bits/pixel, about 40 % of the classes are allocated zero rates. Thus the side information is usually much lower.

ACTCQ encoding

Each class allocated non-zero rate is encoded using the step size derived from the rate allocation. The universal ACTCQ system is used because no codebooks need be stored at the encoder or the decoder. In practice, the TCQ codebooks have to be finite. The choice of number of TCQ codewords depends on the step size, variance, and the GG shape parameter of the source being encoded. An unnecessarily large number of codewords can hinder the adaptation of the source model in arithmetic coding, whereas a smaller number can result in large overload distortion. We used an empirical formula for determining how to truncate the ACTCQ codebook. Use of escape mechanism as described in [20] is also possible.

---

[3]Using both intra- and inter-band dependence results in slightly improved performance as reported in [19].

[4]The mean of the zeroth subband (16 bits) has to be sent to the decoder.

5 Simulation results

In this section we present simulation results for encoding monochrome images 'Lenna', 'Barbara' and 'goldhill'. The images can be obtained via anonymous ftp from "spcot.sanders.com". The directory is "/pub/pnt". All the images are of size 512 × 512 pixels. All the bit rates reported in this section are based on actual encoded file sizes. The side information has been taken into account while calculating the bit rates. The peak signal-to-noise ratio (PSNR) defined as

$$PSNR = 10 \log_{10} \left(\frac{(255)^2}{MSE} \right) \text{dB} \tag{11.16}$$

is used for reporting the coding performance. Various parameters such as number of classes, block sizes, subband decomposition are as described in Section 4.1. Table 11.1 presents the PSNRs for the three images.

TABLE 11.1. PSNR results for the images 'Lenna', 'Barbara' and 'goldhill'.

	PSNR (dB)		
Rate (b/p)	Lenna	Barbara	goldhill
0.20	32.85	26.95	29.70
0.25	33.88	28.03	30.48
0.30	34.79	28.96	31.19
0.40	36.21	30.50	32.33
0.50	37.24	31.94	33.33
0.60	38.13	33.20	34.23
0.70	38.87	34.46	34.99
0.80	39.52	35.48	35.71
0.90	40.14	36.39	36.36
1.00	40.71	37.26	37.05

When compared with the results in [18], [19], for the same filters and subband decomposition, the results in Table 11.1 are lower by up to 0.5 dB. The main reasons for this are

- Use of equally likely classification instead of the maximum gain classification (for reduced complexity at the encoder).
- Sending only a few centroids as side information (for reduced storage requirement at the decoder).

Table 11.2 lists the execution times, including the time required for subband split/merge, for encoding and decoding the 'Lenna' image at rates of 0.25 and 0.5 bits/pixel. The execution times are for a SPARC-20 running at 85 MHz. Since the implementation has not been optimized for execution speed, the execution times in Table 11.2 are provided simply to provide a rough idea as to the complexity of the algorithm. From the table it can be seen that the encoding complexity is asymmetric; the encoder complexity is about 3 times the decoder complexity.

Figure 9 show the test images encoded at bit rate of 0.5 bits/pixel. Additional encoded images can be found at the website. It can be seen that for images encoded at 1.0 bits/pixel there is almost no visual degradation. The subjective quality of the images encoded at 0.5 bits/pixel is also excellent. There are no major objectionable artifacts. Ringing noise and smoothing is apparent for images encoded at 0.25 bits/pixel.

TABLE 11.2. Execution times for encoding and decoding 'Lenna'.

	Execution times (sec.)					
	Encoder			Decoder		
Rate (b/p)	usr	system	total	usr	system	total
0.25	2.552	0.232	2.784	0.733	0.184	0.917
0.50	3.281	0.279	3.560	0.889	0.179	1.068

6 Acknowledgment

This work was supported, in part, by the National Science Foundation under grants NCR-9303868 and NCR-9627815.

7 References

[1] M. Antonini, M. Barlaud, P. Mathieu, and I. Daubechies, "Image coding using wavelet transform," *IEEE Trans. Image Proc.*, vol. IP-1, pp. 205–220, April 1992.

[2] J. L. Arrowwood and M. J. T. Smith, "Exact reconstruction analysis/synthesis filter banks with time varying filters," *Proceedings, IEEE Int. Conf. on Acoust., Speech, and Signal Proc.*, vol. 3, pp. 233-236, April 1993.

[3] K. A. Birney and T. R. Fischer, "On the modeling of DCT and subband image data for compression," *IEEE Trans. Image Proc.*, vol. IP-4, pp. 186-193, February 1995.

[4] W. H. Chen and C. H. Smith, "Adaptive coding of monochrome and color images," *IEEE Trans. Commun.*, vol. COM-25, pp. 1285-1292, November 1977.

[5] P. A. Chou, T. Lookabaugh, and R. M. Gray, "Entropy-constrained vector quantization," *IEEE Trans. Acoustics, Speech and Signal Proc.*, vol. ASSP-37, pp. 31-42, January 1989.

[6] C. Chrysafis and A. Ortega, "Efficient context-based entropy coding for lossy wavelet image compression," *Proceedings, DCC-97*, pp. 241-250, March 1997.

[7] W. C. Chung and M. J. T. Smith, "Spatially-varying IIR filter banks for image coding," *Proceedings, IEEE Int. Conf. on Acoust., Speech, and Signal Proc.*, vol. 5, pp. 570-573, April 1993.

[8] N. Farvardin and J. W. Modestino, "Optimum quantizer performance for a class of non-Gaussian memoryless sources," *IEEE Trans. Inform. Th.*, vol.30, no.3, pp. 485-497, May 1984.

[9] T. R. Fischer and M. Wang, "Entropy-constrained trellis coded quantization," *IEEE Trans. Inform. Th.*, vol.38, no.2, pp. 415-426, March 1992.

[10] G. D. Forney, Jr., "The Viterbi Algorithm," *Proc. IEEE*, vol. 61, pp. 268-278, March 1973.

[11] C. Herley, J. Kovačević, K. Ramchandran, and M. Vetterli, "Time-varying orthonormal tilings of the time-frequency plane," *Proceedings, IEEE Int. Conf. Acoust., Speech and Signal Proc.*, pp. 205-208, vol. 3, April 1993.

[12] H. Jafarkhani, N. Farvardin, and C.-C. Lee, "Adaptive image coding based on the discrete wavelet transform," *Proceedings, Int. Conf. on Image Proc.*, vol. 3, pp. 343-347, November 1994.

[13] N. S. Jayant and Peter Noll, *Digital coding of waveforms*, Prentice-Hall, Englewood Cliffs, New Jersey 1984.

[14] Rajan L. Joshi, Thomas R. Fischer and Roberto H. Bamberger, "Optimum classification in subband coding of images," *Proceedings, Int. Conf. on Image Proc.*, vol. 2, pp. 883-887, November 1994.

[15] Rajan L. Joshi, Valerie J. Crump, and Thomas R. Fischer, "Image subband coding using arithmetic coded trellis coded quantization," *IEEE Trans. Circ. & Syst. Video Tech.*, vol. 5, no. 6, pp. 515-523, December 1995.

[16] R. L. Joshi, T. R. Fischer, and R. R. Bamberger, "Comparison of different methods of classification in subband coding of images," *IS&T/SPIE's symposium on Electronic Imaging: Science & Technology, Technical conference 2418, Still Image Compression*, February 1995.

[17] Rajan L. Joshi and Thomas R. Fischer, "Comparison of generalized Gaussian and Laplacian modeling in DCT image coding," *IEEE Signal Processing Letters*, vol. 2, No. 5, pp. 81-82, May 1995.

[18] R. L. Joshi, "Subband image coding using classification and trellis coded quantization," Ph.D. thesis, Washington State University, Pullman, August 1996.

[19] R. L. Joshi, H. Jafarkhani, J. H. Kasner, T. R. Fischer, N. Farvardin, M. W. Marcellin and R. H. Bamberger, "Comparison of different methods of classification in subband coding of images," To appear in *IEEE Trans. Image Proc.*.

[20] J. H. Kasner and M. W. Marcellin, "Adaptive wavelet coding of images," *Proceedings, Int. Conf. on Image Proc.*, vol. 3, pp. 358-362, November 1994.

[21] James H. Kasner and Michael W. Marcellin, "Universal Trellis Coded Quantization," *Proceedings, Int. Symp. Inform. Th.*, pp. 433, September 1995.

[22] John M. Lervik and Tor A. Ramstad, "Optimal entropy coding in image subband coders," *Proceedings, Int. Conf. on Image Proc.*, vol. 2, pp. 439-444, June 1995.

[23] Scott M. LoPresto, Kannan Ramchandran, and Michael T. Orchard, "Image coding based on mixture modeling of wavelet coefficients and a fast estimation-quantization framework," *Proceedings, DCC-97*, pp. 221-230, March 1997.

[24] S. G. Mallat, "A theory for multiresolution signal decomposition: The wavelet representation," *IEEE Trans. Pattern Anal. Mach. Intel.*, vol. 11, pp. 674-693, July 1989.

[25] M. W. Marcellin and T. R. Fischer, "Trellis coded quantization of memoryless and Gauss-Markov sources," *IEEE Trans. Commun.*, vol.38, no.1, pp. 82-93, January 1990.

[26] M. W. Marcellin, "On entropy-constrained trellis coded quantization," *IEEE Trans. Commun.*, vol. 42, no. 1, pp. 14-16, January 1994.

[27] T. Naveen and John W. Woods, "Subband finite state scalar quantization," *Proceedings, Int. Conf. on Acoust., Speech, and Signal Proc.*, pp. 613-616, April 1993.

[28] Athanasios Papoulis, *Probability, random variables and stochastic processes*, third edition, McGraw-Hill Book Company, 1991.

[29] R. Pasco, *Source coding algorithms for fast data compression*, Ph.D. thesis, Stanford University, 1976.

[30] K. Ramchandran, Z. Xiong, K. Asai and M. Vetterli, "Adaptive transforms for image coding using spatially-varying wavelet packets," *IEEE Trans. Image Proc.*, vol. IP-5, pp. 1197-1204, July 1996.

[31] J. Rissanen, "Generalized Kraft inequality and arithmetic coding," *IBM J. Res. Develop.*, vol. 20, pp. 198-203, May 1976.

[32] A. Said and W. A. Pearlman, "A new fast and efficient image codec based on set partitioning in hierarchical trees," *IEEE Trans. Circ. & Syst. Video Tech.*, vol. 6, no. 3, pp. 243-250, June 1996.

[33] J. M. Shapiro, "Embedded image coding using zerotrees of wavelet coefficients," *IEEE Trans. Signal Proc.*, vol. 41, no. 12, pp. 3445-3462, December 1993.

[34] Y. Shoham and A. Gersho, "Efficient bit allocation for an arbitrary set of quantizers," *IEEE Trans. Acoust., Speech, and Signal Proc.*, vol. ASSP-36, pp. 1445-1453, September 1988.

[35] P. Sriram and M. W. Marcellin, "Image coding using wavelet transforms and entropy-constrained trellis-coded quantization," *IEEE Trans. Image Proc.*, vol. IP-4, pp. 725-733, June 1995.

[36] N. Tanabe and N. Farvardin, "Subband image coding using entropy-coded quantization over noisy channels," *IEEE J. on Select. Areas in Commun.*, pp. 926-942, vol. 10, No. 5, June 1992.

[37] Tse, Yi-tong, *Video coding using global/local motion compensation, classified subband coding, uniform threshold quantization and arithmetic coding*, Ph.D. thesis, University of California, Los Angeles, 1992.

[38] G. Ungerboeck, "Channel coding with multilevel/phase signals," *IEEE Trans. Inform. Th.*, vol. IT-28, pp. 55-67, January 1982.

[39] P. H. Westerink, J. Biemond and D. E. Boekee, "An optimal bit allocation algorithm for sub-band coding," *Proceedings, Int. Conf. on Acoust., Speech, and Signal Proc.*, pp. 757-760, April 1988.

[40] I. H. Witten, R. M. Neal, and J. G. Cleary, "Arithmetic coding for data compression," *Commun. of the ACM*, vol. 30, no. 6, pp. 520-540, June 1987.

[41] J. W. Woods and S. D. O'Neil, "Subband coding of images," *IEEE Trans. on Acoust., Speech, and Signal Proc.*, vol. ASSP-34, pp. 1278-1288, October 1986.

[42] J. W. Woods and T. Naveen, "A filter based allocation scheme for subband compression of HDTV," *IEEE Trans. Image Proc.*, vol. IP-1, pp. 436-440, July 1992.

[43] Zixiang Xiong, Kannan Ramchandran, and Michael T. Orchard, "Space-frequency quantization for wavelet image coding," *IEEE Trans. Image Proc.*, vol. IP-6, pp. 677-693, May 1997.

FIGURE 9. Comparison of original and encoded images. Left column : Original, right column : Encoded at 0.5 bits/pixel

12

Low-Complexity Compression of Run Length Coded Image Subbands

John D. Villasenor
Jiangtao Wen

1 Introduction

Recent work in wavelet image coding has resulted in quite a number of advanced coding algorithms that give extremely good performance. Building on the signal processing foundations laid by researchers including Vetterli [VH92] and in some instances on the idea of zerotree coding introduced by Shapiro [Sha93], algorithms described by Said and Pearlman [SP96], Xiong *et al.* [XRO97], Joshi et al. [JJKFFMB95], and LoPresto *et al.* [LRO97] appear to be converging on performance limits of wavelet image coding. These algorithms have been developed with performance as the primary goal, with complexity viewed as an important, but usually secondary consideration to be addressed after the basic algorithmic framework has been established. This does not imply that that algorithms listed above are overly complex; for example, the algorithm of Said and Pearlman involves low arithmetic complexity. However, the question still arises as to how small the sacrifice in coding performance can be made while reducing the complexity even further.

There is a very interesting and less well explored approach to wavelet image coding in which the problem can be set up as follows: Impose constraints on the arithmetic and addressing complexity, and then work to find an algorithm which optimizes the coding performance within those constraints. While the importance of low arithmetic complexity is universally appreciated, there has been less consideration given to addressing complexity. Despite the increasing capabilities of workstations and PCs which have ample resources for handling large memory structures, there is still an enormous amount of processing which occurs in environments characterized by very limited memory resources. This is especially true in wireless communications devices, in which power limitations and the need for low-cost hardware make dataflow-oriented processing highly desirable. For these environments, at least, this argues against zerotree image coding algorithms, which although they have the significant advantage of offering an embedded structure, employ memory structures which are quite complex. This becomes even more important in coding of data sets such as seismic data [VED96] having more than 2 dimensions, which upon wavelet transformation can produce hundreds of subbands.

The range of complexity-constrained wavelet coding algorithms is far too large to consider in its entirety here. We address a particular framework consisting of a wavelet transform, uniform scalar quantization using a quantizer with a dead zone at the origin, run length coding and finally entropy coding of the run lengths and levels. We focus in particular on low-complexity ways of efficiently coding run-length coded subband data. No adaptation to local statistical variations within subbands is performed, though we do allow adaptation to the characteristics of different subbands. Even this simple coding structure allows a surprisingly varied set of tradeoffs and performance levels. We focus the discussion on approaches suitable for compression of grayscale images to rates ranging from approximately .1 to .5 bits/sample. This range of bits rates was chosen because it is here that the most fertile ground for still image coding research lies. At higher bit rates, almost all reasonable coding algorithms perform quite well, and it is difficult to obtain significant improvement over a well-designed JPEG (DCT) coder. (It should also be pointed at that as Crouse *et al.* [CR97] and others have shown, a properly optimized JPEG implementation can also do extremely well at low bit rates.) At rates below about .1 bits/pixel, the image quality degradations become so severe even in a well-designed algorithm that the utility of the compressed image is questionable. The concepts here can readily be extended to enable color image representation through consideration of the chrominance components, much as occurs in JPEG and MPEG.

2 Large-scale statistics of run-length coded subbands

Wavelet subbands have diverse and highly varying statistical properties, as would be expected from image data that are subjected to multiple combinations of high and low pass filtering. Coding algorithms that look for and exploit local variations in statistics can do extremely well. Such context modeling has a long and successful history in the field of lossless image coding, where, for example, it is the basis for the LOCO-I and CALIC [WSS96, WM97] algorithms. More recently, context modeling has been applied successfully to lossy image coding as well [LRO97]. Context modeling comes inevitably at a cost in complexity. When, as is the case in the techniques we are using, context modeling within subbands is not used, the statistics of one or more subbands in their entirety must be considered, and the coder design optimized accordingly. Since the coders we are using involve scalar quantization and run length coding, it is most relevant to consider the distributions of run lengths and nonzero levels that occur after those operations.

One of the immediate tradeoffs one faces in examining these statistics lies in how to group subbands together. The simplest, and least accurate approach would be to consider many or even all of the subbands from a particular image or images as a single statistical entity. The loss in coding performance from this approach would be fairly large. At the other end of the scale one can consider one subband at a time, and use data from many images and rates to build up a sufficiently large set of data to draw conclusions regarding the pdf of runs and levels. We choose a middle ground and divide the subbands in a wavelet-transformed image into inner, middle, and outer regions. For the moment we will assume a 5-level dyadic wavelet decomposition using the subband numbering scheme shown in Figure 1. We consider subbands 1-7 to be inner subbands, 8-13 to be middle subbands, and 14-16 to be

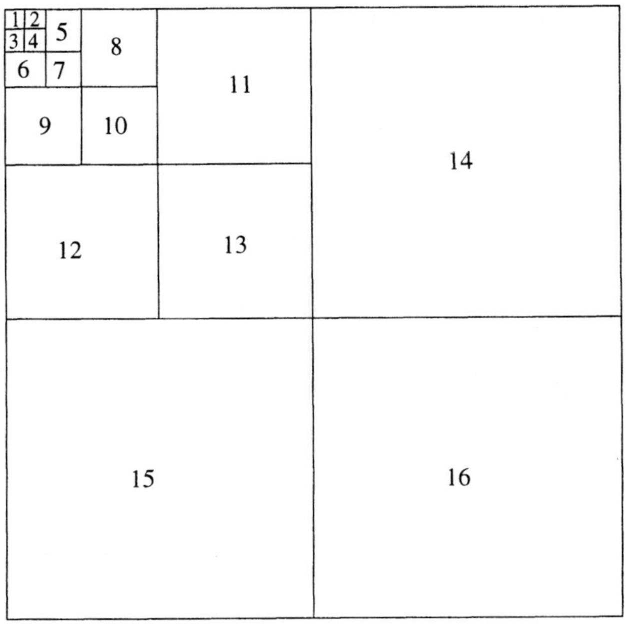

FIGURE 1. 5-level decomposition showing subband numbering scheme.

outer subbands. For the decomposition we will use the 9/7 filter bank pair that has been described in [ABMD92] and elsewhere, though there is relatively little variation of the statistics as different good filter banks are used. The 10/18 filter described in [TVC96] can boost PSNR performance by up to several tenths of a dB relative to the 9/7 filter for some images. However the 9/7 filter has been used for most of the results in the reported literature, and to preserve this consistency we also use it here. The raster scanning directions are chosen (non-adaptively) to maximize the length of zero runs - in other words the scanning used in subbands containing vertical detail was at 90 degrees with respect to the scanning used in "horizontal" subbands. Scanning was performed horizontally for "corner" subbands.

Since we are utilizing a combination of run length coding and entropy coding, we are handling quantized transform coefficients in a way that resembles the approaches used in image coding standards such as JPEG, MPEG, H.263, and the FBI wavelet compression standard, though with the difference that runs and levels are considered separately here. The separate handling of runs and levels represents another tradeoff made in favor of lower complexity. To consider run/level pairs jointly would result in a much larger symbol alphabet, since in contrast with the 8x8 blocks used in the block DCT schemes, wavelet subbands can include thousands of different run length values.

Figure 2 shows the normalized histograms for the run lengths in the inner, middle, and outer subbands obtained from a set of 5 grayscale, 512 x 512 images. The images used where "bridge," "baboon," "boat," "grandma," and "urban." A quantization step size consistent with compression to .5 bpp and uniform across all subbands was used to generate the plots. Also shown in Figure 2 is the curve for a one-sided generalized Gaussian (GG) pdf with a shape parameter of $\nu = .5$ and a variance chosen to give an approximate fit to the histograms. The GG density can be expressed as

$$f_X(x) = C_1 \exp(-C_2|x|^\nu) \tag{12.1}$$

where C_1, C_2, and η are given by

$$C_1 = \frac{\nu\eta(\sigma,\nu)}{2\Gamma(1/\nu)}, \qquad C_2 = [\eta(\sigma,\nu)]^\nu, \qquad \eta(\sigma,\nu) = \frac{1}{\sigma}\left[\frac{\Gamma(3/\nu)}{\Gamma(1/\nu)}\right]^{1/2}.$$

The variable ν is the shape parameter and the standard deviation of the source is σ. When $\nu = 2$ the GG becomes Gaussian and when $\nu = 1$ it is Laplacian. A number of recent authors, including Laroia and Farvardin [LF93], LoPresto *et al.* [LRO97], Calvagno *et al.* [CGMR97], and Birney and Fischer [BF95] have noted the utility of modeling wavelet transformed image data using GG sources with a shape parameter in the range $\nu < 1.0$ By proper choice of ν and σ a good fit can be obtained to most run and level data from quantized image subbands. It is also of interest to understand how the statistics change with the coding rate. Figure 3 illustrates the differences between the run histograms that occur for step sizes corresponding to image coding rates of .25 and .5 bpp respectively. At each coding rate, a single common step size was used for all subbands shown in the figure. Optimizing the step sizes on a per-subband basis gives some coding improvement but has little effect on the shape of the histograms. It is clear from the figures that the form of both the runs and levels in quantized subbands can be well-modeled using a properly parameterized GG. The challenge in coding is to come up with a low-complexity framework to efficiently code data that follow this distribution. As discussed below, structured entropy coding trees can indeed meet these low-complexity/high performance goals.

3 Structured code trees

3.1 Code Descriptions

Structured trees can be easily parameterized and can be used to match a wide range of source statistics. Trees with structure have been used before in image coding. The most widely known example of a structured tree is the family of Golomb-Rice codes [Golomb66, GV75, Rice 79]. Golomb-Rice codes have recently been applied for many applications, including for coding of prediction errors in lossless image coding applications [WSS96]. Golomb-Rice codes are nearly optimal for coding of exponentially distributed non-negative integers, and describe a non-negative integer i in terms of a quotient and a remainder. A Golomb-Rice code with parameter k uses a divisor with value 2^k. The quotient can be arbitrarily large and is expressed using a unary representation; the remainder is bounded by the range $[0, 2^k - 1]$ and is expressed in binary form using k bits. To form the codeword, unary|binary concatenation is often used. For example, for a non-negative discrete source, a Golomb-Rice code with $k = 2$ can represent the number 9 as 00101. The first

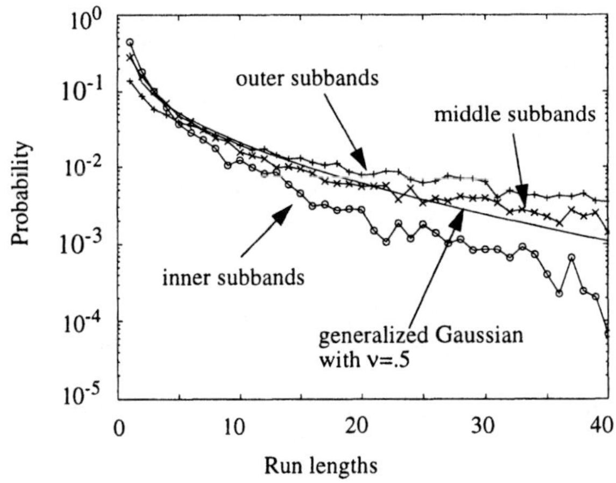

FIGURE 2. Normalized histograms of run lengths in image subbands. The histograms are generated using the combined data from five grayscale test images transfromed using 9/7 filters and the decomposition of the previous figure. Subbands 1-7 are "inner" subbands; 8-13 are "middle" subbands. A generalized Gaussian fit to the data is also shown.

two zeros, terminated by the first one, constitute the unary prefix and identify the quotient of $9/2^2$ as having value 2; the 01 suffix is a 2-bit binary expression of the remainder. Figure 6(a) shows the upper regions of the tree corresponding to Golomb-Rice code with parameter $k = 2$. Since each level in the Golomb-Rice tree contains an identical number (2^k) of code words, it is clear that Golomb-Rice codes are well suited to exponential sources. In general the codeword for integer i in a Golomb-Rice code with parameter k will have length $k + 1 + \lfloor i/2^k \rfloor$.

It is also possible to construct code trees in which the number of codewords of

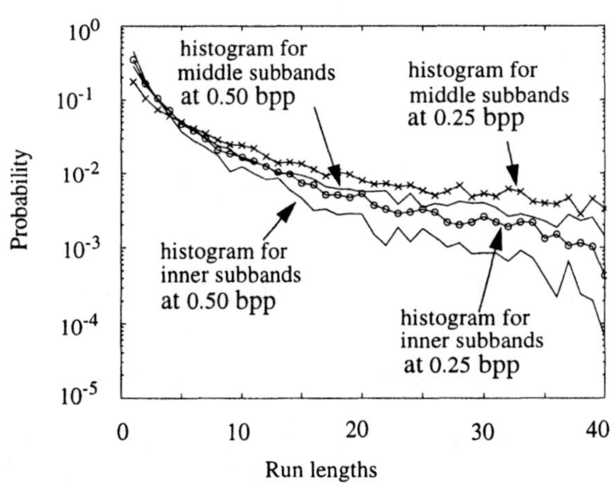

FIGURE 3. Normalized histograms of run lengths in image subbands.

a given length grows geometrically with codeword length. The exp-Golomb codes proposed by Teuhola [Teuhola78] and covered implicitly in the theoretical treatment by Elias [Elias75] are parameterized by k, and have 2^k codewords of shortest length, 2^{k+1} codewords of next shortest length, etc. Because the number of exp-Golomb codewords of given length increases with code length, exp-Golomb codes are matched to pdfs such as GGs with $\nu < 1$ in which the decay rate with respect to exponential lessens with distance from the origin. Though Teuhola's goal in proposing exp-Golomb codes was to provide more robustness for coding of exponential sources with unknown decay rate, we show here that these codes are also very relevant for GG sources. Figure 6(b) shows the upper regions of the tree corresponding to exp-Golomb code for $k = 1$. As with Golomb-Rice codes, exp-Golomb codes support description in terms of a unary|binary concatenation. The unary term is $j + 1$ bits in length, and identifies the codeword as being in level j. The binary term contains $k + j$ bits and selects among the 2^{k+j} codewords in level j. The codeword for integer i (assuming a non-negative discrete monotonically decreasing source) will have length $k + 1 + 2\lfloor\log_2(1 + i/2^k)\rfloor$ for an exp-Golomb code with parameter k.

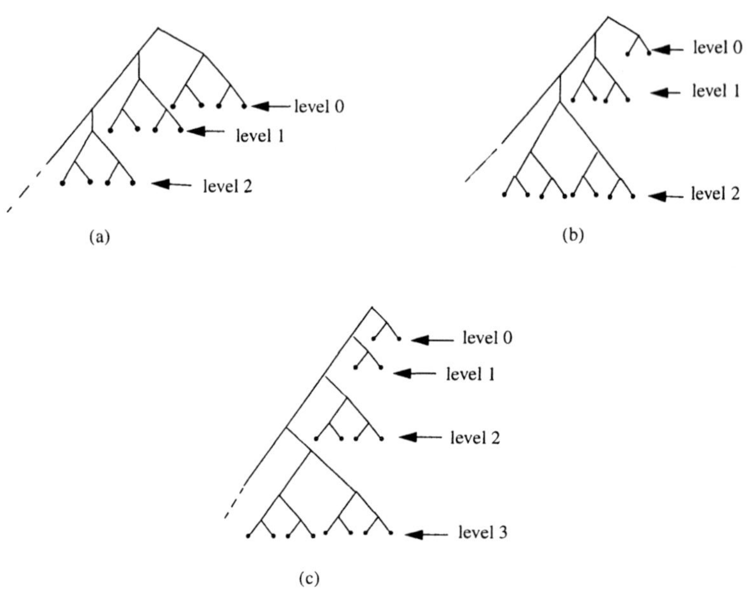

FIGURE 4. Upper regions of coding trees. (a) Golomb-Rice tree with k=2. (b) Exp-Golomb tree with k=1. (c) Hybrid tree.

Figure 5 provides a qualitative illustration of the good match between low-shape parameter GG pdfs, exp-Golomb codes and real data. The figure shows the histogram of run lengths from scalar quantized high-frequency wavelet subbands of the 512 x 512 "Lena" image. Also shown on the figure is the pdf matching the exp-Golomb code (each circle on the figure corresponds to a codeword) with $k = 1$ as well as a GG with shape parameter $\nu = .5$ and variance chosen to fit the exp-

Golomb code. It is clear that both the GG and the exp-Golomb code are better suited to the data than the best Golomb-Rice code, which has codewords as indicated by the "x"s in the figure. The goodness of the fit of the exp-Golomb pdf to the GG pdf is also noteworthy.

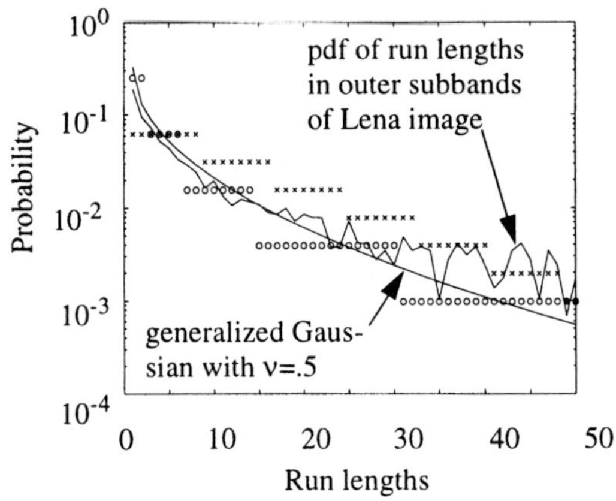

FIGURE 5. pdf matched to k=1 exp-Golomb tree (circles), and pdf matched to the Golomb-Rice tree of k=3 ("x"s), shown with generalized Gaussian pdf, and histogram of run lengths from image data. Both the exp-Golomb tree and the GG pdf match the data more closely than a Golomb-Rice tree.

3.2 Code Efficiency for Ideal Sources

In associating a discrete source with a GG, we use a uniform scalar quantizer having a step size of δ and a deadzone at the origin of width $(1+\alpha)\delta$, $\alpha \geq 0$, as shown in Figure 6. Such uniform scalar quantizers with a deadzone are common in many coding systems. Figure 6 also illustrates three mappings of the quantizer outputs that produce, respectively, positive, non-negative, and two-sided non-zero discrete sources. A full characterization of the discrete source is enabled by specifying the GG shape parameter ν, quantizer step size (normalized by standard deviation) δ/σ, deadzone parameter α, and by identifying which of the three mappings from Figure 6 is used. Positive, non-negative, and two-sided nonzero distributions are all important in run-length coded image data. The run values are restricted to the non-negative integers, or, if runs of 0 are not encoded, to the positive integers. The levels are two-sided, nonzero integers.

We next consider the code efficiency $h/(R+h)$ where h is the source entropy and R is the redundancy. Figure 7 plots the efficency of exp-Golomb codes (with $k=1$) for coding of positive discrete sources. The actual redundancy as opposed to the bounds was used for this and all other efficiency calculations presented here. The horizontal axis gives the normalized quantizer step size. The efficiency for shape parameters of $\nu = .3, .5, .7$, and $.9$ are shown. As might be expected, the efficiency of the codes is better for lower shape parameters, since as $\nu \to 1$ Golomb-Rice codes

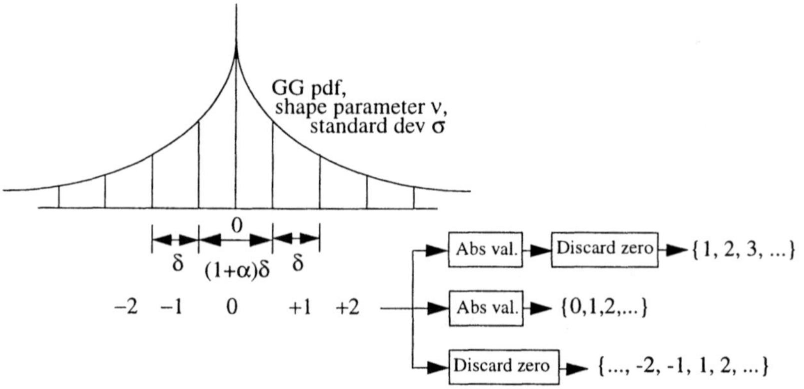

FIGURE 6. Mapping of continuous GG pdf into positive discrete, nonnegative discrete, and two-sided nonzero discrete distributions. The GG is parameterized by ν and σ. The quantizer is uniform with step size δ and a deadzone of width $(1+\alpha)\delta$ at the origin.

often become the optimal codes and therefore become better than exp-Golomb codes.

Figure 8 shows the efficiency that can be achieved when proper choice of code (Golomb-Rice or exp-Golomb) and code parameter k are made. Figure 8(a) shows the efficiency of exp-Golomb codes and Golomb-Rice codes for positive sources for a shape parameter of $\nu = .3$; Figure 8(b) considers $\nu = .7$. The Golomb-Rice codes, which are optimal for exponential distributions ($\nu = 1$) are better at $\nu = .7$ than at $\nu = .3$. Also, in Figure 8(a) it is evident that exp-Golomb codes are more efficient for smaller step sizes while Golomb-Rice codes are better for larger step sizes. In Figure 8(b), Golomb-Rice codes outperform exp-Golomb codes for all step sizes.

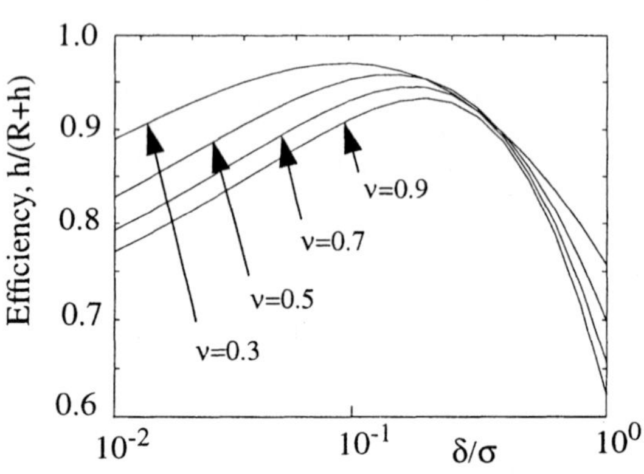

FIGURE 7. Efficiency of k=1 exp-Golomb codes for positive discrete sources derived from a GG, shown as a function of the ratio of quantizer step size δ to standard deviation σ. The quantizer deadzone parameter is $\alpha = .5$

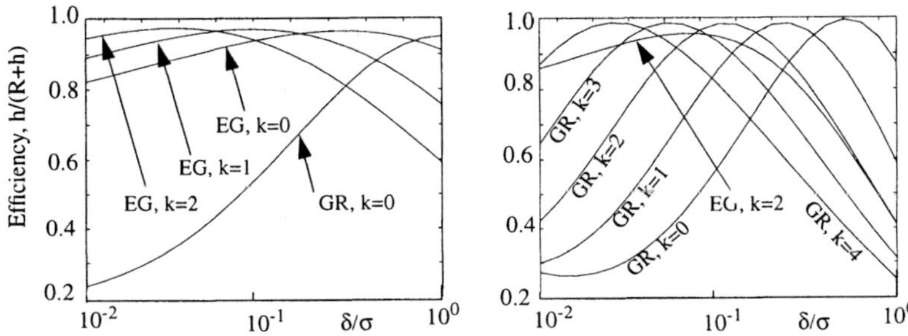

FIGURE 8. Efficiency of Golomb-Rice and exp-Golomb codes for positive discrete sources derived from a GG, showing effects of code choice and code parameter k. (a) shows case for $\nu = .3$. Curves for $\nu = .7$ are given in (b). The quantizer deadzone parameter is $\alpha = .5$.

Among the exp-Golomb codes, codes with a larger k perform better for smaller step sizes. This occurs because as the step size is lowered, the discrete pdf of the source produced at the quantizer output decays more slowly. This results in a better match with exp-Golomb codes with larger k, which have more codewords of shortest length. Efficiency plots for non-negative and two-sided sources are similar in character to Figures 8(a) and 8(b) so are not shown here.

Code efficiency results are summarized in Table 12.1. The table lists the best code choice (exp-Golomb or Golomb-Rice) code parameter k, and corresponding efficiency for a variety of combinations of ν, α, and δ/σ. The deadzone parameter used to generate the table is $\alpha = .5$. The table considers positive, non-negative and two-sided nonzero discrete sources. For the positive and nonnegative sources, optimal (i.e. associating shorter codewords with more probable integers) mapping of integers to tree nodes is used. When the code parameter k of the optimal tree for the positive discrete source is larger than 0, the code parameter for the nonzero discrete source with the same combanition of $(\nu, \alpha, \delta/\sigma)$ is $k + 1$. This shows the optimality (among Golomb-Rice and exp-Golomb codes) when coding two-sided nonzero sources of labeling tree nodes according to absolute value, as with the positive source, and then conveying sign information by adding a bit to each codeword. When $k = 0$ for the positive source, as occurs in several locations in the table, this labeling is not equivalent. When $k = 0$, therefore, it is possible in the optimal labeling for the code for integer m to be of different length than the code for $-m$. The table shows that despite the simplicity of exp-Golomb and Golomb-Rice codes, when a proper code choice is made, it is possible to code quantized GG sources having $\nu < 1$ with efficiencies that typically exceed 90% and in many cases exceed 95%.

4 Application to image coding

There are many different ways to use the coding trees discussed above in image coding. In run-length coded data, it is necessary to ensure that sign information for levels is represented, and that runs are distinguished from levels. Sign information can be most simply (and as noted above, often optimally) handled by adding an

ν	δ/σ	Non-negative Source		Positive Source		Non-zero Source	
		Code, k	Eff.	Code, k	Eff.	Code, k	Eff.
0.3	0.03	EG, 2	.969	EG, 2	.974	EG, 3	.978
	0.1	EG, 0	.972	EG, 1	.968	EG, 2	.975
	0.3	EG, 0	.916	EG, 0	.963	EG, 1	.973
	1.0	GR, 0	.633	GR, 0	.944	GR, 1	.964
0.5	0.03	EG, 3	.971	EG, 3	.971	EG, 4	.975
	0.1	GR, 3	.962	GR, 2	.964	GR, 3	.971
	0.3	GR, 0	.939	GR, 1	.954	GR, 2	.966
	1.0	GR, 0	.801	GR, 0	.921	GR, 1	.952
0.7	0.03	GR, 4	.986	GR, 4	.986	GR, 5	.988
	0.1	GR, 2	.986	GR, 2	.983	GR, 3	.986
	0.3	GR, 1	.975	GR, 1	.970	GR, 2	.978
	1.0	GR, 0	.863	GR, 0	.874	GR, 0	.956
0.9	0.03	GR, 4	.994	GR, 4	.994	GR, 5	.995
	0.1	GR, 2	.989	GR, 2	.988	GR, 3	.991
	0.3	GR, 1	.988	GR, 1	.977	GR, 2	.983
	1.0	GR, 0	.885	GR, 0	.824	GR, 0	.961

TABLE 12.1. Best code choice (among Golomb-rice and exp-Golomb codes) and corresponding code efficiency for quantized GG sources. The quantizer is uniform with deadzone parameter $\alpha = .5$. for each GG shape parameter ν and normalized step size δ/σ, the table gives the best code (GR or EG) and code parameter k, and code efficiency for the resulting discrete source of non-negative integers, the discrete source of positive (one-sided) integers and the discrete source of two-sided non-zero integers.

extra bit after each level that indicates sign. To solve the second problem, we note that there is always a level after a run, but a level can be followed by either a level or a run. This can be handled using a one-bit flag after the codeword for each level that tells the decoder whether the next codeword represents a run or level. This is equivalent to coding zero run-lengths with one bit and adding a bit to the representations of all other run-lengths.

The stack-run coding algorithm described in [TVC96] is one example of an algorithm that indirectly exploits the structured trees described above. Stack-run coding utilizes a 4-ary symbol alphabet to efficiently map runs and levels into a symbol stream that is further compressed using adaptive arithmetic coding. This combines the advantages of the exp-Golomb tree structure with an arithmetic coding framework. The symbol alphabet consists of four symbols, denoted here by "+", "−", "0" and "1", and which have the following meanings: "0" is used to signify a binary bit value of 0 in encoding of levels. "1" is used for binary 1 in levels, but it is not used for the most significant bit (MSB). "+" is used to represent the MSB of a significant positive coefficient; in addition "+" is used for binary 1 in representing run lengths. "−" is used for the MSB of negative significant coefficients, and for binary 0 in representing run lengths.

The symbols take on meanings which are context dependent. For example, in representing level values, the symbols "+" and "−" are used simultaneously to terminate the word (e.g. the sequence of symbols) describing the level, to encode the sign of the level, and to identify the bit plane location of the MSB. The run length representations are terminated by the presence of a "0" or "1", which also conveys the value of the LSB of the next level. An additional gain in efficiency is obtained in representation of the run lengths, which are ordered from LSB to MSB, and represented using the symbols "+" and "−". Since all binary run lengths start with 1, one can omit the final (e.g. MSB) "+" from most run-length representations without loss of information. The only potential problem occurs in representation of a run of length one which would not be expressible if all MSB "+" symbols were eliminated. To circumvent this, it is necessary to retain the MSB "+" symbol for all runs of length $2^k - 1$, where k is an integer. No explicit "end- of-subband" symbol is needed since the decoder can track the locations of coefficients as the decoding progresses. Also, because the symbols "+" and "−" are used to code both the runs and MSB of the levels, a level of magnitude one, which would be represented only by the "+" or "−" that is its MSB, would be indistinguishable from a run. This is handled most easily by simply incrementing by one the absolute value of all levels prior to performing the symbol mapping.

A more careful analysis of the above procedure shows that there are two factors that lead to efficient coding. The first is the use of a mapping that is equivalent to the exp-Golomb tree, which is therefore well matched to the statistics of run-length coded wavelet subbands. A feature of the stack-run symbol mapping is that the appearance of a symbol from the subalphabet ("+","−") can terminate a word described using the subalphabet ("0","1"), or vice versa. The mapping is not a prefix code, although an equivalent prefix code (i.e. the exp-Golomb code) exists. The second factor is the expression of this mapping via a small symbol set which can be efficiently compressed using an adaptive arithmetic coder. There are, in effect, two cascaded entropy coding steps used in stack-run coding – the mapping into symbols, and the further compression of this symbol stream using adaptive arithmetic coding. The arithmetic coding helps to reduce the penalty incurred by the mismatch between the histogram of the subband data and the ideal source pdf corresponding to the symbol mapping.

From a coding theory standpoint, there is no advantage to using the 4-ary stack-run mapping in place of the equivalent binary prefix code. However, there are several practical differences. If the binary prefix code is used, a larger number of contexts will be needed relative to the stack-run mapping. Signed integers can be expressed with 1 less bit if the prefix code is used, but the prefix code will require an additional bit (and corresponding context) to identify the class (e.g. run or level) to which the next word belongs. In addition, the stack-run is parsed differently, with each symbol (e.g. pair of bits) corresponding to one bit in the value being represented. This contrasts with the log/remainder representation that can be used to create the prefix code.

In optimizing the entropy coding step for image data, it is also possible to build hybrid code trees that combine the structures of Golomb-Rice and exp-Golomb codes. Figure 6(c) shows an example of one tree which is particularly well suited to image data. This tree is a exp-Golomb code with an extra "level" added at the top. This alters the initial slope of the pdf to which it is optimized, making it more closely match commonly encountered image data.

5 Image coding results

Table 12.2 gives the PSNR values for the 512 by 512 grayscale "Barbara" and "Lena" images obtained using several of the approaches described above. The filter

Algorithm and filter	Ref	Barbara (dB)			Lena (dB)		
		DWT	.25 bpp	.50 bpp	DWT	.25 bpp	.50 bpp
Structured Tree, 9/7 filter	[WMV96]	A	27.20	30.89	A	33.45	36.60
Stack-run with arithmetic coder, 10/18 filter	[TVC96]	B	28.90	32.66	B	33.87	36.93
Stack-run with arithmetic coder, 9/7 filter	[TVC96]	A	27.32	30.94	A	33.58	36.67
EZW, 9-tap QMF filter	[Shapiro93]	C	26.77	30.53	C	33.17	36.28
Space-Freq Tree , 9/7 filter	[KRO96]	B	29.37	33.13	D	34.35	37.41
EQ	[LRO97]	E	n/a	n/a	E	34.57	37.69
PC-AUTO	[YOY97]	F	n/a	n/a	F	34.46	37.56
Said and Pearlman, 9/7 filter with arithmetic coder	[SP96]	A	27.40	31.25	A	34.11	37.21
Said and Pearlman, 9/7 filter without arithmetic coder	[SP96]	A	27.00	30.70	A	33.68	36.82

TABLE 12.2. Summary of coding results in PSNR. A: 5-level dyadic decomposition. B: 3-level uniform followed by 3-level dyadic of reference signal (73 total subbands). C: 6-level dyadic. D: adaptive. E: 10/18 filter, 4 level dyadic. F: 9/7 filter, 3 level dyadic.

bank and decomposition are as noted in the table. The table also includes comparisons with results for other algorithms. The first is the original zerotree coder introduced by Shapiro [Sha93]. Results from the improvement of zerotree proposed by Said and Pearlman [SP96] also given. This implementation is lower in complexity than the original zerotree algorithm, and gives substantially higher PSNR numbers. In addition, the table includes the results from the adaptive space frequency decomposition described by Xiong *et al.* [XRO97], the adaptive algorithm of Joshi *et al.* [JJKFFMB95], and the recently proposed algorithm by LoPresto *et al.* [LRO97].

The basic message communicated by this table is that when coding efficiency is at a premium, the best approach is to perform a decomposition and/or quantization that is adaptive to the characteristics of the image. As the results from

[JJKFFMB95], [XRO97], and [LRO97] show, this typically gives between 1 and 2 dB of PSNR gain over implementations that are not adaptive. On the other hand, when complexity is a major constraint, an effective organization of the data, as used in the work of Said and Pearlman [SP96] and in the results from this chapter, can lead to simple implementations at a surprisingly low penalty in PSNR. When the structured tree is used with a five-level decomposition (line 1 of the table), the performance is about the same as the Said and Pealrman algorithm without arithmetic coding.

We give some running time information here, though we hasten to add that while operations such as multiplies are relatively easy to tally, the costs of decisions and of different data and memory structures vary strongly with implementation and computation platform. On a Sparc 20 workstation, the CPU time needed for the quantization, run-length and entropy coding of the quantized Lena image wavelet coefficients using the structured code tree (line 1 of Table 12.2) is about 0.12 second at 0.25 bpp and 0.16 second at 0.50 bpp. Note that these times do not include the wavelet transform, which needs to be performed for all of the algorithms in the table. Our implementation has not been optimized for speed, so these numbers should be interpreted with a great deal of caution. It seems safe to conclude that coding using structured trees (or related approaches such as stack-run without arithmentic coding) is of similar complexity to the Said and Pearlman algorithm, and certainly much less complex than many of the other algorithms in the table, which are highly adaptive.

Also, while the embedded coding enabled by zerotree-like approaches is a significant advantage, we believe that the coding performance improvement enabled by exploiting intersubband dependencies is not as great as has often been assumed. This is borne out by comparing our results to the zerotree results of Shapiro and of Said and Pearlman, and also by the high performance of the model based coder (which is not a zerotree algorithm) of LoPresto *et al.*. Certainly, it will always help to exploit any intersubband dependencies that do exist. However, for many images the coding gains that this enables may be smaller than the gains allowed by other techniques such as good local context modeling and efficient representation of run-length coded data.

6 Conclusions

The combination of scalar quantization, run-length, and entropy coding offers a simple and effective way to obtain wavelet image coding results that are within one or two dB in PSNR of the best known algorithms at low bit rates. Run-length coded wavelet subbands have statistics that can be well matched by a family of structured entropy coding trees known as exp-Golomb trees which were first proposed (for other purposes) in [Teuhola78]. Further enhancements to the coding performance can be obtained if a hybrid tree is formed which combines characteristics of a Golomb-Rice code for small integers, and an exp-Golomb code for larger integers.

At the cost of some complexity increase, it is possible to map the entropy-coded representation of the runs and levels into a small symbol alphabet which can then be further compressed using adaptive arithmetic coding. This decreases the efficiency penalty due to mismatch between the data and the pdf for which the coding trees are optimized. One such mapping known as stack-run coding which uses a 4-ary symbol alphabet was described in [TVC96]. Other equivalent mappings that involve different alphabet sizes and numbers of contexts also exist.

The approaches described here are examples of algorithms in which the priority has been placed on minimizing the arithmetic and addressing complexity involved in coding a wavelet transformed image. Algorithms in this class are quite important for the many applications of image coding in which resources for computation and memory access are at a premium. However, there are several disadvantages to the coding framework that should be mentioned. First, the goals of low addressing complexity and embedded coding are difficult to achieve simultaneously. Because we have chosen a raster scan, intrasubband format to minimize the burden on memory accesses, the resulting algorithms are not fully embedded. Some scalability is inherent in the granularity offered by the subband structure itself, though in contrast with zerotree approaches, it is difficult to obtain rate control to single-bit precision. Nevertheless, by proper processing during the wavelet transform and run-length coding stages, we believe it is possible to design relatively simple, single pass rate control algorithms that would deliver a bit rate within 5 or 10% of the target bit rate.

7 References

[VH92] M. Vetterli and C. Herley. "Wavelets and filter banks: Theory and design," *IEEE Trans. on Signal Proc.*, vol. 40, pp. 2207-2232, 1992.

[Shapiro93] J. Shapiro, "Embedded Image Coding Using Zerotrees of Wavelet Coefficients", *IEEE Trans. on Signal Proc.*, vol. 41, no. 12, pp. 3445-3462, December, 1993.

[SP96] A. Said and W.A. Pearlman "A new, fast, and efficient image codec based on set partitioning in hierarchical trees," *IEEE Trans. on Circuits and Systems for Video Technology, June 1996.*

[XRO96] Z. Xiong, K. Ramchandran and M.T. Orchard, "Wavelet packets image coding using space-frequency quantization," submitted to *IEEE Trans. on Image Proc.*, January 1996.

[JJKFFMB95] R.L. Joshi, H. Jafarkhani, J.H. Kasner, T.R. Fischer, N. Farvardin, M.W. Marcellin, and R.H. Bamberger, "Comparison of different methods of classification in subband coding of images," submitted to *IEEE Trans. on Image Proc.*, 1995.

[LRO97] S.M. LoPresto, K. Ramchandran, and M.T. Orchard, "Image Coding basedon Mixture Modeling of Wavelet Coefficients and a Fast Estimation-Quantization Framework," *Proc. of the 1997 IEEE Data Compression Conference*, Snowbird, UT, pp.221-230, March, 1997.

[VED96] J. Villasenor, R.A. Ergas, and P.L. Donoho, "Seismic data compression using high dimensional wavelet transforms," *Proc. of the 1996 IEEE Data Compression Conference*, Snowbird, UT, pp. 396-405, March, 1996.

[CR97] M. Crouse and K. Ramchandran, "Joint thresholding and quantizer selection for transform image coding: Entropy-constrained analysis and applications to baseline JPEG," *IEEE Trans. on Image Proc.*, vol. 6, pp. 285-297, Feb. 1997.

[WSS96] M.J. Weinberger, G. Seroussi, and G. Sapiro, "LOCO-I: A low complexity, context-based lossless image compression algorithm," *Proc. of the 1996 IEEE Data Compression Conference,*, Snowbird, UT, pp. 140-149, April, 1996.

[WM97] X. Wu, N. Memon, "Context-based, adaptive, lossless image coding", *IEEE Trans. on Communications.*, vol. 45, No.4, Arpil 1997.

[ABMD92] M. Antonini, M. Barlaud, P. Mathieu, and I. Daubechies, "Image coding using wavelet transform," *IEEE Trans. on Image Proc.*, vol. 1, pp. 205-220, 1992.

[TVC96] M.J. Tsai, J. Villasenor, and F. Chen, "Stack-run Image Coding," *IEEE Trans. on Circuits and Systems for Video Technology,* vol. 6, pp. 519-521, Oct. 1996.

[LF93] R. Laroia and N. Farvardin, "A structured fixed-rate vector quantizer derived from a variable length scalar quantizer – Part I: Memoryless sources," *IEEE Trans. on Inform. Theory,* vol. 39, pp. 851-867, May 1993.

[CGMR97] G. Calvagno, C. Ghirardi, G.A. Mian, and R. Rinaldo, "Modeling of subband image data for buffer control," *IEEE Trans. on Circuits and Systems for Video Technology,* vol.7, pp.402-408, April 1997.

[BF95] K.A. Birney and T.R. Fischer, "On the modeling of DCT and subband image data for compression," *IEEE Trans. on Image Proc.*, vol. 4, pp. 186-193, February 1995.

[Golomb66] S.W. Golomb, "Run-length encodings," *IEEE Trans. on Inf. Theory,* vol. IT-12, pp. 399-401, July 1966.

[GV75] R.G. Gallager and D.C. Van Voorhis, "Optimal source codes for geometrically distributed integer alphabets," *IEEE Trans. on Inf. Theory,* vol.21, pp.228-230, March 1975.

[Rice 79] R.F. Rice, "Some practical universal noiseless coding techniques," Tech. Rep. JPL-79-22, Jet Propulsion Laboratory, Pasadena, CA, March 1979.

[Teuhola78] J. Teuhola, "A Compression Method for Clustered Bit-Vectors," *Information Processing Letters*, vol. 7, pp. 308-311, October 1978.

[Elias75] P. Elias, "Universal codeword sets and representations of the integers," *IEEE Trans. on Inf. Theory*, Vol. 21, pp. 194-203, March 1975.

Part III

Special Topics in Still Image Coding

13

Fractal Image Coding as Cross-Scale Wavelet Coefficient Prediction

Geoffrey M. Davis

1 Introduction

Fractal image compression techniques, introduced by Barnsley and Jacquin [3], are the product of the study of iterated function systems (IFS)[2]. These techniques involve an approach to compression quite different from standard transform coder-based methods. Transform coders model images in a very simple fashion, namely, as vectors drawn from a wide-sense stationary random process. They store images as quantized transform coefficients. Fractal block coders, as described by Jacquin, assume that "image redundancy can be efficiently exploited through *self-transformability* on a blockwise basis" [16]. They store images as contraction maps of which the images are approximate fixed points. Images are decoded via iterative application of these maps.

In this chapter we examine the connection between the fractal block coders as introduced by Jacquin [16] and transform coders. We show that fractal coding is a form of wavelet subtree quantization. The basis used by the Jacquin-style block coders is the Haar basis. Our wavelet based analysis framework leads to improved understanding of the behavior of fractal schemes. We describe a simple generalization of fractal block coders that yields a substantial improvement in performance, giving results comparable to the best reported for fractal schemes. Finally, our wavelet framework reveals some of the limitations of current fractal coders.

The term "fractal image coding" is defined rather loosely and has been used to describe a wide variety of algorithms. Throughout this chapter, when we discuss fractal image coding, we will be referring to the block-based coders of the form described in [16] and [11].

2 Fractal Block Coders

In this section we describe a generic fractal block coding scheme based on those in [16][10], and we provide some heuristic motivation for the scheme. A more complete overview of fractal coding techniques can be found in [9].

2.1 Motivation for Fractal Coding

Transform coders are designed to take advantage of very simple structure in images, namely that values of pixels that are close together are correlated. Fractal compression is motivated by the observation that important image features, including straight edges and constant regions, are invariant under rescaling. Constant gradients are covariant under rescaling, i.e. rescaling changes the gradient by a constant factor. Scale invariance (and covariance) presents an additional type of structure for an image coder to exploit.

Fractal compression takes advantage of this local scale invariance by using coarse-scale image features to quantize fine-scale features. Fractal block coders perform a vector quantization (VQ) of image blocks. The vector codebook is constructed from locally averaged and subsampled isometries of larger blocks from the image. This codebook is effective for coding constant regions and straight edges due to the scale invariance of these features. The vector quantization is done in such a way that it determines a contraction map from the plane to itself of which the image to be coded is an approximate fixed point. Images are stored by saving the parameters of this map and are decoded by iteratively applying the map to find its fixed point. An advantage of fractal block coding over VQ is that it does not require separate storage of a fixed vector codebook.

The ability of fractal block coders to represent straight edges, constant regions, and constant gradients efficiently is important, as transform coders fail to take advantage of these types of spatial structures. Indeed, recent wavelet transform based techniques that have achieved particularly good compression results have done so by augmenting scalar quantization of transform coefficients with a zerotree vector that is used to efficiently encode locally constant regions.

For fractal block coders to be effective, images must be composed of features at fine scales that are also present at coarser scales up to a rigid motion and an affine transform of intensities. This is the "self-transformability" assumption described by [16]. It is clear that this assumption holds for images composed of isolated straight lines and constant regions, since these features are self-similar. That it should hold when more complex features are present is much less obvious. In Section 4 we use a simple texture model and our wavelet framework to provide a more detailed characterization of "self-transformable" images.

2.2 Mechanics of Fractal Block Coding

We now describe a simple fractal block coding scheme based on those in [16][10]. For convenience we will focus on systems based on dyadic block scalings, but we note that other scalings are possible. Let $\mathcal{I}$ be a $2^N \times 2^N$ pixel grayscale image. Let $\mathbf{B}^J_{K,L}$ be the linear "get-block" operator which when applied to $\mathcal{I}$ extracts the $2^J \times 2^J$ subblock with lower left corner at (K, L). The adjoint of this operator, $(\mathbf{B}^J_{K,L})^*$, is a "put-block" operator that inserts a $2^J \times 2^J$ image block into a $2^N \times 2^N$ all-zero image so that the lower left corner of the inserted block is at (K, L). We will use capital letters to denote block coordinates and lower case to denote individual pixel coordinates. We use a capital Greek multi-index, usually Γ, to abbreviate the block coordinates K, L and a lower-case Greek multi-index to abbreviate pixel coordinates within blocks.

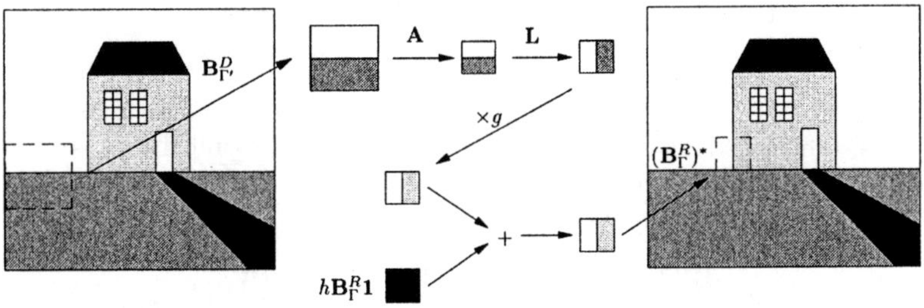

FIGURE 1. We quantize the small range block $\mathbf{B}_{\Gamma}^{R}\mathcal{I}$ on the right using the codebook vector $g\mathbf{LAB}_{\Gamma'}^{D}\mathcal{I} + h\mathbf{B}_{\Gamma}^{R}\mathbf{1}$ obtained from the larger domain block on the left. $\mathbf{A}$ averages and subsamples the block, $\mathbf{L}$ rotates it, multiplication by the gain g modifies the contrast, and the addition of the offset $h\mathbf{B}_{\Gamma}^{R}\mathbf{1}$ adjusts the block DC component.

We partition $\mathcal{I}$ into a set of non-overlapping $2^R \times 2^R$ *range blocks.* The goal of the compression scheme is to approximate each range block with a block from a codebook constructed from a set of $2^D \times 2^D$ *domain blocks*, where $0 < R < D \leq N$. Forming this approximation entails the construction of a contraction map from the image to itself, i.e. from the domain blocks to the range blocks, of which the image is an approximate fixed point. We store the image by storing the parameters of this map, and we recover the image by iterating the map to its fixed point. Iterated function system theory motivates this general approach to storing images, but gives little guidance on questions of implementation. The basic form of the block coder described below is the result of considerable empirical work. In Section 3 we see that this block-based coder arises naturally in a wavelet framework, and in Section 4 we obtain greatly improved coder performance by generalizing these block-based maps to wavelet subtree-based maps.

The range block partition is a disjoint partition of the image consisting of the blocks $\{\mathbf{B}_{K,L}^{R}\mathcal{I} | (K,L) \in \mathcal{R}\}$. Here $\mathcal{R} = \{(2^R m, 2^R n) | 0 \leq m,n < 2^{N-R}\}$. The domain blocks from which the codebook is constructed are drawn from the *domain pool*, the set $\{\mathbf{B}_{K,L}^{D}\mathcal{I} : (K,L) \in \mathcal{D}\}$. A variety of domain pools are used in the literature. A commonly used pool [16] is the set of all unit translates of $2^D \times 2^D$ blocks, $\mathcal{D} = \{(m,n) | 0 \leq m,n < 2^N - 2^D\}$. Some alternative domain pools that we will discuss further are the *disjoint domain pool*, $\mathcal{D} = \{(2^D m, 2^D n) | 0 \leq m,n < 2^{N-D}\}$, a disjoint tiling of $\mathcal{I}$, and the *half-overlapping domain pool*, $\mathcal{D} = \{(2^{D-1}m, 2^{D-1}n) | 0 \leq m,n < 2^{N-D+1}\}$, the union of four disjoint partitions shifted by a half block length in the x or y directions (we periodize the image at its boundaries).

Two basic operators are used for codebook construction. The "average-and-subsample" operator $\mathbf{A}$ maps a $2^J \times 2^J$ image block to a $2^{J-1} \times 2^{J-1}$ block by averaging each pixel in $\mathbf{B}_{\Gamma}^{J}$ with its neighbors and then subsampling. We define $(\mathbf{AB}_{\Gamma}^{J}\mathcal{I})(k,l) = \frac{1}{4}[(\mathbf{B}_{\Gamma}^{J}\mathcal{I})(2k,2l) + (\mathbf{B}_{\Gamma}^{J}\mathcal{I})(2k+1,2l) + (\mathbf{B}_{\Gamma}^{J}\mathcal{I})(2k,2l+1) + (\mathbf{B}_{\Gamma}^{J}\mathcal{I})(2k+1,2l+1)]$ where $\mathbf{B}_{\Gamma}^{J}\mathcal{I}(k,l)$ is the pixel at coordinates (k,l) within the subblock $\mathbf{B}_{\Gamma}^{J}\mathcal{I}$. A second operator is the symmetry operator $\mathbf{L}_k$, $1 \leq k \leq 8$, which maps a square block to one of the 8 isometries obtained from compositions of reflections and 90 degree rotations.

Range block approximation is similar to shape-gain vector quantization[14]. Range blocks are quantized to a linear combination of an element from the codebook and a constant block. The codebook used for quantizing range blocks consists of averaged and subsampled isometries of domain blocks, the set $\mathcal{C} = \{\mathbf{L}_k \mathbf{A}^{D-R} \mathbf{B}^D_\Gamma \mathcal{I} : \Gamma \in \mathcal{D}, 0 \leq k \leq 8\}$. Here $\mathbf{A}^{D-R}$ denotes the operator $\mathbf{A}$ applied $D-R$ times. The contrast of the codewords in $\mathcal{C}$ is adjusted by a gain factor g, and the DC component is adjusted by adding a subblock of the $2^N \times 2^N$ matrix of ones, $\mathbf{1}$, multiplied by an offset factor h. For each range block $\mathbf{B}^R_\Gamma \mathcal{I}$ we have

$$\mathbf{B}^R_\Gamma \mathcal{I} \approx g_\Gamma \mathbf{L}_{P(\Gamma)} \mathbf{A}^{D-R} \mathbf{B}^D_{\Pi(\Gamma)} \mathcal{I} + h_\Gamma \mathbf{B}^R_\Gamma \mathbf{1}. \tag{13.1}$$

Here $\Pi : \mathcal{R} \to \mathcal{D}$ assigns an element from the domain pool to each range element and $P : \mathcal{R} \to \{1 \ldots 8\}$ assigns each range element a symmetry operator index. Ideally the parameters g, h, Π, and P should be chosen so that they minimize the error in the decoded image. The quantization process is complicated by the fact that the codebook used by the decoder is different from that used by the encoder, since the decoder doesn't have access to the original domain blocks. Hence errors made in quantizing range blocks are compounded because they affect the decoder codebook. These additional effects of quantization errors have proven difficult to estimate, so in practice g, h, Π, and P are chosen to minimize the l^2 quantization error. This tactic gives good results in practice.

2.3 Decoding Fractal Coded Images

The approximations for the range blocks (13.1) determine a constraint on the image $\mathcal{I}$ of the form $\mathcal{I} \approx \mathbf{G}\mathcal{I} + \mathcal{H}$. Expanding $\mathcal{I}$ as a sum of range blocks we obtain

$$\begin{aligned} \mathcal{I} &= \sum_{\Gamma \in \mathcal{R}} (\mathbf{B}^R_\Gamma)^* \mathbf{B}^R_\Gamma \mathcal{I} \\ &\approx \sum_{\Gamma \in \mathcal{R}} g_\Gamma (\mathbf{B}^R_\Gamma)^* \mathbf{L}_{P(\Gamma)} \mathbf{A}^{D-R} \mathbf{B}^D_{\Pi(\Gamma)} \mathcal{I} \\ &\quad + \sum_{\Gamma \in \mathcal{R}} h_\Gamma (\mathbf{B}^R_\Gamma)^* \mathbf{B}^R_\Gamma \mathbf{1} \\ &= \mathbf{G}\mathcal{I} + \mathcal{H}. \end{aligned}$$

Provided the matrix $\mathbf{I} - \mathbf{G}$ is nonsingular, there is a unique fixed point solution $\mathcal{I}_{fp}$ satisfying

$$\mathcal{I}_{fp} = \mathbf{G}\mathcal{I}_{fp} + \mathcal{H} \tag{13.2}$$

given by $\mathcal{I}_{fp} = (\mathbf{I} - \mathbf{G})^{-1}\mathcal{H}$. Because $\mathbf{G}$ is a $2^{2N} \times 2^{2N}$ matrix, inverting $\mathbf{I} - \mathbf{G}$ directly is an inordinately difficult task. If (and only if) the eigenvalues of $\mathbf{G}$ are all less than 1 in magnitude, we can find the fixed point solution $\mathcal{I}_{fp}$ by iteratively applying (13.2) to an arbitrary image $\mathcal{I}_0$. Decoding of fractal coded images proceeds by forming the sequence

$$\mathcal{I}_n = \mathbf{G}\mathcal{I}_{n-1} + \mathcal{H} = \mathbf{G}^n \mathcal{I}_0 + \sum_{k=0}^{n-1} \mathbf{G}^k \mathcal{H}. \tag{13.3}$$

Convergence results for this procedure are discussed in detail in [6].

3 A Wavelet Framework

3.1 Notation

The wavelet transform is a natural tool for analyzing fractal block coders since wavelet bases possess the same type of dyadic self-similarity that fractal coders seek to exploit. In particular, the Haar wavelet basis possesses a regular block structure that is aligned with the range block partition of the image. We show below that the maps generated by fractal block coders reduce to a simple set of equations in the wavelet transform domain.

Separable 2-D biorthogonal wavelet bases consist of translates and dyadic scalings of a set of oriented wavelets $\psi_{LH}(x,y)$, $\psi_{HL}(x,y)$, and $\psi_{HH}(x,y)$ together with translates of a scaling function $\phi(x,y)$. We will use the subscript ω to represent one of the three orientations in $\Omega = \{LH, HL, HH\}$. We will limit our attention to symmetrical (or antisymmetrical) bases. The discrete wavelet transform of a $2^N \times 2^N$ image $\mathcal{I}$ expands the image into a linear combination of the basis functions in the set $\mathcal{W}_J = \{\phi^J_{k,l} | 0 \le k,l < 2^J\} \cup \{\psi^j_{\omega,k,l} | \omega \in \Omega; J \le j < N; 0 \le k,l < 2^j\}$. We will use a single lower-case Greek multi-index, usually γ, to abbreviate the orientation and translation subscripts of ϕ and ψ. The coefficients for the basis functions $\phi^j_{k,l}$ and $\psi^j_{\omega,k,l}$ are given by $\langle\tilde{\phi}^j_{k,l}, \mathcal{I}\rangle$ and $\langle\tilde{\psi}^j_{\omega,k,l}, \mathcal{I}\rangle$, respectively, where $\tilde{\phi}^j_{k,l}$ and $\tilde{\psi}^j_{\omega,k,l}$ are dual scaling functions and wavelets.

An important property of wavelet basis expansions, particularly Haar expansions, is that they preserve the spatial localization of image features. For example, the coefficient of the Haar scaling function $\phi^J_{k,l}$ is proportional to the average value of an image in the $2^J \times 2^J$ block of pixels with lower left corner at $2^J k, 2^J l$. The wavelet coefficients associated with this region are organized into three quadtrees. We call this union of three quadtrees a *wavelet subtree*. Coefficients forming such a subtree are shaded in each of the transforms in Figure 2. At the root of a wavelet subtree are the coefficients of the wavelets $\psi^J_{\omega,k,l}$, where $\omega \in \Omega$. These coefficients correspond to the block's coarse-scale information. Each wavelet coefficient $\langle\tilde{\psi}^j_{\omega,k,l}, \mathcal{I}\rangle$ in the tree has four children that correspond to the same spatial location and the same orientation. The children consist of the coefficients of the wavelets of the next finer scale, $\psi^{j+1}_{\omega,2k,2l}$, $\psi^{j+1}_{\omega,2k+1,2l}$, $\psi^{j+1}_{\omega,2k,2l+1}$, and $\psi^{j+1}_{\omega,2k+1,2l+1}$. A wavelet subtree consists of the coefficients of the roots, together with all of their descendents in all three orientations. The scaling function $\phi^J_{k,l}$ is localized in the same region as the subtree with roots given by $\psi^J_{\omega,k,l}$, and we refer to this $\phi^J_{k,l}$ as the scaling function associated with the subtree.

3.2 A Wavelet Analog of Fractal Block Coding

We now describe a wavelet-based analog of fractal block coding introduced in [5]. Fractal block coders approximate a set of $2^R \times 2^R$ range blocks using a set of $2^D \times 2^D$ domain blocks. The wavelet analog of an image block, a set of pixels associated with a small region in space, is a wavelet subtree together with its associated scaling function coefficient. We define a linear "get-subtree" operator $\mathbf{S}^J_{K,L} : \mathbb{R}^{2^{2N}} \to \mathbb{R}^{2^{2(N-J)}-1}$ which extracts from an image the subtree whose root level consists of the coefficients of $\psi^J_{\omega,K,L}$ for all ω. We emphasize that when we discuss wavelet subtrees in

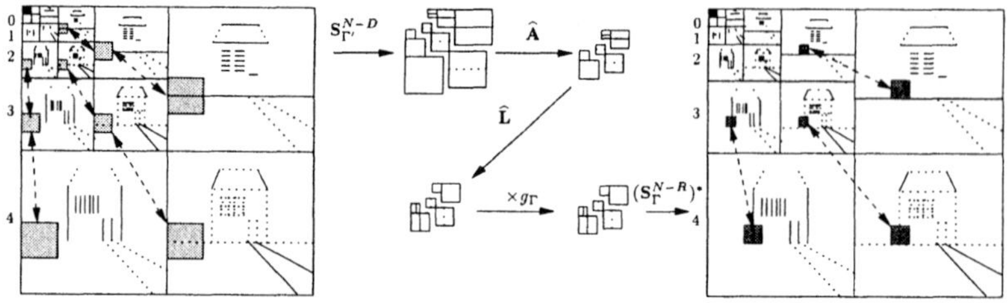

FIGURE 2. We approximate the darkly shaded range subtree $\mathbf{S}_{\Gamma}^{N-R}\mathcal{I}$ using the codebook element $g_{\Gamma}\widehat{\mathbf{L}}\widehat{\mathbf{A}}\mathbf{S}_{\Gamma'}^{N-D}\mathcal{I}$ derived from the lightly shaded domain subtree $\mathbf{S}_{\Gamma'}^{N-D}\mathcal{I}$. $\widehat{\mathbf{A}}$ truncates the finest scale coefficients of the domain subtree and multiplies the coefficients by $\frac{1}{2}$, and $\widehat{\mathbf{L}}$ rotates it. When storing this image we save the coarse-scale wavelet coefficients in subbands of scale 2 and lower, and we save the encodings of all subtrees with roots in scale subband 3. Note that in our usage a subtree contains coefficients from all three orientations, HL, LH, and HH.

this chapter, we will primarily be discussing trees of coefficients of *all 3 orientations* as opposed to more commonly used subtrees of a fixed orientation.

The adjoint of $\mathbf{S}_{K,L}^{J}$ is a "put-subtree" operator which inserts a given subtree into an all-zero image so that the root of the inserted subtree corresponds to the coefficients $\psi_{\omega,K,L}^{J}$. For the Haar basis, subblocks and their corresponding subtrees and associated scaling function coefficients contain identical information, i.e. the transform of a range block $\mathbf{B}_{\Gamma}^{R}\mathcal{I}$ yields the coefficients of subtree $\mathbf{S}_{\Gamma}^{N-R}\mathcal{I}$ and the scaling function coefficient $\langle\tilde{\phi}_{\Gamma}^{N-R},\mathcal{I}\rangle$. For the remainder of this section we will take our wavelet basis to be the Haar basis. The actions of the get-subtree and put-subtree operators are illustrated in Figure 2.

The linear operators used in fractal block coding have simple behavior in the transform domain. We first consider the wavelet analog $\widehat{\mathbf{A}}$ of the average-and-subsample operator $\mathbf{A}$. Averaging and subsampling the finest-scale Haar wavelets sets them to 0. The local averaging has no effect on coarser scale Haar wavelets, and subsampling ψ_{γ}^{j} yields the Haar wavelet at the next finer scale, ψ_{γ}^{j+1}, multiplied by $\frac{1}{2}$. Similarly, averaging and subsampling the scaling function ϕ_{γ}^{j} yields $\frac{1}{2}\phi_{\gamma}^{j+1}$ for $j < N-1$ and 0 for $j = N-1$. The action of the averaging and subsampling operator thus consists of a shifting of coefficients from coarse-scale to fine, a multiplication by $\frac{1}{2}$, and a truncation of the finest-scale coefficients. The operator $\widehat{\mathbf{A}}$ prunes the leaves of a subtree and shifts all remaining coefficients to the next finer scale. The action of $\widehat{\mathbf{A}}$ is illustrated in Figure 2.

For symmetrical wavelets, horizontal/vertical block reflections correspond to a horizontal/vertical reflection of the set of wavelet coefficients within each scale of a subtree. Similarly, 90 degree block rotations correspond to 90 degree rotations of the set of wavelet coefficients within each scale and a switching of the ψ_{LH} coefficients with ψ_{HL} coefficients. Hence the wavelet analogs $\widehat{\mathbf{L}}_k$ of the block symmetry operators $\mathbf{L}_k$ permute wavelet coefficients within each scale. Figure 2 illustrates the action of a symmetry operator on a subtree. Note that the Haar basis is the only orthogonal basis we consider here, since it is the only compactly supported symmetrical wavelet basis. When we generalize to non-Haar bases, we must use biorthogonal bases to obtain both symmetry and compact support.

The approximation (13.1) leads to a similar relation for subtrees in the Haar wavelet transform domain,

$$\mathbf{S}_{\Gamma}^{N-R}\mathcal{I} \approx g_{\Gamma}\widehat{\mathbf{L}}_{P(\Gamma)}\widehat{\mathbf{A}}^{D-R}\mathbf{S}_{\Pi(\Gamma)}^{N-D}\mathcal{I}. \tag{13.4}$$

We refer to this quantization of subtrees using other subtrees as the *self-quantization* of $\mathbf{S}_{\Gamma}^{N-R}\mathcal{I}$. The offset terms h_{Γ} from (13.1) affect only the scaling function coefficients because the left hand side of (13.4) is orthogonal to the subblocks of **1**. Breaking up the subtrees into their constituent wavelet coefficients, we obtain a system of equations for the coefficients of the ψ_{γ}^{j} in $\mathbf{S}_{\Gamma}^{N-R}\mathcal{I}$,

$$\langle\tilde{\psi}_{\gamma}^{j},\mathcal{I}\rangle \approx \frac{g_{\Gamma}}{2^{D-R}}\langle\tilde{\psi}_{\gamma'}^{j-(D-R)},\mathcal{I}\rangle = \frac{g_{\Gamma}}{2^{D-R}}\langle\mathbf{T}\tilde{\psi}_{\gamma}^{j},\mathcal{I}\rangle. \tag{13.5}$$

Here **T** is the map induced by the domain block selection followed by averaging, subsampling, and rotating. We obtain a similar relation for the scaling function coefficients,

$$\begin{aligned}\langle\tilde{\phi}_{\Gamma}^{N-R},\mathcal{I}\rangle &\approx \frac{g_{\Gamma}}{2^{D-R}}\langle\tilde{\phi}_{\Pi(\Gamma)}^{N-D},\mathcal{I}\rangle + h_{\Gamma}\\ &= \frac{g_{\Gamma}}{2^{D-R}}\langle\mathbf{T}\tilde{\phi}_{\Gamma}^{N-R},\mathcal{I}\rangle + h_{\Gamma}\end{aligned} \tag{13.6}$$

From the system (13.5) and (13.6) we see that, roughly speaking, the fractal block quantization process constructs a map from coarse-scale wavelet coefficients to fine. It is important to note that the operator **T** in (13.5) and (13.6) does *not* necessarily map elements of $\mathcal{W}_{N-D}$ to elements of $\mathcal{W}_{N-D}$, since translation of domain blocks by distances smaller than 2^D leads to non-integral translates of the wavelets in their corresponding subtrees.

4 Self-Quantization of Subtrees

4.1 Generalization to non-Haar bases

We obtain a wavelet-based analog of fractal compression by replacing the Haar basis used in (13.5) and (13.6) with a symmetric biorthogonal wavelet basis. This change of basis brings a number of benefits. Bases with a higher number of vanishing moments than the Haar better approximate the K-L basis for the fractional Brownian motion texture model we discuss below, and therefore improve coder performance in these textured regions. Moreover, using smooth wavelet bases eliminates the sharp discontinuities at range block boundaries caused by quantization errors. These artifacts are especially objectionable because the eye is particularly sensitive to horizontal and vertical lines. Figure 4 compares images coded with Haar and smooth spline bases. We see both an increase in overall compressed image fidelity with the spline basis as well as a dramatic reduction in block boundary artifacts.

We store an image $\mathcal{I}$ by storing the parameters in the relations (13.5) and (13.6). We must store one constant h_{Γ} for each scaling function. Image decoding is greatly simplified if we store the scaling function coefficients directly rather than storing the h_Γ's. We then only have to recover the wavelet coefficients when decoding. Also, because we know how quantization errors for the scaling function coefficients will affect the decoded image, we have greater control over the final decoded error than we do with the h_Γ's. We call this modified scheme in which we use (13.4) to quantize wavelet coefficients and we store scaling function coefficients directly the *self-quantization of subtrees* (SQS) scheme.

4.2 Fractal Block Coding of Textures

In section 2.1 we motivated the codebook used by fractal block coders by emphasizing the scale-invariance (covariance) of isolated straight edges, constant regions, and constant gradients. More complex image structures lack this deterministic self-similarity, however. How can we explain fractal block coders' ability to compress images containing complex structures?

We obtain insight into the performance of fractal coders for more complex regions by examining a fractional Brownian motion texture model proposed by Pentland [17]. Flandrin [13] has shown that the wavelet transform coefficients of a fractional Brownian motion process are stationary sequences with a self-similar covariance structure. This means that the codebook constructed from domain subtrees will possess the same second order statistics as the set of range subtrees. Tewfik and Kim [21] point out that for fractional Brownian motion processes with spectral decay like that observed in natural images, wavelet bases with higher numbers of vanishing moments provide much better approximations to the K-L basis than does the Haar basis, so we expect improved performance for these bases. This is borne out in numerical results below.

5 Implementation

5.1 Bit Allocation

Our self-quantization of subtrees scheme possesses a structure similar to that of a number of recently introduced hierarchical wavelet coders. We have two basic methods for quantizing data: we have a set of coarse-scale wavelet coefficients that we quantize using a set of scalar quantizers, and we have a set of range subtrees that we self-quantize using codewords generated from domain subtrees. Given a partition of our data into range subtrees and coarse-scale coefficients, the determination of a near-optimal quantization of each set of data is a straightforward problem. The trick is finding the most effective partition of the data. An elegant and simple algorithm that optimizes the allocation of bits between a set of scalar quantizers and a set of subtree quantizers is described [22]. We use a modified version of this algorithm for our experiments. We refer the reader to [22] and [6] for details.

The source code for our implementation and scripts for generating the data values and images in the figures are available from the web site *http://math.dartmouth.edu/~gdavis/fractal/fractal.html.* The implementation is based on the public domain Wavelet Image Compression Construction Kit, available from *http://math.dartmouth.edu/~gdavis/wavelet/wavelet.html.*

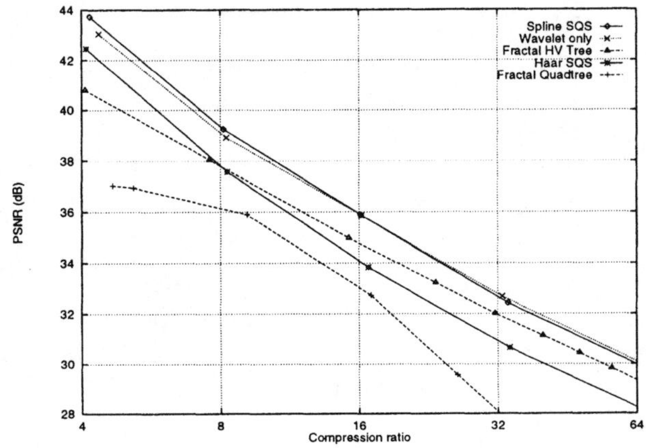

FIGURE 3. PSNR's as a function of compression ratio for the 512 × 512 Lena image using fractal block coding, our self-quantization of subtrees (SQS) scheme, and a baseline wavelet transform coder.

6 Results

6.1 SQS vs. Fractal Block Coders

Figure 3 compares the peak signal to noise ratios of the 512 × 512 Lena image compressed by two fractal block coders, by our self-quantization of subtrees (SQS) scheme, and by a wavelet transform coder. Images compressed at roughly 64:1 by the various methods are shown in Figure 4 to illustrate the artifacts generated by the various methods.

The bottommost line in Figure 3, 'Fractal Quadtree', was produced by the quadtree block coder listed in the appendix of [9]. We used the disjoint domain pool to quantize range blocks with sizes from 4 × 4 to 64 × 64 for this scheme as well as the SQS schemes below.

The next line, 'Haar SQS', was generated by our adaptive SQS scheme using the Haar basis. As we see from the decompressed images in Figure 4, the SQS scheme produces dramatically improved results compared to the quadtree scheme, although both schemes use exactly the same domain pool. Much of this improvement is attributable to the fact that the quadtree coder uses no entropy coding, whereas the SQS coder uses an adaptive arithmetic coder. However, a significant fraction of the bitstream consists of domain block offset indices, for which arithmetic coding is of little help. Much of the gain for SQS is because our improved understanding of how various bits contribute to final image fidelity enables us to partition bits more efficiently between wavelet coefficients and subtrees.

Fisher notes that the performance of quadtree coders is significantly improved by enlarging the domain pool [10]. The third line from the bottom of Figure 3, 'Fractal HV Tree', was produced by a fractal block encoding of rectangular range blocks using rectangular domain blocks [12]. The use of rectangular blocks introduces an additional degree of freedom in the construction of the domain pool and gives increased flexibility to the partitioning of the image. This strategy uses an enormous

FIGURE 4. The top left shows the original 512 × 512 Lena image. The top right image has been compressed at 60.6:1 (PSNR = 24.9 dB) using a disjoint domain pool and the quadtree coder from [9]. The lower left image has been compressed at 68.2:1 (PSNR = 28.0 dB) using our self-quantization of subtrees (SQS) scheme with the Haar basis. Our SQS scheme uses exactly the same domain pool as the quadtree scheme, but our analysis of the SQS scheme enables us to make much more efficient use of bits. The lower right image has been compressed at 65.6:1 (PSNR = 29.9 dB) using a smooth wavelet basis. Blocking artifacts have been completely eliminated.

domain pool. The reconstructed images in [12] show the coding to be of high quality and in fact, the authors claim that their algorithm gives the best results of any fractal block coder in the literature (we note that these results have been since superseded by hybrid transform-based coders such as those of [4] and [18]). The computational requirements for this scheme are quite large due to the size of the domain pool and the increased freedom in partitioning the image. Image encoding times were as high as 46 CPU-hours on a Silicon Graphics Personal IRIS 4D/35. In contrast, the SQS encodings required roughly 90 minutes apiece on a 133 MHz Intel Pentium PC.

The top line in Figure 3, 'Spline SQS', illustrates an alternative method for improving compressed image fidelity: changing bases. As can be seen in Figure

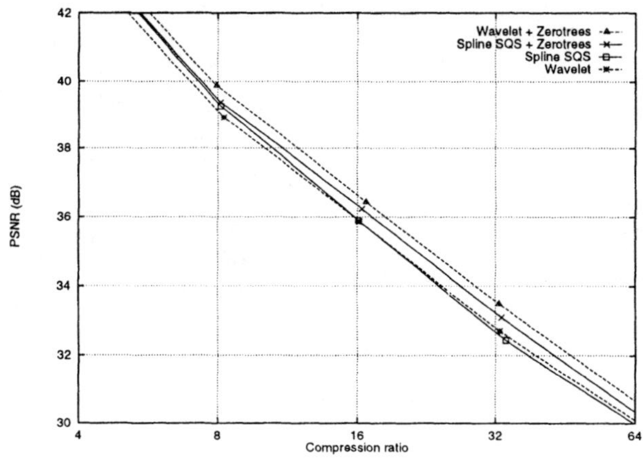

FIGURE 5. Baseline wavelet coder performance vs. self-quantization of subtrees (SQS) with the disjoint domain pool for the 512 × 512 Lena image. PSNR's are shown for both unmodified and zerotree-enhanced versions of these coders. The spline basis of [1] is used in all cases.

4, switching to a smooth basis (in this case the spline variant of [1]) eliminates blocking artifacts. Moreover, our fractional Brownian motion texture model predicts improved performance for bases with additional vanishing moments.

The fourth line in Figure 3 shows the performance of the wavelet transform portion of the SQS coder alone. We see that self-quantization of subtrees yields a modest improvement over the baseline transform coder at high bit rates. This improvement is due largely to the ability of self-quantization to represent zerotrees.

6.2 Zerotrees

Recent work in psychophysics [8] and in image compression [22] shows that energy in natural images is concentrated not only in frequency, but in space as well. Transform coders such as JPEG and our baseline wavelet coder take advantage of this concentration in frequency but not the spatial localization. One consequence of this spatial clustering of high energy, high frequency coefficients is that portions of images not containing edges correspond to wavelet subtrees of low energy coefficients. These subtrees can be efficiently quantized using *zerotrees*, subtrees of all-zero coefficients.

Zerotrees are trivially self-similar, so they can be encoded relatively cheaply via self-quantization. Our experiments reveal that much of fractal coders' effectiveness is due to their ability to effectively represent zerotrees. To see this, we examine the results of incorporating a separate inexpensive zerotree codeword into our codebook. This addition of zerotrees to our codebook results in a modest increase in the performance of our coder. More importantly, it results in a large change in how subtrees are quantized. The first image in Figure 6 shows the range subtrees that are self-quantized when no zerotrees are used. 58% of all coefficients in the image belong to self-quantized subtrees. Self-quantization takes place primarily along locally

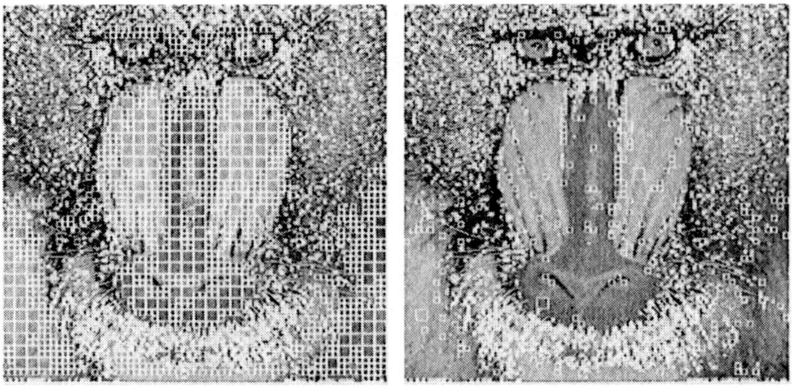

FIGURE 6. The white squares in above images correspond to the self-quantized subtrees used in compressing the 512×512 mandrill image. The squares in the image on the left correspond to the support of the self-quantized subtrees used in a standard self-quantization of subtrees (SQS) scheme with a compression ratio of 8.2:1 (PSNR = 28.0 dB). The squares in the image on the right correspond to the support of the self-quantized subtrees used in a zerotree-enhanced SQS scheme with a compression ratio of 8.4:1 (PSNR = 28.0 dB).

straight edges and locally smooth regions, with some sparse self-quantization in the textured fur. This is consistent with our analysis of the fBm texture model. The second image in Figure 6 shows the self-quantized subtrees in a zerotree-enhanced SQS coder. Only 23% of the coefficients are contained in self-quantized subtrees, and these coefficients are primarily from regions containing locally straight edges. Most of the self-quantized subtrees in the first image can be closely approximated by zerotrees; we obtain similar results for the Barbara and Lena test images.

Adding zerotrees to our baseline wavelet coder leads to a significant performance improvement, as can be seen in Figure 5. In fact, the performance of the wavelet coder with zerotrees is superior to or roughly equivalent to that of the zerotree-enhanced SQS scheme for all images tested. On the whole, once the zerotree codeword is added to the baseline transform wavelet coder, the use of self-quantization actually *diminishes* coder performance.

6.3 Limitations of Fractal Coding

A fundamental weakness of fractal block coders is revealed both in our examination of optimal quantizer density results and in our numerical experiments. The coder possesses no control over the codebook. Codewords are densely clustered around the very common all-zero subtree, but only sparsely distributed in other regions where we need more codebook vectors. This dense clustering of near-zerotrees increases codeword cost but contributes very little to image fidelity.

A number of authors have addressed the problem of codebook inefficiencies by augmenting fractal codebooks. See, for example, the work of [15]. While this codebook supplementation adds codewords in the sparse regions, it does not address the problem of overly dense clustering of code words around zero. At 0.25 bits per pixel, over 80 percent of all coefficients in the 512×512 Lena image are assigned to zerotrees by our zerotree augmented wavelet coder. Hence only about 20 percent of the fractal coder's codewords are significantly different from a zerotree! This redundancy is costly, since when using self-quantization we pay a substantial number of bits to differentiate between these essentially identical zero code words, and we pay roughly 2 bits to distinguish non-zero code words from zero code words. Relatively little attention has been paid to this problem of redundancy. A codebook pruning strategy of Signes [20] is a promising first attempt. Finding a means of effectively addressing these codebook deficiencies is a topic for future research.

Acknowledgments

This work was supported in part by a National Science Foundation Postdoctoral Research Fellowship and by DARPA as administered by the AFOSR under contract DOD F4960-93-1-0567.

7 References

[1] M. Antonini, M. Barlaud, and P. Mathieu. Image Coding Using Wavelet Transform. *IEEE Trans. Image Proc.*, 1(2):205–220, Apr. 1992.

[2] M. F. Barnsley and S. Demko, "Iterated function systems and the global construction of fractals", *Proc. Royal Society of London*, vol. A399, pp. 243–275, 1985.

[3] M. F. Barnsley and A. Jacquin. Application of recurrent iterated function systems to images. *Proc. SPIE*, 1001:122–131, 1988.

[4] K. U. Barthel. Entropy constrained fractal image coding. In Y. Fisher, editor, *Fractal Image Coding and Analysis: a NATO ASI Series Book.* Springer Verlag, New York, 1996.

[5] G. M. Davis, "Self-quantization of wavelet subtrees: a wavelet-based theory of fractal image compression", in *Proc. Data Compression Conference, Snowbird, Utah*, James A. Storer and Martin Cohn, Eds. IEEE Computer Society, Mar. 1995, pp. 232–241.

[6] G. M. Davis. A wavelet-based analysis of fractal image compression. *preprint*, 1996.

[7] D. J. Field. Relations between the statistics of natural images and the response properties of cortical cells. *J. Optical Soc. America A*, 4(12):2379–2394, Dec. 1987.

[8] D. J. Field. What is the goal of sensory coding? *Neural Computation*, 6:559–601, 1994.

[9] Y. Fisher. *Fractal Compression: Theory and Application to Digital Images.* Springer Verlag, New York, 1994.

[10] Y. Fisher. Fractal image compression with quadtrees. In Y. Fisher, editor, *Fractal Compression: Theory and Application to Digital Images.* Springer Verlag, New York, 1994.

[11] Y. Fisher, B. Jacobs, and R. Boss. Fractal image compression using iterated transforms. In J. Storer, editor, *Image and Text Compression*, pages 35–61. Kluwer Academic, 1992.

[12] Y. Fisher and S. Menlove. Fractal encoding with HV partitions. In Y. Fisher, editor, *Fractal Compression: Theory and Application to Digital Images.* Springer Verlag, New York, 1994.

[13] P. Flandrin. Wavelet analysis and synthesis of fractional Brownian motion. *IEEE Transactions on Information Theory*, 38(2):910–917, Mar. 1992.

[14] A. Gersho and R. M. Gray, *Vector Quantization and Signal Compression*, Kluwer Academic, Boston, 1992.

[15] M. Gharavi-Alkhansari and T. Huang. Generalized image coding using fractal-based methods. In *Proc. ICIP*, pages 440–443, 1994.

[16] A. Jacquin. Image coding based on a fractal theory of iterated contractive image transformations. *IEEE Trans. Image Proc.*, 1(1):18–30, Jan. 1992.

[17] A. Pentland. Fractal-based description of natural scenes. *IEEE Transactions on Pattern Analysis and Machine Intelligence*, 6:661–673, 1984.

[18] R. Rinaldo and G. Calvagno. Image coding by block prediction of multiresolution subimages. *IEEE Transactions on Image Processing*, 4(7):909–920, July 1995.

[19] J. Shapiro. Embedded image coding using zerotrees of wavelet coefficients. *IEEE Transactions on Signal Processing*, 41(12):3445–3462, Dec. 1993.

[20] J. Signes. Geometrical interpretation of IFS based image coding. In Y. Fisher, editor, *Fractal Image Coding and Analysis: a NATO ASI Series Book.* Springer Verlag, New York, 1996.

[21] A. H. Tewfik and M. Kim. Correlation structure of the discrete wavelet coefficients of fractional Brownian motion. *IEEE Transactions on Information Theory*, 38(2):904–909, Mar. 1992.

[22] Z. Xiong, K. Ramchandran, and M. T. Orchard. Space-frequency quantization for wavelet image coding. *preprint*, 1995.

14

Region of Interest Compression In Subband Coding

Pankaj N. Topiwala

1 Introduction

We develop a second-generation subband image coding algorithm which exploits the pyramidal structure of transform-domain representations to achieve variable-resolution compression. In particular, regions of interest within an image can be selectively retained at higher fidelity than other regions. This type of coding allows for rapid, very high compression, which is especially suitable for the timely communication of image data over bandlimited channels.

In the past decade, there has been an explosion of interest in the use of wavelets and pyramidal schemes for image compression, beginning with the papers [1], [2], [3]. These and the papers that followed have exploited wavelets and subband coding schemes to obtain lossy image compression with improved quality over DCT-based methods such as JPEG, especially at high compression ratios; see [6] and the proceedings of the Data Compression conferences. Moreover, the pyramidal structure naturally permits other desirable image processing capabilities such as progressive transmission, and browsing. In essence, the lowpass information in the transformed image can serve as an initial "thumbnail" version of the image, with successively higher reconstructions as providing approximations of the full resolution image. The same pyramidal structure can also allow for improved error-correction coding for transmission over noisy channels, due to an effective prioritization of transform data from coarse to fine resolution. In this paper, we exploit the pyramidal structure for another application.

Circumstantial evidence appears to suggest that wavelet still image compression technology is now beginning to mature. Example methods can now provide excellent (lossy) reconstructions at up to about 32:1 compression, e.g., for 512 x 512 grayscale images [8], [9], [10]. In particular, the reported mean-squared error for reconstructions of the well-known Lena image seem to be tightly converging at this compression ratio – PSNRs of 33.17 dB, 34.10 dB, and 34.24 dB, respectively. While low mean-squared error is admittedly an imperfect measure of coding success, that evidence also seems to be corroborated by subjective analysis of sample reconstructions. Further gains in compression, if needed, may now require sacrificing reconstruction fidelity in some portions of an image in favor of others.

Indeed, if regions of interest within an image can be preselected before compression, a prioritized quantization scheme can be constructed for the transform coefficients to achieve region-dependent quality of encoding. Our approach is to subdivide the transform-domain image pyramid into subpyramids, allowing direct access to image regions; designated regions can then be retained at high (or full) fidelity, while sacrificing the background, providing variable-resolution compression. This coding approach can be viewed as a modification of what may be called the standard or first-generation subband image coder: a combination of a subband transformer, a scalar quantizer, and an entropy coder. Each of the recent advances in subband image coding [8], [9], [10] have been achieved by exploiting the interband transform-domain structure, developing what may be called second-generation coding techniques; this paper has been motivated in part by this trend. We remark that [12] also attempts a region-based image coding scheme, but using vector quantization techniques unrelated to the method described here.

2 Error Penetration

Consider a standard two-channel analysis/synthesis filter bank, corresponding, for example, to orthogonal or biorthogonal wavelets; see figure 1. We may choose the filter bank to be perfectly reconstructing, although the approach is also applicable to more general approximately reconstructing quadrature mirror filter (QMF) banks.

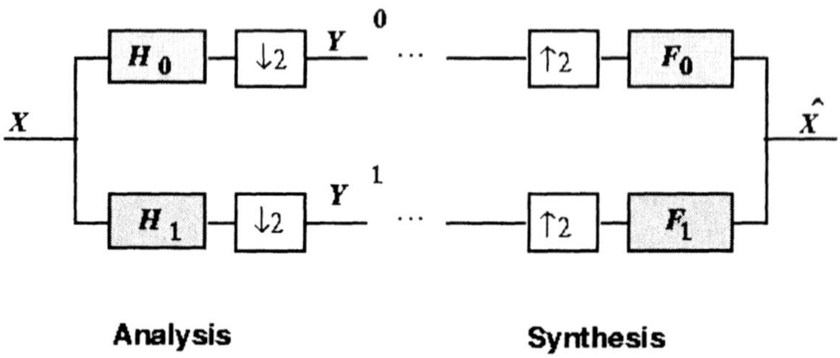

FIGURE 1. A Two-Channel Analysis/Synthesis Filtering Scheme.

In order to achieve variable-resolution coding, and in particular to retain certain regions at high fidelity, our objective is to control the spread of quantization error into such regions. To estimate the spread of quantization error (or "error penetration"), observe first that it depends a priori only on properties of the synthesis filters, and not directly on the analysis filters. This is simply because quantization error is introduced after the analysis stage. Note that convolving a signal with a filter F_i of length T taps replaces a single pixel by the weighted sum of T pixels. Suppose we have an L-level decomposition, and consider the synthesis operation from level L to $L-1$. For error penetration into a region of interest, the filter must straddle the region boundary; in particular, at most $T-2$ taps can enter the region, as in figure 2. We will assume that the region is very large compared to the filter size T. And we make the further simplifying assumption that the region is given by a whole number of pixels in the lowest level L.

The error penetration depth is then essentially $T-2$. In fact, the penetration depth must be odd, so it is $T-2$ if that is odd, and $T-3$ otherwise. This can be expressed as a depth (where [ ] means the integer part of):

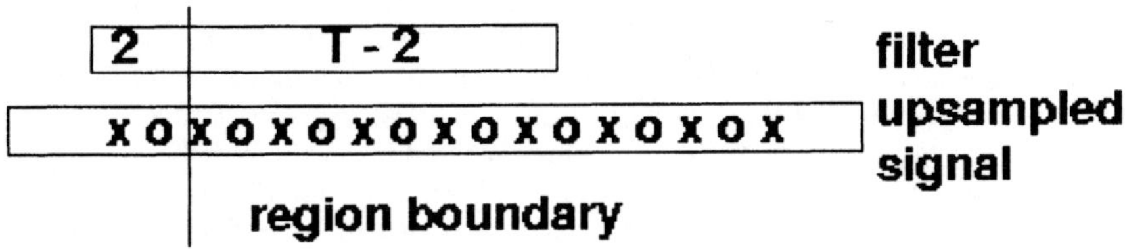

FIGURE 2. Quantization Error Penetration Into A Region Of Interest.

$$A_1 = 2[\frac{T-3}{2}] + 1.$$

Proceeding to level $L - 2$, this gets expanded to twice the amount A_1, plus an additional amount A_1 due to filtering at level $L - 1$. Again, because penetration depth must be odd, we actually get the recursion

$$A_2 = 2A_1 + A_1, \ldots, A_k = 2A_{k-1} + A_1.$$

This recursion can be solved to obtain the formula,

$$A_L = (2^L - 1)(2[\frac{T-3}{2}] + 1).$$

We remark that the error penetration depth grows exponentially in the number of levels, which makes it desirable to limit the number of levels in the decomposition. This is at odds, however, with the requirement of energy compaction for compression, which increases dramatically with the number of levels, so a balance has to be established. The penetration depth is also proportional to the synthesis filter length, so that short synthesis filters are desirable. Thus the 2-tap Haar filter and the 4-tap filter DAUB4 seem to have a preferred role to play here. In fact, one can show that the Haar filter (uniquely) allows perfect separation of background from regions of interest, with no penetration of compression error. Naturally, its energy compaction capability is inferior to that of longer, smoother filters [5], [6]. Among the biorthogonal wavelets, the pairs with asymmetric filter lengths are not necessarily preferred since one of the synthesis filters will then be long. However, one can then selectively control the error penetration more precisely among the two channels.

3 Quantization

An elementary scalar quantization method is outlined here for brevity. More advanced vector quantization techniques can provide further coding advantages [6]; however, the error penetration analysis above would not be directly applicable. A two-channel subband coder applied with L levels transforms an image into a sequence of subbands organized in a pyramid P:

$$(a_{ij}) \to P = (LOW_L, H_L, V_L, D_L, \ldots, H_1, V_1, D_1).$$

Here LOW_L refers to the lowpass subband, and the H, V, and D subbands respectively are the horizontal, vertical, and diagonal subbands at the various levels. In fact, as is well known [8], this is not one pyramid but a multitude of pyramids, one for each pixel in LOW_L. Every partition π of LOW_L gives rise to a sequence of pyramids $P_1, P_2, \ldots, P_\pi$, with $|\pi|$ the number of regions in the partition π. Each of these partitions can be quantized separately, leading to differing levels of fidelity. In particular, image pixels outside regions of interest can be assigned relatively reduced bit rates. A rapid real-time implementation would be to simply apply a low bitrate mask to noncritical regions below a certain level $K < L$ (the inequality is necessary to assure that a nominal resolution is available for all parts of the image). The level K then becomes a control parameter for the background reconstruction fidelity. More sophisticated scalar as well vector quantizers can also be designed, which leverage the tree structure more fully.

Variable quantization by dissecting the image pyramid in this manner allows considerable flexibility in bit allocation, capturing high resolution where it is most desired. Note that this approach is a departure from the usual method of quantizing subband-by-subband [4], [2], [6], [7], while there is commonality with recent tree-based approaches to coding [8], [9], [10]. Indeed, the techniques of [8], [9], [10] can be leveraged to give high-performance variable-resolution coding of image data. Research in this direction is currently ongoing.

4 Simulations

For illustrative purposes, our simulations were performed using modifications of the public-domain (for research) image compression utility EPIC developed at MIT [11]. Figure 4a is an example of a typical 512 x 512 grayscale electro-optical portrait image ("Jay"), with fairly sharp object boundaries and smooth texture. It can be compressed up to 32:1 with high-quality reconstruction by a number of wavelet techniques. Figure 3 shows a standard comparision of wavelet compression to JPEG at 32:1; this compression ratio serves as a benchmark in the field, in that it indicates a breaking point for JPEG, yet a threshold of acceptability for wavelet coding of grayscale images. However, for deeper compression we must now employ alternative strategies; here we have used region of interest compression (ROIC) to achieve a 50:1 ratio. This is a level of compression for which no general image compression technology, including wavelet coding, is typically satisfactory for most applications. Nevertheless, ROIC coding delivers good quality where it is needed; the PSNR for the region of interest (central quarter of image) is 33.8 dB, indicating high image quality. Figure 5 is a much more difficult example for compression: a synthetic aperture radar (SAR) image. Due to the high speckle noise in SAR imagery, compression ratios for this type of image beyond 15:1 may already be unacceptable (e.g., for target recognition purposes). For this example, three regions have been selected, with the largest region being designated for the highest fidelity, and the two smaller regions for moderate fidelity. We achieve 48:1 compression by suppressing most of the noise-saturated background.

5 Acknowledgements

I Thank Chris Heil and Gil Strang for useful discussions.

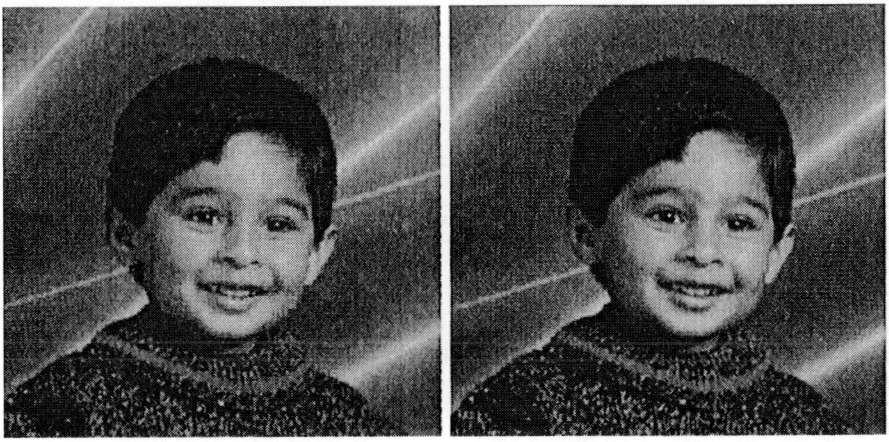

(a) (b)

FIGURE 3. A standard comparison of JPEG to wavelet image compression, on a 512 x 512 grayscale image "Jay", at 0.25 bpp, 32:1 compression: (a) JPEG reconstruction; (b) wavelet reconstruction.

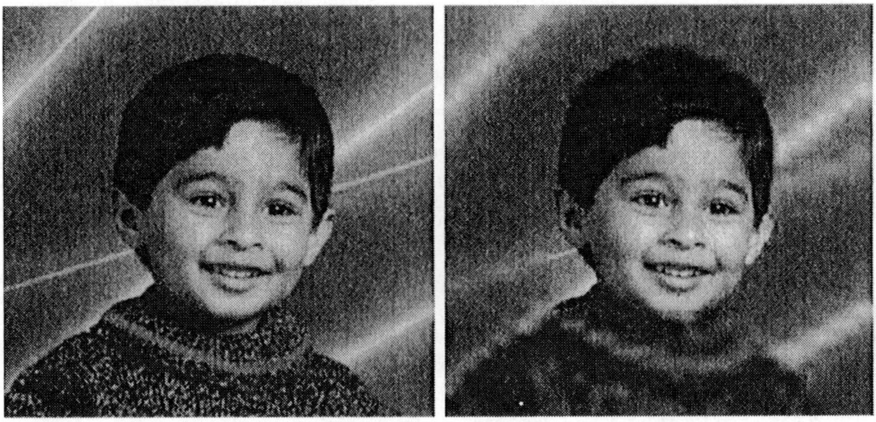

(a) (b)

FIGURE 4. An example application of wavelet-based Region of Interest Compression (ROIC): (a) Original 512 x 512 grayscale image "Jay"; (b) ROIC reconstruction at 50:1, emphasizing the facial portion of the image.

(a) (b)

FIGURE 5. An example application in synthetic aperture radar-based remote sensing for surveillance applications. Here, target detection algorithms naturally lead to the selection of regions of interest, which can then be prioritized for enhanced downlink for further exploitation processing and decision making.

6 References

[1] P. J. Burt and E. H. Adelson, "The Laplacian Pyramid as a Compact Image Code," IEEE Trans. Comm., v. 31, pp. 532-540, 1983.

[2] J. W. Woods and S. D. ONeil, "Subband Coding of Images," IEEE Trans. Acous. Speech Sig. Proc., v. 34, pp. 1278-1288, 1986.

[3] S. Mallat, "A Theory of Multiresolution Signal Decomposition: The Wavelet Representation," IEEE Trans. Pat. Anal. Mach. Intel., v. 11, pp. 674-693, 1989.

[4] N. S. Jayant and P. Noll, Digital Coding of Waveforms, Prentice-Hall, 1984.

[5] I. Daubechies, Ten Lectures on Wavelets, SIAM, 1992.

[6] M. Antonini, M. Barlaud, P. Mathieu, and I. Daubechies, "Image Coding Using Wavelet Transform," IEEE Trans. Image Proc, v. 1, 4, pp. 205-221, April, 1992.

[7] R. DeVore, B. Jawerth, and B. Lucier, "Image Compression Through Wavelet Transform Coding," IEEE Trans. Info. Thry., v. 38, no.2, pp. 719-746, 1992.

[8] J. Shapiro, "Embedded Image Coding Using Zerotrees of Wavelet Coefficients," IEEE Trans. Sig. Proc., v. 41, no.12, pp. 3445-3462, 1993.

[9] M. Smith, "ATR and Compression," Clipping Service Workshop, MITRE Corp., May, 1995.

[10] Z. Xiong, K. Ramchandran, and M. Orchard, "Joint Optimization of Scalar and Tree-Structured Quantization of Wavelet Image Decomposition," Proc. 27th Asilomar Conf. Sig. Sys. Comp., Pacific Grove, Ca, Nov., 1993.

[11] E. Simoncelli and E. Adelson, Efficient Pyramid Image Coder (EPIC), Vision Science Group, Media Lab, MIT, 1989. ftp: whitechapel.media.mit.edu / pub/epic.tar.Z.

[12] S. Fioravanti, "A Region-Based Approach to Image Coding," Proc. Data Comp. Conf., IEEE, p. 515, March 1994.

15

Wavelet-Based Embedded Multispectral Image Compression

Pankaj N. Topiwala

1 Introduction

The rapid pace of change in the earth's environment, as exemplified by urbanization, land-development, deforestation, pollution, global climate change, and many other phenomena have intensified efforts worldwide for monitoring the earth's environment for evidence of the impact of human activity by use of multispectral remote sensing systems. These systems are capable of both high-resolution imaging and spectrometry, simultaneously permitting accurate location and characterization of ground cover materials/compounds through spectrometric analysis. While the exploitation of this type of imagery for environmental studies and a host of other applications is steadily being developed, the downlink and distribution of such voluminous data continues to be problematic. A high-resolution multispectral (or hyperspectral) imager can easily capture hundreds of megabytes per second of highly correlated data. Lossless compression of this type of data, which provides only marginal gains (about 3:1 compression), cannot alleviate the severe downlink and terrestrial transmission bottlenecks encountered; however, lossy compression can permit mass storage and timely transmission, and can add considerable value to the utility of such data by permitting wider collection/distribution. In this paper, we develop a high-performance wavelet-based multispectal image compression algorithm which achieves near-optimal, fully embedded compression. The technology is based on combining optimal spectral decorrelation with embedded wavelet spatial compression in a single, efficient multispectral compression engine.

In the context of lossy coding, the issue of image quality versus bandwidth efficiency becomes important, and requires a detailed analysis in terms of degradations in exploitation as a function of compression. It is interesting to note that, in a different context, the FBI has adopted lossy image compression of fingerprint images (roughly 15:1 compression), despite the severe implications of the exploitation of fingerprint imagery for criminal justice applications [7]. Remarkably, a variety of tests conducted by the FBI has revealed that 15:1 compression of fingerprint images by a wavelet-based algorithm leads to absolutely no degradation in the ability of fingerprint experts to make identifications. Correspondingly, the availability of a high-performance embedded multispectral image coder can facilitate analysis of the tradeoff between compression and image exploitation value in this context; compression ratios can be finely controlled and varied over large ranges without compromising performance, all within a single framework. We have also developed a reduced complexity coder whose performance approximates the embedded coder, but which offers substantial throughput efficiencies.

2 An Embedded Multispectral Image Coder

The nature of multispectral imagery offers unprecedented opportunities to realize high-fidelity lossy compression with strictly limited loss. By definition, a sequence of image bands is available which is pixel-registered and highly correlated both spatially and spectrally; more bands mean greater redundancy, and thus opportunity to reduce bits per pixel - a compression paradise. A survey of a variety of lossless coding techniques for data with over 200 spectral bands shows that advanced techniques still yield only about 3:1 compression typically [24]. On the other hand, lossy compression techniques yielding up to 70:1 compression have been reported in the literature while delivering reconstructions with average PSNRs exceeding 40 dB in each band [1]; [6] reports lossy compression at over 100:1 with almost imperceptible loss in an 80-band system. This MSE (and qualitative visual) analysis suggests that substantial benefits accrue in using high-fidelity lossy compression over lossless; however, a definitive, exploitation- based analysis of the value/limitations of lossy compression is required (although see [19, 22]).

The use and processing of multispectral imagery is in fact closely related to that of natural color imagery, in several ways. First, multispectral imagery, especially with a large number of spectral bands, is difficult to visualize and interpret. The human visual system is well-developed to visualize a three-color space, and quite often, three of the many spectral bands are taken together to form a pseudo-color image. This permits visualizing the loss due to compression, not only as loss in spatial resolution, but in the color space representation (e.g., color bleeding). Second, with proper tuning, the color representation can also be used effectively to highlight features of interest for exploitation purposes (e.g., foliage vs. urban development, pollutants vs. natural resources, etc.). Third, the compression of multispectral imagery, as well, is a generalization of color image compression, which is typically performed by decorrelating in the color space before compressing the bands separately. The decorrelation of RGB data is now well understand and can be formalized by a fixed basis change, for example to YUV space. In general, multispectral bands have a covariance matrix whose diagonalization need not permit a fixed basis change, hence the use of data-dependent KLT methods. In any case, we present an example of color image compression using wavelets to indicate its value (figure 1).

While a global optimization problem in multispectral image compression remains to be formulated and solved, practical compression systems typically divide the problem into two blocks: spectral decorrelation, followed by spatial compression by band (see figure 2). A wide variety of techniques is available for spectral decorrelation, ranging from vector quantization [14] to transform techniques such as the DCT [10] to simple differencing tecniques such as DPCM [1]; however statistically optimal decorrelation is available by application of the Karhunen-Loeve transform or KLT [25, 29, 5, 6]. Spatial compression techniques range from applying vector quantization [4], to transform methods such as the DCT [25] and wavelets [5, 6, 16, 21].

2.1 Algorithm Overview

In this work, we use the following three steps (figure 2) to multispectral image compression:

1. optimal KLT for spectral decorrelation (figure 3),

(a)

(b)

FIGURE 1. Wavelet image coding of a color image. (a) Original NASA sattelite image, 1280 x 854 pixels, 24 bpp; (b) wavelet image compression at 100:1 compression ratio.

2. energy-based bit-allocation, and
3. state-of-the-art embedded wavelet image coding for the eigenimages, e.g., [23, 26].

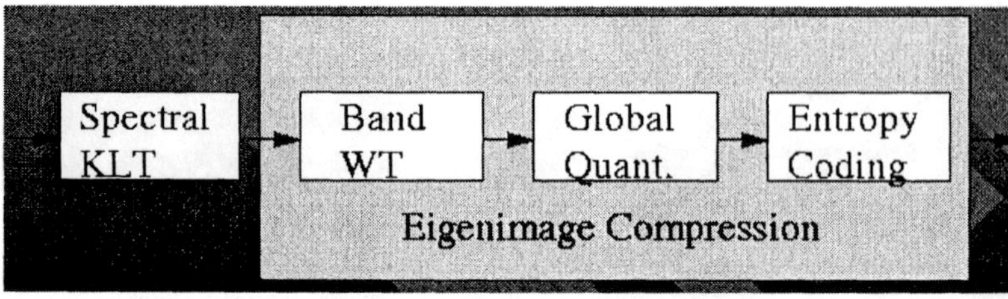

FIGURE 2. Schematic structure of a wavelet-KLT multispectral image compression algorithm.

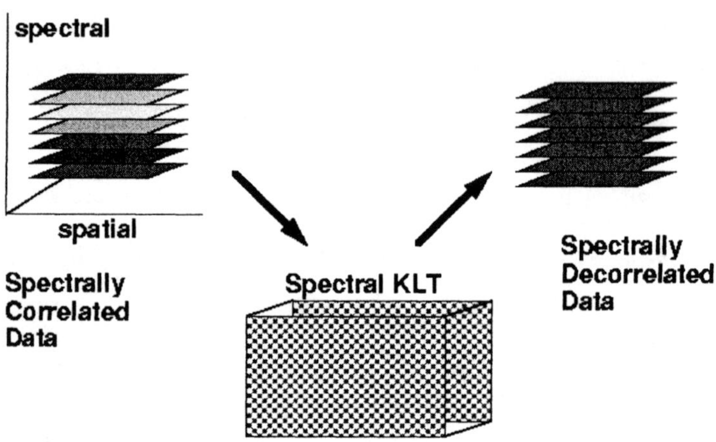

FIGURE 3. Schematic structure of a wavelet-KLT multispectral image compression algorithm.

Steps (1) and (2) now appear to be on their way to being standardized in the literature, and are statistically optimal from the point of view of energetics. An exploitation-based criterion could possibly change (2) but is less likely to change the (statistically optimal) use of (1). On the other hand, the KLT can be prohibitively complex for imaging systems with hundreds of spectral bands, and may be replaced by a spectral transform or predictive coder. In any case, it is mainly step (3), the eigenband coding, on which a number of algorithms significantly differ in their approaches. The method of [23] offers exceptional coding performance at modest complexity, as well as fully embedded coding. An alternative to [23] for eigenimage coding is to use the high-performance low-complexity wavelet image coder of [27, 26], which offers nearly the performance of [23] at less than one-fourth the complexity. Such a system can offer extremely high throughput rates, with real-time operation using DSP chips or custom adaptive computing implementations.

2.2 Transforms

The evidence is nearly unanimous that spectral decorrelation via the Karhunen-Loeve transform [12] is ideally suitable for multispectral image compression application. Note that this arises out of the absence of any specific dominant spectral signatures, giving priority to statistical and energetic considerations. The KLT affords excellent decorrelation of the spectral bands, and can also be conveinently encoded for low to moderate number of bands. Other transforms currently underperform the KLT by a noticeable margin.

Incidentally, if the fact that the KLT is well-suited appears to be a tautology, consider the fact that the same is not directly true in still image coding of grayscale imagery. An illustrative example [15] is one of coding a simple image consisting of a bright square in a dark background. If we consider the process consisting of that image under random translations of the plane, we obtain a stationary process, whose KLT would be essentially a transformation into a basis of sinusoids (we are here considering the two axes independently). Such a basis would be highly inefficient for the coding of any one example image, and indicates that optimality results can be misleading. Even when the KLT (obtained by the sample covariance rather than as a process) can offer optimal MSE performance, MSE may be less important than preserving edge information in imagery, especially if it is to be subjectively interpreted. In multispectral imagery, in the absence of universal exploitation criteria based on preserving specific spectral signatures, optimal spectral decorrelation via the KLT is herein adopted; see figure 3.

Once the KLT in the spectral directions is performed, the resulting eigenimages are essentially continuous-tone grayscale images; these can be encoded by any one of a variety of techniques. We apply two techniques for eigenimage coding: the embedded coder of [23], and the high-performance coder of [27]. Both rely on wavelet decompositions in an octave-band format, using biorthogonal wavelet filters [2, 3]. One practical consideration is that the eigenimages are of variable dynamic range, with some images, especially those corresponding to large eigenvalues, exceeding the dynamic range of the original bands. There are essentially two approaches to dealing with this situation: (1) convert each image to a fixed integer representation (e.g., an 8-bit grayscale image) by a nonlinear pixel transformation, and apply any standard algorithm such as JPEG, as done in [25]; or (2) work in the floating-point domain directly, accomodating the variable dynamic range. We have investigated both approaches, and find that (2) has some modest advantages, but entails slightly higher complexity.

2.3 Quantization

Given a total bitrate budget R, the first task is to divide the bit budget among the eigenbands of the KLT; the next step is to divide the budget for each band into bit allocations for the wavelet transform subbands per image. However, since the KLT is a unitary transform, it is easy to show [28, 26] that the strategy for optimal bit-allocation for all wavelet transform subbands for all images is identical to the formula for bit allocation within a single image. We assume that the KLT, to a first approximation, fully decorrelates the eigenimages; we further assume that the wavelet transform fully decorrelates the pixels in each eigenimage. It is a matter of experimental evidence that each subband in the wavelet transform (except the residual lowpass subband) is approximately Laplace distributed. A more precise fit is often possible among the larger family of Generalized Gaussian distributions, although we will restrict attention to the simpler Laplace family. This distribution has only one scale parameter, which is effectively the standard deviation. If we normalize all variables according to their standard deviation, they become identically distributed. We make the further simplification that all pixels are actually independent (if they were Gaussian distributed, uncorrelated would imply independent).

By our assumptions, pixels in various subbands of the various eigenimages of the KLT are viewed as independent random sources $\{X_1, \ldots, X_k\}$ which are identically distributed when normalized. These possess rate-distortion curves $D_i(R_i)$, and we are given a bitrate budget R. The problem of optimal joint quantization of the random sources then has the solution that all rate-distortion curves must have a constant slope $-\lambda$. λ itself is determined by satisfying the bitrate budget. Again, by using the "high-resolution" approximation, one derives the formula:

$$D_i(R_i) = C_i \sigma_i^2 2^{-2R_i}.$$

Here C_i is a constant depending on the normalized distribution [8] (thus a global constant by our assumptions), and σ_i^2 is the variance of X_i. This approximation leads to the following analytic solution to the bit-allocation problem:

$$R_i = \bar{R} + \frac{1}{2} log_2 \frac{\sigma_i^2}{\rho^2},$$
$$\bar{R} = R/k, \quad \rho^2 = (\prod \sigma_i^2)^{1/K}.$$

This formula applies to all wavelet subbands in all eigenimages. It implies, in particular, that eigenimages with proportionately larger variances get higher bitrates; this is both intuitive and consistent with the literature, e.g., [25]. However, our bitrate assignments may differ in detail.

More advanced approaches can be based on using (a) more precise parametric models of the transform data, and (b) accounting for interpixel, interband, and intereigenband dependencies for enhanced coding performance. However, the marginal return in coding gain that may be available by these techniques is counterbalanced by the corresponding complexity penalties; increasing the operations per pixel is costly as this type of data is very intensive.

2.4 Entropy Coding

The quantized eigenimages are separately entropy coded using an arithmetic coder [31, 17] to generate a compressed symbol stream in our simulations below. The eigenimages can also be combined to produce a globally embedded bitstream; details of that will be reported elsewhere. The current compression engine, in any case, can deliver any compression ratio, by precisely controlling the bit allocations across the eigenimages to achieve a global bit budget. Meanwhile, efforts to extend fully embedded coding to another 3-D realm — video — have been undertaken and are reported later in this volume in a chapter by Pearlman et al. Their techniques can be suitably adapted here to achieve the "interleaved" embedded coding of the various eigenimages.

3 Simulations

For experiments, we use a 3-band SPOT image, and a public-domain 12-band image, both available on the internet. The 3-band SPOT image, with 2 visible bands and one infrared band, is interpretable as a false color image and is presented as a single image. Our simulations present the original false color image and at 50:1 compression, see figure 4. Note that the usual color space transformations do not apply to this unusual color image, hence the use of the KLT. Although the compression is clearly lossy, it does provide reasonable performance despite the spatial and spectral busyness (unfortunately unseen in this dot-dithered grayscale representation, but the full-color images will be available at our web page). In figure 5, we present an original (unrectified) 12-band image which is part of the public-domain software package MultiSpec [18] from Purdue University. Figure 6 presents the image at 50:1 compression, along with an example of compression without prior spectral decorrelation, indicating the value of spectral decorrelation.

As a useful indication of the spectral decorrelation efficiency, we consider two measures: (1) the eigenvalues of the KLT, as shown in figure 7, and (2) the actual variances of the eigenbands after the KLT, as shown in 8. The bitrate needed for a fixed quality reproduction of the multispectral image is proportional to the area under the curve in figure 8; see for example, [25].

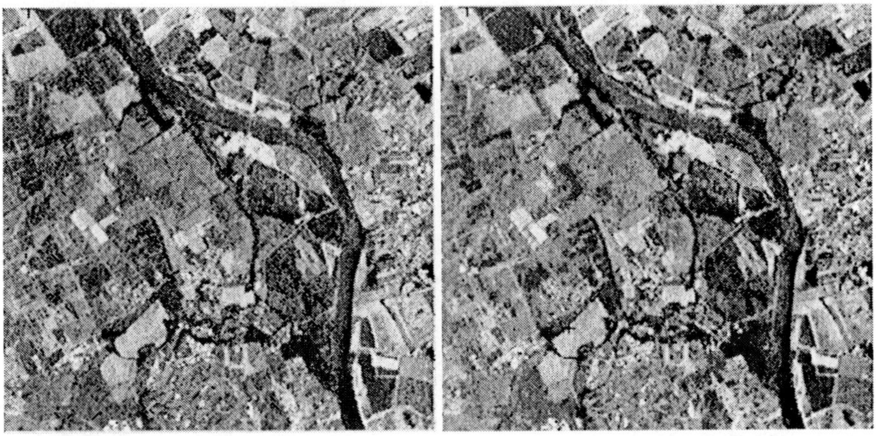

FIGURE 4. A 3-band (visible-infrared) SPOT image, (a) original, and (b) compressed 50:1 by methods discussed.

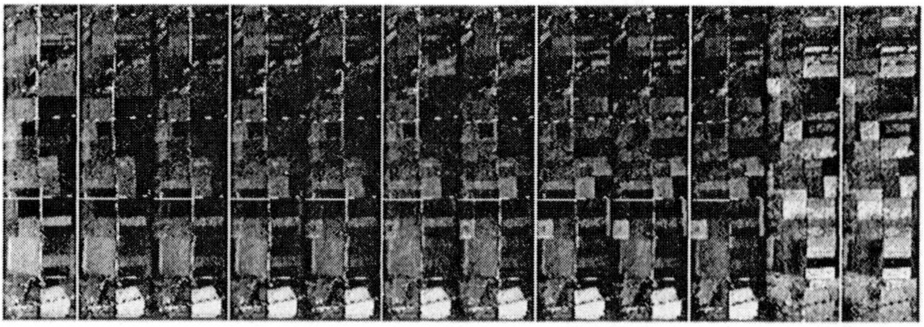

FIGURE 5. A 12-band multispectral image with spectral range 0.4 - 1 μm, with bandwidths in the range 0.03 - 0.2 μm.

FIGURE 6. Multispectral image compression example. (a) band 1 of a 12-band image; (b) multispectral compression at 50:1; (c) single-band compression at 50:1. Spectral decorrelation provides substantial coding gain.

4 References

[1] G. Abousleman, "Compression of Hyperspectral Imagery...," Proc. DCC95 Data Comp. Conf., pp. 322-331, 1995.

[2] M. Antonini et al, "Image Coding Using Wavelet Transform," IEEE Trans. Image Proc., pp. 205-220, April, 1992.

[3] I. Daubechies, Ten Lectures on Wavelets, SIAM, 1992.

[4] S. Budge et al, "An Adaptive VQ and Rice Encoder Algorithm," Im. Comp. Appl. Inn. Work., 1994.

[5] B. Epstein et al, "Multispectral KLT-Wavelet Data Compression...," Proc. DCC92 Data Comp. Conf., pp. 200-208, 1992.

[6] B. Evans et al, "Wavelet Compression Techniques for Hyperspectral Data," Proc. Dcc94 Data Comp. Conf., p. 532, 1994.

[7] "Wavelet Scalar Quantization Fingerprint Image Compression Standard," Criminal Justice Information Services, FBI, March, 1993.

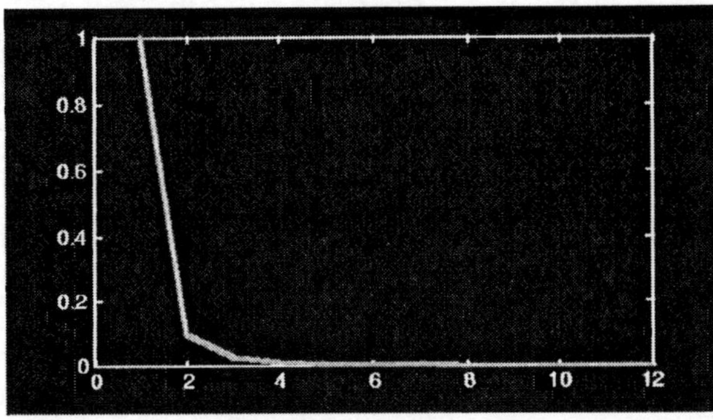

FIGURE 7. The eigenvalues of the KLT for the 12-band image in the previous examples.

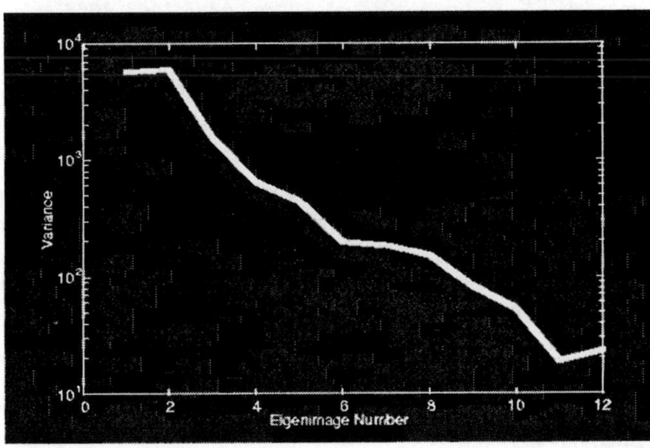

FIGURE 8. The spectral decorrelation efficiency of the KLT can be visualized in terms of the roll-off of the variances of the eigenimages after the transform. The bitrate needed for a fixed quality reconstruction is proportional to the area under this curve.

[8] A. Gersho and R. Gray, *Vector Quantization and Signal Compression*, Kluwer, 1992.

[9] N. Griswold, "Perceptual Coding in the Cosine Transform Domain," Opt. Eng., v. 19, pp. 306-311, 1980.

[10] P. Heller et al, "Wavelet Compression of Multispectral Imagery," Ind. Wkshp. DCC94 Data Comp. Conf., 1994.

[11] J. Huang and P. Schultheiss, "Block Quantization of Correlated Gaussian Variables," IEEE Trans. Comm., CS-11, pp. 289-296, 1963.

[12] N. Jayant and P. Noll, *Digital Coding of Waveforms*, Prentice-Hall, 1984.

[13] W. Pennebaker and J. Mitchell, *JPEG Still Image Data Compression Standard*, Van Nostrand, 1993.

[14] R. Lindsay et al, "Predictive VQ For Landsat Imagery Compression using Cross Spectral Vectors," Im. Comp. Appl. Inn. Work., 1994.

[15] S. Mallat, "Understanding Image Compression Codes," SPIE Aero. Conf., Orlando, FL, April, 1997.

[16] R. Matic et al, "A Comparison of Spectral Decorrelation Techniques...," Im. Comp. Appl. Innov. Wkshp., March, 1994.

[17] A. Moffat et al, "Arithmetic Coding Revisited," Proc. DCC95 Data Comp. Conf. , pp. 202-211, 1995.

[18] http://dynamo.ecn.purdue.edu/ biehl/MultiSpec/

[19] K. Riehl et al, "An Assessment of the Effects of Data Compression on Multispectral Imagery," Proc. Ind. Wksp DCC94 Data Comp. Conf., March, 1994.

[20] Y. Shoham and A. Gersho, "Efficient Bit Allocation for an Arbitrary Set of Quantizers," IEEE Trans. ASSP, vol. 36, pp. 1445-1453, 1988.

[21] J. Shapiro et al, "Compression of Multispectral Landsat Imagery Using the EZW Algorithm," Proc. DCC94 Data Comp. Conf., p. 472, 1994.

[22] S. Shen et al, "Effects of Multispectral Image Compression on Machine Exploitation," Proc. Asil. Conf. Sig., Sys. Comp., Nov. 1993

[23] A. Said and W. Pearlman, "A New, Fast, and Efficient Image Codec Based on Set Partitioning In Hierarchical Trees," IEEE Trans. Cir. Sys. Vid. Tech., June, 1996.

[24] S. Tate, "Band Ordering in Lossless Compression of Multispectral Images," Proc. DCC94 Data Comp. Conf., pp. 311-320, 1994.

[25] J. Saghri et al, "Practical Transform Coding of Multispectral Imagery," IEEE Sig. Proc. Mag., pp. 32-43, Jan. 1995

[26] P. Topiwala, ed., *Wavelet Image and Video Compression*, Kluwer Acad. Publ., June, 1998 (this volume).

[27] P. Topiwala et al, "High-Performance Wavelet Compression Algorithms...," Proc. ImageTech97, April, 1997.

[28] P. Topiwala, "HVS-Motivated Quantization Schemes in Wavelet Image Compression," SPIE Denver, App. Dig. Img. Proc., July, 1996.

[29] V. Vaughn et al, "System Considerations for Multispectral Image Compression Design," IEEE Sig. Proc. Mag., pp. 19-31, Jan. 1995

[30] M. Vetterli and J. Kovacevic, *Wavelets and Subband Coding*, Prentice-Hall, 1995.

[31] I. Witten et al, "Arithmetic Coding for Data Compression," Comm. ACM, IEEE, pp. 520-541, June, 1987.

16 The FBI Fingerprint Image Compression Specification

Christopher M. Brislawn

1 Introduction

The Federal Bureau of Investigation (FBI) has formulated national specifications for digitization and compression of gray-scale fingerprint images. The compression algorithm [Fed93] for the digitized images is based on adaptive uniform scalar quantization of a discrete wavelet transform subband decomposition, a family of techniques referred to as wavelet/scalar quantization (WSQ) methods.

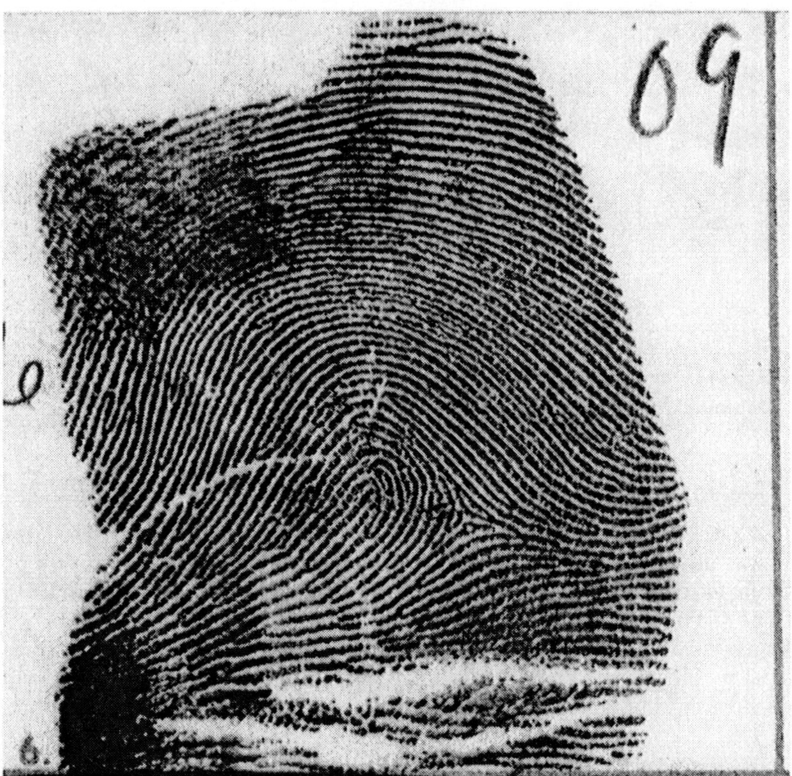

FIGURE 1. Original 500 dpi 8-bit grayscale fingerprint image.

FIGURE 2. Detail of 500 dpi fingerprint image.

DISCLAIMER:

This article is for informative purposes only; the reader is referred to the official FBI document [Fed93] for normative specifications.

1.1 Background

The FBI is faced with the task of converting its criminal fingerprint database, which currently consists of around 200 million inked fingerprint cards, to a digital electronic format. A single card contains 14 separate images: 10 rolled impressions, duplicate (flat) impressions of both thumbs, and simultaneous impressions of all of the fingers on each hand. During the scanning process, fingerprints are digitized at a spatial resolution of 500 dots per inch (dpi), with 8 bits of gray-scale resolution. (Details concerning digitization requirements can be found in the ANSI standard for fingerprint data formats [Ame93].)

A typical scanned fingerprint image is shown in Figure 1, and a zoom of the "core," or center, of the print is shown in Figure 2 to illustrate the level of detail discernable at 500 dpi. Technology is also under development to capture fingerprint

FIGURE 3. Fingerprint image compressed 14.5:1 by the ISO JPEG algorithm.

images by direct optical scanning of the fingertips, without the use of an intervening inked card, thereby eliminating a major source of distortion in the fingerprinting process. However the digital image is obtained, digitization replaces a single fingerprint card with about 10 megabytes of raster image data. This factor, multiplied by the size of the FBI's criminal fingerprint database, plus the addition of some 50,000 new cards submitted to the FBI each work day for background checks, gives some indication of why data compression was deemed necessary for this project.

The first compression algorithm investigated was the ISO JPEG still image compression standard [Ame91]. Figure 3 shows a detail of the image in Figure 1 after JPEG compression and reconstruction; the compression ratio was 14.5. This particular JPEG compression was done using Huffman coding and the JPEG "example" luminance quantization matrix, which is based on human perceptual sensitivity in the DCT domain. Experimentation with customized JPEG quantization matrices by the UK Home Office Police Research Group has demonstrated that the blocking artifacts visible in Figure 3 are inherent in JPEG-compressed fingerprint images at compression ratios above about 10. After some consideration by the FBI, it was decided that the JPEG algorithm was not well-suited to the requirements of the fingerprint application.

FIGURE 4. Fingerprint image compressed 15.0:1 by the FBI WSQ algorithm.

Instead, on the basis of a preliminary technology survey [HP92], the FBI elected to work with Los Alamos National Laboratory to develop a lossy compression method based on scalar quantization of a discrete wavelet transform decomposition. The FBI WSQ algorithm produces archival-quality reconstructed images at compression ratios of around 15. Figure 4 gives an example of WSQ compression at a compression ratio of 15.0; note the lack of blocking artifacts as well as the superior preservation of fine-scale detail when compared to the JPEG image in Figure 3.

What about lossless compression? Based on recent research [Wat93, Miz94], lossless compression of gray-scale fingerprint images appears to be limited to compression ratios of less than 2. By going to a lossy WSQ algorithm, it is possible to get almost an order of magnitude more compression while still providing acceptable image quality. Indeed, testing by the FBI has shown that fingerprint examiners are able to perform their identification tasks successfully using images that have been compressed more than 15:1 using the FBI WSQ algorithm.

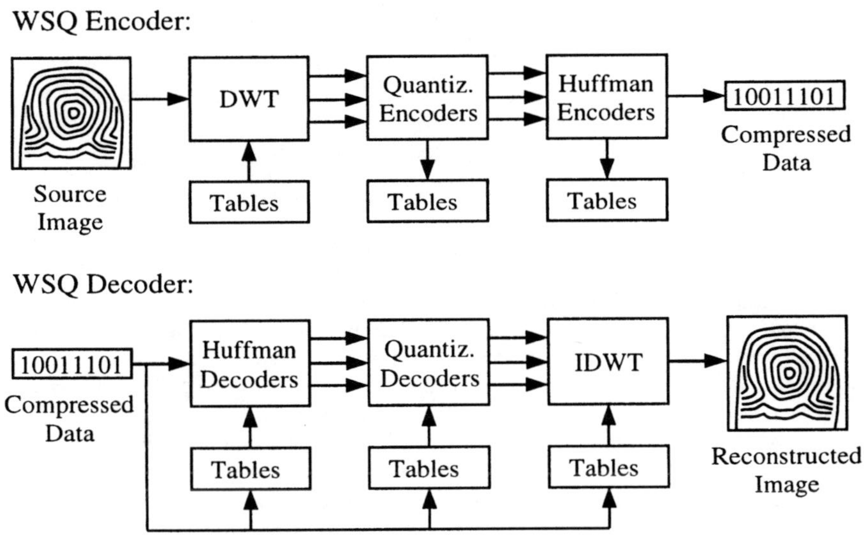

FIGURE 5. Overview of the WSQ algorithm.

1.2 Overview of the algorithm

A high-level overview of the fingerprint compression algorithm is depicted in Figure 5. Encoding consists of three main processes: discrete wavelet transform (DWT) decomposition, scalar quantization, and Huffman entropy coding. The structure of the FBI WSQ compression standard is a specification of the syntax for compressed image data and a specification of a "universal" decoder, which must be capable of reconstructing compressed images produced by any compliant encoder. Several degrees of freedom have been left open in the specification to allow for future improvements in encoding. For this reason, it is most accurate to speak of the specification [Fed93], Part 1: Requirements and Guidelines, as a *decoder* specification, with multiple, independent encoder specifications possible. The first such encoder specification is [Fed93], Part 3: FBI Parameter Settings, Encoder Number One. The next three sections of this article outline the decoding tasks and indicate the level of generality at which they must be implemented by a WSQ decoder. A description of the first-generation FBI WSQ encoder is given in Section 5.

As shown in Figure 5, certain side information, some of it image-dependent, is transmitted in tabular form along with the entropy coded data (in the "interchange format") and extracted by the decoder to enable reconstruction of the compressed image. Specifically, tables are included in the compressed bit stream for transmitting the DWT filter coefficients, the scalar quantizer characteristics, and up to eight image-dependent Huffman coders. In a storage archive environment, this side information can be removed so that compressed data can be stored more economically (the "abbreviated format"), with the environment responsible for providing the necessary tables when the image needs to be decompressed.

The use of "interchange" and "abbreviated" formats was adopted from the JPEG

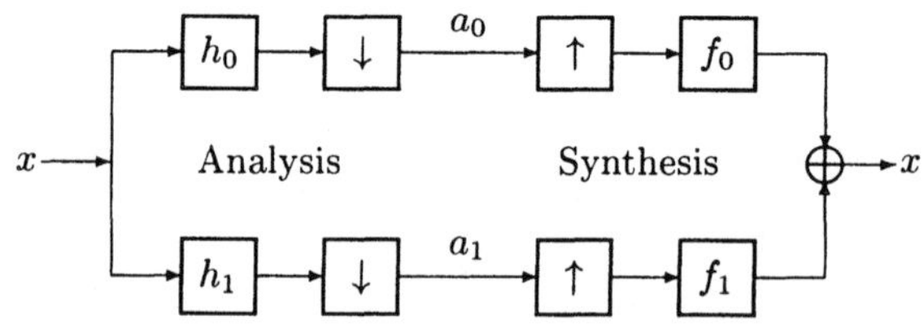

FIGURE 6. Two-channel perfect reconstruction multirate filter bank.

specification [Ame91, PM92], as were many aspects of the syntax for the compressed bit stream. Readers familiar with the JPEG standard will recognize the great similarity between JPEG and FBI WSQ file syntax, based on the use of two-byte markers to delineate the components of the compressed bit stream. This was done to relieve both the writers and the implementors of the FBI specification from the need to master a totally new compressed file format. Since the FBI WSQ file syntax is so close to the JPEG syntax, we will not discuss it at length in this article but will instead refer the reader to the specification [Fed93] for the differences from the JPEG standard, such as the syntax for transmitting wavelet filter coefficients. The FBI specification also borrows heavily from JPEG in other areas, such as the standardization of the procedure for constructing and transmitting adaptive Huffman coders. It should be clear the tremendous debt the FBI WSQ specification owes to the ISO JPEG standard for blazing the way on many issues generic to all image coding standards.

2 The DWT subband decomposition for fingerprints

The spatial frequency decomposition of fingerprint images in the FBI algorithm is obtained using a cascaded two-dimensional (separable) linear phase multirate filter bank [Vai93, VK95, SN96] in a symmetric extension transform (SET) implementation.

2.1 Linear phase filter banks

The FBI standard allows for the use of arbitrary two-channel finite impulse response linear phase perfect reconstruction multirate filter banks (PR MFB's) of the sort shown in Figure 6, with filters up to 32 taps long. It is not required that the filters correspond to an analog scaling function/mother wavelet pair with any particular regularity, but such "wavelet filter banks" are particularly well-suited to algorithms involving multiple levels of filter bank cascade (in Section 2.3 we will see that the fingerprint decomposition involves five levels of cascade in the lowest-frequency portion of the spectrum).

The tap weights are assumed to be floating-point numbers, and while the standard doesn't preclude the use of scaled, integer-arithmetic implementations, compliance testing is done against a direct-form, double-precision floating point implementation, so scaling must be done with care. In the interchange file format, tap weights (for the right halves of the analysis impulse responses) are transmitted as side information along with the lengths of the two filters. Since the filters are linear phase, the decoder can reconstruct the complete analysis filters by reflecting their transmitted halves appropriately (either symmetrically or antisymmetrically, as determined from their lengths and bandpass characteristics).

As we saw in Section 5, the phases of the analysis filters are important for determining the symmetry properties of the subbands produced by the symmetric extension method, so the standard assumes that the analysis filters are applied in a noncausal, minimal phase implementation. (Naturally, there are mathematically equivalent ways to produce the desired phase effects using causal implementations with suitable delays, but the desired symmetry properties are easiest to specify in the minimal phase case.) Specifically, the standard assumes that lowpass WS analysis filters are implemented with a group delay of 0, highpass WS analysis filters with a group delay of -1, and (both) HS/HA analysis filters with a group delay of $-1/2$. In Section 2.2, we'll see that this determines the symmetries of the subbands unambiguously.

In addition, the amplitudes of the transmitted analysis filters are scaled so that the determinant of the alias-component matrix for the analysis bank is $-2z$. With these conventions, corresponding PR synthesis filters can be generated by the decoder using the "anti-aliasing relations"

$$f_0(n) = (-1)^n h_1(n-1) \quad \text{and} \quad f_1(n) = (-1)^{n-1} h_0(n-1)\ . \tag{16.1}$$

Again, delays need to be compensated to obtain zero phase distortion in the reconstructed signal when causal implementations are used; extensive details can be found in [Bri96].

2.2 *Symmetric boundary conditions*

In Chapter 5 we showed how the use of linear phase filters with symmetric boundary extensions allows for the nonexpansive transformation of images of arbitrary sizes, including odd dimensions. This turned out to be a desirable feature in the FBI WSQ specification because there was already in place a standard for digitized fingerprint records [Ame93] that allowed a range of scanning resolutions of anywhere from 500 to 520 dpi. Since inked impressions on a fingerprint card lie in boxes whose physical dimensions must be captured precisely in scanned images, the pixel dimensions of compliantly scanned images are not uniquely specified. For instance, a plain thumbprint (0.875 × 1.875 inches) scanned at 520 dpi measures 455 × 975 pixels. Images cannot be cropped down to "nice" dimensions, however, without discarding part of the required image capture area, and padding out to a *larger* "nice" dimension runs counter to the desire to compress the data as much as possible. Using a nonexpansive symmetric extension transform (SET) to accomodate odd image dimensions sidesteps this dilemma.

Suppose we're filtering a row or column of length N_0 with a linear phase analysis filter, h. Using the language of Chapter 5, the FBI algorithm employs a (1,1)-SET when the length of h is odd and a (2,2)-SET when the length of h is even. The group delay of h is assumed to satisfy the conventions described above in Section 2.1. Under these conditions, the length of the SET output (for both (1,1)-SET's and

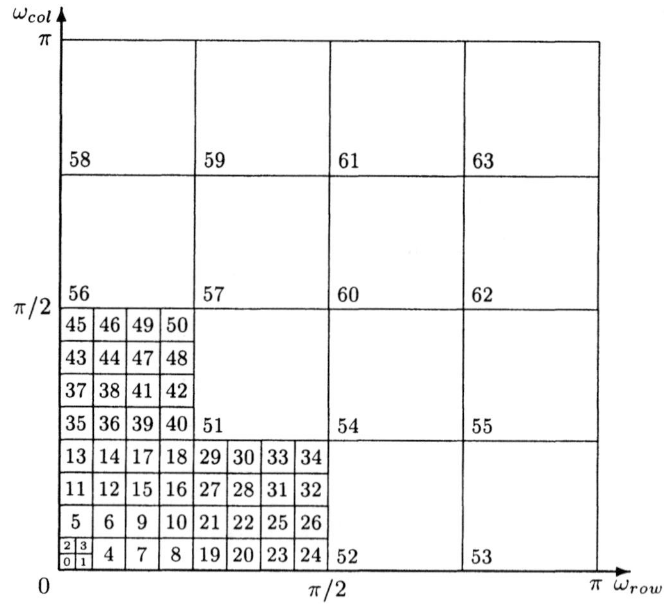

FIGURE 7. Frequency support of DWT subbands in the FBI WSQ specification.

(2,2)-SET's) is $N_0/2$ for both the lowpass and highpass channels when N_0 is even. In case N_0 is odd, the lowpass output will have length $(N_0+1)/2$ and the highpass output will have length $(N_0-1)/2$. These dimensions, together with the symmetric extensions that need to be applied to the transmitted subbands in the synthesis filter bank, are given in Tables 5.1 and 5.2.

These SET algorithms are applied successively to the rows and the columns of the array being transformed, one cascade level at a time. In particular, the analysis filter bank applies either an even-length SET or an odd-length SET at each stage of the cascade based on the length of the input array being filtered at that stage. The synthesis filter bank, by comparison, cannot tell from the size of its input subband whether it is reconstructing an even- or an odd-length vector (i.e., the correct synthesis of a vector of length K doesn't necessarily result in a synthesis output of length $2K$). It is necessary for the WSQ decoder to read the dimensions of the image being decoded and then recreate the logic employed by the encoder to deduce the size of the subbands being synthesized at each stage in the cascade. Only then can it select the correct symmetric synthesis extension from Table 5.1 or Table 5.2.

2.3 Spatial frequency decomposition

The structure of the two-dimensional subband frequency decomposition used on fingerprints is shown in Figure 7. Each of the 64 subbands is positioned in the diagram in the region that it (approximately) occupies in the two-dimensional spectrum. Since the decomposition is nonexpansive, the total number of transform-domain

TABLE 16.1. DWT subband filter cascades.

Band	Cascade						
0	00,00,00,00,00	16	00,00,11,01	32	00,10,01,01	48	00,01,10,01
1	00,00,00,00,10	17	00,00,11,10	33	00,10,01,10	49	00,01,10,10
2	00,00,00,00,01	18	00,00,11,00	34	00,10,01,00	50	00,01,10,00
3	00,00,00,00,11	19	00,10,10,00	35	00,01,01,00	51	00,11
4	00,00,00,10	20	00,10,10,10	36	00,01,01,10	52	10,10
5	00,00,00,01	21	00,10,10,01	37	00,01,01,01	53	10,00
6	00,00,00,11	22	00,10,10,11	38	00,01,01,11	54	10,11
7	00,00,10,10	23	00,10,00,10	39	00,01,11,10	55	10,01
8	00,00,10,00	24	00,10,00,00	40	00,01,11,00	56	01,01
9	00,00,10,11	25	00,10,00,11	41	00,01,11,11	57	01,11
10	00,00,10,01	26	00,10,00,01	42	00,01,11,01	58	01,00
11	00,00,01,01	27	00,10,11,01	43	00,01,00,01	59	01,10
12	00,00,01,11	28	00,10,11,11	44	00,01,00,11	60	11,11
13	00,00,01,00	29	00,10,11,00	45	00,01,00,00	61	11,01
14	00,00,01,10	30	00,10,11,10	46	00,01,00,10	62	11,10
15	00,00,11,11	31	00,10,01,11	47	00,01,10,11	63	11,00

samples in this decomposition is exactly equal to the size of the original image, and the size of each subband in the diagram is proportional to the number of samples it contains. This particular frequency partition is a fixed aspect of the specification and must be used by all encoders.

The exact sequence of row and column filtering operations needed to generate each subband is given in Table 16.1. Each pair of digits, $i\ j$, indicates the row and column filters (respectively) implemented as SET's in a two-dimensional product filtering. Successive levels of (two-dimensional) cascade are separated by commas. Thus, e.g., for band 59 ("01, 10") we apply the lowpass filter to the rows and the highpass filter to the columns in the first level of the decomposition, then in the second level we highpass the rows and lowpass the columns. Note that the location of this subband in the two-dimensional spectrum is a result of spectral inversion in the second-level column filtering operation.

The decomposition shown in Figure 7 was designed specifically for 500 dpi fingerprint images by a combination of spectral analysis and trial-and-error. The fingerprint power spectrum estimate described in the obscure technical report [BB92] shows that the part of the spectrum containing the fingerprint ridge pattern is in the frequency region from approximately $\pi/8$ to about $3\pi/8$. This is the part of the spectrum covered by subbands 7–22, 27–30, 35–42, and a bit of 51. Five levels of frequency cascade were deemed necessary to separate this important part of the spectrum from the DC spike (frequencies below $\pi/32$). Trials by the FBI with different frequency band decompositions then demonstrated that the algorithm benefitted from having fairly fine frequency partitioning in the spectral region containing the fingerprint ridge patterns.

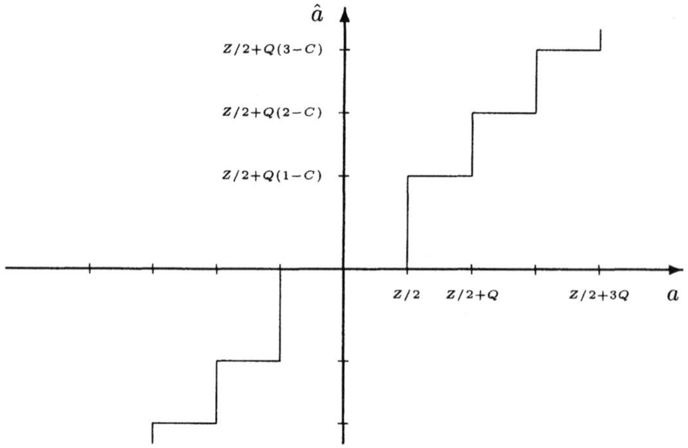

FIGURE 8. WSQ subband quantization characteristic.

3 Uniform scalar quantization

Once the subbands in Figure 7 have been computed, it is necessary to quantize the resulting DWT coefficients prior to entropy coding. This is accomplished via (almost) uniform scalar quantization [JN84, GG92], where the quantizer characteristics may depend on the image and are transmitted as side information in the compressed bit stream.

3.1 Quantizer characteristics

Each of the 64 floating point arrays, $a_k(m,n)$, $k = 0,\ldots,63$, in the frequency decomposition described in Section 2 is quantized to a signed integer array, $p_k(m,n)$, of quantizer indices using an "almost uniform" scalar quantizer with a quantization characteristic for each subband like the one shown in Figure 8.

This quantizer characteristic has several distinctive features. First, the width, Z, of the zero-bin is generally different from the width, Q, of the other bins in the quantizer (hence "almost" uniform scalar quantization). The table of quantizer characteristics transmitted with the compressed bit stream therefore contains two bin widths for each of the 64 quantizers: Z_k, Q_k, $k = 0,\ldots,63$. Second, the output levels of the quantization decoder are determined by the parameter C, $0 \leq C < 1$, which is also transmitted in the quantization table (the value C is held constant across all subbands). Note that if $C = 1/2$ then the reconstructed value corresponding to each quantization bin is the bin's midpoint. Third, the mapping from floating point DWT coefficients to (integer) quantizer bin indices performed by the quantization encoder has no *a priori* bounds on its range; i.e., we do not allow for overload distortion in this quantization strategy but instead code outliers with escape sequences in the entropy coding model.

Mathematical expressions for the quantization encoding and decoding functions are given in equations (16.2) and (16.3), respectively. The expressions $\lceil\cdot\rceil$ and $\lfloor\cdot\rfloor$ denote the functions that round numbers to the next larger and next lower integers.

$$p_k(m,n) = \begin{cases} \left\lfloor \dfrac{(a_k(m,n) - Z_k/2)}{Q_k} \right\rfloor + 1\,, & a_k(m,n) > Z_k/2 \\ 0\,, & -Z_k/2 \le a_k(m,n) \le Z_k/2 \\ \left\lceil \dfrac{(a_k(m,n) + Z_k/2)}{Q_k} \right\rceil - 1\,, & a_k(m,n) < -Z_k/2\,, \end{cases} \quad (16.2)$$

$$\hat{a}_k(m,n) = \begin{cases} (p_k(m,n) - C)Q_k + Z_k/2\,, & p_k(m,n) > 0 \\ 0\,, & p_k(m,n) = 0 \\ (p_k(m,n) + C)Q_k - Z_k/2\,, & p_k(m,n) < 0\,. \end{cases} \quad (16.3)$$

3.2 Bit allocation

The method for determining the bin widths in the WSQ quantizers is not mandated in the FBI decoder specification. Since the decoder does not need to know how the bin widths were determined, the method of designing the 64 quantizers for each image can vary between different encoder versions. Thus, the solution to the quantizer design problem has been left up to the designer of a particular encoder. In Section 5 we will describe how this problem is solved in encoder #1 using a constrained optimal bit allocation approach.

4 Huffman coding

Compression of the quantized DWT subbands is done via adaptive (image-specific) Huffman entropy coding [BCW90, GG92].

4.1 The Huffman coding model

Rather than construct 64 separate Huffman coders for each of the 64 quantized subbands, the specification simplifies matters somewhat by grouping the subbands into a smaller number of blocks and compressing all of the subbands in each block with a single Huffman coder. Encoders may divide the subbands up into anywhere from three to eight different blocks, with a mandatory block boundary between subbands 18 and 19 and another mandatory break between subbands 51 and 52, to facilitate progressive decoding. Although multiple blocks may be compressed using the same Huffman codebook, for simplicity we will describe the design of a WSQ Huffman coder based on a single block of subbands.

Subband boundaries within a single block are eliminated by concatenating the quantized subbands (row-wise, in order of increasing subband index) into a single, one-dimensional array. The quantizer bin indices in each block are then run-length coded for zero-runs (including zero-runs across subband boundaries) and mapped to a finite alphabet of symbols chosen from the coding model in Table 16.2. Encoders may elect to use a proper subset of symbols from Table 16.2, but all FBI WSQ Huffman coders must use symbols taken from this particular alphabet.

TABLE 16.2. Huffman coding model.

Symbol	Value
1	zero run of length 1
2	zero run of length 2
3	zero run of length 3
⋮	⋮
100	zero run of length 100
101	escape for positive 8 bit quantizer bin index
102	escape for negative 8 bit quantizer bin index
103	escape for positive 16 bit quantizer bin index
104	escape for negative 16 bit quantizer bin index
105	escape for 8 bit zero run length
106	escape for 16 bit zero run length
107	quantizer bin index value −73
108	quantizer bin index value −72
109	quantizer bin index value −71
⋮	⋮
180	*Not used. Use symbol 1.*
⋮	⋮
253	quantizer bin index value 73
254	quantizer bin index value 74

Distinct symbols are provided for all zero-runs of length up to 100 and for all signed integers in the interval [−73, 74]. To code outliers, the model provides escape symbols for coding zero-run lengths and signed integer values requiring either 8- or 16-bit representations. For instance, a zero-run of length 127 would be coded as the Huffman codeword for escape symbol #105 followed by the byte 0x7F.

4.2 Adaptive Huffman codebook construction

Symbol frequencies are computed for each block and Huffman codebooks are constructed following the procedures suggested in Annex K of the JPEG standard [Ame91, PM92]. Symbols are encoded and written into entropy coded data segments delineated by special block headers indicating which of the transmitted Huffman coders is to be used for decoding each block. Restart markers for error-control use are optional in the entropy coded bit stream.

The BITS and HUFFVAL lists specified in the JPEG standard are transmitted as side information for each Huffman coder. The BITS list contains 16 integers giving the number of Huffman codewords of length 1, the number of length 2, etc., up to length 16 (codewords longer than 16 bits are precluded by the codebook design algorithm). From this the WSQ decoder reconstructs the tree of Huffman codewords. HUFFVAL lists the symbols associated with the codewords of length 1, length 2, etc., enabling the Huffman decoder to decode the compressed bit stream using the procedures of Annex C in the JPEG standard. The coding model in Table 16.2 is then inverted to map the decoded symbols back to quantizer indices. The decoder is responsible for keeping count of the number of quantizer indices extracted and determining when each subband has been completely decoded.

5 The first-generation fingerprint image encoder

We now outline the details of the first FBI-approved WSQ fingerprint image encoder. As mentioned above, the standard allows for multiple encoders within the framework of the decoder specification, so any or all of the methods described in this section are subject to change in future encoder versions. To help the decoder cope with the possibility of multiple WSQ encoders, the frame header in compressed fingerprint image files identifies the encoder version used on that particular image.

5.1 Source image normalization

Before an image, $I(m,n)$, is decomposed using the DWT, it is first normalized according to the following formula:

$$I'(m,n) = \frac{I(m,n) - M}{R} \quad , \tag{16.4}$$

where M is the image mean and

$$R = \frac{1}{128}\max(I_{max} - M,\, M - I_{min}) \quad .$$

I_{min} and I_{max} are, respectively, the minimum and maximum pixel values in the image $I(m,n)$. The main effect of this normalization is to give the lowest frequency DWT subband a mean of approximately zero. This brings the statistical distribution of quantizer bin indices for the lowest frequency subband more in line with the distributions from the other subbands in the lowest frequency block and facilitates compressing them with the same Huffman codebook.

5.2 First-generation wavelet filters

The analysis filter bank used in the first encoder is a WS/WS filter bank whose impulse responses have lengths of 9 taps (lowpass filter) and 7 taps (highpass filter). These filters were constructed by Cohen, Daubechies, and Feauveau [CDF92, Dau92]; they correspond to a family of symmetric, biorthogonal wavelet functions. The impulse response taps are given in Table 16.3. Subbands 60–63 in the spatial frequency decomposition shown in Figure 7 are not computed or transmitted by encoder #1; the decoder assumes these bands have been quantized to zero.

5.3 Optimal bit allocation and quantizer design

Adaptive quantization of the DWT subbands in encoder #1 is based on optimal adaptive allocation of bits to the subbands subject to a rate-control mechanism in

TABLE 16.3. Analysis filter impulse responses for encoder #1.

Tap	Exact Value	Approx. Value
$h_0(0)$	$-5\sqrt{2}x_1(48\|x_2\|^2 - 16\Re e\, x_2 + 3)/32$	0.85269867900940
$h_0(\pm 1)$	$-5\sqrt{2}x_1(8\|x_2\|^2 - \Re e\, x_2)/8$	0.37740285561265
$h_0(\pm 2)$	$-5\sqrt{2}x_1(4\|x_2\|^2 + 4\Re e\, x_2 - 1)/16$	−0.11062440441842
$h_0(\pm 3)$	$-5\sqrt{2}x_1(\Re e\, x_2)/8$	−0.023849465019380
$h_0(\pm 4)$	$-5\sqrt{2}x_1/64$	0.037828455506995
$h_1(-1)$	$\sqrt{2}(6x_1 - 1)/16x_1$	0.78848561640566
$h_1(-2, 0)$	$-\sqrt{2}(16x_1 - 1)/64x_1$	−0.41809227322221
$h_1(-3, 1)$	$\sqrt{2}(2x_1 + 1)/32x_1$	−0.040689417609558
$h_1(-4, 2)$	$-\sqrt{2}/64x_1$	0.064538882628938

$$x_1 = A + B - 1/6 \quad , \quad x_2 = -\frac{A+B}{2} - 1/6 + i\sqrt{3}(A-B)/2 \quad ,$$

$$A = \left(\frac{-14\sqrt{15} + 63}{1080\sqrt{15}}\right)^{1/3} \quad , \quad B = \left(\frac{-14\sqrt{15} - 63}{1080\sqrt{15}}\right)^{1/3} .$$

the form of a "target" bit rate constraint, r, that is provided by the user. Rather than attempt to reproduce the derivation of the optimal bit allocation formula in this exposition, we will concentrate on simply describing the recipe used to design the bank of subband quantizers and refer the reader to the references [BB93, BB94, BB95, BBOH96] for details.

The formula for optimal bit allocation is based on rate-distortion theory and modelled in terms of subband variances, so a subband variance estimate is made on a central subregion of each DWT subband. The rationale for using a central subregion is to avoid ruled lines, text and handwriting that are typically found near the borders of fingerprint images. The goal of optimal bit allocation is to divide up the target bit rate, r, amongst the DWT subbands in a manner that will minimize a model for overall distortion in the reconstructed image.

Let m_k be the factor by which the kth DWT subband has been downsampled, e.g., $m_{63} = 16$ and $m_4 = 256$. The bit rate to be assigned to the kth subband will be denoted r_k; the target bit rate, r, imposes a constraint on the subband bit rates via the relation

$$r = \sum_k \frac{r_k}{m_k} \quad . \tag{16.5}$$

For encoder #1, an appropriate value of r is specified by the FBI based on imaging system characteristics; $r = 0.75$ bpp can be regarded as a typical value, although certain systems may require higher target bit rates to achieve acceptable image quality. Typical fingerprint images coded at a target of 0.75 bpp tend to achieve about 15:1 compression on average.

The decoder specification allows the encoder to discard some subbands and transmit a bin width of zero ($Q_k = 0$) to signify that no compressed image data is being transmitted for subband k. For instance, this is always done for $60 \leq k \leq 63$ in encoder #1, and may be done for other subbands as well on an image by image basis if the encoder determines that a certain subband contains so little information that it should be discarded altogether. To keep track of the subband bit allocation, let K be the index set of all subbands assigned positive bit rates (in particular, for encoder #1, $K \subset \{0, 1, \ldots, 59\}$). The fraction of DWT coefficients being coded at positive bit rates will be denoted by S, where

$$S = \sum_{k \in K} \frac{1}{m_k} \quad . \tag{16.6}$$

To relate bit rates to quantizer bin widths, we model the data in each subband as lying in some interval of finite extent, specifically, as being contained within an interval spanning 5 standard deviations. This assumption may not be valid in general, but we will not incur overload distortion due to outliers because outliers are coded using escape sequences in the Huffman coding model. Therefore, for the sake of quantizer design we assume that the data lies in the interval $[\mu_k - \gamma\sigma_k\ ,\ \mu_k + \gamma\sigma_k]$, where the loading factor, γ, has the value $\gamma = 2.5$. If we model the average transmission rate for a quantizer with L_k bins by

$$r_k = \log_2 L_k \quad \text{bits/sample}$$

then we obtain a relation connecting bin widths and bit rates:

$$Q_k \;=\; \frac{2\gamma\sigma_k}{L_k} \;=\; \frac{2\gamma\sigma_k}{2^{r_k}} \quad . \tag{16.7}$$

Now we can present a formula for bin widths, Q_k, whose corresponding subband bit rates, r_k, satisfy the constraint (16.5) imposed by r. Let Q'_k denote *relative* bin widths,

$$Q'_k \;=\; \frac{Q_k}{Q_0} \;=\; q \cdot Q_k \quad \text{with } q = Q_0^{-1} \quad , \tag{16.8}$$

which can be regarded as "weights" related to the relative allocation of bits. The parameter q is a constant related in a complicated manner to the absolute overall bit rate of the quantizer system. The weights chosen by the FBI are

$$Q'_k \;=\; \begin{cases} 1\,, & 0 \le k \le 3\,, \\ \frac{10}{A_k \log_e(\sigma_k^2)}\,, & 4 \le k \le 59\,, \end{cases} \tag{16.9}$$

where the constants A_k are given in the encoder specification. Note that the weights, Q'_k, are image-dependent (since σ_k^2 is the variance of the kth subband), which means that this quantization strategy is employing an image-dependent bit allocation.

To achieve effective rate control, it remains to relate the parameter q to the overall target bit rate, r. It can be shown [BB93, BB94] that the value

$$q \;=\; \gamma^{-1} 2^{r/S-1} \left[\prod_{k \in K} \left(\frac{\sigma_k}{Q'_k} \right)^{1/m_k} \right]^{-1/S} \quad . \tag{16.10}$$

produces a quantizer bank corresponding to a bit allocation that satisfies a target bit rate constraint of r bpp. The effectiveness of this rate control mechanism has been documented in practice [BB93] and can be interpreted as an optimal bit allocation with respect to an image-dependent weighted mean-square distortion metric. (The dependence of the distortion metric on the individual image is a consequence of the use of image-dependent weights in (16.9).)

Two cases require special attention. First, to prevent overflow if $\log_e(\sigma_k^2) \approx 0$, the encoder discards any subband for which $\sigma_k^2 < 1.01$ and sets $Q_k = 0$. Second, if $Q_k > 2\gamma\sigma_k$ then the above quantization model implies that $r_k < 0$. Since this is not physically meaningful, we use an iterative procedure (given explicitly in the encoder specification) to determine q. The iterative procedure excludes from the bit allocation those subbands that have theoretically nonpositive bit rates; this will ensure that the overall bit rate constraint, r, is met. Once q has been determined, bin widths are computed and quantization performed for all nondiscarded subbands, including those with theoretically negative bit rates. While we expect that quantization of bands with negative bit allocations will produce essentially zero bits, it may nonetheless happen that a few samples in such bands actually get mapped to nonzero values by quantization and therefore contribute information to the reconstructed image.

For all subbands, the zero bin width, Z_k, is given in terms of Q_k by the formula

$$Z_k = 1.2\,Q_k \quad . \tag{16.11}$$

Although the parameter, C, determining the quantization decoder output levels is not used in the encoding process, it can be changed by the WSQ encoder since it is transmitted as side information in the compressed bit stream. The value of C used with encoder #1 is $C = 0.44$.

5.4 Huffman coding blocks

The subbands produced by encoder #1 are divided into three blocks for Huffman coding, with one Huffman encoder constructed for block 1 (subbands 0 through 18) and a second Huffman encoder constructed for both blocks 2 and 3 (subbands 19–51 and 52–59, respectively). Both Huffman coders make use of the entire symbol alphabet in Table 16.2.

6 Conclusions

Aside from fine-tuning some of the details and parameters, the FBI WSQ algorithm described above has changed little since the first draft of the specification [Fed92] was presented to a technical review conference at the National Institute of Standards and Technology (NIST) in Gaithersburg, MD, on December 8, 1992. All of the main aspects—the use of a cascaded two-channel product filter bank with symmetric boundary conditions, optimal bit allocation and uniform scalar quantization, zero run-length and adaptive Huffman coding—have been in place since late 1992. As it stands, this specification represents the basic, commercially implementable subband image coding technology of that moment in time.

Of course, all standards are, to some extent, deliberately "frozen in time." While we anticipate future improvements in encoder design, within the framework stipulated by the decoder specification, it is necessary to give implementors a well-defined target at which to aim their efforts. That these efforts have been successful is borne out by the results of the FBI's WSQ compliance testing protocol, administered by NIST [BBOH96]. As of mid-1996, *all* of the 28-odd implementations that had been submitted for compliance certification had eventually passed the tests, albeit some after more than one attempt. The next, crucial phase of this project will be to integrate this technology successfully into the day-to-day operations of the criminal justice system.

7 References

[Ame91] Amer. Nat'l. Standards Inst. *Digital Compression and Coding of Continuous-Tone Still Images, Part 1, Requirements and Guidelines*, ISO Draft Int'l. Standard 10918-1, a.k.a. "the JPEG Standard", February 1991.

[Ame93] Amer. Nat'l. Standards Inst. *American National Standard—Data Format for the Interchange of Fingerprint Information*, ANSI/NIST-CSL 1-1993, November 1993.

[BB92] Jonathan N. Bradley and Christopher M. Brislawn. Compression of fingerprint data using the wavelet vector quantization image compression algorithm. Technical Report LA-UR-92-1507, Los Alamos Nat'l. Lab, April 1992. FBI report.

[BB93] Jonathan N. Bradley and Christopher M. Brislawn. Proposed first-generation WSQ bit allocation procedure. In *Proc. Symp. Criminal Justice Info. Services Tech.*, pages C11–C17, Gaithersburg, MD, September 1993. Federal Bureau of Investigation.

[BB94] Jonathan N. Bradley and Christopher M. Brislawn. The wavelet/scalar quantization compression standard for digital fingerprint images. In *Proc. Int'l. Symp. Circuits Systems*, volume 3, pages 205–208, London, June 1994. IEEE Circuits Systems Soc.

[BB95] Jonathan N. Bradley and Christopher M. Brislawn. FBI parameter settings for the first WSQ fingerprint image coder. Technical Report LA-UR-95-1410, Los Alamos Nat'l. Lab, April 1995. FBI report.

[BBOH96] Christopher M. Brislawn, Jonathan N. Bradley, Remigius J. Onyshczak, and Tom Hopper. The FBI compression standard for digitized fingerprint images. In *Appl. Digital Image Process. XIX*, volume 2847 of *Proc. SPIE*, pages 344–355, Denver, CO, August 1996. SPIE.

[BCW90] T. C. Bell, J. G. Cleary, and J. H. Witten. *Text Compression.* Prentice Hall, Englewood Cliffs, NJ, 1990.

[Bri96] Christopher M. Brislawn. Classification of nonexpansive symmetric extension transforms for multirate filter banks. *Appl. Comput. Harmonic Anal.*, 3:337–357, 1996.

[CDF92] Albert Cohen, Ingrid C. Daubechies, and J.-C. Feauveau. Biorthogonal bases of compactly supported wavelets. *Commun. Pure Appl. Math.*, 45:485–560, 1992.

[Dau92] Ingrid C. Daubechies. *Ten Lectures on Wavelets.* Number 61 in CBMS-NSF Regional Conf. Series in Appl. Math., (Univ. Mass.—Lowell, June 1990). Soc. Indust. Appl. Math., Philadelphia, 1992.

[Fed92] Federal Bureau of Investigation. *WSQ Gray-Scale Fingerprint Image Compression Specification*, Revision 1.0, November 1992. Drafted by T. Hopper, C. Brislawn, and J. Bradley.

[Fed93] Federal Bureau of Investigation. *WSQ Gray-Scale Fingerprint Image Compression Specification*, IAFIS-IC-0110v2. Washington, DC, February 1993.

[GG92] Allen Gersho and Robert M. Gray. *Vector Quantization and Signal Compression.* Kluwer, Norwell, MA, 1992.

[HP92] Tom Hopper and Fred Preston. Compression of grey-scale fingerprint images. In *Proc. Data Compress. Conf.*, pages 309–318, Snowbird, UT, March 1992. IEEE Computer Soc.

[JN84] Nuggehally S. Jayant and Peter Noll. *Digital Coding of Waveforms.* Prentice Hall, Englewood Cliffs, NJ, 1984.

[Miz94] Shoji Mizuno. Information-preserving two-stage coding for multilevel fingerprint images using adaptive prediction based on upper bit signal direction. *Optical Engineering*, 33(3):875–880, March 1994.

[PM92] William B. Pennebaker and Joan L. Mitchell. *JPEG Still Image Data Compression Standard.* Van Nostrand Reinhold, New York, NY, 1992.

[SN96] Gilbert Strang and Truong Nguyen. *Wavelets and Filter Banks.* Wellesley-Cambridge, Wellesley, MA, 1996.

[Vai93] P. P. Vaidyanathan. *Multirate Systems and Filter Banks.* Prentice Hall, Englewood Cliffs, NJ, 1993.

[VK95] Martin Vetterli and Jelena Kovačević. *Wavelets and Subband Coding.* Prentice Hall, Englewood Cliffs, NJ, 1995.

[Wat93] Craig I. Watson. NIST special database 9: Mated fingerprint card pairs. Technical report, Nat'l. Inst. Standards Tech., Gaithersburg, MD, February 1993.

17

Embedded Image Coding Using Wavelet Difference Reduction

Jun Tian
Raymond O. Wells, Jr.

ABSTRACT We present an embedded image coding method, which basically consists of three steps, *Discrete Wavelet Transform*, *Differential Coding*, and *Binary Reduction*. Both J. Shapiro's embedded zerotree wavelet algorithm, and A. Said and W. A. Pearlman's codetree algorithm use spatial orientation tree structures to implicitly locate the significant wavelet transform coefficients. Here a direct approach to find the positions of these significant coefficients is presented. The encoding can be stopped at any point, which allows a target rate or distortion metric to be met exactly. The bits in the bit stream are generated in the order of importance, yielding a fully embedded code to successively approximate the original image source; thus it's well suited for progressive image transmission. The decoder can also terminate the decoding at any point, and produce a lower (bit) rate reconstruction image. Our algorithm is very simple in its form (which will make the encoding and decoding very fast), requires no training of any kind or prior knowledge of image sources, and has a clear geometric structure. The image coding results of it are quite competitive with almost all previous reported image compression algorithms on standard test images.

1 Introduction

Wavelet theory and applications have grown explosively in the last decade. It has become a cutting-edge technology in applied mathematics, neural networks, numerical computation, and signal processing, especially in the area of image compression. Due to its good localization property in both the spatial domain and spectral domain, a wavelet transform can handle well transient signals, and hence significant compression ratios may be obtained. Current research on wavelet based image compression (see for example [Sha93, SP96, XRO97], etc) has shown the high promise of this relatively new yet almost mature technology.

In this paper, we present a lossy image codec based on index coding. It contains the following features:

- A discrete wavelet transform which removes the spatial and spectral redundancies of digital images to a large extends.

- Index coding (differential coding and binary reduction) which represents the positions of significant wavelet transform coefficients very efficiently.

- Ordered bit plane transmission which provides a successive approximation of image sources and facilitates progressive image transmission.
- Adaptive arithmetic coding which requires no training of any kind or prior knowledge of image sources.

This paper is organized as follows. Section 2, 3, and 4 explain the discrete wavelet transform, differential coding, and binary reduction, respectively. In Section 5 we combine these three methods together and present the Wavelet Difference Reduction algorithm. Section 6 contains the experimental results of this algorithm. As a result of the algorithm, we discuss synthetic aperture radar (SAR) image compression in Section 7. The paper is concluded in Section 8.

2 Discrete Wavelet Transform

The terminology "wavelet" was first introduced in 1984 by A. Grossmann and J. Morlet [GM84]. An $L^2(\mathbf{R})$ function $\psi(x)$ induces an *orthonormal wavelet system* of $L^2(\mathbf{R})$ if the dilations and translations of $\psi(x)$, $\{\psi_{j,k}(x) = 2^{j/2}\psi(2^j x - k) : j,k \in \mathbf{Z}\}$ constitute an orthonormal basis of $L^2(\mathbf{R})$. We call $\psi(x)$ the *wavelet function.* If $\psi(x)$ is compactly supported, then this wavelet system is associated with a *multiresolution analysis.* A multiresolution analysis consists of a sequence of embedded closed subspaces $V_j \subset L^2(\mathbf{R}), j \in \mathbf{Z}$, satisfying $\overline{\bigcup_{j\in\mathbf{Z}} V_j} = L^2(\mathbf{R})$ and $\bigcap_{j\in\mathbf{Z}} V_j = \{0\}$.Moreover, there should be an $L^2(\mathbf{R})$ function $\phi(x)$ such that $\{\phi_{j,k}(x) = 2^{j/2}\phi(2^j x - k) : k \in \mathbf{Z}\}$ is an orthonormal basis for V_j for all $j \in \mathbf{Z}$. Since $\phi(x) \in V_0 \subset V_1$, we have

$$\phi(x) = \sum_{k\in\mathbf{Z}} a_k \phi(2x - k)\,, \tag{17.1}$$

and $\psi(x)$ and $\phi(x)$ are related by

$$\psi(x) = \sum_{k\in\mathbf{Z}} (-1)^k a_{-k+1} \phi(2x - k)\,. \tag{17.2}$$

We call $\{a_k : k \in \mathbf{Z}\}$ and $\phi(x)$ the *scaling filter* and the *scaling function* of the wavelet system, respectively. Be defining $b_k := (-1)^k a_{-k+1}$, we call $\{b_k : k \in \mathbf{Z}\}$ the *wavelet filter.* As one knows, it is straightforward to define an orthonormal wavelet system by (17.2) from a multiresolution analysis.

For a discrete signal x, the *discrete wavelet transform* (DWT) of x consists of two part, the low frequency part L and the high frequency part H. They can be computed by

$$L_k = 2^{-1/2} \sum_{n\in\mathbf{Z}} a_{n-2k} x_n\,,$$

$$H_k = 2^{-1/2} \sum_{n\in\mathbf{Z}} b_{n-2k} x_n\,.$$

And the *inverse discrete wavelet transform* (IDWT) will give back x from L and H,

$$x_n = 2^{-1/2} \sum_{k\in\mathbf{Z}} (a_{n-2k} L_k + b_{n-2k} H_k)\,.$$

In the case of a *biorthogonal wavelet system* where synthesis filters are different from analysis filters (and consequently synthesis scaling/wavelet functions are different from analysis scaling/wavelet functions), the IDWT is given by

$$x_n = 2^{-1/2} \sum_{k \in \mathbf{Z}} \left(\tilde{a}_{n-2k} L_k + \tilde{b}_{n-2k} H_k \right) ,$$

where $\tilde{a}$ and $\tilde{b}$ are synthesis scaling filter and wavelet filter, respectively.

For more details about discrete wavelet transform and wavelet analysis, we refer to [Chu92, Dau92, Mal89, Mey92, RW97, SN95, VK95, Wic93], etc. The discrete wavelet transform can remove the redundancies of image sources very successfully, and we will take it as the first step in our image compression algorithm.

3 Differential Coding

Differential Coding [GHLR75] takes the difference of adjacent values. This is a quite useful coding scheme when we have a set of integers with monotonically increasing order. For example, if we have an integer set $\mathcal{S}$,

$$\mathcal{S} = \{1, 2, 5, 36, 42\} ,$$

then its *difference set* $\mathcal{S}'$ is

$$\mathcal{S}' = \{1, 1, 3, 31, 6\} .$$

And it's straightforward to get back $\mathcal{S}$ from the difference set $\mathcal{S}'$ by taking the partial sum of $\mathcal{S}'$.

4 Binary Reduction

Binary Reduction is one of the representations of positive binary integers, with the shortest representation length, as described in [Eli75]. It's the binary coding of an integer, with the most significant bit removed. For example, since the binary representation of 19 is 10011, then the binary reduction of 19 is 0011. For the example in Section 3, the binary reduction of $\mathcal{S}'$ will be

$$\mathcal{S}'' = \{, , 1, 1111, 10\} .$$

And we call $\mathcal{S}''$ the *reduced set* of $\mathcal{S}$. Note that there are no coded symbols before the first two commas "," in $\mathcal{S}''$. In practice, one will need some end of message symbol to separate different elements when the reduced set is used, like the comma "," above.

The binary reduction is a reversible procedure by adding a "1" as the most significant bit in the binary representation.

5 Description of the Algorithm

After taking the discrete wavelet transform of an image data, all wavelet transform coefficients will be ordered in such a way that the coefficients at coarser scale will come before the coefficients at finer scale. A typical scanning pattern [Sha93] is indicated in Figure 1. In a wavelet decomposition domain with N scales, the scan begins at LL_N, then HL_N, LH_N, HH_N, at which point it moves on to scale $N-1$, etc. Because of the nature of the discrete wavelet transform, inside each high-low subregion $HL_N, HL_{N-1}, \cdots, HL_1$, the scanning order goes column by column, and inside each low-high subregion $LH_N, LH_{N-1}, \cdots, LH_1$, the scanning order goes row by row. A wavelet transform coefficient x is defined as *significant* with respect to a threshold T if $|x| \geq T$, otherwise x is said to be *insignificant.*

The wavelet transform coefficients are stored in three ordered sets, the set of significant coefficients (**SCS**), the temporary set containing significant coefficients found in a given round (**TPS**), and the set of insignificant coefficients (**ICS**). The initial **ICS** contains all the wavelet transform coefficients with the order shown in Figure 1, while **SCS** and **TPS** are empty. And the initial threshold T is chosen such that $|x_j| < 2T$ for all the wavelet transform coefficients x_j, and for some j_0, $|x_{j_0}| \geq T$. The encoder of this Wavelet Difference Reduction algorithm will output the initial threshold T.

First we have a sorting pass. In the sorting pass, all significant coefficients in **ICS** with respect to T will be moved out and put into **TPS**. Let $\mathcal{S}$ be the indices (in **ICS**) of these significant coefficients. The encoder outputs the reduced set $\mathcal{S}''$ of $\mathcal{S}$. Instead of using "," as the end of message symbol to separate different elements in $\mathcal{S}''$, we will take the signs (either "+" or "−") of these significant coefficients as the end of message symbol. For example, if $\mathcal{S} = \{1, 2, 5, 36, 42\}$, and the signs of these five significant coefficients are "$+ - + + -$", then the encoding output $\mathcal{S}''$ will be "$+ - 1 + 1111 + 10-$". Then update the indexing in **ICS**. For example, if x_3 is moved to **TPS**, then all coefficients after x_3 in **ICS** will have their indices subtracted by 1, and so on.

Right after the sorting pass, we have a refinement pass. In the refinement pass, the magnitudes of coefficients in **SCS** will have an additional bit of precision. For example, if the magnitude of a coefficient in **SCS** is known to be in $[32, 64)$, then it will be decided at this stage whether it is in $[32, 48)$ or $[48, 64)$. And a "0" symbol will indicate it is in the lower half $[32, 48)$, while a "1" symbol will indicate it is in the upper half $[48, 64)$. Output all these refinement values "0" and "1". Then append **TPS** to the end of **SCS**, reset **TPS** to the empty set, and divide T by 2. Another round begins with the sorting pass.

The resulting symbol stream in the sorting pass and refinement pass will be further coded by adaptive arithmetic coding [WNC87]. And the encoding can be stopped at any point, which allows a target rate or distortion metric to be met exactly.

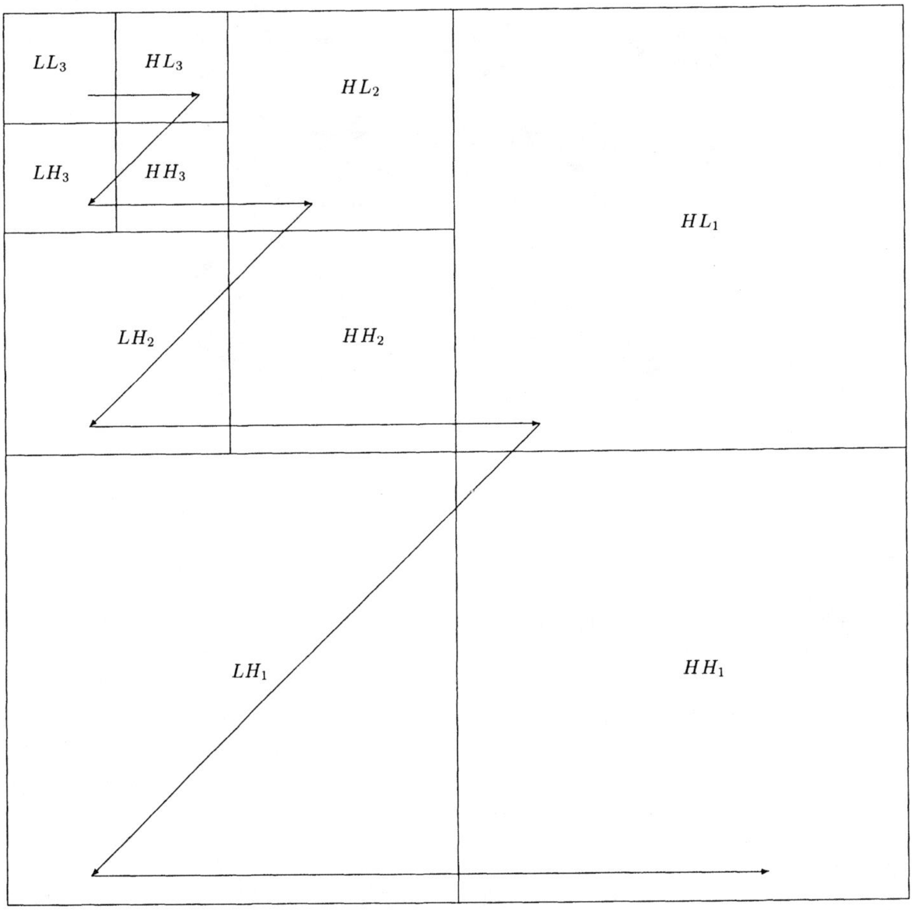

FIGURE 1. Scanning order of the wavelet transform coefficients with 3 scales.

FIGURE 2. Original

6 Experimental Results

Experiments have been done on all the 8 bits per pixel (bpp), grayscale test images, available from ftp://links.uwaterloo.ca:/pub/BragZone/, which include "Barbara", "Goldhill", "Lena" and others. And we used the Cohen-Daubechies-Feauveau 9/7-tap filters (CDF-97) [CDF92] with six scales. The symmetry of CDF-97 allows the "reflection" extension at the images edges. For our purpose, the compression performance is measured by the peak signal to noise ratio

$$\text{PSNR} = 10 \log_{10} \left(\frac{255^2}{\text{MSE}} \right) \text{dB},$$

where MSE is the mean square error between the original image and the reconstructed one. Some other criterion might have been more preferable. However, to make a direct comparison with other coders, PSNR is chosen. And the bit rate is calculated from the actual size of the compressed file.

Our experimental results show that the coding performance of the current implementation of this Wavelet Difference Reduction algorithm is quite competitive with other previous reported image compression algorithms on standard test images. Some early results were reported in [TW96]. Here we include some coding results for the 8 bpp, 512 × 512 grayscale "Lena" image. Figure 2 is the original "Lena" image, and Figure 3 is the one with compression ratio 8:1, using the Wavelet Difference Reduction algorithm, and having PSNR = 40.03 dB. Because the Wavelet Difference Reduction algorithm is an embedded coding scheme, one can actually achieve any given compression ratio.

FIGURE 3. 8:1 Compression, PSNR = 40.03 dB

7 SAR Image Compression

Real-time transmission of sensor data such as synthetic aperture radar (SAR) images is of great importance for both time critical applications such as military search and destroy missions as well as in scientific survey applications. Furthermore, since post processing of the collected data in either application involves search, classification and tracking of targets, the requirements for a "good" compression algorithm is typically very different from that of lossy image compression algorithms developed for compressing still-images. The definition of targets are application dependent and could be military vehicles, trees in the rain forest, oil spills etc.

To compress SAR images, first we take the logarithm of each pixel value in the SAR image data. Then apply the discrete wavelet transform on these numbers. And the following steps are just alternatively index coding and getting refinement values on the wavelet transform coefficients, as described in the Wavelet Difference Reduction algorithm. Adaptive arithmetic coding will compress the symbol stream further.

Since the Wavelet Difference Reduction algorithm locates the position of significant wavelet transform coefficients directly and contains a clear geometric structure, we may process the SAR image data directly in the compressed wavelet domain, for example, speckle reduction. For more details, we refer to [TGW+96].

The test data we used here is a fully polarimetric SAR image of the Lincoln north building in Lincoln, MA, collected by the Lincoln Laboratory MMW SAR. It was preprocessed with a technique known as the polarimetric whitening filer (PWF) [NBCO90]. We apply the Wavelet Difference Reduction algorithm on this PWFed SAR image. In practice the compression ratio can be set to any real number greater

FIGURE 4. PWFed SAR Image of a Building

than 1. Figure 4, 5, 6 and 7 show the SAR image and a sequence of images obtained by compressing the SAR image using the Wavelet Difference Reduction algorithm at the compression ratios 20:1, 80:1, and 400:1. Visually the image quality is still well preserved at the ratio 80:1, which indicates substantial advantage over JPEG compression [Wal91].

8 Conclusions

In this paper we presented an embedded image coding method, the Wavelet Difference Reduction algorithm. Compared with Shapiro's EZW [Sha93], and Said and Pearlman's SPC [SP96] algorithm, one can see that all these three embedded compression algorithm share the same basic structure. More specifically, a generic

FIGURE 5. Decompressed Image at the Compression Ratio 20:1

model including all these three algorithms (and some others) consists of five steps:

1. Take the discrete wavelet transform of the original image.
2. Order the wavelet transform coefficients from coarser scale to finer scale, as in Figure 1. Set the initial threshold T.
3. (Sorting Pass) Find the positions of significant coefficients with respect to T, and move these significant coefficients out.
4. (Refinement Pass) Get the refinement values of all significant coefficients, except those just found in the sorting pass of this round.
5. Divide T by 2 and go to step **3**.

The resulting symbol stream in step 3 and 4 will be further encoded by a lossless data compression algorithm.

FIGURE 6. Decompressed Image at the Compression Ratio 80:1

The only difference among these three algorithms is in Step 3, the sorting pass. In EZW, Shapiro employs the self similarity tree structure. In SPC, a set partitioning algorithm is presented which provides a better tree structure. In our algorithm, we combines the differential coding and binary reduction. The concept of combining the differential coding and binary reduction is actually a fairly general concept and not specific to the wavelet decomposition domain. For example, it can be applied to the Partition Priority Coding (PPC) [HDG92], and one would expect some possible improvement in the image coding results.

Acknowledgments: We would like to thank our colleagues in the Computational Mathematics Laboratory, Rice University, especially C. Sidney Burrus, Haitao Guo and Markus Lang, for fruitful discussion, exchange of code, and lots of feedback. Amir Said explained their SPC algorithm in details to us and made their code

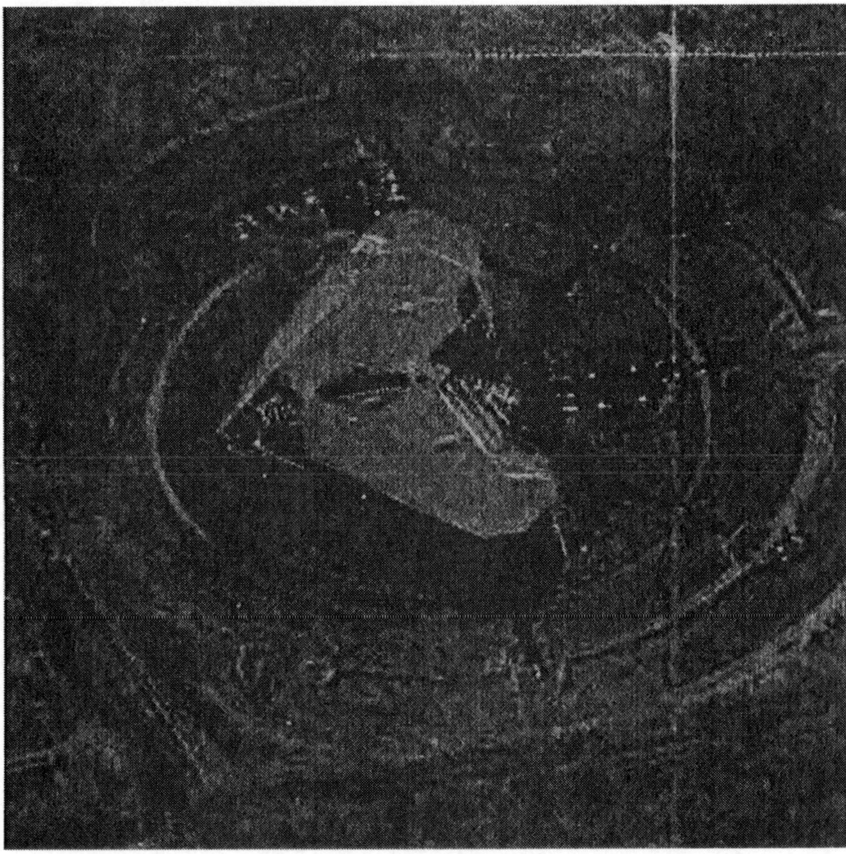

FIGURE 7. Decompressed Image at the Compression Ratio 400:1

available. We would like to take this opportunity to thank him. A special thanks goes to Alistair Moffat for valuable help. Part of the work for this paper was done during our visit at the Center for Medical Visualization and Diagnostic Systems (MeVis), University of Bremen. We would like to thank Heinz-Otto Peitgen, Carl Evertsz, Hartmut Juergens, and all others at Mevis, for their hospitality. This work was supported in part by ARPA and Texas ATP. Part of this work was presented at the IEEE Data Compression Conference, Snowbird, Utah, April 1996 [TW96] and the SPIE's 10th Annual International Symposium on Aerospace/Defense Sensing, Simulation, and Controls, Orlando, Florida, April 1996 [TGW+96].

9 References

[CDF92] A. Cohen, I. Daubechies, and J.-C. Feauveau. Biorthonormal bases of compactly supported wavelets. *Commun. Pure Appl. Math.*, XLV:485–560, 1992.

[Chu92] C. K. Chui. *An Introduction to Wavelets.* Academic Press, Boston, MA, 1992.

[Dau92] I. Daubechies. *Ten Lectures on Wavelets.* SIAM, Philadelphia, PA, 1992.

[Eli75] P. Elias. Universal codeword sets and representations of the integers. *IEEE Trans. Informat. Theory*, 21(2):194–203, March 1975.

[GHLR75] D. Gottlieb, S. A. Hagerth, P. G. H. Lehot, and H. S. Rabinowitz. A classification of compression methods and their usefulness for a large data processing center. In *Nat. Com. Conf.*, volume 44, pages 453–458, 1975.

[GM84] A. Grossmann and J. Morlet. Decomposition of Hardy functions into square integrable wavelets of constant shape. *SIAM J. Math. Anal.*, 15(4):723–736, July 1984.

[HDG92] Y. Huang, H. M. Dreizen, and N. P. Galatsanos. Prioritized DCT for compression and progressive transmission of images. *IEEE Trans. Image Processing*, 1:477–487, October 1992.

[Mal89] S. G. Mallat. Multiresolution approximation and wavelet orthonormal bases of $L^2(\mathbf{R})$. *Trans. AMS*, 315(1):69–87, September 1989.

[Mey92] Y. Meyer. *Wavelets and Operators.* Cambridge University Press, Cambridge, 1992.

[NBCO90] L. M. Novak, M. C. Burl, R. Chaney, and G. J. Owirka. Optimal processing of polarimetric synthetic aperture radar imagery. *Linc. Lab. J.*, 3, 1990.

[RW97] H. L. Resnikoff and R. O. Wells, Jr. *Wavelet Analysis and the Scalable Structure of Information.* Springer-Verlag, New York, 1997. To appear.

[Sha93] J. M. Shapiro. Embedded image coding using zerotrees of wavelet coefficients. *IEEE Trans. Signal Processing*, 41:3445–3462, December 1993.

[SN95] G. Strang and T. Nguyen. *Wavelets and Filter Banks.* Wellesley-Cambridge Press, Wellesley, MA, 1995.

[SP96] A Said and W. A. Pearlman. A new fast and efficient image codec based on set partitioning in hierarchical trees. *IEEE Trans. Circ. Syst. Video Tech.*, 6(3):243–250, June 1996.

[TGW+96] J. Tian, H. Guo, R. O. Wells, Jr., C. S. Burrus, and J. E. Odegard. Evaluation of a new wavelet-based compression algorithm for synthetic aperture radar images. In E. G. Zelnio and R. J. Douglass, editors, *Algorithms for Synthetic Aperture Radar Imagery III*, volume 2757 of *Proc. SPIE*, pages 421–430, April 1996.

[TW96] J. Tian and R. O. Wells, Jr. A lossy image codec based on index coding. In J. A. Storer and M. Cohn, editors, *Proc. Data Compression Conference*. IEEE Computer Society Press, April 1996. (ftp://cml.rice.edu/pub/reports/CML9515.ps.Z).

[VK95] M. Vetterli and J. Kovačević. *Wavelets and Subband Coding*. Prentice Hall, Englewood Cliffs, NJ, 1995.

[Wal91] G. K. Wallace. The JPEG still picture compression standard. *Commun. ACM*, 34:30–44, April 1991.

[Wic93] M. V. Wickerhauser. *Adapted Wavelet Analysis from Theory to Software*. Wellesley, MA, 1993.

[WNC87] I. H. Witten, R. Neal, and J. G. Cleary. Arithmetic coding for data compression. *Commun. ACM*, 30:520–540, June 1987.

[XRO97] Z. Xiong, K. Ramchandran, and M. Orchard. Space-frequency quantization for wavelet image coding. *to appear in IEEE Trans. Image Processing*, 1997.

18

Block Transforms in Progressive Image Coding

Trac D. Tran and Truong Q. Nguyen

1 Introduction

Block transform coding and subband coding have been two dominant techniques in existing image compression standards and implementations. Both methods actually exhibit many similarities: relying on a certain transform to convert the input image to a more decorrelated representation, then utilizing the same basic building blocks such as bit allocator, quantizer, and entropy coder to achieve compression.

Block transform coders enjoyed success first due to their low complexity in implementation and their reasonable performance. The most popular block transform coder leads to the current image compression standard JPEG [1] which utilizes the 8×8 Discrete Cosine Transform (DCT) at its transformation stage. At high bit rates (1 bpp and up), JPEG offers almost visually lossless reconstruction image quality. However, when more compression is needed (i.e., at lower bit rates), annoying blocking artifacts show up because of two reasons: (i) the DCT bases are short, non-overlapped, and have discontinuities at the ends; (ii) JPEG processes each image block independently. So, inter-block correlation has been completely abandoned.

The development of the lapped orthogonal transform [2] and its generalized version GenLOT [3, 4] helps solve the blocking problem to a certain extent by borrowing pixels from the adjacent blocks to produce the transform coefficients of the current block. Lapped transform outperforms DCT on two counts: (i) from the analysis viewpoint, it takes into account inter-block correlation, hence, provides better energy compaction that leads to more efficient entropy coding of the coefficients; (ii) from the synthesis viewpoint, its basis functions decay asymptotically to zero at the ends, reducing blocking discontinuities drastically. However, earlier lapped-transform-based image coders [2, 3, 5] have not utilized global information to their full advantage: the quantization and the entropy coding of transform coefficients are still independent from block to block.

Recently, subband coding has emerged as the leading standardization candidate in future image compression systems thanks to the development of the discrete wavelet transform. Wavelet representation with implicit overlapping and variable-length basis functions produces smoother and more perceptually pleasant reconstructed images. Moreover, wavelet's multiresolution characteristics have created an intuitive foundation on which simple, yet sophisticated, methods of encoding the transform coefficients are developed. Exploiting the relationship between the parent and the offspring coefficients in a wavelet tree, progressive wavelet coders [6, 7, 9] can effectively order the coefficients by bit planes and transmit more significant bits first. This coding scheme results in an embedded bit stream along with many other advantages such as exact bit rate control and near-idempotency (perfect idempotency is obtained when the transform maps integers to integers). In these subband coders, global information is taken into account fully.

From a frequency domain point of view, the wavelet transform simply provides an octave-band representation of signals. The dyadic wavelet transform is analogous to a non-uniform-band lapped transform. It can sufficiently decorrelate smooth images; however, it has problems with images with well-localized high frequency components, leading to low energy compaction. In this appendix, we shall demonstrate that the embedded framework is not only limited to the wavelet transform; it can be utilized with uniform-band lapped transforms as well. In fact, a judicious choice of appropriately-optimized lapped transform coupled with several levels of wavelet decomposition of the DC band can provide much finer frequency spectrum partitioning, leading to significant improvement over current wavelet coders. This appendix also attempts to shed some light onto a deeper understanding of wavelets, lapped transforms, their relation, and their performance in image compression from a multirate filter bank perspective.

2 The wavelet transform and progressive image transmission

Progressive image transmission is perfect for the recent explosion of the internet. The wavelet-based progressive coding approach first introduced by Shapiro [6] relies on the fundamental idea that more important information (defined here as what decreases a certain distortion measure the most) should be transmitted first. Assume that the distortion measure is the mean-squared error (MSE), the transform is paraunitary, and transform coefficients $c_{i,j}$ are transmitted one by one, it can be proven that the mean squared error decreases by $\frac{|c_{i,j}|}{N}$, where N is the total number of pixels [16]. Therefore, larger coefficients should be transmitted first. If one bit is transmitted at a time, this approach can be generalized to ranking the coefficients by bit planes and the most significant bits are transmitted first [8]. The progressive transmission scheme results in an embedded bit stream (i.e., it can be truncated at any point by the decoder to yield the best corresponding reconstructed image). The algorithm can be thought of as an elegant combination of a scalar quantizer with power-of-two stepsizes and an entropy coder to encode wavelet coefficients.

Embedded algorithm relies on the hierachical coefficients' tree structure that we called a *wavelet tree*, defined as a set of wavelet coefficients from different scales that belong in the same spatial locality as demonstrated in Figure 1(a), where the tree in the vertical direction is circled. All of the coefficients in the lowest frequency band make up the *DC band* or the *reference signal* (located at the upper left corner). Besides these DC coefficients, in a wavelet tree of a particular direction, each lower-frequency *parent node* has four corresponding higher-frequency *offspring nodes.* All coefficients below a parent node in the same spatial locality is defined as its *descendents.* Also, define a coefficient $c_{i,j}$ to be *significant* with respect to a given threshold T if $|c_{i,j}| \geq T$, and *insignificant* otherwise. Meaningful image statistics have shown that if a coefficient is insignificant, it is very likely that its offspring and descendents are insignificant as well. Exploiting this fact, the most sophisticated embedded wavelet coder SPIHT can output a single binary marker to represent very efficiently a large, smooth image area (an insignificant tree). For more details on the algorithm, the reader is refered to [7].

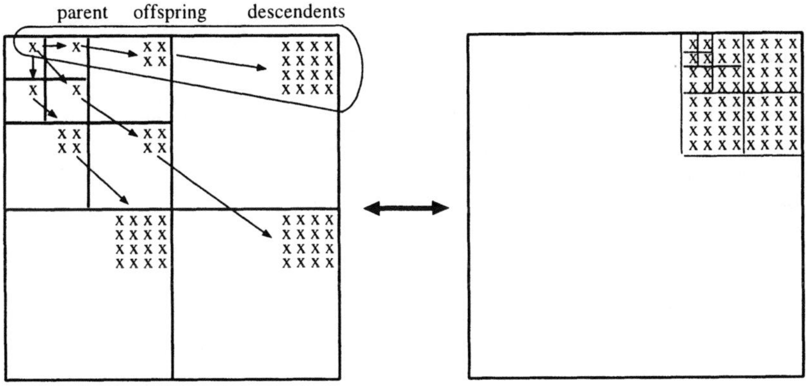

(a) (b)

FIGURE 1. Wavelet and block transform analogy.

Although the wavelet tree provides an elegant hierachical data structure which facilitates quantization and entropy coding of the coefficients, the efficiency of the coder still heavily depends on the transform's ability in generating insignificant trees. For non-smooth images that contain a lot of texture, the wavelet transform is not as efficient in signal decorrelation comparing to transforms with finer frequency selectivity and superior energy compaction. Uniform-band lapped transforms hold the edge in this area.

3 Wavelet and block transform analogy

Instead of obtaining an octave-band signal decomposition, one can have a finer uniform-band partitioning as depicted in Figure 2 (drawn for $M = 8$). The finer frequency partitioning compacts more signal energy into a fewer number of coefficients and generates more insignificant ones, leading to an enhancement in the performance of the zerotree algorithm. However, uniform filter bank also has uniform downsampling (all subbands now have the same size). A parent node does not have four offspring nodes as in the case of the wavelet representation. How would one come up with a new tree structure that still takes full advantage of the inter-scale correlation between block-transform coefficients?

The above question can be answered by investigating an analogy between the wavelet and block transform as illustrated in Figure 1. The parent, the offspring, and the descendents in a wavelet tree cover the same spatial locality, and so are the

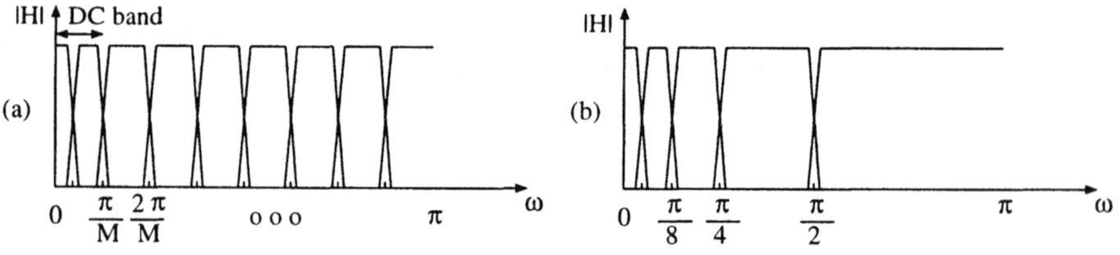

FIGURE 2. Frequency spectrum partitioning of (a) M-channel uniform-band transform (b) dyadic wavelet transform.

coefficients of a transform block. In fact, a wavelet tree in an L-level decomposition is analogous to a 2^L-channel transform's coefficient block. The difference lies at the bases that generate these coefficients. It can be shown that a 1D L-level wavelet decomposition, if implemented as a lapped transform, has the following coefficient matrix:

$$\mathbf{P}_L = \begin{bmatrix} h_0[n] * h_0[\frac{n}{2}] * \cdots * h_0[\frac{n}{2^{L-2}}] * h_0[\frac{n}{2^{L-1}}] \\ h_0[n] * h_0[\frac{n}{2}] * \cdots * h_0[\frac{n}{2^{L-2}}] * h_1[\frac{n}{2^{L-1}}] \\ h_0[n] * h_0[\frac{n}{2}] * \cdots * h_1[\frac{n}{2^{L-2}}] \\ h_0[n] * h_0[\frac{n}{2}] * \cdots * h_1[\frac{n}{2^{L-2}}] \\ \vdots \\ h_1[n] \\ h_1[n] \\ h_1[n] \\ h_1[n] \end{bmatrix}. \quad (18.1)$$

From the coefficient matrix $\mathbf{P}_L$, we can observe several interesting and important characteristics of the wavelet transform through the block transform's prism:

- The wavelet transform can be viewed as a lapped transform with filters of variable lengths. For an L-level decomposition, there are 2^L filters.

- Each basis function has linear phase; however, they do not share the same center of symmetry.

- The block size is defined by the length of the longest filter. If $h_0[n]$ is longer and has length N_0, the longest filter is on top, covering the DC component, and it has a length of $(2^L - 1)(N_0 - 1) + 1$. For the biorthogonal wavelet pair with $h_0[n]$ of length 9 and $h_1[n]$ of length 7 and $L = 3$, the eight resulting basis functions have lengths of $57, 49, 21, 21, 7, 7, 7$, and 7.

- For a 6-level decomposition using the same $9-7$ pair, the length of the longest basis function grows to 505! The huge amount of overlapped pixels explains the smoothness of as well as the complete elimination of blocking artifacts in wavelet-based coders' reconstructed images.

Each block of lapped transform coefficients represents a spatial locality similarly to a tree of wavelet coefficients. Let $\mathcal{O}(i,j)$ be the set of coordinates of all offspring of the node (i,j) in an M-channel block transform $(0 \leq i,j \leq M-1)$, then $\mathcal{O}(i,j)$

can be represented as follows:

$$\mathcal{O}(i,j) = \{(2i,2j),(2i,2j+1),(2i+1,2j),(2i+1,2j+1)\}. \tag{18.2}$$

All $(0,0)$ coefficients from all transform blocks form the DC band, which is similar to the wavelet transform's reference signal, and each of these nodes has only three offsprings: $(0,1)$, $(1,0)$, and $(1,1)$. This is a straightforward generalization of the structure first proposed in [10]. The only requirement here is that the number of channel M has to be a power of two. Figure 3 demonstrates through a simple rearrangement of the block transform coefficients that the redefined tree structure above does possess a wavelet-like multiscale representation. The quadtree grouping of the coefficients is far from optimal in the rate-distortion sense; however, other parent-offspring relationships for uniform-band transform such as the one mentioned in [6] do not facilitate the further usage of various entropy coders to increase the coding efficiency.

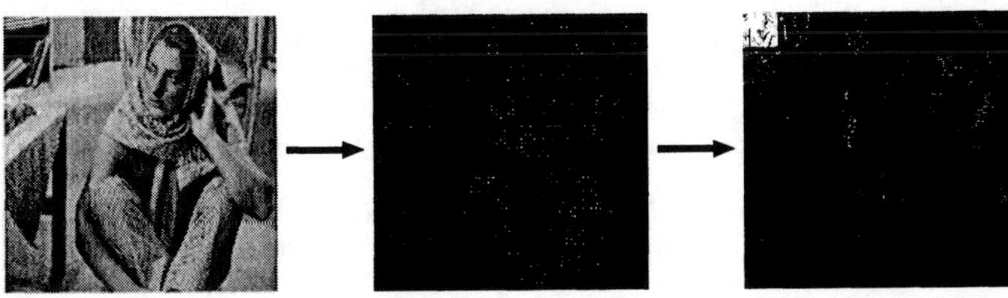

FIGURE 3. Demonstration of the analogy between uniform-band transform and wavelet representation.

4 Transform Design

A mere replacement of the wavelet transform by low-complexity block transforms is not enough to compete with SPIHT as testified in [10, 11]. We propose below several novel criteria in designing high-performance lapped transforms. The overall cost used for transform optimization is a combination of coding gain, DC attenuation, attenuation around the mirror frequencies, weighted stopband attenuation, and unequal-length constraint on filter responses:

$$C_{\text{overall}} = \alpha_1 C_{\text{coding gain}} + \alpha_2 C_{\text{DC}} + \alpha_3 C_{\text{mirror}} + \alpha_4 C_{\text{weighted-stopband}} + \alpha_5 C_{\text{unequal-length}}. \tag{18.3}$$

The first three cost functions are well-known criteria for image compression. Among them, higher coding gain correlates most consistently with higher objective performance (PSNR). Transforms with higher coding gain compact more energy into a fewer number of coefficients, and the more significant bits of those coefficients always get transmitted first. All designs in this appendix are obtained with a version of the generalized coding gain formula in [19]. Low DC leakage and high attenuation near the mirror frequencies are not as essential to the coder's objective performance as coding gain. However, they do improve the visual quality of the reconstructed image [5, 17].

The ramp-weighted stopband attenuation cost is defined as

$$C_{\text{weighted-stopband}} = \sum_{k=1}^{M-1} \int_{\omega \in \Omega_S} |W_k(e^{j\omega})H_k(e^{j\omega})|^2 d\omega,$$

where $W_k(e^{j\omega})$ is a linear function starting with value one at the peak of the frequency response decaying to zero at DC. The frequency weighting forces the highband filters to pick up as little energy as possible, ensuring a high number of insignificant trees. This cost function also helps the optimization process in obtaining higher coding gains.

The unequal-length constraint forces the tails of the high-frequency band-pass filters' responses to have very small values (not necessarily zeroes). The higher the frequency band, the shorter the effective length of the filter gets. This constraint is added to minimize the ringing around strong image edges at low bit rates, a typical characteristic of transforms with long filter lengths. Similar ideas have been presented in [20, 26, 27] where the filters have different lengths. However, these methods restrict the parameter search space severely, leading to low coding gains.

High-performance lapped transforms designed specifically for progressive image coding are presented in Figure 4(c)-(d). Figure 4(a) and (b) show the popular DCT and LOT for comparison purposes. The frequency response and the basis functions of the 8-channel 40-tap GenLOT shown in Figure 4(c) exemplify a well-optimized filter bank: high coding gain and low attenuation near DC for best energy compaction, smoothly decaying impulse responses for blocking artifacts elimination, and unequal-length filters for ringing artifacts suppression.

Figure 3 shows that there still exists correlation between DC coefficients. To decorrelate the DC band even more, several levels of wavelet decomposition can be used depending on the input image size. Besides the obvious increase in the coding efficiency of DC coefficients thanks to a deeper coefficient trees, wavelets provide variably longer bases for the signal's DC component, leading to smoother reconstructed images, i.e., blocking artifacts are further reduced. Regularity objective can be added in the transform design process to produce M-band wavelets, and a wavelet-like iteration can be carried out using uniform-band transforms as well. The complete coder's diagram is depicted in Figure 5.

5 Coding Results

The objective coding results (PSNR in dB) for standard 512×512 Lena and Barbara test images are tabulated in Table 18.1 where several different transforms are used:

- DCT, 8-channel 8-tap filters, shown in Figure 4(a).
- LOT 8-channel 16-tap filters, shown in Figure 4(b).
- GenLOT, 8-channel 40-tap filters, shown in Figure 4(c).
- LOT, 16-channel 32-tap filters, shown in Figure 4(d).

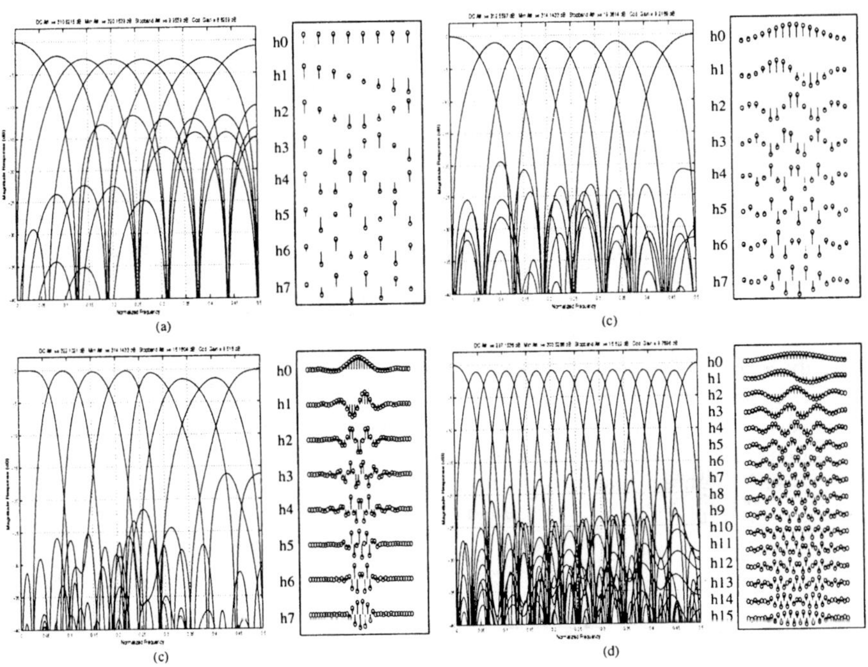

FIGURE 4. Frequency and impulse responses of orthogonal transforms: (a) 8-channel 8-tap DCT (b) 8-channel 16-tap LOT (c) 8-channel 40-tap GenLOT (d) 16-channel 32-tap LOT.

The block transform coders are compared to the best progressive wavelet coder SPIHT [7] and an earlier DCT-based embedded coder [10]. All computed PSNR quotes in dB are obtained from a real compressed bit stream with all overheads included. The rate-distortion curves in Figure 6 and the tabulated coding results in Table 18.1 clearly demonstrate the superiority of well-optimized lapped trans-

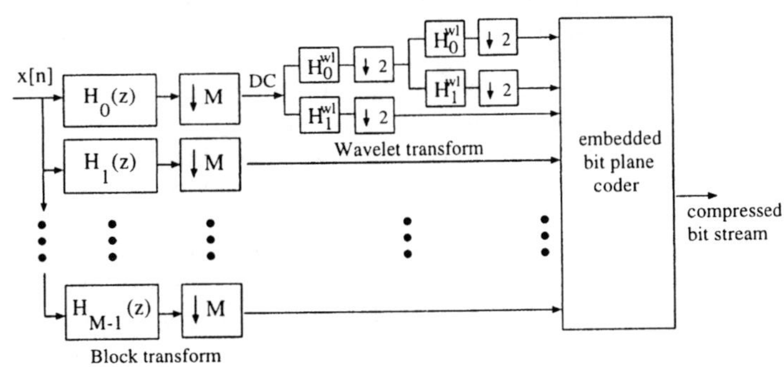

FIGURE 5. Complete coder's diagram.

Lena	Progressive Transmission Coders					
Comp. Ratio	SPIHT (9-7WL)	Xiong et al (DCT)	8 x 8 DCT	8 x 16 LOT	8 x 40 GenLOT	16 x 32 LOT
1:8	40.41	39.62	39.91	40.09	40.43	40.16
1:16	37.21	36.00	36.38	36.75	37.32	36.96
1:32	34.11	32.25	32.90	33.57	34.23	33.87
1:64	31.10	--	29.67	30.48	31.16	30.85
1:100	29.35	--	27.80	28.61	29.31	28.98
1:128	28.38	--	26.91	27.61	28.35	27.99

(a)

Barbara	Progressive Transmission Coders					
Comp. Ratio	SPIHT (9-7WL)	Xiong et al (DCT)	8 x 8 DCT	8 x 16 LOT	8 x 40 GenLOT	16 x 32 GLBT
1:8	36.41	36.10	36.31	37.43	38.08	38.02
1:16	31.40	30.82	31.11	32.70	33.47	33.47
1:32	27.58	26.83	27.28	28.80	29.53	29.70
1:64	24.86	--	24.58	25.70	26.37	26.63
1:100	23.76	--	23.42	24.34	24.95	25.14
1:128	23.35	--	22.68	23.37	24.01	24.09

(b)

TABLE 18.1. Coding results of various progressive coders (a) for Lena (b) for Barbara.

forms over wavelets. For a smooth image like Lena where the wavelet transform can sufficiently decorrelate, SPIHT offers a comparable performance. However, for a highly-textured image like Barbara, the 8 × 40 GenLOT and the 16 × 32 LOT coder can provide a PSNR gain of around 2 dB over a wide range of bit rates. Unlike other block transform coders whose performance dramatically drops at very high compression ratios, the new progressive coders are consistent throughout as illustrated in Figure 6. Lastly, better decorrelation of the DC band provides around 0.3 – 0.5 dB improvement over the earlier DCT embedded coder in [10].

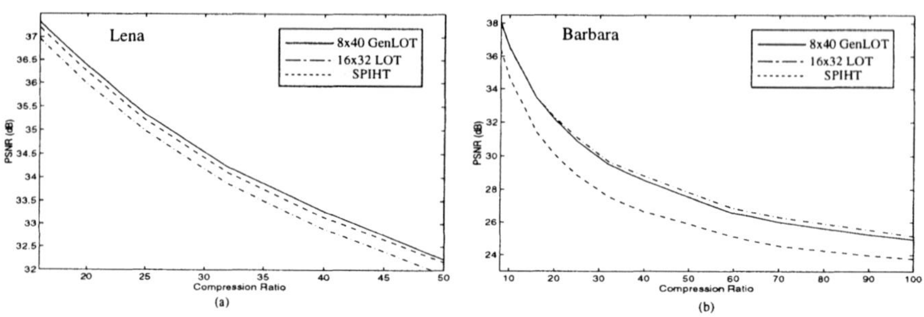

FIGURE 6. Rate-distortion curves of various progressive coders (a) for Lena (b) for Barbara.

Figure 7 - 9 confirm lapped transforms' superiority in reconstructed image quality as well. Figure 7 shows reconstructed Barbara images at 1:32 by various block transforms. Comparing to JPEG, blocking artifacts are already remarkably reduced in the DCT-based coder in Figure 7(a). Blocking is completely eliminated when DCT is replaced by better lapped transforms as shown in Figure 7(c)-(d), and Figure 8. A closer look in Figure 9(a)-(c) (where only 256 × 256 image portions are shown so that artifacts can be more easily seen) reveals that besides blocking elimination, good lapped transform can preserve texture beautifully (the table cloth and the clothes pattern) while keeping the edges relatively clean. The absence of excessive ringing considering the transform's long filters should not come across as a surprise: a glimpse of the time responses of the GenLOT in Figure 4(c) reveals that the high-frequency bandpasses and the highpass filter are very carefully designed – their lengths are essentially under 16-tap. Comparing to SPIHT, the reconstructed images have an overall sharper and more natural look with more defining edges and more evenly reconstructed texture regions. Although the PSNR difference is not as striking in the Goldhill image, the improvement in perceptual quality is rather significant as shown in Figure 9(d)-(f). Even at 1:100, the reconstructed Goldhill image in Figure 8(d) is still visually pleasant: no blocking and not much ringing. More objective and subjective evaluation of block-transform-based progressive coding can be found at *http://saigon.ece.wisc.edu/~waveweb/Coder/index.html.*

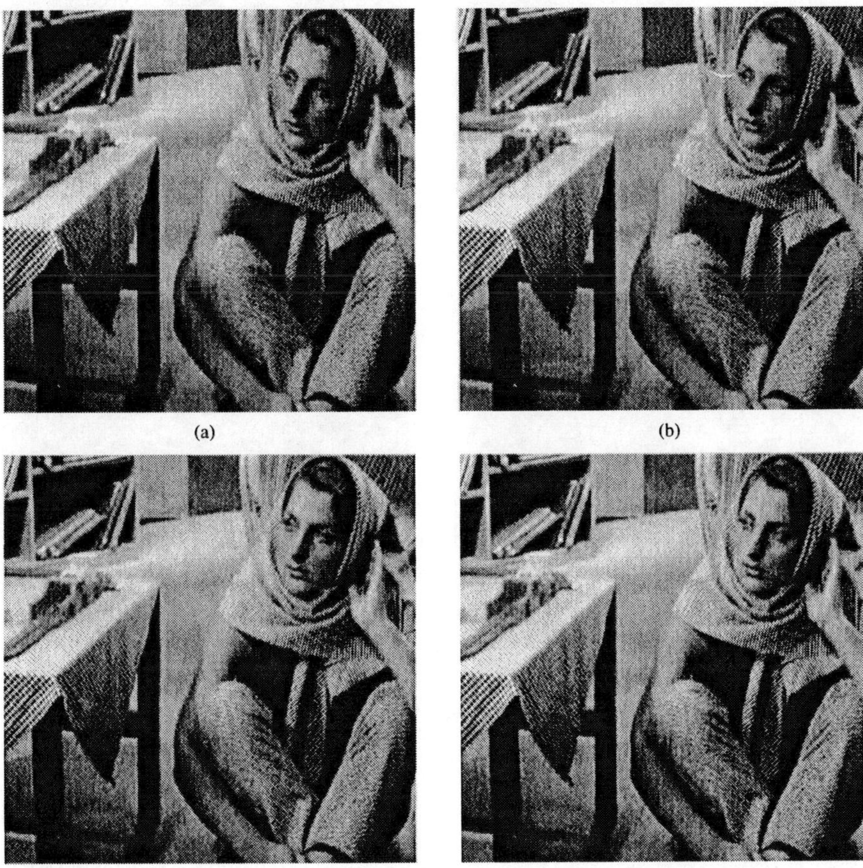

FIGURE 7. Barbara coded at 1:32 by (a) 8×8 DCT (b) 8×16 LOT (c) 8×40 GenLOT (d) 16×32 LOT.

As previously mentioned, the improvement over wavelets keys on the lapped transform's ability to capture and separate localized signal components in the frequency domain. In the spatial domain, this corresponds to images with directional repetitive texture patterns. To illustrate this point, the lapped-transform-based coder is compared against the FBI Wavelet Scalar Quantization (WSQ) standard [23]. When the original 768×768 gray-scale fingerprint image is shown in Figure

FIGURE 8. Goldhill coded by the 8 × 40 GenLOT coder at (a) 1:16, 33.36 dB (b) 1:32, 30.79 dB (c) 1:64, 28.60 dB (d) 1:100, 27.40 dB.

10(a) is compressed at 1 : 13.6 (43366 bytes) by WSQ, Bradley *et al* reported a PSNR of 36.05 dB. Using the 16 × 32 LOT in Figure 4(d), a PSNR of 37.87 dB can be achieved at the same compression ratio. For the same PSNR, the LOT coder can compress the image down to 1 : 19 where the reconstructed image is shown in Figure 10(b). To put this in perspective, the wavelet packet SFQ coder in [22] reported a PSNR of only 37.30 dB at 1:13.6 compression ratio. At 1 : 18.036 (32702 bytes), WSQ's reconstructed image as shown in Figure 10(c) has a PSNR of 34.42 dB while the LOT coder produces 36.32 dB. At the same distortion, we can compress the image down to a compression ratio of 1:26 (22685 bytes) as shown in Figure 10(d). Notice the high perceptual image quality in Figure 10(b) and (d): no visually disturbing blocking and ringing artifacts.

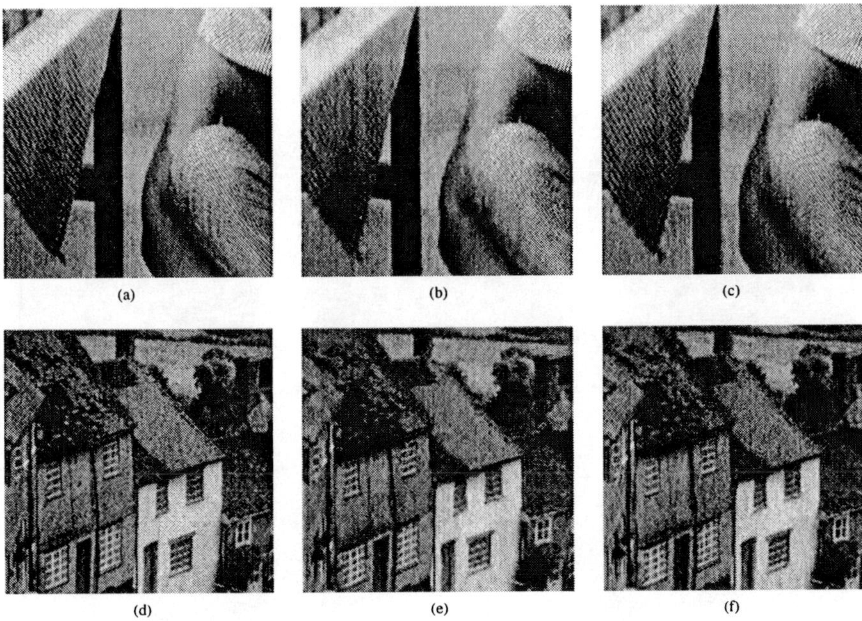

FIGURE 9. Perceptual comparison between wavelet and block transform embedded coder. Zoom-in portion of (a) original Barbara (b) SPIHT at 1:32 (c) 8×40 GenLOT embedded coder at 1:32 (d) original Goldhill (e) SPIHT at 1:32 (c) 8×40 GenLOT embedded coder at 1:32.

6 References

[1] W. B. Pennebaker and J. L. Mitchell, *JPEG: Still Image Compression Standard*, Van Nostrand Reinhold, 1993.

[2] H. S. Malvar, *Signal Processing with Lapped Transforms*, Artech House, 1992.

[3] R. de Queiroz, T. Q. Nguyen, and K. Rao, "The GenLOT: generalized linear-phase lapped orthogonal transform," *IEEE Trans. on SP,* vol. 40, pp. 497-507, Mar. 1996.

[4] T. D. Tran and T. Q. Nguyen, "On M-channel linear-phase FIR filter banks and application in image compression," *IEEE Trans. on SP,* vol. 45, pp. 2175-2187, Sept. 1997.

[5] S. Trautmann and T. Q. Nguyen, "GenLOT – design and application for transform-based image coding," *Asilomar conference*, Monterey, Nov. 1995.

[6] J. M. Shapiro, "Embedded image coding using zerotrees of wavelet coefficients," *IEEE Trans. on SP,* vol. 41, pp. 3445-3462, Dec. 1993.

[7] A. Said and W. A. Pearlman, "A new fast and efficient image codec on set partitioning in hierarchical trees," *IEEE Trans on Circuits Syst. Video Tech.*, vol. 6, pp. 243-250, June 1996.

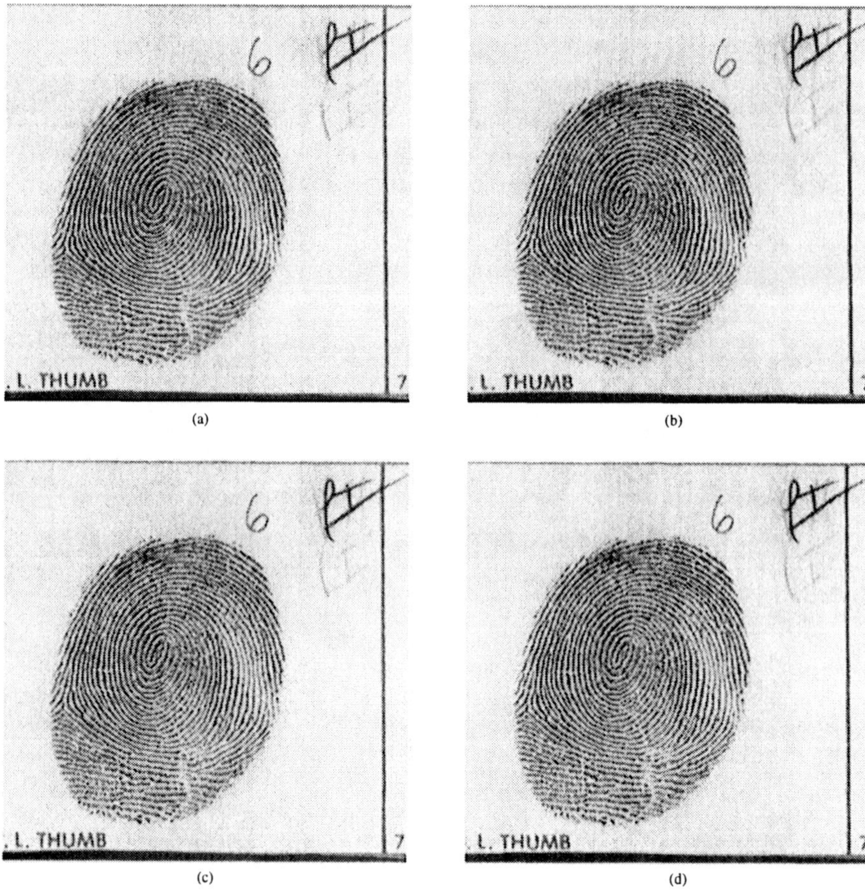

FIGURE 10. (a) original Fingerprint image (589824 bytes) (b) coded by the 16×32 LOT coder at 1:19 (31043 bytes), 36.05 dB (c) coded by WSQ coder at 1:18 (32702 bytes), 34.43 dB (d) coded by the 16×32 LOT coder at 1:26 (22685 bytes), 34.42 dB.

[8] M. Rabbani and P. W. Jones, *Digital Image Compression Techniques,* SPIE Opt. Eng. Press, Bellingham, Washington, 1991.

[9] "Compression with reversible embedded wavelets," RICOH Company Ltd. submission to ISO/IEC JTC1/SC29/WG1 for the JTC1.29.12 work item, 1995. Can be obtained on the World Wide Web, address: *http://www.crc.ricoh.com/CREW.*

[10] Z. Xiong, O. Guleryuz, and M. T. Orchard, "A DCT-based embedded image coder," *IEEE SP Letters*, Nov 1996.

[11] H. S. Malvar, "Lapped biorthogonal transforms for transform coding with reduced blocking and ringing artifacts," *ICASSP*, Munich, April 1997.

[12] T. D. Tran, R. de Queiroz, and T. Q. Nguyen, "The generalized lapped biorthogonal transform," *ICASSP,* Seattle, May 1998.

[13] P. P. Vaidyanathan, *Multirate Systems and Filter Banks*, Prentice Hall, 1993.

[14] G. Strang and T. Q. Nguyen, *Wavelets and Filter Banks*, Wellesley-Cambridge Press, 1996.

[15] M. Vetterli and J. Kovacevic, *Wavelets and Subband Coding*, Prentice-Hall, 1995.

[16] R. A. DeVore, B. Jawerth, and B. J. Lucier, "Image compression through wavelet trasnform coding," *IEEE Trans on Information Theory*, vol. 38, pp. 719-746, March, 1992.

[17] T. A. Ramstad, S. O. Aase, J. H. Husoy, *Subband Compression of Images: Principles and Examples*, Elsevier, 1995.

[18] A. Soman, P. P. Vaidyanathan, and T. Q. Nguyen, "Linear phase paraunitary filter banks," *IEEE Trans. on SP,* V. 41, pp. 3480-3496, 1993.

[19] J. Katto and Y. Yasuda, "Performance evaluation of subband coding and optimization of its filter coefficients," *SPIE Proc. Visual Comm. and Image Proc.*, 1991.

[20] T. D. Tran and T. Q. Nguyen, "Generalized lapped orthogonal transform with unequal-length basis functions," *ISCAS*, Hong Kong, June 1997.

[21] Z. Xiong, K. Ramchandran, and M. T. Orchard, "Space-frequency quantization for wavelet Image Coding," *IEEE Trans. on Image Processing*, vol. 6, pp. 677-693, May 1997.

[22] Z. Xiong, K. Ramchandran, and M. T. Orchard, "Wavelet packet image coding using space-frequency quantization," *submitted to IEEE Trans. on Image Processing*, 1997.

[23] J. N. Bradley, C. M. Brislawn, and T. Hopper, "The FBI wavelet/scalar quantization standard for gray-scale fingerprint Image Compression," *Proc. VCIP*, Orlando, FL, April 1993.

[24] R.L. Joshi, H. Jafarkhani, J.H. Kasner, T.R. Fischer, N. Farvardin, M.W. Marcellin, and R. H. Bamberger, "Comparison of different methods of classification in subband coding of images," submitted to *IEEE Trans. Image Processing,* 1996.

[25] S.M. LoPresto, K. Ramchandran, and M.T. Orchard, "Image coding based on mixture modeling of wavelet coefficients and a fast estimation-quantization framework," *IEEE DCC Proceedings,* pp. 221-230, March 1997.

[26] M. Ikehara, T. D. Tran, and T. Q. Nguyen, "Linear phase paraunitary filter banks with unequal-length filters," *ICIP,* Santa Barbara, Oct. 1997.

[27] T. D. Tran, M. Ikehara, and T. Q. Nguyen, "Linear phase paraunitary filter bank with variable-length Filters and its Application in Image Compression," submitted to *IEEE Trans. on Signal Processing* in Dec. 1997.

Part IV

Video Coding

19

Brief on Video Coding Standards

Pankaj N. Topiwala

1 Introduction

Like the JPEG International Standard for still image compression, there have been several international video coding standards which are now in wide use today. There are a number of excellent texts on these standards, for example [1]. However, for continuity we provide a brief outline of the developments here, in order to provide a background and basis for comparision. Wavelet-based video coding is in many ways still a nascent subject, unlike the wavelet still coding field which is now maturing. Nevertheless, wavelet coding approaches offer a number of desirable features that make them very attractive, such as natural scalability in resolution and rate using the pyramidal structure, along with full embeddedness and perhaps even error-resiliency in certain formulations. In terms of achieving international standard status, the first insertion of wavelets could likely be in the still image coding arena (e.g., JPEG2000); this could then fuel an eventual wavelet standard for video coding. But substantial research remains to be conducted to perfect this line of coding to reach that insertion point.

We note that all the video coding standards we discuss here have associated audio coding standards as well, which we do not enter into. Video coding is a substantially larger market than the still image coding market, and video coding standards in fact predate JPEG. Unlike the JPEG standard, which has no particular target compression ratio, the video coding standards may be divided according to specific target bitrates and applications envisioned.

2 H.261

For example, the International Telecommunications Union-Telecommunications Sector (ITU-T, formerly the CCITT) Recommendation H.261, also known as "$p \times 64$," was developed for video streams of $p \times 64$ Kbps (p = 1 to 30) bandwidth and targetted for visual telephony applications (for example, videoconferencing on ISDN lines). Work on it began as early as December 1984, and was completed in December 1990. Like the JPEG standard, both the 8x8 DCT and Huffman coding play a central role in the decorrelation and coding for the *intra*frame compression, but now there is also *inter*frame prediction and motion-compensation prior to compression.

First, given three different preexisting TV broadcasting standards, already in use for decades: NTSC, PAL, and SECAM, the Common Intermediate Format (CIF) was established. Instead of the 720x480 (NTSC) or 720x576 (PAL) pixel resolutions of the TV standards, the video telephony application envisioned permitted a common reduced resolution of 360x288 pixels (the Quarter CIF (QCIF) format, also used, has resolution 180x144), along with a frame rate of 29.97 frames/sec, consistent with the NTSC standard. One difference to note is that the TV standard have "interlaced" lines in which the odd and even rows of an image are transmitted separately as two "fields" of a single frame, whereas the CIF and QCIF formats are noninterlaced (they are "progressive").

The color coordinates of red-green-blue (RGB), assumed to be 8-bit integers, are first transformed to a format called YCbCr [1]. This is accomplished in two steps: (1) gamma correction, and (2) color conversion. Gamma correction is a process of predistorting the intensity of each color field to compensate for the nonlinearities in the cathode-ray tubes (CRTs) in TVs. When signal voltages are normalized to lie in $[0, 1]$, CRTs are observed to display

$$I_{disp} = I_{rec}^{\gamma}, \quad \gamma = 2.2 \ \ for \ NTSC.$$

As such, it is predistorted to $I_{trans} = I_{signal}^{0.45}$. Next, if the gamma-corrected colors (scaled to $[0, 1]$) are denoted by R', G', B', then the color conversion is given by

$$Y = 0.299R' + 0.587G' + 0.114B'.$$

$$\overline{Y} = 219Y + 16;$$

$$\overline{C_b} = 126(B - Y) + 128;$$

$$\overline{C_r} = 160(R - Y) + 128 :$$

Finally, the Cb and Cr components are subsampled by a factor of 2 in each direction (4:2:2 format). The subsampled color coordinates Cb and Cr are placed at the center of 2x2 luminance Y coordinates.

The various 8x8 blocks in the pictures are organized in four layers, from the full picture layer down to a basic block (8x8). Huffman codes and motion vectors are are constructed for the macroblock layer (16x16 or 4 blocks). Given a new frame, motion estimation is accomplished by considering a macroblock of the current frame and comparing to its various repositionings in the previous frame and selecting the best match. Motion estimation is performed by using the mean-squared error (MSE) criterion, and matches are constructed to single pixel accuracy in H.261 (half-pixel in MPEG-1). A decision to use intra or interframe coding is based on a tradeoff between the variance of the current block and the best MSE match available.

Upon motion compensation decisions, the macroblocks are quantized and Huffman coded. Quantizers are either uniform midrisers (intra DC coefficients only) or midtreaders (all others) with a deadzone [1]. A buffer rate controller is designed to control quantizer binsizes to maintain a given rate, since the DCT-based coding scheme is not itself rate-controlled.

3 MPEG-1

MPEG-1, begun in Ocotober 1988 and completed in November 1994, differs from H.261 is several ways. It was aimed at higher bitrates (e.g., 1.5 Mbps) and targetted for video-on-CD applications to mimic VHS quality. It was the first joint audio/video coding standard in a single framework, and allows either component to dominate (e.g., video equivalent to VHS, or audio equivalent to audio CDs). The global structure of the coder is of the H.261 type: motion-compensation or not, intra or inter, quantize or not, etc. The compression algorithms are also similar to H.261, but motion prediction can also be bidirectional. In fact, the concept of group of pictures (GOP) layer was established, in which an example sequence of frames might look like

$$IBBPBBPBBI...,$$

where I, B, and P are intraframe, bidirectional and (forward) predicted frames. The bidirectional prediction allows for further coding efficiencies at the price of more computation; it also means that the decoder has to buffer enough frames and work backwards in some instances to recover the proper signal order.

The synchronous audio coding structure forms a significant part of the standard. For some applications, further issues relate to packetization for delivery on packet-switched networks (e.g., Internet). In terms of complexity, due to the high cost of motion estimation/compensation, the encoder is substantially more complex than the decoder, which is reflected in the cost of hardware implementations as well.

4 MPEG-2

MPEG-2 was initiated in July 1990 and parts of it attained standard status between 1994 and 1997. It is aimed at even higher bit rates: 2-15 (or even 100) Mbps, and designed to address contribution-quality video coding. The follow-on standard, MPEG-3, was to address HDTV, but was later merged into MPEG-2. This standard was meant to be more generic, include backward compatibility with MPEG-1 for decoding, permit a variety of color video formats, and allow scalability to a variety of bitrates in its domain. It was designed to handle interlaced or progressive data, allow error resilience, and other compatibility/scalability features (such as resolution and SNR scalability) in order to be widely applicable. It has recently been adopted for the video on DVD applications by the major manufacturers, is being used for satellite TV, and is a highly successful standard. The workhorse of the compression is again the 8x8 DCT, but sophisticated motion estimation/compensation makes this a robust standard from the point of view of performance. It may be noted that the complexity of this coder is extremely high, and software-only encoders cannot be conceived to this day (software-only decoders have just been made possible with today's leading high-performance chips). For all topics up to MPEG-2, an excellent general reference is [1]. For an in-depth analysis of the MPEG-1 and MPEG-2 standards, consult [2].

5 H.263 and MPEG-4

MPEG-4 was initiated in November 1993 with the aim of developing very low bitrate video coding (< 64 Kbps), with usable signals down to 10 Kbps. The 10Kbps regime proved to be extremely challenging, and was later dropped from the requirement. The applications envisioned include videophone and videoconferencing, including wireless communications. With this in mind, the key objectives of this standard included (1) improved low bit rate performance, (2) content scalability/accessibility, and (3) SNR scalability.

This standard is due for completion in February 1999. Meanwhile, a near-term draft standard named H.263 was released in April 1995. Again, the compression workhorse is the 8x8 DCT, although the enhanced half-pixel motion estimation and forward/bidirectional prediction used make for excellent temporal decorrelation. The image resolution is assumed to be QCIF or sub-QCIF. Even then, the complexity of this coder means that real-time implementation in software is challenging.

As of this writing (April 1998) the MPEG-4 Standard, which is currently in Final Committee Draft (FCD) phase, permits content-scalability by use of coding techniques that utilize several video object planes (VOPs) [4, 3]. These are a segmentation of the frames into regions that can be later manipulated individually for flexible editing, merging, etc. This approach assumes layered imagery to begin with, and is a priori applicable only to new contributions constructed in this form. Other proposed methods attempt to segment the imagery directly (as opposed to requiring segmented video sequences) in terms of image objects to achieve object-based video coding; this approach may apply equally to existing video content. Note that wavelet still image coding is now explicitly incorporated into the draft standard, and plays a critical role as an enhanced image coding scheme (called *texture coding* in MPEG-4 parlance). As MPEG-4 moves towards completion, these or similar technologies incorporated into the coding structure will give this standard the flexibility to apply to a wide variety of multimedia applications, and lead to interactive video—the promise of video's future. For more information on MPEG, visit *www.mpeg.org*.

6 References

[1] K. Rao and J. Hwang, *Techniques and Standards for Image, Video and Audio Coding*, Prentice-Hall, 1996.

[2] J. Mitchell et al, *MPEG Video Compression Standard*, Chapman and Hall, 1997.

[3] IEEE Signal Processing Magazine, Special Issue on MPEG Audio and Video Coding, Sept., 1997.

[4] IEEE Transactions on Circuits and Systems for Video Technology, Special Issue on MPEG-4, February, 1997.

20

Interpolative Multiresolution Coding of Advanced TV with Subchannels

K. Metin Uz, Martin Vetterli and Didier J. LeGall

1 Introduction[1]

The evolution of the current television standards toward increased quality and realism builds on higher spatial resolution, wider aspect ratio, better chroma resolution, digital audio (CD–quality) and possibly a new scanning format. In addition to the high bandwidth requirements, transmission systems for advanced television face the challenge that the quality has to be maintained throughout the system: for this reason component signals will be preferred over composite and digital representation over analog.

The bandwidth requirements for advanced television (typically more than 1 Gbits/sec) ask for powerful digital compression schemes so as to make transmission and storage manageable. The quality requirements and the high resolution of advanced television material make very high signal to noise ratios necessary. It is therefore required to develop source coding schemes for digital video signals which achieve a compression of an order of magnitude or more at the highest possible quality. Two specific cases of interest are contribution quality advanced television at around 100-140 Mbits/sec (where objective quality has to be nearly perfect to allow post processing like chroma-keying) and distribution quality for the consumer at rates which are 2–5 times lower and where high subjective quality is required. Besides production and distribution of advanced television (ATV), another application of great interest is the coding for digital storage media (e.g. VTR, CD–ROM), where it is desirable to be able to access any segment of the data, as well as to browse the data (i.e. fast forward or reverse search).

Currently, there is an on-going debate between proponents of interlaced and non-interlaced (also called sequential or progressive) scanning formats for ATV. Both formats have their respective advantages: interlaced scanning saves bandwidth and is well matched to current display and camera technology, and non-interlaced scanning is better suited for graphics and movie applications. In this paper, we will thus deal with both scanning formats.

Our approach to the compression problem of ATV is based on the concept of multiresolution (MR) representation of signals. This concept has emerged as a powerful tool both for representation and for coding purposes. Then, we choose a finite

---

memory (or finite impulse response, FIR) scheme for robustness and for fast random access. The resulting FIR multiresolution scheme successfully addresses the following problems of interest in coding, representation and storage of ATV:

- signal decomposition for compression purposes
- representation well suited for fast random access or reverse mode in digital storage devices
- robustness and error recovery
- suitable signal representation for joint source/channel coding
- compatibility with lower resolution representations.

Note that multiresolution decomposition is also called hierarchical or pyramidal decomposition, and associated coding schemes are sometimes called embedded or layered coding methods. In the framework of coding, multiresolution decompositions go back to pyramid coding [1] and subband coding [2], [3], [4], [5], and in applied mathematics, they are related to the theory of wavelets [6], [7]. Note that in this paper, the multiresolution concept will be employed not only for coding but also for motion estimation [8] (as an approximate solution to an optimization problem), a technique also known as hierarchical motion estimation [9], [10].

For robustness purposes, it is advantageous to develop coding schemes with finite memory. This can either be achieved with an inherently finite memory approach, or by periodically restarting a recursive scheme. If a multiresolution decomposition is desired, be it for compatibility purposes or for joint source/channel coding, it turns out that recursive schemes employing a prediction loop, like DPCM or motion compensated hybrid DCT coding [11] have some loss in performance [12]. This is due to the fact that only a suboptimal prediction based on the coarse resolution is possible. In the case of coding for storage media, FIR schemes facilitate easy random access to the data. Additional features such as fast search or reverse playback are provided at almost no extra cost. While error correction is widely used in magnetic media, uncorrectable errors are still unavoidable. The finite memory structure of the scheme assures that errors will not be lasting for more than a few frames.

Among the various possible FIR multiresolution schemes, we choose to develop a three dimensional pyramidal coding scheme which uses 2D spatial interpolation over frames, and motion based interpolation between frames. Note that the temporal interpolation is similar to the proposed MPEG standard [13]. We will justify the reasons for this choice by examining pros and cons in detail, and showing that the resulting coding scheme marries simplicity and compression at high quality. The complexity of the overall scheme is comparable to alternative coding schemes that are considered for high quality ATV applications.

The outline of the paper is as follows. We begin by looking at multiresolution representations for coding, and examine two representative techniques in detail: subband and pyramid coding. We compare these two cases in terms of representation, coding efficiency, and quantization noise. Section 4 describes the spatiotemporal pyramid, a three-dimensional pyramid structure that forms the basis of our coding scheme. In the following section, we focus on the temporal interpolation within the pyramid, describing the multiresolution motion estimation and motion based interpolation procedures in detail. The coding system and simulation results are given in section 6, along with a discussion of relation to the evolving MPEG standard [13]. Finally, we analyze the computational and memory complexity of the proposed scheme.

2 Multiresolution Signal Representations for Coding

The idea of multiresolution is similar to that of successive approximation. A coarse approximation to a signal is refined step by step, until the signal is obtained at the desired resolution. Very similarly, an initial solution to an optimization problem can be refined stepwise, until the full resolution solution is achieved. To get the coarse approximation, as well as to refine this approximation to increase the resolution, one needs operators adapted to the particular resolution change. These can be linear filters, or more sophisticated model based operators. Typical examples are decimation of a signal (fine–to–coarse), which is usually preceded by a low-pass anti–aliasing filter, and upsampling (coarse–to–fine) which is followed by an interpolation filter. We will see that video processing calls for more sophisticated, non-linear operators, such as motion based frame interpolation used to increase time resolution.

Multiresolution approaches are particularly successful when some a priori structure or hierarchy can be found in the problem. A classic approximation technique used in statistical signal processing and waveform coding is the Karhunen-Loeve transform (KLT) [14]. Given a vector process (typically obtained by blocking a stream of samples), one computes a linear transform $\mathbf{T}$ such that $\mathbf{y}_n = \mathbf{T} \cdot \mathbf{x}_n$. The rows of the transform matrix are chosen as the eigenvectors of the autocorrelation matrix $\mathbf{R}$ of $\mathbf{x}_n$ which is symmetric (by stationarity assumption), and therefore $\mathbf{T}$ is unitary. The samples of $\mathbf{y}_n$ are thus decorrelated. Now, the rows of $\mathbf{T}$ can be ordered so that:

$$E[y_n(i)^2] \geq E[y_n(j)^2], \qquad i < j \tag{20.1}$$

That is, the first k coefficients of the KLT are a best k coefficient approximation to the process $\mathbf{x}_n$ in the mean squared error (MSE) sense.

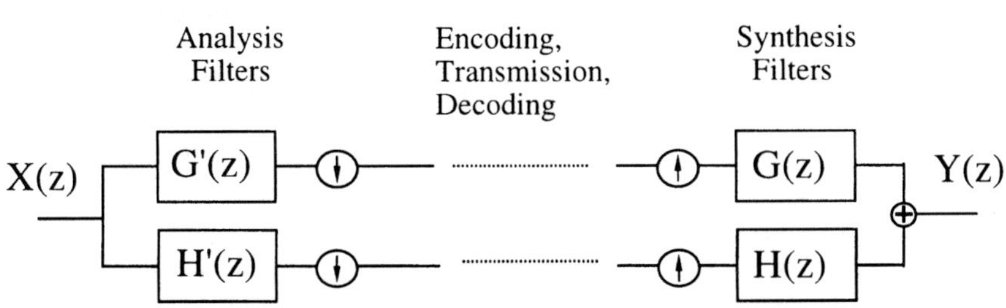

FIGURE 1. Two channel subband coding system

Transform and subband coding, two successful standard image compression techniques when combined with appropriate quantization and entropy coding, can be

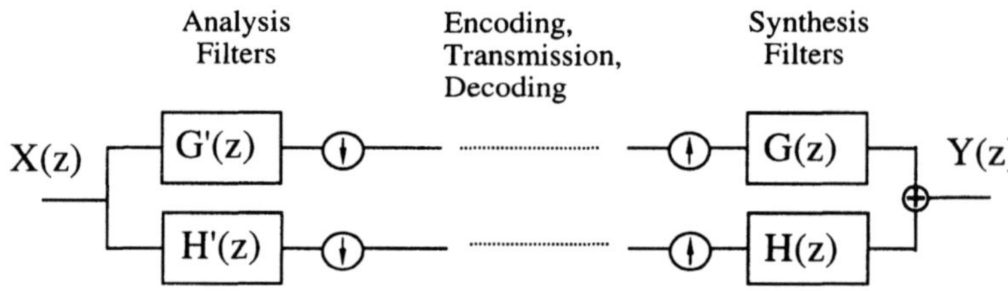

FIGURE 2. Two channel subband coding system.

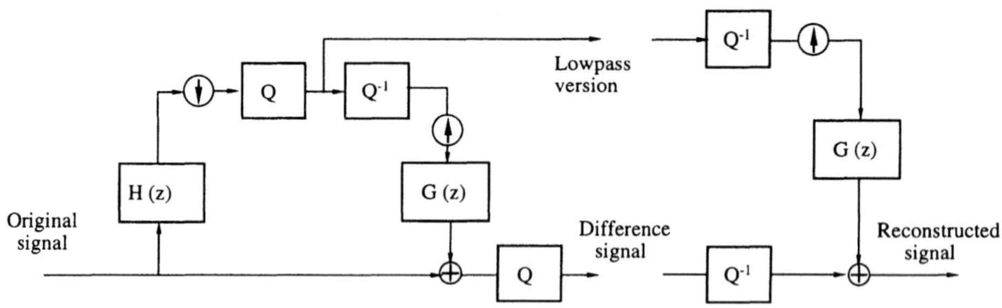

FIGURE 3. One step in a pyramid decomposition and coding. Note that the only source of reconstruction error is the quantizer for the difference signal.

seen as variations on this idea. In terms of multiresolution approximation, their suitability stems from the fact that images typically have a power spectrum that falls off at higher frequencies. Thus, low frequency components of a transform coded picture form a good approximation to the picture. This multiresolution nature of transform coding is used for example in progressive transmission of pictures. Subband coding (see Figure 2) can be seen as a transform with basis vectors extending over more than one block. Constraints are imposed on the analysis and synthesis filterbanks so as to achieve perfect reconstruction [15]. Note that both transform and subband decompositions form a one–to–one mapping, as the number of samples is preserved. In contrast, pyramid decomposition is a redundant representation, since a low resolution version as well as a full resolution difference are derived (see Figure 3). This redundancy, or increase in the number of sample points, becomes negligible as the dimensionality increases, and allows greater freedom in the filter design.

3 Subband and Pyramid Coding

Transform coding and subband coding are very similar decomposition methods, and will thus be discussed together. Below, we will compare and contrast respective advantages and problems of subband schemes versus pyramid schemes. It should be noted that subband decomposition can be viewed as a special case of pyramid decomposition using constrained linear operators in the absence of quantization [16]

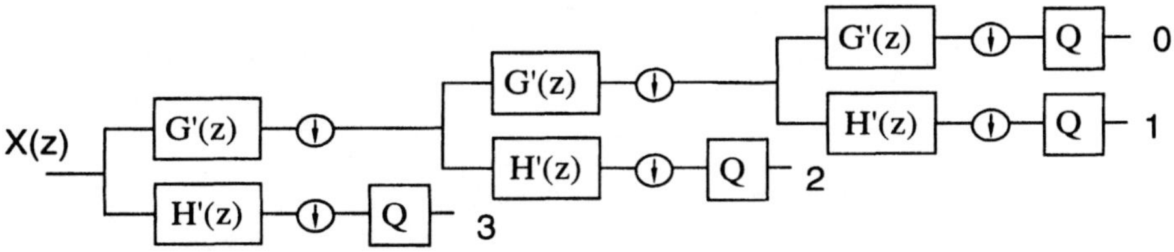

FIGURE 4. Subband analysis filterbank where the band splitting has been iterated three times on the low band, yielding four channels with logarithmic spacing.

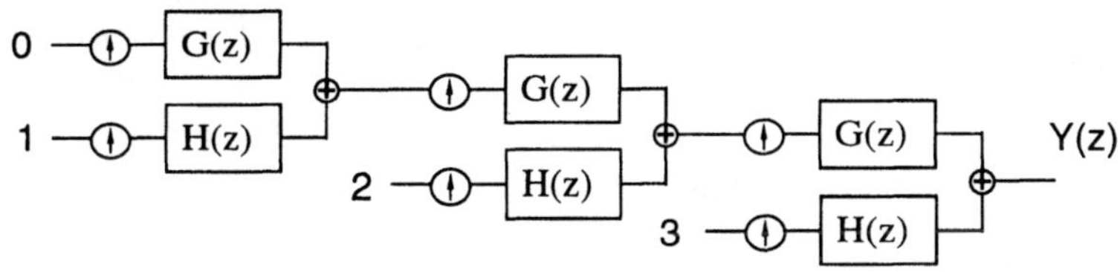

FIGURE 5. The corresponding subband synthesis filterbank.

[7].

3.1 Characteristics of Subband Schemes

The most obvious advantage of subband schemes is the fact that they are critically sampled, that is, there is no increase in the number of samples. The price paid is a constrained filter design and therefore a relatively poor lowpass version as a coarse approximation. This is undesirable if the coarse version is used for viewing in a compatible subchannel or in the case of progressive transmission. Only linear processing is possible in subband coding systems, and any non-linearity has effects which are hard to predict. In particular, quantization noise can produce artifacts in the reconstructed signal which are difficult to foresee.

The problems are linked to the fact that the bound on the maximum error produced in the reconstructed signal because of quantization in the subbands is fairly weak. This is different from the MSE (or l_2 norm of the error), which is conserved since the subband decomposition and reconstruction processes are unitary operators [2]. However, the maximum error is given by the l_∞ norm, which is not conserved by unitary transforms. To make this point clear, let us consider the case of the size N DCT. The first row of the IDCT is equal to $1/\sqrt{N}[1, 1, \ldots, 1]$. If the vector of quantization errors is colinear with this row, the backtransformed vector of errors is equal to $[\sqrt{N}\delta, 0, \ldots, 0]$ (where δ is the quantization step in the transform domain). Thus, all the errors accumulated on the first coefficient! Note that $\sqrt{N}$ is typically

[2]Actually, this is only true for paraunitary filter banks [17], but holds approximately true for most perfect reconstruction filter banks.

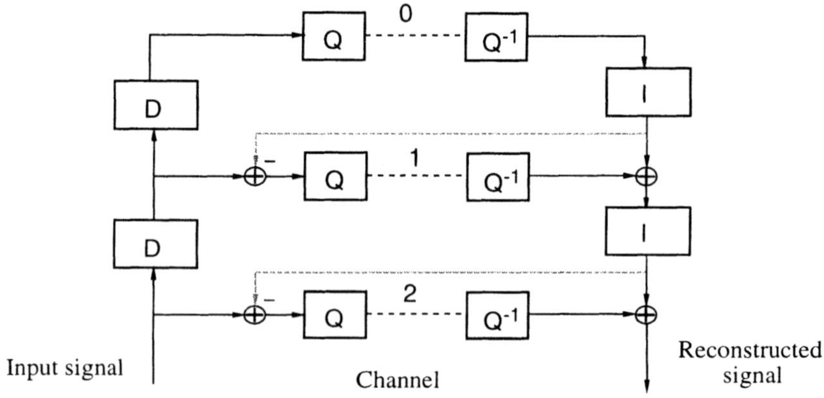

FIGURE 6. Three level pyramid coding, with feed-back of quantization of the high layers into the prediction of the lower ones. "D" and "I" stand for decimation and interpolation operations. Only one source of quantization error remains, namely, that of the highest resolution difference signal.

equal to 8 (corresponding to blocks of 8x8 pels) in image coding applications. Subband schemes behave similarly.

Three dimensional subband coding has been proposed for video coding [7] [19], but has not been able to compete with motion–compensated coding methods in terms of compression efficiency. It seems difficult to naturally include motion compensation within a subband scheme. Motion is a time domain concept, while a subband decomposition leads to a (partly) frequency domain description of the sequence. Thus, motion compensation can be done outside the subband decomposition [20] (in which case SBC replaces DCT for encoding the prediction error) or can be seen as a preprocessing [9] (where only block transforms are used over time). Motion compensation within the bands suffers from accuracy problems [22], and from the fact that independent errors can accumulate after reconstruction [23].

3.2 Pyramid Coding

We have seen that pyramid decomposition is a redundant representation. This redundancy, or increase in the number of sample points, becomes negligible as the dimensionality increases. In a one dimensional system, the increase is upperbounded by $1+1/2+1/4+... < 2$, in two dimensions by $1+1/4+1/16+.. < 4/3$ and in three dimensions, by $1+1/8+1/64... < 8/7$. That is, in the three dimensional case that we will be using, the overhead is less than 15%. At the price of this slight oversampling, one gains complete freedom in the design of the coarse–to–fine and fine–to–coarse resolution change operators, which can be matched to the signal model, and can be non–linear [24]. Constraints in the transform or subband decompositions often result in compromises in the filter quality. If linear filters are chosen to derive the lowpass approximation in a pyramid, it is possible to take very good lowpass filters, and derive visually pleasing coarse versions. Therefore, pyramids can be a better choice when high visual quality must be maintained across a number of scales [25].

We also observe that the l_∞ problem in transform and subband coding case can be avoided in pyramids by quantization noise feedback. A detailed analysis follows in the next section.

3.3 Analysis of Quantization Noise

In this section, we analyze the propagation of quantization noise in pyramid and subband decomposition schemes. We will consider three representative cases: an iterated two-channel subband splitting, and a pyramid with and without error feedback. Simulations are based on actual data from the well known image "Lenna".

A three stage subband analysis bank and the corresponding synthesis bank are depicted in Figures 4 and 5. We assume each band is independently quantized by a scalar quantizer (band quantizer) of equal step size and that quantization noise can be modeled as white. Furthermore, we assume a similar quantizer (with a finer step size) is used following each filter, to model the finite wordlength effects (internal quantizer). For simplicity, we focus on the synthesis bank, although a similar conclusion can be reached for the analysis bank. Let q_i be the noise due to the quantizer for band i, and f_i be the internal quantizer noise due to the synthesis filter pair. Then $E_i(z)$, the z-transform of e_i can be expressed as

$$E_i(z) = E_{i-1}(z^2)G(z) + Q_i(z^2)H(z) + F_i(z) \tag{20.2}$$

(upsampling by 2 means replacing z by z^2 in the z-transform domain [17]) which leads to a final reconstruction error

$$E_N(z) = Q_N(z)\prod_{i=0}^{N-1} G(z^{2^i}) + \sum_{i=1}^{N-1}\left(\prod_{n=1}^{i} G(z^{2^n})\right)\left\{Q_i(z^{2^i})H(z^{2^i}) + F_i(z^{2^i})\right\}. \tag{20.3}$$

We note that the noise spectrum is biased into low frequencies. For an intuitive explanation, consider the signal flow graph corresponding to the analysis–synthesis system. Band N, the subsignal that takes the longest path ($2N$ lowpass filters) covers only $1/2^N$ of the spectrum at the DC end. Therefore, more finite wordlength effects are visible at low frequencies. A numerical simulation was done with the 32-tap Johnston QMF filter pair (C) [26], using 10 bits for the internal quantizers, and 6 bits for the band quantizers. The resulting reconstruction error spectrum is depicted in Figure 7 (a). We should note that in practice, one would choose finer quantizers for the lower bands, partially alleviating the problem, although the accumulation due to finite wordlength effects is unavoidable.

Next, we consider an N stage pyramid without quantization error feedback. We assume that each stage employs a linear scalar quantizer, and that the quantization noise can be modeled as white. For simplicity, we assume the quantizers have equal step size. Let q_i be the noise due to the quantizer at layer i, defined such that coarsest layer is layer 0, and let the reconstruction error at layer i be denoted by e_i. For a linear filter $H(z)$, we can easily analyze the response of the system to the quantization noise by assuming the only input to the decoder is the set $\{q_0, q_1, \ldots q_N\}$. Then

$$E_i(z) = E_{i-1}(z^2)H(z) + Q_i(z) \tag{20.4}$$

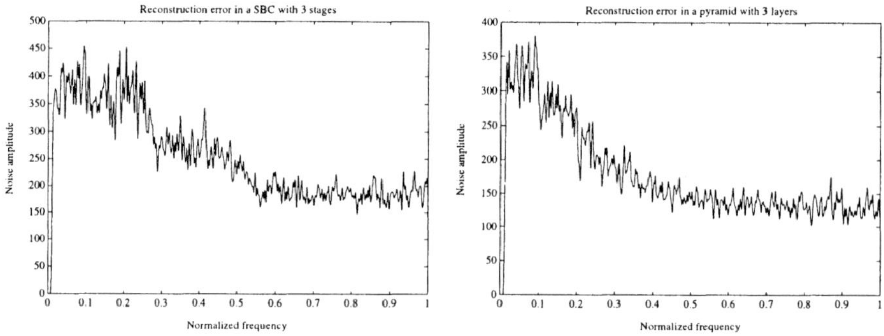

(a) (b)

FIGURE 7. Reconstruction error power spectrum for (a) SBC with three stages; (b) pyramid with three layers without quantization error feedback.

which, by iteration, gives the final reconstruction error as

$$E_N(z) = \sum_{i=0}^{N} \left\{ \prod_{n} H(z^{2^n}) \right\} Q_i(z^{2^i}). \tag{20.5}$$

Qualitatively, it is easy to see how the quantization noise propagates across the layers: The initial error $e_0 = q_0$ is white. Upsampling creates a replica in the spectrum, and h, typically a lowpass filter, attenuates the replica. Thus, e_1 consists of white q_1 plus this lowpass noise. At each layer, previous reconstruction error is "squeezed" into an (approximately) halfband signal, and white noise is added. The results of a numerical simulation using three stages is shown in Figure 7 (b). Here the filters are those used by Burt and Adelson [1], where $a = 0.6$, and the quantizer step size is 4.

Notice that quantization error is hard to control in both cases. Furthermore, the error spectrum is biased toward low frequencies, particularly undesirable since the human visual system (HVS) is much more sensitive to lower frequencies [27]. For comparison, in a pyramid *with* feedback the only source of quantization error is the final quantizer, since prior quantization errors are corrected at each layer. We can thus guarantee a bound δ on the maximum error, and also tailor the quantization to the HVS, by shaping the error spectrum and by using masking based on the original sequence.

4 The Spatiotemporal Pyramid

In this section, we introduce the spatiotemporal pyramid, a multiscale representation of the video signal. It consists of a hierarchy of video signals at increasing temporal and spatial resolutions (see Figure 8). Here, we should stress that the video signal is not a true 3–D signal, but can be modeled as a 2–D signal with an associated displacement field. This fundamental difference between space and time is taken into account in the pyramid by the choice of motion based processing over time. This also justifies the lack of filtering prior to temporal decimation [28].

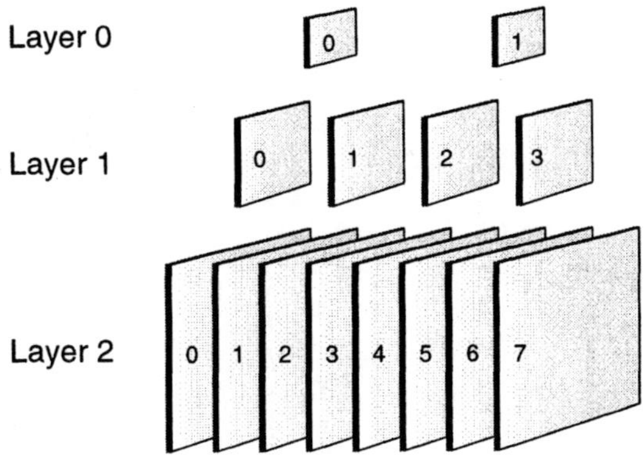

FIGURE 8. The spatiotemporal pyramid structure

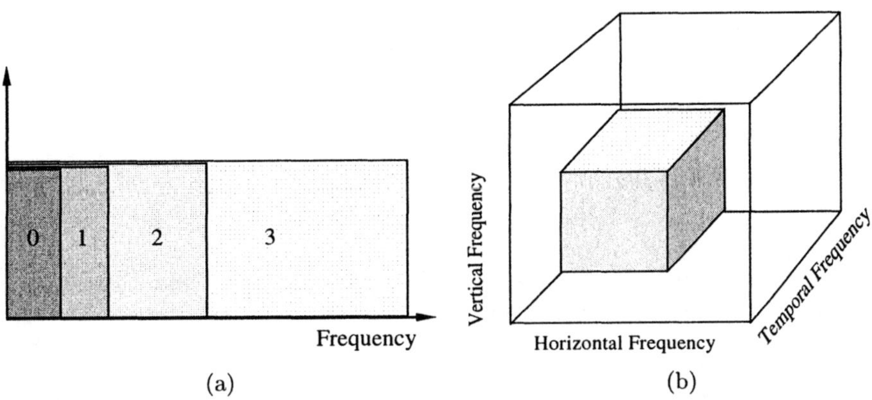

FIGURE 9. The logarithmic frequency division in a pyramid. (a) One dimensional pyramid with four layers. (b) Three dimensional pyramid with two layers. Notice that the spectral "volume" doubles in the first case, but increases eight-fold in the latter.

The structure is formed in a bottom–up manner, starting from the finest resolution, and obtaining a hierarchy of lower resolution versions. Spatially, images are subsampled after anti–aliasing filtering. Temporally, the reduction is achieved by simple frame skipping.

The frequency division obtained with a pyramid is depicted in Figure 9 (b). The decomposition provides a logarithmic scaling in frequency (see Figure 9 (a)), with the bandwidth increasing by a factor of 2 in each dimension down the pyramid. Thus, the spectral "volume" of the signal is increased by a factor of 8 at each coarse–to–fine resolution change (actual scaling factor may depend on the sampling grid).

The encoding is done in a stepwise fashion, starting at the top layer and working down the pyramid in a series of successive refinement steps. At each step, the signal is first spatially interpolated, increasing the spatial resolution by 2 in each dimension (a factor of 4 in the number of samples per frame). Motion based interpolation

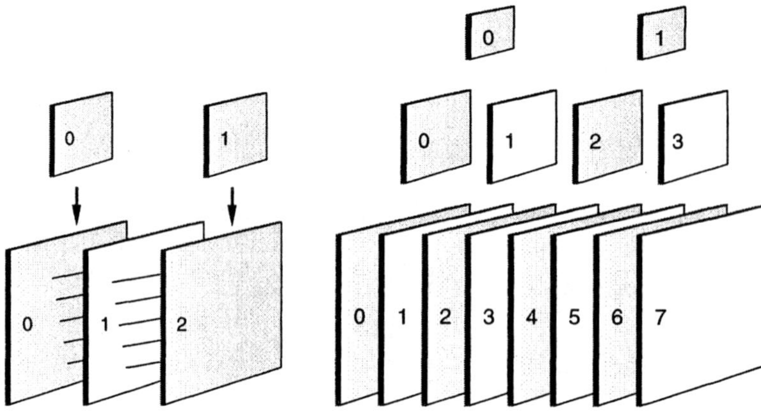

FIGURE 10. Reconstruction of the pyramid (a) One step of coarse-to-fine scale change (b) The reconstructed pyramid. Note that approximately one half of the frames in the structure (shown as shaded) are spatially coded/interpolated.

follows, doubling the temporal resolution and completing the reconstruction of the next layer. We describe the motion-based processing in more detail in the next section, and now focus on some key properties of the scheme.

The structure forms the basis of a finite-memory coding procedure. The frames at a particular layer are based upon the frames directly above them. Note that the dependence graph is in the form of a binary tree, in which the nodes are the individual frames, and the leaves are the frames in the final layer. Therefore any channel error has a finite (and short) duration effect on the output: only the frames with branches passing through the error are affected. With typical parameters, this implies no error can last more than a small fraction of a second. This is in contrast with predictive coders, where the prediction loop has to be restarted frequently to avoid error accumulation.

The encoding procedure is also computationally attractive. Although the scheme provides a 3-D decomposition, the complexity is kept linear by using separable kernels for interpolation. We should note that an optimal algorithm would require complex motion-adaptive spatiotemporal filters, computationally expensive and hard to design. By decoupling space and time, we achieve significant reductions in computation with little penalty, especially in view of our source model: A sequence is formed by images displaced smoothly over time.

The coarse-to-fine scale change step is illustrated in Figure 10 (a). First the spatial resolution is increased, then the temporal interpolation is done based on these new frames at the finer scale. We should note that reversing the order of these operations would cause increased energy in the difference signals to be encoded. In other words, interpolation is statistically more successful over time than over space, so the temporal difference signal has lower energy and thus is easier to compress.

As a side note, we should note that this mismatch problem can be partially alleviated by interpolating more than one frame over time[3]. However, this scheme is not without drawbacks: It becomes harder to maintain the quality between frames, and the effect of a channel error has now longer duration.

We have seen that motion based interpolation is central to the multiresolution approach over time. Therefore we will focus on the motion estimation problem in the next section.

5 Multiresolution Motion Estimation and Interpolation

Motion estimation is based on a multiresolution algorithm: The motion field is initially estimated on a coarse grid, and successively refined until a sufficiently dense motion field is obtained. Here, we are concerned with computing the *apparent* 2-D motion, rather than the *actual* 3-D velocities. However, the motion based interpolation still requires a reasonably accurate estimate, not just a match in the MSE sense.

Hierarchical approaches have been applied to motion estimation problem [9],[29], [30]. The motivation for the algorithm lies in the observation that "typical" scenes frequently contain motion at all scales. Camera movements such as pan and zoom induce global motion, and objects in the scene move with velocities roughly proportional to their sizes.

The structural model, i.e. the relation between the image and motion field, also suggests a MR strategy. Consider a scene consisting of a superposition of sine waves (any frequency domain technique implicitly assumes this). Now consider uniformly displacing this scene in time. Looking at a single sinusoid, the largest displacement that can be computed is limited to less than half its wavelength. With low frequencies, it's hard to resolve small differences in displacement, i.e. precision is reduced. With high spatial frequencies, one gets high resolution at small displacements, but large displacements are ambiguous at best, due to aliasing. However, we assume that all these components move at the *same* velocity, because they belong to the same rigid object. So a coarse–to–fine strategy at estimating motion seems to be a natural choice: Start at a low resolution image, compute a coarse motion field, refine the motion estimate while looking at higher spatial frequencies.

The second argument in favor of a MR technique is the computational complexity. Brute force algorithms such as full search require $O(d^2)$ searches, where d is the maximum allowable displacement, and is typically a fraction of the picture size. Thus, as the picture definition increases, so does d, quadratically increasing the search complexity. In contrast, MR schemes can be designed with roughly logarithmic complexity. So, the MR choice is also a computationally attractive one.

An inherent difficulty in motion compensation is the problem of covered/uncovered areas. In a predictive scheme, one cannot account for the area that has just been uncovered: an object appears on screen for which no motion vector can be computed. Interpolation within an FIR structure elegantly solves this problem: Covered areas are visible in the past, and uncovered areas in the future.

[3]Indeed, the current MPEG proposal [13] for video coding provides two interpolated frames between conventional predicted frames.

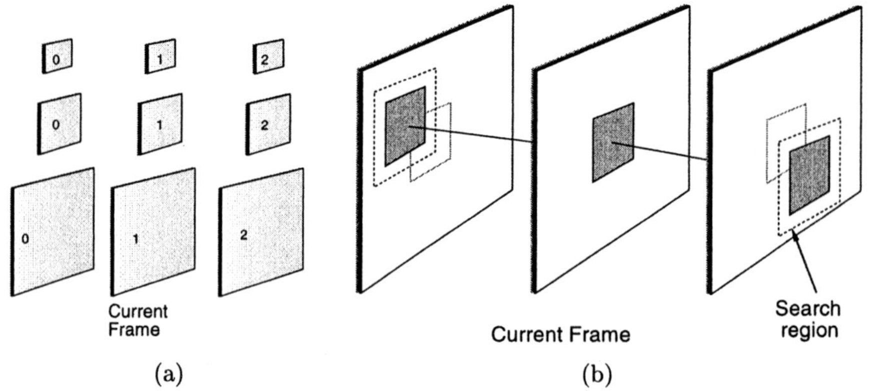

FIGURE 11. Motion estimation starts from a coarse image, and gradually refines the estimate. The search is performed in a symmetric fashion within a window around the previous estimate.

We will present the motion estimation algorithm in two steps. First, we describe a hierarchical "symmetric mode" search similar to that of Bierling [10], but uses the previous and the following frames to compute the motion field. Next, we consider the problem of motion based interpolation, and modify the estimation algorithm for the "best" reconstruction. Essentially, the algorithm consists of three concurrent searches, and a control mechanism to select the best mode. In effect, this also selects the interpolation mode, and sends it as side information.

5.1 Basic Search Procedure

We start with the video signal $I(\mathbf{r}, n)$ where $\mathbf{r}$ denotes the spatial coordinate (x, y), and n denotes the time. The goal is to find a mapping $\mathcal{D}(\mathbf{r}, n)$ that would help reconstruct $I(\mathbf{r}, n)$ from $I(\mathbf{r}, n-1)$ and $I(\mathbf{r}, n+1)$. We assume a restrictive motion model, where the image is assumed to be composed of rigid objects in translational motion on a plane.

$$I(\mathbf{r}, n) = I(\mathbf{r} - \mathcal{D}(\mathbf{r}, n), n-1). \tag{20.6}$$

We also expect homogeneity in time, i.e.

$$I(\mathbf{r}, n) = I(\mathbf{r} + \mathcal{D}(\mathbf{r}, n), n+1). \tag{20.7}$$

Furthermore, we are using a block based scheme, expecting these assumptions are approximately valid for all points within a block b using the same displacement vector $\mathcal{D}_b$. These assumptions are easily justified when the blocks are much smaller than the objects, and temporal sampling is sufficiently dense (we have used 8x8 blocks, but any size down to 4x4 works quite well).

In what follows, we change the notation slightly, omitting the spatial coordinate $\mathbf{r}$ when the meaning is clear, and replacing $I(\mathbf{r}, n)$ by IKn, and $\mathcal{D}(\mathbf{r}, n)$ by $\mathcal{D}_K(n)$.

To compute $\mathcal{D}_K(n)$, we require a hierarchy of lower spatial resolution representations of IKn denoted Ikn, $0 \geq k < K$. Ikn is computed from Ik+1n by first spatially low-pass filtering with half-band filters, reducing the spatial resolution. This filtered

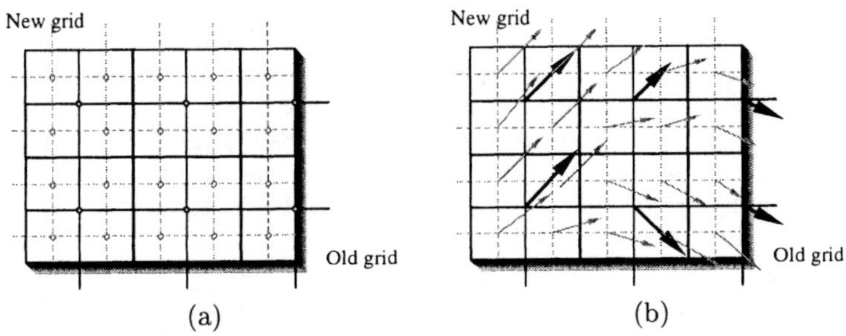

FIGURE 12. Stepwise refinement for motion estimation. (a) The blocks and corresponding grids at two successive scales in the spatial hierarchy. (b) The motion field is resampled in the new finer grid, giving an initial estimate for the next search.

image is then decimated, giving Ikn. Note that this reduces by 4 the number of pels in a frame at each level (see Figure 11).

The search starts at the top level of the spatial hierarchy, I0n. The image Ikn is partitioned into non-overlapping blocks of size MxM. For every block b in the image Ikn, a displacement $\mathcal{D}_b$ is searched to minimize the matching criterion

$$\sum_{r \in b} |I_k(\mathbf{r}, n) - \tilde{I}_k(\mathbf{r}, n)| \tag{20.8}$$

where $\tilde{I}_k(\mathbf{r}, n)$ is the motion based estimate of Ikn computed as

$$\tilde{I}_k(\mathbf{r}, n) = 1/2(I_k(\mathbf{r} - \mathcal{D}_b, n-1) + I_k(\mathbf{r} + \mathcal{D}_b, n+1)) \tag{20.9}$$

Notice that this estimate implies that if a block has moved the distance $\mathcal{D}$ between the previous and the current frame, it is expected to move $\mathcal{D}$ between the current and the following frame. This constitutes the symmetric block based search.

5.2 *Stepwise Refinement*

Going one step down the hierarchy, we want to compute $\mathcal{D}_k(n)$, given that $\mathcal{D}_{k-1}(n)$ is already known. We may think of $\mathcal{D}_k$ as a sampled version of an underlying continuous displacement field. Then $\{\mathcal{D}_k : 0 \leq k \leq K\}$ is a set of multi-resolution samples taken from the underlying displacement field. Therefore, $\mathcal{D}_k$ can be written as

$$\mathcal{D}_k(\mathbf{r}, n) = 2\tilde{\mathcal{D}}_{k-1}(\mathbf{r}, n) + \Delta\mathcal{D}_k(\mathbf{r}, n) \tag{20.10}$$

where $\tilde{\mathcal{D}}_{k-1}$ is the displacement field $\mathcal{D}_{k-1}$ interpolated at the new sampling grid (see Figure 12), and $\Delta\mathcal{D}_k$ is the correction term for the displacement due to increased resolution at level k. The rescaling by 2 arises due to the finer sampling grid for I_k: the distance between two pels in the upper level has now doubled. Thus, we have reduced the problem of computing $\mathcal{D}_k$ into two subtasks which we now describe.

Computing $\tilde{\mathcal{D}}_{k-1}$ on the new sampling grid requires a resampling, or interpolation (this may seem to be a minor point, however the interpolation is crucial for best performance with the constrained search). The discontinuities in the displacement field are often due to the occluding boundaries of moving objects. Therefore, the interpolation may use a (tentative) segmentation of the frame based on the displacement field, or even on the successive frame difference. The segmentation would allow classifying the blocks as belonging to distinct objects, and preventing the "corona effect" around the moving objects. However, we currently use a simple bilinear interpolation for efficiency, moving the burden to computing $\Delta\mathcal{D}_k$.

After the interpolation, every block b has an initial displacement estimate $\hat{\mathcal{D}}_b$, which is equal to $2\tilde{\mathcal{D}}_{k-1}(\mathbf{r}_b, n)$, where $\mathbf{r}_b$ is the coordinate of the center of block b. Now the search for $\Delta\mathcal{D}_k$ is done inside a window centered around $\hat{\mathcal{D}}_b$, i.e

$$\mathcal{D}_b = \hat{\mathcal{D}}_b + \Delta\mathcal{D}, \qquad \Delta\mathcal{D} \in \{(x, y) : -D \leq x, y \leq +D\}. \tag{20.11}$$

Note that the number of displacements searched for a block is constant, while the number of searches increases geometrically down the hierarchy. The procedure is repeated until one obtains the motion field $\mathcal{D}_K(n)$, corresponding to IKn.

Suppose the displacement in the original sequence IKn were limited to $\pm d_{\max}$ pels in each dimension. Then the displacement in the $(K-k)$th level is limited to $\pm d_{\max}/2^k$, since the frames at this level have been k times spatially decimated by 2. In general, K can be chosen such that the displacement at the top level is known to be less than $\pm D$. Given this choice, we can limit the search for each $\mathcal{D}_k(r, n)$ to $\{(x, y) : -D \leq x, y \leq +D\}$. This results in $(2D+1)^2$ tests to compute $\mathcal{D}_b$ for each block. The maximum displacement that can be handled is

$$\sum_{i=0}^{K} 2^i D = (2^{K+1} - 1)D. \tag{20.12}$$

For a typical sequence, D can be 2 or 3, and K can be 3, allowing a maximum displacement of 30 or 45. This constrained window symmetric search algorithm yields a smooth motion field with high temporal and spatial correlation, at the same time allowing large displacements. Three stages in the estimation procedure are shown in Figure ??.

5.3 Motion Based Interpolation

Given frames $I(n-1)$ and $I(n+1)$, and the displacement $\mathcal{D}(n)$, we form $\tilde{I}(n)$, the motion based estimate of $I(n)$ by displaced averaging:

$$\tilde{I}(\mathbf{r}, n) = 1/2(I(\mathbf{r} - \mathcal{D}, n-1) + I(\mathbf{r} + \mathcal{D}, n+1)). \tag{20.13}$$

Here, we use $\mathcal{D} = \mathcal{D}_b$, the displacement of the block containing $\mathbf{r}$, i.e. we operate on a block level. There are other alternatives, including a pel-level resampling of the displacement field $\mathcal{D}(\mathbf{r})$. However, this would significantly increase the complexity of the decoder, which is kept to a minimum by the blockwise averaging scheme. Furthermore, simulations indicate that the time-varying texture distortions caused by pel-level interpolation are visually unpleasant. It is desirable to preserve textures, but it is especially critical to avoid time-varying distortions which tend to be perceptually unacceptable.

We shall use motion based interpolation as the temporal interpolator in a three dimensional pyramid scheme. However, the method, as presented, has some limitations especially when temporal sampling rate is reduced. Consider temporally decimating $I_1(n)$, call it $I_1^1(n)$ (recall that $I_1(n)$ has a reduced spatial resolution). If the original sequence $I_K(n)$ has a frame-to-frame maximum displacement of $d_{\max}$, this property is preserved in $I_1^1(n)$. However, as we go up the spatial hierarchy, the displacements start to deviate significantly from our original assumptions: We can no longer assume that the velocities are constant, nor that time is homogeneous, although we know that the maximum displacement is preserved. To make matters even worse, the area covered (or uncovered) by a moving object increases proportional to the temporal decimation factor. Thus, an increasingly larger ratio of the viewing area is covered/uncovered as we go up in our pyramid structure. In that case, simple averaging by a symmetric motion vector will likely result in double images and other unpleasant artifacts.

For a successful interpolation, we have to be able to (i) handle discontinuities in motion; (ii) interpolate uncovered areas. These are fundamental problems in any motion based scheme; a manifestation of the fact that we are dealing with solid objects in 3-D space, of which we have incomplete observations in the form of a 2-D projection on the focal plane.

To overcome these difficulties, we allow the interpolator to selectively use previous, following, or both frames on a block by block basis. The algorithm is thus modified in the final step, where $\Delta\mathcal{D}_K$ is computed to yield the final displacement $\mathcal{D}_K$. In addition to the symmetric search so far discussed, two other searches are run in parallel: one using the current and previous, and the other using the following and current frames. Then the motion interpolated blocks using the three schemes are compared against the original block. Now the displacement and interpolation mode minimizing the interpolation error is chosen as the final displacement. (In practice, one may weight the errors, favoring the displacement obtained by the symmetric search). Tests performed using different scenes indicate that the scheme works as intended: Symmetric mode is chosen most of the time, with the other two modes being used i) when there is a pan, in which case the covered/uncovered image boundary is interpolated from future/past; ii) around moving object boundaries (for the same reason); iii) when there is irregular motion, i.e. the symmetry assumption no longer holds (see Figure ??). This interpolation technique has originated from an earlier study for MPEG [31], and is similar to the current MPEG proposal [13].

6 Compression for ATV

Compression of ATV signals for digital transmission is a challenging problem. High quality, meaning both high signal–to–noise ratio and near perfect perceptual quality must be maintained at an order of magnitude reduction of bandwidth. Additional constraints such as cost, real–time implementation and suitability to broadcast environment narrow down the number of alternatives even further.

The layers in the pyramid, except for the top one, consist of interpolation errors, i.e. they are mostly bandpass signals with logarithmic spacing. The bands can be modeled to be independent to a good approximation, although this assumption is often violated around edges in the image. Nevertheless, multiresolution decomposition provides a good compromise between efficiency and coding complexity, and facilitates joint source–channel coding. Properties of the HVS can also be exploited in such a decomposition. HVS has been modeled to consist of independent bandpass channels (at least for still images) [27] [32], and distributing error into these bands provides high compression with little unpleasant artifacts. Temporal and spatial masking phenomena [33] can also be used to advantage while maintaining high perceptual quality.

In this section, we apply the multiresolution concepts so far developed to coding for ATV. First we describe how the scheme can be applied for coding interlaced sequences, and then give some simulation results. We show that both scan types can be successfully merged within a multiresolution hierarchy, providing a versatile, compatible representation for the video signal.

6.1 Compatibility and Scan Formats

All existing broadcast TV standards currently employ interlaced–scan, and we can expect it to dominate the ATV scene in the foreseeable future. The inherent bandwidth reduction and efficient utilization has been the major factor in its acceptance. Today's high resolution, low noise cameras rely on interlacing: switching to non–interlaced scan may degrade the performance by as much as 9 dB [34] (for the same number of lines). On the other hand, non–interlaced scan has many desirable features, including less visible artifacts (absence of line flicker), and convenience for digital processing. Furthermore, some sources such as movies and computer generated graphics are available only in non–interlaced format. Both objective and subjective tests indicate that non–interlaced displays are superior [35]. The next generation systems may be required to handle both scan types, which we show can actually be achieved within a multiresolution scheme at little extra cost. An excellent overview of sampling and reconstruction of multidimensional signals can be found in [36].

Approaches to the coding of interlaced television signals have been studied within CCIR's CMTT/2 Expert Group [37], [38]. The CMTT algorithm considers a recursive coder with three prediction modes (intrafield, interfield, or motion–compensated interframe). The work done in the CMTT/2 Expert Group did not consider interpolative or non-causal prediction, however the idea of getting a predictor — forward or backward — from either the nearest reference field or the nearest cosited field is critical in dealing with interlaced video: depending on the amount and the nature of motion either field may be a good candidate for prediction.

In a finite memory scheme, we are faced with a choice: either group two adjacent fields to form a reference frame, or limit the references to one field. The second solution is much more natural in a temporal pyramid. If the the decimation factor is two, the low temporal resolution signal is all of one parity and the signal to be interpolated is of the other parity. Work along those lines has also been performed independently by Wang and Anastassiou and is reported in [39]. In a temporal pyramid with a decimation factor of two, the input signal is thus separated into odd and even fields, as illustrated in Figure 13. The three–dimensional pyramid decomposition is performed as described on the even fields. As also suggested in

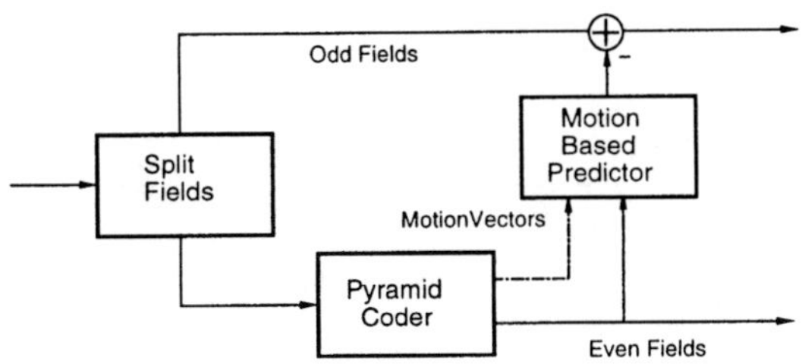

FIGURE 13. Block diagram for coding interlaced input signals

[39], the odd fields are encoded by motion–compensated interfield interpolation. The three interpolation modes previously described can again be used on a block by block basis.

Overhead can be kept low by using the motion information already available at the decoder, or a new temporal interpolation step can be performed, along with transmission of motion vectors and interpolation modes. This is a particularly attractive solution, for it marries the simplicity of non–interlaced sampling inside the spatiotemporal pyramid with the scan compatibility by providing an interlaced signal at the finest resolution. Initial results in this direction look promising, with compression and quality comparable to those obtained with non–interlaced sequences.

6.2 Results

The proposed coding system consists of an entropy coder following the three-dimensional pyramidal decomposition. A discrete cosine transform (DCT) based coder is used to encode the top layer and the subsequent bandpass difference images. The coding algorithm is similar to the JPEG standard [40] [41] for coding still images. DCT is probably not the best choice for coding difference signals, as it approximates the KLT for stationary signals with high autocorrelation [14]. However, it has been widely used in practice, and VLSI implementations make it an attractive choice for inexpensive real–time implementations.

There are several parameters for adjusting the quality versus bandwidth. Each DCT coefficient is quantized by a linear scalar quantizer. Luminance and chrominance components have separate quantizers, with steps adjusted based on MSE and perceptual quality. Typically, chrominance components use about 20 % of the total bandwidth for MSE comparable to the luminance MSE. The bit allocation among the layers is determined by setting the quantizer step size to maintain good visual quality in each layer. Selection of optimal quantizers across the multiresolution hierarchy remains as an interesting problem. We should note that the "optimum" must be based on a joint MR criteria: If the final layer MSE were the criteria, the optimal coder would not have an MR structure except for a restricted class of input signals. Therefore, all bits would be allocated for the final layer for most inputs. Notice that forcing higher layers to zero in a pyramid (effectively allocating no bits) makes the input signal available at the final layer (see Figure 6).

Spatial decimation and interpolation are performed by very simple kernels similar

Layer	Spatial	Temporal	Overall
1	3.06	N/A	0.04
2	2.10	0.55	0.16
3	2.21	0.45	1.33
Total	2.52	0.52	1.53

TABLE 20.1. Bit allocation to various layers for the "MIT" sequence. All numbers are in bits per pixel. The "overall" column indicates the contribution of each layer toward the final bitrate, as summarized in the last row.

to those of Burt and Adelson [1]. The interpolation filter involves two or three pixel averaging (recall that every other sample is zero in the upsampled signal) and is given by $[0.5 - a \quad 0.5 \quad 2a \quad 0.5 \quad 0.5 - a]$ in one-dimensional form. The parameter $0 < a < 0.5$ determines the smoothness of the interpolation kernel, and was chosen as 0.4 for the simulations. This forms a relatively smooth lowpass filter chosen so that the interpolation does not create high frequency error signals that are hard to encode with DCT.

There is also some overhead information associated with each block that has to be coded, most notably the motion vectors. They are differentially coded, with the DPCM loop initialized at the first vector at the upper left of each frame and the predictor is given by the motion vector of the previous block. The differential vectors are coded by a variable length code, typically requiring (on the average) 2 to 5 bits per vector. The horizontal and vertical displacements are coded with a 2–D variable length coder where the less likely events are coded with a prefix and PCM code. The interpolation mode is also encoded, which is one of "backward", "forward", or "averaged". A runlength coder gives satisfactory results, resulting in under a bit per block overhead.

Simulations were performed for several sources. One of the sources is a non-interlaced test sequence produced at MIT that contains a rotating wheel, a subpicture from a movie, and a high detail background with artificially generated zoom and pan. A luminance–chrominance (YCrCb) version is used, with 4:2:2 chrominance subsampling. The picture size is 512 by 512, and 60 frames have been used in the simulation. Blocks of size 8x8 were used throughout, with displacement limited to ± 3 at each stage. The results are summarized in Table 20.1. The second and third columns indicate the average bits per pixel (bpp) used to encode spatial and temporal interpolation errors. Recall that each frame at the top (coarsest) layer and every other frame in the subsequent layers are spatially interpolated. Thus, the "overall" column is computed by averaging the two rates. The last row takes into account the overhead of coarser layers to compute the total rate in terms of bits per pixel in the finest layer.

Subjective quality of each layer was judged to be very good to excellent. No artifacts were visible at normal speed, while some DCT associated artifacts could be seen upon close examination of still frames. The average signal–to–noise (SNR) was 40.2 dB for the luminance component.

"Renata" sequence from RAI and "Table Tennis" from ISO were used for the interlaced coding simulations. A progressive subsequence was obtained by dropping odd fields, and the described algorithm was applied to code this sequence. Then a new set of motion vectors were computed to predict the odd fields from the

Layer	Spatial	Temporal	Overall
1	1.86	N/A	0.01
2	1.64	0.31	0.06
3	1.86	0.75	1.06
Total	2.07	0.76	1.13

TABLE 20.2. Bit allocation to various layers for the interlaced "Renata" sequence. All numbers are in bits per pixel. The "overall" column indicates the contribution of each layer toward the final bitrate, as summarized in the last row. Note that three times as many pixels are temporally interpolated at the last layer.

already encoded even fields and were coded together with the prediction error. The reconstruction procedure consists of the motion–based temporal interpolation we have described. The results of a simulation using the "Renata" sequence are presented in Table 20.2. The picture is 1440 by 1152 pixels, in YCrCb format with 4:2:2 chrominance subsampling. 16 frames have been used for the simulation. Note that unlike the non–interlaced scan case, 3/4 of the frames (or rather fields) have been temporally interpolated in the finest layer. This results in a lower overall bitrate than the previous case, with the "overall" column in the table reflects this weighting. The average SNR was 38.9 dB. The original picture contains a high amount of camera noise, and is therefore difficult to compress while retaining high fidelity. There were no visible artifacts in an informal subjective evaluation.

6.3 Relation to Emerging Video Coding Standards

The technique proposed in this paper has a number of commonalities with the emerging MPEG standard — both are based on motion compensated interpolation — and a usable subset of the techniques presented in this paper could be implemented using a MPEG like syntax. An immediate benefit would be a cost effective implementation of the algorithm with off–the–shelf hardware.

In addition to this basic commonality, the techniques presented in this paper generalize the MPEG approach by introducing spatial hierarchies. Spatial hierarchies are particularly useful when compatibility issues require that a coded bitstream corresponding to a low resolution subsignal be easily obtainable from the higher resolution compressed video signal. They also provide a format suitable for recording on digital media, facilitating fast search and reverse playback.

Finally, this paper proposes a solution to apply the motion compensated interpolation techniques (temporal multi–resolution) to interlaced video signal. The solution fits strictly within the multi resolution, FIR approach and provides an efficient technique to deal with most signals. It is important however to continue investigating the interlaced video problem; and solutions where the two parities — odd and even field — play the same role and where the interpolation can take advantage of the nearest spatial and temporal samples available need to be investigated further.

7 Complexity

In this section, we give a detailed analysis of the computational and memory complexity of the scheme we have presented. In particular, we compare it to conventional techniques such as full search motion estimation and hybrid DCT coding. Also of concern is the decoder simplicity. Highly asymmetric processing schemes are desirable for ATV applications. Encoders are relatively few and can have fairly complex circuitry, even non real–time algorithms can sometimes be acceptable, particularly for recording on digital media. However, decoders have to be simple, inexpensive, and must operate in real time.

Another related issue is compatibility. A low resolution decoder should not have to decode the whole bitstream at full speed to extract the low resolution information. This implies that a hierarchical coding scheme has to be employed.

7.1 Computational Complexity

The major computational burden is in the task of motion estimation. Before we give a complete analysis, let us first look at the major differences from a conventional predictive coding scheme. First, every other frame is interpolated temporally, so motion estimation is used half as often. Second, there are three stages that are coded in the analogous fashion, which brings the total cost to $1+1/8+1/64 = 1.14$ times the cost in the final layer.

We use a three stage search, and at each stage do a constrained search in a 7x7 window, allowing a (differential) displacement of ± 3. Thus, each stage involves 49 comparison operations. The total operation count per block is $49+49/4+49/16 = 64.31$ operations per block.

The factors are due to the fact that a block is split into 4 blocks at each coarse–to–fine step. Thus, there are 4 times less blocks on the second stage compared to the third (and last) stage. There is also an interpolation step at each step, but this is extremely simple, involving 2 or 4 additions per block, and can be safely neglected.

Each operation basically involves a three–step summation with appropriate shifts, as given in equations (20.9) and (20.8). Using 8x8 blocks, each operation thus accounts for 192 additions (assuming shifts by 2 are free). This compares with 128 additions in a conventional scheme using two frames (i.e. using the next frame is only 50% more expensive).

The maximum displacement that can be handled is $3+3\cdot 2+3\cdot 4 = 21$. In comparison, a full search covering the same range would require $(2\cdot 21+1)^2 = 1849$ operations per block, prohibitively expensive with the current technology.

At the last step for each layer, *three* independent searches are performed: two conventional searches involving two frames, and one symmetric search. Thus, the number of operations is actually $49+49/4+49/16$ three–frame operations per block plus $2\cdot 49$ two–frame operations per block.

As a final note, we should point out that the same strategy is used in coding the two coarser layers. Recalling that only half of the frames are motion–compensated, we conclude that the computational complexity of the motion estimation task is on par with the hybrid predictive type algorithms, even when hierarchical ME is used for the latter.

Once the motion vectors are computed, decoding is very simple. Interpolation mode is encoded as a side information, and all the decoder has to do is either use one displaced block from the previous or the following frame (*backward* or *forward* mode), or do a symmetrically displaced averaging (*averaged* mode).

7.2 Memory Requirement

Memory requirement of the algorithm is a critical issue from the hardware realization point of view. In what follows, we use the *frame store* as the basic unit, meaning memory required to hold one frame in the final layer. It is relatively easy to see that we differ from a predictive scheme in two ways (see Figure 10):

1. Three frames are used at a time, compared to two.
2. A complete hierarchy has to be decoded, which, as we have seen, involves 15% overhead.

In the best case, no temporal interpolation is performed, so only 1.15 frames are required for the decoding. For the worst case memory usage, consider frame 1 at layer 2 (last layer) in Figure 10. In order to decode it, the following frames must also be decoded: frames 0 and 1 in layer 0; frames 0, 1, and 2 in layer 1; and frames 0, 1, and 2 in layer 2. The total number of frame stores required is $3+3/4+2/16 = 3.875$ frame stores.

In a recursive scheme, such as the hybrid DCT, only 2 frame stores are required. However, we should emphasize that the pyramid structure allows random access to *any* frame after decoding at most 3.875 units of input. In sharp contrast, a predictive scheme would have to decode all frames since the last restart, which might require as much as 30 frames, even if it is restarted every half second.

To conclude, overall complexity is only slightly higher than the conventional predictive schemes. In return, much faster random access is achieved, and reverse decoding is possible. Note that reverse display requires a prohibitive number of frame stores in the recursive case. The scheme is asymmetric, with a much simpler decoder, which is highly desirable in a broadcast environment. Furthermore, many of the encoding tasks can be run in parallel, making it suitable for real–time implementations.

8 Conclusion and Directions

We have introduced a multiresolution approach to signal representation and coding for advanced television. A three–dimensional pyramid structure has been described where motion information is used for temporal processing, in accordance with our video signal model. Very high subjective quality and SNR of over 40 dB has been achieved in coding of high resolution interlaced and non–interlaced sequences. The scheme provides many advantages in a broadcast environment or for digital storage applications:

- It is an FIR structure, and temporal processing with a short kernel ensures that no errors accumulate over time. Ability to use both the past and the future solves covered/uncovered area problems.

- Fast random access and reverse playback modes are possible.
- Different scan formats can be accommodated.
- Layered and prioritized channels ensure compatibility and graceful degradation.

In the near future at least, we are likely to have several de–facto video standards. Multirate processing will be a key technique for the next generation video codecs. We have seen that the proposed scheme has a number of commonalities with the emerging MPEG standard, and can be seen as a possible evolution path. Multiresolution schemes, both conceptually and as algorithms, can be elegant solutions to a number of problems in coding and representation for advanced television.

Acknowledgements: The authors would like to thank Profs. W. Schreiber and J. Lim of MIT and Dr. M. Barbero of RAI for providing the sequences used in the simulations, and the reviewers for helpful suggestions.

9 References

[1] P. J. Burt and E. H. Adelson, "The laplacian pyramid as a compact image code," *IEEE Transactions on Communications*, vol. 31, pp. 532–540, April 1983.

[2] A. Croisier, D. Esteban, and C. Galand, "Perfect channel splitting by use of interpolation, decimation, tree decomposition techniques," in *Int. Conf. on Information Sciences/Systems*, (Patras), pp. 443–446, August 1976.

[3] R. E. Crochiere and L. R. Rabiner, *Multirate Digital Signal Processing.* Englewood Cliffs: Prentice-Hall, 1983.

[4] M. Vetterli, "Multi–dimensional sub–band coding: Some theory and algorithms," *Signal Processing*, vol. 6, pp. 97–112, February 1984.

[5] J. W. Woods and S. D. O'Neil, "Subband coding of images," *IEEE Trans. Acoust., Speech, Signal Processing*, vol. 34, pp. 1278–1288, October 1986.

[6] S. Mallat, "A theory for multiresolution signal decomposition: The wavelet decomposition," *IEEE Transactions on Pattern Analysis and Machine Intelligence*, vol. 11, pp. 674–693, July 1989.

[7] M. Vetterli and C. Herley, "Wavelets and filter banks: Theory and design." submitted to IEEE Trans. on ASSP.

[8] J. K. Aggarwal and N. Nandhakumar, "On the computation of motion from sequences of images — a review," *Proceedings of the IEEE*, vol. 76, pp. 917–935, August 1988.

[9] F. Glazer, G. Reynolds, and P. Anandan, "Scene matching by hierarchical correlation," in *Proceedings of the IEEE Computer Vision and Pattern Recognition Conference*, (Washington, DC), pp. 432–441, June 1983.

[10] M. Bierling, "Displacement estimation by hierarchical blockmatching," in *Proceedings of the SPIE Conference on Visual Communications and Image Processing,* (Boston), pp. 942–951, November 1988.

[11] K. Shinamura, Y. Hayashi, and F.Kishino, "Variable bitrate coding capable of compensating for packet loss," in *Proceedings of the SPIE Conference on Visual Communications and Image Processing,* (Boston, MA), pp. 991–998, November 1988.

[12] M. Vetterli and K. M. Uz, "Multiresolution techniques with application to HDTV," in *Proc. of the Fourth Int. Colloquium on Advanced Television Systems,* (Ottawa, Canada), pp. 3B2.1–10, June 1990.

[13] Motion Picture Expert Group, ISO/IEC JTC1/SC2/WG8, CCITT SGVIII, "Coded representation of picture and audio information, MPEG video simulation model two," 1990.

[14] N. Jayant and P. Noll, *Digital Coding of Waveforms.* Englewood Cliffs, NJ: Prentice-Hall, 1984.

[15] M. Vetterli, "Filter banks allowing perfect reconstruction," *Signal Processing,* vol. 10, pp. 219–244, April 1986.

[16] M. Vetterli and D. Anastassiou, "An all digital approach to HDTV," in *Proc. ICC-90,* (Atlanta, GA), pp. 881–885, April 1990.

[17] P. P. Vaidyanathan, "Quadrature mirror filter banks, m-band extensions and perfect-reconstruction technique," *IEEE ASSP Magazine*, vol. 4, pp. 4–20, July 1987.

[18] G. Karlsson and M. Vetterli, "Three dimensional sub-band coding of video," in *IEEE International Conference on ASSP,* (New York), pp. 1100–1103, April 1988.

[19] W. F. Screiber and A. B. Lippman, "Single–channel HDTV systems, compatible and noncompatible," in *Signal Processing of HDTV* (L. Chiariglione, ed.), Amsterdam, Netherlands: North-Holland, 1988.

[20] J. W. Woods and T. Naveen, "Subband encoding of video sequences," in *Proc. of the SPIE Conf. on Visual Communications and Image Processing,* (Philadelphia, PA), pp. 724–732, November 1989.

[21] T. Kronander, *Some Aspects of Perception Based Image Coding.* PhD thesis, Dept. of EE, Linköping University, March 1989. No. 203.

[22] M. Pecot, P. J. Tourtier, and Y. Thomas, "Compatible coding of television images," *Image Communication,* vol. 2, October 1990.

[23] H.-M. Hang, R. Leonardi, B. G. Haskell, R. L. Schmidt, H. Bheda, and J. Othmer, "Digital HDTV compression at 44 mbps using parallel motion–compensated transform coders," in *Proceedings of the SPIE Conference on Visual Communications and Image Processing,* vol. 1360, (Lausanne, Switzerland), pp. 1756–1772, October 1990.

[24] D. Anastassiou, "Generalized three-dimensional pyramid coding for HDTV using nonlinear interpolation," in *Proceedings of the Picture Coding Symposium*, (Cambridge, MA), pp. 1.2–1–1.2–2, March 1990.

[25] K. H. Tzou, T. C. Chen, P. E. Fleischer, and M. L. Liou, "Compatible HDTV coding for broadband ISDN," in *Proc. Globecom '88*, pp. 743–749, November 1988.

[26] J. D. Johnston, "A filter family designed for use in quadrature mirror filter banks," in *IEEE International Conference on ASSP*, pp. 291–294, April 1980.

[27] D. Marr, *Vision.* San Fransisco, CA: Freeman, 1982.

[28] K. M. Uz, M. Vetterli, and D. LeGall, "A multiresolution approach to motion estimation and interpolation with application to coding of digital HDTV," in *Proc. ISCAS-90*, (New Orleans, LA), May 1990.

[29] P. J. Burt, "Multiresolution techniques for image representation, analysis, and 'smart' transmission," in *Proceedings of the SPIE Conference on Visual Communications and Image Processing*, (Philadelphia, Pennsylvania), pp. 2–15, November 1989.

[30] M. Bierling and R. Thoma, "Motion compensating field interpolation using a hierarchically structured displacement estimator," *Signal Processing*, vol. 11, pp. 387–404, December 1986.

[31] K. M. Uz and D. J. LeGall, "Motion compensated interpolative coding," in *Proceedings of the Picture Coding Symposium*, (Cambridge, MA), pp. 12.1–1–12.1–3, March 1990.

[32] S. Mallat, "Multifrequency channel decompositions of images and wavelet models," *IEEE Trans. Acoust., Speech, Signal Processing*, vol. 37, pp. 2091–2110, December 1989.

[33] B. Girod, "Eye movements and coding of video sequences," in *Proceedings of the SPIE Conference on Visual Communications and Image Processing*, (Boston, MA), pp. 398–405, November 1988.

[34] R. Schäfer, S. C. Chen, P. Kauff, and M. Leptin, "A comparison of HDTV–standards using subjective and objective criteria," in *Proc. of the Fourth Int. Colloquium on Advanced Television Systems*, (Ottawa, Canada), pp. 3B1.1–17, June 1990.

[35] D. Westerkamp and H. Peters, "Comparison between progressive and interlaced scanning for a future HDTV system with digital data rate reduction," in *Signal Processing of HDTV* (L. Chiariglione, ed.), Amsterdam, Netherlands: North-Holland, 1988.

[36] E. Dubois, "The sampling and reconstruction of time-varying imagery with application in video systems," *Proc. IEEE*, vol. 73, pp. 502–522, April 1985.

[37] CCIR, "Draft new report AD/CMTT on the digital transmission of component–coded television signals at 34 mbit/s and 45 mbit/s." CCIR Documents (1986–90) CMTT/46 and CMTT/116 + Corr 1.

[38] L. Stenger, "Digital coding of television signals — CCIR activities for standardization," *Image Communication*, vol. 1, pp. 29–43, June 1989.

[39] F.-M. Wang and D. Anastassiou, "High-quality coding of the even fields based on the odd fields of interlaced video sequences," *IEEE Trans. on Circuits and Systems*, vol. 38, January 1991.

[40] Joint Photographic Expert Group, ISO/IEC JTC1/SC2/WG8, CCITT SGVIII, "JPEG technical specification, revision 5," January 1990.

[41] A. Ligtenberg and G. K. Wallace, "The emerging JPEG compression standard for continuous tone images: An overview," in *Proceedings of the Picture Coding Symposium*, (Cambridge, MA), pp. 6.1–1–6.1–4, March 1990.

21

Subband Video Coding for Low to High Rate Applications

Wilson C. Chung, Faouzi Kossentini and Mark J. T. Smith

1 Introduction [1]

Over the past several years many effective approaches to the problem of video coding have been proposed [1, 2, 3, 4, 5, 6, 7, 8, 9, 10]. Motion estimation (ME) and compensation (MC) are the cornerstone of most video coding systems presently in vogue and are the basic mechanisms by which temporal redundancies are captured in the current video coding standards. The coding gains achievable by using MC-prediction and residual frame coding are well known. Similarly exploiting spatial redundancy within the video frames plays an important part in video coding. DCT-based schemes as well as subband coding methods have been shown to work well for this application. Such methods are typically applied to the MC-prediction residual and to the individual frames themselves. In the strategy adopted by MPEG, the sequence of video frames is first partitioned into Group of Pictures (GOPs) consisting of N pictures. The first picture in the group (or I-picture) is coded using an intra-picture DCT coder, while the other $N-1$ pictures consist of a combination of P-pictures (forward prediction) and B-pictures (bidirectional pictures), which are coded using a MC-prediction DCT coder [11]. Clearly such an approach has been very effective. However, there are limitations. In particular, the high level of performance is not maintained over a full range of bit rates such as those below 16 kilobits per second (kbps) or those at the HDTV rates. In addition, these standards are limited in the way they exploit the spatio-temporal variations in a video sequence.

In this paper, we introduce a more flexible framework based on a subband representation, motion compensation, and a new class of predictive quantizers. As with MPEG, the I-picture/P-picture structure concept is used in similar spirit. However the coding strategy optimizes spatio-temporal coding within subbands. Thus a unique feature of the proposed approach is that it allows for the flexible allocation of bits between the inter-picture and intra-picture coding components of the coder. This is made possible by applying multistage residual vector quantizers (RVQs) to the subbands and coding the output of the RVQs using a high order conditional

---

[1] Based on "A New Approach to Scalable Video Coding", by Chung, Kossentini and Smith, which appeared in the Proceedings of the Data Compression Conference, Snowbird, Utah, March 1995, ©1995 IEEE

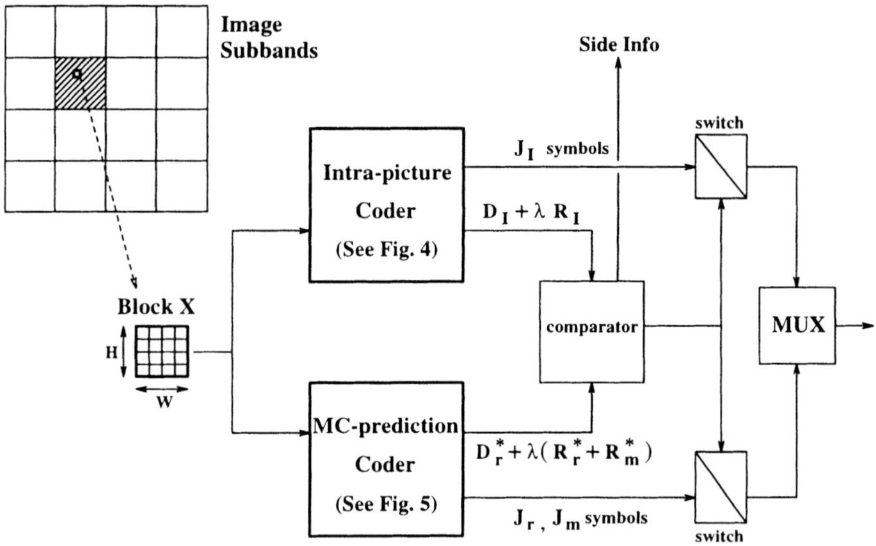

FIGURE 1. Block-level diagram of the proposed video coder.

entropy coding scheme [12]. This coding scheme exploits statistical dependencies between motion vectors in the same subband as well as between those in different subbands of the current and previous frames, simultaneously. The new approach, which we develop next, leads to a fully scalable system, with high quality and reasonable rate-complexity tradeoffs.

2 Basic Structure of the Coder

A block-level description of the framework is shown in Figure 1. Each frame of the input is first decomposed into subbands as illustrated in the figure. Each of the subbands is then decomposed into blocks of size $W \times H$, where each block X is encoded using two coders. One is an intra-picture coder; the other is a MC-prediction coder. The encoder that produces the minimum Lagrangian is selected, and side information is sent to the decoder indicating which of the two coders was chosen.

As will be explained later, both the intra-picture and MC-prediction coders employ high-order entropy coding which exploits dependencies between blocks within and across subbands. Since only one coder is chosen at a time, symbols representing some of the coded blocks in either coder will not be available to the decoder. Thus, both the encoder and decoder must estimate these symbols. More specifically, suppose that the block is intra-picture coded. Then, the coded block is predicted and quantized using the MC-prediction block coder (Figure 2) at both the encoder and decoder. If the block is MC-prediction coded, then the reconstructed block is quantized using the intra-picture block coder (Figure 3), also at both the encoder and decoder. In this case, the additional complexity is relatively small because the

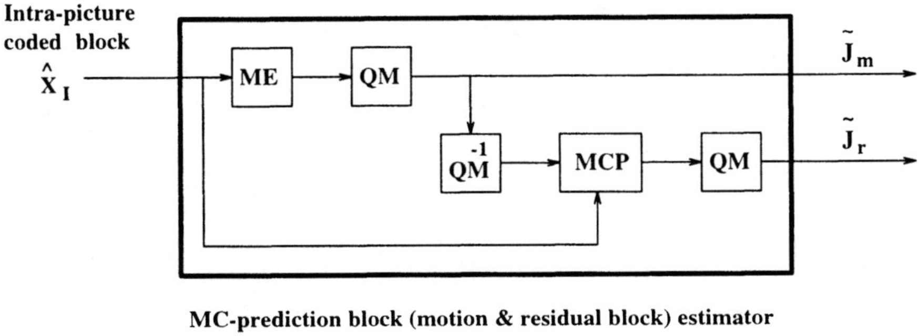

FIGURE 2. MC-prediction block estimator.

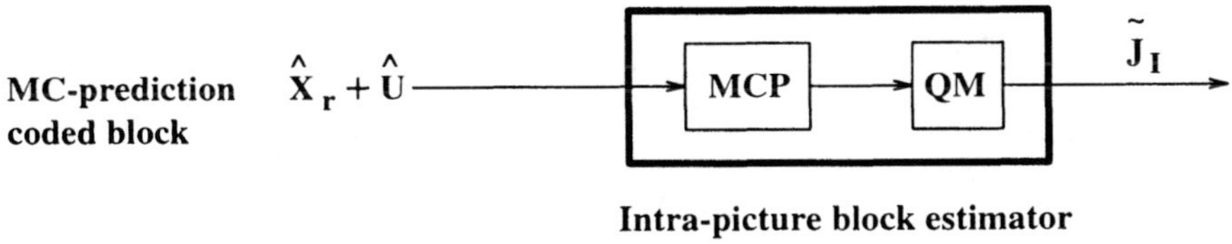

FIGURE 3. Intra-picture block estimator.

the motion-compensated reconstructed blocks are available at both ends, and only intra-picture quantization is performed. In practice, most of the blocks are coded using the MC-prediction coder. Thus, the overall additional complexity required to perform the estimation is relatively small.

Figure 4 shows a typical coder module (CM) structure for the intra-picture coder. An input block is first divided into sub-blocks or vectors. Each is quantized using a fixed-length multistage residual vector quantizer (RVQ) [13] and entropy coded using a binary arithmetic coder (BAC) [14, 15] that is driven by a finite-state machine (FSM) model [16, 17].

Figure 5 shows the structure of the MC-prediction coder. A full-search block matching algorithm (BMA) with the mean absolute distance (MAD) as the cost measure is used to estimate the motion vectors. Since the BMA does not necessarily produce the true motion vectors, we employ a thresholding technique for improving the rate-distortion performance. Let d_{min} be the minimum MAD associated with a block to be coded. Also, let $T > 1.0$ be a threshold that is determined empirically from the statistics of the subband being coded. We then choose all motion vectors with associated MADs d_i that satisfy $\frac{d_i}{d_{min}} < T$. When the motion estimate is accurate, the number M of candidate motion vectors is more likely to have a value of 1. In cases where the video signal undergoes sudden changes (e.g., zoom, occlusion, etc ...) that cannot be accurately estimated using the BMA, a large number of motion vectors can be chosen as candidates, thereby leading to a large complexity. Thus, we limit the number M of candidate motion vectors to a relatively small number M_{max} (e.g., 3 or 4).

In many conventional video subband coders as well as in H.263 and MPEG, pre-

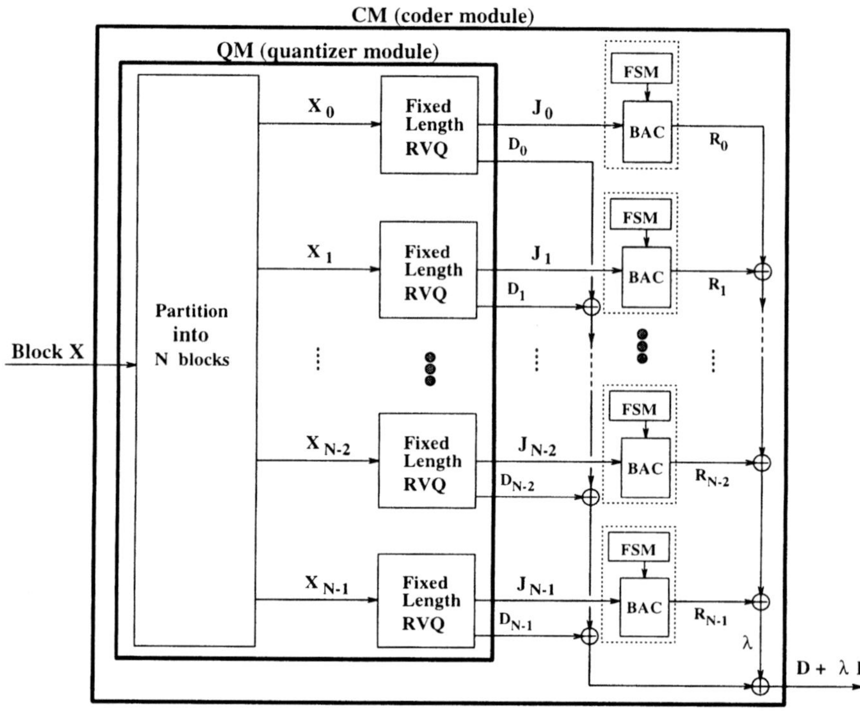

FIGURE 4. A typical coder module (CM) with the corresponding quantizer module (QM), the finite-state machine (FSM) model and the binary arithmetic coder (BAC).

dictive entropy (lossless) coding of motion vectors is employed. Since motion vectors are usually slowly varying, the motion vector bit rate can be reduced further by exploiting dependencies not only between previous motion vectors within and across the subbands but also between the vector coordinates (x- and y-coordinates). For this purpose, we employ a high rate multistage residual 2-dimensional vector quantizer in cascade with a BAC driven by an FSM model that is described later. After the M candidate motion vectors are vector quantized, M MC-prediction reconstructed blocks are generated, and the corresponding residual blocks are computed. The encoding of the residual blocks is done in the same manner as for the original block.

An important part of the encoding procedure is the decision where either the intra-picture or the MC-prediction coder must be chosen for a particular block. Figure 1 shows the encoding procedure. Let R_I and D_I be the rate and distortion associated with intra-picture coding the block X, respectively. Also, let R_M^m, $1 \leq m \leq M$ be the rate required by the 2-dimensional motion coder for the m-th candidate motion vector and (R_r^m, D_r^m) be the rate-distortion pair for the corresponding residual coder. Assuming λ is the Lagrangian parameter that controls the rate-distortion tradeoffs, the intra-picture coder is chosen if

$$D_I + \lambda\, R_I \;<\; D_r^m + \lambda\,(R_r^m + R_M^m), \quad \text{for} \quad m = 1, \ldots, M.$$

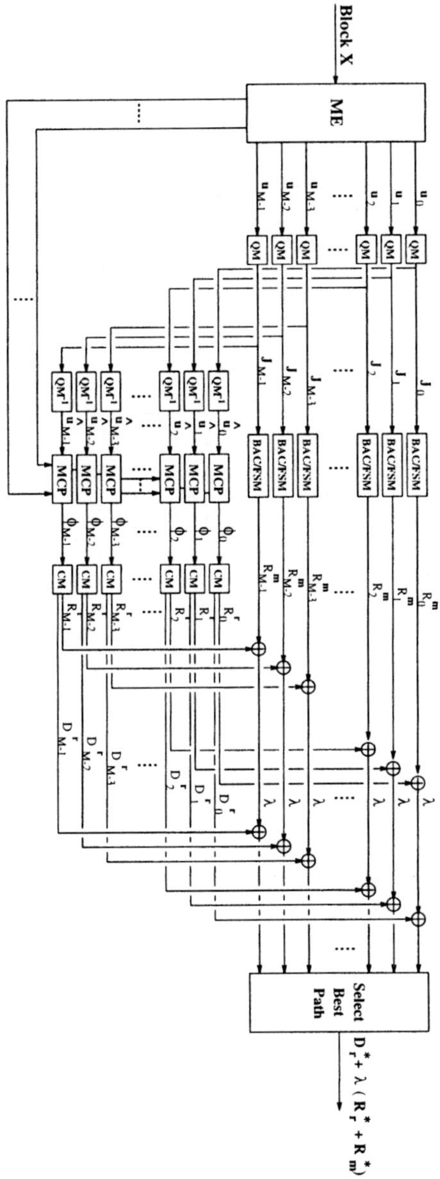

FIGURE 5. A MC-prediction coder.

Otherwise, the MC-prediction coder that leads to the lowest Lagrangian specified by a particular m is chosen.

Statistical modeling for entropy coding the output of the intra-picture, motion vector, and residual coders consists of first selecting, for each subband, N conditioning symbols from a region of support representing a large number of neighboring blocks in the same band as well as other previously coded bands. Then let F be a mapping that is given by

$$u = F(s_1, s_2, \ldots, s_N),$$

where $s_1, s_2, \ldots, s_N$ are the N selected conditioning symbols and $u \in \mathcal{U}$ represents the state of the stage entropy coder. The mapping F converts combinations of realizations of the conditioning symbols to a particular state. For each stage in each subband, a mapping F^* is found such that the conditional entropy $H(J|U)$, where J is the stage symbol random variable and U is the state random variable, are minimized subject to a limit on complexity, expressed in total number of probabilities that must be computed and stored. The tree-based algorithms described in [16, 17] are used to find the best value of N subject to a limit T_1 on the total number of probabilities. The *PNN* algorithm [18], in conjunction with the generalized *BFOS* algorithm [19], is then used to construct tables that represent the best mappings F^* for each stage entropy coder subject to another limit T_2 ($T_2 << T_1$) on the total number of probabilities. Note that the number T_1 controls the tradeoffs between entropy and complexity of the *PNN* algorithm. The output of the RVQs are eventually coded using adaptive binary arithmetic coders (BACs) [14, 15] as determined by the FSM model. BAC coders are very easy to adapt and require a relatively small complexity. They are also naturally suited for this framework because experiments have shown that a stage codebook size of 2 usually leads to a good balance between complexity and performance and provides the highest resolution in a progressive transmission environment.

3 Practical Design & Implementation Issues

In both the intra-picture and MC-prediction coders, the multistage RVQs and associated stage FSM models are designed jointly using an entropy and complexity-constrained algorithm, which is described in [16, 17]. The design algorithm minimizes iteratively the expected distortion $E\{d(X, \hat{X})\}$ subject to a constraint on the overall entropy of the stage models. The algorithm is based on a Lagrangian minimization and employs a Lagrangian parameter λ to control the rate-distortion tradeoffs. The overall FSM model, or the sequence of stage FSM models, enables the design algorithm to optimize jointly the entropy encoders and decoders subject to a limit on complexity. To reduce substantially the complexity of the design algorithm, only independent subband fixed-length encoders and decoders are used. However, the RVQ stage quantizers in each subband are optimized jointly through dynamic M-search [20], and the decoders are optimized jointly using the *Gauss-Seidel* algorithm [13].

To achieve the lowest bit rate, the FSM models used to entropy code the output of the RVQs should be generated on-line. However, this requires a two-pass process where statistics are generated in the first pass, and the modeling algorithm described above is used to generate the conditional probabilities. These probabilities must then be sent to the BAC decoders so that they can track the corresponding encoders. In most cases, this requires a large complexity. Moreover, even by restricting the number of states to be relatively small (such as 8), the side information can be excessive. Therefore, we choose to initialize the encoder with a generic FSM model, which we generate using a training video sequence, and then employ dynamic adaptation [14] to track the local statistics of the video signal.

The proposed coder has many practical advantages, due to both the subband structure and the multistage structure of RVQ. For example, multiresolution transmission can be easily implemented in such a framework. Another example is error correction, where the more probable of the two stage code vectors is selected if an uncorrectable error is detected. Since each stage code vector represents only a small part of the coded vector, this should not significantly affect the video reconstruction quality.

4 Performance

To demonstrate the versatility of the new video coder, we design a low bit rate coder for QCIF ($176 \times 144 \times \times 10Hz$) YUV color video sequences and a high bit rate coder for HDTV video pictures.

For the low bit rate coder, we use the full-search BMA for estimating the motion vectors with the MAD criterion as the matching measure between two blocks. In practice, it is found that the MAD error criterion works well. Our statistical-based BMA uses a block size of 2×2 and a search window area of ± 3 in each direction. Each picture is decomposed into 4 uniform subbands. Motion estimation is performed on all Y luminance, U and V chrominance components. Both the intra-picture and the residual coders use a 1×1 vector size (scalar), and the motion vector coder employs a vector size of 1×2.

To compare the performance of our video coder with the current technology for low bit rate video coding (i.e., below 64 kbits/sec), we used Telenor's software simulation model of the new H.263 standard obtained from *ftp://bonde.nta.no/pub/tmn* [21]. The target bit rate is set to be approximately 16 kbits/sec. Fig. 6(a) and (b) show the bit rate usage and the PSNR coding performance of our coder and Telenor's H.263 for 50 frames of the luminance component of the color test sequence MISS AMERICA. We fixed the PSNR and compared the corresponding bit rates required by both coders. While the average PSNR is approximately 39.4 dB for both coders, the average bit rate for our coder is only 13.245 kbits/sec as opposed to 16.843 kbits/sec for Telenor's H.263. To achieve the same PSNR performance, our coder requires only 78% of the overall bit rate required by Telenor's H.263 video coder.

For the high rate experiments, we use the HDTV test sequence BRITS for comparison, which has a frame size of 720×1280 pixels. Approximately 1.3 Gbps is required for transmission and storage if no compression is performed. Each frame is decomposed into 64 uniform subbands. The full-search BMA algorithm uses a block size of 2×2 and a search area of ± 4 in each direction. The intra-picture and residual coders have a vector size of 2×2 and are designed for each of the

(a) BIT RATE USAGE

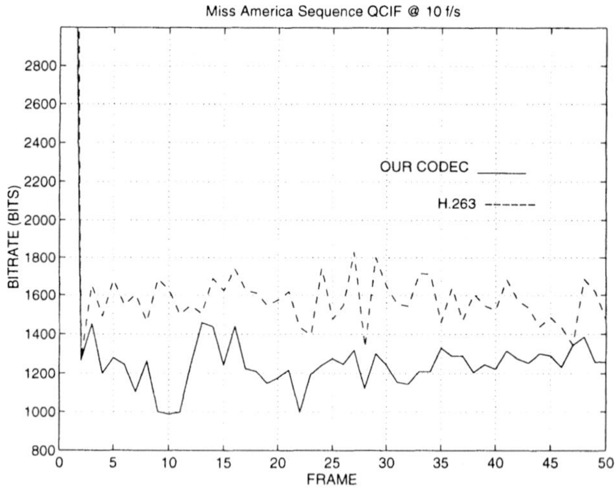

(b) PSNR PERFORMANCE

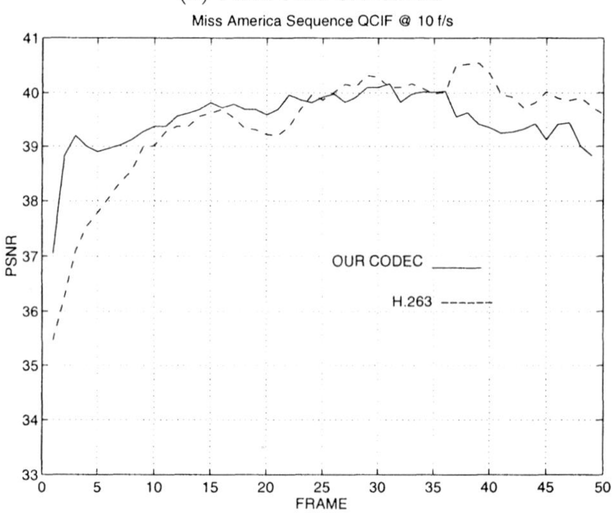

FIGURE 6. The comparison of overall performance — (a) bit rate usage, and (b) PSNR performance, — of our video coder with Telenor's H.263 for the MISS AMERICA QCIF sequence at 10 frames/sec. Only the PSNR of the Y luminance frames are shown in (b).

YUV components. Figure 7(a) shows the bit rate usage and Figure 7(b) shows the PSNR result of our coder in comparison with the MPEG-2 coder for 10 frames of the luminance component of the color video sequence BRITS. The overall average bit rate is approximately 18.0 megabits/sec and the average PSNR is 34.75 dB for the proposed subband coder and 33.70 dB for the MPEG-2 coder. This is typical of the improvement achievable over MPEG-2.

5 References

[1] M. I. Sezan and R. L. Lagendik, *Motion Analysis and Image Sequence Processing.* Boston: Kluwer Academic Publishers, 1992.

[2] W. Li, Y.-Q. Zhang, and M. L. Liou, "Special Issue on Advances in Image and Video Compression," *Proc. of the IEEE*, vol. 83, pp. 135–340, Feb. 1995.

[3] B. Girod, D. J. LeGall, M. I. Sezan, M. Vetterli, and H. Yasuda, "Special Issue on Image Sequence Compression," *IEEE Trans. on Image Processing*, vol. 3, pp. 465–716, Sept. 1994.

[4] K.-H. Tzou, H. G. Musmann, and K. Aizawa, "Special Issue on Very Low Bit Rate Video Coding," *IEEE Trans. on Circuits and Systems for Video Technology*, vol. 4, pp. 213–367, June 1994.

[5] C. T. Chen, "Video compression: Standards and applications," *Journal of Visual Communication and Image Representation*, vol. 4, pp. 103–111, June 1993.

[6] B. Girod, "The efficiency of motion-compensated prediction for hybrid coding of video sequences," *IEEE Journal on Selected Areas in Communications*, vol. 5, pp. 1140–1154, Aug. 1987.

[7] H. Gharavi, "Subband coding algorithms for video applications: Videophone to HDTV-conferencing," *IEEE Trans. on Circuits and Systems for Video Technology*, vol. 1, pp. 174–183, June 1991.

[8] J.-R. Ohm, "Advanced packet-video coding based on layered VQ and SBC techniques," *IEEE Trans. on Circuits and Systems for Video Technology*, vol. 3, pp. 208–221, June 1993.

[9] J. Woods and T. Naveen, "Motion compensated multiresolution transmission of HD video coding using multistage quantizers," in *Proc. IEEE Int. Conf. Acoust., Speech, and Signal Processing*, vol. V, (Minneapolis, MN, USA), pp. 582–585, Apr. 1993.

[10] V. Bhaskaran and K. Konstantinides, *Image and Video Compression Standards: Algorithms and Architecture.* Boston: Kluwer Academic Publishers, 1995.

[11] Motion Picture Experts Group, ISO-IEC JTC1/SC29/WG11/602, "Generic Coding of Moving Pictures and Associated Audio," *Recommendation H.262 ISO/IEC 13818-2, Committe Draft*, Seoul, Korea, Nov. 5, 1993.

(a) BIT RATE USAGE

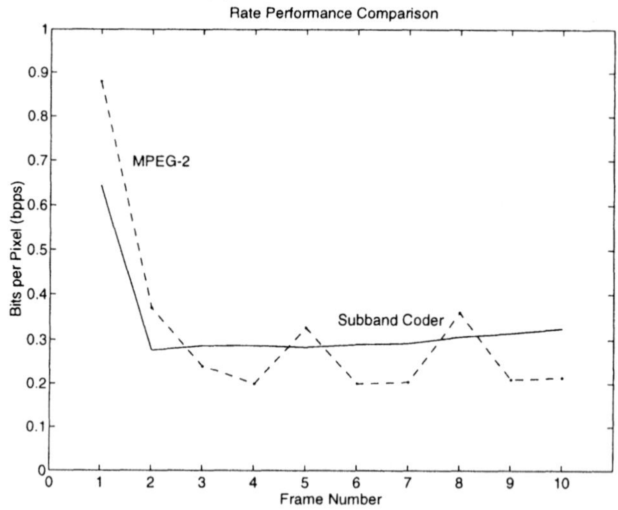

(b) PSNR PERFORMANCE

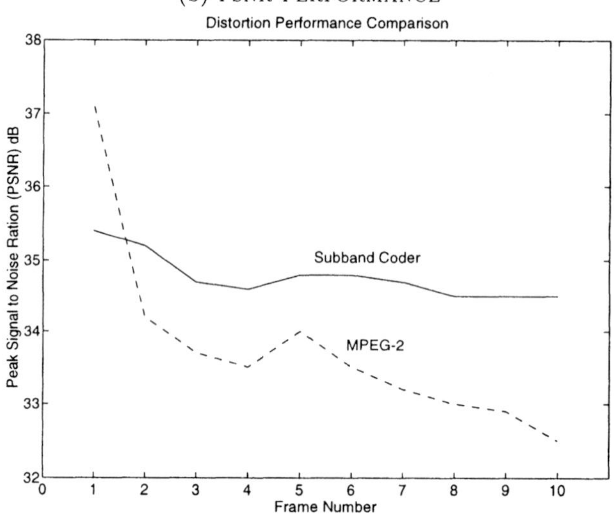

FIGURE 7. *Brits* HDTV sequence. (a) Bit rate usage and (b) PSNR performance for the proposed coder at approximately 18 megabits/sec.

[12] S. M. Lei, T. C. Chen, and K. H. Tzou, "Subband HDTV coding using high-order conditional statistics," *IEEE Journal on Selected Areas in Communications*, vol. 11, pp. 65–76, Jan. 1993.

[13] F. Kossentini, M. Smith, and C. Barnes, "Necessary conditions for the optimality of variable rate residual vector quantizers," *IEEE Trans. on Information Theory*, vol. 41, pp. 1903–1914, Nov. 1995.

[14] G. G. Langdon and J. Rissanen, "Compression of black-white images with arithmetic coding," *IEEE Transactions on Communications*, vol. 29, no. 6, pp. 858–867, 1981.

[15] W. B. Pennebaker, J. L. Mitchel, G. G. Langdon, and R. B. Arps, "An overview of the basic principles of the Q-coder adaptive binary arithmetic coder," *IBM J. Res. Dev.*, vol. 32, pp. 717–726, Nov. 1988.

[16] F. Kossentini, W. Chung, and M. Smith, "A jointly optimized subband coder," *To appear in Transactions on Image Processing*, 1996.

[17] F. Kossentini, W. C. Chung, and M. J. T. Smith, "Subband image coding with jointly optimized quantizers," in *Proc. IEEE Int. Conf. Acoust., Speech, and Signal Processing*, (Detroit, MI, USA), pp. 2221–2224, Apr. 1995.

[18] W. H. Equitz, "A new vector quantization clustering algorithm," *IEEE Trans. on Acoustics, Speech, and Signal Processing*, vol. 37, pp. 1568–1575, Oct. 1989.

[19] E. A. Riskin, "Optimal bit allocation via the generated BFOS algorithm," *IEEE Trans. on Information Theory*, vol. 37, pp. 400–402, Mar. 1991.

[20] F. Kossentini and M. J. T. Smith, "A fast searching technique for residual vector quantizers," *Signal Processing Letters*, vol. 1, pp. 114–116, July 1994.

[21] Telenor Research, "TMN (H.263) encoder / decoder, version 1.4a, *ftp://bonde.nta.no/pub/tmn*," *TMN (H.263) codec*, may 1995.

22

Very Low Bit Rate Video Coding Based on Matching Pursuits

Ralph Neff and Avideh Zakhor
[1]

We present a video compression algorithm which performs well on generic sequences at very low bit rates. This algorithm was the basis for a submission to the November 1995 MPEG4 subjective tests. The main novelty of the algorithm is a matching-pursuit based motion residual coder. The method uses an inner-product search to decompose motion residual signals on an overcomplete dictionary of separable Gabor functions. This coding strategy allows residual bits to be concentrated in the areas where they are needed most, providing detailed reconstructions without block artifacts. Coding results from the MPEG4 Class A compression sequences are presented and compared to H.263. We demonstrate that the matching pursuit system outperforms the H.263 standard in both PSNR and visual quality.

1 Introduction

The MPEG4 standard is being developed to meet a wide range of current and anticipated needs in the area of low bit rate audio-visual communications. Although the standard will be designed to be flexible, there are two main application classes which are currently motivating its development. The first involves real-time two way video communication at very low bit rates. This requires real-time operation at both the encoder and decoder, as well as robustness in noisy channel environments such as wireless networks or telephone lines. The second class involves multimedia services and remote database access. While these applications may not require real-time encoding, they do call for additional functionalities such as object-based manipulation and editing. Both classes of applications are expected to function at bit rates as low as 10 or 24 kbit/s.

The current standard of comparison for real-time two way video communication is the ITU's recommendation H.263 [1]. This system is based on a hybrid motion-compensated DCT structure similar to the two existing MPEG standards. It includes additional features such as an overlapping motion model which improves performance at very low bit rates. There is currently no standard system for doing the object-based coding required by multimedia database applications. However, numerous examples of object oriented coding can be found in the literature [2] [3] [4], and several such systems were evaluated at the MPEG4 subjective tests.

The hybrid motion-compensated DCT structure is a part of nearly all existing video coding standards, and variations of this structure have been proposed for object based coding as well [4] [5]. This popularity can be attributed to the fact that the DCT performs well in a wide variety of coding situations, and because a substantial investment has been made in developing chipsets to implement DCT-based coders. Unfortunately, block-based DCT systems have trouble coding sequences at very low bit rates. At rates below 20 kbit/s, the number of coded DCT coefficients becomes very small, and each coefficient must be represented at a very coarse level of quantization. The resulting coded images have noticeable distortion, and block edge artifacts can be seen in the reconstruction.

To overcome these problems, we propose a hybrid system in which the block-DCT residual coder is replaced with a new coding method which behaves better at low rates. Instead of expanding the motion residual signal on a complete basis such as the DCT, we propose an expansion on a larger, more flexible basis set. Since such an overcomplete basis contains a wider variety of structures than the DCT basis, we expect it to be better able to represent the residual signal using fewer coefficients. The expansion is done using a multistage technique called matching pursuits. This technique was developed for signal analysis by Mallat and Zhang [6], and is related to earlier work in statistics [7] and multistage vector quantization [8]. The matching pursuit technique is general in that it places virtually no restrictions on the choice of basis dictionary. We choose an overcomplete set of separable Gabor functions, which do not contain artificial block edges. We also remove grid positioning restrictions, allowing elements of the basis set to exist at any pixel resolution location within the image. These assumptions allow the system to avoid the artifacts most often produced by low bit rate DCT systems.

The following section reviews the theory and some of the mathematics behind the matching pursuit technique. Section 3 provides a detailed description of the coding system. This includes subsections relating to intraframe coding, motion compensation, matching pursuit residual coding, and rate control. Coding results are presented in Section 4, and conclusions can be found in Section 5.

2 Matching Pursuit Theory

The Matching Pursuit algorithm, as proposed by Mallat and Zhang, [6] expands a signal using an overcomplete dictionary of functions. For simplicity, the procedure can be illustrated with the decomposition of a 1-D time signal. Suppose we want to represent a signal $f(t)$ using basis functions from a dictionary set $\mathcal{G}$. Individual dictionary functions can be denoted as:

$$g_\gamma(t) \in \mathcal{G} \tag{22.1}$$

Here γ is an indexing parameter associated with a particular dictionary element. The decomposition begins by choosing γ to maximize the absolute value of the following inner product:

$$p = < f(t), g_\gamma(t) > \tag{22.2}$$

We then say that p is an expansion coefficient for the signal onto the dictionary function $g_\gamma(t)$. A residual signal is computed as:

$$R(t) = f(t) - p\, g_\gamma(t) \tag{22.3}$$

This residual signal is then expanded in the same way as the original signal. The procedure continues iteratively until either a set number of expansion coefficients are generated or some energy threshold for the residual is reached. Each stage n yields a dictionary structure specified by γ_n, an expansion coefficient p_n, and a residual R_n which is passed on to the next stage. After a total of M stages, the signal can be approximated by a linear function of the dictionary elements:

$$\hat{f}(t) = \sum_{n=1}^{M} p_n\, g_{\gamma_n}(t) \tag{22.4}$$

The above technique has some very useful signal representation properties. For example, the dictionary element chosen at each stage is the element which provides the greatest reduction in mean square error between the true signal $f(t)$ and the coded signal $\hat{f}(t)$. In this sense, the signal structures are coded in order of importance, which is desirable in situations where the bit budget is very limited. For image and video coding applications, this means that the most visible features tend to be coded first. Weaker image features are coded later, if at all. It is even possible to control which types of image features are coded well by choosing dictionary functions to match the shape, scale, or frequency of the desired features.

An interesting feature of the matching pursuit technique is that it places very few restrictions on the dictionary set. The original Mallat and Zhang paper considers both Gabor and wavepacket function dictionaries, but such structure is not required by the algorithm itself. Mallat and Zhang showed that if the dictionary set is at least complete, then $\hat{f}(t)$ will eventually converge to $f(t)$, though the rate of convergence is not guaranteed [6]. Convergence speed and thus coding efficiency are strongly related to the choice of dictionary set. However, true dictionary optimization can be difficult since there are so few restrictions. Any collection of arbitrarily sized and shaped functions can be used with matching pursuits, as long as completeness is satisfied.

There has been much recent interest in using matching pursuits for image processing applications. Bergeaud and Mallat [9] use the technique to decompose still images on a dictionary of oriented Gabor functions. Vetterli and Kalker [10] present an interesting twist on the traditional hybrid DCT video coding algorithm. They use matching pursuits blockwise with a dictionary consisting of DCT basis functions and translated motion blocks. The matching pursuit representation can also be useful for pattern recognition, as demonstrated by Phillips [11]. For many applications, the algorithm can be computationally intensive. For this reason, several researchers have proposed speed enhancements [9], [12], [13].

The system presented in this paper is a hybrid motion compensated video codec in which the motion residual is coded using matching pursuits. This is an enhanced version of the coding system presented in [14] and [15]. The current system allows dictionary elements to have arbitrary sizes, which speeds the search for small scale functions and makes storage more efficient. Larger basis functions are added to increase coding efficiency for scenes involving higher motion or lighting changes. The complete system, including enhancements, is described in the next section.

3 Detailed System Description

This section describes the matching pursuit coding system used to generate the results presented in Section IV. Simplified block diagrams of the encoder and decoder are shown in Figure 1. As can be seen in Figure 1a, original images are first motion compensated using the previous reconstructed image. Here we use the advanced motion model from H.263, as described in Section 3.1. The matching pursuit algorithm is then used to decompose the motion residual signal into coded dictionary functions which are called atoms. The process of finding and coding atoms is described in Section 3.2. Once the motion vectors and atom parameters are found, they can be efficiently coded and sent to the decoder, shown in Figure 1b. This process is governed by a rate control system, which is discussed in Section 3.3. Finally, Section 3.4 describes the scalable subband coder [16] which is used to code the first intra frame.

3.1 Motion Compensation

The motion model used by the matching pursuit system is identical to the advanced prediction model used by TMN [17], which is a specific implementation of H.263 [1]. This section provides a brief description of this model, and also describes how intra blocks are coded when motion prediction fails.

The system operates on QCIF images which consist of 176×144 luminance and 88×72 chrominance components. The luminance portion of the image is divided into 16×16 blocks, and each block is assigned either a single motion vector, a set of four motion vectors, or a set of intra block parameters. The motion search for a block begins with a ± 15 pixel search at integer resolution. The inter/intra decision criterion from TMN is used to decide which prediction mode should be used. If the block is coded in inter mode, then the integer resolution search is refined with a ± 1 half pixel search at half pixel resolution. The block is then split into four 8×8 subblocks and each of these is given a ± 2 half pixel search around the original half-pixel block vector. The block splitting criterion from TMN is used to decide whether one or four vectors should be sent. In either case, vectors are coded differentially using a prediction scheme based on the median of neighboring vector values. Although motion vectors are found using rectangular block matching, the prediction image is formed using the overlapping motion window from H.263. For more details on these methods, consult the H.263 and TMN documents [1],[17].

An H.263 encoder normally codes intra blocks using the DCT. Since we wish to avoid DCT artifacts in the matching pursuit coder, we follow a different approach for coding intra blocks. If a 16×16 block is coded intra, the average pixel intensities from each of the six subblocks are computed. These include four 8×8 luminance subblocks and two 8×8 chrominance subblocks. These six DC intensity values are quantized to five bits each and transmitted as intra block information. The H.263 overlapping motion window is used to smooth the edges of intra blocks and thus

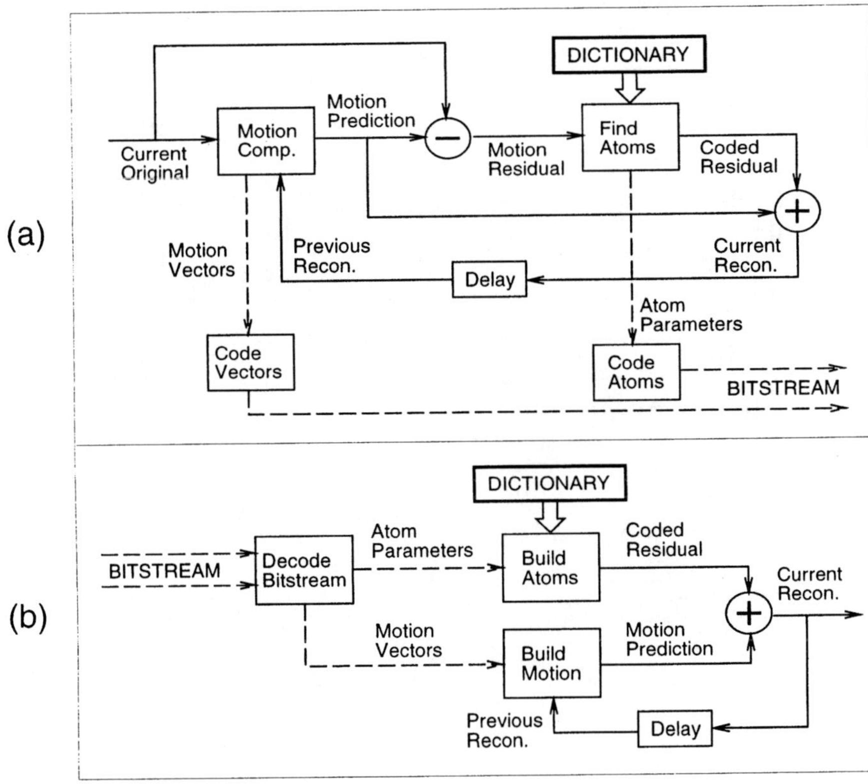

FIGURE 1. (a) Block diagram of encoder. (b) Decoder.

reduce blocking effects in the prediction image. It is assumed that the additional detail coding of intra blocks will be done by the matching pursuit coder during the residual coding stage.

3.2 Matching-Pursuit Residual Coding

After the motion prediction image is formed, it is subtracted from the original image to produce the motion residual. This residual image is coded using the matching pursuit technique introduced in Section 2. To use this method to code the motion residual signal, we must first extend the method to the discrete 2-D domain with the proper choice of a basis dictionary.

The Dictionary Set

The dictionary set we use consists of an overcomplete collection of 2-D separable Gabor functions. We define this set in terms of a prototype Gaussian window:

$$g(t) = \sqrt[4]{2}e^{-\pi t^2} \tag{22.5}$$

FIGURE 2. The 2-D separable Gabor dictionary.

We can then define 1-D discrete Gabor functions as a set of scaled, modulated Gaussian windows:

$$g_{\vec{\alpha}}(i) = K_{\vec{\alpha}} \cdot g(\frac{i - \frac{N}{2} + 1}{s}) \cdot cos(\frac{2\pi\xi(i - \frac{N}{2} + 1)}{16} + \phi) \qquad i \in \{0, 1, ...N{-}1\}; \quad (22.6)$$

Here $\vec{\alpha} = (s, \xi, \phi)$ is a triple consisting respectively of a positive scale, a modulation frequency, and a phase shift. This vector is analogous to the γ parameter of Section 2. The constant $K_{\vec{\alpha}}$ is chosen such that the resulting sequence is of unit norm. If we consider $\mathcal{B}$ to be the set of all such triples $\vec{\alpha}$, then we can specify our 2-D separable gabor functions to be of the following form:

$$G_{\vec{\alpha},\vec{\beta}}(i, j) = g_{\vec{\alpha}}(i) g_{\vec{\beta}}(j) \qquad i, j \in \{0, 1, ...N{-}1\} \quad \vec{\alpha}, \vec{\beta} \in \mathcal{B} \qquad (22.7)$$

In practice, a finite set of 1-D basis functions is chosen and all separable products of these 1-D functions are allowed to exist in the 2-D dictionary set. Table 22.1 shows how the 1-D dictionary used to generate the results of Section 4 can be defined in terms of its 1-D Gabor basis parameters. To obtain this parameter set, a training set of motion residual images was decomposed using a dictionary derived from a much larger set of parameter triples. The dictionary elements which were most often matched to the training images were retained in the reduced set. The dictionary

TABLE 22.1. Dictionary triples and associated sizes that form the 1-D basis set.

k	s_k	ξ_k	ϕ_k	*size* (pixels)
0	1.0	0.0	0	1
1	3.0	0.0	0	5
2	5.0	0.0	0	9
3	7.0	0.0	0	11
4	9.0	0.0	0	15
5	12.0	0.0	0	21
6	14.0	0.0	0	23
7	17.0	0.0	0	29
8	20.0	0.0	0	35
9	1.4	1.0	$\pi/2$	3
10	5.0	1.0	$\pi/2$	9
11	12.0	1.0	$\pi/2$	21
12	16.0	1.0	$\pi/2$	27
13	20.0	1.0	$\pi/2$	35
14	4.0	2.0	0	7
15	4.0	3.0	0	7
16	8.0	3.0	0	13
17	4.0	4.0	0	7
18	4.0	2.0	$\pi/4$	7
19	4.0	4.0	$\pi/4$	7

must remain reasonably small, since a large dictionary decreases the speed of the algorithm.

A visualization of the 2-D basis set is shown in Figure 2. In previously published experiments [14], [15] a fixed size of $N = 16$ pixels was used. In the current system, each 1-D function has an associated size, which allows larger basis functions to be used. An associated size also allows the basis table to be stored more efficiently and increases the search speed, since smaller basis functions are no longer padded with zeros to a fixed size. The size for each 1-D basis element is determined by imposing an intensity threshold on the scaled prototype gaussian window $g(\cdot)$ in Equation 22.6. Larger scale values generally translate to larger sizes, as can be seen in Table 22.1.

Note that in the context of matching pursuits, the dictionary set pictured in Figure 2 forms an overcomplete basis for the residual image. This basis set consists of all of the shapes in the dictionary placed at all possible integer pixel locations in the residual image. To prove that this is an overcomplete set, consider a subset containing only the 1×1 pixel element in the upper left corner of Figure 2. Since this element may be placed at any integer pixel location, it alone forms the standard basis for the residual image. Using all 400 shapes from the Figure produces a much larger basis set. To understand the size of this set, consider that a complete basis set for a 176×144 QCIF image must contain 25344 elements. The block-DCT basis used by H.263 satisfies this criterion, having 64 basis functions to represent each of 396 blocks. Our basis set effectively contains 400 elements for each of the 25344

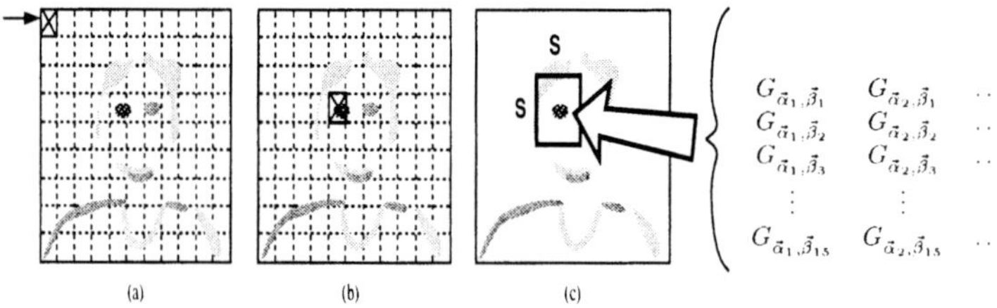

FIGURE 3. Inner product search procedure. (a) Sum-of squares energy search. (b) Largest energy location becomes estimate for exhaustive search. (c) Exhaustive inner product search performed in $S \times S$ window.

locations, for a total of more than 10 million basis elements. Using such a large set allows the matching pursuit residual coder to represent the residual image using fewer coefficients than the DCT. This is a tradeoff, since the cost of representing each element increases with the size of the basis set. Experiments have shown that this tradeoff is extremely favorable at low bit rates [15].

One additional basis function is added at the frame level to allow the coder to more efficiently represent global lighting changes. This is effectively the DC component of the luminance residual frame. It is computed once per frame, and transmitted as a single 8-bit value. This increases coding efficiency on sequences which experience an overall change in luminance with time. Further research is necessary to efficiently model and represent more localized lighting changes.

Finding Atoms

The previous section defined a dictionary of 2-D structures which are used to decompose motion residual images using matching pursuits. A direct extension of the matching pursuit algorithm would require us to examine each 2-D dictionary structure at all possible integer-pixel locations in the image and compute all of the resulting inner products. In order to reduce this search to a manageable level, we make some assumptions about the residual image to be coded. Specifically, we assume that the image is sparse, containing pockets of energy at locations where the motion prediction model was inadequate. If this is true, we can "pre-scan" the image for high-energy pockets. The location of such pockets can be used as an initial estimate for the inner-product search.

The search procedure is outlined in Figure 3. The motion residual image to be coded is first divided into blocks, and the sum of the squares of all pixel intensities is computed for each block, as shown in Figure 3a. This procedure is called "Find Energy." The center of the block with the largest energy value, depicted in Figure 3b, is adopted as an initial estimate for the inner product search. The dictionary is then exhaustively matched to an $S \times S$ window around the initial estimate, as shown in Figure 3c. In practice, we use overlapping 12×12 blocks for the "Find Energy" procedure, and a search window size of $S = 16$.

The exhaustive search can be thought of as follows. Each $N \times N$ dictionary structure is centered at each location in the search window, and the inner product between the structure and the corresponding $N \times N$ region of image data is computed. Fortunately the search can be performed quickly using separable inner

TABLE 22.2. Parameters which define an atom

$\vec{\alpha}, \vec{\beta}$	Best match structure element from dictionary
x, y	Location of best match in residual image
p	Value of largest inner product; Projection of image data at (x, y) onto $G_{\vec{\alpha},\vec{\beta}}(i, j)$

product computations. This technique is shown in Section 3.2.

The largest inner product, along with the corresponding dictionary structure and image location, form a set of five parameters, as shown in Table 22.2. We say that these five parameters define an atom, a coded structure within the image. The atom decomposition of a motion residual image is illustrated in Figure 4. The motion residual from a sample frame of the Hall sequence is shown in Figure 4a. The first five coded atoms are shown in Figure 4b. Note that the most visually prominent features are coded with the first few atoms. Figures 4c and 4d show how the coded residual appears with fifteen and thirty coded atoms, respectively. The final coded residual, consisting of sixty-four atoms, is shown in Figure 4e. The total number of atoms is determined by the rate control system described in Section 3.3. For this example, the sequence was coded at 24 kbit/s. A comparison between the coded residual and the true residual shows that the dominant features are coded but lower energy details and noise are not coded. These remain in Figure 4f, which is the final coding error image. The reconstructed frame is shown in Figure 4g, and a comparison image coded with H.263 at the same bit rate and frame rate is shown in Figure 4h.

The above atom search procedure is also used to represent color information. The "Find Energy" search is performed on each of the three component signals (Y, U, V), and the largest energy measures from the color difference signals are weighted by a constant before comparison with the largest energy block found in luminance. For the experiments presented here, a color weight of 2.5 is used. If either of the weighted color "best energy" measures exceeds the value found in luminance, then a color atom is coded.

Coding Atom Parameters

When the atom decomposition of a single residual frame is found, it is important to code the resulting parameters efficiently. The atoms for each frame are grouped together and coded in position order, from left to right and top to bottom. The positions (x, y) are specified with adaptive Huffman codes derived from the previous ten frames worth of position data. Since position data from previous frames is available at the decoder, no additional bits need be sent to describe the adaptation. One group of codewords specifies the horizontal displacement of the first atom on a position line. A second group of codes indicates the horizontal distance between atoms on the same position line, and also contains 'end-of-line' codes which specify the vertical distance to the next active position line.

The other three atom parameters are coded using fixed Huffman codes. The basis shape is specified by horizontal and vertical components $\vec{\alpha}$ and $\vec{\beta}$, each of which is represented by an index equivalent to k in Table 22.1. Separate code tables

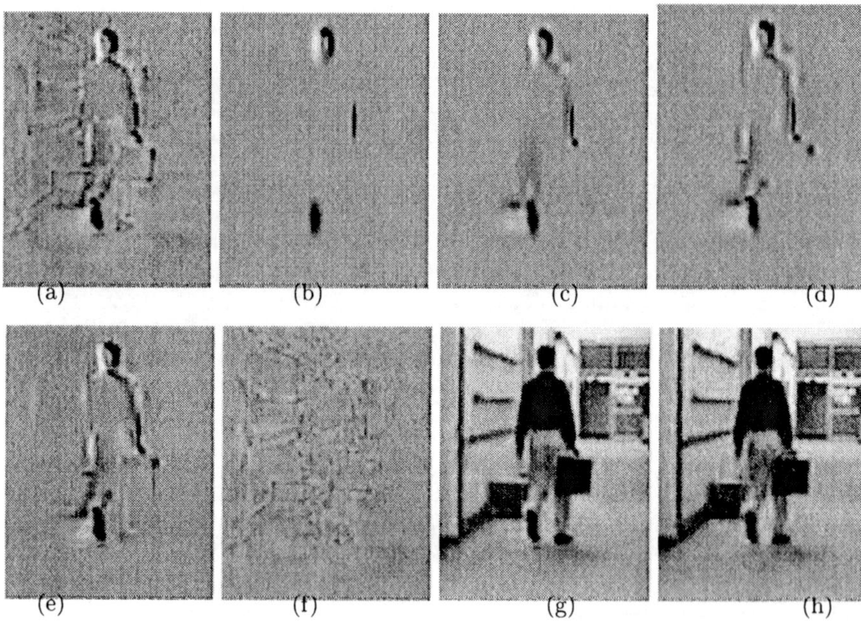

FIGURE 4. Atom decomposition of Hall, frame 50. (a) Motion residual image. (b) First 5 coded atoms. (c) First 15 coded atoms. (d) First 30 coded atoms. (e) All 64 coded atoms. (f) Final coding error. (g) Reconstructed image. (h) Same frame coded by H.263.

are maintained for the horizontal and vertical shape indices. Two special escape codewords in the horizontal Y shape table are used to indicate when a decoded atom belongs to the U or V color difference signals. In all cases, the projection value p is quantized by a linear quantizer with a fixed stepsize, and transmitted using variable length codes.

To obtain the fixed huffman codes described above, it is necessary to gather statistics from previous encoder runs. For this purpose, our training set consists of the five MPEG4 Class A sequences which are listed in Section 4. In each case, we omit the sequence to be coded from the training set. When encoding the Akiyo sequence, for example, the huffman code statistics are summed from previous runs of Container, Hall, Mother and Sean. This technique allows us to avoid the case where an encoded sequence depends on its own training data.

Fast Inner Product Search

The separable nature of the dictionary structures can be used to reduce the inner product search time significantly. Recall that each stage in which an atom is found requires the algorithm to compute the inner product between each 2-D dictionary shape and the underlying image patch at each location in an $S \times S$ search window. To compute the number of operations needed to find an atom, it will first be necessary to introduce some notation describing the basis functions and their associated sizes.

Suppose h and v are scalar indices into Table 22.1 representing the 1-D components of a single 2-D separable basis function. More specifically, h and v are the indices which correspond respectively to $\vec{\alpha}$ and $\vec{\beta}$ in Equation 22.7. The dictionary set can thus be compactly written as

$$G_{h,v}(i,j) = g_h(i)g_v(j) \qquad i \in \{0,...L_h{-}1\} \qquad (22.8)$$

$$j \in \{0,...L_v{-}1\} \qquad (22.9)$$

$$h,v \in \{0,...B{-}1\} \qquad (22.10)$$

where B is the number of 1-D basis elements used to generate the 2-D set, and L_h and L_v are the associated sizes of the horizontal and vertical basis components, respectively.

Suppose we ignore separability. In this case, the inner product between a given 2-D basis element and the underlying image patch can be written as

$$p = \sum_{i=0}^{L_h-1}\sum_{j=0}^{L_v-1} G_{h,v}(i,j)I(i,j) \qquad (22.11)$$

Computing p requires L_hL_v multiply-accumulate operations. Finding a single atom requires this inner product to be computed for each combination of h and v at each location in the $S \times S$ search window. The total number of operations is thus

$$T_{nonsep} = S^2 \sum_{h=0}^{B-1}\sum_{v=0}^{B-1} L_hL_v$$

Using the parameters in Table 22.1 with a search size of $S = 16$, we get 21.8 million multiply-accumulate operations.

If we now assume a separable dictionary, the inner product of Equation 22.11 can be written instead as

$$p = \sum_{i=0}^{L_h-1}\sum_{j=0}^{L_v-1} g_h(i)g_v(j)I(i,j) = \sum_{i=0}^{L_h-1} g_h(i) \sum_{j=0}^{L_v-1} g_v(j)I(i,j)$$

Computing a single 2-D inner product is thus equivalent to taking L_h vertical 1-D inner products, each of length L_v, and then following with a single horizontal inner product of length L_h. This concept is visualized in Figure 5. The atom search requires us to exhaustively compute the inner product at each location using all combinations of h and v. It is thus natural to pre-compute the necessary vertical 1-D inner products with a particular g_v, and then cycle through the horizontal 1-D inner products using all possible g_h. Furthermore, the results from the 1-D vertical pre-filtering also apply to the inner products at adjoining locations in the search window. This motivates the use of a large "prefiltering matrix" to store all the 1-D vertical filtering results for a particular g_v. The resulting search algorithm is shown in Figure 6. The operation count for finding a single atom is thus:

$$T_{sep} = \sum_{v=0}^{B-1}(L_vS(S + L_{max}) + S^2 \sum_{h=0}^{B-1} L_h)$$

TABLE 22.3. Complexity of the H.263 motion search

Search Resolution	Block Size	Search Range	Motion Ops Per Block	Total Blocks
Integer	16×16	± 15	246016	99
Half Pixel	16×16	± 1	2304	99
Half Pixel	8×8	± 2	1600	396

Note that L_{max} is the size of the largest 1-D basis function. Using the values from Table 22.1, 1.7 million multiply-accumulate operations are required to find each atom. This gives a speed improvement factor of about 13 over the general nonseparable case.

An encoding speed comparison between the matching pursuit coder and TMN [17] can be made by counting the operations required to code a single inter-frame using each method. We start with a few simplifying assumptions. First, assume that the TMN encoder is dominated by the motion prediction search, and that the DCT computation can be ignored. In this case coding a TMN frame has a complexity of Z_m, the number of operations needed to perform the motion search for a single frame. The TMN motion search can be broken down into three levels, as shown in Table 22.3. Integer resolution and half-pixel resolution searches are first performed using the full 16×16 pixel block size. A further search using a reduced 8×8 block size is also performed at half-pixel resolution. Each search involves the repetition of a basic operation in which the difference of two pixel intensities is computed and the absolute value of the result is added to an accumulator. Consider this to be a "motion operation." The number of motion operations per block is computed in the fourth column of Table 22.3, and the total number of blocks per frame is shown in the fifth column. From these numbers it is easy to compute $Z_m = 25.22$ million motion operations per frame. Since we are ignoring the DCT computation, this will also be the number of operations needed to encode a single TMN frame.

To model the complexity of the matching pursuit coder, we note that the same motion search is used, and we add the complexity of the atom search. The "Find Energy" search is ignored, since this search can be implemented very efficiently. With this in mind, searching for a single atom requires 1.7 million multiply-accumulate operations as discussed above. The number of coded atoms per frame depends on the number of bits available, as set by the target bit rate. For the comparison which follows, we average the number of atoms coded per frame in the matching pursuit experiments presented in Section 4. By multiplying the average number of coded atoms by 1.7 million operations per atom, we can measure the complexity of the atom search. This value is then added to Z_m to get the total matching pursuit algorithm complexity. For simplicity, we have assumed that a single motion operation takes the same amount of processor time as a single multiply-accumulate. The validity of this assumption depends on the hardware platform, but we have found it to hold reasonably well for the platforms we tested. Table 22.4 shows the theoretical complexity comparison. From this table, it is evident that matching pursuit encoding is several times more complex than H.263, but the complexity difference is reduced at lower bit rates.

TABLE 22.4. Theoretical Encoder Complexity Comparison

Target Bit Rate	Avg. Coded Atoms	H.263 Complexity*	M.P. Complexity*	Complexity Ratio
10 kbit/s	51.18	25.22	112.23	4.45
24 kbit/s	95.00	25.22	186.72	7.40

* Complexity is in millions of operations per frame, where each motion operation or multiply-accumulate is counted as a single operation.

TABLE 22.5. Software Encoder Complexity Comparison

		HP-755 99MHz			Pentium 200MHz		
Sequence	Rate (kbit/s)	H.263 Time*	M.P. Time*	Ratio	H.263 Time*	M.P. Time*	Ratio
Container	10	0.83	4.04	4.86	0.50	1.96	3.94
Mother	10	0.81	2.95	3.64	0.52	1.39	2.69
Container	24	0.79	7.59	9.61	0.52	3.65	7.01
Mother	24	0.81	6.58	8.12	0.52	3.13	6.01

*Time in seconds per coded frame.

To show that this theoretical analysis is reasonable, we provide preliminary software encoding times for some sample runs of the TMN and matching pursuit systems. These are presented in Table 22.5. The times are given as the average number of seconds needed to encode each frame on the given hardware platforms. Times are compared across two platforms. The first is an HP-755 workstation running HP-UX 9.05 at a clock speed of 99 MHz. The second is a Pentium machine running FreeBSD 2.2-SNAP with a clock speed of 200 MHz. The GNU compiler was used in all cases. It can be seen from the final column that the software complexity ratios are fairly close to the theoretical values seen in Table 22.4.

Note that both software encoders employ the full search TMN motion model described in Section 3.1, and that the matching pursuit coder uses the exhaustive local atom search described earlier in this section. Neither encoder has been optimized for speed, and further speed improvements are certainly possible. For example, the exhaustive motion search could be replaced with a reduced hierarchical search. A similar idea could be employed to reduce the atom search time as well. Also, changes in the number of dictionary elements or in the elements themselves can have a large impact on coding speed. Further research will be necessary to exploit these ideas while still maintaining the coding efficiency of the current system.

3.3 Buffer Regulation

For the purpose of comparing the matching pursuit encoder to the TMN H.263 encoder, a very simple mechanism is applied to synchronize rate control between the two systems. The TMN encoder is first run using the target bit rates and frame rates listed in Section 4. The simple rate control mechanism of TMN is used, which adaptively drops frames and adjusts the DCT quantization stepsize in order to meet the target rates [17]. A record is kept of which frames were coded by the TMN encoder, and how many bits were used to represent each frame. The matching pursuit system then uses this record, encoding the same subset of original frames with approximately the same number of bits for each frame. This method eliminates rate control as a variable and allows a very close comparison between the two systems.

The matching pursuit system is thus given a target number of bits to use for each coded frame. It begins by coding the frame header and motion vectors, subtracting the bits necessary to code this information from the target total. The remaining bits become the target residual bit budget, B_r. The encoder also computes the average number of bits each coded atom cost in the previous frame, B_{atom}. An approximation of the number of atoms to code in the current frame is computed as:

$$N_{curr} = \frac{B_r}{B_{atom}}$$

The encoder proceeds with the residual coding process, stopping after N_{curr} atoms have been coded. This approximately consumes B_r bits, and closely matches the target bit rate for the frame. In practice, the number of bits used by the matching pursuit system falls within about one percent of the H.263 bit rate.

3.4 Intraframe Coding

To code the first intra frame, we use the scalable subband coder of Taubman and Zakhor [19]. The source code and documentation for this subband coding system were released to the public in 1994, and can be obtained from an anonymous ftp site [16]. The package is a complete image and video codec which uses spatial and temporal subband filtering along with progressive quantization and arithmetic coding. The resulting system is scalable in both resolution and bit rate. We chose this coder because it produces high quality still images at low bit rates, and because it is free of the blocking artifacts produced by DCT-based schemes. Specifically, it produces much better visual quality than the simple intra-DCT coder of H.263 at low bit rates. For this reason, we use the subband system to code the first frame on both the matching pursuit and H.263 runs presented in the next section.

An illustration of the scalable subband coder is presented in Figure 7, which shows coded versions of Frame 0 of the Hall sequence using both the scalable subband coder and H.263 in intraframe mode. Both images are coded at approximately 15 kbits. The subband version shows more detail and is free of blocking artifacts. The H.263 coded image has very noticeable blocking artifacts. While these could be reduced using post-filtering, such an operation generally leads to a reduction in detail.

4 Results

This section presents results which compare the matching pursuit coding system to H.263. The H.263 results were produced with the publicly available TMN encoder

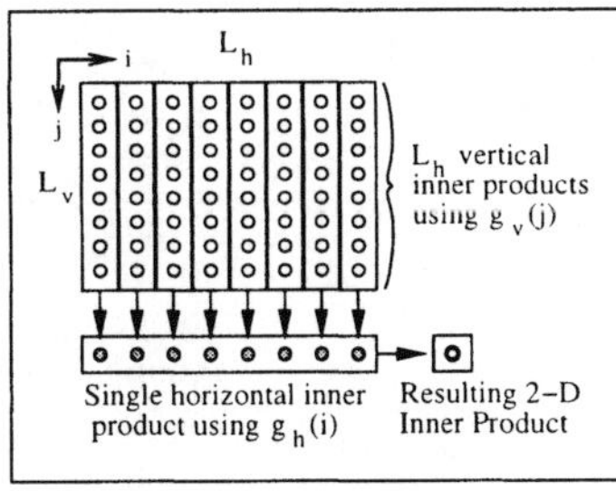

FIGURE 5. Separable computation of a 2-D inner product.

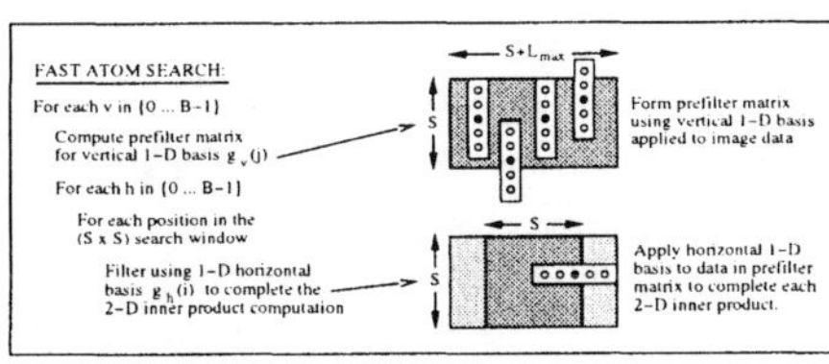

FIGURE 6. The separable inner product search algorithm.

(a) (b)

FIGURE 7. (a) Frame 0 of Hall coded with the H.263 intra mode using 15272 bits. (b) The same frame coded using the subband coder with 14984 bits.

TABLE 22.6. Numerical results from Class A sequences.

Sequence	Bit Rate kbit/s	Frame Rate frame/s	Bits to First Frame	M.P. Total Bits*	M.P. PSNR (dB)	H.263 PSNR (dB)	PSNR Diff. (dB)
Akiyo	10	7.5	7808	100291	34.05	33.82	0.23
Akiyo	24	10	20072	238668	38.79	37.94	0.85
Container	10	7.5	14096	99112	30.36	30.04	0.32
Container	24	10	19880	239203	33.35	33.00	0.35
Hall	10	7.5	12096	99639	30.84	30.63	0.21
Hall	24	10	20096	238062	35.11	34.71	0.40
Mother	10	7.5	8046	99600	32.40	32.22	0.18
Mother	24	10	19264	239602	35.76	35.59	0.17
Sean	10	7.5	9544	99529	30.24	29.68	0.56
Sean	24	10	14088	240454	34.03	33.29	0.74

* Total bits for 10 seconds of video, including first frame.

software [17]. As described earlier, the motion model used by the matching pursuit coder is identical to the model used in the H.263 runs, and the rate control of the two systems is synchronized. Both the matching pursuit and H.263 runs use identical subband-coded first frames. The five MPEG4 Class A test sequences are coded at 10 and 24 kbit/s. The Akiyo, Mother and Sean sequences are suitable test material for video telephone systems. The other two sequences, Container and Hall, are more suited to remote monitoring applications.

Table 22.6 shows some of the numerical results from the ten coding runs. Included are statistics on bit rate, frame rate, and the number of bits used to code the first frame in each case. Average luminance PSNR values for each experiment are also included in the table. It can be seen that the matching pursuit system has a higher average PSNR than H.263 in all the cases. The largest improvement is seen in the 24 kbit/s runs of Akiyo and Sean, which show average gains of .85 dB and .74 dB, respectively. More detail is provided in Figures 8-12, which plot the luminance PSNR against frame number for each experiment. In each case, solid lines are used to represent the matching pursuit runs and dotted lines are used for the H.263 runs. In most of the plots, the matching pursuit system shows a consistent PSNR gain over H.263 for the majority of coded frames. Performance is particularly good for Container, Sean, and the 24 kbit/s runs of Akiyo and Hall. The least improvement is seen on the Mother sequence, which shows a more modest PSNR gain and includes periods where the performance of the two coders is approximately equal. One possible explanation is that the Mother sequence contains intervals of high motion in which both coders are left with very few bits to encode the residual. Since the residual encoding method is the main difference between the two encoders, a near equal PSNR performance can be expected.

The matching pursuit coder shows a visual improvement over H.263 in all the test cases. To illustrate the visual improvement, some sample reconstructed frames are shown in Figures 13-16. In each case, the H.263 coded frame is shown on the left, and the corresponding matching pursuit frame is on the right. Note that all

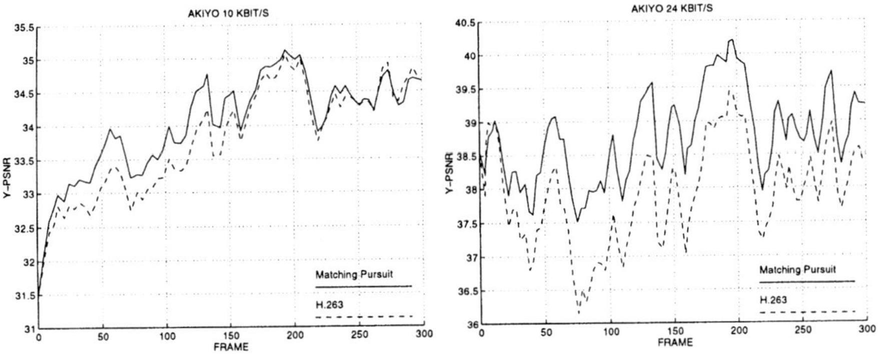

FIGURE 8. PSNR comparison for Akiyo at 10 and 24 kbit/s.

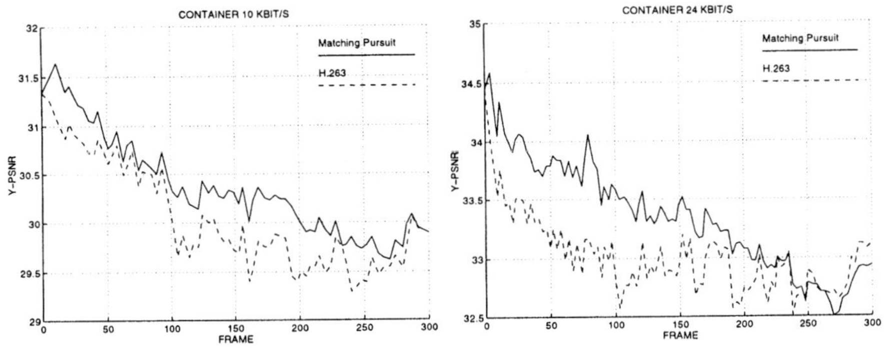

FIGURE 9. PSNR comparison for Container at 10 and 24 kbit/s.

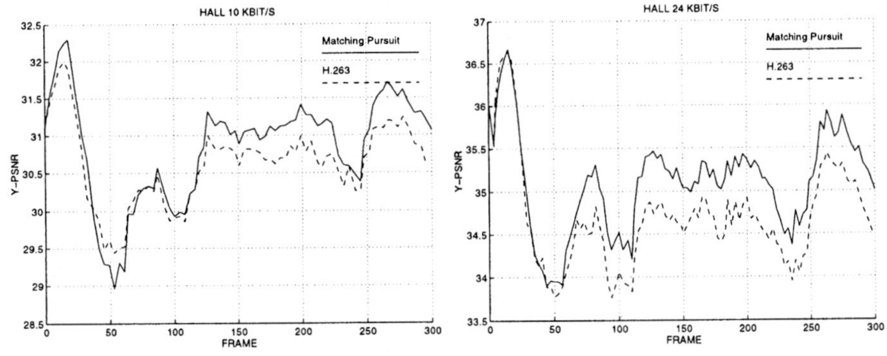

FIGURE 10. PSNR comparison for Hall at 10 and 24 kbit/s.

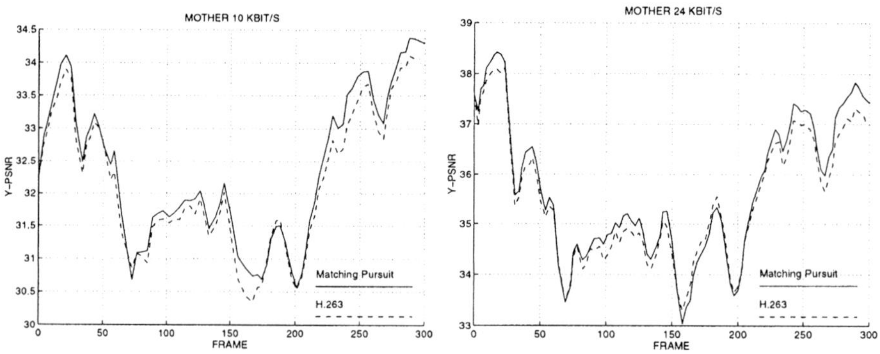

FIGURE 11. PSNR comparison for Mother at 10 and 24 kbit/s.

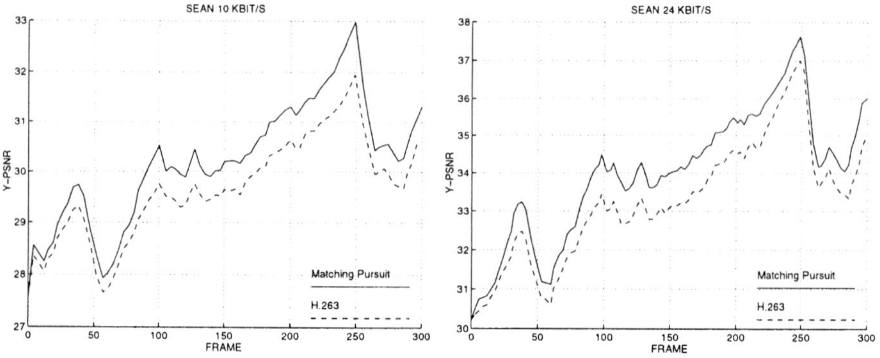

FIGURE 12. PSNR comparison for Sean at 10 and 24 kbit/s.

of the photographs except for Container show enlarged detail regions. Figure 13 shows a sample frame of Akiyo coded at 10 kbit/s. Note that the facial feature details are more clearly preserved in the matching-pursuit coded frame. The H.263 frame is less detailed, and contains block edge artifacts on the mouth and cheeks, as well as noise around the facial feature outlines. Figure 14 shows a frame of Container coded at 10 kbit/s. Both algorithms show a lack of detail at this bit rate. However, the H.263 reconstruction shows noticeable DCT artifacts, including block outlines in the water, and moving high frequency noise patterns around the edges of the container ship and the small white boat. Moving objects in the matching pursuit reconstruction appear less noisy and more natural, and the water and sky are much smoother. Figure 15 shows a frame of Hall coded at 24 kbit/s. Again, the matching pursuit image is more natural looking, and free of high frequency noise. Such noise patterns surround the two walking figures in the H.263 frame, particularly near the legs and briefcase of the foreground figure and around the feet of the background figure. This noise is particularly noticeable when the sequence is viewed in motion. Finally, Figure 16 shows a frame of the Mother sequence coded at 24 kbit/s. The extensive hand motion in this sequence causes the H.263 coded

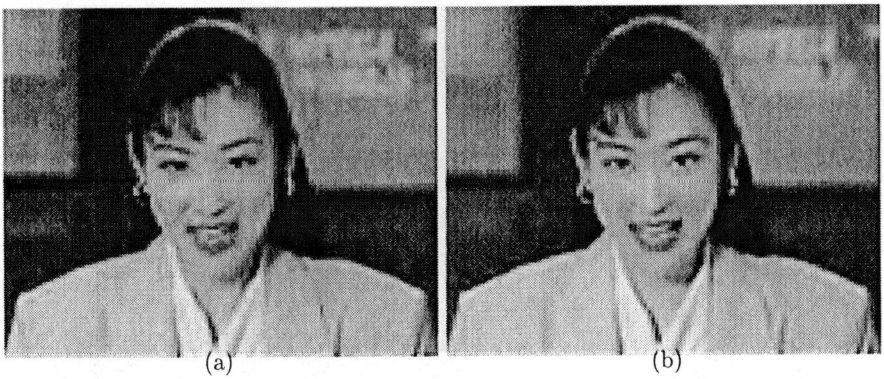

FIGURE 13. Akiyo frame 248 encoded at 10 kbit/s by (a) H.263, (b) Matching Pursuits.

FIGURE 14. Container frame 202 encoded at 10 kbit/s by (a) H.263, (b) Matching Pursuits.

sequence to degrade into a harsh pattern of blocks. The matching pursuit sequence develops a less noticeable blur. Notice also that the matching pursuit coder does not introduce high frequency noise around the moving outline of the mother's face, and the face of the child also appears more natural.

5 Conclusions

When coding motion residuals at very low bit rates, the choice of basis set is extremely important. This is because the coder must represent the residual using only a few coarsely quantized coefficients. If individual elements of the basis set are not well matched to structures in the signal, then the basis functions become visible in the reconstructed images. This explains why DCT based coders produce block edges in smoothly varying image regions and high-frequency noise patterns around the edges of moving objects. Wavelet based video coders can have similar problems at low bit rates. In these coders, the wavelet filter shapes can be seen around moving objects and edges. In both cases, limiting the basis to a complete set has

(a) (b)

FIGURE 15. Hall frame 115 encoded at 24 kbit/s by (a) H.263, (b) Matching Pursuits.

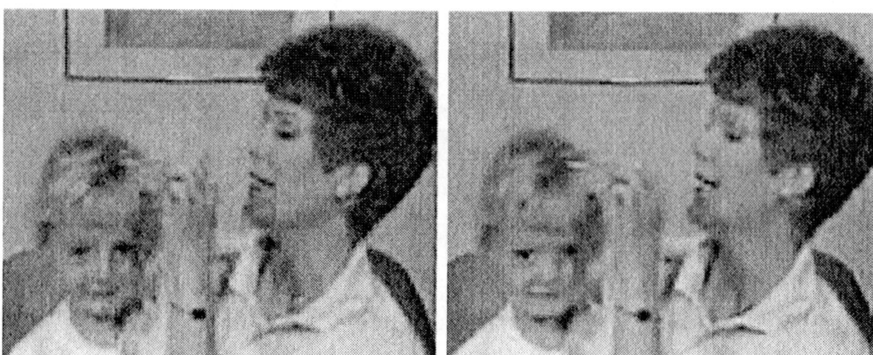

(a) (b)

FIGURE 16. Mother frame 82 encoded at 24 kbit/s by (a) H.263, (b) Matching Pursuits.

the undesirable effect that individual basis elements do not match the variety of structures found in motion residual images.

The matching pursuit residual coder presented in this paper lifts this restriction by providing an expansion onto an overcomplete basis set. This allows greater freedom in basis design. For example, the separable Gabor basis set used to produce the results of this paper contains functions at a wide variety of scales and arbitrary image locations. This flexibility insures that individual atoms are only coded when they are well matched to image structures. The basis set is "invisible" in the sense that it does not impose its own artificial structure on the image representation. When bit rate is reduced, objects coded using matching pursuits tend to lose detail, but the basis structures are not generally seen in the reconstructions. This type of degradation is much more graceful than that produced by hybrid-DCT systems.

Our results demonstrate that a matching-pursuit based residual coder can outperform an H.263 encoder at low bit rates in both PSNR and visual quality. Future research will include increasing the speed of the algorithm and optimizing the basis set to improve coding efficiency.

6 References

[1] ITU-T Study Group 15, Working Party 15/1, Draft Recommendation H.263, December 1995.

[2] H. G. Musmann, M. Hotter and J. Ostermann, "Object-Oriented Analysis Synthesis Coding of Moving Images," *Signal Processing: Image Communication,* October 1989, Vol. 1, No. 2, pp. 117-138.

[3] M. Gilge, T. Engelhardt and R. Mehlan, "Coding of Arbitrarily Shaped Image Segments Based on a Generalized Orthogonal Transform," *Signal Processing: Image Communication,* October 1989, Vol. 1, No. 2, pp. 153-180.

[4] S. F. Chang and D. Messerschmitt, "Transform Coding of Arbitrarily Shaped Image Segments," Proceeedings of 1st ACM International Conference on Multimedia, Anaheim, CA, 1993, pp.83-90.

[5] T. Sikora, "Low Complexity Shape-Adaptive DCT for Coding of Arbitrarily-Shaped Image Segments," *Signal Processing: Image Communication,* Vol. 7, No. 4, November 1995, pp. 381-395.

[6] S. Mallat and Z. Zhang, "Matching Pursuits With Time-Frequency Dictionaries," *IEEE Transactions in Signal Processing,* Vol. 41, No. 12, December 1993, pp. 3397-3415.

[7] J. H. Friedman and W. Stuetzle, "Projection Pursuit Regression," *Journal of the American Statistical Association,* Vol. 76, No. 376, December 1981, pp. 817-823.

[8] L. Wang and M. Goldberg, "Lossless Progressive Image Transmission by Residual Error Vector Quantization," *IEE Proceedings,* Vol. 135, Pt. F, No. 5, October 1988, pp. 421-430.

[9] F. Bergeaud and S. Mallat, "Matching Pursuit of Images," Proceedings of IEEE-SP International Symposium on Time-Frequency and Time-Scale Analysis, October 1994, pp.330-333.

[10] M. Vetterli and T. Kalker, "Matching Pursuit for Compression and Application to Motion Compensated Video Coding," Proceedings of ICIP, November 1994, pp. 725-729.

[11] P. J. Phillips, "Matching Pursuit Filters Applied to Face Identification," Proceedings of SPIE, Vol. 2277, July 1994, pp. 2-11.

[12] H. G. Feichtinger, A. Türk and T. Strohmer, "Hierarchical Parallel Matching Pursuit," Proceedings of SPIE, Vol. 2302, July 1994, pp. 222-232.

[13] T. -H. Chao, B. Lau and W. J. Miceli, "Optical Implementation of a Matching Pursuit for Image Representation," *Optical Engineering,* July 1994, Vol. 33, No. 2, pp. 2303-2309.

[14] R. Neff, A. Zakhor, and M. Vetterli, "Very Low Bit Rate Video Coding Using Matching Pursuits," Proceedings of SPIE VCIP, Vol. 2308, No. 1, September 1994, pp. 47-60.

[15] R. Neff and A. Zakhor, "Matching Pursuit Video Coding at Very Low Bit Rates," IEEE Data Compression Conference, Snowbird, Utah, March 1995, pp. 411-420.

[16] D. Taubman, "Fully Scalable Image and Video Codec," Public software release, available via anonymous ftp from tehran.eecs.berkeley.edu under /pub/taubman/scalable.tar.Z.

[17] ITU Telecommunication Standardization Sector LBC-95, "Video Codec Test Model TMN5." Available from Telenor Research at http://www.nta.no/brukere/DVC/

[18] C. Auyeung, J. Kosmach, M. Orchard and T. Kalafatis, "Overlapped Block Motion Compensation," Proceedings of SPIE VCIP, November 1992, Vol. 1818, No. 2, pp.561-572.

[19] D. Taubman and A. Zakhor, "Multirate 3-D Subband Coding of Video," *IEEE Transactions on Image Processing,* September 1994, Vol. 3, No. 5, pp. 572-88.

23

Object-Based Subband/Wavelet Video Compression

Soo-Chul Han
John W. Woods

ABSTRACT This chapter presents a subband/wavelet video coder using an object-based spatiotemporal segmentation. The moving objects in a video are extracted by means of a joint motion estimation and segmentation algorithm based on a compound Markov random field (MRF) model. The two important features of our technique are the temporal linking of the objects, and the guidance of the motion segmentation with spatial color information. This results in spatiotemporal (3-D) objects that are stable in time, and leads to a new motion-compensated temporal updating and contour coding scheme that greatly reduces the bit-rate to transmit the object boundaries. The object interiors can be encoded by either 2-D or 3-D subband/wavelet coding. Simulations at very low bit-rates yield comparable performance in terms of reconstructed PSNR to the H.263 coder. The object-based coder produces visually more pleasing video with less blurriness and is devoid of block artifacts.

1 Introduction

Video compression to very low bit-rates has attracted considerable attention recently in the image processing community. This is due to the growing list of very low bit-rate applications such as video-conferencing, multimedia, video over telephone lines, wireless communications, and video over the internet.

However, it has been found that standard block-based video coders perform rather poorly at very low bit-rates due to the well-known blocking artifacts. A natural alternative to the block-based standards is object-based coding, first proposed by Musmann *et al* [1]. In the object-based approach, the moving objects in the video scene are extracted, and each object is represented by its shape, motion, and texture. Parameters representing the three components are encoded and transmitted, and the reconstruction is performed by synthesizing each object. Although a plethora of work on the extraction and coding of the moving objects has appeared since [1], few works carry out the entire analysis-coding process from start to finish. Thus, the widespread belief that object-based methods could outperform standard techniques at low bit-rates (or any rates) has yet to be firmly established. In this chapter, we attempt to take the step in that direction with new ideas in both motion analysis and the source encoding. Furthermore, the object-based scheme leads to increased functionalities such as scalability, content-based manipulation, and the combination of synthetic and natural images. This is evidenced by the MPEG-4 standard, which is adopting the object-based approach.

Up to now, several roadblocks have prevented object-based coding systems from outperforming standard block-based techniques. For one thing, extracting the moving objects, such as by means of segmentation, is a very difficult problem in itself due to its ill-posedness and complexity [2]. Next, the gain in improving the motion compensated prediction must outweigh the additional contour information inherent in an object-based scheme. Applying intraframe techniques to encode the contours at each frame has been shown to be inefficient. Finally, it is essential that some objects or regions be encoded in "Intra" mode at certain frames due to lack of information in the temporal direction. This includes uncovered regions due to object movement, new objects that appear in a scene, and objects which undergo complex motion that cannot be properly described by the adopted motion model.

An object-based coder addressing all of the above mentioned issues is presented in this chapter. Moreover, we need to make no *a priori* assumptions about the contents of the video scene (such as constant background, head-and-shoulders only, etc). The extraction of the moving objects is performed by a joint motion estimation and segmentation algorithm based on compound Markov random field (MRF) models. In our approach, the object motion and shape are guided by the spatial color intensity information. This not only improves the motion estimation/segmentation process itself by extracting meaningful objects true to the scene, but it also aids the process of coding the object intensities. The latter because a given object has a certain spatial cohesiveness. The MRF formulation also allows us to temporally link objects, thus creating object volumes in the space-time domain. This helps stabilize the object segmentation process in time, but more importantly, allows the object boundaries to be predicted temporally using the motion information, reducing the boundary coding overhead. With linked objects, uncovered regions and new objects are detected by utilizing both the motion and intensity information.

Object interiors are encoded by either 2-D or 3-D subband/wavelet coding. The 2-D hybrid coding allows objects to be encoded adaptively at each frame, meaning that objects well described by the motion parameters are encoded in "Inter" mode, while those that cannot be predicted in time are encoded in "Intra" mode. This is analogous to P-blocks and I-blocks in the MPEG coding structure, where we now have P-objects and I-objects. Alternatively, the spatiotemporal objects can be encoded by 3-D subband/wavelet coding, which leads to added advantages such as frame-rate scalability and improved rate control [3]. In either case, the subband/wavelet transform must be modified to account for arbitrarily-shaped objects.

2 Joint Motion Estimation and Segmentation

In this section, a novel motion estimation and segmentation scheme is presented. Although the algorithm was specifically designed to meet coding needs as described in the previous section, the end results could very well be applied to other image sequence processing applications. The main objective is to segment the video scene into objects that are undergoing distinct motion, along with finding the parameters

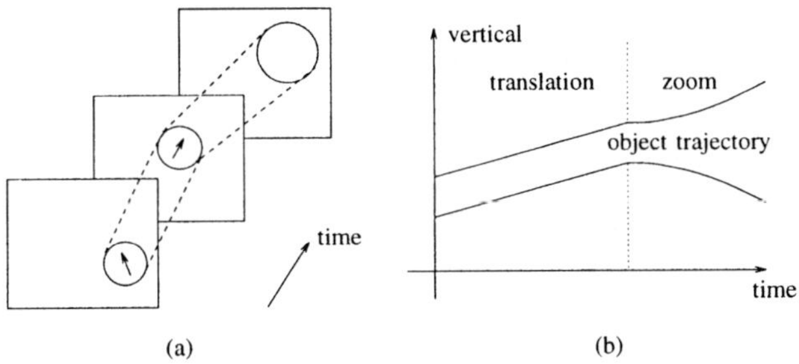

FIGURE 1. The trajectory of a moving ball.

that describe the motion. In Fig. 1(a), the video scene consists of a ball moving against a stationary background. At each frame, we would like to segment the scene into two objects (the ball and background) and find the motion of each. Furthermore, if the objects are linked in time, we can create 3-D objects in space-time as shown in Fig. 1(b). We adopt a Bayesian formulation based on a Markov random field (MRF) model to solve this challenging problem. Our algorithm extends previously published works [4, 5, 6, 7, 8].

2.1 *Problem formulation*

Let $\mathbf{I}^t$ represent the frame at time t of the discretized image sequence. The motion field $\mathbf{d}^t$ represents the displacement between $\mathbf{I}^t$ and $\mathbf{I}^{t-1}$ for each pixel. The segmentation field $\mathbf{z}^t$, consists of numerical labels at every pixel with each label representing one moving object, i.e. $\mathbf{z}^t(\mathbf{x}) = n$ $(n = 1, 2, .., N)$, for each pixel location $\mathbf{x}$ on the lattice Λ. Here, N refers to the total number of moving objects. Using this notation, the goal of motion estimation/segmentation is to find $\{\mathbf{d}^t, \mathbf{z}^t\}$ given $\mathbf{I}^t$ and $\mathbf{I}^{t-1}$.

We adopt a maximum *a posteriori* (MAP) formulation:

$$\{\hat{\mathbf{d}}^t, \hat{\mathbf{z}}^t\} = \arg \max_{\{\mathbf{d}^t, \mathbf{z}^t\}} p(\mathbf{d}^t, \mathbf{z}^t | \mathbf{I}^t, \mathbf{I}^{t-1}), \tag{23.1}$$

which can be rewritten via Bayes rule as

$$\{\hat{\mathbf{d}}^t, \hat{\mathbf{z}}^t\} = \arg \max_{\{\mathbf{d}^t, \mathbf{z}^t\}} p(\mathbf{I}^{t-1} | \mathbf{d}^t, \mathbf{z}^t, \mathbf{I}^t) p(\mathbf{d}^t | \mathbf{z}^t, \mathbf{I}^t) P(\mathbf{z}^t | \mathbf{I}^t). \tag{23.2}$$

Given this formulation, the rest of the work amounts to specifying the probability densities (or the corresponding energy functions) involved and solving.

2.2 *Probability models*

The first term on the right-hand side of (23.2) is the likelihood functional that describes how well the observed images match the motion field data. We model the likelihood functional by

$$p(\mathbf{I}^{t-1} | \mathbf{d}^t, \mathbf{z}^t, \mathbf{I}^t) = Q_l^{-1} exp\{-U_l(\mathbf{I}^{t-1} | \mathbf{d}^t, \mathbf{I}^t)\}, \tag{23.3}$$

which is also Gaussian. Here the energy function

$$U_l(\mathbf{I}^{t-1}|\mathbf{d}^t, \mathbf{I}^t) = \sum_{\mathbf{x}\in\Lambda} (\mathbf{I}^t(\mathbf{x}) - \mathbf{I}^{t-1}(\mathbf{x} - \mathbf{d}^t(\mathbf{x})))^2/2\sigma^2, \tag{23.4}$$

and Q_l is a normalization constant [5, 9].

The *a priori* density of the motion $p(\mathbf{d}^t|\mathbf{z}^t, \mathbf{I}^t)$, enforces prior constraints on the motion field. We adopt a coupled MRF model to govern the interaction between the motion field and the segmentation field, both spatially and temporally. The energy function is given as

$$\begin{aligned} U_d(\mathbf{d}^t|\mathbf{z}^t) = \\ \lambda_1 \sum_{\mathbf{x}} \sum_{\mathbf{y}\in\mathcal{N}_\mathbf{x}} \| \mathbf{d}^t(\mathbf{x}) - \mathbf{d}^t(\mathbf{y}) \|^2 \delta(z^t(\mathbf{x}) - z^t(\mathbf{y})) \\ + \lambda_2 \sum_{\mathbf{x}} \| \mathbf{d}^t(\mathbf{x}) - \mathbf{d}^{t-1}(\mathbf{x} - \mathbf{d}^t(\mathbf{x})) \|^2 \\ - \lambda_3 \sum_{\mathbf{x}} \delta(z^t(\mathbf{x}) - z^{t-1}(\mathbf{x} - \mathbf{d}^t(\mathbf{x}))), \end{aligned} \tag{23.5}$$

where $\delta(\cdot)$ refers to the usual Kronecker delta function, $\|\cdot\|$ is the Euclidean norm in $\mathbf{R}^2$, and $\mathcal{N}_\mathbf{x}$ indicates a spatial neighborhood system at location $\mathbf{x}$. The first two terms of (23.5) are similar to those in [7], while the third term was added to encourage consistency of the object labels along the motion trajectories. The first term enforces the constraint that the motion vectors be locally smooth on the spatial lattice, but only within the same object label. This allows motion discontinuities at object boundaries without introducing any extra variables such as line fields [10]. The second term accounts for causal temporal continuity, under the model restraints (ref. second factor in 23.2) that the motion vector changes slowly frame-to-frame along the motion trajectory. Finally, the third term in (23.5) encourages the object labels to be consistent along the motion trajectories. This constraint allows a framework for the object labels to be linked in time.

Lastly, the third term on the right hand side of (23.2) $P(\mathbf{z}^t|\mathbf{I}^t)$, models our *a priori* expectations about the nature of the object label field. In the temporal direction, we have already modeled the object labels to be consistent along the motion trajectories. Our model incorporates spatial intensity information ($\mathbf{I}^t$) based on the reasonable assumption that object discontinuities coincide with spatial intensity boundaries. The segmentation field is a discrete-valued MRF with the energy function given as

$$U_z(\mathbf{z}^t|\mathbf{I}^t) = \sum_{\mathbf{x}} \sum_{\mathbf{y}\in\mathcal{N}_\mathbf{x}} V_c(z(\mathbf{x}), z(\mathbf{y})|\mathbf{I}^t), \tag{23.6}$$

where we specify the clique function as

$$V_c(z(\mathbf{x}), z(\mathbf{y})|\mathbf{I}^t) = \begin{cases} -\gamma & \text{if } z(\mathbf{x}) = z(\mathbf{y}),\ s(\mathbf{x}) = s(\mathbf{y}) \\ 0 & \text{if } z(\mathbf{x}) = z(\mathbf{y}),\ s(\mathbf{x}) \neq s(\mathbf{y}) \\ +\gamma & \text{if } z(\mathbf{x}) \neq z(\mathbf{y}),\ s(\mathbf{x}) = s(\mathbf{y}) \\ 0 & \text{if } z(\mathbf{x}) \neq z(\mathbf{y}),\ s(\mathbf{x}) \neq s(\mathbf{y}). \end{cases} \tag{23.7}$$

Here, $\mathbf{s}$ refers to the spatial segmentation field that is pre-determined from $\mathbf{I}$. It is important to note that $\mathbf{s}$ is treated as *deterministic* in our Markov model. A simple region-growing method [11] was used in our experiments. According to (23.7), if the spatial neighbors $\mathbf{x}$ and $\mathbf{y}$ belong to the same *intensity-based* object ($s(\mathbf{x}) = s(\mathbf{y})$), then the two pixels are encouraged to belong to the same *motion-based* object. This is achieved by the $\pm\gamma$ terms. On the other hand, if $\mathbf{x}$ and $\mathbf{y}$ belong to *different* intensity-based objects ($s(\mathbf{x}) \neq s(\mathbf{y})$), we do not enforce z to be either way, and hence, the 0 terms in (23.7). This slightly more complex model ensures that the moving object segments we extract have some sort of spatial cohesiveness as well. This is a very important property for our adaptive coding strategy to be presented in Section 4. Furthermore, the above clique function allows the object segmentations to remain stable over time and adhere to what the human observer calls "objects."

2.3 Solution

Due to the equivalence of MRFs and Gibbs distributions [10], the MAP solution amounts to a minimization of the sum of potentials given by (23.4), (23.5), and (23.6). To ease the computation, a two-step iterative procedure [8] is implemented, where the motion and segmentation fields are found in an alternating fashion assuming the other is known. Mean field annealing [9] is used for the motion field estimation, while the object label field is found by the deterministic iterated conditional modes (ICM) algorithm [2]. Furthermore, the estimation is performed on a multiresolution pyramid structure. Thus, crude estimates are first obtained at the top of the pyramid, with successive refinement as we traverse down the pyramid. This greatly reduces the computational burden, making it possible to estimate relatively large motion, and also allowing the motion and segmentation to be represented in a scalable way.

2.4 Results

Simulations for the motion analysis and subsequent video coding were done on three test sequences, *Miss America*, *Carphone*, and *Foreman*, all suitable for low bit-rate applications.

In Fig. 2, the motion estimation and segmentation results for *Miss America* are illustrated. A three-level pyramid was used in speeding up the algorithm, using the two-step iterations as described earlier. The motion field was compared to that obtained by hierarchical block matching. The block size used was 16x16. We can see that the MRF model produced smooth vectors within the objects with definitive discontinuities at the image intensity boundaries. Also, it can be observed that the object boundaries more or less define the "real" objects in the scene.

The temporal linkage of the object segments is illustrated in Fig. 3, which represents a "side view" of the 3-D image sequence, and Fig. 4, which represents a "top view". We can see that our segments are somewhat accurate and follow the true movement. Note that Fig. 3 is analogous to our ideal representation in Fig. 1 (b).

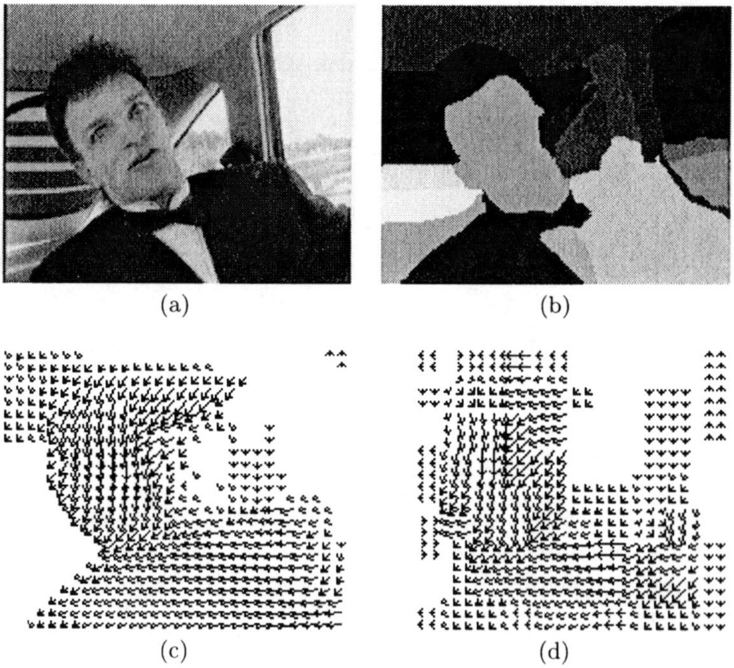

(a) *Carphone* frame no 88 (b) Motion segmentation
motion field by (c) proposed method (d) hierarchical block matching.

FIGURE 2. Motion estimation and segmentation for *Carphone*.

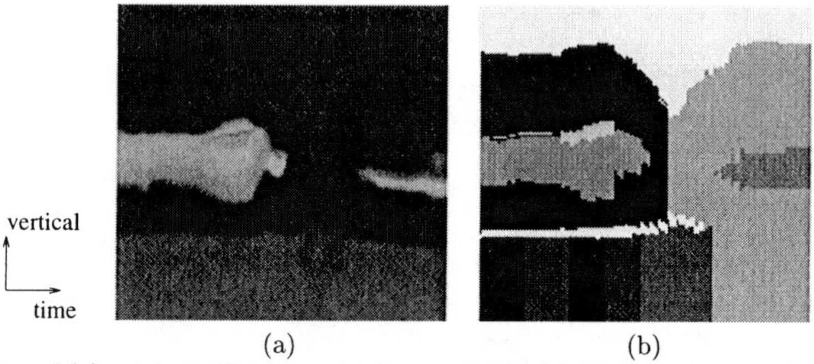

side-view of (a) original *Miss America* frames 0-149 (b) Object segmentation mask

FIGURE 3. Objects in vertical-time space (side-view).

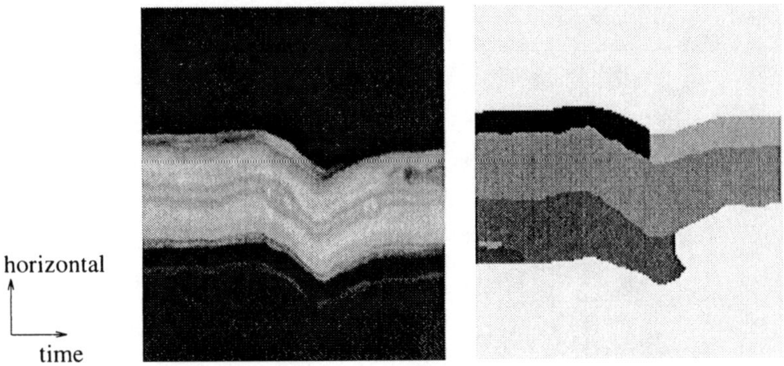

(a) (b)

top-view of (a) original *Miss America* (b) Object segmentation mask

FIGURE 4. Objects in horizontal-time space (top-view).

3 Parametric Representation of Dense Object Motion Field

The motion analysis from the previous section provides us with the boundaries of the moving objects and a dense motion field within each object. In this section, we are interested in efficiently representing and coding the found object information.

3.1 Parametric motion of objects

A 2-D planar surface undergoing rigid 3-D motion yields the following affine velocity field under orthographic projection [2]

$$\begin{aligned} v_x &= a_1 + a_2 x + a_3 y \\ v_y &= a_4 + a_5 x + a_6 y, \end{aligned} \tag{23.8}$$

where $\mathbf{v} = (v_x(x,y), v_y(x,y))$ refers to the apparent velocity at pixel (x,y) in the x and y directions respectively. Thus, the motion of each object can be represented by the six parameters of (23.8). In [12], Wang and Adelson employ a simple least-squares fitting technique, and point out that this process is merely a plane fitting algorithm in the velocity space. In our work, we improve on this method by introducing a matrix **W** so that data with higher confidence can be given more weight in the fitting process. Specifically, we denote the six parameter set of (23.8) by

$$\mathbf{p} = \begin{bmatrix} a_1 & a_4 \\ a_2 & a_5 \\ a_3 & a_6 \end{bmatrix}. \tag{23.9}$$

For a particular object, we can order the M pixels that belong to the object as $(x_1, y_1),...,(x_M, y_M)$, and define the weighting matrix **W** as

$$\mathbf{W} = diag[w_1, w_2, ..., w_M],$$

where $w_i(i = 1, 2, .., M)$ corresponds to the weight at pixel i. Then, the *weighted* least-squares solution for $\mathbf{p}$ is given by

$$\mathbf{p} = \mathbf{\Phi}_W^{-1}\mathbf{V}_W, \tag{23.10}$$

with

$$\mathbf{\Phi}_W \equiv \begin{bmatrix} \sum_i^M w_i^2 & \sum_i^M w_i^2 x_i & \sum_i^M w_i^2 y_i \\ \sum_i^M w_i^2 x_i & \sum_i^M w_i^2 x_i^2 & \sum_i^M w_i^2 x_i y_i \\ \sum_i^M w_i^2 y_i & \sum_i^M w_i^2 y_i x_i & \sum_i^M w_i^2 y_i^2 \end{bmatrix}$$

and

$$\mathbf{V}_W \equiv \begin{bmatrix} \sum_i^M w_i^2 v_x(x_i, y_i) & \sum_i^M w_i^2 v_y(x_i, y_i) \\ \sum_i^M w_i^2 x_i(v_x(x_i, y_i)) & \sum_i^M w_i^2 x_i(v_y(x_i, y_i)) \\ \sum_i^M w_i^2 y_i(v_x(x_i, y_i)) & \sum_i^M w_i^2 y_i(v_y(x_i, y_i)) \end{bmatrix}.$$

We designed the weighting matrix $\mathbf{W}$ based on experimental observations on our motion field data. For one thing, we would like to eliminate (or lessen) the contribution of inaccurate data on the least-squares fitting. This can be done using the displaced frame difference (DFD),

$$\text{DFD}(\mathbf{x}_i) = \mathbf{I}^t(\mathbf{x}_i) - \mathbf{I}^{t-1}(\mathbf{x}_i - d(\mathbf{x}_i)). \tag{23.11}$$

Obviously, we want pixels with higher DFD to have a lower weight, and thus define

$$\alpha_i \equiv 1 - \text{DFD}(\mathbf{x}_i)/\text{MAX\_DFD}, \tag{23.12}$$

where

$$\text{MAX\_DFD} = \max_i \text{DFD}(\mathbf{x}_i).$$

Also, we found that in image regions with almost homogeneous gray level distribution, the mean field solution favored the zero motion vector. This problem was solved by measuring the busyness in the search region of each pixel, represented by a binary weight function

$$\beta_i = \begin{cases} 0 & \text{not busy} \\ 1 & \text{busy} \end{cases}$$

The business is decided by comparing the gray-level variance in the search region of pixel i against a pre-determined threshold. Combining these two effects, the final weight for pixel i is determined as

$$w_i = \lambda\alpha_i + (1 - \lambda)\beta_i \tag{23.13}$$

where $0 < \lambda < 1$ to ensure that $w_i \in [0, 1]$. Fig. 5 images the weights for a selected object in *Miss America*.

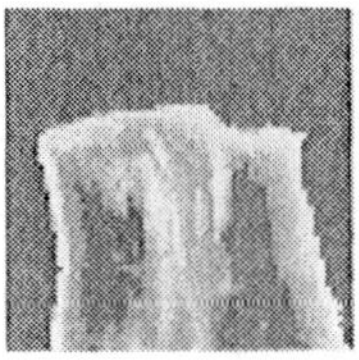

FIGURE 5. Least-squares weights for *Miss America*'s right shoulder: pixels with lighter gray-level values are given more weight.

3.2 Appearance of new regions

To extract meaningful new objects, additional processing was necessary based on least-squares fitting. The basic idea is taken from the "top-down" approach of Musmann *et al* [1], in which regions where the motion parametrization fail are assigned as new objects. However, we begin the process by splitting a given object into subregions using our spatial color segmentator. Then, a subregion is labeled as a new object only if all three of the following conditions are met:

1. The norm difference between the synthesized motion vectors and the original dense motion vectors is large.
2. The prediction error resulting from the split is significantly reduced.
3. The prediction error within the subregion is high without a split.

Because of the smoothness constraint, splitting within objects merely based on affine fit failure, did not produce meaningful objects. The second condition ensures that the splitting process decreases the overall coding rate. Finally, the third condition guards against splitting the object when there is no need to in terms of coding gain.

3.3 Coding the object boundaries

We have already seen that temporally linked objects in an object-based coding environment offer various advantages. However, the biggest advantage comes in reducing the contour information rate. Using the object boundaries from the previous frame and the affine transformation parameters, the boundaries can be predicted with a good deal of accuracy. Some error occurs near boundaries, and the difference is encoded by chain coding. These errors are usually small because the MRF formulation explicitly *makes* them small. It is interesting to note that in [13], this last step of updating is omitted, thus eliminating the need to transmit the boundary information altogether.

4 Object Interior Coding

Two methods of encoding the object interiors have been investigated. One is to code the objects, possibly motion-compensated, at each frame. The other is to explicitly

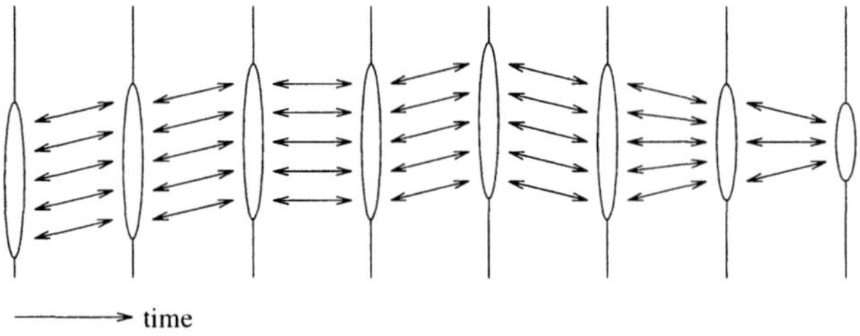

FIGURE 6. Object-based motion-compensated filtering

encode the spatiotemporal objects in the 3-D space-time domain.

4.1 Adaptive Motion-Compensated Coding

In this scheme, objects that can be described well by the motion are encoded by motion compensated predictive (MCP) coding, and those that cannot are encoded in "Intra" mode. This adaptive coding is done independently on each object using spatial subband coding. Since the objects are arbitrarily shaped, the efficient signal extension method proposed by Barnard[14] was applied.

Although motion compensation was relatively good for most objects at most frames, the flexibility to switch to intra-mode (I-mode) in certain cases is quite necessary. For instance, when a new object appears from outside the scene, it cannot be properly predicted from the previous frame. Thus, these new objects must be coded in I-mode. This includes the initial frame of the image sequence, where all the objects are considered new. Even for "continuing" objects, the motion might be too complex at certain frames for our model to properly describe, resulting in poor prediction. Such classification of objects into I-objects and P-objects is analogous to P-blocks and I-blocks in current video standards MPEG and H.263 [15].

4.2 Spatiotemporal (3-D) Coding of Objects

Alternatively, to overcome some of the generic deficiencies of MCP coding [16], the objects can be encoded by object-based 3-D subband/wavelet coding (OB-3DSBC). This is possible because our segmentation algorithm provides objects linked in time. In OB-3DSBC, the temporal redundancies are exploited by motion-compensated 3-D subband/wavelet analysis, where the temporal filtering is performed along the motion trajectories within each object (Figure 6). The object segmentation/motion and covered/uncovered region information enables the filtering to be carried out in a systematic fashion. Up to now, rather *ad-hoc* rules had been employed to solve the temporal region consistency problem in filter implementation [3, 17]. In other words, the object-based approach allows us to make a better decision on where and how to implement the motion-compensated temporal analysis.

Following the temporal decomposition, the objects are further analyzed by spatial subband/wavelet coding. The generalized BFOS algorithm [18] is used in distributing the bits among the spatiotemporal subbands of the objects. The sub-

sequence	bit-rate (kbps)	channel	OB-Coder (PSNR)	H.263 (PSNR)
Miss America (7.5 fps)	8	Y	36.0	35.9
		U	37.5	37.4
		V	35.5	35.3
	16	Y	38.0	38.2
		U	38.2	38.4
		V	37.2	37.6
Carphone (7.5 fps)	24	Y	30.0	30.0
		U	35.6	35.8
		V	36.4	36.4
	32	Y	30.8	31.1
		U	36.1	36.5
		V	36.9	37.2
Foreman (10 fps)	32	Y	30.8	30.7
		U	39.8	36.2
		V	40.7	36.9
	65	Y	32.5	32.5
		U	40.3	37.5
		V	41.4	38.5

TABLE 23.1. PSNR results for OB-MCP coder.

band/wavelet coefficients are encoded by uniform quantization followed by adaptive arithmetic coding.

5 Simulation results

Our proposed object-based coding scheme was applied to three QCIF test sequences, *Miss America*, *Carphone*, and *Foreman*. They were compared to the Telenor research group's H.263 implementation[1]. Table 23.1 gives PSNR results comparing the object-based motion-compensated predictive (OB-MCP) coder and block-based H.263 coder. Table 23.2 shows the performance of our object-based 3-D subband/wavelet coder.

In terms of PSNR the proposed object-based coder is comparable in performance to the conventional technique at very low bit-rates. However, more importantly, the object-based coder produced visually more pleasing video. Some typical reconstructed frames are given in Fig. 7 and Fig. 8. The annoying blocking artifacts that dominate the block-based methods at low bit-rates are not present. Also, the object-based coder gave clearer reconstructed frames with less blurriness.

[1] `http://www.nta.no/brukere/DVC/`

sequence	bit-rate (kbps)	channel	OB-Coder (PSNR)	H.263 (PSNR)
Miss America (15 fps)	20	Y	37.2	37.9
		U	38.9	38.5
		V	37.6	37.4
Carphone (15 fps)	40	Y	32.8	33.4
		U	38.1	38.6
		V	38.9	38.1

TABLE 23.2. PSNR results for OB-3DSBC.

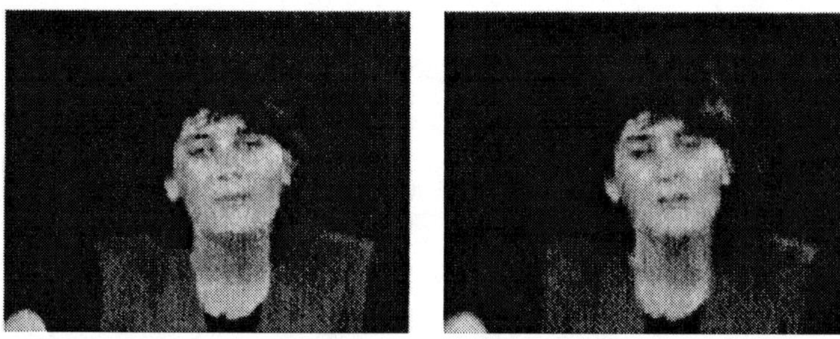

(a) Object-based coder (Frame PSNR = 34.0 dB)

(b) H.263 coder (Frame PSNR = 34.1 dB)

FIGURE 7. Decoded frame 112 for *Miss America* at 8 kbps.

(a) Object-based coder (Frame PSNR = 29.1 dB)

(b) H.263 coder (Frame PSNR = 29.1 dB)

FIGURE 8. Decoded frame 196 for *Carphone* at 24 kbps.

6 Conclusions

We have presented an object-based video compression system with improved coding performance from a visual perspective. Our motion estimation/segmentation algorithm enables the extraction of moving objects that correspond to the true scene. By following the objects in time, the object motion and contour can be encoded efficiently with temporal updating. The interior of the objects are encoded by 2-D subband analysis/synthesis. The objects are encoded adaptively based on the scene contents. No *a priori* assumptions about the image content or motion is needed. We conclude from our results that object-based coding could be a viable alternative to the block-based standards for very low bit-rate applications. Further research on reducing the computation is needed.

7 References

[1] H. Musmann, M. Hotter, and J. Ostermann, "Object-oriented analysis-synthesis coding of moving images," *Signal Processing: Image Communications*, vol. 1, pp. 117–138, Oct. 1989.

[2] A. M. Tekalp, *Digital Video Processing.* Upper Saddle River, NJ: Prentice Hall, 1995.

[3] S. Choi and J. Woods, "Motion-compensated 3-D subband coding of video," *IEEE Trans. Image Process.* submitted for publication, 1996.

[4] D. W. Murray and B. F. Buxton, "Scene segmentation from visual motion using global optimization," *IEEE Trans. Pattern Analysis and Machine Intelligence*, pp. 220–228, Mar. 1987.

[5] J. Konrad and E. Dubois, "Bayesian estimation of motion vector fields," *IEEE Trans. Pattern Analysis and Machine Intelligence*, vol. 14, pp. 910–927, Sept. 1992.

[6] P. Bouthemy and E. François, "Motion segmentation and qualitative dynamic scene analysis from an image sequence," *International Journal of Computer Vision*, vol. 10, no. 2, pp. 157–182, 1993.

[7] C. Stiller, "Object-oriented video coding employing dense motion fields," in *Proc. IEEE Int. Conf. Acoust., Speech, Signal Processing*, vol. V, pp. 273–276, Adelaide, Australia, 1994.

[8] M. Chang, I. Sezan, and A. Tekalp, "An algorithm for simultaneous motion estimation and scene segmentation," in *Proc. IEEE Int. Conf. Acoust., Speech, Signal Processing*, vol. V, pp. 221–224, Adelaide, Australia, 1994.

[9] J. Zhang and G. G. Hanauer, "The application of mean field theory to image motion estimation," *IEEE Trans. Image Process.*, vol. 4, pp. 19–33, 1995.

[10] S. Geman and D. Geman, "Stochastic relaxation, Gibbs distributions, and Bayesian restoration of images," *IEEE Trans. Pattern Analysis and Machine Intelligence*, vol. PAMI-6, pp. 721–741, Nov. 1984.

[11] R. Haralick and L. Shapiro, *Computer and Robot Vision.* Reading, MA: Addison-Wesley Pub. Co., 1992.

[12] J. Wang and E. Adelson, "Representing moving images with layers," *IEEE Trans. Image Process.*, vol. 3, pp. 625–638, Sept. 1994.

[13] Y. Yokoyama, Y. Miyamoto, and M. Ohta, "Very low bit rate video coding using arbitrarily shaped region-based motion compensation," *IEEE Trans. Circuits and Systems for Video Technology,* vol. 5, pp. 500–507, Dec. 1995.

[14] H. J. Barnard, *Image and Video Coding Using a Wavelet Decomposition.* PhD thesis, Delft University of Technology, The Netherlands, 1994.

[15] ITU-T Recommendation H.263, *Video Coding for Low Bitrate Communication,* Nov. 1995.

[16] S. Han and J. Woods, "Three dimensional subband coding of video with object-based motion information," *to be presented at IEEE International Conference on Image Processing,* Oct. 1997.

[17] J. Ohm, "Three-dimensional subband coding with motion compensation," *IEEE Trans. Image Process.*, vol. 3, pp. 559–571, Sept. 1994.

[18] E. Riskin, "Optimal bit allocation via the generalized BFOS algorithm," *IEEE Trans. Inform. Theory,* vol. IT-37, pp. 400–402, Mar. 1991.

24

Embedded Video Subband Coding with 3D SPIHT

William A. Pearlman, Beong-Jo Kim, and Zixiang Xiong

ABSTRACT This chapter is devoted to the exposition of a complete video coding system, which is based on coding of three dimensional (wavelet) subbands with the SPIHT (set partitioning in hierarchical trees) coding algorithm. The SPIHT algorithm, which has proved so successful in still image coding, is also shown to be quite effective in video coding, while retaining its attributes of complete embeddedness and scalability by fidelity and resolution. Three-dimensional spatio-temporal orientation trees coupled with powerful SPIHT sorting and refinement renders 3D SPIHT video coder so efficient that it provides performance superior to that of MPEG-2 and comparable to that of H.263 with minimal system complexity. Extension to color-embedded video coding is accomplished without explicit bit-allocation, and can be used for any color plane representation. In addition to being rate scalable, the proposed video coder allows multiresolution scalability in encoding and decoding in both time and space from one bit-stream. These attributes of scalability, lacking in MPEG-2 and H.263, along with many desirable features, such as full embeddedness for progressive transmission, precise rate control for constant bit-rate (CBR) traffic, and low-complexity for possible software-only video applications, makes the proposed video coder an attractive candidate for for multi-media applications. Moreover, the codec is fast and efficient from low to high rates, obviating the need for a different standard for each rate range.

Keywords: very low bit-rate video, SPIHT, scalability, progressive transmission, embeddedness, multi-media

1 Introduction

The demand for video has accelerated for transmission and delivery across both high and low bandwidth channels. The high bandwidth applications include digital video by satellite (DVS) and the future high-definition television (HDTV), both based on MPEG-2 compression technology. The low bandwidth applications are dominated by transmission over the Internet, where most modems transmit at speeds below 64 kilobits per second (Kbps). Under these stringent conditions, delivering compressed video at acceptable quality becomes a challenging task, since the required compression ratios are quite high. Nonetheless, the current test model standard of H.263 does a creditable job in providing video of acceptable quality for certain applications at low bit rates, but better schemes with increased functionality are actively being sought by the MPEG-4 and MPEG-7 standards committees.

Both MPEG-2 and H.263 are based on block DCT coding of displaced frame differences, where displacements or motion vectors are determined through block-matching estimation methods. Although reasonably effective, these standards lack the functionality now regarded as essential for emerging multimedia applications. In particular, resolution and fidelity (rate) scalability, the capability of progressive transmission by increasing resolution and increasing fidelity, is considered essential for emerging video applications to multimedia. Moreover, if a system is truly progressive by rate or fidelity, then it can presumably handle both the high-rate and low-rate regimes of digital satellite and Internet video, respectively.

In this chapter, we present a three-dimensional (3D) subband-based video coder that is fast and efficient, and possesses the multimedia functionality of resolution and rate scalability in addition to other desirable attributes. Subband coding has been known to be a very effective coding technique. It can be extended naturally to video sequences due to its simplicity and non-recursive structure that limits error propagation within a certain group of frames (GOF). Three-dimensional (3D) subband coding schemes have been designed and applied for mainly high or medium bit-rate video coding. Karlsson and Vetterli [7] took the first step toward 3D subband coding using a simple 2-tap Haar filter for temporal filtering. Podilchuk, Jayant, and Farvardin [15] used the same 3D subband framework without motion compensation. It employed adaptive differential pulse code modulation (DPCM), and vector quantization to overcome the lack of motion compensation.

Kronander [10] presented motion compensated temporal filtering within the 3D SBC framework. However, due to the existence of pixels not encountered by the motion trajectory, he needed to encode a residual signal. Based on the previous work, motion compensated 3D SBC with lattice vector quantization was introduced by Ohm [13]. Ohm introduced the idea for a perfect reconstruction filter with the block-matching algorithm, where 16 frames in one GOF are recursively decomposed with 2-tap filters along the motion trajectory. He then refined the idea to better treat the *connected/unconnected* pixels with arbitrary motion vector field for a perfect reconstruction filter, and extended to arbitrary symmetric (linear phase) QMF's [14]. Similar work by Choi and Woods [5], using a different way of treating the *connected/unconnected* pixels and a sophisticated hierarchical variable size block matching algorithm, has shown better performance than MPEG-2.

Due to the multiresolutional nature of SBC schemes, several scalable 3D SBC schemes have appeared. Bove and Lippman [2] proposed multiresolutional video coding with a 3D subband structure. Taubman and Zakhor [24, 23] introduced a multi-rate video coding system using global motion compensation for camera panning, in which the video sequence was pre-distorted by translating consecutive frames before temporal filtering with 2-tap Haar filters. This approach can be considered as a simplified version of Ohm's [14] in that it treats *connected/unconnected* pixels in a similar way for temporal filtering. However, the algorithm generates a scalable bit-stream in terms of bit-rate, spatial resolution, and frame rate.

Meanwhile, there have been several research activities on embedded video coding systems based on significance tree quantization, which was introduced by Shapiro for still image coding as the embedded zerotree wavelet (EZW) coder [21]. It was later improved through through a more efficient state description in [16] and called improved EZW or IEZW. This two-dimensional (2D) embedded zero-tree (IEZW) method has been extended to 3D IEZW for video coding by Chen and Pearlman [3], and showed promise of an effective and computationally simple video coding system without motion compensation, and obtained excellent numerical and visual results. A 3D zero-tree coding through modified EZW has also been used with good results in compression of volumetric images [11]. Recently, a highly scalable embedded 3D SBC system with *tri-zerotrees* [26] for very low bit-rate environment was reported with coding results visually comparable, but numerically slightly inferior to H.263. Our current work is very much related to previous 3D subband embedded video coding systems in [3, 11, 26]. Our main contribution is to design an even more efficient and computationally simple video coding system with many desirable attributes.

In this chapter, we employ a three dimensional extension of the highly successful set partitioning in hierarchical trees (SPIHT) still image codec [18] and propose a 3D wavelet coding system with features such as complete embeddedness for progressive fidelity transmission, precise rate control for constant bit-rate (CBR) traffic, low-complexity for possible software-only real time applications, and multiresolution scalability. Our proposed coding system is so compact that it consists of only two parts: 3D *spatio-temporal* decomposition and 3D SPIHT coding. An input video sequence is first 3D wavelet transformed with (or without) motion compensation (MC), and then encoded into an embedded bit-stream by the 3D SPIHT kernel. Since the proposed video coding scheme is aimed at both high bit-rate and low bit-rate applications, it will be benchmarked against both MPEG-2 and H.263.

The organization of this chapter follows. Section 2 summarizes the overall system structure of our proposed video coder. Basic principles of SPIHT will be explained in section 3, followed in section 4 by explanations of 3D-SPIHT's attributes of full monochrome and color embeddedness, and multiresolution encoding/decoding scalability. Motion compensation in our proposed video coder is addressed in section 5. Sections 6 and 7 provide implementation details and simulation results. Section 8 concludes the chapter.

2 System Overview

2.1 System Configuration

The proposed video coding system, as shown in Fig. 1 consists primarily of a 3D analysis part with/without motion compensation, and a coding part with 3D SPIHT kernel. As we can see, the decoder has the structure symmetric to that of encoder. Frames in a group of frames, hereafter called GOF, will be first temporally transformed with/without motion compensation. Then, each resulting frame will again be separately transformed in the spatial domain. When motion compensated filtering is performed, the motion vectors are separately lossless-coded, and transmitted over the transmission channel with high priority. With our coding system, there is no complication of a rate allocation, nor is there a feedback loop of prediction error signal, which may slow down the efficiency of the system. With the 3D SPIHT kernel, the preset rate will be allocated over each frame of the GOF automatically according to the distribution of actual magnitudes. However, it is possible to introduce a scheme for bit re-alignment by simply scaling one or more subbands to emphasize or deemphasize the bands so as to artificially control the visual quality of the video. This scheme is also applicable to color planes of video, since it is well known fact that chrominance components are less sensitive than the luminance component to the human observer.

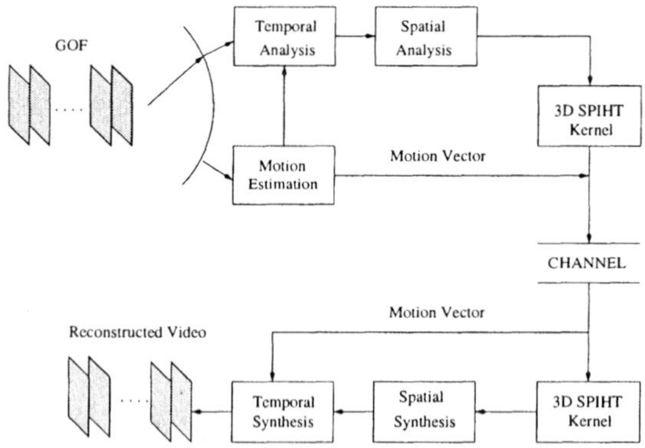

FIGURE 1. System configuration

2.2 3D Subband Structure

A GOF is first decomposed temporally and then spatially into subbands when input to a bank of filters and subsampled. Figure 2 illustrates how a GOF is decomposed into four temporal frequency bands by recursive decomposition of the low temporal subband. The temporal filter here is a one-dimensional (1D) unitary filter. The temporal decomposition will be followed by 2D spatial decomposition with separable unitary filters. As illustrated, this temporal decomposition is the same as in the previous work in [13]. Since the temporal high frequency usually does not contain much energy, previous works [15, 13, 14, 5] apply one level of temporal decomposition. However, we found that with 3D SPIHT, further dyadic decompositions in the temporal high frequency band give advantages over the traditional way in terms of peak signal-to-noise ratio (PSNR) and visual quality. The similar idea of so-called wavelet packet decomposition [26] also reported better visual quality. In addition, there has been some research on optimum wavelet packet image decomposition to optimize jointly the transform or decomposition tree and the quantization [28, 29]. Nonetheless, the subsequent spatial analysis in this work is a dyadic two-dimensional (2D) recursive decomposition of the low spatial frequency subband. The total number of samples in the GOF remains the same at each step in temporal or spatial analysis through the critical subsampling process.

An important issue associated with 3D SBC is the choice of filters. Different filters in general show quite different signal characteristics in the transform domain in terms of energy compaction, and error signal in the high frequency bands [20]. The recent introduction of wavelet theory offers promise for designing better filters for image/video coding. However, the investigation of optimum filter design is beyond of the scope of this paper. We shall employ known filters which have shown good performance in other subband or wavelet coding systems.

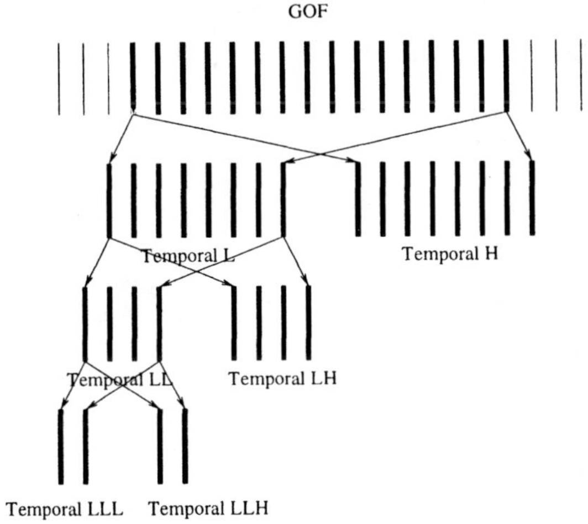

FIGURE 2. Dyadic temporal decomposition of a GOF (group of frames

Fig. 3 shows two templates, the lowest temporal subband, and the highest temporal subband, of typical 3D wavelet transformed frames with the "foreman" sequence of QCIF format (176 × 144). We chose 2 levels of decomposition in the spatial domain just for illustration of the different 3D subband spatial characteristics in the temporal high frequency band. Hence, the lowest spatial band of each frame has dimension of 44 × 36. Each spatial band of the frames is appropriately scaled before it is displayed. Although most of the energy is concentrated in the temporal low frequency, there exists much spatial residual redundancy in the high temporal frequency band due to the object or camera motion. This is the main motivation of further spatial decomposition even in the temporal high subband.

Besides, we can obviously observe not only spatial similarity inside each frame across the different scale, but also temporal similarity between two frames, which will be efficiently exploited by the 3D SPIHT algorithm. When there is fast motion or a scene change, temporal linkages of pixels through trees do not provide any advantage in predicting insignificance (with respect to a given magnitude threshold). However, linkages in trees contained within a frame will still be effective for prediction of insignificance spatially.

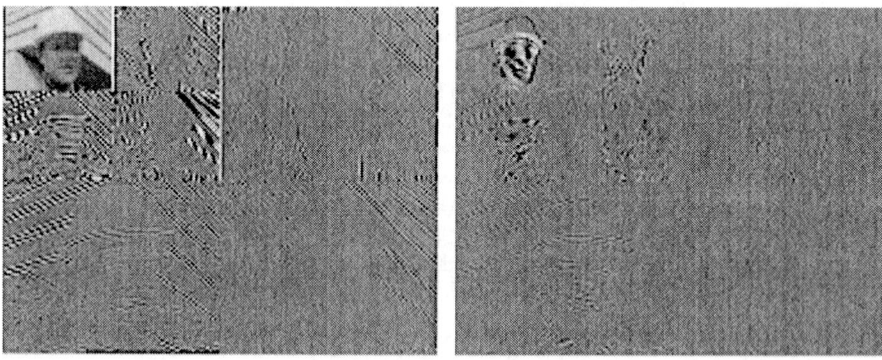

FIGURE 3. Typical lowest and the highest temporal subband frames with 2 levels of dyadic spatial decomposition

3 SPIHT

Since the proposed video coder is based on the SPIHT image coding algorithm [18], the basic principles of SPIHT will be described briefly in this section. The SPIHT algorithm utilizes three basic concepts: (1) coding and transmission of important information first based on the bit-plane representation of pixels; (2) ordered bit-plane refinement; and (3) coding along predefined path/trees called *spatial orientation trees*, which efficiently exploit the properties of a 2D wavelet/transformed image.

SPIHT consists of two main stages, sorting and refinement. In the sorting stage, SPIHT sorts pixels by magnitude with respect to a threshold, which is a power of two, called the level of significance. However, this sorting is a partial ordering, as there is no prescribed ordering among the coefficients with the same level of signicance or highest significant bit. The sorting is based on the significance test of pixels along the spatial orientation trees rooted from the highest level of the pyramid in the 2D wavelet transformed image. Spatial orientation trees were introduced to test significance of groups of pixels for efficient compression by exploiting self-similarity and magnitude localization properties in a 2D wavelet transformed image. In other words, the SPIHT exploits the circumstance that if a pixel in the higher level of pyramid is insignificant, it is very likely that its descendants are insignificant.

Fig. 4 depicts parent-offspring relationship in the spatial orientation trees. In the spatial orientation trees, each node consists of 2×2 adjacent pixels, and each pixel in the node has 4 offspring except at the highest level of pyramid, where one pixel in a node indicated by '*' in this figure does not have any offspring. For practical

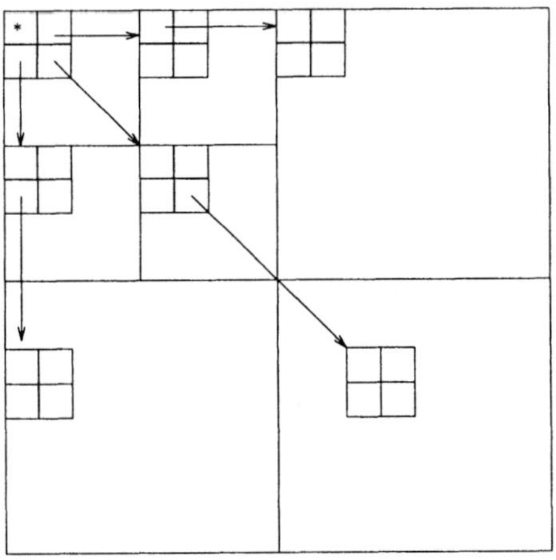

FIGURE 4. *spatial* orientation tree

implementation, SPIHT maintains three linked lists, the list of insignificant pixels (LIP), the list of significant pixels (LSP), the list of insignificant sets (LIS). At the initialization stage, SPIHT initializes the LIP with all the pixels in the highest levels of the pyramid, the LIS with all the pixels in the highest level of pyramid except the pixels which do not have descendants, and the LSP as an empty list. The basic function of the actual sorting algorithm is to recursively partition sets in the LIS to locate individually significant pixels, insignificant pixels, and smaller insignificant sets and move their co-ordinates to the appropriate lists, the LSP, LIP and LIS, respectively. After each sorting stage, SPIHT outputs refinement bits at the current level of bit significance of those pixels which had been moved to the LSP at higher thresholds. In this way, the magnitudes of significant pixels are refined with the bits that decrease the error the most. This process continues by decreasing the current threshold successively by factors of two until the desired bit-rate or image quality is reached. One can refer to [18] for more details.

4 3D SPIHT and Some Attributes

This section introduces the extension of the concept of SPIHT still image coding to 3D video coding. Our main concern is to keep the same simplicity of 2D SPIHT, while still poviding high performance, full embeddedness, and precise rate control.

4.1 Spatio-temporal Orientation Trees

Here, we provide the 3D SPIHT scheme extended from the 2D SPIHT, having the following three similar characteristics: (1) partial ordering by magnitude of the *3D wavelet* transformed video with a *3D set partitioning* algorithm, (2) ordered bit plane transmission of refinement bits, and (3) exploitation of self-similarity across *spatio-temporal* orientation trees. In this way, the compressed bit stream will be completely embedded, so that a single file for a video sequence can provide progressive video quality, that is, the algorithm can be stopped at any compressed file size or let run until nearly lossless reconstruction is obtained, which is desirable in many applications including HDTV.

In the previous section, we have studied the basic concepts of 2D SPIHT. We have seen that there is no constraint to dimensionality in the algorithm itself. Once pixels have been sorted, there is no concept of dimensionality. If all pixels are lined up in magnitude decreasing order, then what matters is how to transmit significance information with respect to a given threshold. In 3D SPIHT, sorting of pixels proceeds just as it would with 2D SPIHT, the only difference being 3D rather than 2D tree sets. Once the sorting is done, the refinement stage of 3D SPIHT will be exactly the same.

A natural question arises as to how to sort the pixels of a three dimensional video sequence. Recall that for an efficient sorting algorithm, 2D SPIHT utilizes a 2D subband/wavelet transform to compact most of the energy to a certain small number of pixels, and generates a large number of pixels with small or even zero value. Extending this idea, one can easily consider a 3D wavelet transform operating on a 3D video sequence, which will naturally lead to a 3D video coding scheme.

On the 3D subband structure, we define a new 3D *spatio-temporal* orientation tree, and its parent-offspring relationships. For ease of explanation, let us review 2D SPIHT first. In 2D SPIHT, a node consists of 4 adjacent pixels as shown in Fig. 4, and a tree is defined such a way that each node has either no offspring (the leaves) or four offspring, which always form a group of 2×2 adjacent pixels. Pixels in the highest levels of the pyramid are tree roots and 2×2 adjacent pixels are also grouped into a root node, one of them (indicated by the star mark in Fig. 4 having no descendants.

A straightforward idea to form a node in 3D SPIHT is to block 8 adjacent pixels with two extending to each of the three dimension, hence forming a node of $2 \times 2 \times 2$ pixels. This grouping is particularly useful at the coding stage, since we can utilize correlation among pixels in the same node. With this basic unit, our goal is to set up trees that cover all the pixels in the 3D *spatio-temporal* domain. To cover all the pixels using trees, we have to impose two constraints except at a node (root node) of the highest level of the pyramid as follows.

1. Each pixel has 8 offspring pixels.

2. Each pixel has only one parent pixel.

With the above constraints, there exists only one reasonable parent-offspring linkage in the 3D SPIHT. Suppose that we have video dimensions of $M \times N \times F$, where M, N, and F are horizontal, vertical, and temporal dimensions of the coding unit or group of frames (GOF). Suppose further that we have l recursive decompositions in both spatial and temporal domains. Thus, we have a root video dimensions of $M_R \times N_R \times F_R$, where $M_R = M/2^l$, $N_R = N/2^l$, and $F_R = F/2^l$. Then, we define three different sets as follows.

Definition 1 *A node represented by a pixel (i,j,k) is said to be a root node, a middle node, or a leaf node according to the following rule.*
If $i < M_R$ and $j < N_R$ and $k < F_R$ then $(i,j,k) \in R$
Else if $i \geq M/2$ and $j \geq N/2$ and $k \geq F/2$, then $(i,j,k) \in L$
Else $(i,j,k) \in M$.
And, the sets R, M, and L represent Root, Middle, and Leaf respectively.

With the above three different classes of a node, there exist three different parent-offspring rules. Let us denote $O(i,j,k)$ as a set of offspring pixels of a parent pixel (i,j,k). Then, we have the following three different parent-offspring relationships depending on a pixel location in the hierarchical tree.

$$
\begin{array}{l}
If\ (i,j,k) \in \mathcal{R}, \\
\mathcal{O}(i,j,k) = \{ \\
(i-1+M_R, j-1+N_R, k-1+F_R), (i+M_R, j-1+N_R, k-1+F_R), \\
(i-1+M_R, j+N_R, k-1+F_R), (i+M_R, j+N_R, k-1+F_R), \\
(i-1+M_R, j-1+N_R, k+F_R), (i+M_R, j-1+N_R, k+F_R), \\
(i-1+M_R, j+N_R, k+F_R), (i+M_R, j+N_R, k+F_R) \\
\} \\
If\ (i,j,k) \in \mathcal{M}, \\
\mathcal{O}(i,j,k) = \{ \\
(2i,2j,2k), (2i+1,2j,2k), (2i,2j+1,2k), \\
(2i+1,2j+1,2k), (2i,2j,2k+1), (2i+1,2j,2k+1), \\
(2i,2j+1,2k+1), (2i+1,2j+1,2k+1) \\
\} \\
If\ (i,j,k) \in \mathcal{L}, \\
\mathcal{O}(i,j,k) = \{\phi\}
\end{array}
\tag{24.1}
$$

One exception as in 2D SPIHT is that one pixel in a root node has no offspring. Fig. 5 depicts the parent-offspring relationships in the highest level of the pyramid, assuming the root dimension is $4 \times 4 \times 2$ for simplicity. S-LL, S-LH, S-HL, and S-HH represent spatial low-low, low-high, high-low, high-high frequency subbands in the vertical and horizontal directions. There is a group (node) of 8 pixels indicated by '*', 'a', 'b', 'c', 'd', 'e', 'f' in S-LL, where pixel 'f' is hidden under pixel 'b'. Every pixel located at '*' position in a root node has no offspring. Each arrow originating from a root pixel pointing to a $2 \times 2 \times 2$ node shows the parent-offspring linkage. In Fig. 5, offspring node 'F' of pixel 'f' is hidden under node 'B' which is offspring node of 'b'. Having defined a tree, the same sorting algorithm can be now applied to the video sequence along the new *spatio-temporal* trees, i.e., set partitioning is now performed in the 3D domain. From Fig. 5, one can see that the trees grow to the order of 8 branches, while 2D SPIHT has trees of order of 4. Hence, the bulk of compression can possibly be obtained by a single bit which represents insignificance of a certain spatio-temporal tree.

The above tree structure required offspring in a $2 \times 2 \times 2$ pixel cube for every parent having offspring. Hence, it appears that there must be the same number of

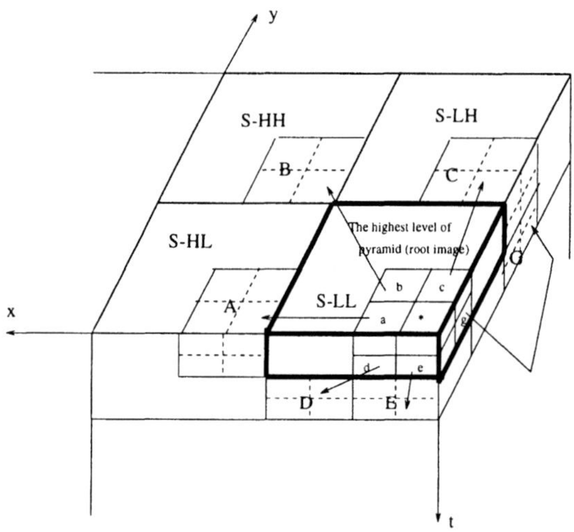

FIGURE 5. *spatio-temporal* orientation tree

decomposition levels in all three dimensions. Therefore, as three spatial decompositions seem to be the minimum for efficient image coding, the same number temporal decompositions forces the GOF size to be a minimum of 16, as SPIHT needs an even number in each dimension in the coarsest scale at the top of the pyramid. Indeed, in previous articles [3, 8], we have reported video coding results using trees with this limitation.

To achieve more flexibility in choosing the number of frames in a GOF, we can break the uniformity in the number of spatial and temporal decompositions and allow *unbalanced trees*. Suppose that we have three levels of spatial decomposition and one level of temporal decomposition with 4 frames in the GOF. Then a pixel with coordinate $(i, j, 0)$ has a longer descendant tree (3 levels) than that of a pixel with coordinate $(i, j, 1)$ (1 level), since any pixel with temporal coordinate of zero has no descendants in the temporal direction. Thus, the descendant trees in the significance tests in the latter terminate sooner than those in the former. This modification in structure can be noted in this case by keeping track of two different kinds of pixels. One pixel has a tree of three levels and the other a tree of one level. The same kind of modification can be made in the case of a GOF size of 8, where there are two levels of temporal decomposition.

With a smaller GOF and removal of structural constraints, there are more possiblities in the choice of filter implementations and the capability of a larger number of decompositions in the spatial domain to compensate for a possible loss of coding performance from reducing the number of frames in the GOF. For example, it would be better to use a shorter filter with short segments of four or eight frames of the video sequence, such as the Haar or S+P [19] filters, which use only integer operations, with the latter being the more efficient.

Since we have already 3D wavelet-transformed the video sequence to set up 3D spatio-temporal trees, the next step is compression of the coefficients into a bitstream. Essentially, it can be done by feeding the 3D data structure to the 3D

SPIHT kernel. Then, 3D SPIHT will sort the data according to the magnitude along the *spatio-temporal orientation trees* (sorting pass), and refine the bit plane by adding necessary bits (refinement pass). At the destination, the decoder will follow the same path to recover the data.

4.2 Color Video Coding

So far, we have considered only one color plane, namely luminance. In this section, we will consider a simple application of the 3D SPIHT to any color video coding, while still retaining full embeddedness, and precise rate control.

A simple application of the SPIHT to color video would be to code each color plane separately as does a conventional color video coder. Then, the generated bit-stream of each plane would be serially concatenated. However, this simple method would require allocation of bits among color components, losing precise rate control, and would fail to meet the requirement of the full embeddedness of the video codec since the decoder needs to wait until the full bit-stream arrives to reconstruct and display. Instead, one can treat all color planes as one unit at the coding stage, and generate one *mixed* bit-stream so that we can stop at any point of the bit-stream and reconstruct the color video of the best quality at the given bit-rate. In addition, we want the algorithm to automatically allocate bits optimally among the color planes. By doing so, we will still keep the claimed full embeddedness and precise rate control of 3D SPIHT. The bit-streams generated by both methods are depicted in the Fig. 6, where the first one shows a conventional color bit-stream, while the second shows how the color embedded bit-stream is generated, from which it is clear that we can stop at any point of the bit-stream, and can still reconstruct a color video at that bit-rate as opposed to the first case.

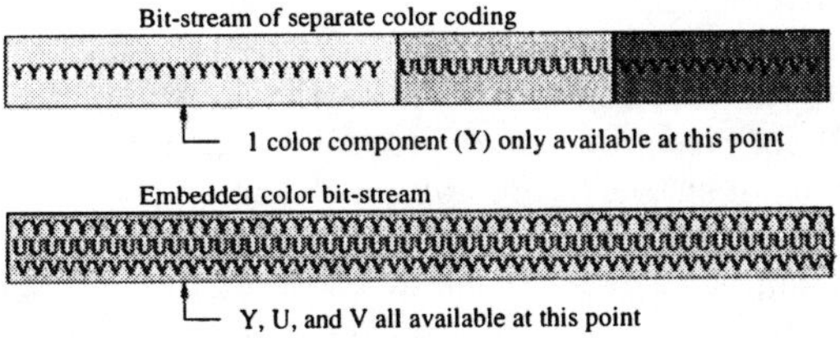

FIGURE 6. bit-streams of two different methods:separate color coding, embedded co lor coding

Let us consider a tri-stimulus color space with luminance Y plane such as YUV, YCrCb, etc. for simplicity. Each such color plane will be separately wavelet transformed, having its own pyramid structure. Now, to code all color planes together, the 3D SPIHT algorithm will initialize the LIP and LIS with the appropriate coordinates of the top level in all three planes. Fig. 7 shows the initial internal structure of the LIP and LIS, where Y,U, and V stand for the coordinates of each root pixel in each color plane. Since each color plane has its own spatial orientation trees, which are mutually exclusive and exhaustive among the color planes, it automatically assigns the bits among the planes according to the significance of the magnitudes of their own coordinates. The effect of the order in which the root pixels of each color plane are initialized will be negligible except when coding at extremely low bit-rate.

YYYYYYYYYYYYYYYYYYYYUUUUUVVVVV

FIGURE 7. Initial internal structure of *LIP* and *LIS*, assuming the U and V planes are one-fourth the size of the Y plane.

4.3 Scalability of SPIHT image/video Coder

Overview

In this section, we address multiresolution encoding and decoding in the 3D SPHIT video coder. Although the proposed video coder naturally gives scalability in rate, it is highly desirable also to have temporal and/or spatial scalabilities for today's many multi-media applications such as video database browsing and multicast network distributions. Multiresolution decoding allows us to decode video sequences at different rate and spatial/temporal resolutions from one bit-stream. Furthermore, a layered bit-stream can be generated with multiresolution encoding, from which the higher resolution layers can be used to increase the spatial/temporal resolution of the video sequence obtained from the low resolution layer. In other words, we achieve full scalability in rate and partial scalability in space and time with multiresolution encoding and decoding.

Since the 3D SPIHT video coder is based on the multiresolution wavelet decomposition, it is relatively easy to add multiresolutional encoding and decoding as functionalities in partial spatial/temporal scalability. In the following subsections, We first concentrate on the simpler case of multiresolutional decoding, in which an encoded bit-stream is assumed to be available at the decoder. This approach is quite attractive since we do not need to change the encoder structure. The idea of multiresolutional decoding is very simple: we partition the embedded bit-stream into portions according to their corresponding spatio-temporal frequency locations, and only decode the ones that contribute to the resolution we want.

We then turn to multiresolutional encoding, where we describe the idea of generating a layered bit-stream by modifying the encoder. Depending on bandwidth availability, different combinations of the layers can be transmitted to the decoder to reconstruct video sequences with different spatial/temporal resolutions. Since the 3D SPIHT video coder is symmetric, the decoder as well as the encoder knows exactly which information bits contribute to which temporal/spatial locations. This makes multiresolutional encoding possible as we can order the original bit-stream into layers, with each layer corresponding to a different resolution (or portion). Although the layered bit-stream is not fully embedded, the first layer is still rate scalable.

Multiresolutional Decoding

As we have seen previously, the 3D SPIHT algorithm uses significance map coding and spatial orientation trees to efficiently predict the insignificance of descendant pixels with respect to a current threshold. It refines each wavelet coefficient successively by adding residual bits in the refinement stage. The algorithm stops when the size of the encoded bit-stream reaches the exact target bit-rate. The final bit-stream consists of significance test bits, sign bits, and refinement bits.

In order to achieve multiresolution decoding, we have to partition the received bit-stream into portions according to their corresponding temporal/spatial location. This is done by putting two flags (one spatial and one temporal) in the bit-stream during the process of *decoding*, when we scan through the bit-stream and mark the portion that corresponds to the temporal/spatial locations defined by the input resolution parameters. As the received bit-stream from the decoder is embedded, this partitioning process can terminate at any point of the bit-stream that is specified by the decoding bit-rate. Fig. 8 shows such a bit-stream partitioning. The dark-gray portion of the bit-stream contributes to low-resolution video sequence, while the light-gray portion corresponds to coefficients in the high resolution. We only decode the dark-gray portion of the bit-stream for a low-resolution sequence and scale down the 3D wavelet coefficients appropriately before the inverse 3D wavelet transformation. We can further partition the dark-gray portion of the bit-stream in Fig. 8 for decoding in even lower resolutions.

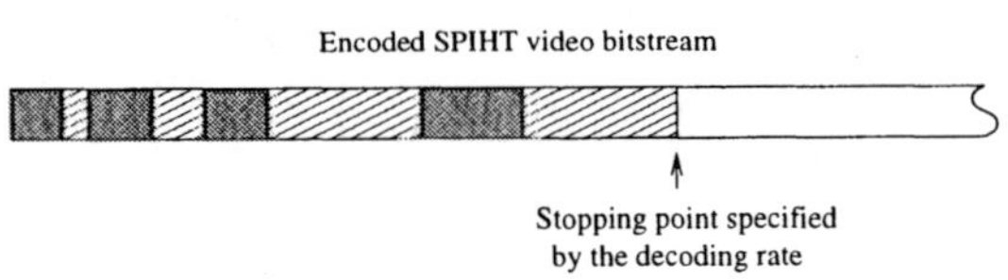

FIGURE 8. Partitioning of the SPIHT encoded bit-stream into portions according to their corresponding temporal/spatial locations

By varying the temporal and spatial flags in decoding, we can obtain different combinations of spatial/temporal resolutions in the encoder. For example, if we encode a QCIF sequence at 24 f/s using a 3-level spatial-temporal decomposition, we can have in the decoder three possible spatial resolutions (176×144, 88×72, 44×36), three possible temporal resolutions (24, 12, 6), and any bit rate that is upper-bounded by the encoding bit-rate. Any combination of the three sets of parameters is an admissible decoding format.

Obvious advantages of scalable video decoding are savings in memory and decoding time. In addition, as illustrated in Fig. 8, information bits corresponding to a specific spatial/temporal resolution are not distributed uniformly over the compressed bit-stream in general. Most of the lower resolution information is crowded at the beginning part of the bit-stream, and after a certain point, most of the bit rate is spent in coding the highest frequency bands which contain the detail of video which are not usually visible at reduced spatial/temporal resolution. What this means is that we can set a very small bit-rate for even faster decoding and browsing applications, saving decoding time and channel bandwidth with negligible degradation in the decoded video sequence.

4.4 Multiresolutional Encoding

The aim of multiresolutional encoding is to generate a layered bit-stream. But, information bits corresponding to different resolutions in the original bit-stream are interleaved. Fortunately, the SPIHT algorithm allows us to keep track of the temporal/spatial resolutions associated with these information bits. Thus, we can change the original encoder so that the new encoded bit-stream is layered in temporal/spatial resolutions. Specifically, multiresolutional encoding amounts to putting into the first (low resolution) layer all the bits needed to decode a low resolution video sequence, in the second (higher resolution) layer those to be added to the first layer for decoding a higher resolution video sequence and so on. This process is illustrated in Fig. 9 for the two-layer case, where scattered segments of the dark-gray (and light-gray) portion in the original bit-stream are put together in the first (and second) layer of the new bit-stream. A low resolution video sequence can be decoded from the first layer (dark-gray portion) alone, and a full resolution video sequence from both the first and the second layers.

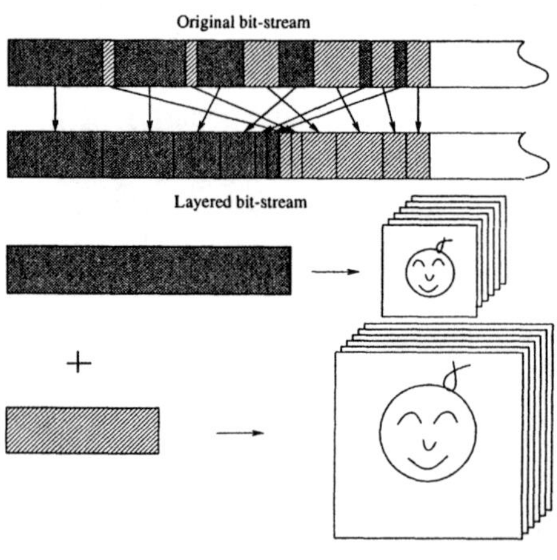

FIGURE 9. A multiresolutional encoder generates a layered bit-stream, from which the higher resolution layers can be use to increase the spatial resolution of the frame obtained from the low resolution layer.

As the layered bit-stream is a reordered version of the original one, we lose overall scalability in rate after multiresolutional encoding. But the first layer (i.e., the dark gray layer in Fig. 9) is still embedded, and it can be used for lower resolution decoding.

Unlike multiresolutional decoding in which the full resolution encoded bit-stream has to be transmitted and stored in the decoder, multiresolutional encoding has the advantage of wasting no bits in transmission and decoding at lower resolution. The disadvantages are that it requires both the encoder and the decoder agree on the resolution parameters and the loss of embeddedness at higher resolution, as mentioned previously.

5 Motion Compensation

When there is a considerable amount of motion either in the form of global camera motion or object motion within a GOF, the PSNR of reconstructed video will fluctuate noticeably due to pixels with high magnitude in the temporal high frequency. On the contrary, if there were a way in which we can measure motion so accurately that we can remove a significant amount of temporal redundancy, there will be not many significant pixels in the temporal high frequency subbands, leading to good compression scheme. Fortunately, as we earlier discussed, typical applications with very low bit-rate deal with rather simple video sequences with slow object and camera motion. Even so, to remove more temporal redundancy and PSNR fluctuation, we attempt to incorporate a motion compensation (MC) scheme as an option. In this section, we digress from the main subject of our proposed video coder, and review some of the previous works on motion compensation. Block matching is the most simple in concept, and has been widely used for many practical video applications due to its smaller hardware complexity. In fact, today's video coding standards such as H.261, H.263, and MPEG 1-2 are all adopting this scheme for simplicity. Hence, we start with a brief introduction of the block matching algorithm, and its variations, followed by some representative motion compensated filtering schemes.

5.1 Block Matching Method

As with other block-based motion estimation methods, the block matching method is based on three simple assumptions: (1) the image is composed of moving blocks, (2) there only exists translational motion, and (3) within a certain block, the motion vector is uniform. Thus, it fails to track more complex motions such as zooming, and rotation of the object, and motion vectors obtained thereby are not accurate enough to describe true object motion.

In the block matching method, the frame being encoded is divided into blocks of size $M \times N$. For each block, we search for a block in the previous frame within a certain search window which gives the smallest measure of closeness (or minimum distortion measure). The motion vector is taken to be the displacement vector between these blocks. Suppose that we have a distortion measure function $D(\cdot)$, which is a function of the displacement dx and dy between the block in the current frame and a block in the previous frame. Then, mathematically, the problem can be formulated as

$$(\hat{dx}, \hat{dy}) = arg \min_{(dx,dy)} D(dx, dy) \tag{24.2}$$

There are several possible measurement criteria including maximum cross-correlation, minimum mean squared error (MSE), minimum mean absolute error (MAD), and maximum matching pel count (MPC). We formally provide the two most popular criteria (MSE, and MAD) just for completeness.

$$MSE(dx, dy) = \frac{1}{MN} \sum_{(x,y)\in S} [I(x, y, t) - I(x - dx, y - dy, t - 1)]^2 \tag{24.3}$$

and

$$MAD(dx, dy) = \frac{1}{MN} \sum_{(x,y)\in S} | I(x,y,t) - I(x-dx, y-dy, t-1) |, \qquad (24.4)$$

where $I(x,y,t)$ denotes the intensity at co-ordinates (x,y,t) and S denotes the block of size $M \times N$.

The advantage of the full search block matching method above is that it guarantees an optimum solution within the search region in terms of the given distortion measure. However, it requires a large number of computations. Suppose that we have $M = N = 8$. Then, it requires 64 differences, and then computation of the sum of the absolute mean square values of each difference. If we have search region of +15 to −15 for $(\hat{dx}, \hat{dy})$, then we need more than 950 comparisons.

There are several ways to reduce computational complexity. One way is to use a larger block. By doing so, we have a fewer number of blocks in the frame (thus, we have a smaller number of motion vectors), while the number of computations per comparison will increase. The drawback with large block size is that it becomes more likely to contain more than one object moving in different directions. Hence, it is more difficult to estimate local object motion. Another way is to reduce the search range, which will reduce number of comparisons for best match search within the search range. However, it will increase the probability of missing a true motion vector. Meanwhile, there are several fast search algorithms such as three-step search, and cross-search. Instead of searching for every possible candidate motion vector, those algorithms evaluate the criterion function only at a predetermined subset of the candidate motion vector locations. Hence, again there is trade-off between speed and performance. For more details of search strategies, one can refer to [25].

Choice of optimum block size is of importance for any block-based motion estimation. As mentioned above, too large a block size fails to meet the basic assumptions in that within a block actual motion vectors may vary. If a block size is too small, a match between the blocks containing two similar gray-level patterns can be established. Because of these two conflicting requirements, the motion vector field is usually not smooth [4]. Hierarchical block matching algorithm, discussed in the next section, has been proposed to overcome these problems, and our proposed video codec employs this method for motion estimation.

5.2 Hierarchical Motion Estimation

Hierarchical representations of image in the form of wavelet transform has been proposed for improved motion estimation. The basic idea of the hierarchical motion estimation is to perform motion estimation at each level successively. The motion vectors estimated at the highest level of pyramid are used for the initial estimates of the motion vectors at the next state of the pyramid to produce motion vector estimates more accurately. At higher resolution levels, we are allowed to use a small search range since we start with a good initial motion vector estimate. Fig. 10 illustrates the procedure of how a motion vector gets refined at each stage. In general, a more accurate and smoother motion vector field (MVF) can be obtained with this method, which results in a lower amount of side information for motion vector coding. Moreover, since the computational burden with full search block

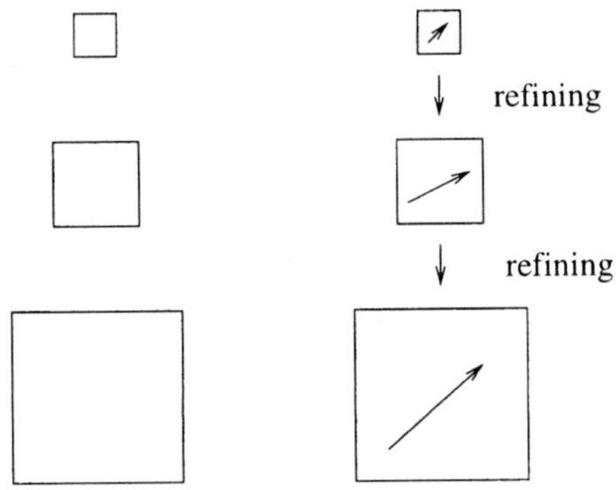

FIGURE 10. Hierarchical motion estimation

matching method is now replaced by a hierarchical wavelet transform plus full search block matching with much smaller search window, we can have faster motion estimation. The additional computational complexity for the wavelet transform is usually much smaller than that of full search block matching with a relatively large search window.

5.3 Motion Compensated Filtering

Motion compensated temporal analysis has been considered to take better advantage of temporal correlation. The basic idea is to perform temporal filtering along the motion trajectory. Although it is conceptually simple and reasonable, it was considered to to be difficult to translate into practice or even impractical [9] due to the complexity of motion in video and the number of connected/unconnected pixels inevitably generated. However, recently there have been some promising results in this area [13, 14, 4], where a heuristic approach has been made to deal with the connected/unconnected pixel problem. Fig. 11 illustrates the problem, and shows solutions to have perfect reconstruction in the presence of unconnected or doubly connected pixels. With Ohm's method [14], for any remaining unconnected pixels, the original pixel value of the current frame is inserted into the temporal low-subband, and the scaled displace frame difference (DFD) into the temporal high-subband. With Choi's method [4], for an unconnected pixel in the reference frame, its original value is inserted into the temporal low-subband with the assumption that unconnected pixels are more likely to be uncovered ones, and for an unconnected one in the current frame, a scaled DFD is taken. In this proposal, we adopted the method in [4] for MC temporal filtering.

One of the main problems in incorporating motion compensation into video coding, especially at very low bit-rate, is the overhead associated with the motion vector. A smaller block size generally increase the amount of overhead information. Too large a block size fails to reasonably estimate diverse motions in the frames. In [4], several different block sizes were used with a hierarchical, variable size block matching method. In that case, additional overhead for the image map for variable sizes of blocks and the motion vectors are needed to be transmitted as side information. Since we can not know the characteristics of the video beforehand, it

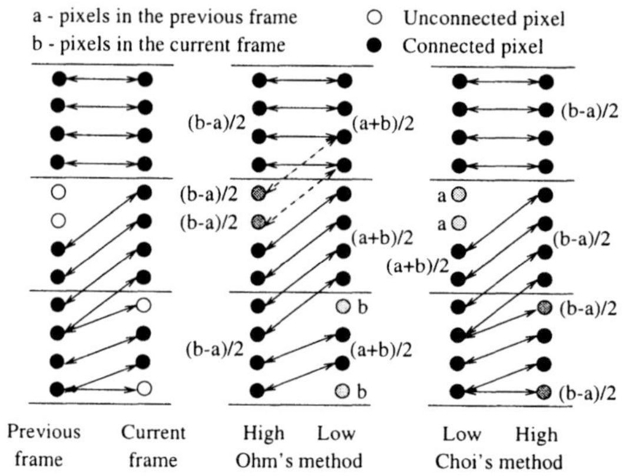

FIGURE 11. MC filtering method in the presence of connected/unconnected pixels

Level	3	2	1
Search Window	±2	±2	±2
Option 1	4	8	16
Option 2	8	16	32

TABLE 24.1. Set of parameters for 3-Level hierarchical block matching

is difficult to choose optimum block size and search window size. However, empirically we provide Table 24.1 for the parameters for our hierarchical block matching method for motion estimation, where there are two options in choosing block size. Motion vectors obtained at a certain level will be scaled up by 2 to be used for initial motion vectors at the next stage.

6 Implementation Details

Performance of video coding systems with the same basic algorithm can be quite different according to the actual implementation. Thus, it is necessary to specify the main features of the implementation in detail. In this section, we will describe some issues such as filter choice, image extension, and modeling of arithmetic coding, that are important in practical implementation.

The S+P [19] and Haar (2 tap) [4, 15] filters are used only for the temporal direction, with the Haar used only when motion-compensated filtering is utilized. The 9/7-tap biorthogonal wavelet filters [1] have the best enegy compaction properties, but having more taps, are used for spatial and temporal decomposition for a size 16 GOF. For GOF sizes of 4 and 8, the S+P filter is used to produce one and two levels of temporal decomposition, respectively. In all cases, three levels of decomposition are performed with the 9/7 biorthogonal wavelet filters.

With these filters, filtering is performed with a convolution operation recursively. Since we need to preserve an even number for dimensions of the highest level of pyramid image of each frame, given some desired number of spatial levels, it is often necessary to extend the frame to a larger image before 2D spatial transformation. For example, we want to have at least 3 levels of spatial filtering for QCIF (176×144) video sequence with 4:2:0 or 4:1:1 subsampled format. For the luminance component, the highest level of the image is 22×18. However, for chrominance components, it will be 11×9 which is not appropriate for the coding stage. To allow 3 levels of decomposition, a simple extrapolation scheme is used to make dimension of chrominance component 96×80 which results in 12×10 of root image after 3 levels of decomposition. Generally, when extending the image by artificially augmenting its boundary can cause some loss of performance. However, extending the image in a smooth fashion (since we do not want to generate artificial high frequency coefficients), the performance loss is expected to be minimal.

After the 3D subband/wavelet transformation is completed, the 3D SPIHT algorithm is applied to the resulting multiresolution pyramid. Then, the output bitstream is further compressed with an arithmetic encoder [27]. To increase the coding efficiency, groups of $2 \times 2 \times 2$ coordinates were kept together in the list. Since the amount of information to be coded depends on the number of insignificant pixels m in that group, we use several different adaptive models, each with 2^m symbols, where $m \in \{1,2,3,4,5,6,7,8\}$, to code the information in a group of 8 pixels. By using different models for the different number of insignificant pixels, each adaptive model contains better estimates of the probabilities conditioned to the fact that a certain number of adjacent pixels are significant or insignificant.

Lastly, when motion compensated temporal filtering is applied, we will need to code motion vectors. When $GOF = 16$, we have 8 first level MVFs, 4 second level MVFs, and 2 third level MVF, resulting in total 14 MVFs to code. In the experiment, we have found that MVFs from different levels have quite different statistical characteristics. In general, more unconnected pixels are generated at higher levels of temporal decomposition. Furthermore, horizontal and vertical motion are assumed to be independent of each other. Thus, each motion vector component is separately coded conditioned to the level of decomposition. For the chrominance components, the motion vector obtained from the luminance component will be used with an appropriate scale factor.

7 Coding results

7.1 *The High Rate Regime*

The first ask is to test the 3D SPIHT codec in the high-rate regime with two monochrome SIF sequences, 'table tennis' and 'football', having a frame size of 352×240 and frame rate of 30 frames per second (fps). Generally, a larger GOF is expected to provide better rate-distortion performance, which is possibly counterbalanced by possibly unacceptable latency (delay), especially in interactive video applications [24], from the temporal filtering and coding. For these SIF sequences with their frame size and rate, we dispensed with motion-compensated filtering and chose GOF sizes of 4, 8, and 16. With GOF's of 4 and 8 frames, the S+P transform [17] is used for 1-level and 2-level temporal decomposition respectively, while with GOF of 16 frames, the 9/7 bi-orthogonal wavelet filters are used. As mentioned previously, we always performed three levels of spatial decomposition with the 9/7 wavelet filters.

From frame 1 to 30 of 'table tennis', the camera is almost still, and the only movements are the bouncing ping pong ball, and hands of the player, where, as shown in Fig. 12, the GOF of 16 outperforms the smaller size of GOF. Then, the camera begins to move away (or zoom out from frame 31 - 67), and the whole body of the player shows up, where GOF size does not seem to affect the performance very much. In this region, MPEG-2 obviously catches up with the camera zooming better than 3-D SPIHT. Then, there occurs a sudden scene change at frame 68 (a new player begins to show up). Note the interesting fact that the GOF of 16 is not affected very much by the sudden scene change since the scene change occurs in the middle of 16 frame segment of frames 65 - 80, which are simultaneously processed by 3-D SPIHT, while GOF of 4 suffers severely from the scene change of frame 68 which is the last frame of 4 frame segment of frames 65 - 68. With several simulations, we found that larger size of GOF is generally less vulnerable or even shows a good adaptation to a sudden scene change. After the frame 68, the sequence stays still, and there is only small local motion (swinging hands), which is effectively coded by 3-D SPIHT. 3-D SPIHT more quickly tracks up to high PSNR in the still region compared with MPEG-2 in the Fig. 12, where MPEG-2 shows lower PSNR in that region.

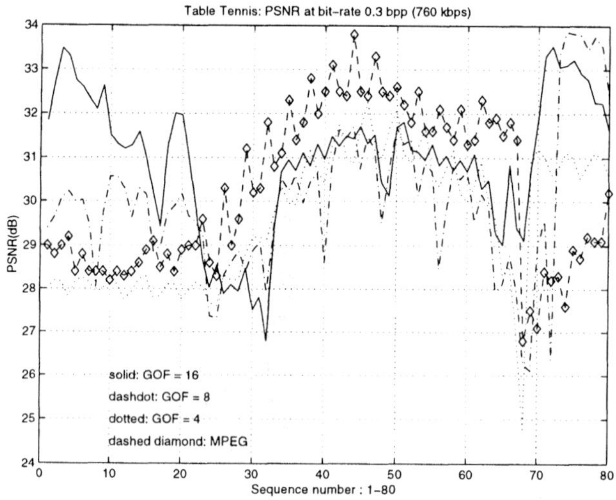

FIGURE 12. Comparison of PSNR with different GOF (4, 8, 16) on table tennis sequence at 0.3 bpp (760 kbits)

In Figure 13, showing the PSNR versus frame number of "football", we again see that GOF of 16 outperforms the smaller GOFs. It has been surprising that 3-D SPIHT is better than MPEG-2 for all different GOF sizes. It seems that the conventional block matching motion compensation of MPEG-2 to find translational movement of blocks in the frame does not work well in this sequence, where there is much more complex local motion everywhere and not much global camera movement. This similar phenomenon has also been reported by Taubman and Zahkor [24]. Average PSNRs with different GOFs are shown in Table 24.2 for the perfor-

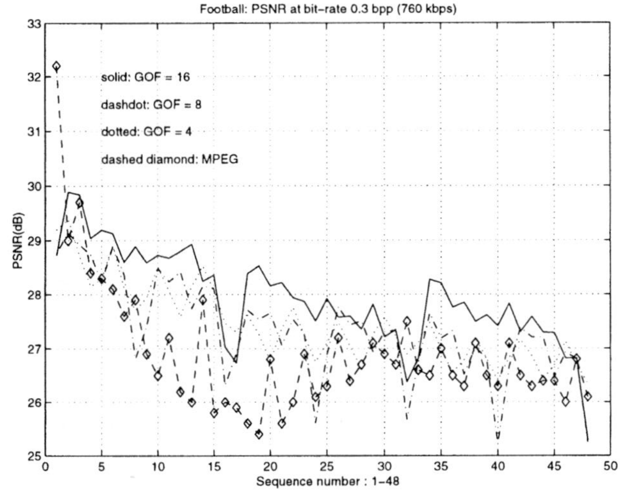

FIGURE 13. Comparison of PSNR with different GOF (4, 8, 16) on football sequence at 0.3 bpp (760 kbits)

mance comparison, in which we can again observe that GOF = 16 gives the best average PSNR performance at the cost of longer coding delay and more memory. However, the gain for more frames in a GOF is not much in the "football" sequence with its complex motion and stationary camera.

To demonstrate the visual performances with different GOFs, reconstructed frames of the same frame number (8) at bit-rate of 0.2 bpp (or 507 Kbps) are shown in Fig. 14, where GOF = 16 performs the best. With GOF = 4, 3-D SPIHT still shows the visual quality comparable to or better than MPEG-2. Overall, 3-D SPIHT suffers from blurs in some local regions, while MPEG-2 suffers from both blurs and blocking artifacts as can be shown in Fig. 14, especially when there exists significant amount of motion.

Sequence	GOF=4(dB)	GOF=8 (dB)	GOF=16 (dB)	MPEG-2 (dB)
tennis	29.59	30.04	30.95	30.27
football	27.45	27.36	27.94	26.90

TABLE 24.2. Coding results with different GOFs at 0.3 bpp (760 kbits) (Average PSNR in dB)

Although it might have been expected that 3-D SPIHT would degrade in performance from motion in the sequence, we have seen that 3-D SPIHT is working relatively well for local motion, even better than MPEG-2 with motion compensation on the "football" sequence. Even with the similar average PSNR, better visual quality of reconstructed video was observed. As in most 3-D subband video coders [3], one can observe that PSNR dips at the GOF boundaries partly due to the object motion and partly due to the boundary extension for the temporal filtering with GOF = 16. However, they are not visible in the reconstructed video. With smaller number of GOF such as 4 and 8, less fluctuation in frame-by-frame PSNR was observed.

7.2 The Low Rate Regime

We now turn to the low-rate regime and test the 3D SPIHT codec with color video QCIF sequences with a frame size of 176 × 144 and frame rate of 10 fps. Because filtering and coding are so much faster per frame than with the larger SIF frames, a GOF size of 16 was selected for all these experiments. In this section, we shall provide simulation results and compare the proposed video codec with H.263 in various aspects such as objective and subjective performance, system complexity, and performance at different camera and object motion. The latest test model number of H.263, tmn2.0 (test model number 2.0), was downloaded from the public domain (ftp://bonde.nta.no/pub/tmn/). As in H.263, the 3D SPIHT video codec is a fully implemented software video codec.

We tested three different color video sequences: "Carphone", "Mother and Daughter", and "Hall Monitor". These video sequences cover a variety of object and camera motions. "Carphone" is a representative sequence for video-telephone application. This has more complicated object motion than the other two, and camera is assumed to be stationary. However, the background outside the car window changes very rapidly. "Mother and Daughter" is a typical head-and-shoulder sequence with relatively small object motion and camera is also stationary. The last sequence "Hall Monitor" is suitable for a monitoring application. The background is always fixed and some objects (persons) appear and then disappear.

All the tests were performed at the frame-rate of 10 fps by coding every third frame. The successive frames of this downsampled sequence are now much less correlated than in the original sequence. Therefore, the action of temporal filtering is less effective for further decorrelation and energy compaction. For a parallel comparison, we first run with H.263 at a target bit-rates (30k and 60k) with all the advanced options (-D -F -E) on. In general, H.263 does not give the exact bit-rate and frame-rate due to the buffer control. For example, H.263 produced actual bit-rate and frame-rate of 30080 bps and 9.17 fps for the target bit-rate and frame-rate of 30000 bps and 10 fps. With our proposed video codec, the exact target bit rates can be obtained. However, since 3D SPIHT is fully embedded video codec, we only needed to decode at the target bit-rates with one bit-stream compressed at the larger bit-rate.

Figs. 15, 16, and 17 show frame-by-frame PSNR comparison among the luminance frames coded by MC 3D SPIHT, 3D SPIHT, and H.263. From the figures, 3D SPIHT is 0.89 – 1.39 dB worse than H.263 at the bit-rate of 30k. At the bit-rate of 60k, less PSNR advantage of H.263 over 3D SPIHT can be observed except the "Hall Monitor" sequence for which 3D SPIHT has a small advantage over H.263. In comparing MC 3D SPIHT with 3D SPIHT, MC 3D SPIHT generally gives more stable PSNR fluctuation than 3D SPIHT without MC, since MC reduces large magnitude pixels in temporal high subbands resulting in more uniform rate allocations over the *GOF*. In addition, a small advantage of MC 3D SPIHT over 3D SPIHT was obtained for the sequences with translational object motion. However, for the "Hall Monitor" sequence, where 3D SPIHT outperforms MC 3D SPIHT in terms of average PSNR, side information associated with motion vectors seemed to exceed the gain provided by motion compensation. In terms of visual quality, H.263, 3D SPIHT, and MC 3D SPIHT showed very competitive performance as shown in Figs. 18, 19, and 20 although our proposed video coders have lower PSNR values at the bit rate of 30 kbps. In general, H.263 preserves edge information of objects better than 3D SPIHT, while 3D SPIHT exhibits blurs in some local regions. However, as we can observe in Fig. 18, H.263 suffers from blocking artifacts around the mouth of the person and background outside the car window. In overall, 3D SPIHT and MC 3D SPIHT showed comparable results in terms of average PSNR and visual quality of reconstructed frames. As in most 3D subband video coders, one can observe that the PSNR dips at the GOF boundaries, partly due to object motion and partly due to the boundary extension for temporal filtering. However, they are not manifested visibly in the reconstructed video. Smaller GOF sizes of 4 and 8 have been used with shorter filters, such as Haar and S+P, for temporal filtering. Consistent with the SIF sequences, the results are not significantly different, but slightly inferior in terms of average PSNR. Again, the fluctuation in PSNR from frame to frame is smaller with the shorter GOF's.

As we discussed earlier section, scalability in video coding is very important for many multi-media applications. Fig. 21 shows first frames of the decoded "Carphone" sequence at different spatial/temporal resolutions from the same encoded bit-stream. Note the improved quality of the lower spatial/temporal resolutions. It is due to the fact that the SPIHT algorithm extracts first the bits of largest significance in the largest magnitude coefficients, which reside in the lower spatial/temporal resoltuions. At low bit rates, the low resolutions will receive most of the bits and be reconstructed fairly accurately, while the higher resolutions will receive relatively few bits and be reconstructed rather coarsely.

7.3 Embedded Color Coding

Next, we provide simulation results of color-embedded coding of 4:2:0 (or 4:1:1) chrominance subsampled QCIF (176 × 144) sequence. Note that not many video coding schemes perform reasonably well in both high bit-rate and low bit-rate. To demonstrate embeddedness of 3-D color video coder, we reconstructed video at bit-rates of 30 Kbps and 60 Kbps, and frame-rate of 10 fps, as before, by coding

Bit-rate	30 Kbps (Y U V) dB	60 Kbps (Y U V) dB
3-D SPIHT	32.95 38.15 40.68	37.95 40.41 42.38
H.263	33.84 37.79 40.03	37.50 39.81 41.65

TABLE 24.3. Coding Results of "Hall monitor" sequence (frames 0 - 285) at bit-rates of 30 Kbps and 60 Kbps, and frame-rate of 10 fps

	3D-SPIHT	MPEG-2
Actual time/frame	2.44 sec	8.63 sec
Relative time/frame	1	3.5

TABLE 24.4. Coding is performed on Sun-sparc 10 machine. Total 48 frames were coded at bit-rate of 0.3 bpp (760 Kbps).

every third frame, with the same color-embedded bit-stream compressed at higher bit-rate 100 Kbps. In this simulation, we remind the reader that 16 frames were chosen for a GOF, since required memory and computational time are reasonable with QCIF sized video. The average PSNRs coding frame 0 – 285 (96 frames coded) are shown in the Table 24.3, in which we can observe that 3-D SPIHT gives average PSNRs of Y component slightly inferior to those of H.263 and average PSNRs of U and V components slightly better than those of H.263. However, visually 3-D SPIHT and H.263 show competitive performances as shown in Fig. 20. Overall, color 3-D SPIHT still preserves precise rate control, self-adjusting rate allocation according to the magnitude distribution of the pixels in all three color planes, and full embeddedness.

7.4 Computation Times

Now, we assess the computational complexity in terms of the run times of the stages of transformation, encoding, decoding, and search for maximum magnitudes. First, absolute and relative encoding times per frame are shown in Table 24.4 for 3D SPIHT and MPEG-2 in coding of 48 frames of a SIF sequence at the bit rate of 0.3 bpp or 760 Kbps. We observe 3-D SPIHT is 3.5 times faster than MPEG-2. For QCIF sequences, 3D SPIHT is 2.53 times faster than H.263, but with motion compensation is 1.2 times slower, as shown in Table 24.5. In motion-compensated 3D SPIHT, most of the execution time is spent in hierarchical motion estimation.

	3D SPIHT	MC 3D SPIHT	H.263
Relative complexity	1	3.09	2.53

TABLE 24.5. Comparison of 3D SPIHT, MC 3D SPIHT, and H.263 relative computational complexity for encoding 0 - 285 "Carphone" sequence at 10 fps, and 30 kbps. Complexity is computed on SUN SPARC 20 machine.

A more detailed breakdown of computation times was undertaken for the various stages in the encoding and decoding of QCIF 4:2:0 or 4:1:1 YUV color sequences. Table 24.6 shows run times of these stages on a SUN SPARC 20 for 96 frames of the 'Carphone' sequence, specifically every third frame of frames 0–285 (10 fps), encoded at 30 Kbps. Note that 57 % of the encoding time and 78 % of the decoding time are spent on wavelet transformation and its inverse, respectively. Considering that the 3D SPIHT coder is not fully optimized and there have been recent advances in fast filter design, 3D SPIHT shows promise of becoming a real-time software-only video codec, as seen from the actual coding and decoding times in Table 24.6.

Functions	time in sec	relative time (%)
3D wavelet transform	16.22	56.95
3D SPIHT compression	1.27	4.46
Maximum magnitude computation	6.16	21.63
I/O	4.83	16.96
Total time	28.48	100.00

Functions	time in sec	relative time (%)
3D inverse-wavelet transform	14.60	78.37
3D SPIHT decompression	0.86	4.62
I/O	3.17	17.01
Total time	18.63	100.00

TABLE 24.6. System complexity of 3D SPIHT encoder/decoder in terms of time. Coding time is based on Sun SPARC 20. The video "Carphone" (0 - 285) was coded at 10 fps, and 30 kbps. Total time is for encoding/decoding 96 frames

In order to quantitatively assess the time saving in decoding video sequences progressively by increasing resolution, we give the total decoder running time on a SUN SPARC 20, including inverse wavelet transform, and I/O in Table 24.7 on a SUN SPARC 20. From Table 24.7, we observe that significant computational time saving can be obtained with the multiresolutional decoding scheme. In particular, when we decoded at half spatial/temporal resolution, we obtained more than 5 times faster decoding speed than full resolution decoding. A significant amount of that time saving results from fewer levels of inverse wavelet transforms on smaller images.

	Spatial full (%)	Spatial half (%)
Temporal full	100	31.52
Temporal half	62.76	19.39

TABLE 24.7. Relative decoding time in % for the "Carphone" sequence at different temporal and spatial resolutions. Full resolution time is 18.63 second for decoding frame 0 – 285 at 10 fps and 30 kbps. (Hence, total number of frames decoded is 96)

8 Conclusions

We have presented an embedded subband-based video coder, which employs the SPIHT coding algorithm in 3D spatio-temporal orientation trees in video coding, analagous to the 2D spatial orientation trees in image coding. The video coder is fully embedded, so that a variety of monochrome or color video quality can thus be obtained with a single compressed bit stream. Since the algorithm approximates the original sequence successively by bit plane quantization according to magnitude comparisons, a precise rate control and self-adjusting rate allocations are achieved. In addition, spatial and temporal scalability can be easily incorporated into the system to meet various types of display parameter requirements.

It should be pointed out that a channel error affecting a significance test bit will lead to irretrievable loss of synchronism between the encoder and decoder and render the remainder of the bit stream undecodable. There are independent efforts now under way to provide more resilience to channel errors for the SPIHT bit stream and to protect it with powerful error-correcting codes [6, 12, 22]. Such techniques must be used when transmitting the compressed video over a noisy channel. Transmission in a noisy environment was considered beyond the scope of this chapter.

The proposed video coding is so efficient that, even without motion compensation, it showed superior or comparable results, visually and measurably, to that of MPEG-2 and H.263 for the sequences tested. Without motion compensation, the temporal decorrelation obtained from subbanding along the frames is more effective for the higher 30 f/s frame rate sequences of MPEG2 than for the 10 f/s sequences of QCIF. The local motion-compensated filtering used with QCIF provided more uniformity in rate and PSNR over a group of frames, which was hard to discern visually. For video sequences in which there is considerable camera pan and zoom, a global motion compensation scheme, such as that proposed in [24], will probably be required to maintain coding efficiency. However, one of the attractive features of 3D SPIHT is its speed and simplicity, so one must be careful not to sacrifice these features for minimal gains in coding efficiency.

9 References

[1] M. Antonini, M. Barlaud, P. Mathieu, and I. Daubechies. Image Coding Using Wavelet Transform. *IEEE Transactions on Image Processing*, 1:205–220, 1992.

[2] V. M. Bove and A. B. Lippman. Scalable Open-Architecture Television. *SMPTE J.*, pages 2–5, Jan. 1992.

[3] Y. Chen and W. A. Pearlman. Three-Dimensional Subband Coding of Video Using the Zero-Tree Method. *Visual Communications and Image Processing '96, Proc. SPIE 2727*, pages 1302–1309, March 1996.

[4] S. J. Choi. *Three-Dimensional Subband/Wavelet Coding of Video with Motion Compensation.* PhD thesis, Rensselaer Polytechnic Institute, 1996.

[5] S. J. Choi and J. W. Woods. Motion-Compensated 3-D Subband Coding of Video. *Submitted to IEEE Transactions on Image Processing*, 1997.

[6] C. D. Creusere. A New Method of Robust Image Compression Based on the Embedded Wavelet Algorithm. *IEEE Trans. on Image Processing*, 6(10):1436–1442, Oct. 1997.

[7] G. Karlsson and M. Vetterli. Three Dimensional Subband Coding of Video. *Proc. ICASSP*, pages 1100–1103, April 1988.

[8] B-J. Kim and W. A. Pearlman. An Embedded Wavelet Video Coder Using Three-Dimensional Set Partitioning in Hierarchical Trees (SPIHT). *Proc. IEEE Data Compression Conference*, pages 251–260, March 1997.

[9] T. Kronander. *Some Aspects of Perception Based Image Coding.* PhD thesis, Linkeoping University, Linkeoping, Sweden, 1989.

[10] T. Kronander. New Results on 3-Dimensional Motion Compensated Subband Coding. *Proc. PCS-90*, Mar 1990.

[11] J. Luo, X. Wang, C. W. Chen, and K. J. Parker. Volumetric Medical Image Compression with Three-Dimensional Wavelet Transform and Octave Zerotree Coding. *Visual Communications and Image Processing'96, Proc. SPIE 2727*, pages 579–590, March 1996.

[12] H. Man, F. Kossentini, and M. J. T. Smith. Robust EZW Image Coding for Noisy Channels. *IEEE Signal Processing Letters*, 4(8):227–229, Aug. 1997.

[13] J. R. Ohm. Advanced Packet Video Coding Based on Layered VQ and SBC Techniques. *IEEE Transactions on Circuit and System for Video Technology*, 3(3):208–221, June 1993.

[14] J. R. Ohm. Three-Dimensional Subband Coding with Motion Compensation. *IEEE Transactions on Image Processing*, 3(5):559–571, Sep. 1994.

[15] C. I. Podilchuk, N. S. Jayant, and N. Farvardin. Three-Dimensional Subband Coding of Video. *IEEE Transactions on Image Processing*, 4(2):125–139, Feb. 1995.

[16] A. Said and W. A. Pearlman. Image Compression Using the Spatial-Orientation Tree. *Proc. IEEE Intl. Symp. Circuits and Systems*, pages 279–282, May 1993.

[17] A. Said and W. A. Pearlman. Reversible Image Compression via Multiresolution Representation and Predictive coding. *Proc. SPIE*, 2094:664–674, Nov. 1993.

[18] A. Said and W. A. Pearlman. A New Fast and Efficient Image Codec Based on Set Partitioning in Hierarchical Trees. *IEEE Transactions on Circuits and Systems for Video Technology*, 6:243–250, June 1996.

[19] A. Said and W. A. Pearlman. An Image Multiresolution Representation for Lossless and Lossy Compression. *IEEE Transactions on Image Processing*, 5(9):1303–1310, Sep. 1996.

[20] K. Sayood. *Introduction to Data Compression.* Morgan Kaufman Publisher, Inc, 1996. page 285–354.

[21] J. M. Shapiro. An Embedded Wavelet Hierarchical Image Coder. *Proceedings IEEE International Conference on Acoustics, Speech, and Signal Processing (ICASSP), San Francisco*, pages IV 657–660, March 1992.

[22] P. G. Sherwood and K. Zeger. Progressive Image Coding for Noisy Channels. *IEEE Signal Processing Letters*, 4(7):189–191, July 1997.

[23] D. Taubman. *Directionality and Scalability in Image and Video Compression.* PhD thesis, University of California, Berkeley, 1994.

[24] D. Taubman and A. Zakhor. Multirate 3-D Subband Coding of Video. *IEEE Transactions on Image Processing*, 3(5):572–588, Sep. 1994.

[25] A. M. Tekalp. *Digital Video Processing.* Prentice Hall, Inc, 1995.

[26] J. Y. Tham, S. Ranganath, and A. A. Kassim. Highly Scalable Wavelet-Based Video Codec for Very Low Bit-rate Environment. *To appear in IEEE Journal on Selected Area in Communications*, 1998.

[27] I. H. Witten, R. M. Neal, and J. G. Cleary. Arithmetic Coding for Data Compression. *Communications of the ACM*, 30(6):520–540, June 1987.

[28] Z. Xiong, K. Ramchandran, and M. T. Orchard. Wavelet Packet Image Coding Using Space-Frequency Quantization. *Submitted to IEEE Transactions on Image Processing*, 1996.

[29] Z. Xiong, K. Ramchandran, and M. T. Orchard. Space-Frequency Quantization for Wavelet Image Coding. *IEEE Transactions on Image Processing*, 6:677–693, May 1997.

FIGURE 14. Comparison of visual performance of 3-D SPIHT with $GOF = 16$ (top), $GOF = 4$ (middle), and MPEG-2 (bottom) at bit-rate of 0.2 bpp

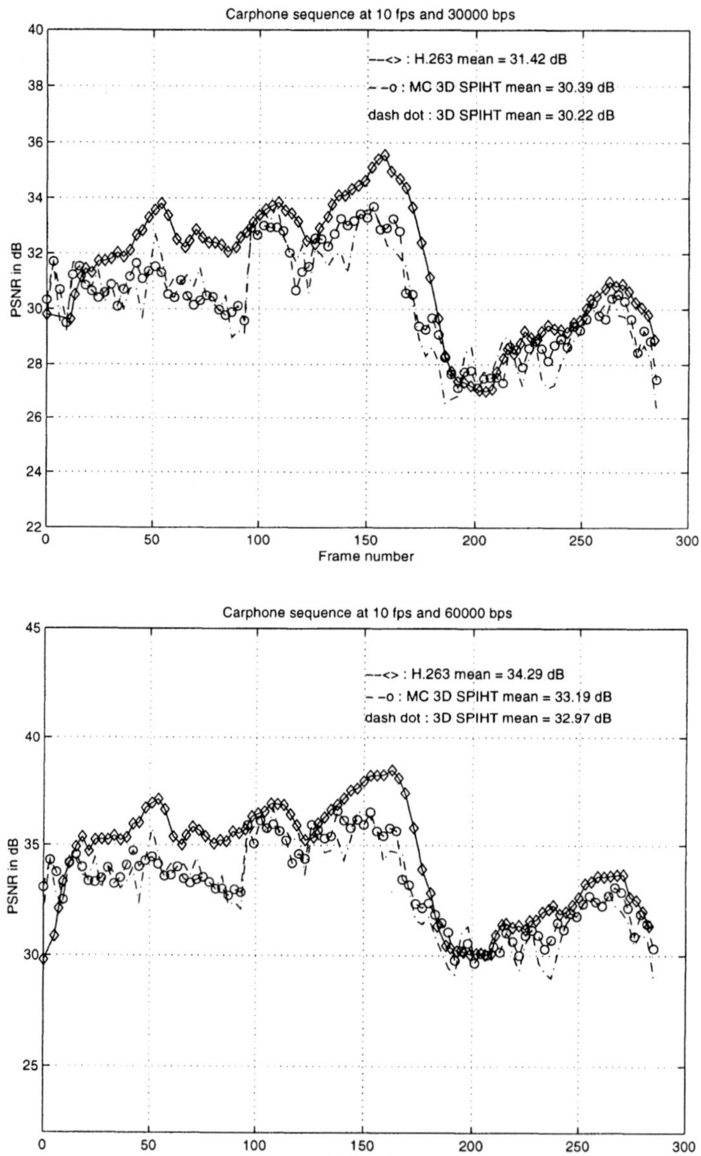

FIGURE 15. Frame-by-frame PSNR comparison of 3D SPIHT, MC 3D SPIHT, and H.263 at 30 and 60 kbps and 10 fps with "Carphone" sequence

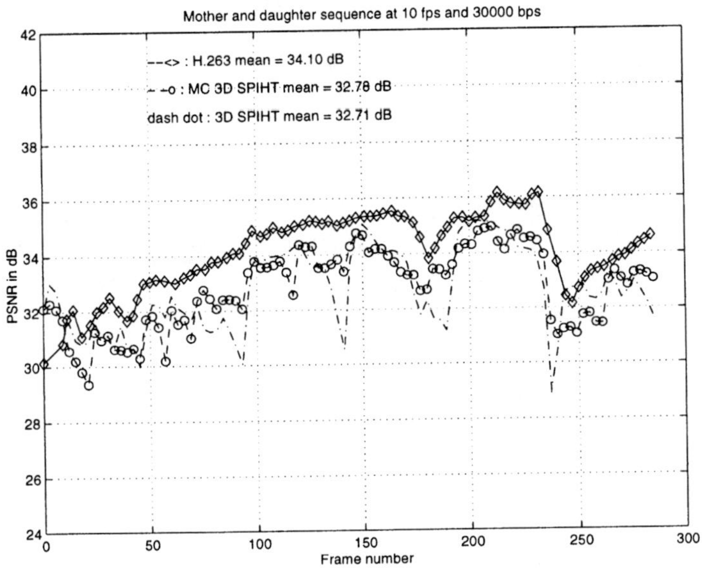

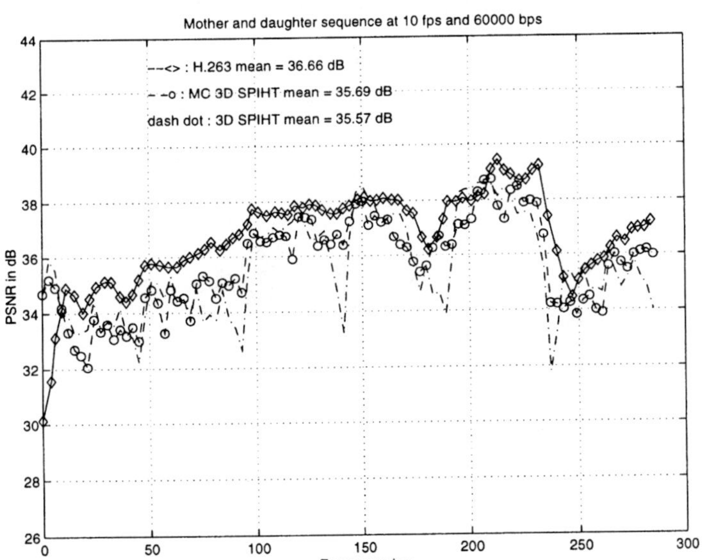

FIGURE 16. Frame-by-frame PSNR comparison of 3D SPIHT, MC 3D SPIHT, and H.263 at 30 and 60 kbps and 10 fps with "Mother and Daughter" sequence

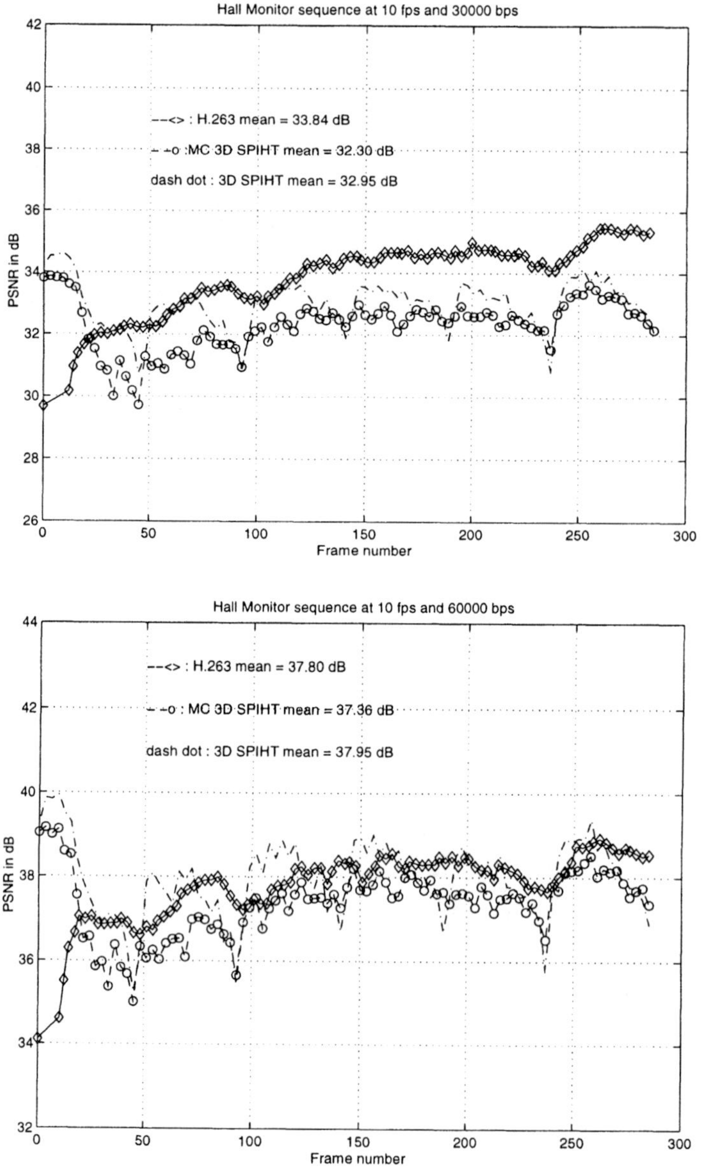

FIGURE 17. Frame-by-frame PSNR comparison of 3D SPIHT, MC 3D SPIHT, and H.263 at 30 and 60 kbps and 10 fps with "Hall Monitor" sequence

FIGURE 18. The same reconstructed frames at 30 kbps and 10 fps (a)Top-left: original "Carphone" frame 198 (b)Top-right: H.263 (c)Bottom-left: 3D SPIHT (d)Bottom-right: MC 3D SPIHT

FIGURE 19. The same reconstructed frames at 30 kbps and 10 fps (a)Top-left: original "Mother and Daughter" frame 102 (b)Top-right: H.263 (c)Bottom-left: 3D SPIHT (d)Bottom-right: MC 3D SPIHT

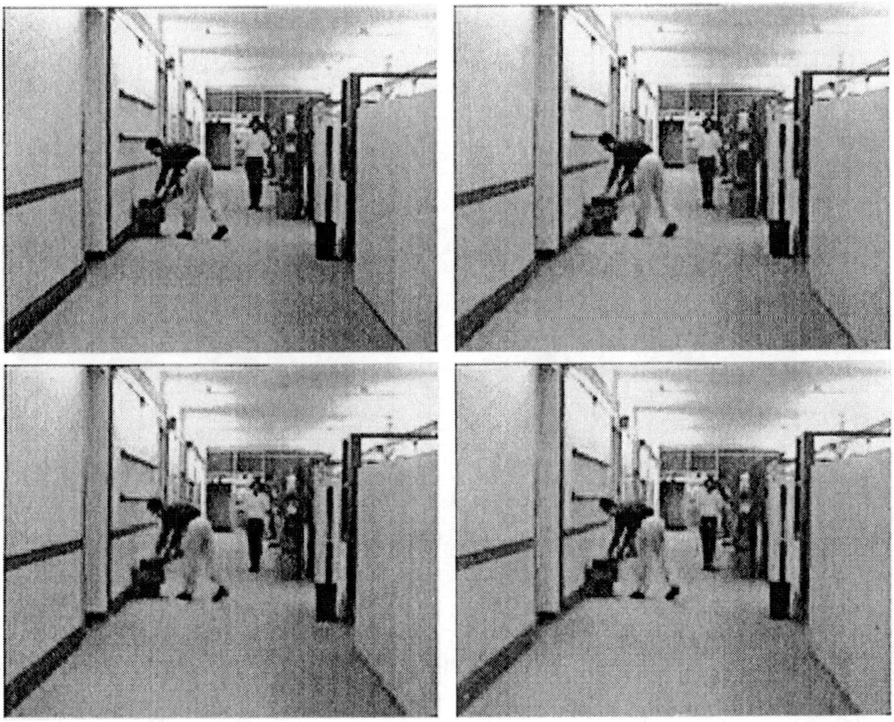

FIGURE 20. The same reconstructed frames at 30 kbps and 10 fps (a)Top-left: original "Hall Monitor" frame 123 (b)Top-right: H.263 (c)Bottom-left: 3D SPIHT (d)Bottom-right: MC 3D SPIHT

FIGURE 21. Multiresolutional decoded frame 0 of "Carphone" sequence with the embedded 3D SPIHT video coder (a)Top-left: spatial half resolution (88 x 72 and 10 fps) (b)Top-right: spatial and temporal half resolution (88 × 72 and 5 fps) (c)Bottom-left: temporal half resolution (176 × 144 and 5 fps) (d)Bottom-right: full resolution (176 × 144 and 10 fps)

Appendix A. Wavelet Image and Video Compression — The Home Page

Pankaj N. Topiwala

1 Homepage For This Book

Since this is a book on the state of the art in image processing, we are especially aware that no mass publication of the digital images produced by our algorithms could adequately represent them. That is why we are setting up a home page for this book at the publishers' site. This home page will carry the actual digital images and video sequences produced by the host of algorithms considered in this volume, giving direct access for comparision with a variety of other methodologies.

Look for it at

http://www.wkap.com

2 Other Web Resources

We'd like to list just a few useful web resources for wavelets and image/video compression. Note that these lead to many more links, effectively covering these topics.

1. www.icsl.ucla.edu/ ipl/psnr_results.htm Extensive compression PSNR results.
2. ipl.rpi.edu:80/SPIHT/ Home page of the SPIHT coder, as discussed in the book.
3. www.cs.dartmouth.edu/ gdavis/wavelet/wavelet.html. Wavelet image compression toolkit, as discussed in the book.
4. links.uwaterloo.ca/bragzone.base.html. A compression brag zone.
5. www.mat.sbg.ac.at/ uhl/wav.html. A compendium of links on wavelets and compression.
6. www.mathsoft.com/wavelets.html. A directory of wavelet papers by topic.
7. www.wavelet.org. Home page of the Wavelet Digest, an on-line newsletter.
8. www.c3.lanl.gov/ brislawn/FBI/FBI.html Home page for the FBI Fingerprint Compression Standard, as discussed in the book.
9. www-video.eecs.berkeley.edu/ falcon/MP/mp_intro.html Home page of the matching pursuit video coding effort, as discussed in the book.

10. saigon.ece.wisc.edu/ waveweb/QMF.html. Homepage of the Wisconsin wavelet group, as discussed in the book.

11. www.amara.com/current/wavelet.html. A compendium of wavelet resources.

Appendix B. The Authors

Pankaj N. Topiwala is with the Signal Processing Center of Sanders, A Lockheed Martin Company, in Nashua, NH. pnt@sanders.com

Christopher M. Brislawn is with the Computer Research and Applications Division of the Los Alamos National Laboratory, Los Alamos, NM. brislawn@lanl.gov

Alen Docef, Faouzi Kossentini, and Wilson C. Chung are with the School of Engineering, and Mark J. T. Smith is with the Office of the President of Georgia Institute of Technology. Contact: mark.smith@carnegie.gatech.edu

Jerome M. Shapiro is with Aware, Inc. of Bedford, MA. shapiro@aware.com

William A. Pearlman is with Center for Image Processing Research in the Electrical and Computer Engineering Department of Rensselaer Polytechnique Institute, Troy, NY; Amir Said is with Iterated Systems of Atlanta, GA. Contact: pearlman@ecse.rpi.edu

Zixiang Xiong, Kannan Ramchandran, and Michael T. Orchard are with the Departments of Electrical Engineering of the University of Hawaii, Honolulu, HI, the University of Illinois, Urbana, IL, and Princeton University, Princeton, NJ, respectively. zx@lena.eng.hawaii.edu; kannan@ifp.uiuc.edu; orchard@ee.princeton.edu

Rajan Joshi is with the Imaging Science Division of Eastman Kodak Company, Rochester, NY, and Thomas R. Fischer is with the School of Electrical Engineering and Computer Science at Washington State University in Pullman, WA. rjoshi@kodak.com; fischer@eecs.wsu.edu

John D. Villasenor and Jiangtao Wen are with the Integrated Circuits and Systems Laboratory of the University of California, Los Angeles, CA. villa@icsl.ucla.edu; gwen@icsl.ucla.edu

Geoffrey Davis is with the Mathematics Department of Dartmouth College, Hanover, NH. davis@cs.dartmouth.edu

Jun Tian and Raymond O. Wells, Jr. are with the Computational Mathematics Laboratory of Rice University, Houston, TX. juntian@math.rice.edu; wells@rice.edu

Trac D. Tran and Truong Q. Nguyen are with the Departments of Electrical and Computer Engineering at the Univeristy of Wisconsin, Madison, WI, and Boston University, Boston, MA, respectively. tran@saigon.ece.wisc.edu; nguyent@bu.edu

K. Metin Uz and Didier J. LeGall are with C-Cube Microsystems of Milpitas, CA, and Martin Vetterli is with Communication Systems Division of the Ecole

Polytechnique Federale de Lausanne, Lausanne, Switzerland, and the Department of Electrical Engineering and Computer Science, University of California, Berkeley, CA. metin@c-cube.com; didier@c-cube.com; martin.vetterli@de.epfl.ch, martin@eecs.berkeley.edu

Wilson C. Chung, Faouzi Kossentini are with the School of Engineering, and Mark J. T. Smith is with the Office of the President of Georgia Institute of Technology. Contact: mark.smith@carnegie.gatech.edu

Ralph Neff and Avideh Zakhor are with the Department of Electrical Engineering and Computer Science at the University of California, Berkeley, CA.
falcon@eecs.berkeley.edu; avz@eecs.berkeley.edu

Soo-Chul Han and John W. Woods are with the Center for Image Processing Research in the ECSE Department of Rennselaer Polytechnique Institute, Troy, NY. sooch@ipl.rpi.edu; woods@ecse.rpi.edu

William A. Pearlman, Beong-Jo Kim, and Zixiang Xiong were at the Center for Image Processing Research in the ECSE Department of Rensselaer Polytechnique Institute, Troy, NY. Xiong is now at the University of Hawaii. Contact: pearlman@eecs.rpi.edu

Index